DIGITAL CONCEPTS AND APPLICATIONS

SECOND EDITION

Amin R. Ismail
Victor M. Rooney

University of Dayton

SAUNDERS COLLEGE PUBLISHING

Harcourt Brace College Publishers

Fort Worth   Philadelphia   San Diego   New York   Orlando   Austin   San Antonio
Toronto   Montreal   London   Sydney   Tokyo

Requests for permission to make copies of any part of the work should be mailed to Copyrights and Permissions Department, Harcourt Brace & Company, 8th Floor, Orlando, Florida 32887.

Text Typeface: Times Roman and Caslon
Compositor: General Graphic Services
Acquisitions Editor: Emily Barrosse
Managing Editor: Carol Field
Project Editor: Laura Shur
Copy Editor: Jennifer Holness-Harze
Manager of Art and Design: Carol Bleistine
Art Director: Robin Milicevic
Cover Designer: Vincent Ceci
Text Artwork: Grafacon
Director of EDP: Tim Frelick
Production Manager: Joanne Cassetti
Marketing Manager: Monica Wilson

Cover Credit: Brett Froomer / The Image Bank

Printed in the United States of America

Digital Concepts and Applications, 2/e

ISBN: 0-03-097196-9

Library of Congress Catalog Card Number: 93-084988

3456 032 987654321

To my wife, Loretta, and granddaughters,
Aarika, Alicia, and Melissa.

To my parents, Rashid and Salma,
my grandfather, H. H. Ismail, and grandmother, Bernadette Carvalho.

THE SAUNDERS COLLEGE PUBLISHING SERIES IN ELECTRONICS TECHNOLOGY

PREFACE

This text covers the concepts and applications of digital circuits and is intended for college courses in two- and four-year programs in electronics technology, electrical engineering technology, and computer technology. The material in the book may also be well suited for digital circuit courses in computer science and some engineering programs. The prerequisites for this book are a basic knowledge of algebra and elementary electrical concepts; for later chapters a familiarity with basic electronics is beneficial but not required. The material covered in this book can be used in either a one- or two-semester course in digital circuit concepts and applications, depending on the depth of coverage.

Approach

The emphasis of this text is an understanding of digital circuit concepts and applications of those concepts. A good understanding of the elementary concepts of digital circuits is vital to understanding the design of various application circuits. Therefore, much of this book is dedicated to building a strong conceptual foundation for use in the analysis and synthesis (design) of digital circuits. Application circuits covered in this book are first introduced by designing the circuit. Then equivalent commercially available circuits (in the form of ICs) are investigated through analysis and compared and contrasted with the circuits designed earlier. This technique makes it easier to understand the function and operation of a digital application circuit.

Organization

The text deals with eight distinct areas of study — numbering systems, combinational logic, sequential and register-transfer logic, memory and I/O systems, digital computer and microprocessor architecture, programmable logic devices (PLDs), logic technology, and Analog/Digital interfacing.

Chapter 1 introduces numbering systems with an emphasis on the binary numbering system. Because the decimal numbering system is most familiar, it is used to explain some of the elementary concepts of other numbering systems. Also included is a detailed look at the octal and hexadecimal numbering systems, both of which play an important role in digital computers. Conversions between numbering systems and concepts of binary arithmetic and binary codes are also included. The chapter is written so that the sequence in which the different numbering systems are covered can be altered very easily to suit the instructor and course organization.

Chapters 2 through 4 cover the basic concepts of combinational logic and their application circuits that make up a digital computer and other digital devices. The elementary concepts of logic are introduced first, followed by the basic logic gates, equations, truth tables, and simple circuits. Next the mathematics of logic — Boolean algebra — is covered in detail along with techniques for analyzing and synthesizing logic circuits. Finally, these concepts are applied to the operation and design of specific application circuits such as adders, subtracters, comparators, etc., and their hardware implementations in the form of integrated circuits. Techniques for implementing and troubleshooting combinational logic circuits in the laboratory are also included in these chapters.

Sequential logic is covered in Chapters 5 and 6, and register-transfer logic is covered in Chapter 7. These chapters introduce the elementary concepts of sequential logic circuits and the building block of all sequential logic — the flip-flop. The operation of the different types of flip-flops is covered in detail, along with their applications in counters, registers, and register files. Also included is an analysis of the various integrated circuits that implement the basic flip-flops, counters, and registers in hardware as well as procedures to troubleshoot sequential logic circuits. Chapter 6 is oriented toward the design, analysis, and applications of various counters. Instructors who do not want to cover detailed design can omit the design procedures in Section 6–4 and use the circuits designed as classroom examples. Chapter 7 deals with the various types of registers and their applications.

The study of sequential logic and register files leads naturally to the design and operation of memory systems. The fundamental characteristics of memory are investigated in Chapter 8, along with the development of memory design concepts and procedures. These concepts and procedures are then used in Chapter 9 in the design of memory systems.

The architecture of a digital computer integrates four areas of study — numbering systems, combinational logic, sequential logic, and memory and I/O systems. Due to their popularity, the microprocessor and microcomputer are used as models for the architecture of a typical digital computer. Chapter 10 covers this material and investigates the basic components and circuits that make up a digital computer. This treatment includes a detailed look at the CPU and microprocessor as well as the relationship between computer hardware and software. Techniques for troubleshooting such systems are also included.

Chapter 11 is a study of a related application of logic circuits — programmable logic devices (PLDs) that are frequently used today to replace many types of combinational and sequential logic circuits. Because of the increasing use of PLDs in industry, this chapter includes a comprehensive study of their concepts, architecture, and applications. Programming with the industry-standard PALASM4 software is also covered in detail.

The technology of logic circuits is covered in Chapter 12, including the electronics that makes up the basic logic gates. The chapter begins with the application of the BJT and MOSFET as electronic switches and their use in simple switching circuits. Next, the various electronic circuits that make up logic gates are analyzed. The switching and loading characteristics of the two major families of logic circuits (MOS and bipolar) are also investigated. Laboratory procedures for troubleshooting the logic circuits of different families are included.

Chapter 13 covers the operation of various interface circuits. Circuits that interface various serial and parallel peripheral I/O devices to a digital computer are examined with their applications. Also covered are the design and characteristics of such circuits as analog-to-digital (ADC) and digital-to-analog (DAC) converters, which provide the link between the analog and digital worlds.

For convenience, the book also contains several appendices. Powers of 2 are listed in Appendix A to assist in various calculations. Appendix B contains a detailed description and explanation of the ANSI/IEEE 91-1984 logic symbols. However, because of their popularity, the distinctive shape symbols of the old standard have been retained for most of the logic circuits discussed in the book. Appendix C contains the manufacturer's data sheets of all the TTL integrated circuits used in the book, although selected portions of these data sheets have been reproduced throughout the text when necessary. A glossary of commonly used terms and definitions is included in Appendix D. Appendix E contains the answers to all odd-numbered end-of-chapter problems.

Flexibility

Many academic programs introduce digital circuits early in their curriculum to allow for a larger selection of more advanced courses in microprocessors and related fields. The material in this book is organized to accommodate such programs because it can be covered without a prerequisite course in semiconductor devices. For instance, Chapter 2 includes a section that deals with the electrical characteristics of logic gates for those instructors who would like their students to be able to interpret data sheets early in their course; this section can be safely omitted with no adverse consequences. Some computer science programs may choose to omit the coverage of digital electronics entirely (Chapter 12) and possibly even the subsections that deal with integrated circuit logic, electrical and switching characteristics, and troubleshooting. The material in Chapters 1 through 11 is organized so that it can be covered with or without a basic knowledge of electronics or electricity. The electrical and switching characteristics of logic gates, however, are introduced early to complement the discussions on troubleshooting. Because this material is sectioned off, it also may be omitted if desired. The internal electronic construction of logic gates is not covered until Chapter 12.

Because many educators prefer a less detailed coverage of the formal procedures for synthesis and analysis of sequential logic circuits, Chapter 6 includes these procedures in separate sections so that they may be omitted if desired with no adverse effects. Finally, the book can be used strictly as a digital text for courses that deal with digital circuits (no microprocessors) only. For such courses we recommend coverage of Chapters 1 through 7 and Chapter 11 (PLDs).

Changes in the second edition

Chapter 1 was reorganized to cover the concepts of numbering systems simultaneously, again using the decimal numbering system as a model for the others. The chapter first covers counting in various numbering systems and then looks at conversions and arithmetic operations between the various numbering systems with emphasis on binary numbers.

Chapters 2 through 4 have been enhanced with more practical examples of combinational logic circuits.

Chapter 5 now includes a detailed study of gates with Schmitt trigger inputs, and an in-depth look at the astable and monostable multivibrators. Additional examples of practical sequential logic applications have also been added.

Chapter 8 was revised to be more focused on the preliminary concepts of semiconductor memories and the design of memory systems. Subject matter that was not essential was removed. As a result, any discussion of microprocessors was postponed until Chapter 10.

Chapter 10 was revised to reflect the concepts of segmented memory, which will provide a greater understanding of some of the more advanced microprocessors. The coverage of isolated and memory-mapped I/O was moved to this chapter.

Chapter 11 was revised to include extensive coverage of PLD programming using PALASM4 software. The coverage of PLDs now includes the concepts of PLD architecture, PLD software, and detailed explanations of how to use that software. Examples of PLD usage have been updated to include programming examples. These examples include both combinational and sequential logic designs. In addition to PLD programming, PLD simulation has also been added.

Chapter 13 now incorporates commercially available DAC's and ADC's. Examples showing applications of these IC's are given.

Learning aids

The organization of each chapter is designed to enhance the learning process. Every chapter begins with an introduction that includes a list of chapter objectives and a chapter outline. This organization allows the reader to quickly identify the content of each chapter and to review the material thoroughly. The material is broken up into logical sections and contains an abundant supply of examples and illustrations to enhance the discussions and reinforce the concepts. Review questions are included at the end of every section and can be used as a self-test to determine if the concepts are understood. An end-of-chapter summary lists the key concepts and terms covered in the chapter in an easy-to-read form. At the end of every chapter are several problems and exercises marked by section, as well as troubleshooting case studies.

The following is a summary of the pedagogical and instructional features that are included in each chapter:

- **Chapter-opening photograph** of general interest relates to the subject matter. This helps to capture the student's interest in the material.

- **Chapter outline** identifies the sections and topics covered in the chapter.

- **List of objectives** assists in identifying material covered in the chapter.

- **Introduction** leads the student into the chapter's contents.

- **Two-color illustrations** make it easier to understand and analyze complex digital circuits and diagrams.

- **Review questions** at the end of each section are provided as a self-test.

- **Examples** and their solutions explain the various concepts and applications introduced in each chapter.

- **Troubleshooting case studies** introduce ''real world'' examples of practical faults in digital circuits and procedures for correcting them.

- **Key terms** and **definitions** are identified throughout the text using italics for easy reference.

- **Chapter summaries** include a compiled list of key terms, definitions, and concepts covered in the chapter in an easy-to-read format.

- **End-of-chapter problems, questions,** and **case studies** are organized by section for easy reference.

Ancillary package

A comprehensive ancillary package supports the textbook. The following is a list of its components that can be used to assist greatly both the instructor and the student in preparing and studying the material covered in the book.

Instructor's manual with transparency masters and test bank

This manual contains the complete solutions and answers to all of the end-of-chapter problems. Also included is a competition comparison chart and a chart to coordinate the individual ancillary components with the material covered in the textbook. Transparency masters for approximately 100 tables and illustrations from the textbook are included to assist with class lectures. The test bank includes approximately 40 test items per chapter for use in preparing quizzes, tests, and exams. These test items are variations on the review questions and end-of-chapter problems from the textbook and include solutions.

Laboratory manual

Approximately 30 class-tested lab experiments related to the material covered in the textbook are included in this manual. Lab experiments are organized with an introduction, lab objectives, equipment, and procedures section. Results and answers obtained from a lab experiment can be filled in by the student on separate pages and detached from the manual for submission to the instructor.

Instructor's laboratory manual

This manual contains the material in the laboratory manual along with answers and expected results for each experiment. Instructions on preparing the laboratory experiment are also included.

Acknowledgments

We would like to thank all the people who were involved with this project, particularly:

Gerald Schickman, Miami Dade Community College
Rich Pfile, Indiana University-Purdue University
Ron McKeen, Ferris State University
Russell English, San Diego Mesa College
Charles Swain, Rochester Institute of Technology
Brian Goodman, Chippewa Valley Technical College
Nazar Karzay, Indiana Vocational Technical College

We appreciate their help, encouragement, and honest criticisms during the review process.

We would also like to thank our Department Chairman, Joseph Farren, for allowing us to use the "raw" manuscript in our Digital Circuits course and for arranging our schedules so that we could teach the course on a regular basis. We would like to acknowledge the dedication and hard work of all the people at Saunders who were involved with the production of the book: Barbara Gingery, Electronics Technology Editor; Maureen Iannuzzi, Project Editor; Alexa Barnes, Developmental Editor; Laura Shur, Editorial Assistant; and Teri Stratford, photo researcher.

Last, but certainly not the least, we would like to thank Loretta Rooney and Rhonda Copley for helping us prepare the manuscript, and the many students at the University of Dayton who helped iron out the wrinkles in the manuscript.

A. R. I.
V. M. R.
November 1993

CONTENTS

1

NUMBERING SYSTEMS

OBJECTIVES

The objectives of this chapter are to:

* Introduce the basic concepts that relate to the decimal and other numbering systems

* Apply the decimal numbering system concepts to the octal, hexadecimal, and binary numbering systems

* Explore the procedures for performing arithmetic using only the process of addition

* Study techniques of converting between different numbering systems

* Study the various procedures for performing binary arithmetic

* Investigate the different types of binary codes and their applications

The abacus is often considered to be the first computer invented. It was used to assist in calculations and was constructed with pebbles or, in Latin, "calculus," which is where the word "calculate" comes from (The Bettman Archive).

1.1
Introduction

Numbering systems have been in existence for thousands of years and have provided us with the means to assign value, determine quantity, calculate distances, and measure sizes. Besides serving as a very important tool in people's everyday lives, numbering systems play an important part in the design and operation of digital circuits.

The decimal numbering system has been in existence for over 2000 years and is simply based on a series of ten unique digits that were originally used to represent the ten fingers of the hand (hence the use of the term "digit," derived from the latin word "digitus," meaning finger). This implies that a numbering system does not have to be based on ten digits but can be based on a larger or smaller number. There are a few other numbering systems that are more suitable in the study of digital computers and their associated circuits. Therefore, it is important to be able to understand some of the basic concepts of numbering systems so that these concepts can be applied to other numbering systems that are important in the study of digital circuits. Because the decimal numbering system is the most familiar, this chapter will begin by introducing some of the basic numbering system concepts using the decimal numbering system as a model and then those concepts will be applied to three other numbering systems—octal, hexadecimal, and binary. Once we have developed a good understanding of our own (decimal) numbering system and two other closely related numbering systems (octal and hexadecimal) we shall then study the binary numbering system. This study will include the binary counting sequence; conversions between binary numbers and decimal, octal, and hexadecimal numbers; and representation of binary numbers, binary arithmetic, and binary codes. Many of the concepts and procedures developed in this chapter will then be applied to the design and operation of various digital circuits to be covered in subsequent chapters.

1.2
Numbering system concepts

The decimal numbering system is based on a set of ten digits, each digit representing a value. These ten digits in the order of value can be listed as follows:

0	(zero)
1	(one)
2	(two)
3	(three)
4	(four)
5	(five)
6	(six)
7	(seven)
8	(eight)
9	(nine)

The first digit 0 (zero) has the lowest value (no value) and the tenth digit 9 (nine) has the highest value. If there is a need to have larger values represented by the decimal numbering system, then *combinations* of the basic ten digits can be used. Therefore, after 9, the next highest values would be

10	(ten)
11	(eleven)

.

.

.

etc.

Note, for example, that the numbers 10 (ten) and 11 (eleven) are not digits but combinations of digits that are assigned increasing values. If values continue to be assigned to these combinations a *count* is obtained. If a count of all possible decimal digits taken singly is obtained, then a count of ten is achieved:

$$0, 1, 2, 3, 4, 5, 6, 7, 8, 9$$

If a count of all possible numbers taking two digits at a time is obtained, a count of 100 is achieved

$$00, 01, 02, 03, \ldots, 10, 11, 12, \ldots, 20, 21, 22, \ldots, 97, 98, 99$$

That is, there are 100 combinations of digits between 0 and 99 when the combinations are taken two at a time.

Similarly, there are 1000 combinations of digits between 0 and 999 when the combinations are taken three at a time:

$$000, 001, 002, \ldots, 010, 011, \ldots, 098, 099, 100, 101, \ldots, 999$$

In general, for any numbering system the number of combinations of n digits can be determined by using the following relationship:

$$B^n$$

where B is the *base* of the numbering system and n is the number of digits to be combined. Because the decimal numbering system is a base-ten numbering system (i.e., there are ten digits in the numbering system), the number of combinations of n decimal digits is

$$10^n.$$

Thus, there are 10^2 or 100 combinations of two digits, 10^3 or 1000 combinations of three digits, 10^4 or 10,000 combinations of four digits, and so on. When these combinations are organized into an orderly sequence of increasing values, a count is obtained.

Example 1.1

How many combinations would there be of five decimal digits and what would the largest value in the count be?

Solution

The number of combinations would be 10^5 or 100,000. The count would begin at 00000 and end at 99999. Therefore, the largest value in the count would be 99999.

The octal numbering system

The octal numbering system is a *base-eight* numbering system. That is, there are only eight digits in the numbering system. The word "octal" is derived from the latin word "octa" or eight.

0	(zero)
1	(one)
2	(two)
3	(three)
4	(four)
5	(five)
6	(six)
7	(seven)

The digits 8 (eight) and 9 (nine) do not exist in this system. Therefore, to count in octal, follow the same procedure as for decimal system — start combining digits after reaching a count of 7

$$\ldots 7, 10, 11, 12, \ldots, 15, 16, 17, 20, 21, \ldots, 75, 76, 77, 100$$

If a count of all possible combinations of two octal digits were to be examined, the largest number would be 77. That is, the count would end at 77. Using the relationship introduced earlier, B^n, the total number of combinations for two octal digits would be

$$8^2$$

or

$$64.$$

Therefore, there are 64 (decimal) counts between the octal numbers 00 and 77. The value of the octal number 77 would actually be 63 (in decimal). Thus, octal numbers are always larger than decimal numbers representing the same value; this is because the octal numbering system has fewer digits. Table 1.1 lists the relationship between the ten decimal digits and their octal equivalents.

A note on terminology — if the decimal combination 10 is referred to as "ten," and the combination 11 as "eleven," etc., should the octal combination 10 be referred to as "ten" and the octal combination 11 as "eleven," or should these combinations be referred to as "eight" and "nine"? The answer is neither! The octal combination 10 should be referred to as "one-zero," 11 as "one-one," 12 as "one-two," and so on, because the terms "eight," "nine," "ten," "eleven," etc. are used to describe decimal digits and combinations that do not exist or are not the same in the octal numbering system. Furthermore, we have now been introduced to a numbering system that uses the same digits as the decimal numbering system

Table 1.1 Decimal digits and their octal equivalents

Decimal	Octal
0	0
1	1
2	2
3	3
4	4
5	5
6	6
7	7
8	10
9	11

although not all similar digits and combinations in both systems have the same value. Therefore, it is important to distinguish between octal numbers and decimal numbers. The notation used to identify numbers in different numbering systems is to append to the number a *subscript* that identifies its base. For example, the following are octal numbers:

$$645_8, 7364510_8, 7_8, 13007_8$$

and the following are decimal numbers:

$$2746_{10}, 92648_{10}, 25347_{10}, 4_{10}$$

At this stage it is important to be consistent in identifying numbers in different bases to eliminate confusion when dealing with different numbering systems. For example, if a number is specified as 72645, we would not know if it were an octal number or a decimal number. However, the number 223648 is definitely a decimal number because it contains the digit 8, which is not part of the octal set. Still, it is recommended that the base of the numbering system be identified when dealing with numbers from now on.

Because the decimal numbering system is the most frequently used, the subscript base specifier (10) is often deleted when identifying decimal numbers and assume decimal as the *default*. Thus, if a number such as 17263 is specified (without any base specifier) it will be assumed to be decimal.

Example 1.2

Obtain an octal count from 275_8 to 310_8. If the count were continued, what would be the largest three-digit octal number obtained?

Solution

275, 276, 277, 300, 301, 302, 303, 304, 305, 306, 307, 310. If the count were continued, the largest three-digit octal number obtained would be 777.

The hexadecimal numbering system

The hexadecimal numbering system is a *base-sixteen* numbering system. That is, there are 16 digits in the numbering system. The word "hexadecimal" is derived from the latin word "hexa", or six, because it is used to describe a numbering system that has six additional digits over the decimal numbering system. Because the decimal numbering system uses the graphical symbols 0, 1, 2, 3, 4, 5, 6, 7, 8, 9 to represent the 10 basic digits, six "new" graphical symbols must be used to represent the additional six hexadecimal digits. However, instead of designing new graphical symbols for these additional six digits, the letters A, B, C, D, E, F are used. Note, however, that when dealing with hexadecimal numbers these are no longer considered to be letters of the alphabet but are now treated as digits in exactly the same way as are the basic 10 digits 0 through 9.

0	(zero)
1	(one)
2	(two)
3	(three)
4	(four)
5	(five)
6	(six)
7	(seven)
8	(eight)
9	(nine)

A	(ay)
B	(bee)
C	(see)
D	(dee)
E	(ee)
F	(ef)

To count in hexadecimal, follow the same procedure as for the decimal and octal systems — start combining digits after a count of F is reached.

$$\ldots F, 10, 11, 12, \ldots, 19, 1A, 1B, 1C, 1D, 1E, 1F, 20, 21, \ldots,$$
$$99, 9A, \ldots, 9F, A0, A1, \ldots, FD, FE, FF, 100$$

If a count of all possible combinations of two hexadecimal digits were to be examined, the largest number would be FF. That is, the count would end at FF. Using the relationship introduced earlier, B^n, the total number of combinations for two hexadecimal digits would be

$$16^2$$

or

$$256.$$

Therefore, there are 256 (decimal) counts between the hexadecimal numbers 00 and FF. The value of the hexadecimal number FF would actually be 255 (in decimal). Thus, hexadecimal numbers are always smaller than decimal numbers representing the same value because the hexadecimal numbering system has more digits. Table 1.2 lists the relationship between the 16 hexadecimal digits and their decimal equivalents.

As before, the subscript 16 must be appended to a number if that number is considered to be hexadecimal. For example, the following are hexadecimal numbers:

$$6A45_{16}, 7364510_{16}, F7_{16}, 13007_{16}.$$

Table 1.2   Hexadecimal digits and their decimal equivalents

Hexadecimal	Decimal
0	0
1	1
2	2
3	3
4	4
5	5
6	6
7	7
8	8
9	9
A	10
B	11
C	12
D	13
E	14
F	15

Example 1.3

Obtain a hexadecimal count from 197_{16} to $1A2_{16}$. If the count were continued, what would be the largest three-digit hexadecimal number obtained?

Solution

197, 198, 199, 19A, 19B, 19C, 19D, 19E, 19F, 1A0, 1A1, 1A2. If the count wre continued, the largest three-digit hexadecimal number obtained would be FFF.

The binary numbering system

The binary numbering system is the most important numbering system in the study of digital circuits and their applications. It provides the basis for all digital computer circuits, and is a natural numbering system for the description of the logic circuits to be studied later in this book.

The binary numbering system is a *base-two* numbering system. That is, there are only two digits in the numbering system. The word "binary" is used to describe something constituted of two things or parts, and when used in this context describes a numbering system that is composed of only two digits, 0 and 1. No other digits exist in the binary numbering system. Therefore, to count in binary, the same procedure would be followed as in decimal — digits are combined after a count of 1 is reached

$$0, 1, 10, 11, 100, 101, 110, 111, 1000,$$
$$1001, 1010, 1011, 1100, 1101, 1110, 1111$$

Counting from 0, the next count is 1. However, note that after 1, the next count is 10 because the digits 2 through 9 do not exist in the binary numbering system. The next count is 11. However, after 11 the next count is 100 because the numbers 12 through 99 do not exist in the binary numbering system. The count then proceeds to 101, 110, 111, and so on.

To examine a count of all possible combinations of four binary digits, the largest number would be 1111. That is, the count would end at 1111. Using the relationship introduced earlier, B^n, the total number of combinations for four binary digits would be

$$2^4$$

or

$$16.$$

Therefore, there are 16 (decimal) counts between the binary numbers 0000 and 1111. The value of the binary number 1111 would actually be 15 (in decimal). Thus, binary numbers are always larger than decimal numbers representing the same value because the binary numbering system has only two digits. Table 1.3 lists the relationship between the ten decimal digits and their binary equivalents. Table 1.3 shows that binary numbers are often operated on in groups. A single binary digit (0 or 1) is called a *bit*, a group of 4 bits is called a *nibble*, a group of 8 bits is called a *byte*, and a group of 16 bits is called a *word*. Since the largest number in Table 1.3 is 9, and the binary number used to represent 9 requires 4 bits, leading zeros are added to the other numbers to make them 4 bits long. Of course, the addition of leading zeros has absolutely no effect on the values of the numbers.

The binary counting sequence is of particular importance for it follows a definite pattern that is important in the design and analysis of digital circuits. Table 1.4 lists all possible combinations of 4 bits in the counting sequence, along with their equivalent decimal, hexadecimal, and octal digits or numbers.

Table 1.4 shows that the count of all possible combinations of 4 bits follows a pattern similar to the counting sequence in any other numbering system. That is, the *least significant digit* (LSD) cycles through every possible digit in sequence whereas the other digits in positions to the left increment once every time a cycle from the digit at the right is completed. Because there are only two possible digits (bits) in the binary numbering system, the *least significant bit* (LSB) toggles back and forth between a 0 and a 1. The next bit toggles once for every cycle of the LSB. The next bit position toggles once for every cycle of the bit to its right or once for every two cycles of the LSB. Finally, the *most significant bit* (MSB) toggles once for every cycle of the bit to its right or once for every four cycles of the LSB.

Table 1.4 also illustrates two very important characteristics of the octal and hexadecimal numbering systems. Note that every 4-bit binary combination can be uniquely represented by a hexadecimal digit. That is, there is one hexadecimal digit available for every 4-bit binary number. This also applies to octal digits and 3-bit binary numbers. Every digit in the octal set can uniquely represent every 3-bit binary combination; there is one octal digit for every 3-bit binary number. As will be seen later in this section, this perfect representation makes conversion among hexadecimal, octal, and binary numbers very simple. Unfortunately, the relationship between binary and decimal is not that perfect. Note in Table 1.4 that even though every decimal digit can be represented by a unique 4-bit binary number, the set of decimal digits cannot represent all 4-bit combinations because there are 16 combinations and only 10 decimal digits. Furthermore, even though a decimal digit can be used to

Table 1.3 Decimal digits and their binary equivalents

Decimal	Binary
0	0000
1	0001
2	0010
3	0011
4	0100
5	0101
6	0110
7	0111
8	1000
9	1001

Table 1.4 Binary, decimal, hexadecimal, and octal equivalence table

Binary	Hexadecimal	Decimal	Octal
0000	0	0	0
0001	1	1	1
0010	2	2	2
0011	3	3	3
0100	4	4	4
0101	5	5	5
0110	6	6	6
0111	7	7	7
1000	8	8	10
1001	9	9	11
1010	A	10	12
1011	B	11	13
1100	C	12	14
1101	D	13	15
1110	E	14	16
1111	F	15	17

represent each unique 3-bit combination, all the digits in the decimal set cannot be represented by 3-bit combinations because there are only eight combinations. Therefore 3-bit or 4-bit numbers are not a perfect fit for the decimal digits.

As before, the subscript 2 must be appended to a number if that number is considered to be binary. For example, the following are binary numbers:

$$1110_2, 1010110_2, 10_2, 10001_2.$$

Example 1.4

Obtain a binary count from 10101_2 to 11101_2. If the count were continued, what would be the largest 5-bit binary number obtained?

Solution

10101, 10110, 10111, 11000, 11001, 11010, 11011, 11100, 11101.

If the count were continued, the largest 5-bit binary number obtained would be 11111.

Binary numbers tend to be very long. For example, the binary representation of the decimal number 60,000 would be a 16-bit binary number, 1110101001100000. It is not uncommon to encounter binary numbers this size (and even larger) during the study of digital circuits and computers. Therefore, a more convenient means of representing these large binary numbers is needed to somewhat ease the task of working with them. At first thought, decimal numbers would be the ideal choice to represent binary numbers because they are shorter in size and because the decimal numbering system is the most familiar of all numbering systems. Binary numbers could be converted to decimal, manipulated, and then reconverted to binary. However, the decimal numbering system is not very well suited to representing binary numbers, and conversions are relatively more time consuming and complicated. As will be seen later in this section, the octal and hexadecimal numbering systems are the ideal choices for representing binary numbers.

Review Questions

1. What is meant by the base of a numbering system?
2. How many digits do each of the following numbering systems have — octal, hexadecimal, and binary?
3. What is the largest possible four-digit number in each of the four numbering systems?
4. Compare the four numbering systems, decimal, octal, hexadecimal, and binary, in terms of their representation of a value (such as 75).

1.3

Conversions to decimal

A decimal number such as 6375 is simply a combination of the four decimal digits 6, 3, 7, and 5. However, the place, or *position,* of each digit in the number is of significance. For example, even though the number 7653 has the same digits as 6375, the two numbers have different values. Each digit in a decimal number is assigned a *positional weight*. For example:

$$\begin{array}{cccc}
\text{number} \longrightarrow & 6 & 3 & 7 & 5 \\
& \uparrow & \uparrow & \uparrow & \uparrow \\
\text{positional weight} \longrightarrow & 10^3 & 10^2 & 10^1 & 10^0
\end{array}$$

In the number 6375, the digit 5 has a weight of 10^0, or 1, the digit 7 has a weight of 10^1, or 10, the digit 3 has a weight of 10^2, or 100, and the digit 6 has a weight of 10^3, or 1000. We can express the value of the number 6375 as follows:

$$\begin{array}{llll}
6 \times 10^3 = 6 \times 1000 = & 6000 \\
3 \times 10^2 = 3 \times 100 & = & 300 \\
7 \times 10^1 = 7 \times 10 & = & 70 \\
5 \times 10^0 = 5 \times 1 & = & \underline{5} + \\
& & 6375
\end{array}$$

Thus each digit in a decimal number is assigned a position (beginning with 0) and a weight that is a power of 10. The rightmost digit of any number is the LSD and the leftmost digit is called the *most significant digit* (MSD). The *least significant digit* (LSD) always has a position of 0 associated with it whereas the position of the MSD will depend on the size of the number. Therefore for the number 6375

$$\begin{array}{cccc}
& 6 & 3 & 7 & 5 \\
& \uparrow & \uparrow & \uparrow & \uparrow \\
\text{position:} & 3 & 2 & 1 & 0 \\
\text{MSD} & \rule{0pt}{0pt}\uparrow & & \uparrow & \text{LSD}
\end{array}$$

the LSD is 5 (position 0) and the MSD is 6 (position 3).

Example 1.5

Determine the value of the number 90,812 in terms of its positional weights.

Solution

$$\begin{array}{llll}
9 \times 10^4 = 9 \times 10{,}000 = & 90{,}000 \\
0 \times 10^3 = 0 \times 1000 & = & 0\,000 \\
8 \times 10^2 = 8 \times 100 & = & 800 \\
1 \times 10^1 = 1 \times 10 & = & 10 \\
2 \times 10^0 = 2 \times 1 & = & \underline{2} + \\
& & 90{,}812
\end{array}$$

In general, for a number in any numbering system, the value of any digit in the number is given by the expression

$$D \times B^p$$

where D is the digit, B is the base of the numbering system, and p is the position of the digit. This relationship will be applied to other numbering systems to convert their numbers to decimal equivalents.

Conversions from octal to decimal

Converting a number from one numbering system to another is simply a process by which the value of a number in one system (or base) is converted to its equivalent value in another system (or base). For example, with reference to Table 1.1, the value of the octal number 11 is 9 in decimal. Therefore, the decimal equivalent of 11_8 is 9_{10}. Extending the count in Table 1.1 to 77_8 and 63_{10}, the octal and decimal equivalents can be obtained. However, this is a very inconvenient way of performing conversions, especially with very large numbers.

Recall from our discussion earlier in this section that the value of a decimal number could be obtained by taking the sum-of-products of each digit and its positional weight. The positional weight of each digit was given by the relationship B^p, where B is the base of the numbering system and p is the position of the digit. This technique can also be applied to the octal numbering system to determine the (decimal) values of octal numbers. For example, consider the number 76251_8. The value of this number in decimal would be

$$(7 \times 8^4) + (6 \times 8^3) + (2 \times 8^2) + (5 \times 8^1) + (1 \times 8^0)$$
$$= (7 \times 4096) + (6 \times 512) + (2 \times 64) + (5 \times 8) + (1 \times 1)$$
$$= 28672 + 3072 + 128 + 40 + 1$$
$$= 31913_{10}.$$

Again, it takes a smaller decimal number (31913) to represent the same value as a larger octal number (76251).

Example 1.6

A woman represents herself as being 102_8 years old. What is her actual age? That is, what is her age in decimal?

Solution

$$102_8 = (1 \times 8^2) + (0 \times 8^1) + (2 \times 8^0)$$
$$= 64 + 0 + 2$$
$$= 66_{10}$$

The woman is thus 66 years old.

Example 1.7

Convert the following numbers from octal to decimal:
(a) 6351_8 (b) 24073_8 (c) 791_8

Solutions

(a) $6351_8 = 6 \times 8^3 + 3 \times 8^2 + 5 \times 8^1 + 1 \times 8^0$
$\qquad\quad = 6 \times 512 + 3 \times 64 + 5 \times 8 + 1 \times 1$
$\qquad\quad = 3072 + 192 + 40 + 1$
$\qquad\quad = 3305_{10}$

(b) $24073_8 = 2 \times 8^4 + 4 \times 8^3 + 0 \times 8^2 + 7 \times 8^1 + 3 \times 8^0$
$\qquad\qquad = 2 \times 4096 + 4 \times 512 + 0 \times 64 + 7 \times 8 + 3 \times 1$
$\qquad\qquad = 8192 + 2048 + 0 + 56 + 3$
$\qquad\qquad = 10299_{10}$

(c) No solution possible. 791_8 is not a valid octal number because 9 is not an octal digit.

Conversions from hexadecimal to decimal

To convert from hexadecimal to decimal, apply the same concepts studied earlier in this section. The procedure used is similar to the octal-to-decimal conversion procedure. To convert a number from hexadecimal to decimal, take the sum-of-products of each digit and its positional weight. The positional weight of each digit in the hexadecimal numbering system is 16^p, where p is the position of each digit in the number to be converted. For example, consider the number 6251_{16}. The value of this number in decimal would be

$$
\begin{aligned}
(6 \times 16^3) &+ (2 \times 16^2) + (5 \times 16^1) + (1 \times 16^0) \\
&= (6 \times 4096) + (2 \times 256) + (5 \times 16) + (1 \times 1) \\
&= 24576 + 512 + 80 + 1 \\
&= 25169_{10}.
\end{aligned}
$$

Again, it takes a larger decimal number (25169) to represent the same value as a smaller hexadecimal number (6251).

Example 1.8

A woman represents herself as being 23_{16} years old. What is her actual age? That is, what is her age in decimal?

Solution

$$
\begin{aligned}
23_{16} &= (2 \times 16^1) + (3 \times 16^0) \\
&= 32 + 3 \\
&= 35_{10}
\end{aligned}
$$

The woman is 35 years old.

Example 1.9

Convert the following numbers from hexadecimal to decimal
(a) $6A51_{16}$ (b) $240F3_{16}$ (c) BAD_{16}

Solutions

(a) $6A51_{16} = 6 \times 16^3 + A \times 16^2 + 5 \times 16^1 + 1 \times 16^0$
$= 6 \times 4096 + A \times 256 + 5 \times 16 + 1 \times 1$
(*Note:* Because A has a decimal value of 10, A $\times$ 256 is evaluated as 10×256.)
$= 24576 + 2560 + 80 + 1$
$= 27217_{10}$

(b) $240F3_{16} = 2 \times 16^4 + 4 \times 16^3 + 0 \times 16^2 + F \times 16^1 + 3 \times 16^0$
$= 2 \times 65{,}536 + 4 \times 4096 + 0 \times 256 + F \times 16 + 3 \times 1$

(*Note:* Because F has a decimal value of 15, F $\times$ 16 is evaluated as 15×16.)
$= 131072 + 16384 + 0 + 240 + 3$
$= 147699_{10}$

(c) $BAD_{16} = B \times 16^2 + A \times 16^1 + D \times 16^0$
$= 11 \times 256 + 10 \times 16 + 13 \times 1$
$= 2816 + 160 + 13$
$= 2989_{10}$

Conversions from binary to decimal

To convert from binary to decimal, apply the concepts studied earlier. The procedure used is similar to the octal-to-decimal and hexadecimal-to-decimal conversion procedures. To convert a number from binary to decimal, take the sum-of-products of each bit and its positional weight. The positional weight of each bit in the binary numbering system is 2^p, where p is the position of each bit in the number to be converted. For example, consider the number 11011_2. The value of this number in decimal would be

$$(1 \times 2^4) + (1 \times 2^3) + (0 \times 2^2) + (1 \times 2^1) + (1 \times 2^0)$$
$$= (1 \times 16) + (1 \times 8) + (0 \times 4) + (1 \times 2) + (1 \times 1)$$
$$= 16 + 8 + 0 + 2 + 1$$
$$= 27_{10}$$

Again, it takes a smaller decimal number (27) to represent the same value as a larger binary number (11011).

Example 1.10

A man represents himself as being 11000_2 years old. What is his actual age? That is, what is his age in decimal?

Solution

$$11000_2 = (1 \times 2^4) + (1 \times 2^3) + (0 \times 2^2) + (0 \times 2^1) + (0 \times 2^0)$$
$$= 16 + 8 + 0 + 0 + 0$$
$$= 24_{10}$$

The man is 24 years old.

Table 1.5 Powers of 2 from 0 to 15

n	2^n
0	1
1	2
2	4
3	8
4	16
5	32
6	64
7	128
8	256
9	512
10	1024
11	2048
12	4096
13	8192
14	16,384
15	32,768

The conversions from binary to decimal can be greatly simplified by simply adding up the powers of 2 for bit positions that are 1's. For example, consider the binary number 10100110_2.

number:	1	0	1	0	0	1	1	0
	↑	↑	↑	↑	↑	↑	↑	↑
position:	7	6	5	4	3	2	1	0
weight:	2^7	2^6	2^5	2^4	2^3	2^2	2^1	2^0
	128	64	32	16	8	4	2	1

To convert the number 10100110_2 to decimal, simply add up all the weights corresponding to the 1 bits (because if the bit is 0 then the product of the bit and the weight is 0)

$$128 + 32 + 4 + 2 = 166_{10}$$

Table 1.5 lists the powers of 2 for bit positions 0 through 15 as a convenience in the conversion of binary numbers up to 16 bits long. Appendix A lists all powers of 2 from 0 to 63.

Example 1.11

Convert the following numbers from binary to decimal using Table 1.5:
(a) 1001010010_2 (b) 1111_2 (c) 01001011_2

Solutions

(a)

1	0	0	1	0	1	0	0	1	0
↑			↑		↑			↑	
512			64		16			2	

$$512 + 64 + 16 + 2 = 594_{10}$$

(b)
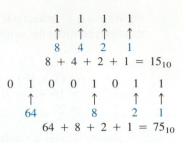

$$8 + 4 + 2 + 1 = 15_{10}$$

(c)   0   1   0   0   1   0   1   1

           ↑   ↑   ↑ ↑

      64   8   2 1

$$64 + 8 + 2 + 1 = 75_{10}$$

Review Questions

1. Why are the decimal numbers 23 and 32 different in value when they both have the same digits?

2. How are positional weights used to convert a number from octal, hexadecimal, and binary, to decimal? Give an example.

1.4
Conversions from decimal

A technique can be developed for working the positional weight system in reverse. That is, given a decimal number that represents a value, the digits that make up the number can be obtained by a series of repetitive divisions by the base of the numbering system. For example, to determine the digits that make up the decimal number 125, first divide 125 by 10

$$125 \div 10 = 12 \text{ with a remainder of } 5$$

Note that the remainder 5 is the LSD.
Next we take the result 12 and again divide by 10

$$12 \div 10 = 1 \text{ with a remainder of } 2$$

Note that the remainder 2 is the next digit.
Because 1 cannot be divided any further, it is the final remainder that is the MSD.
    To illustrate this procedure further, consider the following example.

Example 1.12

Determine the digits that make up the decimal number 8475 by using the repetitive division procedure.

Solution

10	8475	5	⟵ remainder
10	847	7	⟵ remainder
10	84	4	⟵ remainder
	8		⟵ remainder

The digits that make up the decimal number 8475 are given by the remainders (reading bottom to top) 8, 4, 7, 5.

This procedure can also be applied to the other numbering systems to facilitate conversions from decimal to numbers in other numbering systems.

Conversions from decimal to octal

To convert from decimal to octal, the process is similar to the one discussed earlier in this section. That is, obtain the individual digits that make up the number by repetitively dividing the decimal number by 8 (the base of the octal numbering system). The remainders for each successive division will make up the octal number. The following example illustrates this procedure.

Example 1.13

Convert the decimal number (a) 3305 to octal, (b) 10299 to octal.

Solutions

(a)

8	3305	1	⟵ remainder
8	413	5	⟵ remainder
8	51	3	⟵ remainder
	6		⟵ remainder

Reading the remainders from bottom to top, the octal equivalent is 6351_8.

(b)

8	10299	3	⟵ remainder
8	1287	7	⟵ remainder
8	160	0	⟵ remainder
8	20	4	⟵ remainder
	2		⟵ remainder

Reading the remainders from bottom to top, the octal equivalent is 24073_8.

Conversion from decimal to hexadecimal

The process of repetitive division can also be used to convert from decimal to hexadecimal. That is, obtain the individual digits that make up the number by repetitively dividing the decimal number by 16 (the base of the hexadecimal numbering system). The remainders for each successive division will make up the hexadecimal number. The following example illustrates this procedure.

Example 1.14

Convert the decimal number (a) 3305 to hexadecimal, (b) 10299 to hexadecimal.

Solutions

(a)

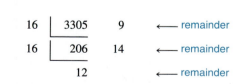

16	3305	9	⟵ remainder
16	206	14	⟵ remainder
	12		⟵ remainder

Note: The hexadecimal digit C represents the decimal number 12, and the hexadecimal digit E represents the decimal number 14. Therefore, the last remainder 12 will be expressed as the hexadecimal digit C and the second remainder 14 will be expressed as the hexadecimal digit E.

Reading the remainders from bottom to top, the hexadecimal equivalent is $CE9_{16}$.

(b)

$$
\begin{array}{r|l l}
16 & 10299 & 11 \quad \longleftarrow \text{remainder} \\
16 & 643 & 3 \quad \longleftarrow \text{remainder} \\
16 & 40 & 8 \quad \longleftarrow \text{remainder} \\
 & 2 & \quad \longleftarrow \text{remainder}
\end{array}
$$

Reading the remainders from bottom to top, the hexadecimal equivalent is $283B_{16}$.

Conversions from decimal to binary

To convert from decimal to binary, the process is similar to the decimal-to-octal and decimal-to-hexadecimal conversions covered earlier. That is, obtain the individual digits that make up the number by repetitively dividing the decimal number by 2 (the base of the binary numbering system). The remainders for each successive division will make up the binary number. The following example illustrates this procedure.

Example 1.15

Convert the decimal number (a) 33 to binary, (b) 624 to binary.

Solutions

(a)

$$
\begin{array}{r|l l}
2 & 33 & 1 \quad \longleftarrow \text{remainder} \\
2 & 16 & 0 \quad \longleftarrow \text{remainder} \\
2 & 8 & 0 \quad \longleftarrow \text{remainder} \\
2 & 4 & 0 \quad \longleftarrow \text{remainder} \\
2 & 2 & 0 \quad \longleftarrow \text{remainder} \\
 & 1 & \quad \longleftarrow \text{remainder}
\end{array}
$$

Reading the remainders from bottom to top, the binary equivalent is 100001_2.

(b)

$$
\begin{array}{r|l l}
2 & 624 & 0 \quad \longleftarrow \text{remainder} \\
2 & 312 & 0 \quad \longleftarrow \text{remainder} \\
2 & 156 & 0 \quad \longleftarrow \text{remainder} \\
2 & 78 & 0 \quad \longleftarrow \text{remainder} \\
2 & 39 & 1 \quad \longleftarrow \text{remainder} \\
2 & 19 & 1 \quad \longleftarrow \text{remainder} \\
2 & 9 & 1 \quad \longleftarrow \text{remainder} \\
2 & 4 & 0 \quad \longleftarrow \text{remainder} \\
2 & 2 & 0 \quad \longleftarrow \text{remainder} \\
 & 1 & \quad \longleftarrow \text{remainder}
\end{array}
$$

Reading the remainders from bottom to top, the binary equivalent is 1001110000_2.

Review Questions

1. In using the repetitive division procedure to convert from decimal to any one of the three numbering systems, what should the number to be converted be divided by?

2. Discuss the prodedure used to convert a number from decimal to any one of the three other numbering systems — octal, hexadecimal, and binary. Give examples.

1.5
Representation of binary numbers

Digital computers and many digital circuits are designed to process and store binary numbers. As discussed earlier, binary numbers representing relatively small values tend to be quite large in size. Because one often has to deal with these binary numbers during digital design, it is convenient to have an easy means of representing these large binary numbers. The designer can work with the representation of the binary number rather than the binary number itself, and switch back and forth between the binary number and its representation as needed. Decimal numbers could be used to represent binary numbers. For example, the binary number 10100001 could be represented by its decimal equivalent, 161. However, this involves a complicated conversion process between decimal and binary. For instance, it would be quite difficult to switch back and forth between binary numbers such as 1010100101010110110 and their decimal representations. Hexadecimal and octal numbers are more suitable for representing binary numbers. This section examines the manner in which conversions can be easily performed between binary and hexadecimal and octal numbers. This section concludes by examining how decimal numbers can be represented by binary numbers for storage in digital computers.

Binary-to-octal and octal-to-binary conversions

It was mentioned in Section 1.4 that the hexadecimal and octal numbering systems are ideal for representing groups of binary numbers because the digits in each system can perfectly represent all possible combinations of 3 bits (octal) or all possible combinations of 4 bits (hexadecimal) (*see* Table 1.4). The capability of the hexadecimal and octal numbering systems to perfectly represent groups of binary bits also makes conversions very simple. For example, consider the binary number, 100111010011. If this number is broken up into groups of 3 bits, starting at the LSB, and the octal digit that represents each group is substituted, the octal equivalent of the binary number will be obtained. That is, binary was converted to octal. Table 1.6 lists the eight octal digits and their 3-bit binary equivalents.

Table 1.6 Binary representation of the octal digits

Binary	Octal
000	0
001	1
010	2
011	3
100	4
101	5
110	6
111	7

$$\underbrace{1\ 0\ 0}_{4}\quad \underbrace{1\ 1\ 1}_{7}\quad \underbrace{0\ 1\ 0}_{2}\quad \underbrace{0\ 1\ 1}_{3}$$

Therefore, 100111010011_2 is equivalent to 4723_8.

To verify that the conversion is valid, convert the binary number 100111010011_2 to decimal

$$100111010011_2 = 1 + 2 + 16 + 64 + 128 + 256 + 2048$$
$$= 2515_{10}$$

and then convert the octal number 4723_8 to decimal

$$4723_8 = 4 \times 8^3 + 7 \times 8^2 + 2 \times 8^1 + 3 \times 8^0$$
$$= 2515_{10}.$$

Because both numbers yield the same decimal equivalent, the two numbers are equivalent.

Example 1.16

Convert the following binary numbers to octal:
(a) 110001010111   (b) 10100110   (c) 1   (d) 0001011000

Solutions

(a)
$$\underbrace{1\ 1\ 0}_{6}\ \underbrace{0\ 0\ 1}_{1}\ \underbrace{0\ 1\ 0}_{2}\ \underbrace{1\ 1\ 1}_{7}$$

Therefore, $110001010111_2 = 6127_8$.

(b)

leading zero added

$$\longrightarrow \underbrace{0\ 1\ 0}_{2}\ \underbrace{1\ 0\ 0}_{4}\ \underbrace{1\ 1\ 0}_{6}$$

Therefore, $10100110_2 = 246_8$.

(c) $1_2 = 1_8$

(d)
$$\underbrace{0\ 0\ 1}_{1}\ \underbrace{0\ 1\ 1}_{3}\ \underbrace{0\ 0\ 0}_{0}$$

Therefore, $0001011000_2 = 130_8$.

To convert from octal to binary, the process is simply reversed. That is, substitute the 3-bit binary code for each octal digit. For instance, *see* Example 1.17.

Example 1.17

Convert the following octal numbers to binary:
(a) 73645   (b) 0347   (c) 64580   (d) 2341034   (e) 10101

Solutions

(a)
$$\begin{array}{ccccc} 7 & 3 & 6 & 4 & 5 \\ \overbrace{111} & \overbrace{011} & \overbrace{110} & \overbrace{100} & \overbrace{101} \end{array}$$

Therefore, $73645_8 = 111011110100101_2$.

(b)
$$\begin{array}{cccc} 0 & 3 & 4 & 7 \\ \overbrace{000} & \overbrace{011} & \overbrace{100} & \overbrace{111} \end{array}$$

Therefore, $0347_8 = 11100111_2$.

(c) Solution not possible—8 is not an octal digit!

(d)
$$\begin{array}{ccccccc} 2 & 3 & 4 & 1 & 0 & 3 & 4 \\ \overbrace{010} & \overbrace{011} & \overbrace{100} & \overbrace{001} & \overbrace{000} & \overbrace{011} & \overbrace{100} \end{array}$$

Therefore, $2341034_8 = 10011100001000011100_2$.

(e)
$$1 \quad 0 \quad 1 \quad 0 \quad 1$$
$$001 \quad 000 \quad 001 \quad 000 \quad 001$$

Therefore, $10101_8 = 1000001000001_2$.

Binary-to-hexadecimal and hexadecimal-to-binary conversions

Table 1.7   Binary representation of the hexadecimal digits

Binary	Hexadecimal
0000	0
0001	1
0010	2
0011	3
0100	4
0101	5
0110	6
0111	7
1000	8
1001	9
1010	A
1011	B
1100	C
1101	D
1110	E
1111	F

For the same reasons stated earlier, the same techniques can be applied to convert from binary to hexadecimal and from hexadecimal to binary. However, because there are 16 hexadecimal digits and because this set perfectly represents all possible combinations of 4 bits, as shown in Table 1.7, the conversion must be performed with 4-bit groups. For example, consider the binary number, 100111010011. If this number is broken up into groups of 4 bits, starting at the LSB, and the hexadecimal digit that represents each group is substituted, the hexadecimal equivalent of the binary number is obtained. That is, binary was converted to hexadecimal.

$$1\ 0\ 0\ 1 \quad 1\ 1\ 0\ 1 \quad 0\ 0\ 1\ 1$$
$$9 \qquad\quad D \qquad\quad 3$$

Therefore, 100111010011_2 is equivalent to $9D3_{16}$.

To verify that the conversion is valid, convert the binary number 100111010011_2 to decimal

$$100111010011_2 = 1 + 2 + 16 + 64 + 128 + 256 + 2048$$
$$= 2515_{10}$$

and then convert the hexadecimal number $9D3_{16}$ to decimal

$$9D3_{16} = 9 \times 16^2 + D \times 16^1 + 3 \times 16^0$$
$$= 2515_{10}.$$

Since both numbers yield the same decimal equivalent, the two numbers are equivalent.

Example 1.18

Convert the following binary numbers to hexadecimal:
(a) 110001010111   (b) 10100110   (c) 1   (d) 0001011000

Solutions

(a)
$$1\ 1\ 0\ 0 \quad 0\ 1\ 0\ 1 \quad 0\ 1\ 1\ 1$$
$$C \qquad\quad 5 \qquad\quad 7$$

Therefore, $110001010111_2 = C57_{16}$.

(b)
$$1\ 0\ 1\ 0 \quad 0\ 1\ 1\ 0$$
$$A \qquad\quad 6$$

Therefore, $10100110_2 = A6_{16}$.

(c)
$$1_2 = 1_{16}$$

(d)
$$0\ 0\ \underbrace{0\ 1\ 0\ 1}_{5}\ \underbrace{1\ 0\ 0\ 0}_{8}$$

Therefore, $0001011000_2 = 58_{16}$.

To convert from hexadecimal to binary, the process is simply reversed. That is, substitute the 4-bit binary code for each hexadecimal digit (*see* Example 1.19).

Example 1.19

Convert the following hexadecimal numbers to binary:
(a) 93E45   (b) 0347   (c) 645G0   (d) 2341034   (e) FACE

Solutions

(a)
$$\underbrace{9}_{1001}\ \underbrace{3}_{0011}\ \underbrace{E}_{1110}\ \underbrace{4}_{0100}\ \underbrace{5}_{0101}$$

Therefore, $93E45_{16} = 10010011111001000101_2$.

(b)
$$\underbrace{0}_{0000}\ \underbrace{3}_{0011}\ \underbrace{4}_{0100}\ \underbrace{7}_{0111}$$

Therefore, $0347_{16} = 1101000111_2$.

(c) Solution not possible — G is not a hexadecimal digit!

(d)
$$\underbrace{2}_{0010}\ \underbrace{3}_{0011}\ \underbrace{4}_{0100}\ \underbrace{1}_{0001}\ \underbrace{0}_{0000}\ \underbrace{3}_{0011}\ \underbrace{4}_{0100}$$

Therefore, $2341034_{16} = 1000110100000100000011010 0_2$.

(e)
$$\underbrace{F}_{1111}\ \underbrace{A}_{1010}\ \underbrace{C}_{1100}\ \underbrace{E}_{1110}$$

Therefore, $FACE_{16} = 1111101011001110_2$.

Octal-to-hexadecimal and hexadecimal-to-octal conversions

The simplest way to convert between the octal and hexadecimal numbering systems is to perform a binary conversion as an intermediate step. That is, first convert from one base to binary, and then regroup the bits and convert to the other base. To illustrate this procedure, consider the following examples.

Example 1.20

Convert the following octal numbers to hexadecimal:
(a) 51   (b) 64320   (c) 100011

Solutions

(a)
$$\underbrace{5}_{101}\ \underbrace{1}_{001}$$

Regroup the number 101001_2 into groups of 4 bits and add two leading zeros
$$\underbrace{0010}_{2}\ \underbrace{1001}_{9}$$

Therefore, $51_8 = 29_{16}$.

(b)

$$\begin{array}{ccccc} 6 & 4 & 3 & 2 & 0 \\ \overbrace{110} & \overbrace{100} & \overbrace{011} & \overbrace{010} & \overbrace{000} \end{array}$$

Regroup the number 110100011010000_2 into groups of 4 bits and add one leading zero

$$\begin{array}{cccc} \underbrace{0110} & \underbrace{1000} & \underbrace{1101} & \underbrace{0000} \\ 6 & 8 & D & 0 \end{array}$$

Therefore, $64320_8 = 68D0_{16}$.

(c)

$$\begin{array}{cccccc} 1 & 0 & 0 & 0 & 1 & 1 \\ \overbrace{001} & \overbrace{000} & \overbrace{000} & \overbrace{000} & \overbrace{001} & \overbrace{001} \end{array}$$

Regroup the number 001000000000001001_2 into groups of 4 bits and discard the two leading zeros

$$\begin{array}{cccc} \underbrace{1000} & \underbrace{0000} & \underbrace{0000} & \underbrace{1001} \\ 8 & 0 & 0 & 9 \end{array}$$

Therefore, $100011_8 = 8009_{16}$.

Example 1.21

Convert the following hexadecimal numbers to octal:
(a) 29 (b) 68D0 (c) 8009

Solutions

(a)

$$\begin{array}{cc} 2 & 9 \\ \overbrace{0010} & \overbrace{1001} \end{array}$$

Regroup the number 00101001_2 into groups of 3 bits and discard the two leading zeros

$$\begin{array}{cc} \underbrace{101} & \underbrace{001} \\ 5 & 1 \end{array}$$

Therefore, $29_{16} = 51_8$.

(b)

$$\begin{array}{cccc} 6 & 8 & D & 0 \\ \overbrace{0110} & \overbrace{1000} & \overbrace{1101} & \overbrace{0000} \end{array}$$

Regroup the number 0110100011010000_2 into groups of 3 bits and discard one leading zero

$$\begin{array}{ccccc} \underbrace{110} & \underbrace{100} & \underbrace{011} & \underbrace{010} & \underbrace{000} \\ 6 & 4 & 3 & 2 & 0 \end{array}$$

Therefore, $68D0_{16} = 64320_8$.

(c)

$$\begin{array}{cccc} 8 & 0 & 0 & 9 \\ \overbrace{1000} & \overbrace{0000} & \overbrace{0000} & \overbrace{1001} \end{array}$$

Regroup the number 1000000000001001_2 into groups of 3 bits and add two leading zeros

$$\begin{array}{cccccc} \underbrace{001} & \underbrace{000} & \underbrace{000} & \underbrace{000} & \underbrace{001} & \underbrace{001} \\ 1 & 0 & 0 & 0 & 1 & 1 \end{array}$$

Therefore, $8009_{16} = 100011_8$.

The main purpose of the octal and hexadecimal numbering system is in the *representation* of binary numbers. Because it is easy to switch back and forth between binary and octal/hexadecimal numbers, there is a convenient way of represent these rather large binary numbers. For example, instead of working with numbers such as

$$1000101001001111100101001_2$$

equivalents such as $8A4F29_{16}$ or 42447451_8 are more convenient to "carry around" and are less prone to error. However, it is important to remember that these numbers are only convenient representations of binary numbers. Using decimal numbers to represent binary numbers is possible but is time-consuming and involves a lot of arithmetic. The selection of either the octal or hexadecimal representation of binary numbers is a matter of choice. However, some digital circuit applications that process binary numbers are more suited to octal representation whereas others are better suited to hexadecimal representation.

Binary coded decimal

At this point it is important to note that the substitution technique used to convert from octal and hexadecimal to binary and vice versa cannot be used for decimal-to-binary and binary-to-decimal conversions. As previously stated, this is because the set of decimal digits cannot perfectly represent all possible combinations of 3 or 4 bits. There are ten decimal digits, whereas there are 2^3 or 8 combinations of 3 bits, and 2^4 or 16 combinations of 4 bits. It may, however, appear to work in some cases, but the results are erroneous. For example, consider the binary number, 10010111. Using the substitution technique

$$\underbrace{1\ 0\ 0\ 1}_{9}\ \underbrace{0\ 1\ 1\ 1}_{7}$$

or the number 97, which is actually a hexadecimal number. The decimal equivalent of 10010111

$$10010111_2 = 1 + 2 + 4 + 16 + 128 = 151$$

is 151 and not 97.

Furthermore, observe that there absolutely is no way to convert a number such as 11001110_2 to decimal using the substitution technique, because there is no *single* decimal digit that can represent the binary combinations 1100 and 1110. In fact, there are no decimal digits to represent any of the six combinations, 1010, 1011, 1100, 1101, 1110, 1111. Using two decimal digits to represent each combination is not permitted because it will not be possible to reconvert the number to its original value

Table 1.8 Binary representation of the decimal digits (BCD)

Binary	Decimal
0000	0
0001	1
0010	2
0011	3
0100	4
0101	5
0110	6
0111	7
1000	8
1001	9

$$\underbrace{1\ 1\ 0\ 0}_{12}\ \underbrace{1\ 1\ 1\ 0}_{14}$$

does not match

$$\underbrace{1}_{0001}\ \underbrace{2}_{0010}\ \underbrace{1}_{0001}\ \underbrace{4}_{0100}$$

This substitution technique can, however, be used to *represent* decimal digits (but not to *convert* from decimal to binary) in binary form. When a 4-bit binary number is substituted for a decimal digit, this 4-bit binary number is called a *binary coded decimal* (BCD) number. Table 1.8 lists the 10 decimal digits and their BCD values. The following examples illustrate the procedures used for converting from decimal to BCD and from BCD to decimal.

Example 1.22

Convert the following decimal numbers to their BCD representations:
(a) 63547　(b) 12　(c) 10101　(d) 0398

Solutions

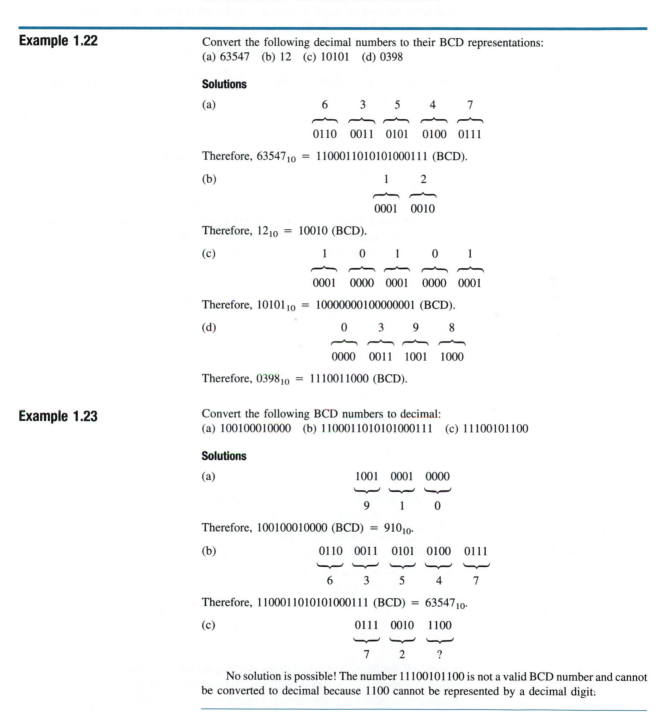

(a)

6	3	5	4	7
0110	0011	0101	0100	0111

Therefore, 63547_{10} = 1100011010101000111 (BCD).

(b)

1	2
0001	0010

Therefore, 12_{10} = 10010 (BCD).

(c)

1	0	1	0	1
0001	0000	0001	0000	0001

Therefore, 10101_{10} = 10000000100000001 (BCD).

(d)

0	3	9	8
0000	0011	1001	1000

Therefore, 0398_{10} = 1110011000 (BCD).

Example 1.23

Convert the following BCD numbers to decimal:
(a) 100100010000　(b) 1100011010101000111　(c) 11100101100

Solutions

(a)

1001	0001	0000
9	1	0

Therefore, 100100010000 (BCD) = 910_{10}.

(b)

0110	0011	0101	0100	0111
6	3	5	4	7

Therefore, 1100011010101000111 (BCD) = 63547_{10}.

(c)

0111	0010	1100
7	2	?

No solution is possible! The number 11100101100 is not a valid BCD number and cannot be converted to decimal because 1100 cannot be represented by a decimal digit.

Digital computers use two techniques to store decimal numbers in memory. The first involves converting a decimal number to binary and then storing the binary representation of the decimal number. An alternative technique involves converting the decimal number to BCD (using the substitution technique) and then storing the BCD representation of the decimal number. The former technique is more complicated because it involves a lot of arithmetic manipulation, the latter technique is preferable because it involves only a simple substitution process. For example, a digital computer can store the decimal number 7395 as

$$1110011100011$$

after converting from decimal to binary using the repetitive division procedure, or it can store the number as

$$111001110010101$$

after converting from decimal to BCD using the substitution procedure.

There are advantages and disadvantages in both procedures. BCD numbers generally require more storage, that is, they are usually larger than their binary equivalents. Performing arithmetic on BCD numbers is more complicated as will be seen in Section 1.9. However, conversions to and from BCD and decimal are much simpler than conversions to and from binary and decimal.

This section has introduced the concepts of binary numbers and their relationship to decimal, octal, and hexadecimal numbers. The binary numbering system may seem very cumbersome to work with and manipulate because of its limited set of digits. However, this binary set lends itself well to the operation of the digital circuits (many of which are designed to manipulate binary numbers) to be examined in the rest of this book. The octal and hexadecimal numbering systems are well suited to representing binary numbers because the conversions back and forth are very simple. As far as the decimal numbering system is concerned — we live in a decimal world and must therefore be able to switch back and forth between the decimal numbering system and the other numbering systems when dealing with digital circuits that interface to the "real world."

Review Questions

1. What is the purpose of the octal and hexadecimal numbering systems?
2. Why is the decimal numbering system not a good system for representing binary numbers?
3. Why can't we use the substitution procedure to convert from decimal to binary and binary to decimal?
4. What is the difference between the binary representation of decimal numbers and the BCD representation of decimal numbers?

1.6
Addition

Almost all mathematical calculations can be performed through the use of the four basic arithmetic operations — addition, subtraction, multiplication, and division. Furthermore, addition is particularly important because it can be used to perform the other three arithmetic operations. For example, two numbers can be subtracted by adding a negative number to a positive number or vice versa. Multiplication can be performed by repetitive addition and division by repetitive subtractions. Many digital computers perform all arithmetic operations through the process of addition.

Therefore, the techniques by which these operations are conducted are worth investigating.

Decimal addition

The simple process of adding decimal numbers is often taken for granted in everyday life. However, because the concepts of the decimal numbering system and decimal arithmetic operations will be applied to various other numbering systems, it is important to identify certain procedures and terms used when performing these operations.

Consider the addition of two decimal numbers, 7632 and 6489. The addition is performed by adding each pair of digits in both numbers starting with the LSDs, obtaining a sum of the two LSDs, keeping the LSD of this sum as the LSD of the final sum, and carrying over the MSD of the sum for addition to the next pair of digits. The process is then repeated for the pairs that follow until all pairs of digits have been added. This procedure can be illustrated in a less complicated manner as follows:

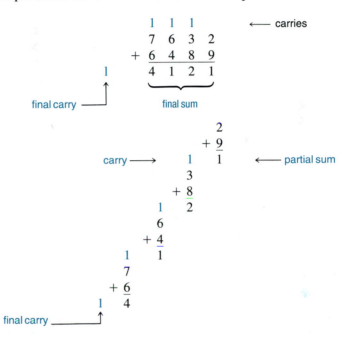

Because the addition of 2 four-digit numbers has been performed, the final sum is considered to be the four-digit number 4121 with a final carry of 1.

Binary addition

The binary addition procedure is similar to addition in any other numbering system. The only thing to remember is that there are only two digits (bits) in this numbering system, and therefore the results obtained when adding these bits are quite different. Table 1.9 lists the sums obtained by adding all possible combinations of 2 bits.

Table 1.9 Sums of all possible combinations of 2 bits

0	+	0	=	0	0
0	+	1	=	0	1
1	+	0	=	0	1
1	+	1	=	1	0

Table 1.10   Sums of all possible combinations of 3 bits

0	+	0	+	0	=	0	0
0	+	0	+	1	=	0	1
0	+	1	+	0	=	0	1
0	+	1	+	1	=	1	0
1	+	0	+	0	=	0	1
1	+	0	+	1	=	1	0
1	+	1	+	0	=	1	0
1	+	1	+	1	=	1	1

Observe in Table 1.9 that the sum (result) of the two numbers has been extended to 2 bits because the largest sum is 10 (2). Also note that $1 + 1$ is equal to 2_{10}, which is 10_2. Table 1.10 extends the arithmetic performed in Table 1.9 to 3 bits. Again we can see in Table 1.10, that because the largest sum is 11 (3), the result of adding $1 + 1 + 1$, only 2 bits are needed to represent the results. Note in Tables 1.9 and 1.10 that in order to list all possible combinations of 2 or 3 bits simply count in binary. Also, the MSB of the sum is referred to as the *carry bit*.

Example 1.24

Construct a table similar to Tables 1.9 and 1.10 that will list the sums of all possible combinations of 4 bits.

Solution

Because the largest sum will be $1 + 1 + 1 + 1$ or 100 (4) it is necessary to extend the result to 3 bits.

$$0 + 0 + 0 + 0 = 0\ 0\ 0$$
$$0 + 0 + 0 + 1 = 0\ 0\ 1$$
$$0 + 0 + 1 + 0 = 0\ 0\ 1$$
$$0 + 0 + 1 + 1 = 0\ 1\ 0$$
$$0 + 1 + 0 + 0 = 0\ 0\ 1$$
$$0 + 1 + 0 + 1 = 0\ 1\ 0$$
$$0 + 1 + 1 + 0 = 0\ 1\ 0$$
$$0 + 1 + 1 + 1 = 0\ 1\ 1$$
$$1 + 0 + 0 + 0 = 0\ 0\ 1$$
$$1 + 0 + 0 + 1 = 0\ 1\ 0$$
$$1 + 0 + 1 + 0 = 0\ 1\ 0$$
$$1 + 0 + 1 + 1 = 0\ 1\ 1$$
$$1 + 1 + 0 + 0 = 0\ 1\ 0$$
$$1 + 1 + 0 + 1 = 0\ 1\ 1$$
$$1 + 1 + 1 + 0 = 0\ 1\ 1$$
$$1 + 1 + 1 + 1 = 1\ 0\ 0$$

So far we have looked at single-bit addition. The next step is to extend this process to the addition of multibit binary numbers. Recall from our discussion at the beginning of this section that to add decimal numbers, simply add the LSDs of each number, obtain a single-digit sum and carry, and then add the carry to the next set of digits in sequence, continuing this process to the MSD. The same procedure is used to add multibit binary numbers. For example, consider the addition of the binary number 1001 (9) and 0101 (5) to produce a result of 1110 (14 or E_{16})

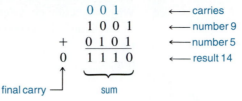

Starting with the LSB, $1 + 1$ produces 10, or 0 with a carry of 1. This carry is then added to the next position, $1 + 0 + 0$, to obtain 01 (or 1 with a carry of 0). The carry along with the next 2 bits in position are then added $0 + 0 + 1$, to obtain a result of 01 (or 1 with a carry of 0). Finally, this carry is then added to the two MSBs, $0 + 1 + 0$, to obtain the MSB of the result, 1, and a final carry of 0. The largest number being added is 9 (1001), which is a 4-bit number. Therefore, the addition has been extended to 4 bits. However, the result is a 5-bit number when the final carry is included. For the addition of two 4-bit numbers the result will never be greater than 5 bits. To illustrate this, consider the addition of the two largest possible 4-bit numbers 1111 (15 or F_{16}),

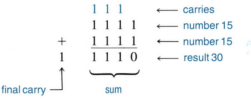

The result of the addition of 1111 (15) and 1111 (15) produces the binary number 11110 (30). The following examples further illustrate the technique of binary addition.

Example 1.25

Perform the following additions in binary arithmetic:
(a) $7 + 2$ (b) $20 + 32$ (c) $3E_{16} + 75_{16}$ (d) $324_8 + 106_8$

Solutions

(a)
```
                    1 0      ← carries
         7          1 1 1
       + 2        + 0 1 0
         9        1 0 0 1
```

(b)
```
                  0 0 0 0 0    ← carries
        20        0 1 0 1 0 0
      + 32      + 1 0 0 0 0 0
        52      0 1 1 0 1 0 0
```

(c)
```
                  1 1 1 1 0 0    ← carries
        3E        0 1 1 1 1 1 0
      + 75      + 1 1 1 0 1 0 1
        B3      1 0 1 1 0 0 1 1
```

(d)
```
                  1 0 0 0 1 0 0    ← carries
       324        1 1 0 1 0 1 0 0
     + 106      + 0 1 0 0 0 1 1 0
       432      1 0 0 0 1 1 0 1 0
```

Review Questions

1. Why is addition considered to be the most important arithmetic operation as far as digital computers are concerned?

2. Discuss the similarities and differences between binary and decimal addition.

1.7

Subtraction

It was stated earlier that addition is perhaps the most important arithmetic operation performed by a digital computer because the other three operations (subtraction, multiplication, and division) are often performed through addition, or a series of high-speed repetitive additions. This section will examine the procedure to subtract two numbers without performing a conventional subtraction but by adding a positive number to a negative number. Again, because the decimal numbering system is most familiar, it will be used as a model to examine the procedure to perform this subtraction. The concepts learned will then be applied to the binary numbering system.

Decimal subtraction —
9's complement

The *9's complement* technique of subtracting decimal numbers allows us to subtract by adding a positive number to a negative number and vice versa. To apply this technique of subtraction, first define the 9's complement of a decimal number.

The *9's complement* of a decimal number can be used to represent a negative number. It is obtained by subtracting each digit in the number from 9. For example, the 9's complement of the number 2 is 7, and the 9's complement of the number 6 is 3. A *signed* negative decimal number can be represented as an *unsigned* 9's complement representation. For example, the 9's complement representation of the number -5 is 4. Here 4 is not a positive number but is an unsigned representation of the number -5. The unsigned number 4 and the signed number -5 are *equivalent* and *complementary*. It can be seen that it is very easy to confuse negative unsigned numbers represented in 9's complement form with positive numbers. The 9's complement of multidigit decimal numbers can be obtained by subtracting each digit in the number from 9.

Example 1.26

Represent the following signed negative numbers in 9's complement form:
(a) -1 (b) -8 (c) -4 (d) -9 (e) -72 (f) -4218

Solutions

(a) $9 - 1 = 8$
Therefore, the 9's complement representation of -1 is 8.

(b) $9 - 8 = 1$
Therefore, the 9's complement representation of -8 is 1.

(c) $9 - 4 = 5$
Therefore, the 9's complement representation of -4 is 5.

(d) $9 - 9 = 0$
Therefore, the 9's complement representation of -9 is 0.

(e) $99 - 72 = 27$
Therefore, the 9's complement representation of -72 is 27.

(f) $9999 - 4218 = 5781$
Therefore, the 9's complement representation of -4218 is 5781.

The word "complement" in the term "9's complement" implies that the same operation can be applied in reverse. For example, if the 9's complement of 6 is 3, then the 9's complement of 3 is 6. Similarly, if the 9's complement of 72 is 27, then the 9's complement of 27 is 72. This allows one to very easily switch back and forth between a signed negative number and an unsigned negative number in 9's complement representation. For example, the number 38,743 is the 9's complement representation of some negative number. To obtain its actual (or signed) value, calculate its 9's complement

$$\begin{array}{r} 99{,}999 \\ -\ 38{,}743 \\ \hline 61{,}256 \end{array}$$

Therefore, the actual value of the number is $-61{,}256$.

Example 1.27

The following is a list of negative numbers expressed in 9's complement form. Determine the signed representation of each number.
(a) 937,569 (b) 8 (c) 54 (d) 734,958,457

Solutions

(a) 9's complement of 937,569 is $999{,}999 - 937{,}569 = 62{,}430$.
Therefore, the signed number is $-62{,}430$

(b) 9's complement of 8 is $9 - 8 = 1$
Therefore, the signed number is -1.

(c) 9's complement of 54 is $99 - 54 = 45$
Therefore, the signed number is -45.

(d) 9's complement of 734,958,457 is
$999{,}999{,}999 - 734{,}958{,}457 = 265{,}041{,}542$
Therefore, the signed number is $-265{,}041{,}542$.

To obtain the difference between two numbers using the 9's complement subtraction technique, the following steps are performed:

1. Obtain the 9's complement representation of the negative number.

2. Add the 9's complement representation of the negative number (from step 1) to the positive number.

3. Add the final carry obtained in step 2 to the final sum obtained by adding the two numbers; this is known as the *end-around* carry. The sum obtained will be the final result (the difference between the two numbers).

To illustrate the procedure used in 9's complement arithmetic, consider the following examples:

Example 1.28

Perform the following subtractions using the 9's complement arithmetic technique:
(a) $6 - 4 = 2$ (b) $9 - 8 = 1$ (c) $42 - 39 = 3$
(d) $125 - 67 = 58$

Solutions

(a) The 9's complement representation of -4 is 5.

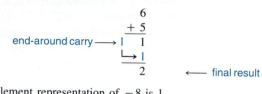

(b) The 9's complement representation of -8 is 1.

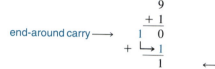

(c) The 9's complement representation of -39 is 60.

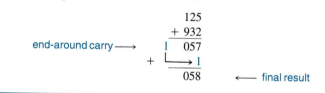

(d) In this problem because the largest number being operated on is a three-digit number, assume that both numbers are three digits long. Thus, $125 - 067 = 058$. The 9's complement representation of -067 is 932.

$$
\begin{array}{r}
125 \\
+ \ 932 \\
\hline
\end{array}
$$

end-around carry $\longrightarrow$ 1   057
$+ \ \ \llcorner\!\!\rightarrow 1$
058    $\longleftarrow$ final result

In all the subtractions in Example 1.28 the results were positive numbers and every addition produced an end-around carry of 1 that was added to the sum to obtain the final result. This is true in all cases of 9's complement arithmetic that yield positive results. However, when the 9's complement technique is performed on operations that yield negative results, the carry will always be 0, and the result will always be in unsigned 9's complement representation. Because the 9's complement representation of a negative number is unsigned, the carry is the only means by which the final result can be identifed as being positive or negative. Furthermore, if the result is negative, the result must be recomplemented to obtain its signed value. To illustrate this fact, consider the following examples.

Example 1.29

Perform the following subtractions using the 9's complement arithmetic technique:
(a) $4 - 6 = -2$ (b) $8 - 9 = -1$ (c) $39 - 42 = -3$
(d) $67 - 125 = -58$ (e) $7 - 7 = 0$

Solutions

(a) The 9's complement representation of -6 is 3.

end-around carry ⟶
(negative result)

⟵ final result in 9's complement representation

The 9's complement of 7 is 2. Therefore the signed result is -2.

(b) The 9's complement representation of -9 is 0.

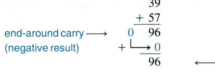

end-around carry ⟶
(negative result)

⟵ final result in 9's complement representation

The 9's complement of 8 is 1. Therefore, the signed result is -1.

(c) The 9's complement representation of -42 is 57.

end-around carry ⟶
(negative result)

⟵ final result in 9's complement representation

The 9's complement of 96 is 3. Therefore, the signed result is -3.

(d) Because the largest number being operated on is a three-digit number, assume that both numbers are three digits long. Thus, $067 - 125 = -058$.
The 9's complement representation of -125 is 874.

end-around carry ⟶
(negative result)

⟵ final result in 9's complement representation

The 9's complement of 941 is 58. Therefore, the signed result is -58.

(e) The 9's complement representation of -7 is 2.

```
        7
      + 2
    0   9
  + └→ 0
        9
```

end-around carry ⟶
(negative result)

⟵ final result in 9's complement representation

The 9's complement of 9 is 0. Therefore, the signed result is -0. Because a sign is not generally associated with the digit 0, the negative sign can be dropped in this case.

Note in all the above examples that even though the end-around carry is 0 in all the cases, it is still added in so that the same procedure is followed for both types of results. In Chapter 4 we will see that the circuit that exactly implements this procedure will always add in the end-around carry regardless of its value.

In all the examples worked so far, it can be seen that the 9's complement technique can be used to obtain the difference between two numbers without having to perform a conventional subtraction, simply by adding two numbers (ignoring the fact that a conventional subtraction was performed to obtain the 9's complement representation of the negative number). The advantage of performing subtraction in this manner will be seen later in this section and in Chapter 4.

Decimal subtraction — 10's complement

The 9's complement technique is just one of two techniques that may be used to obtain the difference between two numbers through the process of addition. The 10's complement is another technique that may be used to obtain the same results.

The 10's complement of a decimal number can also be used to represent a negative number. It is obtained by subtracting the number from 10. For example, the 10's complement of the number 2 is 8, and the 10's complement of the number 6 is 4. A signed negative decimal number can be represented as an unsigned 10's complement representation. For example, the 10's complement representation of the number -4 is 6. Here 6 is not a positive number but is an unsigned representation of the number -4. Another way of determining the 10's complement of a number is to obtain its 9's complement and then add 1 to it; this is the preferred way of obtaining the 10's complement because it is more applicable to the technique used in binary arithmetic (to be examined later) and in digital computer circuits.

Example 1.30

Represent the following signed negative numbers in 10's complement form:
(a) -1 (b) -8 (c) -4 (d) -9 (e) -72 (f) -4218

Solutions

(a) The 9's complement representation of -1 is 8.
Therefore, its 10's complement representation is $8 + 1$, or 9.

(b) The 9's complement representation of -8 is 1.
Therefore, its 10's complement representation is $1 + 1$, or 2.

(c) The 9's complement representation of -4 is 5.
Therefore, its 10's complement representation is $5 + 1$, or 6.

(d) The 9's complement representation of -9 is 0.
Therefore, its 10's complement representation is $0 + 1$, or 1.

(e) The 9's complement representation of -72 is 27.
Therefore, its 10's complement representation is $27 + 1$, or 28.

(f) The 9's complement representation of -4218 is 5781.
Therefore, its 10's complement representation is $5781 + 1$, or 5782.

As mentioned before, the word "complement" in the term "10's complement" implies that the same operation can be applied in reverse. For example, if the 10's complement of 6 is 4, then the 10's complement of 4 is 6. Similarly, if the 10's complement of 72 is 28, then the 10's complement of 28 is 72. We can also switch back and forth between a signed negative number and an unsigned negative number in 10's complement representation. For example, assume that the number 38,743 is the 10's complement representation of some negative number. To obtain its actual (or signed) value, calculate its 10's complement

$$
\begin{array}{r}
99{,}999 \\
-\ 38{,}743 \\
\hline
61{,}256 \\
+\ \phantom{61{,}25}1 \\
\hline
61{,}257
\end{array}
$$

Therefore, the actual value of the number is $-61{,}257$.

Example 1.31

The following is a list of negative numbers expressed in 10's complement form. Determine the signed representation of each number.
(a) 937,569 (b) 8 (c) 54 (d) 734,958,457

Solutions

(a) 999,999 − 937,569 = 62,430 (9's complement)
    62,430 + 1 (10's complement)
    Therefore, the signed number is −62,431.

(b) 9 − 8 = 1 (9's complement)
    1 + 1 (10's complement)
    Therefore, the signed number is −2.

(c) 99 − 54 = 45 (9's complement)
    45 + 1 (10's complement)
    Therefore, the signed number is −46.

(d) 999,999,999 − 734,958,457 = 265,041,542 (9's complement)
    265,041,542 + 1 (10's complement)
    Therefore, the signed number is −265,041,543.

To obtain the difference between two numbers using the 10's complement subtraction technique, do the following:

1. Obtain the 10's complement representation of the negative number.

2. Add the 10's complement representation of the negative number (from step 1) to the positive number.

3. Ignore the final carry obtained from step 2, but interpret the meaning of the carry. That is, a carry of 1 will indicate a positive result, and a carry of 0 will indicate a negative result. The sum obtained from step 2 will be the final result (the difference between the two numbers). If the result is negative it will be represented in unsigned 10's complement form.

To illustrate the procedure used in 10's complement arithmetic, consider the following examples.

Example 1.32

Perform the following subtractions using the 10's complement arithmetic technique:
(a) 6 − 4 = 2 (b) 9 − 8 = 1 (c) 42 − 39 = 3 (d) 125 − 67 = 58
(e) 7 − 7 = 0

Solutions

(a) The 10's complement representation of −4 is 6.

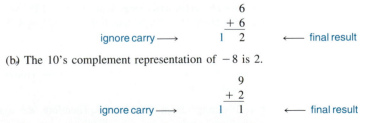

(b) The 10's complement representation of −8 is 2.

(c) The 10's complement representation of -39 is 61.

$$42$$
$$+\ 61$$

ignore carry $\longrightarrow$ 1   03 $\longleftarrow$ final result

(d) Because the largest number being operated on is a three-digit number, assume that both numbers are three digits long. Thus, $125 - 067 = 058$.
The 10's complement representation of -067 is 933.

$$125$$
$$+\ 933$$

ignore carry $\longrightarrow$ 1   058 $\longleftarrow$ final result

(e) The 10's complement representation of -7 is 3.

$$7$$
$$+\ 3$$

ignore carry $\longrightarrow$ 1   0 $\longleftarrow$ $\longleftarrow$ final result

Example 1.33

Perform the following subtractions using the 10's complement arithmetic technique:
(a) $4 - 6 = -2$ (b) $8 - 9 = -1$ (c) $39 - 42 = -3$
(d) $67 - 125 = -58$

Solutions

(a) The 10's complement representation of -6 is 4.

$$4$$
$$+\ 4$$

ignore carry $\longrightarrow$ 0   8 $\longleftarrow$ final result in 10's complement representation
(negative result)

The 10's complement of 8 is 2. Therefore, the signed result is -2.

(b) The 10's complement representation of -9 is 1.

$$8$$
$$+\ 1$$

ignore carry $\longrightarrow$ 0   9 $\longleftarrow$ final result in 10's complement representation
(negative result)

The 10's complement of 9 is 1. Therefore, the signed result is -1.

(c) The 10's complement representation of -42 is 58.

$$39$$
$$+\ 58$$

ignore carry $\longrightarrow$ 0   97 $\longleftarrow$ final result in 10's complement representation
(negative result)

The 10's complement of 97 is 3. Therefore, the signed result is -3.

(d) Because largest number being operated on is a three-digit number, assume that both numbers are three digits long. Thus, $067 - 125 = -058$.
The 10's complement representation of -125 is 875.

$$067$$
$$+\ 875$$

ignore carry $\longrightarrow$ 0   942 $\longleftarrow$ final result in 10's complement representation
(negative result)

The 10's complement of 942 is 58. Therefore, the signed result is -58.

Binary subtraction– 1's complement

The 1's complement technique for binary subtraction is similar to the 9's complement technique for decimal subtraction discussed earlier in this section. The 1's complement of a binary number is obtained by subtracting each bit in the number from 1. However, observe that,

$$1 - 1 = 0 \quad \text{and} \quad 1 - 0 = 1$$

indicating that the 1's complement of 1 is 0 and the 1's complement of 0 is 1. Therefore, a more direct (or easier) approach to obtaining the 1's complement of a binary number is to *invert* each bit in the number. The term "invert" refers to the process of changing a 0 to a 1, and a 1 to a 0. For example, the 1's complement of the binary number, 10010110 is 01101001. Note that this completely eliminates the need for subtraction! Also, the process of obtaining the inverse of a bit can be performed very easily by a digital circuit called an inverter (*see* Chapter 2). To subtract two binary numbers using the 1's complement technique, do the following:

1. Obtain the 1's complement representation of the negative number.

2. Add the positive number to the 1's complement representation of the negative number.

3. Add the end-around (final) carry obtained from step 2 to the sum obtained from step 2.

4. The final sum is the difference between the two numbers.

5. If the end-around carry is a 1, then the result is positive. If the end-around carry is 0, then the result is negative and is represented in unsigned 1's complement form; it must be recomplemented in order to obtain its signed representation.

The following examples illustrate this technique.

Example 1.34

Subtract the following numbers using 1's complement binary arithmetic:
(a) $9 - 5$ (b) $85 - 32$ (c) $F3_{16} - 8B_{16}$ (d) $63_8 - 12_8$

Solutions

(a) $9 - 5 = 4$
$9 = 1001$
$5 = 0101$
Therefore, $-5 = 1010$ (in 1's complement representation).
Note that 5 (101) must be extended to 4 bits (0101) because the largest number being operated on, 9 (1001), is 4 bits long.

$$
\begin{array}{r}
1\ 0\ 0\ 1 \\
+\ 1\ 0\ 1\ 0 \\
\hline
1 \quad 0\ 0\ 1\ 1 \\
+ \quad \longrightarrow 1 \\
\hline
0\ 1\ 0\ 0 \\
\end{array}
$$

The result is 0100 or 4.

(b) $85 - 32 = 53$
$85 = 1010101$
$32 = 0100000$
Therefore, $-32 = 1011111$ (in 1's complement representation).

$$
\begin{array}{r}
1\ 0\ 1\ 0\ 1\ 0\ 1 \\
+\ 1\ 0\ 1\ 1\ 1\ 1\ 1 \\
\hline
1 \quad 0\ 1\ 1\ 0\ 1\ 0\ 0 \\
+ \quad \longrightarrow 1 \\
\hline
0\ 1\ 1\ 0\ 1\ 0\ 1 \\
\end{array}
$$

The result is 0110101 or 53.

(c) $F3_{16} - 8B_{16} = 68_{16}$
F3 = 11110011
8B = 10001011
Therefore, $-8B = 01110100$ (in 1's complement representation).

$$
\begin{array}{r}
1\ 1\ 1\ 1\ 0\ 0\ 1\ 1 \\
+\ 0\ 1\ 1\ 1\ 0\ 1\ 0\ 0 \\
\hline
1\quad 0\ 1\ 1\ 0\ 0\ 1\ 1\ 1 \\
+\ \lfloor\!\!\longrightarrow 1 \\
\hline
0\ 1\ 1\ 0\ 1\ 0\ 0\ 0
\end{array}
$$

The result is 01101000 or 68_{16}.

(d) $63_8 - 12_8 = 51_8$
63 = 110011
12 = 001010
Therefore, $-12 = 110101$ (in 1's complement representation).

$$
\begin{array}{r}
1\ 1\ 0\ 0\ 1\ 1 \\
+\ 1\ 1\ 0\ 1\ 0\ 1 \\
\hline
1\quad 1\ 0\ 1\ 0\ 0\ 0 \\
+\ \lfloor\!\!\longrightarrow 1 \\
\hline
1\ 0\ 1\ 0\ 0\ 1
\end{array}
$$

The result is 101001 or 51_8.

Example 1.35

Subtract the following numbers using 1's complement binary arithmetic:
(a) $5 - 9$ (b) $32 - 85$ (c) $8B_{16} - F3_{16}$ (d) $12_8 - 63_8$

Solutions

(a) $5 - 9 = -4$
5 = 0101
9 = 1001
Therefore, $-9 = 0110$ (in 1's complement representation).
Note that 5 (101) must be extended to 4 bits (0101) because the largest number being operated on, 9 (1001), is 4 bits long.

$$
\begin{array}{r}
0\ 1\ 0\ 1 \\
+\ 0\ 1\ 1\ 0 \\
\hline
0\quad 1\ 0\ 1\ 1 \\
+\ \lfloor\!\!\longrightarrow 0 \\
\hline
1\ 0\ 1\ 1
\end{array}
$$

Because the carry is 0, the result is 1011, or -4, in 1's complement representation. The signed result is the 1's complement of 1011, or -0100.

(b) $32 - 85 = -53$
32 = 0100000
85 = 1010101
Therefore, $-85 = 0101010$ (in 1's complement representation).

$$
\begin{array}{r}
0\ 1\ 0\ 0\ 0\ 0\ 0 \\
+\ 0\ 1\ 0\ 1\ 0\ 1\ 0 \\
\hline
0\quad 1\ 0\ 0\ 1\ 0\ 1\ 0 \\
+\ \lfloor\!\!\longrightarrow 0 \\
\hline
1\ 0\ 0\ 1\ 0\ 1\ 0
\end{array}
$$

Because the carry is 0, the result is 1001010, or -53, in 1's complement representation. The signed result is the 1's complement of 1001010, or -0110101.

(c) $8B_{16} - F3_{16} = -68_{16}$
$8B = 10001011$
$F3 = 11110011$
Therefore, $-F3 = 00001100$ (in 1's complement representation).

$$
\begin{array}{r}
1\ 0\ 0\ 0\ 1\ 0\ 1\ 1 \\
+\ 0\ 0\ 0\ 0\ 1\ 1\ 0\ 0 \\
\hline
0\quad 1\ 0\ 0\ 1\ 0\ 1\ 1\ 1 \\
+\ \rule{0pt}{0pt}\longrightarrow 0 \\
\hline
1\ 0\ 0\ 1\ 0\ 1\ 1\ 1
\end{array}
$$

Because the carry is 0, the result is 10010111, or -68, in 1's complement representation. The signed result is the 1's complement of 10010111, or -01101000.

(d) $12_8 - 63_8 = -51_8$
$12 = 001010$
$63 = 110011$
Therefore, $-63 = 001100$ (in 1's complement representation).

$$
\begin{array}{r}
0\ 0\ 1\ 0\ 1\ 0 \\
+\ 0\ 0\ 1\ 1\ 0\ 0 \\
\hline
0\quad 0\ 1\ 0\ 1\ 1\ 0 \\
+\ \rule{0pt}{0pt}\longrightarrow 0 \\
\hline
0\ 1\ 0\ 1\ 1\ 0
\end{array}
$$

Because the carry is 0, the result is 010110, or -51, in 1's complement representation. The signed result is the 1's complement of 010110, or -101001.

Binary subtraction — 2's complement

The 2's complement technique for binary subtraction is similar to the 10's complement technique for decimal subtraction discussed earlier in this section. The 2's complement of a binary number is obtained by adding 1 to the 1's complement of a number. For example, if the 1's complement of the number 10010110 is 01101001, then the 2's complement of the number 10010110 will be

$$
\begin{array}{ll}
01101001 & \text{1's complement} \\
+\qquad\quad 1 & \text{plus 1} \\
\hline
01101010 & \text{2's complement}
\end{array}
$$

To subtract two binary numbers using the 2's complement technique, do the following:

1. Obtain the 2's complement representation of the negative number.
2. Add the positive number to the 2's complement representation of the negative number.
3. Discard the final carry.
4. The final sum is the difference between the two numbers.
5. If the final carry is a 1, then the result is positive. If the final carry is 0, then the result is negative and is represented in unsigned 2's complement form; it must be recomplemented in order to obtain its signed representation.

The following examples illustrate this technique.

Example 1.36

Subtract the following numbers using 2's complement binary arithmetic:
(a) $9 - 5$ (b) $85 - 32$ (c) $F3_{16} - 8B_{16}$ (d) $63_8 - 12_8$

Solutions

(a) $9 - 5 = 4$
$9 = 1001$
$5 = 0101$
Therefore, $-5 = 1010 + 1 = 1011$ (in 2's complement representation).

$$\begin{array}{r} 1\ 0\ 0\ 1 \\ +\ 1\ 0\ 1\ 1 \\ \hline 1\quad 0\ 1\ 0\ 0 \end{array}$$

discard $\longrightarrow$
carry

The result is 0100 or 4.

(b) $85 - 32 = 53$
$85 = 1010101$
$32 = 0100000$
Therefore, $-32 = 1011111 + 1 = 1100000$ (in 2's complement representation).

$$\begin{array}{r} 1\ 0\ 1\ 0\ 1\ 0\ 1 \\ +\ 1\ 1\ 0\ 0\ 0\ 0\ 0 \\ \hline 1\quad 0\ 1\ 1\ 0\ 1\ 0\ 1 \end{array}$$

discard $\longrightarrow$
carry

The result is 0110101 or 53.

(c) $F3_{16} - 8B_{16} = 68_{16}$
$F3 = 11110011$
$8B = 10001011$
Therefore, $-8B = 01110100 + 1 = 01110101$ (in 2's complement representation).

$$\begin{array}{r} 1\ 1\ 1\ 1\ 0\ 0\ 1\ 1 \\ +\ 0\ 1\ 1\ 1\ 0\ 1\ 0\ 1 \\ \hline 1\quad 0\ 1\ 1\ 0\ 1\ 0\ 0\ 0 \end{array}$$

discard $\longrightarrow$
carry

The result is 01101000 or 68_{16}.

(d) $63_8 - 12_8 = 51_8$
$63 = 110011$
$12 = 001010$
Therefore, $-12 = 110101 + 1 = 110110$ (in 2's complement representation).

$$\begin{array}{r} 1\ 1\ 0\ 0\ 1\ 1 \\ +\ 1\ 1\ 0\ 1\ 1\ 0 \\ \hline 1\quad 1\ 0\ 1\ 0\ 0\ 1 \end{array}$$

discard $\longrightarrow$
carry

The result is 101001 or 51_8.

Example 1.37

Subtract the following numbers using 2's complement binary arithmetic:
(a) $5 - 9$ (b) $32 - 85$ (c) $8B_{16} - F3_{16}$ (d) $12_8 - 63_8$

Solutions

(a) $5 - 9 = -4$
$5 = 0101$
$9 = 1001$
Therefore, $-9 = 0110 + 1 = 0111$ (in 2's complement representation).

$$\begin{array}{r} 0\ 1\ 0\ 1 \\ +\ 0\ 1\ 1\ 1 \\ \hline 0\quad 1\ 1\ 0\ 0 \end{array}$$

discard $\longrightarrow$
carry

Because the carry is 0 the result is 1100, or -4, in 2's complement representation. The signed result is the 2's complement of 1100, or -0100.

(b) $32 - 85 = -53$
$32 = 0100000$
$85 = 1010101$
Therefore, $-85 = 0101010 + 1 = 0101011$ (in 2's complement representation).

$$
\begin{array}{r}
0\ 1\ 0\ 0\ 0\ 0\ 0 \\
+\ 0\ 1\ 0\ 1\ 0\ 1\ 1 \\
\hline
0 \quad 1\ 0\ 0\ 1\ 0\ 1\ 1
\end{array}
$$
discard ⟶ carry

Because the carry is 0 the result is 1001011, or -53, in 2's complement representation. The signed result is the 2's complement of 1001011, or -0110101.

(c) $8B_{16} - F3_{16} = -68_{16}$
$8B = 10001011$
$F3 = 11110011$
Therefore, $-F3 = 00001100 + 1 = 00001101$ (in 2's complement representation).

$$
\begin{array}{r}
1\ 0\ 0\ 0\ 1\ 0\ 1\ 1 \\
+\ 0\ 0\ 0\ 0\ 1\ 1\ 0\ 1 \\
\hline
0 \quad 1\ 0\ 0\ 1\ 1\ 0\ 0\ 0
\end{array}
$$
discard ⟶ carry

Because the carry is 0 the result is 10011000, or -68, in 2's complement representation. The signed result is the 2's complement of 10011000, or -01101000.

(d) $12_8 - 63_8 = -51_8$
$12 = 001010$
$63 = 110011$
Therefore, $-63 = 001100 + 1 = 001101$ (in 2's complement representation).

$$
\begin{array}{r}
0\ 0\ 1\ 0\ 1\ 0 \\
+\ 0\ 0\ 1\ 1\ 0\ 1 \\
\hline
0 \quad 0\ 1\ 0\ 1\ 1\ 1
\end{array}
$$
discard ⟶ carry

Because the carry is 0 the result is 010111, or -51, in 2's complement representation. The signed result is the 2's complement of 010111, or -101001.

Review Questions

1. Discuss the procedure used to represent a negative number without a sign.

2. In 9's and 10's complement subtraction, how is the end-around carry interpreted?

3. Discuss the similarity between 9's complement decimal subtraction and 1's complement binary subtraction.

4. Discuss the similarity between 10's complement decimal subtraction and 2's complement binary subtraction.

5. Discuss how to identify the sign of the result of a binary subtraction.

1.8
Multiplication and division

Just as the process of subtracting two numbers through the process of addition was examined in Section 1.7, this section examines how to multiply and divide numbers through the processes of repetitive addition and subtraction, respectively. Again the concepts of repetitive addition and subtraction are best understood by using a familiar numbering system (decimal) and then applying the concepts learned to the binary numbering system.

Decimal multiplication and division

Two different techniques for calculating the difference between two numbers using only the process of addition were discussed in the previous section. Now the techniques to perform multiplication and division through the processes of repetitive addition and subtraction, respectively, will be examined.

The process of multiplying two numbers is simply a process by which one number, the *multiplicand*, is repetitively added to itself a certain number of times (specified by the *multiplier*) to obtain a *product*. For example, to perform the following multiplication:

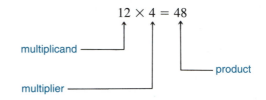

add the multiplicand, 12, four (multiplier = 4) times to itself to obtain a total sum of 48 (product)

$$
\begin{array}{r}
12 \\
12 \\
12 \\
+\ 12 \\
\hline
48
\end{array}
$$

The longhand technique of multiplication can also be used, but the repetitive addition technique is more applicable to how a digital computer would perform multiplication using the binary numbering system (to be discussed later in this section).

Dividing two decimal numbers can be accomplished by repetitive subtraction, and because subtraction can be accomplished by addition (9's or 10's complement subtraction), the entire division process can be done through addition. For example, consider dividing the number 24 (*dividend*) by 3 (*divisor*) to obtain a result of 8 (*quotient*).

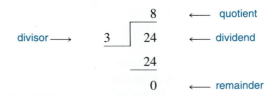

The operation can also be performed by repetitively subtracting the divisor (3) from the dividend (24) until a remainder of 0 or a remainder that is less than the divisor (3) is obtained. The quotient will be a count of the number of subtractions that have taken place

		quotient (tally)
dividend:	24	
divisor:	− 3	1
remainder:	21	
divisor:	− 3	1
remainder:	18	
divisor:	− 3	1
remainder:	15	
divisor:	− 3	1
remainder:	12	
divisor:	− 3	1
remainder:	9	
divisor:	− 3	1
remainder:	6	
divisor:	− 3	1
remainder:	3	
divisor:	− 3	+ 1
final remainder ⟶	0	8 ⟵ quotient

To illustrate this procedure with numbers that yield a remainder when divided, consider the following example.

Example 1.38

Divide the number 96 by 9 using the repetitive subtraction procedure:

$$
\begin{array}{r}
10 \quad \longleftarrow \text{ quotient} \\
9 \overline{)96} \quad \longleftarrow \text{ dividend} \\
90 \\
\hline
6 \quad \longleftarrow \text{ remainder}
\end{array}
$$

divisor ⟶

Solution

		quotient (tally)
dividend:	96	
divisor:	− 9	1
remainder:	87	
divisor:	− 9	1
remainder:	78	
divisor:	− 9	1
remainder:	69	
divisor:	− 9	1
remainder:	60	
divisor:	− 9	1
remainder:	51	
divisor:	− 9	1
remainder:	42	
divisor:	− 9	1
remainder:	33	
divisor:	− 9	1
remainder:	24	
divisor:	− 9	1
remainder:	15	
divisor:	− 9	+ 1
final remainder ⟶	6	10 ⟵ quotient
(less than 9)		

Binary multiplication

Two techniques are used to perform binary multiplication. The first is the process of repetitive additions similar to the procedure discussed earlier in this section in which the multiplicand is added to itself a certain number of times specified by the multiplier. The second is a technique that uses a "shift and add" scheme to perform the multiplication. Both techniques will now be investigated.

The simplest technique of binary multiplication is the repetitive addition procedure. For example, to multiply the binary number 1001 (9) by the binary number 101 (5), add the number 1001 to itself 5 (101) times.

```
        9              1 0 0 1
    +   9              1 0 0 1   +
       18            1 0 0 1 0
    +   9              1 0 0 1   +
       27            1 1 0 1 1
    +   9              1 0 0 1   +
       36          1 0 0 1 0 0
    +   9              1 0 0 1   +
       45          1 0 1 1 0 1
```

The final result is 101101 or 45_{10}.

A more practical way to perform binary multiplication is to apply some of the concepts of longhand multiplication. For example, multiply the two binary numbers 1001 and 101 using conventional long multiplication techniques

```
          1 0 0 1        ⟵ multiplicand
      ×     1 0 1        ⟵ multiplier
          1 0 0 1
        0 0 0 0
      1 0 0 1
      1 0 1 1 0 1        ⟵ product
```

The final result is 101101 or 45_{10}.

To perform the multiplication, just shift and add the multiplicand. Both these operations are easily performed by digital circuits. The following are examples of binary multiplication.

Example 1.39

Multiply the following numbers using the shift-and-add binary multiplication technique (longhand multiplication).
(a) 15 × 12   (b) $3A_{16} \times 2$ (c) $10_8 \times 10_8$

Solutions

(a) 15 × 12 = 180
    In binary, 1111 × 1100 = 10110100.

```
                  1 1 1 1
              ×   1 1 0 0
                  0 0 0 0
                0 0 0 0
            1 1 1 1
          1 1 1 1
          1 0 1 1 0 1 0 0
```

The final result is 10110100 or 180_{10}.

(b) $3A_{16} \times 2 = 74_{16}$
   In binary, $111010 \times 10 = 1110100$.

$$
\begin{array}{r}
1\ 1\ 1\ 0\ 1\ 0 \\
\times \qquad\quad 1\ 0 \\
\hline
0\ 0\ 0\ 0\ 0\ 0 \\
1\ 1\ 1\ 0\ 1\ 0 \\
\hline
1\ 1\ 1\ 0\ 1\ 0\ 0
\end{array}
$$

The final result is 1110100 or 74_{16}.

(c) $10_8 \times 10_8 = 100_8$
   In binary, $1000 \times 1000 = 1000000$.

$$
\begin{array}{r}
1\ 0\ 0\ 0 \\
\times\ 1\ 0\ 0\ 0 \\
\hline
0\ 0\ 0\ 0 \\
0\ 0\ 0\ 0 \\
0\ 0\ 0\ 0 \\
1\ 0\ 0\ 0 \\
\hline
1\ 0\ 0\ 0\ 0\ 0\ 0
\end{array}
$$

The final result is 1000000 or 100_8.

Binary division

Binary division is performed by repetitive subtractions. The technique of division by repetitive subtractions was described earlier in this section with reference to the decimal numbering system. Recall that this process simply involves repetitively subtracting the divisor from the dividend while keeping a count of how many subtractions have taken place. When the remainder of the subtractions reach a value that is less than the divisor or equal to zero, the division is complete and the quotient is equal to the count of the number of subtractions that have taken place. For example, to divide the binary number 11110 (30) by the binary number 101 (5), perform the division by adding the 2's complement of the divisor 101 to the quotient repetitively until a remainder of zero is obtained. It is preferable to use 2's complement subtraction because this does not involve adding in the end-around carry each time. The following examples illustrate the procedure.

Example 1.40

Divide the number 30 by 5 using the repetitive binary subtraction procedure.

Solution

The dividend is $30_{10} = 11110_2$.
The divisor is $5 = 101_2$.
Because the divisor will be subtracted from the dividend, obtain the 2's complement of 5 (101) extended to 5 bits (00101) because the dividend 30 (11110) is a 5-bit number.
2's complement of $00101 = 11010 + 1 = 11011$.

dividend:	30	1 1 1 1 0	quotient (tally)
divisor:	− 5	+ 1 1 0 1 1	1
remainder:	25	1 1 1 0 0 1	
divisor:	− 5	+ 1 1 0 1 1	1
remainder:	20	1 1 0 1 0 0	
divisor:	− 5	+ 1 1 0 1 1	1
remainder:	15	1 0 1 1 1 1	

divisor:	$-$ 5	$+$ 1 1 0 1 1	1
remainder:	10	1 0 1 0 1 0	

divisor:	$-$ 5	$+$ 1 1 0 1 1	1
remainder:	5	1 0 0 1 0 1	

divisor:	$-$ 5	$+$ 1 1 0 1 1	1 $+$
final $\longrightarrow$	0	1 0 0 0 0 0	6
remainder			↑
(less than 5)			└ quotient

Example 1.41

Divide the number 10 by 3 using the repetitive binary subtraction procedure.

Solution

The dividend is $10_{10} = 1010_2$.
The divisor is $3 = 11_2$.
Because the divisor will be subtracted from the dividend, obtain the 2's complement of 3 (11) extended to 4 bits (0011) because the dividend 10 (1010) is a 4-bit number.
2's complement of $0011 = 1100 + 1 = 1101$.

dividend:	10	1 0 1 0	quotient (tally)
divisor:	$-$ 3	$+$ 1 1 0 1	1
remainder:	7	1 0 1 1 1	

divisor:	$-$ 3	$+$ 1 1 0 1	1
remainder:	4	1 0 1 0 0	

divisor:	$-$ 3	$+$ 1 1 0 1	1 $+$
final $\longrightarrow$	1	1 0 0 0 1	3
remainder			↑
(less than 3)			└ quotient

Review Questions

1. Discuss the procedure used to perform multiplication using repetitive addition.

2. Discuss the procedure used to perform division using repetitive subtraction.

1.9
BCD arithmetic

Section 1.5 introduced the representation of decimal numbers in a form known as BCD. These BCD numbers were made up of 4-bit binary codes that represented each digit in a decimal number. Recall that BCD numbers are *not* the binary equivalent of decimal numbers, they are merely a representation. In Section 1.5 it was stated that a digital computer uses one of two techniques to store a decimal number in memory — it either converts the number into its binary equivalent before storage or it converts the number to its BCD representation before storage.

If the digital computer stores decimal numbers as their binary equivalents, arithmetic performed on these binary numbers is simple and can be done using the standard binary addition procedure discussed earlier. The result can then be reconverted from binary to decimal before retrieval. For example, if the two decimal numbers 27 and 34 were to be added, they would first be converted to binary, 011011 and 100010, respectively, and then added as follows:

$$
\begin{array}{rl}
27 & 0\ 1\ 1\ 0\ 1\ 1 \\
+\ 34 & +\ 1\ 0\ 0\ 0\ 1\ 0 \\
\hline
61 & 1\ 1\ 1\ 1\ 0\ 1 \\
\end{array}
$$

The binary sum obtained is 111101 that when reconverted to decimal is 61, the desired result.

However, if the digital computer stored the two numbers, 27 and 34, in BCD representation, 00100111 and 00110100, respectively, the result would be incorrect if these two numbers were added using the standard binary addition procedure used in the previous case

$$
\begin{array}{rl}
27 & 0\ 0\ 1\ 0\ 0\ 1\ 1\ 1 \\
+\ 34 & +\ 0\ 0\ 1\ 1\ 0\ 1\ 0\ 0 \\
\hline
5? & 0\ 1\ 0\ 1\ 1\ 0\ 1\ 1 \\
& \underbrace{}_{5}\ \underbrace{}_{?}
\end{array}
$$

Note that the result of the addition yields not only the wrong MSD, but also an invalid BCD number 1011. This is because binary addition takes into account the six "missing" combinations 1010, 1011, 1100, 1101, 1110, and 1111 that are invalid in BCD. To compensate for these six digits during addition, an operation known as a *decimal adjust* must be performed on the result of a binary addition of two BCD numbers. The rules for BCD addition and for performing this decimal adjust operation are as follows:

1. Add the two BCD numbers using regular binary addition.

2. Check each group of 4 bits (nibble) of the result. If the value of the nibble is greater than 9 (1001), add 6 (0110) to the nibble. Also add 6 if there was a carry-out generated from the nibble when the two BCD numbers were added.

Apply these rules to the addition of the two numbers, 27 and 34, attempted earlier as follows:

$$
\begin{array}{rl}
27 & 0\ 0\ 1\ 0\ 0\ 1\ 1\ 1 \\
+\ 34 & +\ 0\ 0\ 1\ 1\ 0\ 1\ 0\ 0 \\
\hline
5? & 0\ 1\ 0\ 1\ 1\ 0\ 1\ 1 \\
6 & +\qquad\qquad 0\ 1\ 1\ 0 \\
\hline
61 & 0\ 1\ 1\ 0\ 0\ 0\ 0\ 1 \\
\end{array}
$$

Note that no carry-out was generated out of the least significant nibble or the most significant nibble. However, because the least significant nibble has a value greater than 9, 6 was added to the least significant nibble to obtain the desired result 61_{10}.

The following examples of BCD addition illustrate other variations in the use of the rules for BCD addition.

Example 1.42

Perform the addition of the following decimal numbers using BCD addition:
(a) 59 + 39 (b) 98 + 89 (c) 75 + 56 (d) 3928 + 4165

Solutions

(a) 59 + 39 = 98

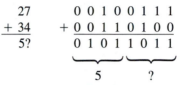

$$
\begin{array}{rl}
59 & 0\ 1\ 0\ 1\quad 1\ 0\ 0\ 1 \\
+\ 39 & +\ 0\ 0\ 1\ 1\quad 1\ 0\ 0\ 1 \\
\hline
?? & 1\ 0\ 0\ 1\quad 0\ 0\ 1\ 0 \\
\end{array}
$$

A carry-out was generated from the least significant nibble. Therefore, add 6 to the least significant nibble

```
    ??        1 0 0 1   0 0 1 0
  +  6      +             0 1 1 0
    98        1 0 0 1   1 0 0 0
```

The final sum is 10011000 (BCD) or 98_{10}.

(b) 98 + 89 = 187

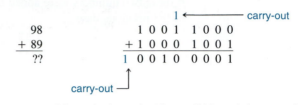

```
    98        1 0 0 1   1 0 0 0
  + 89      + 1 0 0 0   1 0 0 1
    ??      1 0 0 1 0   0 0 0 1
```

A carry-out was generated from the least significant nibble and the most significant nibble. Therefore, add 6 to the least significant nibble and the most significant nibble

```
     ??       1 0 0 1 0   0 0 0 1
   + 66     +   0 1 1 0   0 1 1 0
    187       1 1 0 0 0   0 1 1 1
```

The final sum is 110000111 (BCD) or 187_{10}.

(c) 75 + 56 = 131

```
    75        0 1 1 1   0 1 0 1
  + 56      + 0 1 0 1   0 1 1 0
    ??        1 1 0 0   1 0 1 1
```

The least significant nibble and the most significant nibble are both greater than 9. Therefore, add 6 to both the least significant and most significant nibbles

```
     ??       1 1 0 0   1 0 1 1
   + 66     + 0 1 1 0   0 1 1 0
    131     1 0 0 1 1   0 0 0 1
```

The final sum is 100110001 (BCD) or 131_{10}.

(d) 3928 + 4165 = 8093

```
    3928     0 0 1 1   1 0 0 1   0 0 1 0   1 0 0 0
  + 4165   + 0 1 0 0   0 0 0 1   0 1 1 0   0 1 0 1
    ????     0 1 1 1   1 0 1 0   1 0 0 0   1 1 0 1
```

The least significant nibble and the third nibble (from the right) are greater than 9. Therefore, 6 must be added to both these nibbles

```
    ????     0 1 1 1   1 0 1 0   1 0 0 0   1 1 0 1
  + 0606   +            0 1 1 0             0 1 1 0
    8093     1 0 0 0   0 0 0 0   1 0 0 1   0 0 1 1
```

The final sum is 1000000010010011 (BCD) or 8093_{10}.

Review Questions

1. Why can't decimal numbers represented in BCD be added in the same manner as decimal numbers represented in binary?

2. Why is a decimal adjust required during BCD addition?

3. Discuss the procedure used to add two BCD numbers.

1.10
Binary codes

At this point it is apparent that many of the digital circuits to be covered later in this book will operate on binary numbers. Besides performing arithmetic on binary numbers, many digital circuits interpret binary numbers as instructions, whereas others interpret binary numbers as information. Information transfer between digital devices plays an important part in digital communications.

In a digital communication scheme, if two devices want to transfer information between themselves (i.e., communicate) they do so by transmitting and receiving binary numbers. In much the same way as words make up the vocabulary of a spoken language, these binary numbers make up the ''vocabulary'' of a basic digital communications scheme that allows information to be transferred. The information transferred between two digital communicating devices are binary numbers that represent all the symbols necessary to accomplish the intelligible transfer of information. Thus, each letter in the English alphabet (as well as some international alphabets), each digit in our numbering system, commonly used symbols for punctuation, and other necessary graphical characters are assigned unique binary numbers (codes), which are then transferred back and forth in a digital communications system.

A good example of a simple digital communicating system is the keyboard connected to a computer. When a key is pressed on the keyboard, a device known as an *encoder* inside the keyboard detects which key was pressed and sends a unique binary number (code) to the computer. The computer receives the code and *decodes* it to determine which key was pressed. Standardization of these codes is extremely important. That is, if the keyboard transmits the code 1000001 when the ''A'' key is pressed, the computer should be able to identify that this code corresponds to the letter *A* and not any other character. Just as in a spoken language, all persons involved must speak the same language to prevent a communications breakdown, in a digital communications system, all devices must use the same code assignments for the symbols used in the communications vocabulary.

Over the years there have been several different sets of codes used in digital communications. However, the most popular and universally accepted code is the American Standard Code for Information Interchange (ASCII). Table 1.11 lists the ASCII codes and the symbols they represent.

Observe in Table 1.11 that even though each ASCII code is represented by an 8-bit number, because the MSB of each code is 0, the ASCII code is really a 7-bit code. Because there are 2^7 or 128 possible combinations of 7-bit binary numbers, the ASCII code can represent a maximum of 128 graphic and nongraphic symbols. Graphic ASCII symbols (codes 20_{16} through $7F_{16}$) can be displayed or printed, for example, letters, numbers, and punctuation. Nongraphic ASCII symbols (codes 00_{16} through 19_{16}) cannot be printed or displayed and are used to control the transmission of information or activate certain special features on the communicating devices. For example, when the *RETURN* (or *ENTER*) key is pressed on an ASCII keyboard (a keyboard that generates ASCII codes) no graphic character is actually displayed on the screen, instead the digital computer simply advances the cursor to the next line. Most of the nongraphic codes are generated through the use of the *CTRL* (*CONTROL*) key on an ASCII keyboard. Table 1.11 lists only the function abbreviations of these nongraphic codes; however, the definitions and key sequences for these nongraphic ASCII codes are shown in Table 1.12.

An extension of the ASCII code developed by IBM and called the extended ASCII code uses 8 bits, thus allowing for up to 2^8 or 256 symbols to be represented (*see* Table 1.13). The symbols assigned to the codes 00000000 through 01111111 are exactly the same as the basic ASCII code. The symbols assigned to the codes 10000000 through 11111111 make up the extended portion of the set and are used to represent various graphical characters and symbols.

Table 1.11   The ASCII Code

Hex	Binary	Symbol	Hex	Binary	Symbol	Hex	Binary	Symbol	Hex	Binary	Symbol
00	00000000	NUL	20	00100000	SP	40	01000000	@	60	01100000	
01	00000001	SOH	21	00100001	!	41	01000001	A	61	01100001	a
02	00000010	STX	22	00100010	"	42	01000010	B	62	01100010	b
03	00000011	ETX	23	00100011	#	43	01000011	C	63	01100011	c
04	00000100	EOT	24	00100100	$	44	01000100	D	64	01100100	d
05	00000101	ENQ	25	00100101	%	45	01000101	E	65	01100101	e
06	00000110	ACK	26	00100110	&	46	01000110	F	66	01100110	f
07	00000111	BEL	27	00100111	'	47	01000111	G	67	01100111	g
08	00001000	BS	28	00101000	(	48	01001000	H	68	01101000	h
09	00001001	HT	29	00101001	)	49	01001001	I	69	01101001	i
0A	00001010	LF	2A	00101010	*	4A	01001010	J	6A	01101010	j
0B	00001011	VT	2B	00101011	+	4B	01001011	K	6B	01101011	k
0C	00001100	FF	2C	00101100	,	4C	01001100	L	6C	01101100	l
0D	00001101	CR	2D	00101101	−	4D	01001101	M	6D	01101101	m
0E	00001110	SO	2E	00101110	.	4E	01001110	N	6E	01101110	n
0F	00001111	SI	2F	00101111	/	4F	01001111	O	6F	01101111	o
10	00010000	DLE	30	00110000	0	50	01010000	P	70	01110000	p
11	00010001	DC1	31	00110001	1	51	01010001	Q	71	01110001	q
12	00010010	DC2	32	00110010	2	52	01010010	R	72	01110010	r
13	00010011	DC3	33	00110011	3	53	01010011	S	73	01110011	s
14	00010100	DC4	34	00110100	4	54	01010100	T	74	01110100	t
15	00010101	NAK	35	00110101	5	55	01010101	U	75	01110101	u
16	00010110	SYN	36	00110110	6	56	01010110	V	76	01110110	v
17	00010111	ETB	37	00110111	7	57	01010111	W	77	01110111	w
18	00011000	CAN	38	00111000	8	58	01011000	X	78	01111000	x
19	00011001	EM	39	00111001	9	59	01011001	Y	79	01111001	y
1A	00011010	SUB	3A	00111010	:	5A	01011010	Z	7A	01111010	z
1B	00011011	ESC	3B	00111011	;	5B	01011011	[	7B	01111011	{
1C	00011100	FS	3C	00111100	<	5C	01011100	\	7C	01111100	|
1D	00011101	GS	3D	00111101	=	5D	01011101	]	7D	01111101	}
1E	00011110	RS	3E	00111110	>	5E	01011110	^	7E	01111110	~
1F	00011111	US	3F	00111111	?	5F	01011111	_	7F	01111111	DEL

Many digital communicating devices transfer binary codes with *parity*. Parity generating and checking is a scheme by which any errors that occur during transfer of these binary codes can be detected. There are two types of parity used during the transfer of binary codes — even and odd. To illustrate the concept of even and odd parity, consider the basic ASCII codes shown in Table 1.11. All 128 codes listed in Table 1.11 can be classified as those that have *even parity* and those that have *odd parity*. The codes that have an even number of 1 bits have even parity, whereas the codes that have an odd number of 1 bits have odd parity. For example, the following codes have even parity:

$$\text{``!''} \quad 010\ 0001 \text{— two 1-bits}$$
$$\text{``M''} \quad 100\ 1101 \text{— four 1-bits}$$
$$\text{``w''} \quad 111\ 0111 \text{— six 1-bits}$$

whereas the following have odd parity:

$$\text{``@''} \quad 100\ 0000 \text{— one 1-bit}$$
$$\text{``Q''} \quad 101\ 0001 \text{— three 1-bits}$$
$$\text{``O''} \quad 100\ 1111 \text{— five 1-bits}$$

Table 1.12 Nongraphic ASCII code definitions and key sequences

Code	Key	Definition	Code	Key	Definition
NUL	CTRL @	Null	DLE	CTRL P	Data Link Escape
SOH	CTRL A	Start of Heading	DC1	CTRL Q	Direct Control 1
STX	CTRL B	Start Text	DC2	CTRL R	Direct Control 2
ETX	CTRL C	End Text	DC3	CTRL S	Direct Control 3
EOT	CTRL D	End of Transmission	DC4	CTRL T	Direct Control 4
ENQ	CTRL E	Enquiry	NAK	CTRL U	Negative Acknowledge
ACK	CTRL F	Acknowledge	SYN	CTRL V	Synchronous idle
BEL	CTRL G	Bell	ETB	CTRL W	End Transmission Block
BS	CTRL H	Backspace	CAN	CTRL X	Cancel
HT	CTRL I	Horizontal Tab	EM	CTRL Y	End of Medium
LF	CTRL J	Line Feed	SUB	CTRL Z	Substitute
VT	CTRL K	Vertical Tab	ESC	CTRL [	Escape
FF	CTRL L	Form Feed	FS	CTRL \	Form Separator
CR	CTRL M	Carriage Return	GS	CTRL]	Group Separator
SO	CTRL N	Shift Out	RS	CTRL ^	Record Separator
SI	CTRL O	Shift In	US	CTRL _	Unit Separator

In an even parity transfer of information, the transmitter will add an additional bit (most significant 8 bit) to each 7-bit ASCII code called the *parity bit*. This bit would be set to a 1 if the code had odd parity so as to make it even or set to a 0 if the code had even parity. Thus, all codes transmitted would have even parity by forcing the parity bit to a 1 or a 0. The receiver would be conditioned to check for even parity. That is, when the code is received the receiver checks to see if the total number of 1 bits is even. If the total number of 1 bits is even, the code is transmitted properly, otherwise the code is corrupted during transfer. An odd parity data communications system basically works in the same manner except that the parity bit of each code transmitted is set to a 1 or a 0 so as to make the total number of 1 bits an odd number. The receiver then counts the number of 1 bits received and signals an error if this number is not odd.

Example 1.43

Determine the ASCII codes for the following characters if they were transmitted with no parity, even parity, and odd parity:
(a) "5"   (b) "$"   (c) "E"   (d) ">"

Solutions

(a) "5" = 35_{16} = 011 0101
With no parity, 011 0101
With even parity, 0011 0101
With odd parity, 1011 0101

(b) "$" = 24_{16} = 010 0100
With no parity, 010 0100
With even parity, 0010 0100
With odd parity, 1010 0100

(c) "E" = 45_{16} = 100 0101
With no parity, 100 0101
With even parity, 1100 0101
With odd parity, 0100 0101

(d) ">" = $3E_{16}$ = 011 1110
With no parity, 011 1110
With even parity, 1011 1110
With odd parity, 0011 1110

The binary codes discussed in this section are used mainly for information transfers as well as for representing the alphabets that make up our spoken language. Textual matter is usually stored in a computer's memory in ASCII representation. This also includes the other graphical symbols that are used in everyday writing.

Table 1.13 The IBM extended ASCII code

Hex	Binary	Symbol	Hex	Binary	Symbol	Hex	Binary	Symbol	Hex	Binary	Symbol
80	10000000	Ç	A0	10100000	á	C0	11000000	∟	E0	11100000	α
81	10000001	ü	A1	10100001	í	C1	11000001	⊥	E1	11100001	β
82	10000010	é	A2	10100010	ó	C2	11000010	┬	E2	11100010	Γ
83	10000011	â	A3	10100011	ú	C3	11000011	├	E3	11100011	π
84	10000100	ä	A4	10100100	ñ	C4	11000100	─	E4	11100100	Σ
85	10000101	à	A5	10100101	Ñ	C5	11000101	┼	E5	11100101	σ
86	10000110	å	A6	10100110	a	C6	11000110	╞	E6	11100110	μ
87	10000111	ç	A7	10100111	o	C7	11000111	╟	E7	11100111	γ
88	10001000	ê	A8	10101000	¿	C8	11001000	╚	E8	11101000	Φ
89	10001001	ë	A9	10101001	⌐	C9	11001001	╔	E9	11101001	θ
8A	10001010	è	AA	10101010	¬	CA	11001010	╩	EA	11101010	Ω
8B	10001011	ï	AB	10101011	½	CB	11001011	╦	EB	11101011	δ
8C	10001100	î	AC	10101100	¼	CC	11001100	╠	EC	11101100	∞
8D	10001101	ì	AD	10101101	¡	CD	11001101	═	ED	11101101	φ
8E	10001110	Ä	AE	10101110	≪	CE	11001110	╬	EE	11101110	ε
8F	10001111	Å	AF	10101111	≫	CF	11001111	╧	EF	11101111	∩
90	10010000	É	B0	10110000	▓	D0	11010000	╨	F0	11110000	≡
91	10010001	æ	B1	10110001	▓	D1	11010001	╤	F1	11110001	±
92	10010010	Æ	B2	10110010	▓	D2	11010010	╥	F2	11110010	≥
93	10010011	ô	B3	10110011	│	D3	11010011	╙	F3	11110011	≤
94	10010100	ö	B4	10110100	┤	D4	11010100	╘	F4	11110100	⌠
95	10010101	ò	B5	10110101	╡	D5	11010101	╒	F5	11110101	⌡
96	10010110	û	B6	10110110	╢	D6	11010110	╓	F6	11110110	÷
97	10010111	ù	B7	10110111	╖	D7	11010111	╫	F7	11110111	≈
98	10011000	ÿ	B8	10111000	╕	D8	11011000	╪	F8	11111000	°
99	10011001	Ö	B9	10111001	╣	D9	11011001	┘	F9	11111001	•
9A	10011010	Ü	BA	10111010	║	DA	11011010	┌	FA	11111010	·
9B	10011011	¢	BB	10111011	╗	DB	11011011	█	FB	11111011	√
9C	10011100	£	BC	10111100	╝	DC	11011100	▄	FC	11111100	ⁿ
9D	10011101	¥	BD	10111101	╜	DD	11011101	▌	FD	11111101	²
9E	10011110	₧	BE	10111110	╛	DE	11011110	▐	FE	11111110	■
9F	10011111	ƒ	BF	10111111	┐	DF	11011111	▀	FF	11111111	

Symbolic representation is therefore another application of the binary numbering system without which the processing of information by digital computers cannot be accomplished.

Review Questions

1. What are binary codes used for?
2. Why is there a need for a standard code?
3. What is the purpose of parity?

Summary

- The decimal numbering system is a base-ten numbering system and has ten digits.

- A count in any numbering system is simply an ordered sequence of combinations of digits.

- The octal numbering system is a base-eight numbering system and has a set of only eight digits.

- The hexadecimal numbering system is a base-16 numbering system and has a set of 16 digits.

- The binary numbering system is a base-two numbering system and has only two digits.

- The binary numbering system is used by all digital circuits.

- A single binary digit is called a bit, a group of 4 bits make a nibble, 2 nibbles make up a byte, and 2 bytes make up a word.

- The octal and hexadecimal numbering systems are ideal for representing binary numbers.

- Binary coded decimal numbers represent decimal digits.

- The BCD value of a decimal number is only a representation of the decimal number and not its binary equivalent (value).

- The BCD system is a convenient means of representing decimal numbers because it does not require complicated conversions and can be accomplished through the substitution procedure.

- Almost all mathematical operations can be reduced to the four basic arithmetic operations—addition, subtraction, multiplication, and division.

- Subtraction, multiplication, and division can be accomplished through various techniques of addition.

- The 9's and 10's complement of a decimal number can be used to represent a negative number without a sign.

- The 9's complement of a number is obtained by subtracting each digit in the number from 9.

- The 10's complement of a number can be obtained by adding 1 to its 9's complement.

- The 9's and 10's complement subtraction procedures can be used to subtract two decimal numbers through the process of addition.

- The final carry in 9's and 10's complement subtraction can be used to identify a negative or positive result.

- The end-around carry is the final carry that is added to the sum in 9's complement arithmetic and discarded in 10's complement arithmetic.

- Binary subtraction is performed through the 1's and 2's complement techniques.

- The 1's complement of a binary number is obtained by reversing (inverting) each bit in the number.

- The 2's complement of a binary number is obtained by adding 1 to its 1's complement.

- The 1's and 2's complement form can be used to represent a negative binary number without a sign.

- The final carry in 1's and 2's complement subtraction can be used to identify a negative or positive result.

- The end-around carry is the final carry that is added to the sum in 1's complement arithmetic and discarded in 2's complement arithmetic.

- The addition of BCD numbers involves an extra step called decimal adjust to account for the six combinations that cannot be represented by any decimal digit.

- Binary codes are used in digital communications for transferring information.

- The ASCII code is a standard set of 7-bit binary numbers that represent the various symbols used in information transfer.

- The extended ASCII code simply adds another 128 combinations and symbols to the basic ASCII code by extending it to 8 bits.

- Parity generation and checking are done to ensure that data transmission errors can be detected.

- Two types of parity schemes are used—even and odd.

- Binary numbers transmitted with even parity have an even number of 1 bits whereas binary numbers transmitted with odd parity have an odd number of 1 bits.

Problems

Section 1.2   Numbering system concepts

1. How many combinations would there be of 8 decimal digits, and what would the largest value in the count be?

2. Obtain an octal count sequence for the following decimal sequences:
   a. 0 to 10
   b. 100 to 108
   c. 25 to 35
   d. 55 to 66

3. Obtain a hexadecimal count sequence for the following decimal sequences:
   a. 0 to 12
   b. 100 to 108
   c. 25 to 35
   d. 245 to 257

4. Obtain a binary count sequence for the following sequences:
   a. 0 to 12
   b. 100 to 108
   c. 25_{16} to 30_{16}
   d. 245_8 to 260_8

Section 1.3   Conversions to decimal

5. Determine the value of the number 17,563 in terms of its positional weights.

6. Convert the following numbers from octal to decimal:
   a. 1234_8
   b. 03776_8
   c. 812_8
   d. 10_8
   e. 7777_8
   f. 3057672_8

7. Convert the following numbers from hexadecimal to decimal:
   a. 5691_{16}
   b. $1B6A9_{16}$
   c. $FACE_{16}$
   d. 10_{16}
   e. $FFFF_{16}$
   f. 16789_{16}

8. Convert the following numbers from binary to decimal using Table 1.5:
   a. 1101101101_2
   b. 1010_2
   c. 10110101_2
   d. 100021101_2
   e. 100001_2
   f. 11110_2

Section 1.4   Conversions from decimal

9. Determine the digits that make up the decimal number 9140 by using the repetitive division procedure.

10. Convert the following decimal numbers to octal:
    a. 9731
    b. 18325
    c. 64
    d. 8

11. Convert the following decimal numbers to hexadecimal:
    a. 2916
    b. 12087
    c. 256
    d. 16

12. Convert the following decimal numbers to binary:
    a. 10
    b. 987
    c. 801
    d. 76

Section 1.5   Representation of binary numbers

13. Convert the following binary numbers to octal:
    a. 101010101010
    b. 11100101
    c. 10110
    d. 0110110
    e. 0000001
    f. 11
    g. 101000000

14. Convert the following octal numbers to binary:
    a. 63543
    b. 7777
    c. 06354
    d. 1110110
    e. 53692
    f. 1546
    g. 2645

15. Convert the following binary numbers to hexadecimal:
    a. 111101011001
    b. 01011011
    c. 101
    d. 10010111101
    e. 11111101111111
    f. 010000001
    g. 0000101010010

16. Convert the following hexadecimal numbers to binary:
    a. ABCDE
    b. 6F62B
    c. 27549
    d. 73AC645
    e. BA32
    f. 100101
    g. 88

17. Convert the following octal numbers to hexadecimal:
    a. 77
    b. 10111
    c. 352440
    d. 6235213
    e. 10321
    f. 12

18. Convert the following hexadecimal numbers to octal:
    a. AC
    b. 6345
    c. 10010
    d. 7DA64
    e. 02365
    f. F

19. Convert the following decimal numbers to their BCD representations:
    a. 90017
    b. 65
    c. 63515
    d. 1010
    e. 5
    f. 236021

20. Convert the following BCD numbers to decimal:
    a. 010110011000
    b. 0100010010101001000
    c. 10100101110
    d. 100100010101
    e. 0100100010000010
    f. 00010

21. The number 2103 is a base-four number. Convert this number to:
    a. decimal
    b. octal
    c. hexadecimal
    d. binary

22. Convert the following numbers to their base-four equivalents:
    a. 75
    b. 634_8
    c. $F9_{16}$
    d. 10101001_2

Section 1.6   Addition

23. Perform the following additions in binary arithmetic:
    a. $9 + 7$
    b. $56 + 95$
    c. $F5_{16} + 10_{16}$
    d. $107_8 + 632_8$

24. Perform the following additions in binary arithmetic:
    a. $1010 + 0110$
    b. $0001 + 1111$
    c. $10110 + 110111$
    d. $111110 + 11011$

Section 1.7   Subtraction

25. Represent the following signed negative numbers in 9's complement form:
    a. -4
    b. -12
    c. -6
    d. -640
    e. -98
    f. -4657

26. The following is a list of negative numbers expressed in 9's complement form. Determine the signed representation of each number.
    a. 734,658
    b. 238
    c. 12
    d. 948,505,734
    e. 7

27. Perform the following subtractions using the 9's complement arithmetic technique:
    a. $2 - 4 = -2$
    b. $3 - 8 = -5$
    c. $24 - 87 = -63$
    d. $43 - 205 = -162$
    e. $1 - 643 = -642$

28. Represent the following signed negative numbers in 10's complement form:
    a. -3
    b. -25
    c. -7
    d. -10

    e. -99
    f. -7310

29. The following is a list of negative numbers expressed in 10's complement form. Determine the signed representation of each number.
    a. 671,239
    b. 9
    c. 22
    d. 819,735,021
    e. 912

30. Perform the following subtractions using the 10's complement arithmetic technique:
    a. $4 - 8 = -4$
    b. $1 - 7 = -6$
    c. $9 - 19 = -10$
    d. $7 - 512 = -505$

31. Subtract the following numbers using 1's complement binary arithmetic:
    a. $7 - 3$
    b. $92 - 21$
    c. $E7_{16} - 6D_{16}$
    d. $172_8 - 65_8$

32. Subtract the following numbers using 2's complement binary arithmetic:
    a. $8 - 3$
    b. $43 - 12$
    c. $EF_{16} - AB_{16}$
    d. $42_8 - 33_8$

Section 1.8   Multiplication and division

33. Divide the number 128 by 14 using the repetitive subtraction procedure.

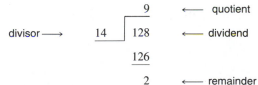

34. Multiply the following numbers using the shift-and-add binary multiplication technique (longhand multiplication):
    a. 10×9
    b. $2C_{16} \times 3_{16}$
    c. $17_8 \times 13_8$

35. Divide the number 70 by 14 using the repetitive binary subtraction procedure.

36. Divide the number 32 by 10 using the repetitive binary subtraction procedure.

Section 1.9   BCD arithmetic

37. Perform the addition of the following decimal numbers using BCD addition.
    a. $23 + 56$
    b. $48 + 12$

c. 90 + 77

d. 7364 + 1238

38. Add the following BCD numbers:

a. 1001 + 0011

b. 1000 + 1000

c. 0101 + 0011

d. 1011 + 1000

Section 1.10 Binary codes

39. Determine the ASCII codes for the following characters if they were transmitted with no parity, even parity, and odd parity:

a. "W"

b. "9"

c. "@"

d. ETX

40. Determine the ASCII codes for the following characters if they were transmitted with no parity, even parity, and odd parity:

a. "4"

b. "="

c. BEL

d. "H"

2

BASIC LOGIC GATES AND CIRCUITS

OBJECTIVES

The objectives of this chapter are to:

• Introduce the basic concepts of logic circuits, their operation, and their analysis

• Introduce the general operation of a logic gate and its hardware implementation in integrated circuits

• Examine the operation of the four basic logic gates — the buffer, the inverter, the *AND* gate, and the *OR* gate

• Introduce the distinctive shape symbols and the ANSI/IEEE uniform shape symbols for each gate

• Introduce the various integrated circuits that implement the four basic gates in hardware

• Examine the operation of the *NAND* and *NOR* gates, and introduce their integrated circuit implementations

• Introduce simple logic circuits that are constructed with the basic gates and develop a technique for analyzing and constructing these circuits

• Examine the procedure to set up logic circuit hardware

• Study the electrical characteristics of a logic gate

• Study the various troubleshooting procedures to test and debug logic gates and circuits

Digital circuit boards are usually made up of many logic circuits in the form of ICs. The photograph shows a person troubleshooting such a circuit board
(Photo courtesy Hewlett-Packard Company).

2.1
Introduction

Logic circuits are the building blocks of all digital computers. The operation of these circuits is described by the principles of logic rather than the principles of conventional mathematics. Logic circuits are also known as *switching circuits* because the basic circuit element of early logic circuits was the electromechanical switch known as the *relay*. Also, the characteristics of logic circuits to switch from one state to another have led to continued use of the term to describe modern logic circuits. Because digital computers are made up of various logic circuits, and logic circuits are often used to process digits and numbers, these circuits are often referred to as digital circuits.

The basic element of modern logic circuits is the gate which electronically implements the various logic primitives or functions that are used in the design of logic circuits. Gates and logic circuits accept binary numbers as input, process or manipulate these numbers, and produce binary numbers as output. These logic circuits can be broadly classified into two categories—*combinational* logic circuits and *sequential* logic circuits. The outputs of a combinational logic circuit (discussed also in Chapters 3 and 4) are dependent only on the inputs of the circuit. However, the outputs of a sequential logic circuit (discussed in Chapters 5 through 7) are dependent not only on its inputs but also on the past history of the outputs; that is, sequential logic circuits include a memory element, or feedback.

This chapter introduces the basic logic gates that will be used in all logic circuits. The standard distinctive shape symbols as well as the new American National Standards Institute/Institute of Electrical and Electronics Engineers (ANSI/IEEE) uniform shape symbols are introduced with each gate. Because of their popularity, however, the distinctive shape symbols will be used in most of the logic circuits discussed. A complete explanation of the ANSI/IEEE logic symbols is included in Appendix B. The principle of operation of each gate along with the actual hardware implementation of the gate in the form of an *integrated circuit* (IC) will then be

studied. Also included in this chapter is an introduction to the electrical characteristics of logic gates and the interpretation of data sheets for digital integrated circuits. The application of the basic gates in simple logic circuits is then examined along with procedures to set up the logic circuit hardware. Chapter 3 continues the study to include procedures to design and analyze logic circuits. At the end of this chapter various procedures to troubleshoot the basic logic gates and circuits in a laboratory environment will be examined. These procedures can then be applied to many of the other circuits included in the rest of this book.

2.2
Basic logic concepts

This section introduces the basic concepts used in the analysis and synthesis of logic gates and circuits. These concepts are developed using simple switching circuit analogies and later in this chapter they are applied to specific gates and circuits. It should therefore be noted that the switching circuits developed in this section are not representative of the circuits to be covered in the rest of the book but are simply conceptual models that are used to illustrate the principles of logic and the operation of logic circuits.

   The basic element of a switching circuit is a switch, which is used to control the flow of current through a circuit. It is a two-state device that can exist in only one of two possible states — ON or OFF. Figure 2.1 illustrates these two possible states. In Figure 2.1a the switch is shown with its contacts open, thus preventing the passage of current though the switch; this will be referred to as the OFF state. In Figure 2.1b, the switch is now closed and allows the passage of current through it; this state will be referred to as the ON state.

Figure 2.1

The two states of a switch.

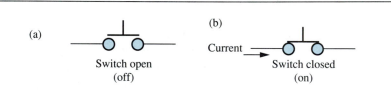

Consider a simple switching circuit such as the one shown in Figure 2.2. The circuit consists of a battery (or some other power supply) connected in series with a switch and a lamp (light bulb). Assume that the battery maintains a fixed voltage and current for the circuit. The switch and the lamp have been identified with the variables S and L, respectively. The lamp, L, is also a two-state device, that is, the lamp can either be ON or OFF. For both the lamp and the switch there are only two possible states — ON or OFF, with nothing in between these two states. Note in Figure 2.2 that the state of the lamp, L, is dependent on the state of the switch, S. However, the state of the switch, S is independent of the state of the lamp, L. In other words, turning the switch ON or OFF will cause the lamp to turn ON or OFF, but turning the lamp ON or OFF will not cause the switch to turn ON or OFF. The switch is therefore the controlling factor, whereas the lamp is the controlled factor. In the study of logic circuits, L is called the *dependent variable* and S is called the *independent variable*.

   To describe the operation of the circuit shown in Figure 2.2, a table can be

Figure 2.2

A simple switching circuit.

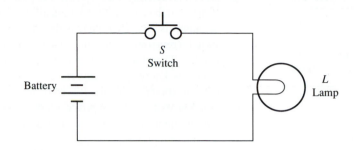

Table 2.1 Truth table for Figure 2.2

S	L
OFF	OFF
ON	ON

constructed that lists all possible states of the dependent variable, *L*, as a function of the independent variable, *S*. Such a table is called a *truth table* (*see* Table 2.1).

We can see in Table 2.1 that the state of the lamp, *L*, corresponds exactly to the state of the switch, *S*. That is, if the switch is ON then the lamp is ON, and if the switch is OFF then the lamp is OFF. The circuit in Figure 2.2 and the truth table in Table 2.1 can also be described by the logic equation

$$L = S$$

which when expressed in words simply means that the state of the lamp is the same as (equal to) the state of the switch. Observe that the dependent variable is always placed on the left-hand side of the logic equation.

The operation of the circuit shown in Figure 2.2 can be determined by simple logical deduction. Hence, such a circuit is often referred to as a *logic circuit* because the operation of such circuits is often governed by the simple rules of logic. For reasons mentioned in Section 2.1, these circuits are also known as *digital circuits*. All digital circuits are two-state circuits, that is, the variables can only have one of two possible states. The counterpart of a digital circuit is often referred to as an *analog circuit*. For example, if the circuit in Figure 2.2 were modified such that instead of the switch a variable resistor were used to control the state of the lamp, then the lamp would have several states of intensity between being completely ON and completely OFF; the circuit could no longer be classified as a digital circuit but would now be considered an analog circuit.

To illustrate the concepts of logic circuits, truth tables, and logic equations further, consider the circuit shown in Figure 2.3.

In Figure 2.2, the switch was placed in series with the lamp so that the switch directly controlled the current flowing through the lamp. In Figure 2.3, the switch is placed in parallel with the lamp (i.e., across the lamp) so that the switch controls the flow of current away from the lamp. Of course, this circuit should not be experimen-

Figure 2.3

An inverting circuit.

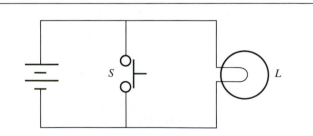

Table 2.2 Truth table for
Figure 2.3

S	L
OFF	ON
ON	OFF

tally implemented without a suitable current-limiting power supply because the switch effectively short circuits the power supply. The logic of the circuit is apparent. If the switch is open (OFF), then all the current flows from the battery to the lamp and the lamp is ON. However, if the switch is closed, all the current flows through the switch (remember, current always follows the path of least resistance) and the lamp is OFF. Therefore, the state of the lamp is the inverse of the state of the switch. The circuit is known as an *inverting circuit* or more simply as an *inverter* and its truth table is shown in Table 2.2. Note in Table 2.2 that the state of the lamp, L, at any given instant is the opposite of the state of the switch, S. L is referred to as the complement of S, and S as the complement of L. In the study of logic circuits, the word "complement" is identified with inversion rather than with equivalence. The logic equation for the circuit of Figure 2.3 and its truth table is

$$L = NOT\, S$$

which when expressed in words means that the state of L is not the state of S. Because there are only two possible states, the word "NOT" identifies the inverse state. For example, if $S = $ ON,

$$\text{then } L = NOT\, S = NOT\, \text{ON} = \text{OFF}$$

if $S = $ OFF,

$$\text{then } L = NOT\, S = NOT\, \text{OFF} = \text{ON}.$$

Instead of using the word "NOT" to identify the inverse state, a horizontal bar over the letter of the variable can be used. Thus, the same logic equation can be represented as

$$L = \overline{S}$$

The equation is read as "L equals S not" or "L equals not S" and identifies the variable L and its complement, S.

Example 2.1

The variable A is the complement of the variable B, and the variable B is the complement of the independent variable C. Construct a truth table for the three variables, A, B, and C. What conclusion can be drawn from the relationship between the variables A and C?

Solution

Because B is the complement of C

$$B = \overline{C}$$

Because A is the complement of B

$$A = \overline{B}$$

Thus, looking at all possible states of the independent variable C and determining the states of A and B from the preceding equations, the following truth table can be constructed:

C	B	A
ON	OFF	ON
OFF	ON	OFF

From the truth table, it can be concluded that the state of A is the same as the state of C. Therefore,

$$A = C.$$

This conclusion can also be reached mathematically as follows:

If $B = \overline{C}$

Then $\overline{B} = C$

But $A = \overline{B}$

Therefore, $A = C$.

So far, circuits that had one dependent variable and one independent variable have been examined. The study of logic concepts will now be extended to circuits that contain more than one independent variable. Figure 2.4 expands the circuit of Figure 2.2 to include two switches, A and B, connected in series with the lamp, L. As before, note that the state of L is dependent on the states of both A and B, but not vice versa. Therefore, L is the dependent variable and A and B are the independent variables.

Figure 2.4

An *AND* circuit.

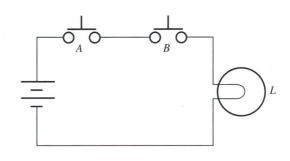

The operation of the circuit in Figure 2.4 can be described in terms of the conditions necessary for the lamp to turn ON. This is known as *positive logic* and will be used throughout this book. If the operation of the circuit was described in terms of the conditions necessary to turn the lamp OFF, this would be known as *negative logic*. The lamp, L, will be ON if both switches A and B are ON. The circuit is known as an *AND* circuit because the dependent variable L will be in the ON state if the independent variables A and B are both in the ON state. If any of the independent variables are OFF, the dependent variable will be OFF. The truth table shown in Table 2.3 illustrates this relationship.

Note in Table 2.3 that we have looked at all possible combinations of the independent variables A and B, and for each combination, determined the state of the dependent variable L. Also, note that if the binary digit 1 were substituted for ON and a 0 for OFF, we would have a binary count of all possible combinations of 2 bits in the columns representing the two independent variables A and B. This format for constructing truth tables will be used for all the logic circuits to follow.

The logic equation for the circuit in Figure 2.4 and its truth table is expressed as follows:

$$L = A \; AND \; B$$

However, in the study of logic circuits, the word "AND" is replaced with the symbol "·" which will be used to represent the *AND* function. Thus,

$$L = A \cdot B$$

The equation is read as L equals A and B. The concept of the *AND* function can be extended to more than one independent variable as shown in the following example.

Table 2.3 Truth table for Figure 2.4

A	B	L
OFF	OFF	OFF
OFF	ON	OFF
ON	OFF	OFF
ON	ON	ON

Example 2.2

Construct a truth table for the dependent variable X for the circuit shown in Figure 2.5, and determine the logic equation for the circuit.

Figure 2.5

AND circuit with three switches.

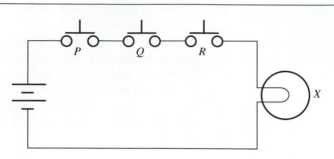

Solution

Regardless of the number of switches in the circuit, the lamp X will be ON, only if all three switches are ON. Therefore, the truth table for the circuit is as follows:

P	Q	R	X
OFF	OFF	OFF	OFF
OFF	OFF	ON	OFF
OFF	ON	OFF	OFF
OFF	ON	ON	OFF
ON	OFF	OFF	OFF
ON	OFF	ON	OFF
ON	ON	OFF	OFF
ON	ON	ON	ON

The logic equation for the circuit is

$$X = P \cdot Q \cdot R$$

Again, observe in Example 2.2 that the binary counting sequence has been used for all possible combinations of 3 bits in order to obtain all possible states of the three independent variables P, Q, and R. Also, note that the logic of the *AND* function is the same irrespective of the number of independent variables; therefore, the logic can be extended to as many variables as needed.

The circuit in Figure 2.6 represents another type of logic function that is commonly used in digital circuits. The circuit is known as an *OR* circuit because either switch A or switch B must be ON in order for the lamp L to turn ON. The truth table for the *OR* circuit is shown in Table 2.4.

Table 2.4 illustrates that if any of the independent variables is ON, the dependent variable will be ON.

The logic equation for the circuit in Figure 2.6 and its truth table is expressed as follows:

$$L = A \; OR \; B$$

However, in the study of logic circuits, the word "OR" is replaced with the symbol "+" which will be used to represent the *OR* function. Thus,

$$L = A + B$$

The equation is read as L equals A or B.

Table 2.4 Truth table for Figure 2.6

A	B	L
OFF	OFF	OFF
OFF	ON	ON
ON	OFF	ON
ON	ON	ON

Figure 2.6

An *OR* circuit.

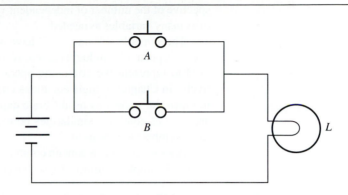

The concept of the *OR* function can be extended to more than one independent variable as shown in the following example.

Example 2.3

Construct a truth table for the dependent variable *X* for the circuit shown in Figure 2.7, and determine the logic equation for the circuit.

Figure 2.7

OR circuit with three switches.

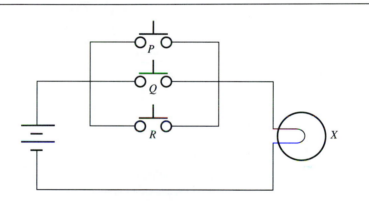

Solution

Regardless of the number of switches in the circuit, the lamp *X* will be ON if any of the three switches are ON. Therefore, the truth table for the circuit is as follows:

P	*Q*	*R*	*X*
OFF	OFF	OFF	OFF
OFF	OFF	ON	ON
OFF	ON	OFF	ON
OFF	ON	ON	ON
ON	OFF	OFF	ON
ON	OFF	ON	ON
ON	ON	OFF	ON
ON	ON	ON	ON

The logic equation for the circuit is

$$X = P + Q + R$$

As before, Example 2.3 shows that the logic of the *OR* function is the same irrespective of the number of independent variables; therefore, the logic can be extended to as many variables as needed.

The symbols ''·'' and ''+'' have been used to represent the *AND* and *OR* functions, respectively, in logic equations. It is apparent that these are the same symbols used to represent the arithmetic operations of multiplication and addition, respectively. In Chapter 3, logic equations containing these functions will be manipulated in much the same way as algebraic expressions are manipulated, and because there are a lot of algebraic similarities between the arithmetic and logical functions, the same symbols will be used.

A logic circuit can contain combinations of the various logic functions discussed earlier. Example 2.4 illustrates such a circuit and its analysis.

Example 2.4

Construct a truth table for the dependent variable *Z* for the circuit shown in Figure 2.8, and determine the logic equation for the circuit.

Figure 2.8

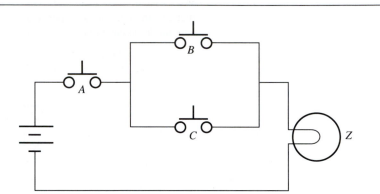

Solution

The logic of the circuit can be described as follows. The lamp *Z* will be ON if and only if switch *A* is ON, and either switch *B* or *C* is ON. This logic can be expressed as a logic equation

$$Z = A \cdot (B + C) \tag{2.1}$$

The logic of the circuit can also be described as follows. The lamp *Z* will be ON if and only if switch *A* AND switch *B* are ON, OR switch *A* AND switch *C* are ON. This logic can also be expressed as an equivalent logic equation

$$Z = (A \cdot B) + (A \cdot C) \tag{2.2}$$

A truth table can now be constructed that looks at all possible states of *A*, *B*, and *C* and for each combination the state of *Z* can be determined as follows:

A	*B*	*C*	*Z*
OFF	OFF	OFF	OFF
OFF	OFF	ON	OFF
OFF	ON	OFF	OFF
OFF	ON	ON	OFF
ON	OFF	OFF	OFF
ON	OFF	ON	ON
ON	ON	OFF	ON
ON	ON	ON	ON

In Example 2.4 two equivalent logic equations were obtained for the circuit in Figure 2.8. Equations 2.1 and 2.2 both described the logic of the circuit in different forms. A logic equation can also be obtained from the truth table of the circuit by analyzing the last three conditions in the table. Note that the last three combinations cause the lamp *Z* to turn ON. Expressed in words, this simply means that in order for the lamp *Z* to turn ON, one of the following three conditions have to be satisfied:

1. Switch *A* must be ON, switch *B* must be OFF, and switch *C* must be ON.

or

2. Switch *A* must be ON, switch *B* must be ON, and switch *C* must be OFF.

or

3. Switch *A* must be ON, switch *B* must be ON, and switch *C* must be ON.

Because an OFF state can be expressed as the complement of the ON state by placing a bar over a variable, these conditions can be expressed in the form of the following logic equation that is equivalent to Equations 2.1 and 2.2.

$$Z = (A \cdot \overline{B} \cdot C) + (A \cdot B \cdot \overline{C}) + (A \cdot B \cdot C) \tag{2.3}$$

In Chapter 3, using various laws defined by a field of logical mathematics known as *Boolean algebra*, it will be proved that Equations 2.1, 2.2, and 2.3 are logically equivalent.

Review Questions

1. Why is a switching circuit considered to be a two-state device?
2. What are the two types of variables that exist in a switching circuit and how do they differ from each other?
3. What is the difference between a digital and an analog circuit?
4. What does an inverter do?
5. What are the conditions necessary for the dependent variable of an *AND* circuit to be ON?
6. What are the conditions necessary for the dependent variable of an *OR* circuit to be ON?

2.3
The logic gate

At this point the three basic logic functions known as the logic inversion (*NOT*), the logic *AND*, and the logic *OR* have been examined. Simple switching circuits have been used to illustrate the concepts of logic functions and the equations and tables that describe them. However, these switching circuits are only conceptual models and are not used in modern digital devices. Instead, we use circuits that operate on the same principles discussed previously but have different configurations. These circuits are known as *logic gates* or more simply, *gates*.

Figure 2.9 illustrates the general model that represents all logic gates. A logic gate can be viewed as a ''black box'' that has one or more inputs and a single output. The inputs and output to the box are 1's and 0's (binary numbers). The gate accepts binary numbers at its input and produces a 1 or a 0 at its output depending on its function. The output of the gate is therefore considered to be the dependent variable because it depends on the binary combinations at the input(s) of the gate (independent variables). Like the switching circuits discussed earlier, the operation of a gate

Figure 2.9

The model of a logic gate.

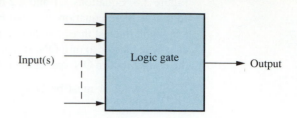

can be described by its truth table and logic equation that relate the dependent and independent variables.

Within the box is an electronic circuit (to be studied in Chapter 12) that makes the gate function in the appropriate manner. At this stage, view the gate as a black box with signals going in and coming out, with its inner workings essentially invisible. The binary digits 1 and 0 are applied to the gate's input(s) and measured at the output in the form of voltages. Typically $+5$ volts represents a logic 1 (also referred to as a logic *high level*) and 0 volts (ground potential) represents a logic 0 (also referred to as a logic *low level*). The details of the electrical characteristics of these gates will be examined in Section 2.9 and Chapter 12.

In the previous section a truth table was obtained for a switching circuit by examining the state of the dependent variable (lamp) for all possible combinations of the states of the independent variables (switches). These combinations followed a binary counting sequence. The truth tables that describe the operation of logic gates follow the same format. That is, look at all possible combinations of the independent variables (inputs) by listing a binary count. Recall from Chapter 1 that the number of binary combinations of n bits is given by the relationship 2^n. Therefore, the truth table for a single input gate would have 2^1 or two entries

Input
0
1

The truth table for a two-input gate would have 2^2 or four entries

Inputs	
0	0
0	1
1	0
1	1

The truth table for a three-input gate would have 2^3 or eight entries

Inputs		
0	0	0
0	0	1
0	1	0
0	1	1
1	0	0
1	0	1
1	1	0
1	1	1

And the truth table for a four-input gate would have 2^4 or 16 entries

Inputs			
0	0	0	0
0	0	0	1
0	0	1	0
0	0	1	1
0	1	0	0
0	1	0	1
0	1	1	0
0	1	1	1
1	0	0	0
1	0	0	1
1	0	1	0
1	0	1	1
1	1	0	0
1	1	0	1
1	1	1	0
1	1	1	1

IC logic

A logic gate is usually packaged in the form of an IC. One of the most popular and most accepted series of digital ICs is the 74xxx transistor-transistor-logic (TTL) series. These ICs provide the digital circuit designer with a wide variety of gates and application circuits that can be used in most applications. Each IC in the series is given a two- or three-digit number ranging from 00 to (currently) 670 that identifies the function of the IC. The numbers are prefixed with the digits 74 (or 54) to identify

Figure 2.10

14- and 16-pin ICs.
(a) Pictorial view; (b) pin layout (top view).

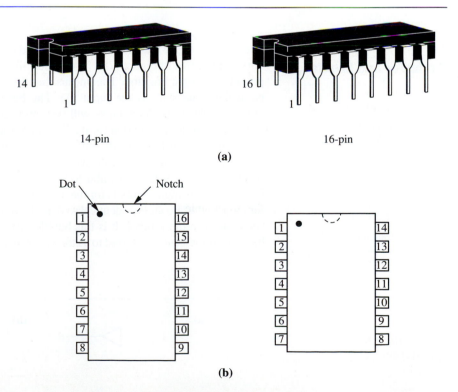

14-pin   16-pin

(a)

(b)

the series, therefore the series is referred to as the 74xxx series, where xxx identifies a specific IC in the series. More will be said about this series in Chapter 12.

ICs are manufactured in a variety of sizes that are classified according to the number of *pins* or connections. Figure 2.10a shows typical IC packages that are used for logic gates in 14- and 16-pin sizes. Figure 2.10b includes the pin layout when each IC (commonly referred to as a *chip*) is viewed from above. A notch on one end of the IC usually identifies the top. The pins are numbered sequentially starting from the upper left-hand side and proceeding down to the end of one side. The count is then continued from the bottom right-hand side to the top right-hand side of the IC. Many ICs include a small dot on the upper left-hand side to identify pin 1. Generally, two pins on an IC are reserved for applying power to the internal circuitry: $+5$ volts and ground potential. These two pins are labeled V_{CC} and GND (ground), respectively, and are usually (but not always) assigned to pins 14 and 7, respectively, in a 14-pin IC, and pins 16 and 8, respectively, in a 16-pin IC. The other pins on the IC allow access to the input(s) and output(s) of the gate(s) in the IC.

The next few sections in this chapter will examine the various gates that implement the basic functions discussed in this section. The operation and analysis of each gate examined will also include the specific 74xxx IC(s) that implement the function of the gate in hardware (i.e., electronically). Appendix C includes the complete data sheets for the ICs discussed in the book, which include the pin configurations, symbols, and electrical characteristics.

Review Questions

1. How is a logic gate similar to a simple switching circuit?
2. What form do the dependent and independent variables of a logic gate take — conceptually and electronically?
3. How are pins identified on an IC?
4. What is the purpose of the V_{CC} and GND pins on an IC?

2.4

The buffer and inverter

Table 2.5   Truth table and logic equation for the buffer

A	B
0	0
1	1

Logic equation: $B = A$

The *buffer* is a gate that has only one input and one output. The logic symbols for the buffer are shown in Figure 2.11. The buffer, like the switching circuit in Figure 2.2, provides no logic function. This means that the output of the buffer (B) will always be at the same state as the input (A). The buffer, however, does provide a very important electronic function as will be seen in Section 2.9 and in Chapter 12. The truth table and logic equation for the buffer are shown in Table 2.5. The similarity in logic can be seen between the buffer and the switching circuit of Figure 2.2.

The *inverter*, also referred to as the *NOT* gate, implements logical inversion. The function of the inverter is similar to the function of the switching circuit shown in Figure 2.3. The logic symbols for the inverter are shown in Figure 2.12. The symbols for the inverter are almost identical to that of a buffer except that a "bubble" (○) has been added to the symbol. It is this bubble that indicates that a logic inversion has been performed and is referred to as the *negation indicator*. The ANSI/IEEE version

Figure 2.11

Logic symbols for the buffer. (a) Distinctive shape symbol; (b) ANSI/IEEE Std. 91-1984 symbol.

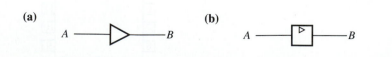

(a)

$A \longrightarrow\!\!\!\!\triangleright\!\!\!\!\longrightarrow B$

(b)

$A \longrightarrow\!\boxed{\triangleright}\!\longrightarrow B$

Figure 2.12

Logic symbols for the inverter. (a) Distinctive shape symbol; (b) ANSI/ IEEE Std. 91-1984 symbol.

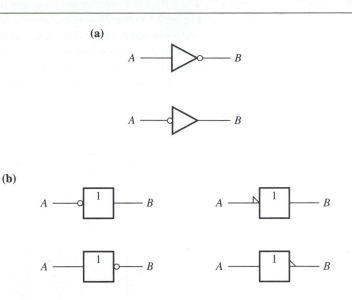

(a)

(b)

Table 2.6 Truth table and logic equation for the inverter

A	B
0	1
1	0

Logic equation: $B = \overline{A}$

of the symbol may also represent inversion by means of the *polarity indicator*, which is shown as a triangle (◁) at either the input or the output of the gate. The negation indicator will be used throughout this textbook to represent inversion.

The output of the inverter (B) is always equal to the complement of the input (A). The truth table and logic equation for the inverter shown in Table 2.6 illustrate its operation.

Because the logic state at the input of an inverter is complemented, two inverters connected in series will simply function logically as a buffer. The following example illustrates this fact:

Example 2.5

Construct a truth table for the dependent variable X for the circuit shown in Figure 2.13, and determine the logic equation for the circuit.

Figure 2.13

Cascaded inverters.

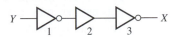

Solution

Y	Output of gate 1	Output of gate 2	X
0	1	1	0
1	0	0	1

Therefore, from the truth table, the logic equation for the circuit is

$$X = Y$$

IC logic

The buffer and inverter are packaged in several 74xxx ICs, two of which are shown in Figure 2.14. The 7404 is known as a *hex inverter* IC because it contains six (hexa) inverters in one package. The inputs to the inverters are accessed through pins 1, 3, 5, 9, 11, and 13, whereas the outputs are derived from pins 2, 4, 6, 8, 10, and 12, respectively. Pins 7 and 14 are used to supply power to the IC. Similarly, the 7407 is known as a "hex buffer" and is used in much the same way as the 7404.

Figure 2.14

Pin configuration of the 7404 inverter and the 7407 buffer.

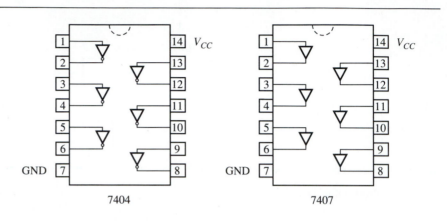

7404   7407

Review Questions

1. Logically, what is the difference between a buffer and an inverter?
2. In the symbols that represent an inverter, what actually identifies the process of logical inversion?

2.5

The *AND* gate

The *AND* gate implements the logic *AND* function discussed in Section 2.2. The function of the *AND* gate is similar to the operation of the *AND* circuit shown in Figure 2.4. This means that the output of the *AND* gate (dependent variable) will be a logic 1 if and only if all inputs (independent variables) are at the logic 1 state. The symbols for the *AND* gate are shown in Figure 2.15. *AND* gates can have several inputs but they must have at least two inputs to properly implement the *AND* function. The truth tables and logic equations for the two-input and three-input *AND* gates shown in Figure 2.15 are given in Tables 2.7 and 2.8, respectively.

Figure 2.15

Logic symbols for the *AND* gate. (a) Distinctive shape symbol; (b) ANSI/IEEE Std. 91-1984 symbol.

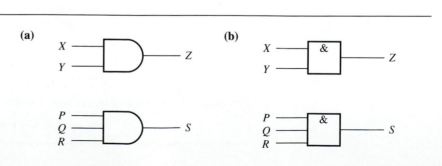

X	Y	Z
0	0	0
0	1	0
1	0	0
1	1	1

Table 2.7 Truth table and logic equation for a two-input *AND* gate

Logic equation: $Z = X \cdot Y$

P	Q	R	S
0	0	0	0
0	0	1	0
0	1	0	0
0	1	1	0
1	0	0	0
1	0	1	0
1	1	0	0
1	1	1	1

Table 2.8 Truth table and logic equation for a three-input *AND* gate

Logic equations: $S = P \cdot Q \cdot R$

Example 2.6

The five-input *AND* gate shown in Figure 2.16 has one of its inputs permanently tied high (logic 1 state). Obtain the truth table and the logic equation for the gate. What would happen to the gate if one of its inputs were permanently tied low (logic 0 state)?

Figure 2.16

A five-input *AND* gate used as a four-input *AND* gate.

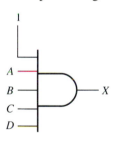

Solution

Because one of the five inputs is permanently tied to the logic 1 state, the five-input *AND* gate now functions like a four-input *AND* gate. This is because the output of the gate (X) now depends on whether the states of the four remaining inputs (A, B, C, D) are all 1s, or if any one of the inputs is a 0.

Therefore, its truth table and logic equation is

A	B	C	D	X
0	0	0	0	0
0	0	0	1	0
0	0	1	0	0
0	0	1	1	0
0	1	0	0	0
0	1	0	1	0
0	1	1	0	0
0	1	1	1	0
1	0	0	0	0
1	0	0	1	0
1	0	1	0	0
1	0	1	1	0
1	1	0	0	0
1	1	0	1	0
1	1	1	0	0
1	1	1	1	1

Logic equation: $X = A \cdot B \cdot C \cdot D$

In any *AND* gate, if any one of the inputs is a logic 0, the *AND* gate is "disabled" and the output will be a logic 0 and will not be affected by the states of the other inputs.

IC logic

The *AND* gate is available in several IC packages as shown in Figure 2.17. The 7408 is known as the ''quad 2-input *AND* gate'' and contains four (quad) 2-input *AND* gates in one IC. The 7411 is known as the ''triple 3-input *AND* gate'' and contains three (triple) 3-input *AND* gates in one IC. The 7421 is known as the ''dual 4-input *AND* gate'' and contains two (dual) 4-input *AND* gates in one IC. As before, observe that power is supplied to pins 7 (GND) and 14 (V_{CC}) on each IC.

Figure 2.17

Pin configurations of various *AND* gate ICs.

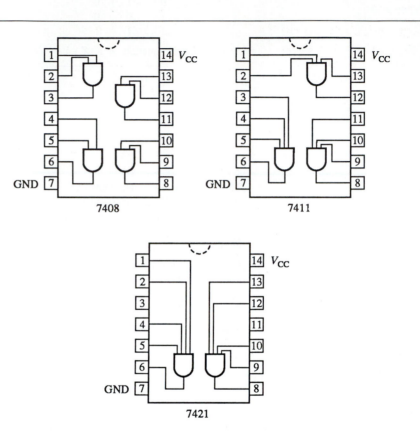

Review Questions

1. What is the difference between an *AND* gate and an inverter?
2. Is it possible to make a four-input *AND* gate function logically like a buffer? If yes, how can it be done?
3. How can we connect two 2-input *AND* gates to function as a three-input *AND* gate?

2.6

The *OR* gate

The *OR* gate implements the logic *OR* function discussed in Section 2.2. The function of the *OR* gate is similar to the operation of the *OR* circuit shown in Figure 2.6. This means that the output of the *OR* gate (dependent variable) will be a logic 1 if any of its inputs (independent variables) are at the logic 1 state. The symbols for the *OR* gate are shown in Figure 2.18. *OR* gates can have several inputs but they must have at least two inputs to properly implement the *OR* function. The truth tables and logic equations for the two- and three-input *OR* gates shown in Figure 2.18 are given in Tables 2.9 and 2.10, respectively.

Figure 2.18

Logic symbols for the *OR* gate. (a) Distinctive shape symbol; (b) ANSI/IEEE Std. 91-1984 symbol.

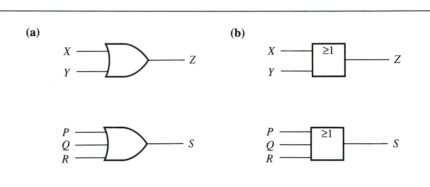

Table 2.9 Truth table and logic equation for a two-input *OR* gate

X	Y	Z
0	0	0
0	1	1
1	0	1
1	1	1

Logic equation: $Z = X + Y$

Table 2.10 Truth table and logic equation for a three-input *OR* gate

P	Q	R	S
0	0	0	0
0	0	1	1
0	1	0	1
0	1	1	1
1	0	0	1
1	0	1	1
1	1	0	1
1	1	1	1

Logic equation: $S = P + Q + R$

Example 2.7

The five-input *OR* gate shown in Figure 2.19 has one of its inputs permanently tied low (logic 0 state). Obtain the truth table and the logic equation for the gate. What would happen to the gate if one of its inputs were permanently tied high (logic 1 state)?

Solution

Because one of the five inputs is permanently tied to the logic 0 state, the five-input *OR* gate now functions like a four-input *OR* gate. This is because the output of the gate (*X*) now depends on whether the states of the four remaining inputs (*A, B, C, D*) are all 0's, or if any one of the inputs is a 1.

Figure 2.19

A five-input *OR* gate used as a four-input *OR* gate.

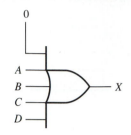

Therefore, its truth table and logic equation is

A	B	C	D	X
0	0	0	0	0
0	0	0	1	1
0	0	1	0	1
0	0	1	1	1
0	1	0	0	1
0	1	0	1	1
0	1	1	0	1
0	1	1	1	1
1	0	0	0	1
1	0	0	1	1
1	0	1	0	1
1	0	1	1	1
1	1	0	0	1
1	1	0	1	1
1	1	1	0	1
1	1	1	1	1

Logic equation: $X = A + B + C + D$

In any *OR* gate, if any one of the inputs is a logic 1, the output will be a logic 1 and will not be affected by the states of the other inputs.

IC logic

Figure 2.20

Pin configuration of a 7432 *OR* gate IC.

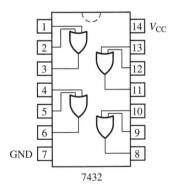

7432

The *OR* gate is only available in one IC package as shown in Figure 2.20. The 7432 is known as the "quad 2-input *OR* gate" and contains four (quad) 2-input *OR* gates in one IC. Having only one IC that provides only two-input *OR* gates is not a limitation in any way. As will be seen in the next section and in Chapter 3, there are several procedures that can be used to emulate the function of *OR* gates having more than two inputs, or to construct *OR* gates with several inputs.

1. Compare and contrast the *AND* and *OR* gates.
2. Is it possible to make a four-input *OR* gate function logically like a buffer? If yes, how can it be done?
3. How could we connect two 2-input *OR* gates to function like a three-input *OR* gate?

2.7

The *NAND* and *NOR* gates

There are two other basic logic gates that are frequently used in digital circuits — the *NAND* and *NOR* gates. Both these gates are actually combinations of *AND* gates, *OR* gates, and inverters but are treated as discrete logic gates.

The *NAND* gate is a combination of an *AND* gate and a *NOT* gate (inverter). Hence, the letter *N* in the word "*NAND*" indicates that the *AND* function is inverted. This inversion is done at the output as shown in Figure 2.21.

Figure 2.21a and b show the symbols that represent two- and three-input *NAND*

Figure 2.21

The *NAND* gate. (a)
Distinctive shape symbol;
(b) ANSI / IEEE Std. 91-
1984 symbol; (c) equivalent
circuit.

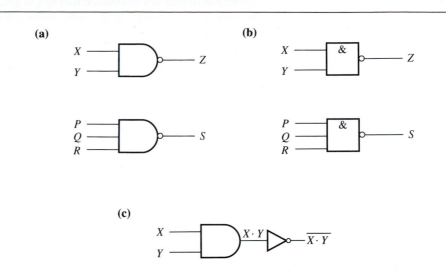

gates. As before, these symbols can be modified to represent more inputs. The symbol for the *NAND* gate is similar to the symbol of an *AND* gate except that the output line has a bubble symbol. Recall from Section 2.3 that the bubble symbol indicates inversion, and therefore this means that the output of the gate is internally inverted. The circuit in Figure 2.21c represents a two-input *NAND* gate as an *AND* gate with an inverter connected to its output. The output of a *NAND* gate therefore is exactly the opposite of the output of an *AND* gate. This means that the output of a *NAND* gate (dependent variable) will be a logic 0 if and only if all inputs (independent variables) are at the logic 1 state. The truth tables and logic equations for the two- and three-input NAND gates shown in Figure 2.21 are given in Tables 2.11 and 2.12, respectively.

Placement of the bar over the entire term on the right-hand side of each logic equation indicates that the *AND* function is inverted to produce the *NAND* function.

Table 2.11   Truth table and logic equation for a two-input *NAND* gate

X	Y	Z
0	0	1
0	1	1
1	0	1
1	1	0

Logic equation: $Z = \overline{X \cdot Y}$

Table 2.12   Truth table and logic equation for a three-input *NAND* gate

P	Q	R	S
0	0	0	1
0	0	1	1
0	1	0	1
0	1	1	1
1	0	0	1
1	0	1	1
1	1	0	1
1	1	1	0

Logic equation: $S = \overline{P \cdot Q \cdot R}$

The *NOR* gate is a combination of an *OR* gate and a *NOT* gate (inverter). Hence, the letter *N* in the word ''*NOR*'' indicates that the *OR* function is inverted. Like the *NAND* gate, this inversion is done at the output as shown in Figure 2.22.

Figure 2.22a and b show the symbols that represent two- and three-input *NOR* gates. As before, these symbols can be modified to represent more inputs. The symbol for the *NOR* gate is very similar to the symbol for an *OR* gate except that the output line has a bubble symbol. Again, because the bubble symbol indicates inversion, this means that the output of the gate is internally inverted. The circuit in Figure 2.21c represents a two-input *NOR* gate as an *OR* gate with an inverter connected to its output. The output of a *NOR* gate therefore is exactly the opposite of the output of an *OR* gate. This means that the output of the *NOR* gate (dependent variable) will be a logic 0 if any of its inputs (independent variables) are at the logic 1 state. The truth tables and logic equations for the two- and three-input *NOR* gates shown in Figure 2.22 are given in Tables 2.13 and 2.14, respectively.

Figure 2.22

The *NOR* gate. (a) Distinctive shape symbol; (b) ANSI/IEEE Std. 91-1984 symbol; (c) equivalent circuit.

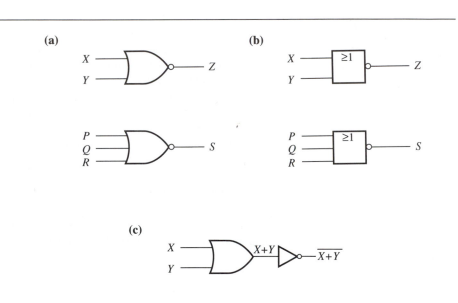

Table 2.13 Truth table and logic equation for a two-input *NOR* gate

X	Y	Z
0	0	1
0	1	0
1	0	0
1	1	0

Logic equation: $Z = \overline{X + Y}$

Table 2.14   Truth table and logic equation for a three-input *NOR* gate

P	Q	R	S
0	0	0	1
0	0	1	0
0	1	0	0
0	1	1	0
1	0	0	0
1	0	1	0
1	1	0	0
1	1	1	0

Logic equation: $S = \overline{P + Q + R}$

As in the *NAND* gate, the placement of the bar over the entire term on the right-hand side of each logic equation indicates that the *OR* function is inverted to produce the *NOR* function.

Example 2.8

Figure 2.23

A five-input *NAND* gate used as a four-input *NAND* gate.

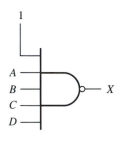

The five-input *NAND* gate shown in Figure 2.23 has one of its inputs permanently tied high (logic 1 state). Obtain the truth table and the logic equation for the gate. What would happen to the gate if one of its inputs were permanently tied low (logic 0 state)?

Solution

Because one of the five inputs is permanently tied to the logic 1 state, the five-input *NAND* gate now functions like a four-input *NAND* gate. This is because the output of the gate (X) now depends on whether the states of the four remaining inputs (A, B, C, D) are all 1's, or if any one of the inputs is a 0.

Therefore, its truth table and logic equation is

A	B	C	D	X
0	0	0	0	1
0	0	0	1	1
0	0	1	0	1
0	0	1	1	1
0	1	0	0	1
0	1	0	1	1
0	1	1	0	1
0	1	1	1	1
1	0	0	0	1
1	0	0	1	1
1	0	1	0	1
1	0	1	1	1
1	1	0	0	1
1	1	0	1	1
1	1	1	0	1
1	1	1	1	0

Logic equation: $X = \overline{A \cdot B \cdot C \cdot D}$

In any *NAND* gate, if any one of the inputs is a logic 0, the output will be a logic 1 and will not be affected by the states of the other inputs.

Example 2.9

The five-input *NOR* gate shown in Figure 2.24 has one of its inputs permanently tied low (logic 0 state). Obtain the truth table and the logic equation for the gate. What would happen to the gate if one of its inputs were permanently tied high (logic 1 state)?

Solution

Because one of the five inputs is permanently tied to the logic 0 state, the five-input *NOR* gate now functions like a four-input *NOR* gate. This is because the output of the gate (X) now depends on whether the states of the four remaining inputs (A, B, C, D) are all 0's, or if any one of the inputs is a 1.

Figure 2.24

A five-input *NOR* gate used
as a four-input *NOR* gate.

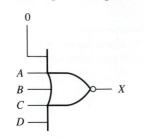

Therefore, its truth table and logic equation is

A	B	C	D	X
0	0	0	0	1
0	0	0	1	0
0	0	1	0	0
0	0	1	1	0
0	1	0	0	0
0	1	0	1	0
0	1	1	0	0
0	1	1	1	0
1	0	0	0	0
1	0	0	1	0
1	0	1	0	0
1	0	1	1	0
1	1	0	0	0
1	1	0	1	0
1	1	1	0	0
1	1	1	1	0

Logic equation: $X = \overline{A + B + C + D}$

In any *NOR* gate, if any one of the inputs is a logic 1, the output will be a logic 0 and will not be affected by the states of the other inputs.

The *NAND* and *NOR* gates are very useful in logic circuits because they can greatly reduce the complexity of a logic circuit. The techniques used to do this will be examined in the next chapter. *NAND* and *NOR* gates can also be used as *AND* and *OR* gates, respectively, by inverting the outputs. Figure 2.25 shows the circuits required to do this.

Figure 2.25

(a) *AND* gate; (b) *OR* gate.

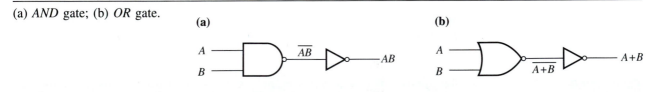

IC logic

The 74xxx series includes a variety of ICs that package the *NAND* and *NOR* gates. Figure 2.26 shows the pinouts of several *NAND* ICs — the 7400 quad 2-input *NAND* gate, the 7410 triple 3-input *NAND* gate, the 7420 dual 4-input *NAND* gate, and the 7430 eight-input *NAND* gate.

There are two ICs that package the *NOR* gates. These are shown in Figure 2.27 — the 7402 quad 2-input *NOR* gate, and the 7427 triple 3-input NOR gate.

Figure 2.26

Pin configurations of various *NAND* gate ICs.

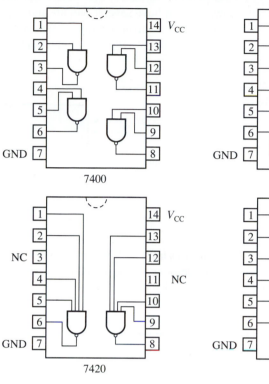
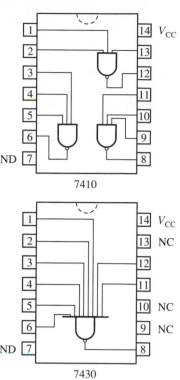

Figure 2.27

Pin configurations of various *NOR* gate ICs.

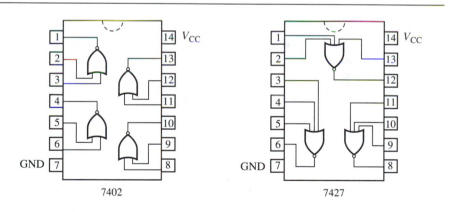

Review Questions

1. Is it possible to make a four-input *NAND* gate or *NOR* gate function as an inverter? If yes, how can it be done?
2. What is the significance of the bubble in the logic symbols that are used to represent the *NAND* and *NOR* gates?
3. Is it possible to connect two 2-input *NOR* gates to function like a three-input *NOR* gate?
4. Is it possible to connect two 2-input *NAND* gates to function like a three-input *NAND* gate?

2.8
Logic circuits

Like logic gates, logic circuits can be viewed as "black boxes" with binary inputs (independent variables) and binary outputs (dependent variables). Logic circuits, however, can have several outputs and are made up of the various logic gates discussed in this chapter and other gates that will be introduced in following chapters. The basic model of a logic circuit is shown in Figure 2.28.

Figure 2.28

The model of a logic circuit.

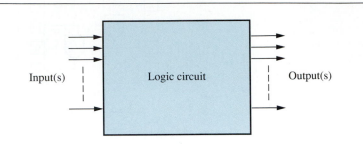

Just as a truth table was used to describe the operation of a logic gate, so can it be used to describe the operation of a logic circuit. Similarly, a logic equation can also be used to relate the output of a logic circuit to its inputs. To illustrate this fact, consider the following examples.

Example 2.10

Obtain the logic equations and truth tables for the circuits shown in Figure 2.29a and b.

Figure 2.29

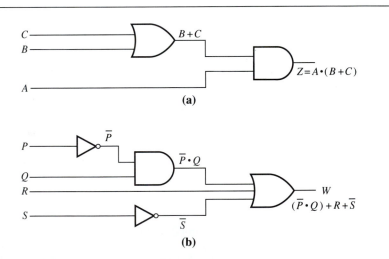

Solution

(a) In Figure 2.29a, the logic equation for the circuit was obtained by tracing the independent variables A, B, and C through each gate in the circuit. Because the input to the *OR* gate is B and C, the output of the *OR* gate is $B + C$. One of the inputs of the *AND* gate is A, while the other input is connected to the output of the *OR* gate, $B + C$. Therefore, the output of the *AND* gate and the logic equation for the entire circuit is

$$Z = A \cdot (B + C)$$

The term $B + C$ is grouped in parentheses to indicate that it is evaluated before the *AND* function. Enclosing the term $B + C$ in parentheses allows the term to be treated as if it were a single variable. The truth table for the logic equation can be obtained by analyzing each gate in the circuit separately as follows:

A	*B*	*C*	*B* + *C*	**Z** *A* · (*B* + *C*)
0	0	0	0	0
0	0	1	1	0
0	1	0	1	0
0	1	1	1	0
1	0	0	0	0
1	0	1	1	1
1	1	0	1	1
1	1	1	1	1

To obtain the output of the *OR* gate in terms of the variables *B* and *C*, "OR" (apply the *OR* function) the columns for *B* and *C* using the basic truth table for an *OR* gate. To obtain the output of the *AND* gate (the final output of the circuit), "AND" (apply the *AND* function) the columns for *A* and *B* + *C* using the basic truth table for an *AND* gate.

(b) In Figure 2.29b apply the same procedure as for Figure 2.29a to determine the logic equation for the circuit

$$W = (\overline{P} \cdot Q) + R + \overline{S}$$

Again, the term $(\overline{P} \cdot Q)$ is grouped together because it is evaluated before the *OR* function and is treated like a single variable at the input of the *OR* gate. As before, the truth table can be determined by tracing all possible combinations of the independent variables *P*, *Q*, *R*, and *S* through each gate in the circuit.

P	*Q*	*R*	*S*	$\overline{P}$	$\overline{S}$	$\overline{P} \cdot Q$	**W** $(\overline{P} \cdot Q) + R + \overline{S}$
0	0	0	0	1	1	0	1
0	0	0	1	1	0	0	0
0	0	1	0	1	1	0	1
0	0	1	1	1	0	0	1
0	1	0	0	1	1	1	1
0	1	0	1	1	0	1	1
0	1	1	0	1	1	1	1
0	1	1	1	1	0	1	1
1	0	0	0	0	1	0	1
1	0	0	1	0	0	0	0
1	0	1	0	0	1	0	1
1	0	1	1	0	0	0	1
1	1	0	0	0	1	0	1
1	1	0	1	0	0	0	0
1	1	1	0	0	1	0	1
1	1	1	1	0	0	0	1

To obtain the $\overline{P}$ and $\overline{S}$ columns simply invert the *P* and *S* columns. The $(\overline{P} \cdot Q)$ column is obtained by applying the *AND* function to the $\overline{P}$ and *Q* columns. The last column represents the final output *W* and is obtained by applying the *OR* function to the *R*, $\overline{S}$, and $(\overline{P} \cdot Q)$ columns.

If a logic circuit has more than one output (dependent variable), then the circuit is described by a series of logic equations, each one representing a single output. Each logic equation can be treated independently in order to obtain the truth table for the circuit. The following example illustrates this procedure.

Example 2.11   Obtain the logic equation and truth table for the circuit shown in Figure 2.30.

Figure 2.30

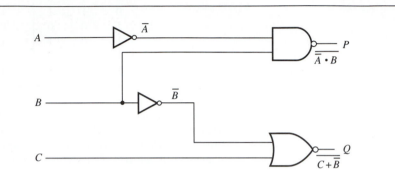

Solution

The outputs of the logic circuit are described by the following logic equations:

$$P = \overline{\overline{A} \cdot B} \tag{2.4}$$

$$Q = \overline{C + \overline{\overline{B}}} \tag{2.5}$$

The truth table of the circuit is obtained by independently determining the values of P and Q for all possible combinations of A, B, and C. That is, think of the circuit in Figure 2.30 as being made up of two circuits, one that is described by Equation 2.4, and the other by Equation 2.5.

A	B	C	$\overline{A}$	$\overline{B}$	P $\overline{A} \cdot B$	Q $C + \overline{B}$
0	0	0	1	1	1	0
0	0	1	1	1	1	0
0	1	0	1	0	0	1
0	1	1	1	0	0	0
1	0	0	0	1	1	0
1	0	1	0	1	1	0
1	1	0	0	0	1	1
1	1	1	0	0	1	0

It is also possible to work backward and draw the circuit for a given logic equation. The order in which each term in the equation is drawn is the same as the order in which each term is evaluated, with terms in parentheses treated as single variables. Example 2.12 illustrates these procedures for a few simple logic equations. More examples will follow in Chapter 3 after the various mathematical techniques of manipulating these equations have been introduced.

Example 2.12

Draw the logic circuits for the following logic equations:
(a) $T = (A \cdot B) + C$
(b) $X = \overline{P} \cdot Q \cdot (R + \overline{Q})$
(c) $F = (\overline{X + Y + Z}) + (W \cdot \overline{X})$

Solutions

Figure 2.31

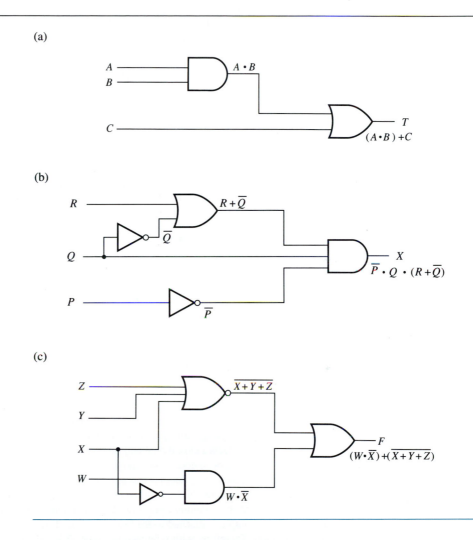

(a)

(b)

(c)

The following example illustrates a practical application of a logic circuit that is used to control a garage door opener system. The circuit used in the example does have its limitations and deficiencies and will need additional circuitry (to be examined in Chapter 5) if it were actually implemented. It does, however, show how logic gates can be used in circuits to make "decisions" about certain conditions.

Example 2.13

The logic circuit shown in Figure 2.32 is used to control the movement (up or down) of a garage door. To detect the current position of the garage door, limit switches LS1 and LS2 are used. Both limit switches are spring loaded momentary single-pole double-throw (SPDT) switches whose normal position maintains continuity between pole 1 and the common pole.

Figure 2.32

Logic circuit for garage door
controller.

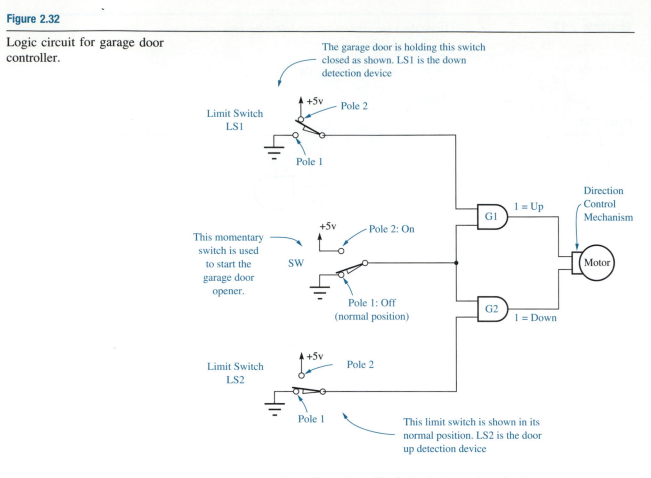

Note: The position of the limit switches are shown for the garage
door in the down position

An actuator button built into the switch will move the contacts to connect pole 2 to the
common pole (forced position) when pressed.

Figure 2.33 illustrates the positioning of the limit switches. There are two limit switches
for detecting garage door position. Limit switch LS1 is positioned at the bottom of the door
travel and detects a closed door and limit switch LS2 is positioned at the top of the door travel
and detects an open door. When the garage door is completely down (Fig. 2.33a) the down tab
attached to the garage door presses against the actuator button of the limit switch LS1 mounted
on the garage door sill. LS1 is then held in the forced position resulting in continuity between
pole 2 and the common pole (Fig. 2.32). Similarly, when the garage door is opened (Fig.
2.33b), an up-tab attached to the top of the door forces the up-limit switch LS2 in the forced
position. Therefore, with reference to Figure 2.32 when the door is down, LS1 will apply a
logic 1 to gate G1 and LS2 will apply a logic 0 to gate G2. Similarly, when the door is up, LS1
will apply a logic 0 to gate G1 and LS2 will apply a logic 1 to gate G2.

The logic circuit shown in Figure 2.32 illustrates how the control of the garage door
movement is accomplished. The logic levels produced by *AND* gates G1 and G2 control the
rotation of a motor to move the garage door up or down as follows:

G2 output	G1 output	Motor rotation
0	0	Off
0	1	Door moves UP
1	0	Door moves DOWN
1	1	Fault

Figure 2.33

Positioning of limit switches for garage door controller.

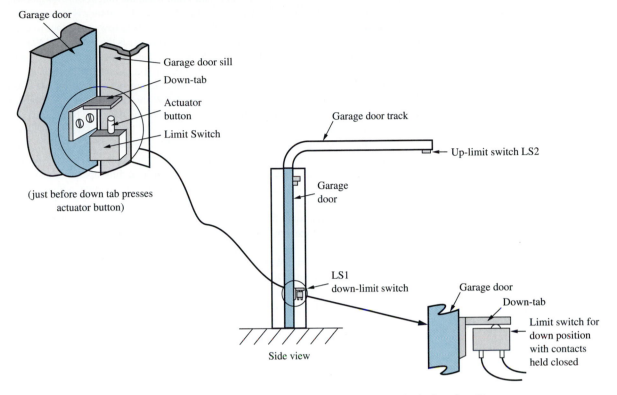

(a) Garage door in the down position — down limit switch LS1 in the forced position

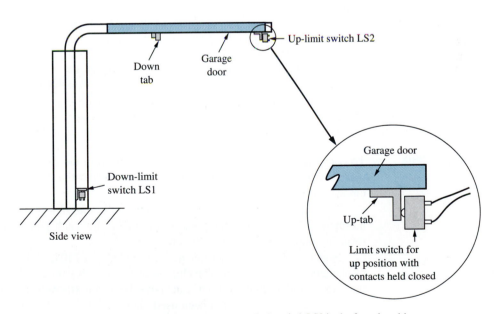

(b) Garage door in the up position — up limit switch LS2 in the forced position

Examining the truth table shown above, the first row states that when the outputs of both *AND* gates G1 and G2 are logic 0's, the garage door motor is not powered and the door remains in its present position. The second row of the table states that when the output of *AND* gate G2 is a logic 0 and the output of *AND* gate G1 is a logic 1, the garage door motor is powered to rotate to move the door in the UP direction (open the door). The third row states that to close the garage door, the output of *AND* gate G1 must be a logic 0 whereas the output of *AND* gate G2 must be a logic 1. The last row of the truth table is an invalid condition (the door cannot be opened and closed at the same time) and can only occur if a circuit fault exists.

The logic levels produced by *AND* gates G1 and G2 are determined by the two limit switches LS1 and LS2 and a momentary SPDT switch SW shown in Figure 2.32. Switch SW is used to move the door up or down. The actual direction of the door will be dictated by limit switches LS1 and LS2. The normal position of SW is at pole 1 (OFF). Switch SW is used to enable or disable *AND* gates G1 and G2. When switch SW is pressed (ON) a logic 1 is applied to *AND* gates G1 and G2, thus enabling both gates. In its normal position switch SW applies a logic 0 to gates G1 and G2, thus disabling both gates and inhibiting operation of the motor.

If the door is down (as indicated by the position of the switches shown in Figure 2.32) and switch SW is pressed, both inputs of *AND* gate G1 are at a logic 1 state and therefore the output of G1 is a logic 1. At the same time the output of G2 is a logic 0 because limit switch LS2 applies a logic 0 to one of the inputs of G2. As indicated in the truth table, the door then moves UP.

If door is up and switch SW is pressed, this time both inputs of gate G2 are at a logic 1 state and therefore the output of *AND* gate G2 is a logic 1. At the same time the output of *AND* gate G1 is a logic 0 because limit switch LS1 applies a logic 0 to G1 when the door is UP. As indicated in the truth table, the door moves DOWN.

In order for this circuit to work, certain assumptions must be made. Assume that when the output of G1 or G2 goes momentarily from a logic 0 to a logic 1, it triggers some sort of latching mechanism (such as a relay) in the motor's direction control circuit to keep the motor powered in the appropriate direction even when the output of G1 or G2 returns to the logic 0 state. This is because the instant the door starts moving from its completely up or down position, the limit switches LS1 or LS2 will apply logic 0's to the inputs of G1 and G2, thus disabling the two gates. According to the truth table, the motor should turn off when this occurs. However, the latching mechanism should keep the motor running. It must also be assumed that another set of limit switches will turn off the motor when the door reaches its terminal positions. A potential problem with this circuit is that if the door's initial position is between its completely down or up position, it could never be moved with this circuit. Chapter 5 will examine other logic devices that will solve some of these problems. This circuit will then be completely redesigned to obtain a more practical and usable version of the garage door controller.

Constructing logic circuits with ICs

To construct a logic circuit in the laboratory, the various 74xxx ICs that implement the basic logic gates are used. For example, consider the construction of the logic circuit shown in Figure 2.34.

The first step would be to draw a schematic diagram that identifies the chip numbers (74xxx IC numbers) and pin numbers for each gate in the circuit. The resulting diagram is shown in Figure 2.35.

Note in Figure 2.35 that three ICs are needed — a 7404, a 7408, and a 7432 — to implement the logic circuit. The chip number for each gate and the pin numbers for the inputs and output of each gate have been identified. Because all the gates available in each IC have not been used, there will be five ''spare'' inverters in the 7404, two ''spare'' *AND* gates in the 7408, and three ''spare'' *OR* gates in the 7432.

The next step would be to draw a layout and wiring diagram that shows how the ICs are laid out on a breadboard and wired together. This diagram is shown in Figure

Figure 2.34

Logic circuit to be
constructed.

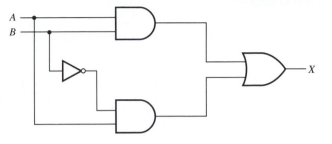

2.36 and represents the actual logic circuit that would be set up in the laboratory. Generally, before the circuit is set up it is good practice to wire-up the V_{CC} and GND pins of all the ICs to the 5-volt power supply as shown in Figure 2.36.

It is important to remember that any unused inputs to a gate that is being used in a logic circuit must be *conditioned* to a logic high or logic low as appropriate. For example, if a three-input *AND* gate is being used to function as a two-input *AND* gate, then the unused third input must be tied to a logic high (V_{CC}) to disable it (see Fig. 2.16). Similarly, if a three-input *NOR* gate is being used to function as a two-input *NOR* gate, then the unused third input must be tied to a logic low (GND) to disable it (see Fig. 2.24). Never assume that because an unconnected input has no voltage on it, it is equivalent to having a logic low; a logic low is a voltage — ground potential! In reality, the inputs to a 74xxx TTL logic gate are internally pulled up to a logic high and therefore if left unconnected will assume the logic high state; however, this is not recommended because unconditioned (or floating) inputs have a tendency to pick up noise that can affect the operation of the logic circuit.

The outputs of two or more logic gates should never be connected together, especially if one output is at a logic high state and another is at a logic low state. Logically, the level at this connection (node) is undefined because there are two logic levels applied to it, and electrically, this sort of connection could cause damage to the outputs of the gates because we are effectively shorting a 5-volt potential to ground through the output of the gate. There are two exceptions to this rule that will be examined in Chapters 4 and 12.

Figure 2.35

Schematic diagram.

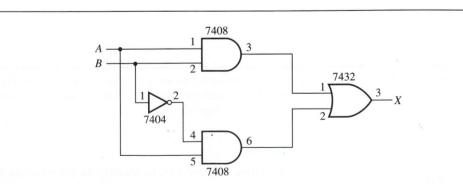

Figure 2.36

Layout and wiring diagram.

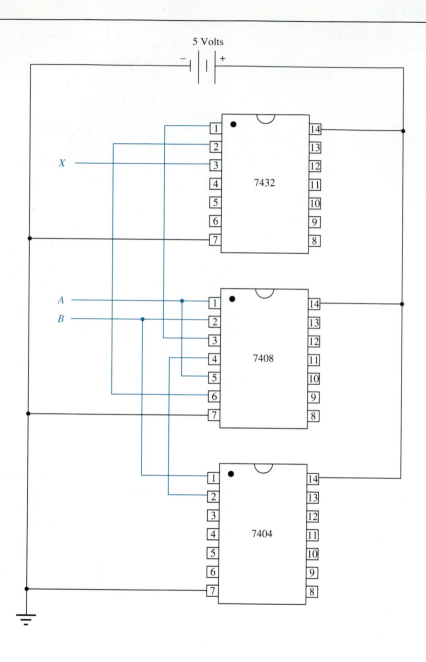

Example 2.14

Draw a schematic diagram for the logic circuit shown in Figure 2.30 showing the chip and pin numbers for each gate. Also draw a layout and wiring diagram that represents the laboratory set-up of the logic circuit.

Solution

See Figure 2.37 and 2.38, respectively, for the schematic and layout diagrams.

Figure 2.37

Schematic diagram.

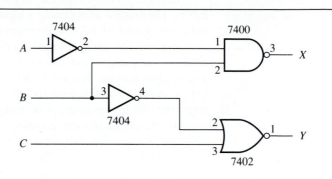

Figure 2.38

Layout and wiring diagram

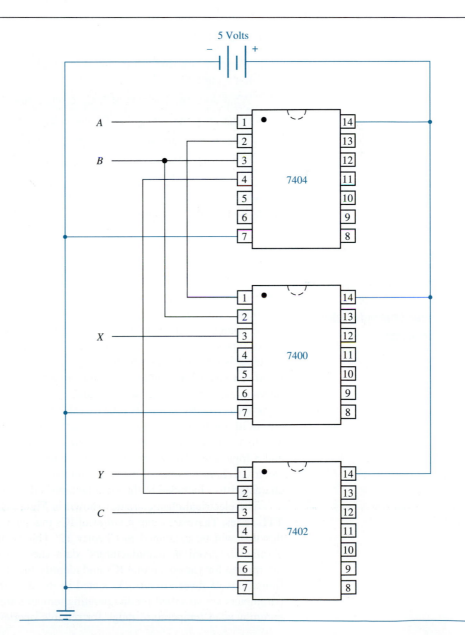

This section has examined simple logic circuits, their equations, and their truth tables. The process of obtaining a logic equation and truth table from a given circuit is known as *analysis* because these are the components of a logic circuit that describe its operation. The study of digital circuits also involves the *synthesis* of logic circuits. Synthesis (the opposite of analysis) involves the putting together or design of logic circuits, given a specification, truth table, or logic equation. This section has included simple examples of both analysis and synthesis of logic circuits. However, in order to study the different procedures of analysis and synthesis of logic circuits, and apply them to the design of more complex circuits, we must first study the mathematical concepts of logic that will allow us to understand the basic laws of logic functions and their interrelationships. This and other techniques of designing and analyzing digital circuits will be discussed in Chapter 3.

Review Questions

1. What are the similarities and differences between a basic logic gate and a logic circuit?
2. What does the analysis of logic circuits involve?
3. What does the synthesis of logic circuits involve?
4. What is the purpose of drawing a schematic diagram for a logic circuit?
5. How does a layout and wiring diagram differ from the schematic diagram of a logic circuit?
6. Why is it important to condition all unconnected inputs of a logic gate to the high or low state?

2.9

Electrical characteristics of logic gates

The previous sections showed that the inputs and outputs of logic gates are the binary numbers 0 and 1 that are represented by the voltages 0 and $+5$ volts, respectively. These voltages correspond to the power supply potentials for the IC, which for the 74xxx series are referred to as GND and V_{CC}, respectively. In electrical terminology, a logic 1 is referred to as a high voltage level (or simply a "high") and a logic 0 is referred to as a low voltage level (or simply a "low"). Because of the effects of internal resistances within the gates and loading effects, an output may not necessarily produce exactly $+5$ volts for a logic high and 0 volts for a logic low. Because the output of a gate may have to drive several inputs, the input of a logic gate must be able to accept a range of voltages that are acceptable for interpretation of a logic high and a logic low. These are some of the gate specifications that are included in the electrical characteristics of the IC manufacturer's data sheets. Some of the important characteristics included in the manufacturer's data sheets are shown in Figure 2.39.

The electrical characteristics shown in Figure 2.39 are for the standard 74xxx TTL series. There are other *families* of ICs that have different electrical characteristics that will be examined in Chapter 12. This section will examine some of the parameters given in manufacturers' data sheets so that the data sheets can be interpreted for various digital ICs and identify the characteristics, requirements, and limitations of these circuits. As stated in the data sheet shown in Figure 2.39, all parameters are specified for an operating temperature range of 0 to 70°C. This is the recommended temperature range for reliable operation of the IC.

Figure 2.39

Manufacturer's data sheet showing the electrical characteristics of the 74xxx ICs.

Recommended operating conditions

Parameter	Minimum	Typical	Maximum	Units
Supply voltage (V_{CC})	4.75	5.0	5.25	V
Operating free-air temperature range	0	25	70	°C
HIGH-level output current (I_{OH})			−400	mA
LOW-level output current (I_{OL})			16	mA

Electrical characteristics over operating temperature range (unless otherwise noted)

Parameter	Minimum	Typical[b]	Maximum	Units	Test conditions[a]
HIGH-level input voltage (V_{IH})	2.0			V	
LOW-level input voltage (V_{IL})			0.8	V	
HIGH-level output voltage (V_{OH})	2.4	3.4		V	V_{CC} = min, I_{OH} = 0.4 mA, V_{IN} = 0.8 V
LOW-level output voltage (V_{OL})		0.2	0.4	V	V_{CC} = min, I_{OL} = 16 mA, V_{IN} = 2.0 V
HIGH-level input current (I_{IH})			40	mA	V_{CC} = max, V_{IN} = 2.4 V
LOW-level input current (I_{IL})			−1.6	mA	V_{CC} = max, V_{IN} = 0.4 V
Short-circuit output current[c] (I_{OS})	− 18		− 55	mA	V_{CC} = max
Total supply current with outputs high (I_{CCH})		4.0	8.0	mA	V_{CC} = max
Total supply current with outputs low (I_{CCL})		12	22	mA	V_{CC} = max

Switching characteristics (T_A = 25°C)

Parameter	Minimum	Typical	Maximum	Units	Test conditions
Propagation delay time, LOW-to-HIGH output (t_{PLH})		11	22	ns	V_{CC} = 5.0 V, C_{LOAD} = 15 pF, R_{LOAD} = 400 ohms
Propagation delay time, HIGH-to-LOW output (t_{PHL})		7.0	15	ns	

NOTES:

[a]For conditions shown as min or max, use the appropriate value specified under recommended operating conditions for the applicable device type.

[b]Typical limits are at V_{CC} = 5.0 V, 25°C.

[c]Not more than one output should be shorted at a time. Duration of short not to exceed 1 s.

Power requirements

The data sheet shown in Figure 2.39 includes the power supply voltage range for V_{CC}. This indicates that the voltage applied to power the IC can be in the range 4.75 to 5.25 volts but is typically 5 volts.

The total amount of current drawn by the IC is given by two parameters, I_{CCH} and I_{CCL}. I_{CCL} is the total amount of current drawn from the power supply V_{CC} when all the outputs are in the low state. I_{CCH} is the total amount of current drawn from the power supply V_{CC} when all the outputs are in the high state. Note that I_{CCL} is usually higher than I_{CCH} (the reason for this will be seen later).

Therefore the average current drawn from V_{CC} is

$$I_{CC} = \frac{I_{CCH} + I_{CCL}}{2}$$

and the average power dissipated by the IC is

$$P_{AVG} = I_{CC} \times V_{CC}.$$

Input characteristics

When a logic high is applied to the input of a gate, the voltage can be in the range of 2 to 5 volts. This is specified by the parameter V_{IH} (the IH stands for "high input voltage"), which is a minimum of 2 volts. This means that any voltage less than 2 volts will not be interpreted as a logic high by the gate. Because a positive voltage is applied to the input of the gate, the gate input represents a load and draws current. This current is typically 40 μA when the positive input voltage is 2.4 volts. This input current at the logic high state is referred to as I_{IH}. The relationship between the input voltage and I_{IH} is shown in Figure 2.40a. The reason why 2.4 volts is chosen for the input voltage will be seen later.

Figure 2.40

Input characteristics of a 74xxx logic gate.

Logic high input

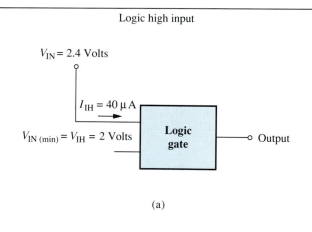

(a)

Logic low input

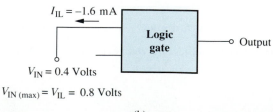

(b)

When a logic low is applied to the input of a gate, current flows out of the input. By convention, this is identified as a negative current. A low is generally 0 volts or ground potential but can be a maximum of 0.8 volts as specified by the parameter V_{IL} in the data sheet. This means that a logic gate will interpret any voltage in the range 0 to 0.8 volts as a logic low but any voltage greater than 0.8 volts will not be considered a logic low. In Figure 2.40b, when 0.4 volts is applied to the input of the gate (the reason for picking this value will be seen later) the current flowing out of the gate, I_{IL}, is -1.6 mA as specified in the data sheet.

Thus, the input of a logic gate will not respond correctly to any voltage between 0.8 and 2 volts because this area does not define a logic high or a logic low state. A logic gate should therefore never be operated in this undefined area.

Values for the input voltages and currents V_{IH}, V_{IL}, I_{IH}, and I_{IL} are given in the data sheet for the 74xxx TTL ICs shown in Figure 2.39.

Output characteristics

The input characteristics of a logic gate have been examined. The output characteristics will now be examined.

When the output of a logic gate is at a logic high its voltage can be anywhere from 2.4 to 5 volts. This is specified by the parameter V_{OH}. The manufacturer guarantees that V_{OH} will be a minimum of 2.4 volts. The value of the output voltage will change depending on how much current is drawn from the output. Typically, the output of a logic gate will be connected to (drive) the inputs of other logic gates and therefore the amount of current that is drawn from an output in the logic high state will depend on the number of loads (or inputs) attached to the output. Figure 2.41 illustrates the relationship between the output voltage, V_{OUT}, and the output current, I_{OUT}. If there is no load attached to the output of the gate, then the current flowing out of the gate is 0 and V_{OUT} has a value very close to 5 volts as shown in Figure 2.41a. When a load is attached to the output it draws current which causes V_{OUT} to decrease (due to internal gate resistances). The greater the loading, the more the

Figure 2.41

Output characteristics of a 74xxx logic gate at the high state.

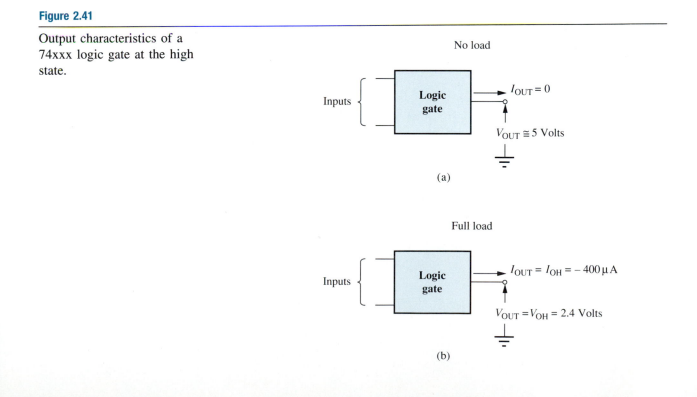

(a)

(b)

current drawn and the smaller the value of V_{OUT}. However the output should not be loaded to a point where V_{OUT} reaches a value less than 2.4 volts (V_{OH}). While maintaining $V_{OUT} = V_{OH}$ at 2.4 volts, the maximum amount of current that can be drawn from the output of the gate is $-400\ \mu A$ (by convention, the negative sign indicates that the current flow is out of the gate) as shown in Figure 2.41b. This current is known as I_{OH}. Any further increase in I_{OUT} will cause the output voltage to drop below V_{OH} (2.4 volts). The current I_{OH} is known as *source current*.

When the output of a logic gate is at a logic low its voltage can be anywhere from 0 to 0.4 volts. This is specified by the parameter V_{OL}. The manufacturer guarantees that V_{OL} will be a maximum of 0.4 volts. The value of the output voltage will change depending on how much current is drawn from the load attached to the output. Typically, the output of a logic gate will be connected to (drive) the inputs of other logic gates and therefore the amount of current that is drawn from an input in the logic low state will depend on the number of loads (or inputs) attached to the output.

When the output of a logic gate is at the logic low state it draws current from the load. This current flows into the output of the logic gate. If no load is attached to the output as shown in Figure 2.42a, then the current is 0 and the output voltage V_{OUT} is very close to 0 because the internal resistance of the gate does not affect the output voltage. If a load is attached to the output of the gate when at the logic low state, the flow of current into the gate's output causes a voltage drop inside the gate and increases the output voltage. The greater the loading, or the larger the number of loads, the larger the output voltage. However, the output should not be loaded to a point where V_{OUT} reaches a value greater than 0.4 volts. While maintaining V_{OUT} at 0.4 volts (V_{OL}), the maximum amount of current that can be drawn from the load(s) is 16 mA and is known as I_{OL}. Any further increase in I_{OUT} will cause the output voltage to increase above V_{OL} (0.4 volts). The current I_{OL} is known as *sink current*.

Figure 2.42

Output characteristics of a
74xxx logic gate at the low
state.

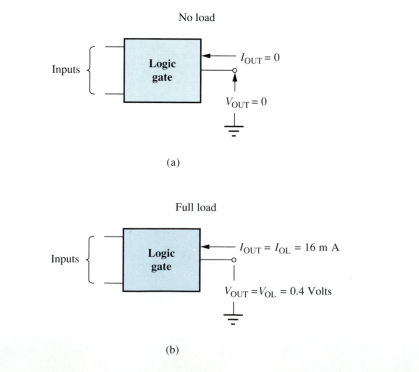

(a)

(b)

Therefore, a logic output sources current when in the high state and sinks current when in the low state.

Values for the input voltages and currents V_{OH}, V_{OL}, I_{OH}, and I_{OL} are given in the data sheet for the 74xxx TTL ICs shown in Figure 2.39. Note that the data sheet also lists the short circuit output current I_{OS}. This is the total amount of current that the output can source when short circuited; of course the output voltage will be 0 volts.

Fan-in and fan-out

It has been shown so far that 40 μA of current (I_{IH}) will flow into the input of a logic gate when a logic high is applied to it, and -1.6 mA of current (I_{IL}) will flow out of the input of a logic gate when a logic low is applied to it. These currents represent 1 *unit load* (UL) to the output of a logic circuit.

Because an output can source -400 μA of current in the high state (I_{OH}), the total number of inputs that an output can drive in the high state will be

$$\frac{I_{OH}}{I_{IH}} = \frac{400\,\mu A}{40\,\mu A} = 10.$$

Thus, an output can drive ten inputs or 10 ULs when in the high state. This number is referred to as the *fan-out (high)* of an output and it is illustrated in Figure 2.43a. The addition of any more loads will cause the output voltage to decrease below V_{OH} and therefore could place the voltage in the undefined area between a logic high and a logic low.

Because an output can sink 16 mA of current in the low state (I_{OL}), the total number of inputs that an output can drive in the low state will be

$$\frac{I_{OL}}{I_{IL}} = \frac{16\,mA}{1.6\,mA} = 10.$$

Thus, an output can also drive ten inputs or 10 ULs when in the low state. This number is referred to as the *fan-out (low)* of an output and it is illustrated in Figure 2.43b. Again, the addition of any more loads will cause the output voltage to increase above V_{OL} and therefore could place the voltage in the undefined area between a logic low and a logic high.

Figure 2.43

(a) Fan-out in the high state;
(b) Fan-out in the low state.

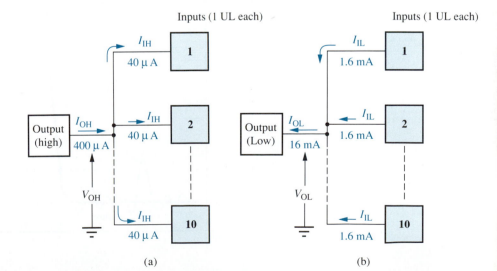

(a)   (b)

Therefore the fan-out of a standard 74xxx TTL logic gate is 10. This means that the output can safely drive ten inputs (or 10 ULs). If an attempt is made to make the output drive more than this number, unpredictable results could occur. Note that the term *fan-in* is sometimes used to refer to a unit load because these inputs are usually connected to the outputs (that have fan-outs). Most standard 74xxx TTL logic gates have a fan-out of 10, but some gates in this family, such as buffers and open-collector ICs, have larger fan-outs; these will be examined in Chapter 12. Similarly, most standard 74xxx TTL logic gates have inputs that represent 1 UL; there are, however, variations of these ICs that have inputs that represent more and even less than 1 UL.

Noise margin

When the input and output characteristics of logic gates are examined, particularly the voltages, the following observations can be made during worst case conditions — that is, when the gate is fully loaded and operating at its thresholds. Assume that the output of a 74xxx TTL gate is connected to ten inputs so that it is fully loaded (fan-out = 10).

The input of a logic gate requires a minimum voltage of 2 volts for a logic high — V_{IH}. The output of a logic gate when at the high state and fully loaded will produce a voltage of 2.4 volts — V_{OH}. There is, therefore, a margin of 0.4 volts when the gate is operating under worst case conditions. If the output voltage drops below this margin of 0.4 volts, it will no longer be considered a logic high by the inputs of the gates connected to it.

The input of a logic gate requires a maximum voltage of 0.8 volts for a logic low — V_{IL}. The output of a logic gate when at the low state and fully loaded will produce a voltage of 0.4 volts — V_{OL}. Again, there is a margin of 0.4 volts when the gate is operating under worst case conditions. If the output voltage rises above this

Figure 2.44

Effects of noise on a logic level.

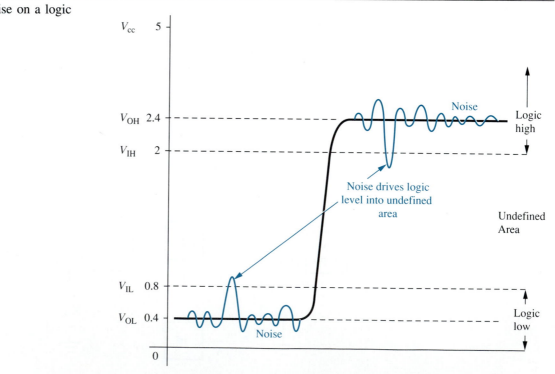

margin of 0.4 volts, it will no longer be considered a logic low by the inputs of the gates connected to it.

Figure 2.44 illustrates the effects of noise on logic gates operating under worst case conditions. Assuming that the logic high level is right at V_{OH} (2.4 volts) and the logic low level is right at V_{OL} (0.4 volts), any electrical noise that is picked up over the interconnecting wires will "ride" on these logic levels and will not affect the operation of the gates as long as the amplitude of the noise is less than 0.4 volts. If the amplitude of the noise exceeds 0.4 volts (peak voltage), then the logic level could be driven into the undefined area.

This value of 0.4 volts is known as the *noise margin* because it establishes the amount of noise that can exist on a logic level under worst case conditions. The noise margin is therefore defined as the difference between worst case output voltage and worst case input voltage.

Thus the noise margin in the high state is

$$V_{NH} = V_{OH} - V_{IH} = 2.4 - 0.4 = 0.4 \text{ volts}$$

and the noise margin in the low state is

$$V_{NL} = V_{IL} - V_{OL} = 0.8 - 0.4 = 0.4 \text{ volts}.$$

Propagation delay

When the input of a gate changes states and this change causes the output to change state, the change of state at the output is not instantaneous. There is a certain amount of delay time associated with the output. This delay time is known as *propagation delay*.

Figure 2.45 shows the effects of propagation delay on an inverter both under ideal conditions and actual conditions. Ideally, when the input of the inverter switches from a low (L) to a high (H), the output of the inverter will instantaneously switch from a high to a low. However, in practice there is a delay associated with a logic level that switches from a low to a high (*rise time*) and a delay associated with a logic level that switches from a high to a low (*fall time*). The propagation delay time is the time between a change of state that occurs at the input and a change of state that occurs at the output. Manufacturer data sheets often specify two propagation delay times as shown in Figure 2.39:

t_{pHL} = propagation delay time when the output changes from high to a low state

t_{pLH} = propagation delay time when the output changes from low to a high state

These two values are usually equal, but if they are not then the average propagation delay time (t_{pd}) is

$$t_{pd} = \frac{t_{pHL} + t_{pLH}}{2}$$

and the worst case propagation delay time is the larger of the two values.

Propagation delays are an indication of the switching speed of the gate. A slow gate has a longer propagation delay time and therefore has a slow switching speed whereas a fast gate has a smaller propagation delay time and a fast switching speed. The effects of gate propagation delays are usually felt on sequential circuits and will be examined in Chapter 5.

Figure 2.45

Propagation delay in a logic gate.

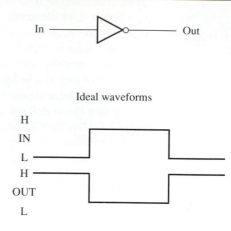

Ideal waveforms

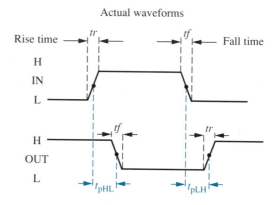

Actual waveforms

Review Questions

1. Explain why the current drawn from the power supply V_{CC} is more when all the outputs are at the low state.

2. Define the range of voltages that are recognized as a logic high and a logic low at the input of a gate.

3. Describe the currents that flow at the input of a gate when a logic high and a logic low are applied to it.

4. Define the range of voltages that are produced by the output of a gate when at the high state and at the low state.

5. Describe the terms *source* and *sink currents* with reference to the output of a gate.

6. Define the terms *fan-out* and *fan-in*.

7. Define *noise margin*.

8. What is propagation delay? What effect does it have on the speed of a gate?

2.10
Troubleshooting logic gates and circuits

Section 2.8 examined the procedures to set up a logic circuit with ICs using the layout and wiring diagram. Once the circuit is constructed on a breadboard there is often a good possibility that the circuit will not work as intended due to one of the following reasons:

1. Faulty power supply
2. Defective IC
3. Wiring mistake

This section will deal with the procedures used to identify and correct such faults in a logic circuit. These procedures are known as *troubleshooting procedures*.

Troubleshooting equipment

One of the advantages in troubleshooting digital circuits over analog circuits is that because digital circuits are two-stated we only have to determine if the inputs and outputs of a circuit are at a logic 1 (high) or a logic 0 (low) state. To troubleshoot analog circuits, one needs to measure continuously varying voltages and currents that have to be within fairly close tolerances. This characteristic of digital circuits allows testing and troubleshooting of logic circuits without the need for expensive equipment and lengthy procedures. The most commonly used equipment for troubleshooting digital circuits can be listed as follows in the order of importance:

1. Logic probe
2. Logic pulser
3. Current tracer
4. Voltmeter and ammeter
5. Oscilloscope
6. Logic analyzer
7. Signature analyzer

This section will examine the use of the logic probe, logic pulser, current tracer, voltmeter, and ammeter in troubleshooting combinational logic circuits. Later chapters will discuss the application of the oscilloscope, logic analyzer, and signature analyzer for the dynamic troubleshooting of various application circuits and sequential logic circuits.

The logic probe can be used to detect 90% of all problems with simple combinational logic circuits. The logic probe simply indicates whether a logic level is at the logic high state, logic low state, or changing states. The logic probe consists of a thin needle-like probe tip connected to a housing that has three *light emitting diodes* (LEDs). A set of leads is used to supply power (usually 5 volts) to the logic probe. To determine the logic level at a particular pin of an IC, touch the probe tip to the pin and observe the LEDs. One LED indicates a logic high state, the other indicates a logic low state, and the third indicates if the logic state at the pin is changing rapidly from a high to a low and vice versa (pulsing). If neither the high LED nor the low LED illuminate, then the logic level is neither high nor low but is floating, or in a high-impedance state (more will be said about this state in Chapter 4). Many logic probes may also have a switch that is used to select between the various families of logic ICs.

The logic pulser, which is similar in construction to the logic probe, is used to inject a pulse into a logic circuit. The logic pulser has a pushbutton switch that applies a short-duration pulse to a circuit node. This pulse causes a node that is at a logic high state to momentarily go low and a node that is at a logic low state to momentarily go high. The logic pulser can be used to inject a pulse into a circuit node without the need to isolate the node from the rest of the circuit. It cannot produce a voltage pulse on a node that is directly shorted to ground or V_{CC}; it can, however, produce a current pulse at such a node.

The current tracer is often used in conjunction with the logic pulser to detect a changing current in a wire or connection between nodes. The current tracer is also similar in construction to the logic probe but contains an insulated tip that can sense the presence of a magnetic field caused by a changing current. An LED or lamp indicates the presence of current flow in the circuit.

Voltmeters and ammeters are useful instruments for measuring logic levels to ensure that the voltages and currents are within the ranges specified for the IC. The oscilloscope and logic analyzer are generally used to examine dynamic logic states in sequential logic circuits where logic levels are constantly changing. The signature analyzer is a useful tool in quickly pinpointing a faulty component in a circuit without having to go through elaborate diagnostic procedures. More will be said about the logic analyzer and signature analyzer in Chapters 8 and 9.

Causes of faults in ICs

Digital ICs often contain several gates in one package. Often a single gate may malfunction whereas the other gates operate properly, or in some cases all the gates in the IC may malfunction due to an internal problem with the power supply. A single gate may not work as it should because of a number of problems related to the internal circuitry of that gate. Typical problems include

1. Damage of internal components such as transistors and resistors due to overloading or overdriving outputs and inputs
2. Inputs or outputs that are internally shorted to V_{CC} or ground
3. Inputs or outputs that are internally open circuited

Because it is impossible to repair any of the above three situations, the IC that exhibits one or more of these symptoms is generally replaced. It is therefore fruitless to devote much troubleshooting time in determining the reason for the defective operation of the gate or IC. Instead, it is beneficial to concentrate on troubleshooting logic circuits to narrow the problem down to a defective IC that can simply be replaced.

Troubleshooting procedures

The first step in any troubleshooting procedure is to make sure that each IC in the circuit is powered up properly. One of the most common faults in a logic circuit is a defective power supply, or, in cases where the circuit is being breadboarded, unconnected or misconnected power supply pins on the IC(s). In cases where no logic levels are detectable in the circuit it is obvious that an inoperable power supply is at fault. In all other cases, a marginally operating power supply could cause erratic operation of the circuit. Power supply potential should be checked with a voltmeter to make sure that it is between the recommended 4.75- to 5.25-volt range. The closer the voltage to 5 volts the better because (depending on the quality of the power supply) the voltage could change a lot when the outputs of gates in the logic circuit switch to the logic low state. If the power supply voltages are acceptable, then the next step is to try to isolate the fault to a defective IC assuming that there are no wiring errors. If the circuit is being breadboarded, then the wiring should be checked for proper connections.

Consider a logic circuit such as the one shown in Figure 2.46 that contains a fault. The expected values for the output of the circuit and the recorded values are shown in the truth table in Table 2.15.

Figure 2.46

Logic circuit containing a fault.

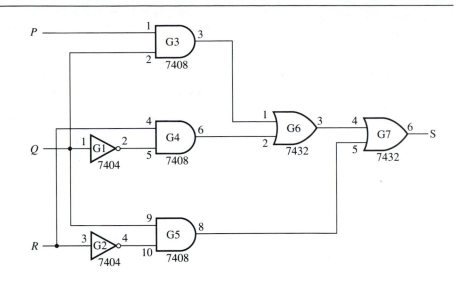

Table 2.15

P	Q	R	Expected S	Recorded S	
L	L	L	L	L	
L	L	H	H	H	
L	H	L	H	H	
L	H	H	L	L	
H	L	L	L	L	
H	L	H	H	H	
H	H	L	H	H	
H	H	H	H	L	←—fault

Table 2.15 shows that the last combination does not produce the expected output of a logic high; instead, it produces a logic low. All other combinations work as expected. To troubleshoot this circuit, apply this combination (111) to the inputs of the circuit. The logic levels that we can expect from the output of each gate is shown in Figure 2.47. Using the logic probe, verify that each gate produces the correct output as shown in Figure 2.47. However, the logic probe indicates that the output of gate G3 is at the logic low (L) state when it should really be at a logic high (H) state. Because the output of G3 is connected to the input of G6, the fault has been isolated to one of two possible causes

1. The output of gate G3 is shorted to GND, thereby pulling the node down to the logic low state.

2. The input of gate G6 is shorted to GND, thereby pulling the node down to the logic low state.

To isolate the defective gate, disconnect (if possible) the connection between the output of G3 and the input of G6 and measure the logic level at the output of G3 as shown in Figure 2.48. If the output of G3 now indicates a logic high level (Fig. 2.48a), the defective gate is G6 due to an internally shorted input to ground. If the

Figure 2.47

Identifying the fault in a
circuit by using the logic
probe.

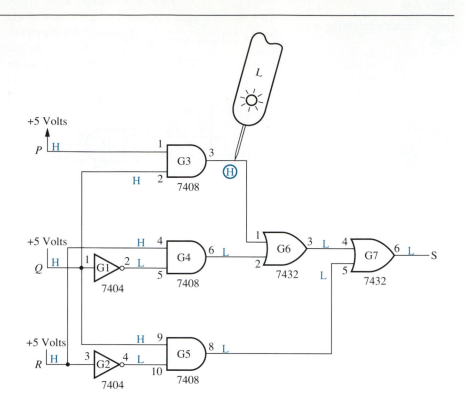

Figure 2.48

Isolation of the fault in the circuit using a logic probe.

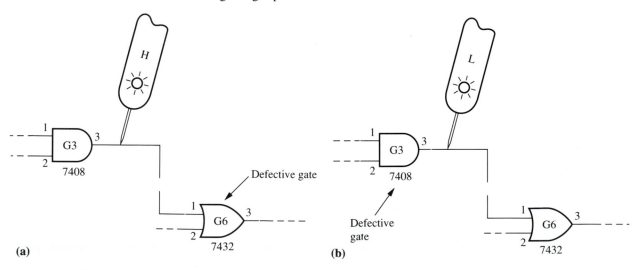

(a)

(b)

output of G3 is still a logic low, then gate G3 is defective (Fig. 2.48b) due to an
internally shorted output to ground.

If it is not possible to disconnect the connection between the output of G3 and
the input of G6 (such as in a printed circuit board), then the current pulser and current
tracer can be used to isolate the defective gate. The pulser is used to inject a pulse of

current into the node while the current tracer is used to determine if the current pulse flows into the input of G6 or into the output of G3. The current will always flow in the direction of the short. In Figure 2.49a the pulser is used to inject a pulse of current into the node and the tracer is placed near the input of G6. If the tracer detects the current pulse, then the input of G6 is shorted to ground and therefore G6 is the defective gate. In Figure 2.49b the tracer is placed near the output of G3. If the tracer detects the current pulse, the output of G3 is shorted to ground and therefore G3 is the defective gate.

The problem has now been isolated to a single gate and can be corrected by replacing the IC.

The following example illustrates another case study of a similar procedure used to troubleshoot a logic circuit.

Figure 2.49

Isolation of the fault in the circuit using a pulser and current tracer.

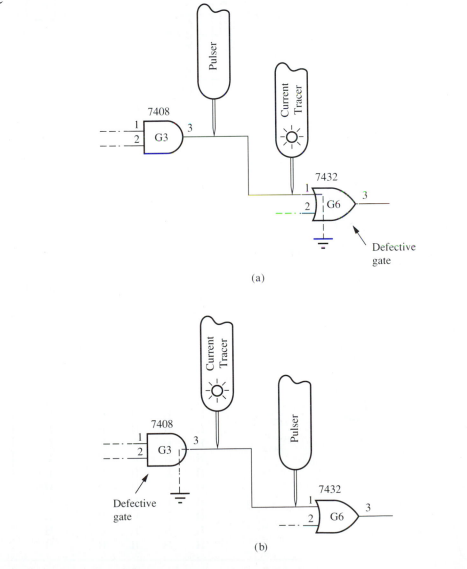

Example 2.15

The circuit shown in Figure 2.50 has a fault as shown in Table 2.16. The recorded output of the circuit indicates that the output remains at a logic high state and does not respond to any combination applied to the input. After making sure that the ICs have proper power supply, determine which one of the gates in the circuit has a fault. The best approach is to begin at the output of the circuit because if the output of G7 is shorted to V_{CC} it would remain high regardless of its input levels.

Figure 2.50

Circuit containing a fault.

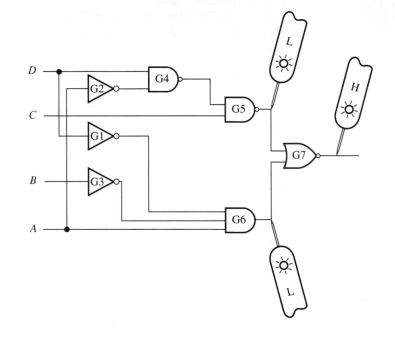

Figure 2.51

Testing a gate for proper operation.

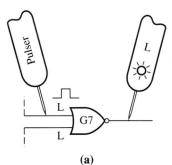

(a)

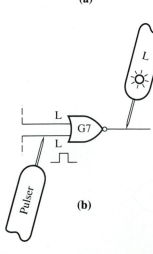

(b)

Table 2.16

A	B	C	D	Expected X	Recorded X
L	L	L	L	L	H
L	L	L	H	L	H
L	L	H	L	H	H
L	L	H	H	L	H
L	H	L	L	L	H
L	H	L	H	L	H
L	H	H	L	H	H
L	H	H	H	L	H
H	L	L	L	L	H
H	L	L	H	L	H
H	L	H	L	L	H
H	L	H	H	H	H
H	H	L	L	L	H
H	H	L	H	L	H
H	H	H	L	H	H
H	H	H	H	H	H

An examination of the inputs of G7 with the logic probe indicates that both inputs are at the low state. This could account for the output of G7 being high (because if both inputs of a *NOR* gate are low the output is high). To make sure that the inputs of G7 are not shorted to ground, inject a high going pulse (a pulse that changes from a low to a high state) to one input using the pulser and examine the output with a logic probe as shown in Figure 2.51a. The same procedure is repeated with the other input of G7 as shown in Figure 2.51b. In both cases the output goes low when the pulse is applied indicating that gate G7 is not at fault.

It is quite unlikely that the outputs of G5 and G6 are shorted to ground because the pulser would not have been able to produce a voltage pulse on the inputs of G7 if this were the case. To see if it is possible to change the output of G5 to a logic high, ground (apply a logic low) the *C* input as shown in Figure 2.52a; the logic probe attached to the output of G5 still indicates a logic low. This indicates that gate G5 is defective due to an internal malfunction because it should have produced a logic high at its output.

To see if the output of gate G6 changes to a logic high, the *A* input is connected high, the *B* input is connected low, and the *D* input is connected low so that all the inputs of G6 are high as shown in Figure 2.52b. The output of gate G6 remains low. However, the output of G3

Figure 2.52

Isolation of the fault.

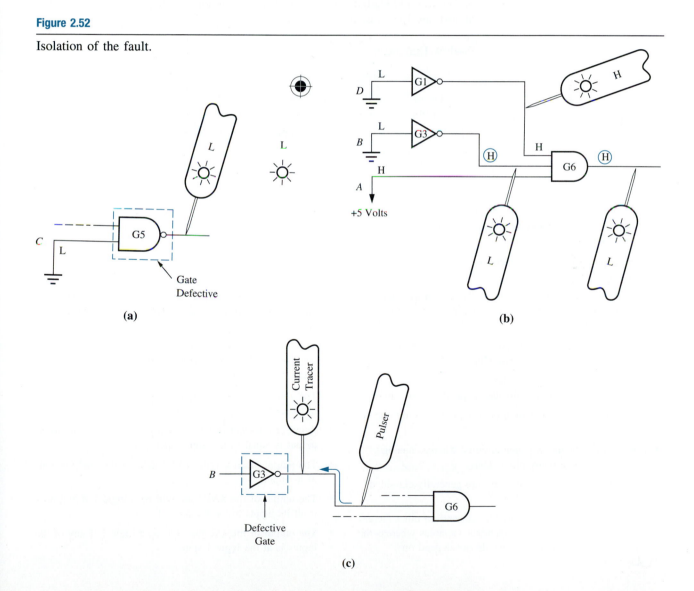

(a)

(b)

(c)

indicates a low state when it should be high. Use of the current tracer as shown in Figure 2.52c indicates that the output of G3 is shorted to ground.

Replacement of G5 (7400 IC) and G3 (7404 IC) produces recorded results that match the expected results of Table 2.16.

The troubleshooting procedures examined in this section provide a systematic approach to isolating and correcting faults in a logic circuit. A major portion of troubleshooting time is spent on isolation of the fault. Correction of the fault is usually as simple as replacement of an IC or in some cases removal of an external short such as a solder bridge. The logic probe is a versatile tool to isolate most common problems. However, the gates being tested with a logic probe generally have to be isolated from the rest of the circuit to make measurements. The pulser / current tracer combination is excellent for in-circuit tests of logic gates. Because most combinational logic circuits are similar to the ones studied in this chapter, the procedures and equipment introduced in this section can be applied to troubleshoot almost any logic circuit.

Review Questions

1. What is troubleshooting? What does it involve?
2. Describe the use of the logic probe, the pulser, and the current tracer to troubleshoot logic circuits.
3. What are the typical faults that could exist in a logic gate?

Summary

- Logic circuits are also called digital circuits and switching circuits.

- The basic element of a modern logic circuit is a device known as the logic gate.

- A gate is a device that electronically implements the various logic functions that are used in the design of logic circuits.

- Gates are implemented in hardware in the form of an electronic IC also known as a chip.

- Logic circuits are broadly classified into two categories — combinational logic circuits and sequential logic circuits.

- The variables in a switching circuit can have only one of two states.

- Digital circuits are two-state circuits, whereas analog circuits can have many more states.

- The variables in a digital circuit are generally classified as dependent or independent.

- The states of dependent variables depend on other factors such as the state of the independent variables whereas the states of independent variables do not depend on anything.

- Logical inversion is a process or function that reverses the state of a variable.

- The complement of a variable is its opposite state.

- A logic gate can have several inputs (independent variables) and one output (dependent variable) whose value depends on the binary combinations at the inputs and the function of the gate.

- The inputs and outputs of a gate are represented conceptually by binary numbers and electronically by voltages.

- The 74xxx TTL series of ICs is a family of ICs that implement various logic gates and digital circuits.

- A buffer does not provide any logic function — the logic output is equal to the logic input.

- The output of an inverter (*NOT* gate) is the complement of its input.

- The output of the *AND* gate will be a logic 1 if and only if all its inputs are at the logic 1 state.

- The output of the *OR* gate will be a logic 1 if any of its inputs is at the logic 1 state.

- *AND* and *OR* gates can have any number of inputs but must have at least two.

- The output of the *NAND* gate will be a logic 0 if and only if all its inputs are at the logic 1 state.

- The output of the *NOR* gate will be a logic 0 if any of its inputs is at the logic 1 state.

- *NAND* and *NOR* gates can have any number of inputs but must have at least two.

- Logic circuits are constructed from logic gates and can have several inputs and outputs.

- Truth tables and logic equations are used to describe the operation of logic gates and circuits.

- The analysis of logic circuits involves determining the operation of the circuit by obtaining its truth table and logic equation.

- The synthesis of logic circuits involves the design of the circuit given its logic equation, truth table, or a specification.

- In electrical terminology, a logic 1 is referred to as a high and a logic 0 is referred to as a low.

- A logic output sinks current in the low state and sources current in the high state.

- A UL describes the amount of loading an input puts on an output and is also known as fan-in.

- Fan-out is a measure of how many inputs (ULs) an output can drive under worst case conditions.

- Noise margin is the difference between worst case output voltage and worst case input voltage.

- Propagation delay is the amount of delay that occurs from the time the input of a gate changes states to the time the output changes states.

- The propagation delay of a gate is a direct indication of the switching speed of the gate; a slow gate has a long propagation delay time whereas a fast gate has a short propagation delay time.

Problems

Section 2.2   Basic logic concepts

1. Obtain a truth table and logic equation for each of the switching circuits shown in Figure 2.53.

2. Obtain a truth table and logic equation for each of the switching circuits shown in Figure 2.54.

Figure 2.53

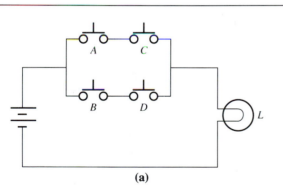

(a)

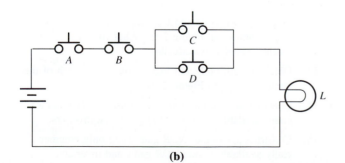

(b)

Figure 2.54

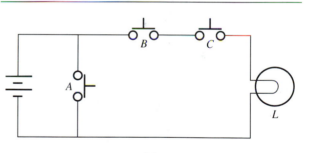

(a)

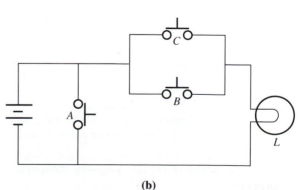

(b)

Figure 2.55

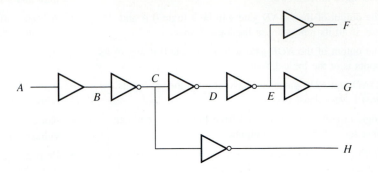

3. Describe (in words) the conditions necessary for the lamp *L* in Figure 2-53a to turn ON.

4. Describe (in words) the conditions necessary for the lamp *L* in Figure 2-53b to turn ON.

Section 2.4   The buffer and inverter

5. Trace through the circuit shown in Figure 2.55 and determine the logic states of *B* through *H* for all possible values of the independent variable *A*.

6. From the circuit shown in Figure 2.55, determine the logic equations for the dependent variables *B* through *H* in terms of the independent variable *A*.

Section 2.5   The *AND* gate

7. Obtain a logic equation and truth table for the circuit shown in Figure 2.56. What can be concluded about this circuit?

Figure 2.56

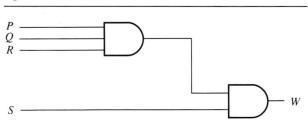

8. Construct a 16-input *AND* gate using only three-input *AND* gates.

Section 2.6   The OR gate

9. Obtain a logic equation and truth table for the circuit shown in Figure 2.57. What can you conclude about this circuit?

10. Construct a 16-input *OR* gate using only three-input *OR* gates.

Figure 2.57

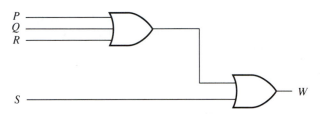

Section 2.7   The NAND and NOR gates

11. Obtain a truth table for each of the circuits shown in Figure 2.58. What conclusions can be reached about these circuits?

Figure 2.58

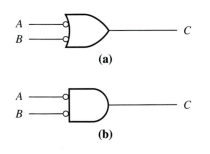

(a)

(b)

12. Obtain a truth table and logic equation for each of the circuits shown in Figure 2.59. What conclusions can be reached about these circuits?

13. Obtain a truth table and logic equation for each of the circuits shown in Figure 2.60.

14. Construct a 16-input *NAND* gate using only commercially available (74xxx) *AND* gates and inverters.

15. Construct a 16-input *NOR* gate using only commercially available (74xxx) *OR* gates and inverters.

Figure 2.59

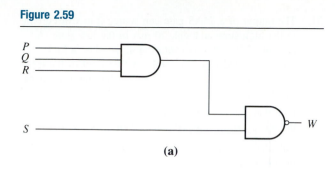

(a)

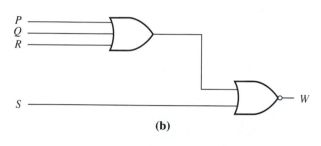

(b)

Figure 2.60

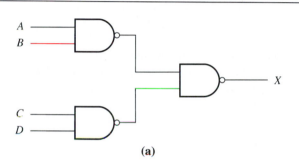

(a)

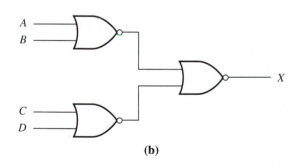

(b)

Figure 2.61

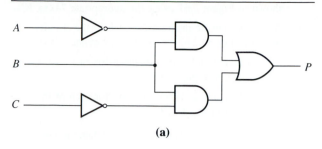

(a)

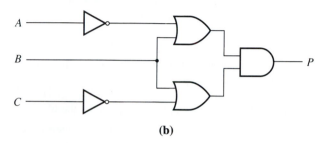

(b)

Figure 2.62

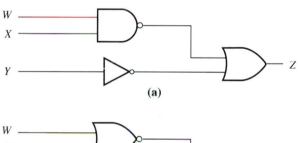

(a)

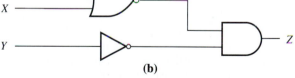

(b)

Section 2.8   Logic circuits

16. Analyze each of the circuits shown in Figure 2.61 and obtain their truth tables.

17. Analyze each of the circuits shown in Figure 2.62 and obtain their truth tables.

18. Obtain the logic equations for each of the circuits shown in Figure 2.61.

19. Obtain the logic equations for each of the circuits shown in Figure 2.62.

20. Draw the logic circuits that represent the following logic equations:
    a. $A = \bar{P} \cdot (Q + \bar{R}) \cdot S$
    b. $Z = (A \cdot \bar{B} \cdot C) + (\bar{A} \cdot \bar{B} \cdot \bar{C})$
    c. $T = \overline{X + Y} + \overline{Y + Z}$
    d. $W = \bar{A} \cdot B \cdot C \cdot \bar{D}$

21. Draw the logic circuits that represent the following logic equations:
    a. $X = \bar{W} + \bar{Z} + (\bar{P} \cdot Z)$
    b. $Z = \overline{A \cdot B \cdot C \cdot D}$
    c. $W = \overline{P + (Q \cdot R)}$
    d. $A = (X \cdot Y) + \bar{X} + \bar{Y}$

22. Draw a schematic diagram for each of the circuits shown in Figure 2.61 using the appropriate 74xxx ICs.

23. Draw a schematic diagram for each of the circuits shown in Figure 2.62 using the appropriate 74xxx ICs.

24. Draw a layout and wiring diagram for each of the circuits shown in Figure 2.61.

25. Draw a layout and wiring diagram for each of the circuits shown in Figure 2.62.

Section 2.9 Electrical characteristics of logic gates

26. An IC belonging to the 74LS family of logic ICs has the following electrical parameters:
    $I_{CCH} = 16$ mA, $I_{CCL} = 18$ mA, and $V_{CC} = 5$ volts
    Calculate the average power dissipated by the IC.

27. An IC belonging to the 74S family of logic ICs has the following electrical parameters:
    $I_{CCH} = 7$ mA, $I_{CCL} = 10$ mA, and $V_{CC} = 5$ volts
    Calculate the average power dissipated by the IC.

28. The 74LS family of logic ICs can detect at their inputs all voltages in the range 2 to 5 volts as a logic high and all voltages in the range 0 to 0.8 volts as a logic low. The outputs are guaranteed to produce a voltage in the range 2.7 to 5 volts in the high state and a voltage of 0 to 0.5 volts in the low state. Determine the voltages V_{OH}, V_{OL}, V_{IH}, V_{IL} for this family.

29. The 74AS family of logic ICs is guaranteed to produce a minimum of 3 volts at its output when at the logic high state and a maximum of 0.5 volts when at the logic low state. The minimum amount of voltage required at the input for a logic high is 2 volts and the maximum amount of voltage at the input for a logic low is 0.8 volts. Determine the voltages V_{OH}, V_{OL}, V_{IH}, and V_{IL} for this family.

30. The output of a 74LS gate can source 0.4 mA of current in the high state and sink 8 mA in the low state. A current of 20 μA flows into the input when a logic high is applied to it and 400 μA flows out of the input when a logic low is applied to it. Determine the currents I_{IL}, I_{IH}, I_{OL}, and I_{OH}.

31. The output of a 74AS gate can source 2 mA of current in the high state and sink 20 mA in the low state. A current of 20 μA flows into the input when a logic high is applied to it, and 500 μA flows out of the input when a logic low is applied to it. Determine the current I_{IL}, I_{IH}, I_{OL}, and I_{OH}.

32. Calculate the fan-out of the 74LS family of ICs from the electrical parameters obtained from Problem 30.

33. Calculate the fan-out of the 74AS family of ICs from the electrical parameters obtained from Problem 31.

34. Because 1 UL is defined with reference to the standard 74xxx TTL family as being an input that draws 40 μA in the high state and supplies 1.6 mA in the low state, determine how many ULs the input of a 74LS gate corresponds to with reference to the parameters obtained from Problem 30.

35. Because 1 UL is defined with reference to the standard 74xxx TTL family as being an input that draws 40 μA in the high state and supplies 1.6 mA in the low state, determine how many ULs the input of a 74AS gate corresponds to with reference to the parameters obtained from Problem 31.

36. Calculate the noise margin of the 74LS family of ICs from the electrical parameters obtained in Problem 28.

37. Calculate the noise margin of the 74AS family of ICs from the electrical parameters obtained in Problem 29.

38. A 74LS05 inverter IC has the following propagation delay times: $t_{pLH} = 11$ ns and $t_{pHL} = 8$ ns. Draw a diagram of the input and output waveform showing these delays, and calculate the average propagation delay time, t_{pd}.

39. A 74L02 *NOR* gate IC has the following propagation delay times: $t_{pLH} = 25$ ns and $t_{pHL} = 21$ ns. Draw a diagram of the input and output waveform showing these delays, and calculate the average propagation delay time, t_{pd}. Assume that one of the inputs to the *NOR* gate is permanently tied low.

Troubleshooting

Section 2.10 Troubleshooting logic gates and circuits

40. The output of gate G2 in Figure 2.63 has an internal open circuit. What would a logic probe indicate at the output of G2?

41. What would the output of the circuit (Z) shown in Figure 2.63 be if the output of G2 was an open circuit and the three inputs to the circuit (A, B, and C) are logic lows.

42. If the output of gate G2 (refer to Fig. 2.63) is an open circuit, discuss procedures that could be used to detect this fault.

Figure 2.63

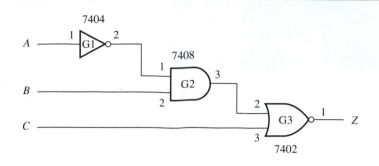

43. The output of gate G3 in Figure 2.64 is always at the logic low state and is not affected by any change in logic levels at the inputs *A* and *B*. Where could the possible fault lie?

Figure 2.64

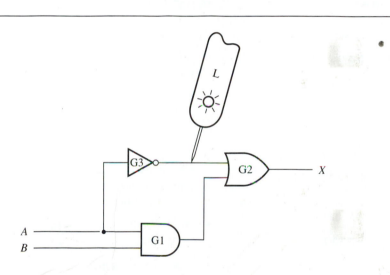

44. Discuss a troubleshooting procedure that could be used to identify the defective gate in Problem 43 without having to disconnect the connection between G3 and G2.

45. Discuss how a logic probe alone could be used to isolate the fault in the malfunctioning circuit described in Problem 43.

3

THE ANALYSIS AND SYNTHESIS OF LOGIC CIRCUITS

OBJECTIVES

The objectives of this chapter are to:

- Study the various techniques of analyzing logic circuits to obtain their logic equations and truth tables

- Introduce the concepts of minterms and maxterms

- Design logic circuits using the sum of minterms (sum-of-products) and product of maxterms (product-of-sums) forms

- Introduce and verify the basic laws of Boolean algebra

- Apply the laws of Boolean algebra to the simplification of logic equations and circuits

- Apply DeMorgan's law to convert from *AND-OR* logic circuits to *NAND* and *NOR* circuits

- Study the Karnaugh map techniques for simplifying logic equations and circuits

The English mathematician, George Boole (1815.1864) was the developer of Boolean Algebra, the mathematics of logic, which is the basis of all digital computers in existence today (The Bettman Archive).

3.1 Introduction

The study of digital circuits includes two areas that provide its foundations — the analysis and synthesis of logic circuits. Techniques of analyzing logic circuits can be used to determine their function and to assist in electronic troubleshooting procedures. The synthesis of logic circuits encompasses techniques to design logic circuits for specific applications. The procedures for analysis and synthesis of logic circuits have opposite paths because the objective of synthesis is to obtain a circuit, given its specification, whereas the objective of analysis is to obtain its function (and indirectly its specifications), given the circuit.

The analysis and synthesis of logic circuits are based on the elementary principles and functions of logic circuits discussed in Chapter 2. As have been shown, the behavior of gates and circuits can be described by mathematical logic equations. The manipulation of these equations also plays a vital part in the analysis and synthesis of logic circuits, and includes the study of Boolean algebra, a branch of discrete mathematics. Using Boolean algebraic techniques, procedures for designing and analyzing logic circuits can be greatly simplified.

This chapter continues the study of logic circuits introduced in Chapter 2 to include other techniques of analysis and synthesis of logic circuits. The mathematics of Boolean algebra along with its application in design and analysis procedures is also included in this chapter. The material covered in this chapter then serves as a foundation for the study of the various application circuits that are designed and analyzed in Chapter 4.

3.2 Analysis of logic circuits

Chapter 2 introduced simple logic circuits and procedures for analyzing them in order to obtain their logic equations and truth tables. This section continues the study of the analysis of similar logic circuits using more formal techniques. A good understanding of the basic logic gates and their functions is an important prerequisite to the study of these techniques.

To obtain the logic equation for any logic circuit, trace the basic logic functions through each gate in the circuit and then build the logic equation from these components (*see* Fig. 3.1).

To obtain the logic equation for the circuit in Figure 3.1, begin at the inputs of the circuit and trace the independent variables through each gate. For example, because *A* and *B* are the inputs of the *AND* gate, the output of the *AND* gate is

$$A \cdot B.$$

Similarly, the output of the inverter is

$$\overline{B}.$$

Because the inputs to the *OR* gate are connected to the outputs of the *AND* gate and the inverter, the output of the *OR* gate and the output of the circuit is

$$(A \cdot B) + \overline{B}.$$

Also, because the term $(A \cdot B)$ is treated like a single variable that makes up one of the inputs of the *OR* gate, the term is enclosed in parentheses. Section 3.4 will show that this is not always necessary. The parentheses enclosing the *AND* function also indicates that the *AND* function is evaluated before the *OR* function, that is, the *AND* gate will appear before the *OR* gate in the logic circuit. The logic equation for the circuit in Figure 3.1 is

$$X = (A \cdot B) + \overline{B}.$$

Once the logic equation for the circuit has been determined, one of two techniques can be used to obtain its truth table. The first technique involves the application of the basic logic functions (*AND, OR, NOT*) to the logic equations. This is done by breaking down the logic equation into its basic logic components, obtaining the logic states of each one of these components as a function of the independent variables, and then determining the logic states of the dependent variable by combining these logic components appropriately (*see* Table 3.1). The first two columns in Table 3.1 list all possible combinations of the independent variables, *A* and *B*. Because the logic equation for the circuit is made up of the two basic components or terms, $\overline{B}$ and $(A \cdot B)$, there is a column for each one of these terms. Finally, because the logic equation for the circuit is simply the logical *OR* of these two terms, the last column represents the dependent variable *X* that represents the logic equation for the circuit.

Figure 3.1

Trace to determine the logic equation of a circuit.

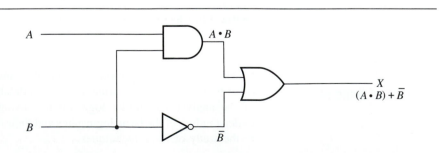

Table 3.1 Truth table analysis for Figure 3.1

A	B	$\bar{B}$	$A \cdot B$	X $(A \cdot B) + \bar{B}$
0	0			
0	1			
1	0			
1	1			

To complete the truth table, determine the logic states in each column starting with the leftmost column. The column for $\bar{B}$ is obtained by complementing the column for B. The column for $A \cdot B$ is obtained by *AND*ing (applying the *AND* function to) the A and B columns. Finally, the column for $(A \cdot B) + \bar{B}$ (the dependent variable X) is obtained by "*OR*ing" (applying the *OR* function to) the columns for $\bar{B}$ and $(A \cdot B)$. The completed truth table is shown in Table 3.2.

An alternative technique for obtaining the truth table of a logic circuit is to apply all possible binary combinations to the inputs of the circuit and trace these logic states through the circuit. Figure 3.2 shows the logic states at the outputs of each gate in the circuit of Figure 3.1 for all possible combinations of inputs. The construction of the truth table for each trace is also included in Figure 3.2. Note that it is not necessary to obtain the logic equation for the circuit first, as was the case with the previous technique of analysis.

The technique of obtaining a truth table for a logic circuit by tracing through all possible combinations of the inputs can often be quite time-consuming, especially when the circuit is fairly complex and has more than two input variables. However, in many logic circuits this process of analysis can be accelerated by recognizing certain configurations and applying the general rules of the basic *AND, OR,* and *NOT* functions. For example, the circuit in Figure 3.3 has four inputs and therefore we must trace through 16 combinations to obtain the truth table. However, note that the output of the circuit, W, is actually the output of the *AND* gate, G4, and if any of the inputs of an *AND* gate is a logic 0, its output will be a logic 0 regardless of the states of the other inputs. Because one of the inputs of the *AND* gate, G4, is connected to the input variable A, through the inverter, G3, the output of the *AND* gate G4 (which is also the output of the circuit) will be a 0 whenever the variable A is a 1. Therefore,

Table 3.2 Completed truth table for Figure 3.1

A	B	$\bar{B}$	$A \cdot B$	X $(A \cdot B) + \bar{B}$
0	0	1	0	1
0	1	0	0	0
1	0	1	0	1
1	1	0	1	1

Figure 3.2

Tracing logic states through a circuit.

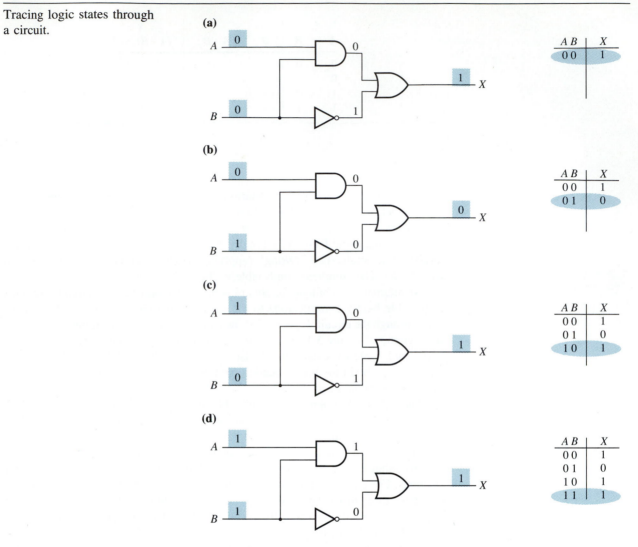

Figure 3.3

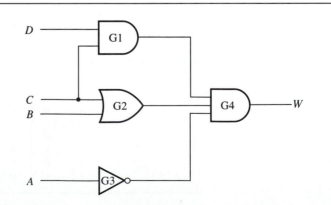

8 of the 16 truth table entries can be completed as follows:

A	B	C	D	W
0	0	0	0	
0	0	0	1	
0	0	1	0	
0	0	1	1	
0	1	0	0	
0	1	0	1	
0	1	1	0	
0	1	1	1	
⟶ 1	0	0	0	0
⟶ 1	0	0	1	0
⟶ 1	0	1	0	0
⟶ 1	0	1	1	0
⟶ 1	1	0	0	0
⟶ 1	1	0	1	0
⟶ 1	1	1	0	0
⟶ 1	1	1	1	0

Similarly, apply the same rule to the other inputs of the *AND* gate, G4. Because the second input of the *AND* gate, G4, is connected to the output of the *OR* gate, G2, this input will be a logic 0 if and only if both *B* and *C* are at the logic 0 state. Thus the output *W* will be a 0 if *B* and *C* are both 0's. Therefore, all remaining truth table combinations can be filled in where *B* and *C* are both at the logic 0 states

A	B	C	D	W
⟶ 0	0	0	0	0
⟶ 0	0	0	1	0
0	0	1	0	
0	0	1	1	
0	1	0	0	
0	1	0	1	
0	1	1	0	
0	1	1	1	
1	0	0	0	0
1	0	0	1	0
1	0	1	0	0
1	0	1	1	0
1	1	0	0	0
1	1	0	1	0
1	1	1	0	0
1	1	1	1	0

Finally, the third input of the *AND* gate, G4, will be a logic 0 if either *D* or *C* is at the logic 0 state because *D* and *C* are *AND*ed together by G1. All remaining combinations where D or C is at the logic 0 state can be completed:

A	B	C	D	W
0	0	0	0	0
0	0	0	1	0
→ 0	0	1	0	0
0	0	1	1	
→ 0	1	0	0	0
→ 0	1	0	1	0
→ 0	1	1	0	0
0	1	1	1	
1	0	0	0	0
1	0	0	1	0
1	0	1	0	0
1	0	1	1	0
1	1	0	0	0
1	1	0	1	0
1	1	1	0	0
1	1	1	1	0

Because we have now looked at all possible conditions necessary to produce a 0 at one or more of the inputs of the *AND* gate, G4, the remaining conditions will produce logic 1s on all the inputs of G4; this can be verified by tracing the remaining two combinations through the circuit. The completed truth table is

A	B	C	D	W
0	0	0	0	0
0	0	0	1	0
0	0	1	0	0
0	0	1	1	1
0	1	0	0	0
0	1	0	1	0
0	1	1	0	0
0	1	1	1	1
1	0	0	0	0
1	0	0	1	0
1	0	1	0	0
1	0	1	1	0
1	1	0	0	0
1	1	0	1	0
1	1	1	0	0
1	1	1	1	0

Thus, either of these techniques can be used to obtain the truth table for a logic circuit. The first procedure obtains the truth table from the logic equation derived from the circuit, whereas the second procedure obtains the truth table directly from the circuit by tracing through various logic combinations. The decision as to which procedure to use depends on the configuration of the circuit and its complexity. However, both procedures can be used to check the results of the analysis. The following examples illustrate both techniques of analysis for various circuits.

Example 3.1

The logic equation for the circuit shown in Figure 3.4 can be obtained by tracing the logic functions through each gate in the circuit. The logic function outputs of each gate are also shown in the circuit.

Figure 3.4

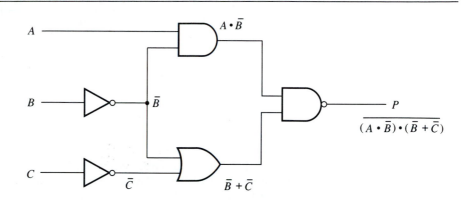

The output of the *AND* gate is

$$A \cdot \bar{B}.$$

and the output of the *OR* gate is

$$\bar{B} + \bar{C}.$$

Because these two terms (functions) are fed into the input of the *NAND* gate, they are treated like single variables (grouped in parentheses) and the logic equation for the circuit is

$$P = \overline{(A \cdot \bar{B}) \cdot (\bar{B} + \bar{C})}.$$

The truth table for the circuit can be obtained by breaking down its logic equation into its basic logic components as shown in Table 3.3.

The truth table for the circuit in Figure 3.4 can also be obtained by tracing through all possible binary combinations of the input variables *A*, *B*, and *C*. However, to simplify the process of analysis, certain judgments can be made based on the rules of the basic logic functions.

Because the output of the circuit is a *NAND* gate, if any of its inputs is at the logic 0 state, the output will be a logic 1. Because one of the inputs of the *NAND* gate is connected to the

Table 3.3 Truth table for Figure 3.4

A	B	C	$\bar{B}$	$\bar{C}$	$A \cdot \bar{B}$	$\bar{B} + \bar{C}$	P $\overline{(A \cdot \bar{B}) \cdot (\bar{B} + \bar{C})}$
0	0	0	1	1	0	1	1
0	0	1	1	0	0	1	1
0	1	0	0	1	0	1	1
0	1	1	0	0	0	0	1
1	0	0	1	1	1	1	0
1	0	1	1	0	1	1	0
1	1	0	0	1	0	1	1
1	1	1	0	0	0	0	1

output of the *OR* gate (whose inputs are connected to *B* and *C* through inverters), it can be concluded that the output of the *OR* gate (also the input of the *NAND* gate) will be a logic 0 if and only if both *B* and *C* are at the logic 1 states. Thus, the output, *P*, will be a logic 1 if both *B* and *C* are at the logic 1 states. Therefore, all entries in the following truth table that satisfy this condition can be completed

A	B	C	P
0	0	0	
0	0	1	
0	1	0	
→ 0	1	1	1
1	0	0	
1	0	1	
1	1	0	
→ 1	1	1	1

It can also be concluded that the remaining input of the *NAND* gate will be a logic 0 if either *A* is a logic 0 or if *B* is a logic 1, because this input is connected to an *AND* gate whose inputs are *A* and $\overline{B}$. Therefore, all remaining entries in the truth table where *P* will be a logic 1 can be completed

A	B	C	P
→ 0	0	0	1
→ 0	0	1	1
→ 0	1	0	1
0	1	1	1
1	0	0	
1	0	1	
→ 1	1	0	1
1	1	1	1

Because all conditions necessary for the variable *P* to be a logic 1 have been satisfied, the remaining combinations for *A*, *B*, and *C* will cause the output of the circuit to be a logic 0. This can be verified by tracing the remaining two combinations (100, 101) through the circuit. The final truth table is shown in Table 3.4.

It can be seen that the truth tables shown in Tables 3.3 and 3.4 are identical with respect to the dependent variable *P*.

Table 3.4   Truth table for Figure 3.4

A	B	C	P
0	0	0	1
0	0	1	1
0	1	0	1
0	1	1	1
1	0	0	0
1	0	1	0
1	1	0	1
1	1	1	1

Example 3.2

The logic equation for the circuit shown in Figure 3.5 can be obtained by tracing the logic functions through each gate in the circuit. The logic function outputs of each gate are also shown in the circuit.

The inputs to the four-input *OR* gate, G7, are

$$(\overline{P \cdot Q}), \ (\overline{Q + R}), \ \overline{R}, \ (\overline{R + S}).$$

As before, the terms in parentheses have been grouped because each term is treated as a single variable to obtain the final logic equation for the circuit

$$Z = (\overline{P \cdot Q}) + (\overline{Q + R}) + \overline{R} + (\overline{R + S})$$

The truth table for the circuit can be obtained by breaking down its logic equation into its basic logic components as shown in Table 3.5.

The truth table for the circuit in Figure 3.5 can also be obtained by tracing through all possible binary combinations of the input variables *P*, *Q*, *R*, and *S*. However, to simplify the

Figure 3.5

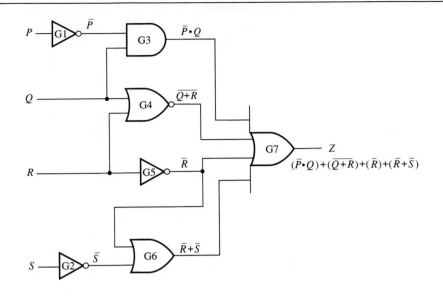

Table 3.5 Truth table for Figure 3.5

P	Q	R	S	$\bar{P}$	$\bar{S}$	$\bar{P} \cdot Q$	$\overline{Q + R}$	$\bar{R}$	$\bar{R} + \bar{S}$	Z
0	0	0	0	1	1	0	1	1	1	1
0	0	0	1	1	0	0	1	1	1	1
0	0	1	0	1	1	0	0	0	1	1
0	0	1	1	1	0	0	0	0	0	0
0	1	0	0	1	1	1	0	1	1	1
0	1	0	1	1	0	1	0	1	1	1
0	1	1	0	1	1	1	0	0	1	1
0	1	1	1	1	0	1	0	0	0	1
1	0	0	0	0	1	0	1	1	1	1
1	0	0	1	0	0	0	1	1	1	1
1	0	1	0	0	1	0	0	0	1	1
1	0	1	1	0	0	0	0	0	0	0
1	1	0	0	0	1	0	0	1	1	1
1	1	0	1	0	0	0	0	1	1	1
1	1	1	0	0	1	0	0	0	1	1
1	1	1	1	0	0	0	0	0	0	0

process of analysis, certain judgments can be made based on the rules of the basic logic functions.

Because the output of the circuit is the *OR* gate, G7, if any of its inputs are at the logic 1 state, the output will be a logic 1. Because one of the inputs of G7 is connected to the output of the *AND* gate, G3 (whose inputs are connected to $\bar{P}$ and Q), it can be concluded that the output of the *AND* gate, G3 (also the input of the *OR* gate), will be a logic 1 if and only if P is at the logic 0 state and Q is a logic 1. Thus, the output, Z, will be a logic 1 if P is a logic 0 and Q is a logic 1. Therefore, all entries in the following truth table that satisfy this condition can be completed

P	Q	R	S	Z
0	0	0	0	
0	0	0	1	
0	0	1	0	
0	0	1	1	
→ 0	1	0	0	1
→ 0	1	0	1	1
→ 0	1	1	0	1
→ 0	1	1	1	1
1	0	0	0	
1	0	0	1	
1	0	1	0	
1	0	1	1	
1	1	0	0	
1	1	0	1	
1	1	1	0	
1	1	1	1	

The output of the *NOR* gate, G4, also feeds into the input of the *OR* gate, G7. This *NOR* gate will produce a logic 1 at its output if and only if both Q and R are at the logic 0 state. Therefore, the output of the circuit, Z, will be a logic 1 when both Q and R are logic 0's.

P	Q	R	S	Z
→ 0	0	0	0	1
→ 0	0	0	1	1
0	0	1	0	
0	0	1	1	
0	1	0	0	1
0	1	0	1	1
0	1	1	0	1
0	1	1	1	1
→ 1	0	0	0	1
→ 1	0	0	1	1
1	0	1	0	
1	0	1	1	
1	1	0	0	
1	1	0	1	
1	1	1	0	
1	1	1	1	

The third input of the *OR* gate, G7, is connected to R through the inverter, G5. Therefore, the output of the circuit, Z, will be a logic 1 if R is a logic 0

P	Q	R	S	Z
0	0	0	0	1
0	0	0	1	1
0	0	1	0	
0	0	1	1	
0	1	0	0	1
0	1	0	1	1
0	1	1	0	1
0	1	1	1	1
1	0	0	0	1
1	0	0	1	1
1	0	1	0	
1	0	1	1	
⟶ 1	1	0	0	1
⟶ 1	1	0	1	1
1	1	1	0	
1	1	1	1	

Finally, note that the fourth input of the *OR* gate, G7, is connected to another *OR* gate, G6, whose output will be a logic 1 if either *R* or *S* is at the logic 0 state (because *R* and *S* are inverted before being connected to the input of this *OR* gate)

Table 3.6 Truth table for Figure 3.5

P	Q	R	S	Z
0	0	0	0	1
0	0	0	1	1
0	0	1	0	1
0	0	1	1	0
0	1	0	0	1
0	1	0	1	1
0	1	1	0	1
0	1	1	1	1
1	0	0	0	1
1	0	0	1	1
1	0	1	0	1
1	0	1	1	0
1	1	0	0	1
1	1	0	1	1
1	1	1	0	1
1	1	1	1	0

P	Q	R	S	Z
0	0	0	0	1
0	0	0	1	1
⟶ 0	0	1	0	1
0	0	1	1	
0	1	0	0	1
0	1	0	1	1
0	1	1	0	1
0	1	1	1	1
1	0	0	0	1
1	0	0	1	1
⟶ 1	0	1	0	1
1	0	1	1	
1	1	0	0	1
1	1	0	1	1
⟶ 1	1	1	0	1
1	1	1	1	

Because all conditions necessary for the output of the circuit, *Z*, to be a logic 1 have been satisfied, the remaining combinations for *P*, *Q*, *R*, and *S* (0011, 1011, 1111) will cause the output of the circuit to be a logic 0. This can be verified by tracing the remaining three combinations through the circuit. The final truth table is shown in Table 3.6.

It can be seen that the truth table shown in Tables 3.5 and 3.6 are identical with respect to the dependent variable *Z*.

This section has examined procedures for analyzing various logic circuits. The objective of analysis is to determine the function or operation of a logic circuit by obtaining its truth table, or logic equation, or both. The analysis of logic circuits plays an important role in the study of digital circuits because these procedures also allow us to simplify the design of existing logic circuits and to troubleshoot logic circuits. The analysis procedures can also be applied to check the proper design of logic circuits, as will be seen in the next section.

Review Questions

1. What is the purpose of analyzing logic circuits?
2. What does the analysis of a logic circuit involve?
3. Describe the two techniques used in the analysis of logic circuits.
4. Why are parentheses used in some of the logic equations obtained from a circuit?

3.3
Synthesis of logic circuits

The synthesis of logic circuits involves designing or putting together logic circuits from a given specification, truth table, or logic equation. Synthesis, therefore, is the opposite of analysis. Chapter 2 examined several simple examples of synthesizing logic circuits from logic equations. This section will investigate the procedures for obtaining logic equations from truth tables; these logic equations will then be used to produce the logic circuits that represent the equations and truth tables.

Before learning the techniques for designing logic circuits, certain characteristics of truth tables must be identified. Table 3.7 includes all possible binary combinations of two independent variables, X and Y. The first column (labeled "number") in the table lists the decimal equivalent of each combination.

Table 3.7   Two-variable minterms and maxterms

Number	X	Y	Minterms	Maxterms
0	0	0	$\overline{X} \cdot \overline{Y}$	$X + Y$
1	0	1	$\overline{X} \cdot Y$	$X + \overline{Y}$
2	1	0	$X \cdot \overline{Y}$	$\overline{X} + Y$
3	1	1	$X \cdot Y$	$\overline{X} + \overline{Y}$

The fourth column in Table 3.7 lists the minterms for each combination. Each minterm in the table will produce a logic 1 when its corresponding combination is applied to it, and a logic 0 for all other combinations. This can be verified by drawing the logic circuits for each of the minterms shown in Table 3.7 and then applying all possible combinations to each circuit. Figure 3.6 illustrates these circuits.

Figure 3.6a represents the circuit for the minterm, $\overline{X} \cdot \overline{Y}$. The output of the circuit will be a logic 1 only when the combination 00 is applied to it. All other combinations will produce a logic 0 at the output. Similarly, the circuit in Figure 3.6b represents the circuit for the minterm $\overline{X} \cdot Y$, whose output will be a logic 1 if and only if the combination 01 is applied to it. Figure 3.6c and d show the circuits for the remaining two minterms $X \cdot \overline{Y}$ and $X \cdot Y$, whose outputs will be a logic 1 for the combinations 10 and 11, respectively.

Figure 3.6

Minterm circuits.

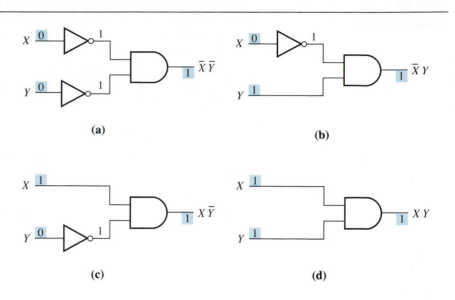

(a) (b) (c) (d)

A *minterm* is therefore a logic expression (or term) that will produce a logic 1 for one and only one combination. Note that this combination very closely resembles the minterm. For example, in Table 3.7, note that the variables in each minterm are barred (complemented) if its value in its corresponding combination is a logic 0

minterm:	$\overline{X} \cdot \overline{Y}$	$\overline{X} \cdot Y$	$X \cdot \overline{Y}$	$X \cdot Y$
combination:	0 0	0 1	1 0	1 1

A minterm can therefore be used to detect a certain combination by producing a logic 1 for that combination only, and a logic 0 for all other combinations. Minterms are often referred to by number, which is the decimal (or hexadecimal) equivalent of the combination that the minterm represents. For example, with reference to Table 3.7, minterm 0 is $\overline{X} \cdot \overline{Y}$, minterm 1 is $\overline{X} \cdot Y$, and so on. Because we use the ''·'' to represent the multiplication operator in conventional algebra, minterms are often referred to as *logical products*.

A *maxterm* is the complement of a minterm, a fact that will be proved in the next section. A maxterm will produce a logic 0 for one and only one combination. The fifth column in Table 3.7 lists the maxterms for all possible combinations of X and Y. Figure 3.7 shows the four circuits that represent the maxterms along with the unique combinations that produce a logic 0 for each circuit.

Observe in Figure 3.7a that if any combination other than a 00 is applied to the inputs of the circuit, the output will be a logic 1. Therefore, the circuit will produce a logic 0 if and only if the combination applied to its inputs is 00. Thus, the maxterm $X + Y$ represents the combination 00. Similarly, the circuits in Figure 3.7b, c, and d represent the maxterms $X + \overline{Y}$, $\overline{X} + Y$, and $\overline{X} + \overline{Y}$, respectively, which correspond to the combinations 01, 10, and 11, respectively.

A maxterm is therefore a logic expression (or term) that will produce a logic 0 for one and only one combination. Note that this combination very closely resembles the maxterm. For example, in Table 3.7 note that the variables in each maxterm are barred (complemented) if its value in its corresponding combination is a logic 1

maxterm:	$X + Y$	$X + \overline{Y}$	$\overline{X} + Y$	$\overline{X} + \overline{Y}$
combination:	0 0	0 1	1 0	1 1

Figure 3.7

Maxterm circuits.

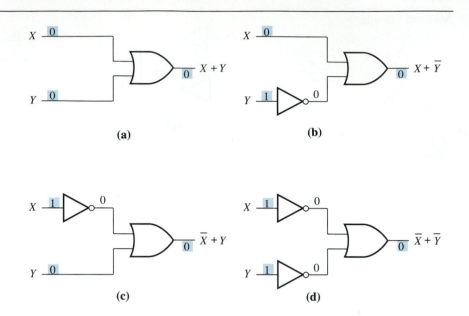

(a)   (b)

(c)   (d)

Like a minterm, a maxterm can also be used to detect a certain combination by producing a logic 0 for that combination only, and a logic 1 for all other combinations. Maxterms can also be referred to by a number that is the decimal (or hexadecimal) equivalent of the combination that the maxterm represents. For example, with reference to Table 3.7, maxterm 0 is $X + Y$, maxterm 1 is $X + \bar{Y}$, and so on. Because "+" is used to represent the addition operator in conventional algebra, maxterms are often referred to as *logical sums*.

Minterms and maxterms for combinations of any size can be obtained. Tables 3.8 and 3.9 list the minterms and maxterms for all possible combinations of 3 and 4 bits, respectively.

The logic equation for a truth table can be obtained in two forms—sum-of-products and product-of-sums. For example, consider the truth table shown in Table 3.10. Because the combinations 01 and 10 produce logic 1's for the dependent variable X, the logic equation for the circuit can be obtained by *OR*ing the minterms for these two combinations

$$X = (\bar{A} \cdot B) + (A \cdot \bar{B}) \tag{3.1}$$

Table 3.8   Three-variable minterms and maxterms

Number	X	Y	Z	Minterms	Maxterms
0	0	0	0	$\bar{X} \cdot \bar{Y} \cdot \bar{Z}$	$X + Y + Z$
1	0	0	1	$\bar{X} \cdot \bar{Y} \cdot Z$	$X + Y + \bar{Z}$
2	0	1	0	$\bar{X} \cdot Y \cdot \bar{Z}$	$X + \bar{Y} + Z$
3	0	1	1	$\bar{X} \cdot Y \cdot Z$	$X + \bar{Y} + \bar{Z}$
4	1	0	0	$X \cdot \bar{Y} \cdot \bar{Z}$	$\bar{X} + Y + Z$
5	1	0	1	$X \cdot \bar{Y} \cdot Z$	$\bar{X} + Y + \bar{Z}$
6	1	1	0	$X \cdot Y \cdot \bar{Z}$	$\bar{X} + \bar{Y} + Z$
7	1	1	1	$X \cdot Y \cdot Z$	$\bar{X} + \bar{Y} + \bar{Z}$

Table 3.9   Four-variable minterms and maxterms

Number	W	X	Y	Z	Minterms	Maxterms
0	0	0	0	0	$\overline{W} \cdot \overline{X} \cdot \overline{Y} \cdot \overline{Z}$	$W + X + Y + Z$
1	0	0	0	1	$\overline{W} \cdot \overline{X} \cdot \overline{Y} \cdot Z$	$W + X + Y + \overline{Z}$
2	0	0	1	0	$\overline{W} \cdot \overline{X} \cdot Y \cdot \overline{Z}$	$W + X + \overline{Y} + Z$
3	0	0	1	1	$\overline{W} \cdot \overline{X} \cdot Y \cdot Z$	$W + X + \overline{Y} + \overline{Z}$
4	0	1	0	0	$\overline{W} \cdot X \cdot \overline{Y} \cdot \overline{Z}$	$W + \overline{X} + Y + Z$
5	0	1	0	1	$\overline{W} \cdot X \cdot \overline{Y} \cdot Z$	$W + \overline{X} + Y + \overline{Z}$
6	0	1	1	0	$\overline{W} \cdot X \cdot Y \cdot \overline{Z}$	$W + \overline{X} + \overline{Y} + Z$
7	0	1	1	1	$\overline{W} \cdot X \cdot Y \cdot Z$	$W + \overline{X} + \overline{Y} + \overline{Z}$
8	1	0	0	0	$W \cdot \overline{X} \cdot \overline{Y} \cdot \overline{Z}$	$\overline{W} + X + Y + Z$
9	1	0	0	1	$W \cdot \overline{X} \cdot \overline{Y} \cdot Z$	$\overline{W} + X + Y + \overline{Z}$
10	1	0	1	0	$W \cdot \overline{X} \cdot Y \cdot \overline{Z}$	$\overline{W} + X + \overline{Y} + Z$
11	1	0	1	1	$W \cdot \overline{X} \cdot Y \cdot Z$	$\overline{W} + X + \overline{Y} + \overline{Z}$
12	1	1	0	0	$W \cdot X \cdot \overline{Y} \cdot \overline{Z}$	$\overline{W} + \overline{X} + Y + Z$
13	1	1	0	1	$W \cdot X \cdot \overline{Y} \cdot Z$	$\overline{W} + \overline{X} + Y + \overline{Z}$
14	1	1	1	0	$W \cdot X \cdot Y \cdot \overline{Z}$	$\overline{W} + \overline{X} + \overline{Y} + Z$
15	1	1	1	1	$W \cdot X \cdot Y \cdot Z$	$\overline{W} + \overline{X} + \overline{Y} + \overline{Z}$

Note that because the two minterms have been *OR*ed, the variable X will only be a 1 when either the combination 01 or the combination 10 is applied to the circuit.

Because minterms are known as logical products and because we have taken the logical sum of these products, this equation is said to be in sum-of-products (SOP) form. The circuit for Equation 3.1 is shown in Figure 3.8.

It can be proved that Equation 3.1 represents Table 3.10 by analyzing the circuit in Figure 3.8 as follows:

Table 3.10

A	B	X
0	0	0
0	1	1
1	0	1
1	1	0

A	B	$\overline{A}$	$\overline{B}$	$\overline{A} \cdot B$	$A \cdot \overline{B}$	X
0	0	1	1	0	0	0
0	1	1	0	1	0	1
1	0	0	1	0	1	1
1	1	0	0	0	0	0

It is also possible to obtain the logic equation and circuit for the truth table in Table 3.10 by *AND*ing the maxterms for combinations that produce logic 0s for the

Figure 3.8

Circuit obtained from sum-of-products equation.

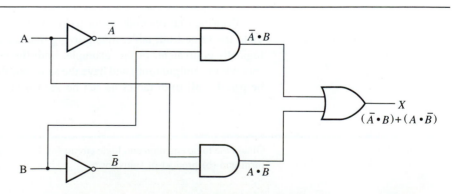

dependent variable X. Because the two combinations 00 and 11 produce logic 0's, the logic equation for the circuit could also be expressed as

$$X = (A + B) \cdot (\overline{A} + \overline{B}). \qquad (3.2)$$

Note that because the two maxterms have been *AND*ed, the variable X will only be a 0 when either the combination 00 or the combination 11 is applied to the circuit.

Because maxterms are known as logical sums and because we have taken the logical product of these sums, this equation is said to be in product-of-sums (POS) form. The circuit for Equation 3.2 is shown in Figure 3.9.

Figure 3.9

Circuit obtained from product-of-sums equation.

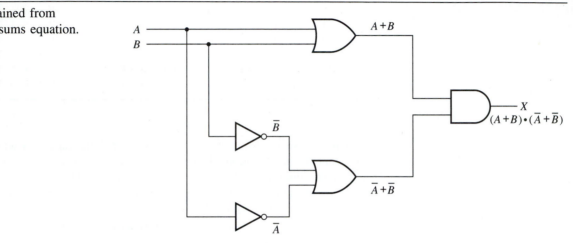

It can be proved that Equation 3.2 represents Table 3.10 by analyzing the circuit in Figure 3.9 as follows:

A	B	$\overline{A}$	$\overline{B}$	$A + B$	$\overline{A} + \overline{B}$	X
0	0	1	1	0	1	0
0	1	1	0	1	1	1
1	0	0	1	1	1	1
1	1	0	0	1	0	0

Both the sum-of-products form (Eq. 3.1) and the product-of-sums form (Eq. 3.2) yield the same truth table, thus proving that the circuits of Figures 3.8 and 3.9 are logically equivalent. In this example, both the circuits are the same in complexity (but not in configuration) and have the same number of gates. However, this may not be true for all truth tables as can be seen in the following examples.

Example 3.3

Obtain the logic equation and logic circuit for the following truth table in (a) sum-of-products form and (b) product-of-sums form.

P	Q	R	S
0	0	0	1
0	0	1	0
0	1	0	0
0	1	1	0
1	0	0	1
1	0	1	1
1	1	0	1
1	1	1	1

Solution

(a) The logic equation for the truth table in sum-of-products form is

$$S = (\overline{P} \cdot \overline{Q} \cdot \overline{R}) + (P \cdot \overline{Q} \cdot \overline{R}) + (P \cdot \overline{Q} \cdot R) + (P \cdot Q \cdot \overline{R}) + (P \cdot Q \cdot R). \tag{3.3}$$

The logic circuit for Equation 3.3 is shown in Figure 3.10.

(b) The logic equation for the truth table in product-of-sums form is

$$S = (P + Q + \overline{R}) \cdot (P + \overline{Q} + R) \cdot (P + \overline{Q} + \overline{R}) \tag{3.4}$$

The logic circuit for Equation 3.4 is shown in Figure 3.11.

The equations obtained in Example 3.3 (Eq. 3.3 and 3.4) are logically equivalent and therefore the circuits in Figures 3.10 and 3.11 are also logically equivalent. However, the circuit in Figure 3.10 is more complex in construction and requires many more gates than the circuit in Figure 3.11.

Figure 3.10

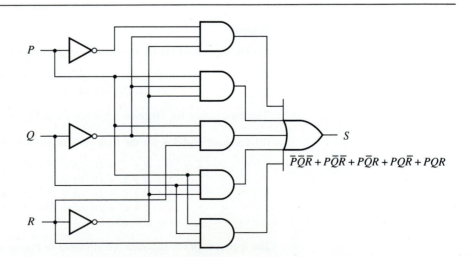

$$\overline{P}\overline{Q}\overline{R} + P\overline{Q}\overline{R} + P\overline{Q}R + PQ\overline{R} + PQR$$

Figure 3.11

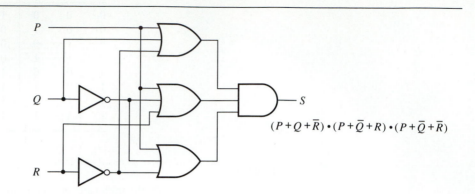

$$(P+Q+\bar{R}) \cdot (P+\bar{Q}+R) \cdot (P+\bar{Q}+\bar{R})$$

Example 3.4

Obtain the logic equation and logic circuit for the following truth table in (a) sum-of-products form and (b) product-of-sums form.

A	B	C	D	X
0	0	0	0	0
0	0	0	1	1
0	0	1	0	0
0	0	1	1	1
0	1	0	0	0
0	1	0	1	0
0	1	1	0	1
0	1	1	1	0
1	0	0	0	0
1	0	0	1	1
1	0	1	0	1
1	0	1	1	0
1	1	0	0	0
1	1	0	1	0
1	1	1	0	0
1	1	1	1	1

Solution

(a) The logic equation for the truth table in sum-of-products form is

$$X = (\bar{A} \cdot \bar{B} \cdot \bar{C} \cdot D) + (\bar{A} \cdot \bar{B} \cdot C \cdot D) + (\bar{A} \cdot B \cdot C \cdot \bar{D}) + (A \cdot \bar{B} \cdot \bar{C} \cdot D)$$
$$+ (A \cdot \bar{B} \cdot C \cdot \bar{D}) + (A \cdot B \cdot C \cdot D). \tag{3.5}$$

The logic circuit for Equation 3.5 is shown in Figure 3.12.

(b) The logic equation for the truth table in product-of-sums form is

$$X = (A + B + C + D) \cdot (A + B + \bar{C} + D) \cdot (A + \bar{B} + C + D) \cdot$$
$$(A + \bar{B} + C + \bar{D}) \cdot (A + \bar{B} + \bar{C} + \bar{D}) \cdot$$
$$(\bar{A} + B + C + D) \cdot (\bar{A} + B + \bar{C} + \bar{D}) \cdot$$
$$(\bar{A} + \bar{B} + C + D) \cdot (\bar{A} + \bar{B} + C + \bar{D}) \cdot$$
$$(\bar{A} + \bar{B} + \bar{C} + D) \tag{3.6}$$

The logic circuit for Equation 3.6 is shown in Figure 3.13.

Figure 3.12

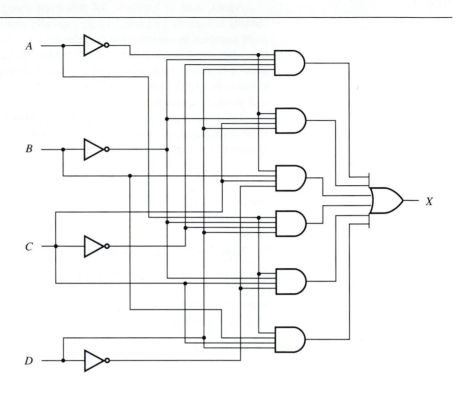

Figure 3.13

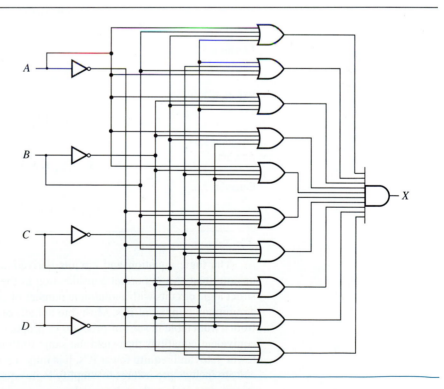

Again, note in Example 3.4 that even though Equations 3.5 and 3.6 and the circuits in Figures 3.12 and 3.13 are logically equivalent, the circuit in Figure 3.13 is more complex in construction than the circuit in Figure 3.12. It is therefore apparent that when the dependent variable in a truth table has more logic 0's than logic 1's, the sum-of-products form will yield a simpler circuit. However, when the dependent variable in a truth table has more logic 1's than logic 0's, the product-of-sums form will produce a simpler circuit.

The logic equations derived so far can also be expressed in a more convenient short-form notation in terms of the minterm and maxterm numbers. For example, Equation 3.1 can be expressed as

$$X = \sum_{m} 1, 2 = m_1 + m_2.$$

where the Greek letter Σ (sigma) represents a sum and the letter m signifies the sum of minterms numbered 1 and 2. m_1 and m_2 represent the minterms $(\overline{A} \cdot B)$ and $(A \cdot \overline{B})$, respectively.

Equation 3.2 is represented in short-form notation as follows:

$$X = \prod_{M} 0, 3 = M_0 + M_3$$

where the Greek letter Π (pi) represents a product and the letter M signifies the product of maxterms numbered 0 and 3. M_0 amd M_3 represent the maxterms $(A + B)$ and $(\overline{A} + \overline{B})$, respectively.

The short-form notation is particularly useful for representing very long and complicated logic equations as can be seen in the following example.

Example 3.5

The short-form notation for Equations 3.3 through 3.6 can be obtained by determining the appropriate minterm and maxterm numbers from their truth tables that produce 1's and 0's, respectively. Note that the minterm and maxterm numbers are shown in decimal, but can also be expressed in hexadecimal.

Equation 3.3:

$$S = \sum_{m} 0, 4, 5, 6, 7$$

Equation 3.4:

$$S = \prod_{M} 1, 2, 3$$

Equation 3.5:

$$X = \sum_{m} 1, 3, 6, 9, 10, 15$$

Equation 3.6:

$$X = \prod_{M} 0, 2, 4, 5, 7, 8, 11, 12, 13, 14$$

The logic equation and circuits derived in this section are considered to be unsimplified. An important consideration in the design of digital circuits is to construct logic circuits with a minimum number of components (gates) and consequently a minimum number of ICs. Most (but not all) of the logic equations derived directly from a truth table in sum-of-products or product-of-sums form can be simplified into equivalent equations that yield the same truth table and produce circuits that have fewer gates and require fewer ICs. It is important to extend the synthesis of logic circuits to include procedures to simplify logic equations and circuits. The remainder of this chapter deals with such procedures.

Review Questions

1. What is the difference between the procedures of analysis and synthesis of logic circuits?
2. What does the synthesis of a logic circuit involve?
3. What are minterms and maxterms?
4. What two forms can the equations derived from a truth table take?
5. What is the purpose of having a short-form notation for expressing logic equations?

3.4
Boolean algebra

As shown in the previous sections and in Chapter 2, logic equations that represent logic circuits very closely resemble conventional algebraic equations. In fact, these logic equations contain some of the same operators ("+" and "·") that are used in conventional algebra because logic equations and expressions can be manipulated in much the same way as conventional algebraic equations and expressions. The mathematics of manipulating logic equations and expressions is called *Boolean algebra*, which was developed by the English mathematician George Boole around 1847. Boolean algebra, or the "algebra of logic," is a natural system for representing logic circuits even though it was developed much before the invention of electronic switching circuits. This section will study the various laws, theorems, and postulates of Boolean algebra that will allow us to manipulate and simplify logic equations and expressions so that the resulting circuits can be constructed with a minimum number of gates and ICs.

In Boolean algebra there are certain assumptions that must be made based on the basic logic functions discussed previously—*AND*, *OR*, and *NOT*. These assumptions are known as postulates and will be described for each one of the three basic logic functions.

There are two postulates for the *NOT* function

$$\overline{1} = 0 \quad \text{and} \quad \overline{0} = 1$$

The two postulates simply describe the operation of the inverter and read "*NOT* 1 is equal to 0" and "*NOT* 0 is equal to 1." In other words, just as a variable can be complemented, so can a logic state because a variable actually represents a logic state.

A double inverted variable will always produce the variable. Thus,

$$\overline{\overline{X}} = X.$$

There are four postulates for the *AND* function, often referred to as *logical multiplication* because the *AND* operator " · " is also the multiplication operator in conventional mathematics.

$$0 \cdot 0 = 0 \qquad 0 \cdot 1 = 0$$
$$1 \cdot 0 = 0 \qquad 1 \cdot 1 = 1$$

Note that the four *AND* postulates are derived from the truth table of the *AND* gate, which states that "any logic level" *AND*ed with a 0 will produce a 0.

There are also four postulates for the *OR* function, often referred to as *logical addition* because the *OR* operator " + " is also the addition operator in conventional mathematics.

$$0 + 0 = 0 \qquad 0 + 1 = 1$$
$$1 + 0 = 1 \qquad 1 + 1 = 1$$

The four *OR* postulates are derived from the truth table of the *OR* gate, which simply states that "any logic level" *OR*ed with a 1 will produce a 1.

In conventional arithmetic and algebra, certain rules of precedence are used to evaluate expressions. For example:

$$3 \cdot 2 + 5 = 11$$

because 3 and 2 are multiplied first and then the result (6) is added to 5 for a final result of 11. Therefore, the multiplication function has a higher precedence than the addition function. To add 2 and 5 and multiply the result (7) by 3 for a final result of 21, parentheses are used to force precedence of evaluation in the expression as follows:

$$3 \cdot (2 + 5) = 21$$

The same rules of precedence are used for the logical operators *AND* ("·") and *OR* ("+"), as for the arithmetic operators *MULTIPLY* ("·") and *ADD* ("+"). That is, in a logic expression containing both types of operators, the *AND* function is always evaluated first. The same rule applies to the construction of logic circuits. Thus, the following logic equation:

$$X = A \cdot B + C \tag{3.7}$$

is the same as

$$X = (A \cdot B) + C \tag{3.8}$$

because the *AND* function has a higher precedence than the *OR* function. In the logic circuit for Equation 3.7, *A* and *B* would be *AND*ed first and then *OR*ed with *C* as shown in Figure 3.14. Parentheses can be used to force precedence of evaluation. For example, in Equation 3.7, to *OR B* and *C* first and then *AND* the result with *A*, the logic equation would be

$$X = A \cdot (B + C). \tag{3.9}$$

Note in Equation 3.9 that $B + C$ must be enclosed in parentheses to force precedence because the *OR* function has a lower precedence than the *AND* function. The circuit for Equation 3.9 is shown in Figure 3.15. Equation 3.9 is *not* logically equivalent to Equation 3.7.

Terms that are enclosed in parentheses can be treated as single variables. For example, if the term $(B + C)$ is to be represented in Equation 3.9 by the variable, *P*, then the equation becomes

$$X = A \cdot P$$

which identifies the output gate (outermost gate) of the circuit as an *AND* gate.

The *NAND* and *NOR* functions have the same precedence as terms enclosed in

Figure 3.14

Circuit for Equation 3.7.

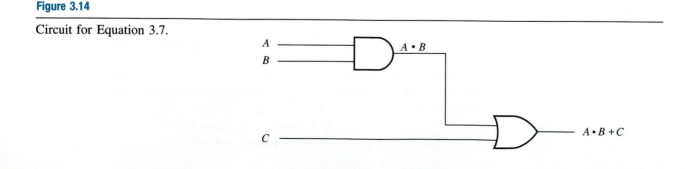

Figure 3.15

Circuit for Equation 3.9.

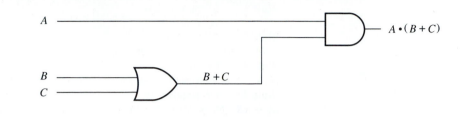

parentheses. Thus, the expression

$$\overline{A \cdot B} + \overline{A + B}$$

is equivalent to

$$\overline{(A \cdot B)} + \overline{(A + B)}.$$

The *NOT* operator always has the highest precedence. That is, inversions are always done first, before the *AND* and *OR* functions.

As in conventional algebra, for convenience, the "·" operator is often dropped when dealing with logic expressions. Thus, the following logic equation

$$Z = P \cdot \overline{Q} \cdot R + \overline{P} \cdot Q \cdot \overline{R}$$

can also be written as

$$Z = P\overline{Q}R + \overline{P}Q\overline{R}.$$

The latter form will be used for most of the logic equations that follow.

Example 3.6

In each of the following logic equations, determine the value of the dependent variable for the following values of the independent variables:

$$A = 0 \quad B = 1 \quad C = 1 \quad D = 0$$

(a) $X = A\overline{B} + AD$ (b) $Y = A(B + \overline{C})$ (c) $Z = B(\overline{A} + \overline{C})C$
(d) $P = 1 + \overline{D}B + \overline{A}$

Solutions

(a) $X = A \cdot \overline{B} + A \cdot D$
$= 0 \cdot \overline{1} + 0 \cdot 0$
$= 0 \cdot 0 + 0 \cdot 0$
$= 0 + 0$
$= 0$

(b) $Y = A \cdot (B + \overline{C})$
$= 0 \cdot (1 + \overline{1})$
$= 0 \cdot (1 + 0)$
$= 0 \cdot 1$
$= 0$

(c) $Z = B \cdot (\overline{A} + \overline{C}) \cdot C$
$= 1 \cdot (\overline{0} + \overline{1}) \cdot 1$
$= 1 \cdot (1 + 0) \cdot 1$
$= 1 \cdot 1 \cdot 1$
$= 1$

(d) $P = 1 + \overline{D} \cdot B + \overline{A}$
$= 1 + \overline{0} \cdot 1 + \overline{0}$
$= 1 + 1 \cdot 1 + 1$
$= 1 + 1 + 1$
$= 1$

There are ten basic Boolean algebra theorems that allow logic equations to be simplified and the number of gates in logic circuits to be minimized. These ten theorems or rules are known as the *laws of equivalence* and are listed in Table 3.11. Each rule listed in Table 3.11 has two interpretations, the *OR* interpretation (on

Boolean algebra laws of equivalence

Rule 1	Idempotent law	
	$A + A = A$	$A \cdot A = A$
Rule 2	Associative law	
	$(A + B) + C = A + (B + C)$	$(A \cdot B) \cdot C = A \cdot (B \cdot C)$
Rule 3	Commutative law	
	$A + B = B + A$	$A \cdot B = B \cdot A$
Rule 4	Distributive law	
	$A + (B \cdot C) = (A + B) \cdot (A + C)$	$A \cdot (B + C) = (A \cdot B) + (A \cdot C)$
Rule 5	Identity law	
	$A + 0 = A$	$A \cdot 1 = A$
Rule 6	Dominance law	
	$A + 1 = 1$	$A \cdot 0 = 0$
Rule 7	Complementarity law	
	$A + \overline{A} = 1$	$A \cdot \overline{A} = 0$
Rule 8	Absorbtion law	
	$A + (A \cdot B) = A$	$A \cdot (A + B) = A$
Rule 9	DeMorgan's law	
	$\overline{A + B} = \overline{A} \cdot \overline{B}$	$\overline{A \cdot B} = \overline{A} + \overline{B}$
Rule 10	Miscellaneous	
	$\overline{A} + (A \cdot B) = \overline{A} + B$	$A + (\overline{A} \cdot B) = A + B$

the left-hand side) and the *AND* interpretation (on the right-hand side). The remainder of this section will deal with the explanation and verification of each rule.

The idempotent law

The idempotent law can be verified by constructing a circuit for each interpretation as shown in Figure 3.16. In Figure 3.16a a two-input *OR* gate has both its inputs connected to the independent variable, A. The output of the circuit therefore is $A + A$. If $A = 0$, then the output of the *OR* gate is 0, and if $A = 1$, then the output of the *OR* gate is 1. Therefore, the output of the *OR* gate will always be at the same state as A. The following truth table confirms this fact:

A	$A + A = A$
0	$0 + 0 = 0$
1	$1 + 1 = 1$

Similarly, in Figure 3.16b, if $A = 0$, then the output of the *AND* gate is 0, and if $A = 1$, then the output of the *AND* gate is 1. The idempotent law can also be verified by the following truth table using the Boolean postulates:

A	$A \cdot A = A$
0	$0 \cdot 0 = 0$
1	$1 \cdot 1 = 1$

Figure 3.16

Circuits to represent idempotence.

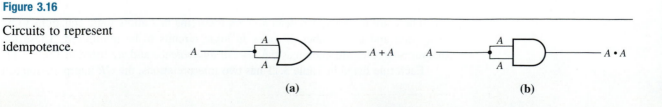

(a)   (b)

The idempotent law can be extended to more than two occurrences. Thus,

$$A \cdot A \cdot A = A \quad \text{and} \quad A + A + A = A$$
$$A \cdot A \cdot A \cdot A = A \quad \text{and} \quad A + A + A + A = A$$
etc.

The associative law

The associative law specifies that there are no rules of precedence when evaluating logical expressions that contain the same operators. Therefore, the circuits in Figure 3.17a through c are all equivalent. This can also be proved by applying the Boolean postulates to the logic expressions as shown in Table 3.12.

Similarly, the associative law can also be applied to the *AND* function, and therefore the circuits shown in Figure 3.17d through f are equivalent. Table 3.13 verifies the associative law for the *AND* function.

Observe that the associative laws can be applied to extend the number of inputs in *AND* and *OR* gates. For example, the circuits in Figure 3.17a and b are equivalent to the three-input *OR* gate in Figure 3.17c. Similarly, the circuits in Figure 3.17d and e are equivalent to the three-input *AND* gate in Figure 3.17f.

Figure 3.17

Circuits to represent association.

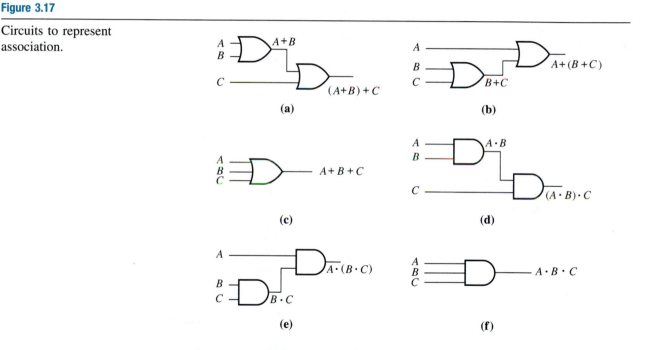

Table 3.12

A	B	C	A + B	B + C	A + (B + C)	(A + B) + C
0	0	0	0	0	0	0
0	0	1	0	1	1	1
0	1	0	1	1	1	1
0	1	1	1	1	1	1
1	0	0	1	0	1	1
1	0	1	1	1	1	1
1	1	0	1	1	1	1
1	1	1	1	1	1	1

Table 3.13

A	B	C	$A \cdot B$	$B \cdot C$	$A \cdot (B \cdot C)$	$(A \cdot B) \cdot C$
0	0	0	0	0	0	0
0	0	1	0	0	0	0
0	1	0	0	0	0	0
0	1	1	0	1	0	0
1	0	0	0	0	0	0
1	0	1	0	0	0	0
1	1	0	1	0	0	0
1	1	1	1	1	1	1

The commutative law

The commutative law states that the position of the operands of a logical operator has no bearing on the logic of a logic function. This means that the function of a logic gate will be the same regardless of the order in which the inputs are applied, as can be seen in Figure 3.18a and b.

Figure 3.18

Circuits to represent commutation.

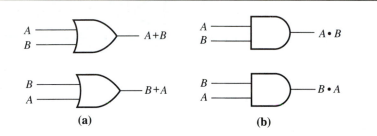

(a)   (b)

The distributive law

The distributive law is one of the most important laws because it is used frequently in the reduction of logic equations. Figure 3.19a and b show the two circuits that represent the left-hand and right-hand side of the *OR* interpretation of the distributive law. The two circuits can be proved to be equivalent by analyzing the two circuits and ob-

Figure 3.19

Circuits to represent distribution.

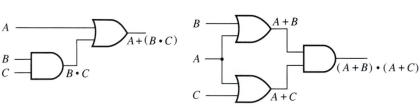

(a)   (b)

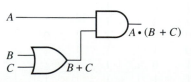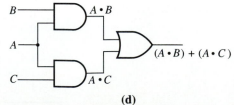

(c)   (d)

taining a truth table for the outputs. The results of the analysis are shown in Table 3.14.

Similarly, Figure 3.19c and d are equivalent as shown by the analysis in Table 3.15.

Table 3.14

A	B	C	$A + (B \cdot C)$	$(A + B) \cdot (A + C)$
0	0	0	0	0
0	0	1	0	0
0	1	0	0	0
0	1	1	1	1
1	0	0	1	1
1	0	1	1	1
1	1	0	1	1
1	1	1	1	1

Table 3.15

A	B	C	$A \cdot (B + C)$	$(A \cdot B) + (A \cdot C)$
0	0	0	0	0
0	0	1	0	0
0	1	0	0	0
0	1	1	0	0
1	0	0	0	0
1	0	1	1	1
1	1	0	1	1
1	1	1	1	1

The identity law

The identity law can be verified by constructing the circuit for each interpretation as shown in Figure 3.20. Note in Figure 3.20a that because one of the inputs of the *OR* gate is tied permanently to a logic 0, if $A = 1$, then the output of the gate will be a 1, but if $A = 0$, the output will be a 0. Thus, the state of the output is equal to the state of A. The following truth table illustrates this analysis:

A	$A + 0 = A$
0	$0 + 0 = 0$
1	$1 + 0 = 1$

Figure 3.20

Circuits to represent identity.

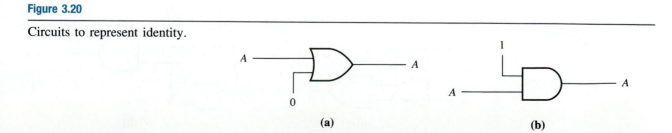

(a)   (b)

The identity law when applied to the *AND* function produces the same result. Figure 3.20b shows that because one of the inputs to the *AND* gate is tied permanently to a logic 1, the output will be a logic 1 if $A = 1$, and a logic 0 if $A = 0$. Therefore, the state of the output of the *AND* gate is equal to the state of A. This is illustrated in the following truth table:

A	$A \cdot 1 = A$
0	$0 \cdot 1 = 0$
1	$1 \cdot 1 = 1$

The identity law is very useful when it is necessary to disable certain inputs of the *AND* and *OR* gates. For example, in Figure 3.21a a four-input *OR* gate has been effectively converted to a three-input *OR* gate by disabling one of the inputs by tying it at the logic 0 state. Similarly, Figure 3.21b reduces the number of effective inputs in a four-input *AND* gate to three.

Figure 3.21

Reducing the number of effective inputs in a gate. (a) *OR* gate; (b) *AND* gate.

(a)   (b)

The dominance law

The dominance law describes the basic characteristics of the *AND* and *OR* functions (and gates). That is, if any of the inputs of an *OR* gate is a logic 1, the output of the gate will be a logic 1 regardless of the states of the other inputs. Similarly, if any of the inputs of an *AND* gate is a logic 0, the output of the gate will be a logic 0 regardless of the states of the other inputs. The circuits that represent the dominance laws are shown in Figure 3.22 and the following truth table verifies the laws:

A	$A + 1 = 1$	$A \cdot 0 = 0$
0	$0 + 1 = 1$	$0 \cdot 0 = 0$
1	$1 + 1 = 1$	$1 \cdot 0 = 0$

The identity and the dominance laws can be applied to a two-input gate to effectively pass through or block out the logic level of a variable.

Figure 3.22

Circuits to represent dominance.

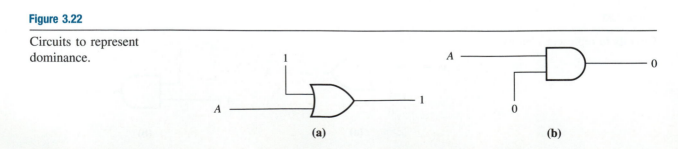

(a)   (b)

The complementarity law

The complementarity law is used frequently in the simplification of logic circuits. Figure 3.23 includes the two circuits that represent this law. Note in Figure 3.23a that the output of the *OR* gate will always be a logic 1 because one input is the complement of the other and therefore for any value of *A*, one of the inputs will always be a logic 1. Similarly, in Figure 3.23b, the output of the *AND* gate will always be a logic 0 because the logic states at the two inputs will always be complementary, and therefore for any value of *A*, one of the inputs will always be a logic 0. The following truth table verifies the complementarity law:

A	$\bar{A}$	$A + \bar{A} = 1$	$A \cdot \bar{A} = 0$
0	1	$0 + 1 = 1$	$0 \cdot 1 = 0$
1	0	$1 + 0 = 1$	$1 \cdot 0 = 0$

Figure 3.23

Circuits to represent complementarity.

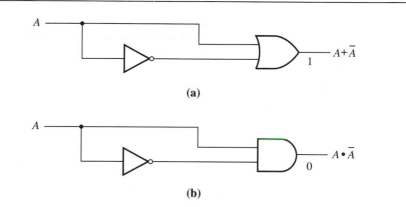

(a)

(b)

The absorbtion law

The circuits that represent the absorbtion laws are shown in Figure 3.24. Note in Figure 3.24a that if *A* is a logic 1, then the output of the *OR* gate will be a logic 1, and if *A* is a logic 0, the output of the *OR* gate will be a logic 0 because both its inputs are at the logic 0 state; therefore, the output of the circuit is equal to the state of *A*. In Figure 3.24b if *A* is a logic 0, then the output of the *AND* gate is a logic 0, and if *A* is a logic 1,

Figure 3.24

Circuits to represent absorbtion.

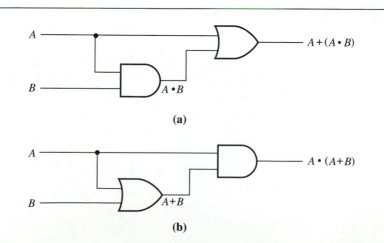

(a)

(b)

Table 3.16

A	B	$A \cdot B$	$A + B$	$A + (A \cdot B)$	$A \cdot (A + B)$
0	0	0	0	0	0
0	1	0	1	0	0
1	0	0	1	1	1
1	1	1	1	1	1

the output of the *AND* gate will be a logic 1 because both its inputs are at logic 1 state; therefore, the output of the circuit is equal to the state of *A*. The truth table shown in Table 3.16 verifies the absorbtion law.

DeMorgan's law

DeMorgan's law (also known as DeMorgan's theorem) relates the three basic functions—*AND*, *OR*, and *NOT*—to the *NAND* and *NOR* functions. Its importance in changing the configuration of a circuit will be seen in Section 3.6.

DeMorgan's law states that the complement of the logical sum of two or more variables equals the logical product of the complements of the variables. Conversely, the complement of the logical products of two or more variables equals the logical sum of the complements of the variables.

Figure 3.25 includes the four circuits that represent DeMorgan's law. Figure 3.25a is a simple two-input *NOR* gate, while Figure 3.25b is its equivalent circuit constructed with an *AND* gate and inverters. Recall that the output of a *NOR* gate will be a logic 0 if any of its inputs is a logic 1. This logic also applies to the circuit in Figure 3.25b because if either *A* or *B* is a logic 1, the output of the *AND* gate will be a logic 0. The truth table shown in Table 3.17 verifies DeMorgan's law for the circuits shown in Figure 3.25a and b.

Similarly, Figure 3.25c is a simple two-input *NAND* gate, whereas Figure 3.25d is its equivalent circuit constructed with an *OR* gate and inverters. Recall that the output of a *NAND* gate will be a logic 1 if any of its inputs is a logic 0. This logic also applies to the circuit in Figure 3.25d because if either *A* or *B* is a logic 0, the output of the *OR* gate will be a logic 1. The truth table shown in Table 3.18 verifies DeMorgan's law for the circuits shown in Figure 3.25c and d.

Figure 3.25

Circuits to represent
DeMorgan's law.

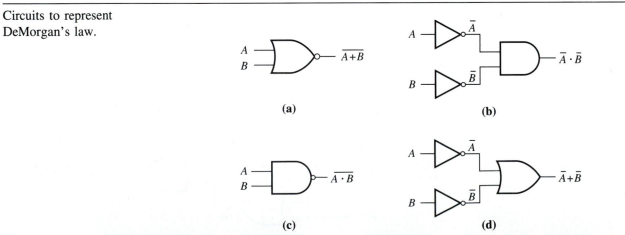

(a)

(b)

(c)

(d)

Table 3.17

A	B	$\overline{A}$	$\overline{B}$	$\overline{A + B}$	$\overline{A} \cdot \overline{B}$
0	0	1	1	1	1
0	1	1	0	0	0
1	0	0	1	0	0
1	1	0	0	0	0

Table 3.18

A	B	$\overline{A}$	$\overline{B}$	$\overline{A \cdot B}$	$\overline{A} + \overline{B}$
0	0	1	1	1	1
0	1	1	0	1	1
1	0	0	1	1	1
1	1	0	0	0	0

DeMorgan's law can also be used to convert minterms to their corresponding maxterms and vice versa. Recall that minterms and maxterms are complements of each other. For example, minterm no. 3 in Table 3.8 can be converted to its corresponding maxterm (no. 3) as follows:

minterm no. 3: $\qquad \overline{X} \cdot Y \cdot Z$

applying DeMorgan's law: $\qquad \overline{\overline{X} \cdot Y \cdot Z}$

$$= \overline{\overline{X}} + \overline{Y} + \overline{Z}$$

maxterm no. 3: $\qquad = X + \overline{Y} + \overline{Z}$

Similarly, maxterm no. 3 can be converted to its complementary minterm (no. 3) as follows:

maxterm no. 3: $\qquad X + \overline{Y} + \overline{Z}$

applying DeMorgan's law: $\qquad \overline{X + \overline{Y} + \overline{Z}}$

$$= \overline{X} \cdot \overline{\overline{Y}} \cdot \overline{\overline{Z}}$$

minterm no. 3: $\qquad \overline{X} \cdot Y \cdot Z$

DeMorgan's law allows the symbols for the *NAND* and *NOR* gates to be represented in terms of *AND* and *OR* gates. For example, because a *NAND* gate is equivalent to an *OR* gate with inverted inputs, it can also be represented as

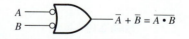

$$\overline{A} + \overline{B} = \overline{A \bullet B}$$

Recall that the bubble in a logic symbol indicates inversion.

Similarly, a *NOR* gate is equivalent to an *AND* gate with inverted inputs and can be represented as

$$\overline{A} \bullet \overline{B} = \overline{A + B}$$

Miscellaneous rules

The two miscellaneous rules listed in Table 3.11 are not part of the Boolean algebra laws but are useful in the reduction of logic equations. The circuits that represent these rules are shown in Figure 3.26. According to the first rule, the circuits in Figure 3.26a and b are equivalent, and according to the second rule, the circuits in Figure 3.26c and d are equivalent. The verification of the two rules is shown in Tables 3.19 and 3.20 by obtaining truth tables for each set of circuits through analysis.

Figure 3.26

Circuits to represent the miscellaneous rules.

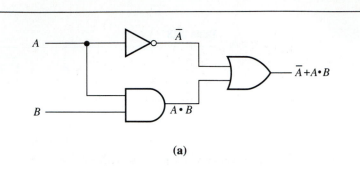

(a)

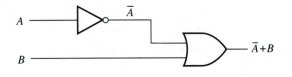

(b)

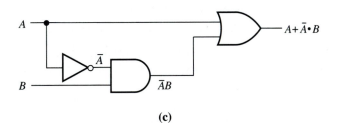

(c)

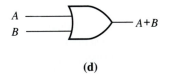

(d)

Table 3.19

A	B	$\overline{A}$	$A \cdot B$	$\overline{A} + (A \cdot B)$	$\overline{A} + B$
0	0	1	0	1	1
0	1	1	0	1	1
1	0	0	0	0	0
1	1	0	1	1	1

Table 3.20

A	B	$\overline{A}$	$\overline{A} \cdot B$	$A + (\overline{A} \cdot B)$	$A + B$
0	0	1	0	0	0
0	1	1	1	1	1
1	0	0	0	1	1
1	1	0	0	1	1

This section has dealt with the basic postulates, laws, theorems, and rules of Boolean algebra. It was stated earlier that one of the applications of Boolean algebra is in the simplification (or reduction) of logic equations to yield logic circuits with a minimum number of gates. The applications of these laws in such procedures will be seen in the next section.

Review Questions

1. What is Boolean algebra? What is its purpose in the study of digital circuits?

2. What is meant by logical multiplication and addition?

3. What is the purpose of using parentheses in logic expressions?

4. What are the Boolean algebra laws of equivalence used for?

5. Are the *NAND* and *NOR* functions associative? That is, can the *AND* and *OR* gates in the circuits of Figure 3.17 be replaced with *NAND* and *NOR* gates, respectively?

3.5
Boolean simplification

The reduction of logic equations and minimization of logic circuits are important procedures in the design of digital circuits because a circuit that contains fewer gates can usually be implemented with fewer ICs, and consequently at a lower cost and smaller size. The objective, therefore, in the design of logic circuits is to obtain a required circuit with as few gates and ICs as possible. In Section 3.3, a procedure was developed to obtain logic equations from a given truth table. It was stated that these equations were unsimplified and in most cases could be reduced to equivalent equations that produced much simpler circuits. This section deals with the application of Boolean algebra in the reduction of logic circuits.

To illustrate the process of Boolean simplification, consider the truth table shown in Table 3.21. The unsimplified logic equation can be obtained for the circuit in sum-of-products form as

$$Z = \overline{X}\,\overline{Y} + X\overline{Y} \tag{3.10}$$

Table 3.21

X	Y	Z
0	0	1
0	1	0
1	0	1
1	1	0

However, by inspection of Table 3.21 we can see that the dependent variable, Z, is the complement of the independent variable Y. Therefore, the *simplified* logic equation for the circuit is

$$Z = \overline{Y}. \tag{3.11}$$

In most cases the simplified logic equation cannot be obtained so easily by inspection, and the rules of Boolean algebra (see Table 3.11) must be applied to reduce the equation to its simplest form.

To begin the reduction of Equation 3.10, first apply the distributive law, to factor out the $\overline{Y}$

$$Z = \overline{Y}(\overline{X} + X)$$

The complementarity law can now be applied to the term $(\overline{X} + X)$ to obtain the following equation

$$Z = \overline{Y} \cdot 1$$

Finally, apply the identity law to obtain the final simplified equation

$$Z = \overline{Y}$$

When the Boolean algebra laws can no longer be applied, the equation is in its simplest form.

It is also possible to apply the laws of Boolean algebra to reduce the unsimplified product-of-sums equation for Table 3.21

$$
\begin{aligned}
Z &= (X + \overline{Y})(\overline{X} + \overline{Y}) \\
Z &= \overline{X}(X + \overline{Y}) + \overline{Y}(X + \overline{Y}) && \text{(distributive)} \\
Z &= \overline{X}X + \overline{X}\,\overline{Y} + \overline{Y}X + \overline{Y}\,\overline{Y} && \text{(distributive)} \\
Z &= 0 + \overline{X}\,\overline{Y} + \overline{Y}X + \overline{Y}\,\overline{Y} && \text{(complementarity)} \\
Z &= \overline{X}\,\overline{Y} + \overline{Y}X + \overline{Y}\,\overline{Y} && \text{(identity)} \\
Z &= \overline{X}\,\overline{Y} + \overline{Y}X + \overline{Y} && \text{(idempotent)} \\
Z &= \overline{Y}(\overline{X} + X) + \overline{Y} && \text{(distributive)} \\
Z &= \overline{Y} \cdot 1 + \overline{Y} && \text{(complementarity)} \\
Z &= \overline{Y} + \overline{Y} && \text{(identity)} \\
Z &= \overline{Y} && \text{(idempotent)}
\end{aligned}
$$

It can be proved that the simplified and unsimplified logic equations and their circuits are equivalent by constructing truth tables for each equation or circuit and comparing their truth tables. For example, the unsimplified Equation 3.3 (from Section 3.3) can be reduced as follows:

$$
\begin{aligned}
S &= \overline{P}\,\overline{Q}\,\overline{R} + P\overline{Q}\,\overline{R} + P\overline{Q}R + PQ\overline{R} + PQR \\
S &= \overline{Q}\,\overline{R}(\overline{P} + P) + P\overline{Q}R + PQ(\overline{R} + R) && \text{(distributive)} \\
S &= \overline{Q}\,\overline{R}1 + P\overline{Q}R + PQ1 && \text{(complementarity)} \\
S &= \overline{Q}\,\overline{R} + P\overline{Q}R + PQ && \text{(identity)} \\
S &= \overline{Q}(\overline{R} + PR) + PQ && \text{(distributive)} \\
S &= \overline{Q}(\overline{R} + P) + PQ && \text{(miscellaneous)} \\
S &= \overline{Q}\,\overline{R} + \overline{Q}P + PQ && \text{(distributive)} \\
S &= \overline{Q}\,\overline{R} + P(\overline{Q} + Q) && \text{(distributive)} \\
S &= \overline{Q}\,\overline{R} + P1 && \text{(complementarity)} \\
S &= \overline{Q}\,\overline{R} + P && \text{(identity)} \\
S &= \overline{Q}\,\overline{R} + P && \text{(3.12)}
\end{aligned}
$$

It can be proved that the reduced Equation 3.12 is equivalent to the unsimplified Equation 3.3 by constructing its truth table (Table 3.22). The truth table shown in Table 3.22 is identical to the truth table for Equation 3.3 (see Example 3.3).

The circuit for Equation 3.12 is shown in Figure 3.27. This logic circuit is considerably simpler than the circuit in Figure 3.10, requiring a total of only four gates as compared to nine gates for Figure 3.10. Futhermore, the IC implementation of Figure 3.10 would require four ICs—two 7411's for the three-input *AND* gates,

Table 3.22

P	Q	R	$\overline{Q}$	$\overline{R}$	$\overline{Q}\overline{R}$	S
0	0	0	1	1	1	1
0	0	1	1	0	0	0
0	1	0	0	1	0	0
0	1	1	0	0	0	0
1	0	0	1	1	1	1
1	0	1	1	0	0	1
1	1	0	0	1	0	1
1	1	1	0	0	0	1

Figure 3.27

Figure 3.10 reduced by applying Boolean algebra.

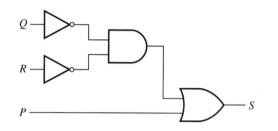

one 7432 for the *OR* gate, and one 7404 for the inverters. The circuit in Figure 3.27 would only require three ICs — one 7408 for the two-input *AND* gate, one 7432 for the two-input *OR* gate, and one 7404 for the inverters.

Example 3.7

Simplify Equation 3.5 using Boolean algebra and draw the logic circuit for the simplified equation. Compare the simplified circuit with the unsimplified circuit (Fig. 3.12) in terms of the number of gates and ICs and verify that the two circuits are equivalent.

Solution

$$X = \overline{A}\overline{B}\overline{C}D + \overline{A}\overline{B}CD + \overline{A}BC\overline{D} + A\overline{B}C\overline{D} + A\overline{B}C\overline{D} + ABCD$$
$$X = \overline{A}\overline{B}\overline{C}D + \overline{A}\overline{B}CD + A\overline{B}\overline{C}D + \overline{A}BC\overline{D} + A\overline{B}C\overline{D} + ABCD \qquad \text{(commutative)}$$
$$X = \overline{A}\overline{B}D\,(\overline{C} + C) + A\overline{B}\overline{C}D + \overline{A}BC\overline{D} + A\overline{B}C\overline{D} + ABCD \qquad \text{(distributive)}$$
$$X = \overline{A}\overline{B}D1 + A\overline{B}\overline{C}D + \overline{A}BC\overline{D} + A\overline{B}C\overline{D} + ABCD \qquad \text{(complementarity)}$$
$$X = \overline{A}\overline{B}D + A\overline{B}\overline{C}D + \overline{A}BC\overline{D} + A\overline{B}C\overline{D} + ABCD \qquad \text{(identity)}$$
$$X = \overline{B}D\,(\overline{A} + A\overline{C}) + \overline{A}BC\overline{D} + A\overline{B}C\overline{D} + ABCD \qquad \text{(distributive)}$$
$$X = \overline{B}D\,(\overline{A} + \overline{C}) + \overline{A}BC\overline{D} + A\overline{B}C\overline{D} + ABCD \qquad \text{(miscellaneous)}$$
$$X = \overline{A}\overline{B}D + \overline{B}\overline{C}D + \overline{A}BC\overline{D} + A\overline{B}C\overline{D} + ABCD \qquad \text{(distributive)}$$

The circuit in Figure 3.12 has a total of 11 gates as compared to the simplified circuit shown in Figure 3.28 consisting of ten gates. To implement the circuit in Figure 3.12, six ICs are needed — two 7432's, three 7421's, and one 7404. To implement the circuit in Figure 3.28, five ICs are needed — one 7432, three 7421's, and one 7404.

It can be proved that the simplified circuit in Figure 3.28 is equivalent to the unsimplified

Figure 3.28

Figure 3.12 reduced by
applying Boolean algebra.

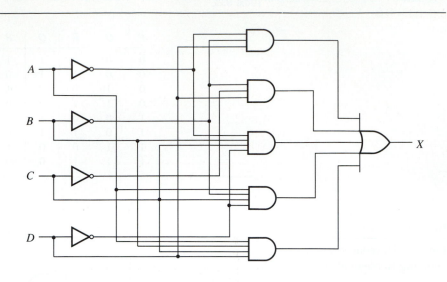

circuit in Figure 3.12 by constructing a truth table for the circuit in Figure 3.28 and comparing
it to the truth table for the circuit in Figure 3.12 (shown in Example 3.4):

A	B	C	D	X
0	0	0	0	0
0	0	0	1	1
0	0	1	0	0
0	0	1	1	1
0	1	0	0	0
0	1	0	1	0
0	1	1	0	1
0	1	1	1	0
1	0	0	0	0
1	0	0	1	1
1	0	1	0	1
1	0	1	1	0
1	1	0	0	0
1	1	0	1	0
1	1	1	0	0
1	1	1	1	1

It can be seen that because the two tables are identical, their circuits and equations are
equivalent.

The rules of Boolean algebra do not always work in the simplification of logic
equations and circuits. There are some logic equations that cannot be simplified by
applying the laws of Boolean algebra. There are others that even when simplified
will not provide much of a benefit in terms of gate and IC reduction. However, in
most cases Boolean algebra can be used to provide the designer with a circuit that
contains the smallest number of gates and ICs. The following examples illustrate the
use of Boolean algebra in the reduction of different types of logic equations.

Example 3.8

Simplify the following logic equations using Boolean algebra:

(a) $H = \overline{X}\overline{Y} + Y\overline{Z} + X\overline{Y}\overline{Z} + X\overline{Y}Z + X(Y + \overline{X}) + X$

(b) $P = \overline{A}\overline{B}\overline{C} + \overline{A}B\overline{C} + \overline{A}BC + A\overline{B}\overline{C} + A\overline{B}C + AB\overline{C}$

(c) $X = R(P + \overline{P}Q + \overline{Q})(\overline{Q} + \overline{R}P)$

(d) $Z = (A\overline{B}\overline{C} + A\overline{B}C + ABC + AB\overline{C})(A + B)$

(e) $W = \overline{\overline{AC} + \overline{AD}} + \overline{\overline{B}C}$

(f) $A = P + \overline{P}Q\overline{R} + \overline{Q + R}$

Solutions

(a)

$H = \overline{X}\overline{Y} + Y\overline{Z} + X\overline{Y}\overline{Z} + X\overline{Y}Z + X(Y + \overline{X}) + X$
$H = \overline{X}\overline{Y} + Y\overline{Z} + X\overline{Y}(\overline{Z} + Z) + X(Y + \overline{X}) + X$ (distributive)
$H = \overline{X}\overline{Y} + Y\overline{Z} + X\overline{Y}1 + X(Y + \overline{X}) + X$ (complementarity)
$H = \overline{X}\overline{Y} + Y\overline{Z} + X\overline{Y} + X(Y + \overline{X}) + X$ (identity)
$H = Y\overline{Z} + \overline{X}\overline{Y} + X\overline{Y} + X(Y + \overline{X}) + X$ (commutative)
$H = Y\overline{Z} + \overline{Y}(\overline{X} + X) + X(Y + \overline{X}) + X$ (distributive)
$H = Y\overline{Z} + \overline{Y}1 + X(Y + \overline{X}) + X$ (complementarity)
$H = Y\overline{Z} + \overline{Y} + X(Y + \overline{X}) + X$ (identity)
$H = \overline{Z} + \overline{Y} + X(Y + \overline{X}) + X$ (miscellaneous)
$H = \overline{Z} + \overline{Y} + XY + X\overline{X} + X$ (distributive)
$H = \overline{Z} + \overline{Y} + XY + 0 + X$ (complementarity)
$H = \overline{Z} + \overline{Y} + XY + X$ (identity)
$H = \overline{Z} + \overline{Y} + X$ (absorbtion)

(b)

$P = \overline{A}\overline{B}\overline{C} + \overline{A}B\overline{C} + \overline{A}BC + A\overline{B}\overline{C} + A\overline{B}C + AB\overline{C}$
$P = \overline{A}\overline{C}(\overline{B} + B) + \overline{A}BC + A\overline{B}(\overline{C} + C) + AB\overline{C}$ (distributive)
$P = \overline{A}\overline{C} + \overline{A}BC + A\overline{B} + AB\overline{C}$ (complementarity / identity)
$P = \overline{A}(\overline{C} + BC) + A(\overline{B} + B\overline{C})$ (distributive)
$P = \overline{A}(\overline{C} + B) + A(\overline{B} + \overline{C})$ (miscellaneous)
$P = \overline{A}\overline{C} + \overline{A}B + A\overline{B} + A\overline{C}$ (distributive)
$P = \overline{A}\overline{C} + A\overline{C} + A\overline{B} + \overline{A}B$ (associative)
$P = \overline{C}(\overline{A} + A) + A\overline{B} + \overline{A}B$ (distributive)
$P = \overline{C} + A\overline{B} + \overline{A}B$ (complementarity / identity)

(c)

$X = R(P + \overline{P}Q + \overline{Q})(\overline{Q} + \overline{R}P)$
$X = R(P + Q + \overline{Q})(\overline{Q} + \overline{R}P)$ (miscellaneous)
$X = R(P + 1)(\overline{Q} + \overline{R}P)$ (complementarity)
$X = R1(\overline{Q} + \overline{R}P)$ (dominance)
$X = R(\overline{Q} + \overline{R}P)$ (identity)
$X = R\overline{Q} + R\overline{R}P$ (distributive)
$X = R\overline{Q} + 0P$ (complementarity)
$X = R\overline{Q} + 0$ (dominance)
$X = R\overline{Q}$ (identity)

(d)

$Z = (A\overline{B}\overline{C} + A\overline{B}C + ABC + AB\overline{C})(A + B)$
$Z = (A\overline{B}(\overline{C} + C) + AB(C + \overline{C}))(A + B)$ (distributive)
$Z = (A\overline{B} + AB)(A + B)$ (complementarity / identity)
$Z = (A(\overline{B} + B))(A + B)$ (distributive)
$Z = A(A + B)$ (complementarity / identity)
$Z = A$ (absorbtion)

(e)

$$W = \overline{\overline{AC} + \overline{AD}} + \overline{\overline{B}C}$$
$$W = \overline{\overline{AC}} \cdot \overline{\overline{AD}} + \overline{\overline{B}C} \qquad \text{(DeMorgan's law)}$$
$$W = (\overline{A} + \overline{C}) \cdot (\overline{A} + \overline{\overline{D}}) + (\overline{\overline{B}} + \overline{C}) \qquad \text{(DeMorgan's law)}$$
$$W = (\overline{A} + \overline{C}) \cdot (\overline{A} + D) + (B + \overline{C}) \qquad (\overline{\overline{X}} = X)$$
$$W = \overline{A}(\overline{A} + D) + \overline{C}(\overline{A} + D) + (B + \overline{C}) \qquad \text{(distributive)}$$
$$W = \overline{A}\overline{A} + \overline{A}D + \overline{C}\overline{A} + \overline{C}D + (B + \overline{C}) \qquad \text{(distributive)}$$
$$W = \overline{A} + \overline{A}D + \overline{C}\overline{A} + \overline{C}D + (B + \overline{C}) \qquad \text{(idempotent)}$$
$$W = \overline{A} + \overline{C}\overline{A} + \overline{C}D + (B + \overline{C}) \qquad \text{(absorbtion)}$$
$$W = \overline{A} + \overline{C}D + (B + \overline{C}) \qquad \text{(absorbtion)}$$
$$W = \overline{A} + \overline{C}D + B + \overline{C} \qquad \text{(associative)}$$
$$W = \overline{C} + \overline{C}D + B + \overline{A} \qquad \text{(commutative)}$$
$$W = \overline{C} + B + \overline{A} \qquad \text{(absorbtion)}$$

(f)

$$A = P + \overline{P}Q\overline{R} + \overline{Q} + R$$
$$A = P + Q\overline{R} + \overline{Q} + R \qquad \text{(miscellaneous)}$$
$$A = P + Q\overline{R} + \overline{QR} \qquad \text{(DeMorgan's law)}$$
$$A = P + \overline{R}(Q + \overline{Q}) \qquad \text{(distributive)}$$
$$A = P + \overline{R} \qquad \text{(complementarity / identity)}$$

This section has examined the application of the Boolean algebra laws, theorems, and postulates in the simplification of logic equations and circuits. In many of the examples of Boolean simplification it can be seen that the number of gates and ICs can be greatly reduced by obtaining an equivalent simplified logic equation. However, it is difficult to determine when a logic equation has been simplified to a point where no further simplification is possible. As a general rule, one can consider a logic equation completely simplified when no other rule can be applied to further simplify it. However, there is always a possibility that the simplest form may not be obtained because the application of a particular law or rule is not apparent. In Section 3.7 a technique will be examined that (if used properly) will allow one to obtain the simplest possible form using a more direct approach. Section 3.6 will also deal with a technique of reducing the size of some logic circuits by applying DeMorgan's law.

Review Questions

1. What is the objective of Boolean simplification?
2. When is a logic equation considered to be completely simplified?
3. Can all logic equations be simplified using Boolean algebra?

3.6
NAND and *NOR* logic

The unsimplified sum-of-products and product-of sums equations that were synthesized in Section 3.3 and their simplified equivalents were all implemented using *AND* gates, *OR* gates, and inverters. These circuits are therefore known as *AND-OR-INVERT* circuits or more simply *AND-OR* logic. The hardware implementation of these circuits, no matter how simple, required a minimum of three ICs if the circuit contained *AND* gates, *OR* gates, and inverters, because each IC usually contains gates having the same function.

NAND and *NOR* gates have a special characteristic that allow them to replace *AND-OR* logic circuits with circuits that contain only *NAND* or *NOR* gates. In some circumstances, the number of ICs required to construct a circuit can be reduced if only *NAND* gates or *NOR* gates are used. *NAND* gate circuits have another advantage in that the manufacturing cost of *NAND* gates is lower than other gates because fewer electronic components are used to make up a *NAND* gate than any other 74xxx gate; this also means that *NAND* gate circuits are more compact than those of any other gate. *NAND* gate circuits are therefore preferable in the integration of complex logic circuits on an IC. This section will therefore deal with the techniques of converting logic equations from *AND-OR* logic to *NAND* and *NOR* logic to further minimize the number of ICs required to implement the circuit rather than the number of gates in the circuit.

To illustrate the universal properties of the *NAND* and *NOR* gates, first consider the circuits shown in Figure 3.29. The circuit shown in Figure 3.29a is an inverter that is constructed from a two-input *NAND* gate. It can be proved that the circuit functions as an inverter by applying the idempotent law

$$A \cdot A = A \qquad (3.13)$$

Figure 3.29

(a) *NAND* gate used as an inverter; (b) *NOR* gate used as an inverter.

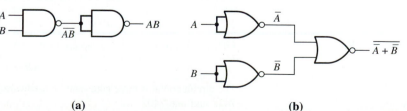

$A \qquad A \cdot A = \overline{A}$

(a)

$A \qquad \overline{A+A} = \overline{A}$

(b)

If the left-hand side of Equation 3.13 is inverted, then the right-hand side must also be inverted to balance the equation

$$\overline{A \cdot A} = \overline{A} \qquad (3.14)$$

The left-hand side of Equation 3.14 represents a two-input *NAND* gate with both its inputs connected to the same independent variable A, and the right-hand side of Equation 3.14 represents the complement of A.

Similarly, Equation 3.15 proves by idempotence that the circuit in Figure 3.29b is an inverter that is constructed from a *NOR* gate.

$$\overline{A + A} = \overline{A} \qquad (3.15)$$

NAND and *NOR* gates can also be used with more than two inputs to function as inverters by tying all inputs together.

Figure 3.30 shows circuits that implement the *AND* function. Figure 3.30a is a circuit that functions as an *AND* gate but is constructed exclusively with *NAND* gates. Observe that the output of a two-input *NAND* gate can be inverted to obtain the *AND* function. To implement the *AND* function using only *NOR* gates is more

Figure 3.30

(a) *AND* gate using only *NAND* gates; (b) *AND* gate using only *NOR* gates.

A
$B \qquad \overline{AB} \qquad AB$

(a)

$A \qquad \overline{A}$
$B \qquad \overline{B} \qquad \overline{\overline{A} + \overline{B}}$

(b)

complex, as can be seen in Figure 3.30b. The output of the circuit is

$$\overline{\overline{A} + \overline{B}}. \tag{3.16}$$

Applying DeMorgan's theorem to Equation 3.16 gives

$$\overline{\overline{A}} \cdot \overline{\overline{B}}. \tag{3.17}$$

Equation 3.17 further simplifies into the equation for an *AND* gate

$$A \cdot B. \tag{3.18}$$

Note that a *NAND* gate can also be implemented using only *NOR* gates by inverting the output of Figure 3.30b. As stated before, these circuits can be extended to have more than two inputs.

Figure 3.31 shows circuits that implement the *OR* function. Figure 3.31b is a circuit that functions as an *OR* gate but is constructed exclusively with *NOR* gates. Observe that the output of a two-input *NOR* gate has been inverted to obtain the *OR* function. To implement the *OR* function using only *NAND* gates is more complex as can be seen in Figure 3.31a. The output of the circuit is

$$\overline{\overline{A} \cdot \overline{B}}. \tag{3.19}$$

Figure 3.31

(a) *OR* gate using only *NAND* gates; (b) *OR* gate using only *NOR* gates.

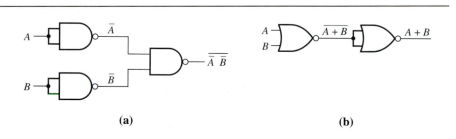

(a) (b)

Applying DeMorgan's theorem to Equation 3.19 gives

$$\tag{3.20}$$
$$\overline{\overline{A}} + \overline{\overline{B}}.$$

Equation 3.20 further simplifies into the equation for an *OR* gate

$$A + B \tag{3.21}$$

Note that a *NOR* gate can also be implemented using only *NAND* gates by inverting the output of Figure 3.31a. As stated before, these circuits can also be extended to have more than two inputs.

By inspecting the circuits shown in Figures 3.29 through 3.31 it may appear that the *NAND* and *NOR* logic implementation of the basic *AND*, *OR*, and INVERT functions would increase the complexity of a logic circuit rather than reduce it. However, this is not the case, as can be seen in the following examples.

Example 3.9

Figure 3.32a is a circuit that represents the *AND-OR* logic implementation of the expression

$$AB + CD. \tag{3.22}$$

The circuit contains three gates and its hardware implementation will contain two ICs — one 7432 and one 7408.

Figure 3.32

(a) *AND-OR*
implementation; (b) *NAND*
implementation.

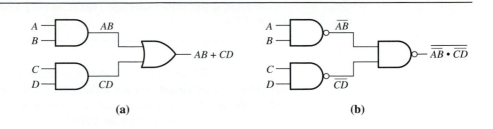

(a)   (b)

Figure 3.32b is the *NAND* logic implementation of the circuit shown in Figure 3.32a. The logic expression for the circuit is

$$\overline{\overline{AB} \cdot \overline{CD}}. \tag{3.23}$$

It can be proved that the two circuits are equivalent by applying DeMorgan's theorem to Equation 3.23

$$\overline{\overline{AB}} + \overline{\overline{CD}} \tag{3.24}$$

Simplifying Equation 3.24 further gives

$$AB + CD. \tag{3.25}$$

Because Equations 3.25 and 3.22 are identical, the two circuits are equivalent.

Even though the circuit in Figure 3.32b has the same number of gates as the circuit in Figure 3.32a, it will require only one 7400 IC as compared to the two-IC implementation of Figure 3.32a.

Example 3.10

Figure 3.33a is a circuit that represents the *AND-OR* logic implementation of the expression

$$(A + B) \cdot (C + D) \tag{3.26}$$

The circuit contains three gates and its hardware implementation will contain two ICs — one 7408 and one 7432.

Figure 3.33b is the *NOR* logic implementation of the circuit shown in Figure 3.33a. The logic expression for the circuit is

$$\overline{\overline{A + B} + \overline{C + D}}. \tag{3.27}$$

The two circuits can be proved to be equivalent by applying DeMorgan's theorem to Equation 3.27

$$\overline{\overline{A + B}} \cdot \overline{\overline{C + D}} \tag{3.28}$$

Simplifying Equation 3.28 further gives

$$(A + B) \cdot (C + D) \tag{3.29}$$

Because Equations 3.29 and 3.26 are identical, the two circuits are equivalent.

Figure 3.33

(a) *AND-OR*
implementation; (b) *NOR*
implementation.

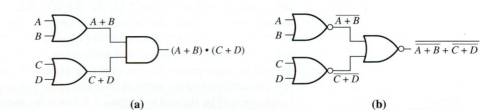

(a)   (b)

Even though the circuit in Figure 3.33b has the same number of gates as the circuit in Figure 3.33a, it will require only one 7402 IC as compared to the two-IC implementation of Figure 3.33a.

Examples 3.9 and 3.10 have shown us that *AND-OR* circuits when converted to *NAND* and *NOR* circuits do not reduce the number of gates in the circuit but can reduce the number of ICs required to implement the circuit in hardware. Furthermore, it can be seen that when an *AND-OR* circuit is in sum-of-products form it is easier to convert the circuit to *NAND* logic, and when the *AND-OR* circuit is in product-of-sums form it is easier to convert it to *NOR* logic. The techniques used to convert *AND-OR* logic circuits to *NAND* and *NOR* logic will now be investigated.

Consider the following *AND-OR* logic equation in sum-of-products form:

$$X = \overline{A}B + AC \tag{3.30}$$

The circuit for Equation 3.30 is made up of four gates and will require three ICs (7404, 7408, 7432) as shown in Figure 3.34. This circuit can be converted to *NAND* logic by applying the following procedure to Equation 3.30.

Figure 3.34

AND-OR circuit for Equation 3.30.

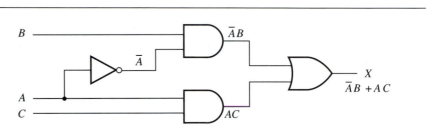

Both sides of the equation are first inverted

$$\overline{X} = \overline{\overline{A}B + AC} \tag{3.31}$$

DeMorgan's law can then be applied to the right-hand side of the equation

$$\overline{X} = \overline{\overline{A}B} \cdot \overline{AC} \tag{3.32}$$

Finally, to restore the equation to its original value in terms of the variable X, invert both sides again

$$\overline{\overline{X}} = \overline{\overline{\overline{A}B} \cdot \overline{AC}} \tag{3.33}$$

The final *NAND* equation is

$$X = \overline{\overline{\overline{A}B} \cdot \overline{AC}}. \tag{3.34}$$

The circuit for Equation 3.34 is shown in Figure 3.35. Like its equivalent *AND-OR* circuit shown in Figure 3.34, the *NAND* circuit also has four gates but only requires one IC (7400) for its hardware implementation.

To convert the *AND-OR* circuit in Figure 3.34 to *NOR* logic, Equation 3.30 must first be converted into product-of-sums form. Converting from sum-of-products to product-of-sums can be done by applying various Boolean algebra rules to Equation 3.30, but the procedure involves trial and error and is time-consuming. A preferred technique will be studied in Section 3.7. For now, use the product-of-sums equation

Figure 3.35

NAND circuit for Equation 3.30.

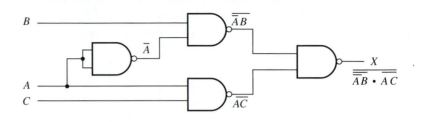

without being concerned about its derivation. Equation 3.35 is the product-of-sums equivalent for Equation 3.30.

$$X = (A + B) \cdot (\overline{A} + C) \tag{3.35}$$

Apply the same procedure as before to convert Equation 3.35 to *NOR* logic. Both sides of the equation are inverted

$$\overline{X} = \overline{(A + B) \cdot (\overline{A} + C)} \tag{3.36}$$

Next, DeMorgan's law is applied to the right-hand side of Equation 3.36

$$\overline{X} = \overline{(A + B)} + \overline{(\overline{A} + C)} \tag{3.37}$$

Finally, both sides of the equation are inverted again to restore the equation to its original value in terms of *X*

$$\overline{\overline{X}} = \overline{\overline{(A + B)} + \overline{(\overline{A} + C)}} \tag{3.38}$$

The final *NOR* equation is

$$X = \overline{\overline{(A + B)} + \overline{(\overline{A} + C)}}. \tag{3.39}$$

The circuit for Equation 3.39 is shown in Figure 3.36. Like its equivalent *AND-OR* circuit shown in Figure 3.34 and its equivalent *NAND* circuit in Figure 3.35, the *NOR* circuit also has four gates. However, like the *NAND* circuit, it only requires one IC (7402) for its hardware implementation. It can be proved by truth tables that the three circuits are equivalent as shown in Table 3.23.

It should be noted that before the previously discussed procedures are applied to *AND-OR* logic equations (in sum-of-products or product-of-sums form) they should be simplified completely using either the Boolean algebra laws or through the Karnaugh map (K-map) procedure to be discussed in the next section. To convert an *AND-OR* equation to *NAND* logic the equation must be in sum-of-products form, and to convert an *AND-OR* equation to *NOR* logic the equation must be in product-of-

Figure 3.36

NOR circuit for Equation 3.30.

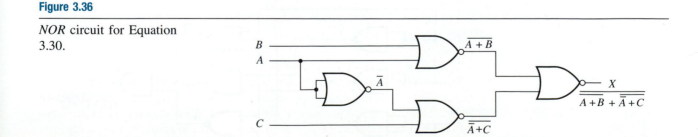

Table 3.23

A	B	C	X (Figure 3.34)	X (Figure 3.35)	X (Figure 3.36)
0	0	0	0	0	0
0	0	1	0	0	0
0	1	0	1	1	1
0	1	1	1	1	1
1	0	0	0	0	0
1	0	1	1	1	1
1	1	0	0	0	0
1	1	1	1	1	1

sums form. As was stated earlier in this section, conversions between logic equations in sum-of-products and product-of-sums form will be examined in the next section.

Example 3.11

Convert the following simplified *AND-OR* logic equations to *NAND* logic and draw the *NAND* circuit for each equation:

(a) $T = XY\bar{Z} + \bar{X}\bar{Y} + \bar{X}Z$

(b) $X = P(Q + \bar{R}S)$

(c) $Z = (\bar{A} + B)C + A(\bar{B} + C)$

Solutions

(a) $T = XY\bar{Z} + \bar{X}\,\bar{Y} + \bar{X}Z$

  Inverting both sides

$$\bar{T} = \overline{XY\bar{Z} + \bar{X}\bar{Y} + \bar{X}Z}$$

  Applying DeMorgan's law

$$\bar{T} = \overline{XY\bar{Z}} \cdot \overline{\bar{X}\bar{Y}} \cdot \overline{\bar{X}Z}$$

  Inverting both sides again

$$\bar{\bar{T}} = \overline{\overline{XY\bar{Z}} \cdot \overline{\bar{X}\bar{Y}} \cdot \overline{\bar{X}Z}}$$

  The final *NAND* equation is

$$T = \overline{\overline{XY\bar{Z}} \cdot \overline{\bar{X}\bar{Y}} \cdot \overline{\bar{X}Z}}. \tag{3.40}$$

The circuit for Equation 3.40 is shown in Figure 3.37.

(b) $X = P(Q + \bar{R}S)$

  First, convert the equation to sum-of-products form using the distributive law

$$X = PQ + P\bar{R}S$$

Figure 3.37

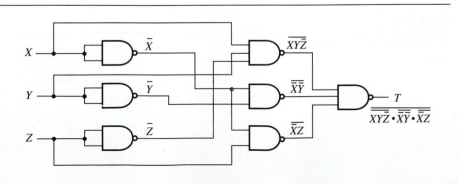

Both sides of the equation are now inverted

$$\overline{X} = \overline{PQ} + \overline{P}\overline{R}S$$

Applying DeMorgan's law

$$\overline{X} = \overline{PQ} \cdot \overline{P\overline{R}S}$$

Inverting both sides again

$$\overline{\overline{X}} = \overline{\overline{PQ} \cdot \overline{P\overline{R}S}}$$

The final *NAND* equation is

$$X = \overline{\overline{PQ} \cdot \overline{P\overline{R}S}}. \qquad (3.41)$$

The circuit for Equation 3.41 is shown in Figure 3.38.

Figure 3.38

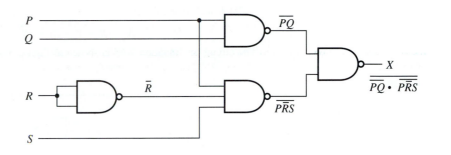

(c) $Z = (\overline{A + B})\,C + A\,(\overline{B + C})$

First, convert the equation to sum-of-products form using DeMorgan's law

$$Z = (\overline{A} \cdot \overline{B})\,C + A\,(\overline{B} \cdot \overline{C})$$

$$Z = \overline{A}\overline{B}C + A\overline{B}\overline{C}$$

Inverting both sides of the equation

$$\overline{Z} = \overline{\overline{A}\overline{B}C + A\overline{B}\overline{C}}$$

Applying DeMorgan's law

$$\overline{Z} = \overline{\overline{A}\overline{B}C} \cdot \overline{A\overline{B}\overline{C}}$$

Inverting both sides again

$$\overline{\overline{Z}} = \overline{\overline{\overline{A}\overline{B}C} \cdot \overline{A\overline{B}\overline{C}}}$$

The final *NAND* equation is

$$Z = \overline{\overline{\overline{A}\overline{B}C} \cdot \overline{A\overline{B}\overline{C}}}. \qquad (3.42)$$

The circuit for Equation 3.42 is shown in Figure 3.39.

Figure 3.39

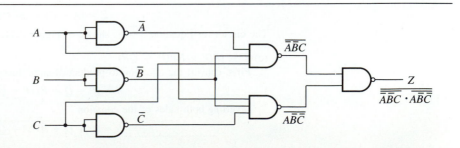

Example 3.12

Convert the following simplified *AND-OR* logic equations to *NOR* logic and draw the *NOR* circuit for each equation.

(a) $A = (\overline{X} + Y + \overline{Z})(X + Y)(X + Z)$

(b) $Z = A + B\overline{C}$

(c) $S = (\overline{PQ} + R)(P + \overline{QR})$

Solutions

(a) $A = (\overline{X} + Y + \overline{Z})(X + Y)(X + Z)$

Inverting both sides

$$\overline{A} = \overline{(\overline{X} + Y + \overline{Z})(X + Y)(X + Z)}$$

Applying DeMorgan's law

$$\overline{A} = \overline{(\overline{X} + Y + \overline{Z})} + \overline{(X + Y)} + \overline{(X + Z)}$$

Inverting both sides again

$$\overline{\overline{A}} = \overline{\overline{(\overline{X} + Y + \overline{Z})} + \overline{(X + Y)} + \overline{(X + Z)}}$$

The final *NOR* equation is

$$A = \overline{\overline{(\overline{X} + Y + \overline{Z})} + \overline{(X + Y)} + \overline{(X + Z)}}. \tag{3.43}$$

The circuit for Equation 3.43 is shown in Figure 3.40.

Figure 3.40

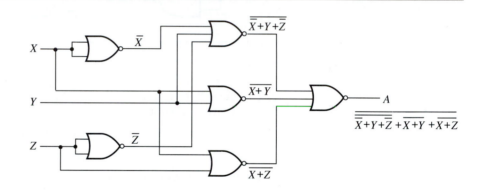

(b) $Z = A + B\overline{C}$

First, convert the equation to product-of-sums form using the distributive law

$$Z = (A + B) \cdot (A + \overline{C})$$

Both sides of the equation are now inverted

$$\overline{Z} = \overline{(A + B) \cdot (A + \overline{C})}$$

Applying DeMorgan's law

$$\overline{Z} = \overline{(A + B)} + \overline{(A + \overline{C})}$$

Inverting both sides again

$$\overline{\overline{Z}} = \overline{\overline{(A + B)} + \overline{(A + \overline{C})}}$$

The final *NOR* equation is

$$Z = \overline{\overline{(A + B)} + \overline{(A + \overline{C})}}. \tag{3.44}$$

The circuit for Equation 3.44 is shown in Figure 3.41.

(c) $S = (\overline{PQ} + R)(P + \overline{QR})$

First, convert the equation to product-of-sums form using DeMorgan's law

$$S = (\overline{P} + \overline{Q} + R)(P + \overline{Q} + \overline{R})$$

Inverting both sides of the equation

$$\overline{S} = \overline{(\overline{P} + \overline{Q} + R)(P + \overline{Q} + \overline{R})}$$

Figure 3.41

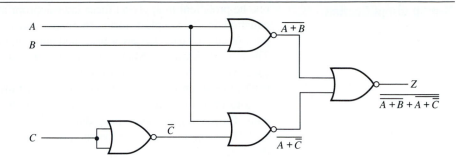

Applying DeMorgan's law

$$\overline{S} = \overline{(\overline{\overline{P} + \overline{Q} + R})} + \overline{(P + \overline{Q} + \overline{R})}$$

Inverting both sides again

$$\overline{\overline{S}} = \overline{\overline{(\overline{\overline{P} + \overline{Q} + R})} + \overline{(P + \overline{Q} + \overline{R})}}$$

The final *NOR* equation is

$$S = \overline{\overline{(\overline{\overline{P} + \overline{Q} + R})} + \overline{(P + \overline{Q} + \overline{R})}}. \tag{3.45}$$

The circuit for Equation 3.45 is shown in Figure 3.42.

Figure 3.42

Review Questions

1. What is an *AND-OR* logic circuit? How does it differ from a *NAND* and *NOR* logic circuit?

2. What are the advantages of *NAND* and *NOR* circuits over *AND-OR* circuits?

3. Why are *NAND* and *NOR* gates considered to have a "universal" property?

4. What form should an *AND-OR* logic equation be in before it is converted to *NAND* logic?

5. What form should an *AND-OR* logic equation be in before it is converted to *NOR* logic?

3.7
K-map simplification

The simplification of logic equations using the laws of Boolean algebra requires that one be proficient in applying these laws appropriately to obtain an equivalent logic equation in its simplest form. In many cases, Boolean simplification must be performed on a trial-and-error basis and consequently can be quite time-consuming. The K-map (Karnaugh Map) simplification technique is a more direct and mechanical technique that (if used properly) can obtain the simplest logic equation from the truth table of an unsimplified logic equation. Furthermore, the K-map can also be used to obtain simplified logic equations in both sum-of-products and product-of-sums form using the same procedures.

The K-map is a modified version of a truth table. Like a truth table, it maps the dependent variable as a function of the independent variables, but its arrangement is such that one can tell by inspection if simplification is possible or not, and if simplification is possible its configuration allows us to obtain the simplest possible logic equation. K-maps are generally classified according to the number of independent variables in the logic equation, circuit, or truth table. Two-variable, three-variable, and four-variable K-maps are generally used. K-maps with more than four variables are possible but can get very complicated and other techniques must be used. This section will examine the configuration of these K-maps and their use in the simplification of logic equations. This technique will be used in the simplification of logic equations for most of the examples covered in the rest of this book. Therefore, it is important that its usage be thoroughly understood.

Two-variable K-maps

The general configuration of a two-variable K-map is shown in Figure 3.43 to represent the independent variables A and B and the dependent variable X. The K-map is made up of four cells (corresponding to the four combinations of two independent variables, $2^2 = 4$). For reference purposes only, each cell has been numbered from 0 to 3 to identify the binary combination of A and B that identifies the cell (assuming that A is the MSB and B is the LSB). Therefore, cell no. 0 corresponds to $A = 0$ and $B = 0$, cell no. 1 corresponds to $A = 0$ and $B = 1$, cell no. 2 corresponds to $A = 1$ and $B = 0$, and cell no. 3 corresponds to $A = 1$ and $B = 1$. The bits that will be placed in each cell will be the values of the dependent variable X. For example, Figure 3.44 shows a two-variable truth table and its corresponding K-map configuration. Note in Figure 3.44 that the value inside each cell is the value of the dependent variable Z, and the coordinates of each cell are the values of the independent variables P and Q.

The short-form notation for writing logic equations that was developed in Section 3.3 is especially useful for filling in K-maps from a truth table. For example, in Figure 3.44, the unsimplified logic equation for the truth table in short-form notation is

$$Z = \sum_m 1, 2, 3.$$

Figure 3.43

General configuration of a two-variable K-map.

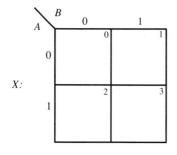

Figure 3.44

A two-variable truth table and its K-map configuration.

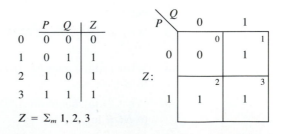

To fill in the K-map, simply place 1's in cells 1, 2, and 3 to correspond to the minterms in the equation.

Before the logic equation can be obtained from the K-map, the *adjacency* of two cells in a K-map must be identified. Two cells are said to be *adjacent* if *only one independent variable changes in going from one cell to the other*. For example, in Figure 3.43, cell no. 0 and cell no. 1 are adjacent because the variable *B* changes from a 0 to a 1 (or a 1 to a 0) but the variable *A* remains the same (0) in both cells. Similarly, cell no. 1 and cell no. 3 are adjacent because *A* changes from a 0 to a 1 (or a 1 to a 0) but the variable *B* remains the same (1) in both cells. Note that cell no. 1 and cell no. 2 are not adjacent because both variables *A* and *B* change in going from one cell to the other. In going from cell no. 1 to cell no. 2, *B* changes from a 1 to a 0 and *A* changes from a 0 to a 1. Similarly, cell no. 0 and cell no. 3 are not adjacent.

There are two methods of using a K-map. The first involves combining all the cells containing 1's (1-cells) to obtain a simplified sum-of-products equation, and the other involves combining all the cells containing 0's (0-cells) to obtain the product-of-sums equation. Note that this procedure is similar to the procedure used to obtain the unsimplified sum-of-products and product-of-sums equation from a truth table. The sum-of-products procedure will be examined first and the concepts developed will be appled to the product-of-sums procedure discussed at the end of this section.

Figure 3.45a shows a K-map representation of a truth table containing two 1's for the dependent variable *P*. Observe that these two 1's are located in cells that are not adjacent, meaning that simplification is not possible. Therefore, we take the sum of the minterms corresponding to these two 1's. The unsimplified equation is

$$P = \overline{X}\,\overline{Y} + XY.$$

Similarly, in Figure 3.45b, the two 1-cells are not adjacent and therefore no simplification is possible. The unsimplified logic equation for the K-map is

$$P = X\overline{Y} + \overline{X}Y.$$

If a K-map contains a single 1-cell, the equation will contain only the minterm for that cell.

Simplification is only possible in K-maps that have logic 1's located in cells that are adjacent. For example, in Figure 3.46a, because there are two adjacent 1-cells, and because the variable *X* remains the same (1) in both cells, the logic equation for the K-map is

$$P = X.$$

Figure 3.45

Nonadjacent 1-cells.

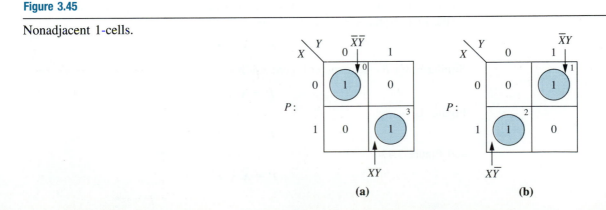

(a) (b)

Figure 3.46

Groups of two adjacent
1-cells.

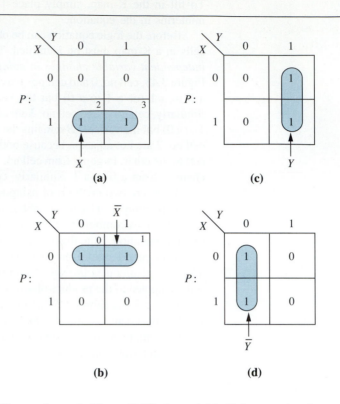

(a) (c)

(b) (d)

In the K-map shown in Figure 3.46b the variable X also remains the same in both cells but its value is 0. Therefore, the logic equation for the K-map is

$$P = \overline{X}.$$

In Figure 3.46c, the variable Y remains constant in both adjacent 1-cells. Therefore, the equation is

$$P = Y.$$

Similarly, in Figure 3.46d, the variable $\overline{Y}$ remains constant in both adjacent 1-cells (because its value is 0). Therefore, the equation for the K-map is

$$P = \overline{Y}.$$

In K-maps where there are more than one set of adjacent 1's, *OR* the variables obtained from each set. For example, in Figure 3.47a because adjacent cells no. 1 and no. 3 produce the variable Y, and adjacent cells no. 2 and no. 3 produce the variable X, the logic equation for the K-map is

$$P = X + Y.$$

Similarly, for the K-maps in Figure 3.47b

$$P = \overline{X} + Y$$

Figure 3.47c

$$P = X + \overline{Y}$$

and Figure 3.47d

$$P = \overline{X} + \overline{Y}.$$

Figure 3.47

Multiple groups of two adjacent 1-cells.

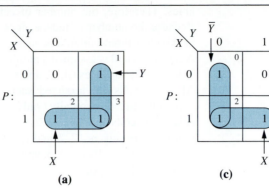

(a)

(c)

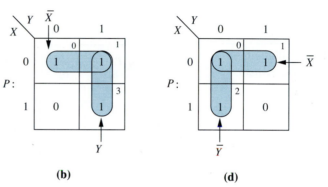

(b)

(d)

Figure 3.48

(a) K-map with all 0-cells;
(b) K-map with all 1-cells.

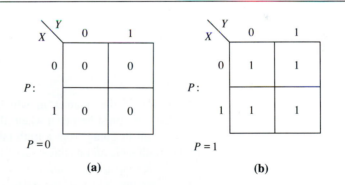

(a)

(b)

If a K-map contains all 0's as shown in Figure 3.48a, it is obvious that the dependent variable has a value of 0 permanently and does not depend on the states of the independent variables. Similarly, in Figure 3.48b the dependent variable P has a value of 1 regardless of the independent variables X and Y.

Two-variable K-maps are not always beneficial in the simplification of logic equations because in most cases the simplified equation for a two-variable truth table can be obtained by inspection or by applying some elementary Boolean algebra rules. However, the advantage of using K-maps becomes evident when they are used to obtain simplified equations for three- and four-variable truth tables.

Three-variable K-maps

Three-variable K-maps map a dependent variable against three independent variables. Therefore, the number of cells in a three-variable K-map will be 2^3 or 8. Because the number of independent variables is an odd number, there are two ways in which a three-variable K-map can be drawn as shown in Figure 3.49a and b. Again, each cell is assigned a number corresponding to the binary combination of the independent variables (A, B, and C), assuming that A is the MSB and C is the LSB. Also, the cells are arranged such that in going from one cell to another, both vertically and horizontally, only one variable changes. Therefore, cell no. 0 is adjacent to cells no. 1, 4, and 2; cell no. 1 is adjacent to cells no. 3, 5, and 0; cell no. 3 is adjacent to cells no. 2, 7, and 1; cell no. 2 is adjacent to cells no. 0, 3, and 6; cell no. 4 is adjacent to cells no. 0, 5, and 6, and so on. As in two-variable K-maps, cells that are diagonally opposite each other are not adjacent because more than one variable changes in going from one cell to another. Thus, cells no. 4 and 1 are not adjacent, cells no. 5 and 3 are not adjacent, and so on.

Figure 3.49

Two different configurations of a three-variable K-map.

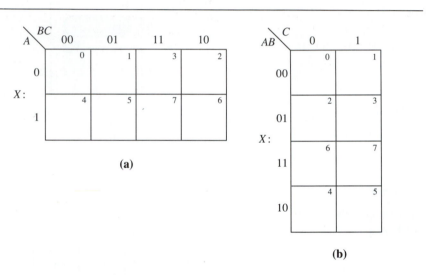

(a)

(b)

Because of the manner in which the cells in a three-variable K-map are numbered, care must be taken when filling in the values of the dependent variable into each cell in going from a truth table to a K-map. Figure 3.50 shows a three-variable truth table and its three-variable K-map representations. The values in each cell of the K-map are the values of the dependent variable Z, and the coordinates of each cell are the values of the independent variables P, Q, and R.

We previously saw that the short-form notation for writing logic equations was useful for filling in K-maps from a truth table. In Figure 3.50, the unsimplified logic equation for the truth table in short-form notation is

$$Z = \sum_m 0, 3, 4, 5, 7.$$

To fill in the K-map, simply place 1's in cells 0, 3, 4, 5, and 7 to correspond to the minterms in the equation.

As stated earlier, there are two methods used to obtain a simplified equation from a K-map — combining cells containing 1's into adjacent groups to obtain a simplified sum-of-products equation, and combining cells containing 0's into adjacent groups to obtain a simplified product-of-sums equation. The first method will be

Figure 3.50

A three-variable truth table and its K-map.

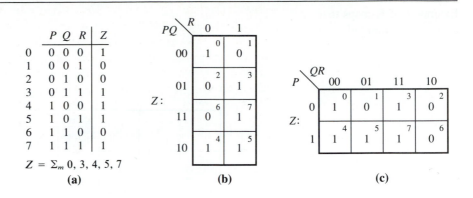

	P Q R	Z
0	0 0 0	1
1	0 0 1	0
2	0 1 0	0
3	0 1 1	1
4	1 0 0	1
5	1 0 1	1
6	1 1 0	0
7	1 1 1	1

$Z = \Sigma_m\ 0, 3, 4, 5, 7$

(a)   **(b)**   **(c)**

examined now and then the concepts will be applied to the second method at the end of this section.

Again, if a K-map has nonadjacent cells containing 1's, then no simplification is possible. Figure 3.51 illustrates such a K-map. In cases such as this the logic equation is simply the sum of the minterms for each cell

$$P = \overline{A}\,\overline{B}\,\overline{C} + \overline{A}BC + A\overline{B}C$$

Figure 3.51

Nonadjacent 1-cells.

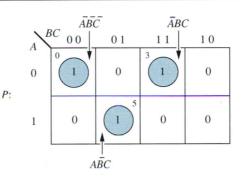

Figure 3.52 illustrates other K-map configurations in which no simplification is possible.

K-maps that have cells containing 1's that are adjacent in groups of two will produce simplified logic equations. For example, in Figure 3.53, K-maps with five cells containing 1's are shown. Figure 3.53a and b illustrate the same K-map, but the adjacent 1-cells have been grouped differently, that is, cells no. 0 and 4 and cells no. 3 and 7 have been combined into two groups. Because cell no. 1 is adjacent to both cell no. 0 and cell no. 3, it can be combined with cell no. 0 (as shown in Fig. 3.53a) or with cell no. 3 (as shown in Fig. 3.53b).

With reference to the group made up of cell no. 0 and cell no. 4, note that the variables B and C remain constant (the variable A changes). Similarly, the variables B and C remain constant in the group made up of cells no. 3 and 7. Because the values of B and C correspond to 00 and 11 for each group, respectively, the common variables are $\overline{B}\,\overline{C}$ for adjacent cells no. 0 and 4 and BC for adjacent cells no. 3 and 7.

Because the simplified sum-of-products equation will be obtained from the

Figure 3.52

Examples of K-maps that cannot be simplified.

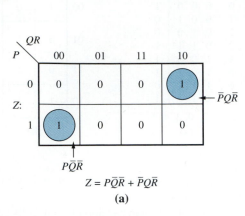

$$Z = P\bar{Q}\bar{R} + \bar{P}Q\bar{R}$$

(a)

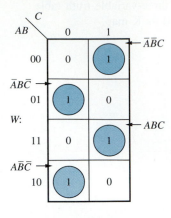
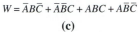

$$W = \bar{A}B\bar{C} + \bar{A}\bar{B}C + ABC + A\bar{B}\bar{C}$$

(c)

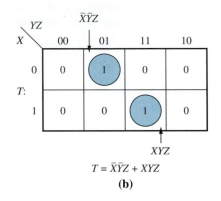

$$T = \bar{X}\bar{Y}Z + XYZ$$

(b)

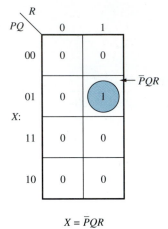

$$X = \bar{P}QR$$

(d)

Figure 3.53

Groups of two adjacent 1-cells.

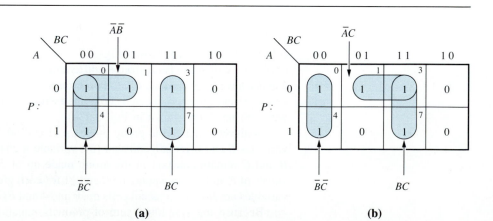

(a)   **(b)**

K-map, each group will produce the logical product of the variables that remain constant. The logical sum of these products will then be taken to obtain the final logic equation.

In Figure 3.53a, the group made up of adjacent cells no. 0 and 1 yields the term

$$\overline{AB}$$

because the variable C changes, but A and B remain constant and their values are 0. The simplified logic equation for the K-map in Figure 3.53a can now be obtained as

$$P = \overline{BC} + \overline{AB} + BC. \tag{3.46}$$

In Figure 3.53b, the group made up of adjacent cells no. 1 and 3 yields the term

$$\overline{A}C$$

because the variable B changes but A and C remain constant at values of 0 and 1, respectively. The simplified logic equation for the K-map in Figure 3.53b can now be obtained as

$$P = \overline{BC} + \overline{A}C + BC. \tag{3.47}$$

Equations 3.46 and 3.47 are logically equivalent and can be considered the same as far as simplification is concerned because both equations are made up of three terms, each term containing two variables. Combining 1-cells into adjacent groups of two produces terms that have only two variables. If, for example, 1-cells were not combined into groups, each cell would have to be taken individually. Therefore, the following logic equation is obtained

$$P = \overline{ABC} + \overline{AB}C + \overline{A}BC + A\overline{BC} + ABC. \tag{3.48}$$

It is apparent that Equation 3.48 is unsimplified and is more complex than Equations 3.46 and 3.47 because it contains five terms, each term having three variables.

To further emphasize the importance of combining 1-cells into groups, obtain the equation for either of the K-maps in Figure 3.53 by not including cell no. 1 in any group

$$P = \overline{BC} + \overline{A}BC + BC \tag{3.49}$$

Note that Equation 3.49 is logically equivalent to Equations 3.46 and 3.47, but is more complex because it contains one term with three variables. We can prove that Equation 3.49 is not completely simplified by using Boolean algebra

$$
\begin{aligned}
P &= \overline{BC} + \overline{A}BC + BC \\
P &= \overline{B}(\overline{C} + \overline{A}C) + BC && \text{(distributive)} \\
P &= \overline{B}(\overline{C} + \overline{A}) + BC && \text{(miscellaneous)} \\
P &= \overline{BC} + \overline{AB} + BC && \text{(distributive)}
\end{aligned}
$$

The equation obtained is the same as Equation 3.46. A different approach to the Boolean simplification could be used to obtain Equation 3.47

$$
\begin{aligned}
P &= \overline{BC} + \overline{A}BC + BC \\
P &= \overline{BC} + C(\overline{A}B + B) && \text{(distributive)} \\
P &= \overline{BC} + C(\overline{A} + B) && \text{(miscellaneous)} \\
P &= \overline{BC} + \overline{A}C + BC && \text{(distributive)}
\end{aligned}
$$

Therefore, to obtain the simplest possible logic equation an attempt must be made to include a 1-cell in a group. 1-cells that are not adjacent to any other 1-cell must be taken alone (i.e., as minterms). However, all 1-cells must be accounted for in the final equation either as groups or taken individually. Overlapping groups are possible as can be seen in Figure 3.53a and b. However, if all 1-cells have been accounted for either alone or in adjacent groups, then it would not be appropriate to combine 1-cells into other groups again. For example, in Figure 3.53a combining cell no. 1 and cell no. 3 into another group would not be appropriate because these two cells have already been included in other groups; if these two cells were combined again, the logic equation obtained (Equation 3.50) would have an extra term although it would be logically equivalent to Equation 3.46

$$P = \overline{BC} + \overline{AB} + BC + \overline{A}C \qquad (3.50)$$

It is possible for K-maps to have four adjacent cells containing 1's (*see* Fig. 3.54). Note in Figure 3.54 that because cell no. 1, 3, 7, and 5 are adjacent (in that order), only one variable, R, remains constant in that group of four cells. Therefore, the term produced by the group of four cells is R. The logic equation for the K-map in Figure 3.54 is

$$Z = R + \overline{P}Q. \qquad (3.51)$$

K-map with a group of four
adjacent 1-cells.

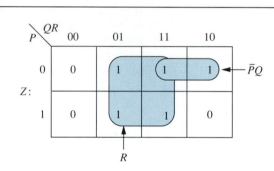

It should now be apparent that the larger the group of adjacent 1-cells, the smaller the resulting term, and consequently the simpler the resulting logic equation. In a three-variable K-map the first attempt should be to try to find groups of four adjacent 1-cells, then groups of two adjacent 1-cells, and finally single 1-cells. Figure 3.55 illustrates other K-map configurations and their logic equations.

Four-variable K-maps

Four-variable K-maps have 2^4 or 16 cells, each cell representing a unique combination of the four independent variables. As stated earlier, the value in each cell corresponds to the value of the dependent variable. Figure 3.56 shows the general configuration of a four-variable K-map that maps the dependent variable X against the independent variables A, B, C, and D. For convenience, each cell has been numbered with numbers that correspond to the binary values of the variables A, B, C, and D (assuming that A is the MSB and D is the LSB). The cells are arranged in the same manner as a three-variable K-map so that only one variable changes in going from one adjacent cell to another. As before, cells located diagonally opposite each other are not adjacent. Cells in all rows and columns are adjacent as well as cells in each corner of the K-map (0, 2, 10, 8).

As stated earlier, care must be taken when filling in the cells of the K-map with the values of the dependent variable from a truth table, because the cells are not arranged in the same order as the entries in the truth table. Figure 3.57 shows a four-variable truth table with its K-map representation.

We previously saw that the short-form notation for writing logic equations was useful for filling in two- and three-variable K-maps from a truth table. In Figure 3.57,

Figure 3.55

Examples of three-variable
K-map simplification.

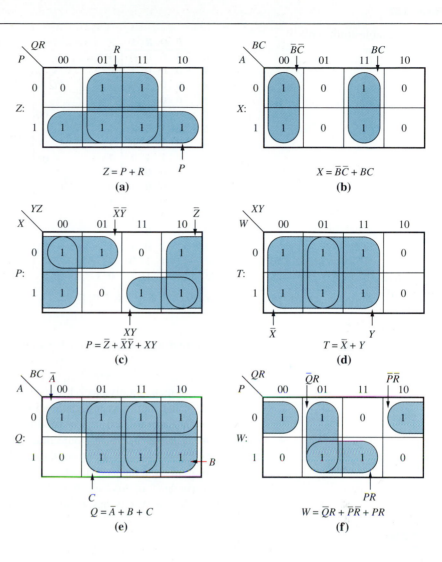

Figure 3.56

General configuration of a
four-variable K-map.

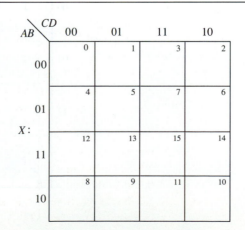

Figure 3.57

A four-variable truth table
and its K-map.

	P	Q	R	S	Z
0	0	0	0	0	1
1	0	0	0	1	0
2	0	0	1	0	1
3	0	0	1	1	1
4	0	1	0	0	0
5	0	1	0	1	1
6	0	1	1	0	1
7	0	1	1	1	1
8	1	0	0	0	1
9	1	0	0	1	0
10	1	0	1	0	1
11	1	0	1	1	0
12	1	1	0	0	1
13	1	1	0	1	1
14	1	1	1	0	1
15	1	1	1	1	0

Z:

PQ \ RS	00	01	11	10
00	1 (0)	0 (1)	1 (3)	1 (2)
01	0 (4)	1 (5)	1 (7)	1 (6)
11	1 (12)	1 (13)	0 (15)	1 (14)
10	1 (8)	0 (9)	0 (11)	1 (10)

$$Z = \Sigma_m\ 0, 2, 3, 5, 6, 7, 8, 10, 12, 13, 14$$

the unsimplified logic equation for the truth table in short-form notation is

$$Z = \sum_m 0, 2, 3, 5, 6, 7, 8, 10, 12, 13, 14.$$

To fill in the K-map, simply place 1's in cells 0, 2, 3, 5, 6, 7, 8, 10, 12, 13, and 14 to correspond to the minterms in the equation.

To obtain the simplified sum-of-products equation from a four-variable K-map, combine the adjacent 1-cells in groups that are as large as possible. In a four-variable K-map, the largest possible group contains eight adjacent 1-cells as shown in Figure 3.58. A group of eight adjacent 1-cells (cell numbers 0, 1, 4, 5, 12, 13, 8, 9) will only have one variable ($\overline{C}$ in this case) that remains constant. A group of eight adjacent 1-cells will yield a term that has a single variable. The next largest group in a four-

Figure 3.58

Combining various groups
of 1-cells in a four-variable
K-map.

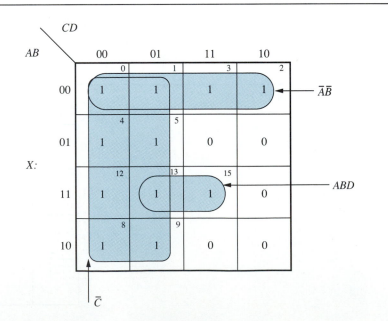

variable K-map contains four cells. In Figure 3.58, cells no. 0, 1, 3, and 2, make up a group of four adjacent 1-cells that yield the term $\overline{A}\overline{B}$. Finally, the smallest group contains two adjacent 1-cells. In Figure 3.58, cell no. 13 and cell no. 15 make up a group of two adjacent 1-cells that yield the term ABD. As stated earlier, 1-cells that are not adjacent to any other 1-cells must be taken singly, as minterms. There are no single 1-cells in Figure 3.58. The simplified logic equation for the K-map in Figure 3.58 can be obtained by taking the logical sum of the simplified terms (products) produced by each group of adjacent 1-cells:

$$X = \overline{C} + \overline{A}\overline{B} + ABD \qquad (3.52)$$

Observe that the larger the group of adjacent 1-cells, the smaller its corresponding term in the final logic equation. Figure 3.59 illustrates various other four-variable K-map configurations and the simplified logic equations obtained from them.

Figure 3.59

Examples of four-variable K-map simplification.

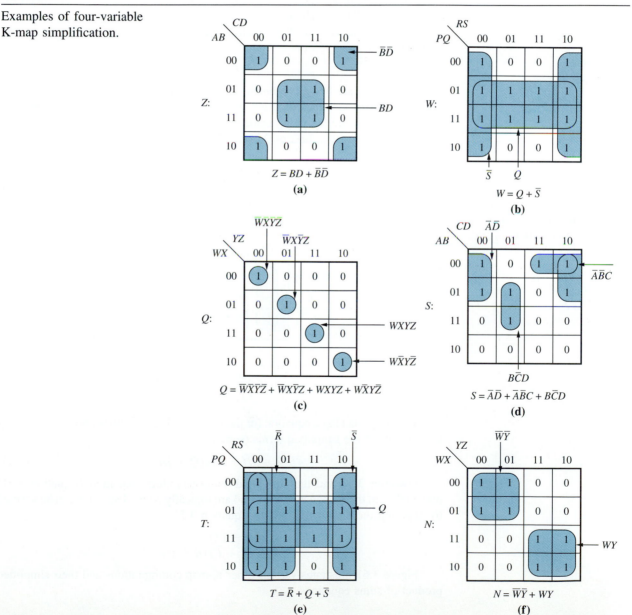

Product-of-sums simplification

Section 3.3 showed that it was possible to obtain an unsimplified logic equation from a truth table in one of two forms — sum-of-products or product-of-sums. This was done by obtaining the logical sum of the minterms (sum-of-products) for which the dependent variable was a 1 or the logical product of the maxterms (product-of-sums) for which the dependent variable was a 0. Earlier in this section, groups of adjacent 1-cells were combined in K-maps to obtain simplified products (terms) that were logically added together to produce the final simplified logic equation. Similarly, groups of adjacent 0-cells in K-maps can also be combined to obtain simplified sums (terms) that are logically multiplied together to produce the final simplified logic equation. Recall that for minterms, a variable that represented a 0 was barred and a variable that represented a 1 was unbarred, but for maxterms a barred variable represented a 1 and an unbarred variable represented a 0.

To illustrate the procedure for obtaining a simplified product-of-sums equation from a K-map, consider the K-map shown in Figure 3.54. For convenience, the K-map has been redrawn and shown in Figure 3.60 with the two groups of adjacent 0-cells identified. The variables P and R remain constant in the group made up of cells no. 4 and 6, and because their values are 1 and 0, respectively, the term that corresponds to this group is

$$(\overline{P} + R).$$

Because the 0's are being combined, the term is a sum rather than a product with variables that have values of 0 left unbarred and variables that have values of 1 barred. Similarly, for the group made up of cells no. 0 and 4, the variables Q and R remain constant with values of 0. Therefore, this group yields the term

$$(Q + R).$$

Figure 3.60

Combining 0-cells in a K-map simplification.

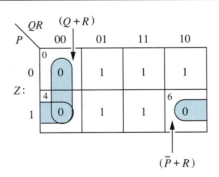

The simplified logic equation for the K-map in Figure 3.60 is obtained by taking the product of the simplified terms (sums)

$$Z = (\overline{P} + R) \cdot (Q + R) \tag{3.53}$$

Equation 3.53 and the simplified sum-of-products equation (Equation 3.51) obtained from the K-map in Figure 3.54 are logically equivalent. This can be verified by applying the distributive law to Equation 3.51

$$Z = R + \overline{P}Q$$
$$Z = (R + \overline{P})(R + Q)$$

Figure 3.61 illustrates various other K-map configurations and their simplified product-of-sums equations.

Figure 3.61

Examples of K-map simplification to yield equations in product-of-sums form.

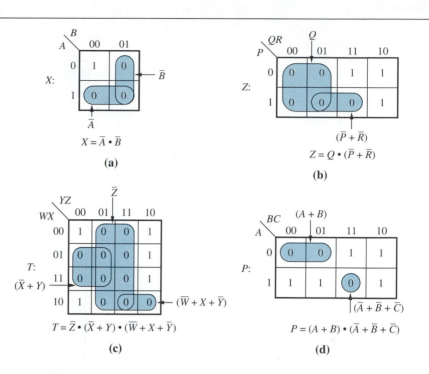

$$X = \overline{A} \cdot \overline{B}$$

(a)

$$Z = Q \cdot (\overline{P} + \overline{R})$$

(b)

$$T = \overline{Z} \cdot (\overline{X} + Y) \cdot (\overline{W} + X + \overline{Y})$$

(c)

$$P = (A + B) \cdot (\overline{A} + \overline{B} + \overline{C})$$

(d)

Note that the procedure for obtaining the simplified product-of sums equation from a K-map is very similar to the procedure for obtaining the simplified sum-of-products equation, except that everything is reversed. K-maps also provide a very convenient means of converting a logic equation from simplified sum-of-products form to simplified product-of-sums form and vice versa. To do this, obtain a truth table for the given equation, construct a K-map, and then solve the K-map to obtain the equation in its complementary form. Example 3.17 illustrates this procedure.

Example 3.13

Convert the following equations into their complementary forms:
(a) $Z = A\overline{B} + \overline{A}C$

(b) $X = (P + Q)(R + \overline{S})$

Solutions

(a) To convert

$$Z = A\overline{B} + \overline{A}C$$

into product-of-sums form, the equation's truth table shown in Table 3.24 is first obtained. Next, the truth table values are entered into the K-map shown in Figure 3.62. Note that we could also go directly from the equation to a K-map because a K-map is effectively a truth table.

From the K-map in Figure 3.62, the simplified product-of-sums equation can be obtained

$$Z = (A + C)(\overline{A} + \overline{B})$$

(b) To convert

$$X = (P + Q)(R + \overline{S})$$

Table 3.24

A	B	C	Z
0	0	0	0
0	0	1	1
0	1	0	0
0	1	1	1
1	0	0	1
1	0	1	1
1	1	0	0
1	1	1	0

Figure 3.62

Using a K-map to convert from sum-of-products to product-of-sums form.

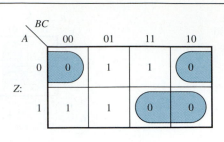

Figure 3.63

Using a K-map to convert from product-of-sums to sum-of-products form.

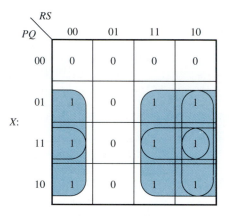

Table 3.25

P	Q	R	S	X
0	0	0	0	0
0	0	0	1	0
0	0	1	0	0
0	0	1	1	0
0	1	0	0	1
0	1	0	1	0
0	1	1	0	1
0	1	1	1	1
1	0	0	0	1
1	0	0	1	0
1	0	1	0	1
1	0	1	1	1
1	1	0	0	1
1	1	0	1	0
1	1	1	0	1
1	1	1	1	1

into sum-of-products form, the equation's truth table shown in Table 3.25 is first obtained. Next, the truth table values are entered into the K-map shown in Figure 3.63.

From the K-map in Figure 3.63, the simplified sum-of-products equation can be obtained

$$Z = PR + QR + P\bar{S} + Q\bar{S}$$

This section has examined an important technique used in the simplification of logic equations and circuits — the K-map technique. The K-map simplification is not designed to be a complete replacement of the Boolean algebra techniques of simplification but is a quick and effective substitute in many cases. Unlike Boolean simplification, the K-map approach can directly indicate whether simplification is possible and if so provide the simplest logic equation. Boolean algebra can still be an effective means of reducing a logic circuit even beyond the capabilities of a K-map. Besides, the function of a K-map is based on the laws of Boolean algebra, and therefore a good understanding of the laws of Boolean algebra is an important prerequisite to the efficient use of a K-map. The techniques of simplification examined in this chapter will be applied to many of the circuits that will be designed and analyzed in subsequent chapters; the specific simplification procedure used however, will always be the one that best fits the application.

Review Questions

1. In what way is a K-map similar to a truth table?
2. What determines adjacency between cells in a K-map?
3. What is the number of cells in a K-map based on?
4. How does one recognize that there is no simplification possible in a K-map?
5. What are the largest group of adjacent 1- and 0-cells that can exist in two-, three-, and four-variable K-maps?
6. What are the differences between the sum-of-products and the product-of-sums procedures of using a K-map?

3.8

IC logic and troubleshooting applications

Many of the rules of Boolean algebra introduced in this chapter can be (and have been) applied to the construction of IC logic circuits and the troubleshooting of logic gates and circuits.

For example, in Figure 3.64a, the operation of an *AND* gate can be tested by applying the identity law ($A \cdot 1 = A$). One input of the *AND* gate is tied high while the other input is pulsed low. If the output pulses low, then the gate is functioning properly. If the output indicates a permanent high or a low (shorted output) or if there is no logic level detected (open output), then the gate is defective. Similarly, in Figure 3.64b, the operation of the *OR* gate can be tested by applying the identity law ($A + 0 = A$). One input of the *OR* gate is tied low while the other input is pulsed high. If the output pulses high, then the gate is functioning properly. If the output indicates a permanent high or a low (shorted output) or if there is no logic level detected (open output), then the gate is defective. The same procedures can also be applied to test *NAND* and *NOR* gates.

In isolating various faults in logic gates and circuits, the dominance law can generally be applied. For example, if the output of an *AND* gate is stuck at the logic low

Figure 3.64

Application of the identity law in troubleshooting a logic circuit.

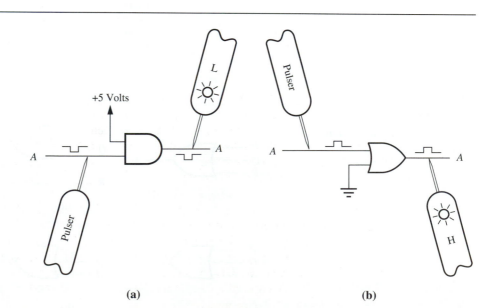

(a)   (b)

state as shown in Figure 3.65a and if it is determined that the output is not shorted to ground, then by logical deduction the dominance law shows that one of the inputs could be internally shorted to ground ($A \cdot 0 = 0$). Similarly, in Figure 3.65b if the output of the *OR* gate is stuck at the logic high state and if it is determined that the output is not shorted to V_{CC}, then by logical deduction the dominance law shows that one of the inputs could be internally shorted to V_{CC} ($A + 1 = 1$).

There are many instances when the design of a logic circuit requires a type of logic gate that may not be available in the form of an IC. For example, if a logic circuit requires a three-input *OR* gate, which is not available in the form of an IC, the associative law can be applied and the circuit implemented using two 2-input *OR* gates from a 7432 IC as shown in Figure 3.66a. Or one of the Boolean postulates (double negation) could be applied and a 7427 three-input *NOR* gate could be used along with a 7404 inverter as shown in Figure 3.66b. DeMorgan's law, which is used extensively to convert from *NAND* and *NOR* logic to *AND-OR* logic and vice versa, can greatly reduce the gate and IC count (and consequently the cost) of a logic circuit. Examples of this were seen in Section 3.6. In the design and implementation of logic circuits, many spare gates are left over in the ICs that make up the circuit. These gates can often be efficiently utilized through the proper application of Boolean algebra to produce a more efficient and compact circuit.

Figure 3.65

Application of the dominance law in troubleshooting a logic circuit.

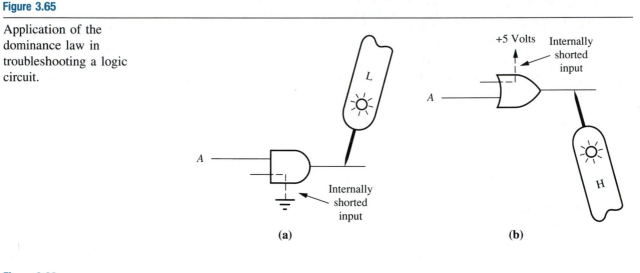

(a)

(b)

Figure 3.66

Use of Boolean algebra in the implementation of logic circuits.

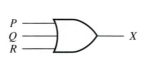

X

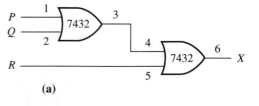

(a)

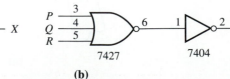

(b)

The technique of analysis introduced in Section 3.2 is used extensively to troubleshoot logic circuits. It allows the outputs of a circuit to be determined by tracing each combination through the circuit from inputs to outputs. For example, to troubleshoot the circuit shown in Figure 3.1, construct a table similar to the one shown in Table 3.2 and record the measured logic levels at the outputs of each gate. The recorded data table can then be compared to the expected data table (Table 3.2) and any discrepancy in logic level can be attributed to a specific gate or a pair of gates. Example 3.14 illustrates this procedure.

Example 3.14

The following table lists the recorded data obtained from Figure 3.67 after a logic probe was used to record the logic outputs of each gate for all possible combinations of the input. The circuit has a fault in it. Identify the fault in the circuit using this data.

| | | | | Outputs | |
A	*B*	*C*	*G1*	*G2*	*G3*
L	L	L	H	H	H ←—fault
L	L	H	H	H	L ←—fault
L	H	L	H	H	H
L	H	H	H	H	L
H	L	L	L	L	H
H	L	H	L	L	H
H	H	L	L	L	H
H	H	H	L	L	H

Figure 3.67

Circuit containing a fault.

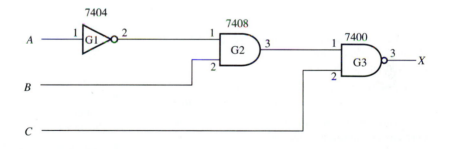

To determine the fault in the circuit, the circuit must be analyzed by tracing through all possible combinations and determining the output of each gate for each combination. The resulting table (expected results) is shown below for the logic equation $X = \overline{\overline{AB} \cdot C}$

| | | | | Outputs | |
A	*B*	*C*	$\overline{A}$	$\overline{A}B$	$\overline{\overline{A}B} \cdot C$
L	L	L	H	L	H ←—fault
L	L	H	H	L	H ←—fault
L	H	L	H	H	H
L	H	H	H	H	L
H	L	L	L	L	H
H	L	H	L	L	H
H	H	L	L	L	H
H	H	H	L	L	H

Note that the first two entries in the above table of expected results (highlighted) do not match the first two (corresponding) entries in the table of recorded results. The fault cannot be in the output of G3 because the low output is produced because both of its inputs are in the high state. Besides, the output of G3 is correct for all other combinations. Because the output of G1 appears to be normal, it would appear that gate G2 has a fault possibly because pin 2 is internally shorted to V_{CC}.

It is important to have a good knowledge of the rules of Boolean algebra when it comes to troubleshooting logic gates and circuits. A basic understanding of the analysis procedures is also important because it enables quick and efficient narrowing down of the fault (or faults) in a logic circuit to a specific gate or IC.

Review Questions

1. Explain the ways in which some of the laws of Boolean algebra can be applied to the construction of logic circuits with ICs.
2. Discuss some of the applications of Boolean algebra in troubleshooting logic circuits.
3. How can we apply the procedure for analyzing logic circuits to troubleshooting?

Summary

-   The analysis of a logic circuit involves procedures to obtain its truth table, logic equation, or both to determine the function of the circuit.
-   The synthesis of a logic circuit involves procedures to obtain (or design) a logic circuit from a truth table, logic equation, or specification.
-   Minterms are logical products that produce a 1 for one and only one combination of the independent variables.
-   Maxterms are logical sums that produce a 0 for one and only one combination of the independent variables.
-   Minterms and maxterms are complements of each other.
-   Logic equations derived from a truth table can be expressed in two equivalent forms — sum-of-products (SOP) and product-of-sums (POS).
-   The sum-of-products and product-of-sums equations obtained directly from a truth table are generally unsimplified.
-   Boolean algebra is also known as the algebra of logic.
-   Boolean algebra is used in the simplification of logic equations and circuits.
-   The objective of Boolean simplification is to reduce the number of gates and ICs required to implement the circuit in hardware.
-   A logic equation is considered to be completely simplified when no rule of Boolean algebra can be used to reduce it any further.

-   Some logic equations cannot be simplifed using Boolean algebra.
-   *AND-OR* logic circuits are made up of *AND* and *OR* gates and inverters.
-   *NAND* logic circuits are made up of *NAND* gates only.
-   *NOR* logic circuits are made up of *NOR* gates only.
-   *NAND* and *NOR* logic circuits can greatly reduce the number of ICs required to implement a circuit in hardware.
-   Any of three basic functions — *AND, OR,* and *NOT* — can be implemented with *NAND* and *NOR* gates.
-   The K-map is a type of truth table that maps the dependent variable against all possible combinations of the independent variables.
-   Adjacent cells in a K-map are cells in which only one independent variable changes in going from one cell to another.
-   The cells in a K-map are organized such that all cells located in rows and columns are adjacent whereas cells that are located diagonally opposite each other are not.
-   The number of cells in a K-map is equal to the number of combinations of the independent variables.
-   The value entered in a K-map cell is the value of the dependent variable whereas the coordinates of the cell are the combination of the independent variables.

- A simplified sum-of-products equation can be obtained from a K-map by combining all 1-cells (cells containing logic 1's).

- A simplified product-of-sums equation can be obtained from a K-map by combining all 0-cells (cells containing logic 0's).

- No simplification is possible if 1-cells or 0-cells are not adjacent in the K-map.

- Adjacent cells in a K-map can be combined in groups of eight, four, and two, with each group producing a simplified term.

- The larger the group of adjacent cells, the smaller the resulting term.

- Cells that are not adjacent to any other cells must be taken as minterms (1-cells) or maxterms (0-cells).

- The K-map can also be used to obtain an equivalent simplified product-of-sums equation, given a sum-of-products equation, and vice versa.

Problems

Section 3.2   Analysis of logic circuits

1. Obtain the logic equations for the circuits shown in Figure 3.68.

2. Obtain the logic equations for the circuits shown in Figure 3.69.

Figure 3.68

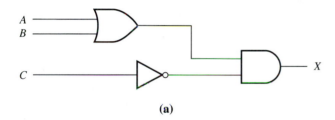

(a)

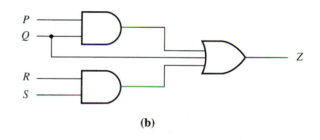

(b)

Figure 3.69

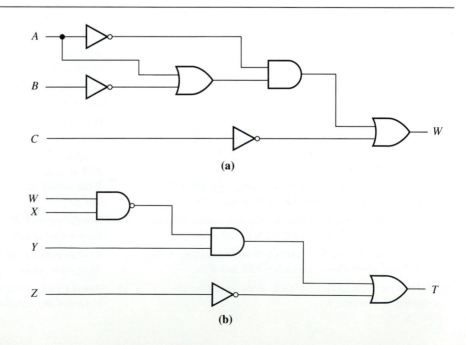

(a)

(b)

3. Obtain the truth tables for the circuits shown in Figure 3.68 by tracing through all possible combinations of the input (independent) variables.

4. Obtain the truth tables for the circuits shown in Figure 3.69 by tracing through all possible combinations of the input (independent) variables.

5. Obtain the truth tables for the circuits shown in Figure 3.68 by breaking down their logic equations into their basic functions and constructing truth tables like the ones shown in Tables 3.2, 3.3, etc.

6. Obtain the truth tables for the circuits shown in Figure 3.69 by breaking down their logic equations into their basic functions and constructing truth tables like the ones shown in Tables 3.2, 3.3, etc.

Section 3.3   Synthesis of logic circuits

7. Derive the unsimplified sum-of-products equations from the truth tables shown in Tables 3.26 and 3.27.

Table 3.26

X	Y	Z
0	0	1
0	1	1
1	0	0
1	1	0

Table 3.27

P	Q	R	S
0	0	0	0
0	0	1	1
0	1	0	0
0	1	1	0
1	0	0	1
1	0	1	1
1	1	0	0
1	1	1	1

8. Derive the unsimplified sum-of-products equations from the truth tables shown in Tables 3.28 and 3.29.

9. Derive the unsimplified product-of-sums equations from the truth tables shown in Tables 3.26 and 3.27.

10. Derive the unsimplified product-of-sums equations from the truth tables shown in Tables 3.28 and 3.29.

11. Draw the logic circuit for each of the equations obtained in Problem 7.

Table 3.28

A	B	C	D	X
0	0	0	0	1
0	0	0	1	1
0	0	1	0	1
0	0	1	1	1
0	1	0	0	0
0	1	0	1	1
0	1	1	0	0
0	1	1	1	1
1	0	0	0	0
1	0	0	1	0
1	0	1	0	0
1	0	1	1	0
1	1	0	0	0
1	1	0	1	1
1	1	1	0	0
1	1	1	1	1

Table 3.29

P	Q	R	S
0	0	0	1
0	0	1	1
0	1	0	1
0	1	1	0
1	0	0	1
1	0	1	1
1	1	0	1
1	1	1	0

12. Draw the logic circuit for each of the equations obtained in Problem 8.

13. Draw the logic circuit for each of the equations obtained in Problem 9.

14. Draw the logic circuit for each of the equations obtained in Problem 10.

15. If the column for the dependent variable S in Table 3.27 were all 1's what would the logic equation be? What would the logic equation for S be if the column had all 0's?

16. Rewrite each of the equations obtained in Problem 8 in short-form notation.

17. Rewrite each of the equations obtained in Problem 7 in short-form notation.

18. Rewrite each of the equations obtained in Problem 10 in short-form notation.

19. Rewrite each of the equations obtained in Problem 9 in short-form notation.

Section 3.4 Boolean algebra

20. In each of the following logic equations, determine the value of the dependent variable for the following values of the independent variable:

$$X = 0 \quad Y = 0 \quad Z = 1$$

   a. $T = \overline{Y} + Z\overline{X} + ZY$
   b. $A = Y(\overline{Y} + X)\overline{Z}$
   c. $C = Z\overline{X} + (Y + 1)$
   d. $S = \overline{X}YZ + X(Y \cdot 0)$
   e. $P = X(\overline{Y} + Z + \overline{X})$

21. In each of the following logic equations, determine the value of the dependent variable for the following values of the independent variable:

$$A = 0 \quad B = 1 \quad C = 0 \quad D = 0$$

   a. $X = A\overline{B}(C + \overline{B}) + \overline{A}C(D + A)$
   b. $Y = B + DC + B\overline{D}C$
   c. $Z = ABCD(0 + A\overline{B}CD)$
   d. $Q = \overline{A + B + \overline{CD}}$
   e. $J = \overline{A} + BC + D$

22. Determine the Boolean algebra law that is applicable in the following logic equations:
   a. $PQ(RS + \overline{R}S) = PQRS + PQ\overline{R}S$
   b. $\overline{AB + CD} = \overline{AB} \cdot \overline{CD}$
   c. $XY + XYZ = XY$
   d. $(P + Q) + (\overline{P + Q}) = 1$

23. Prove by truth tables that the right-hand side of each of the equations given in Problem 22 is equal to the left-hand side.

Section 3.5 Boolean simplification

24. Simplify the equations obtained from Problem 8 using Boolean algebra and verify that the simplified equation is equivalent to the unsimplified equation. Compare the unsimplified and simplified equations in terms of the number of logic gates and ICs required to implement them.

25. Simplify the equations obtained from Problem 7 using Boolean algebra and verify that the simplified equation is equivalent to the unsimplified equation. Compare the unsimplified and simplified equations in terms of the number of logic gates and ICs required to implement them.

26. Simplify the equations obtained from Problem 10 using Boolean algebra and verify that the simplified equation is equivalent to the unsimplified equation. Compare the unsimplified and simplified equations in terms of the number of logic gates and ICs required to implement them.

27. Simplify the equations obtained from Problem 9 using Boolean algebra and verify that the simplified equation is equivalent to the unsimplified equation. Compare the unsimplified and simplified equations in terms of the number of logic gates and ICs required to implement them.

28. Simplify the following logic expressions using Boolean algebra:
   a. $\overline{AB} + \overline{A}B + A\overline{B}$
   b. $\overline{PQR} + \overline{P}Q + P\overline{Q} + \overline{P}R$
   c. $(P + Q + R)(P + \overline{Q} + R)$
   d. $\overline{A}BC + \overline{AB}C + ABC$
   e. $X\overline{Y} + \overline{WX}Y + XY + W\overline{X}Y$
   f. $\overline{PR} + P\overline{QS} + \overline{P}QR\overline{S}$

29. Simplify the following logic expressions using Boolean algebra:
   a. $A(A + B) + B + \overline{A}(A + AB)$
   b. $\overline{A} + AB + A + A\overline{B}$
   c. $\overline{PQR + PR}$
   d. $(P + R) \cdot (\overline{P} + R)$
   e. $\overline{A} + B + \overline{C} + \overline{(A + B) \cdot (\overline{A} + D)} + A\overline{BCD}$
   f. $\overline{(\overline{P} + Q + S)} \cdot (P + Q + S)$

Section 3.6 *NAND* and *NOR* logic

30. Convert each of the simplified sum-of-products equations obtained from Problem 24 to *NAND* logic by applying DeMorgan's theorem. Then verify that the *NAND* logic equations obtained are equivalent to their original *AND-OR* logic equations.

31. Convert each of the simplified sum-of-products equations obtained from Problem 25 to *NAND* logic by applying DeMorgan's theorem. Then verify that the *NAND* logic equations obtained are equivalent to their original *AND-OR* logic equations.

32. Construct the logic circuits for each of the *NAND* equations obtained in Problem 30 and compare these circuits with their *AND-OR* equivalent circuits in terms of the number of gates and ICs required to implement them.

33. Construct the logic circuits for each of the *NAND* equations obtained in Problem 31 and compare these circuits with their *AND-OR* equivalent circuits in terms of the number of gates and ICs required to implement them.

34. Convert the following simplified logic equations to *NAND* logic and draw the *NAND* logic circuit for each equation:
   a. $Z = \overline{P}Q + P(\overline{Q} + RS)$
   b. $X = \overline{A}B + AB + BC$
   c. $G = \overline{Y} + \overline{W}Z + WX\overline{Z}$
   d. $F = \overline{A\overline{B}} + \overline{DC}$
   e. $A = \overline{P} + \overline{R} + \overline{R}(P + QS)$

35. Convert the following simplified logic equations to *NOR* logic and draw the *NOR* logic circuit for each equation:
    a. $Z = (\overline{P} + Q) \cdot (P + \overline{R})$
    b. $T = \overline{X} + YZ$
    c. $S = A \cdot (B + \overline{D}) \cdot (B + C + D)$
    d. $W = \overline{\overline{PQR} \cdot \overline{P\overline{Q}S}}$
    e. $Z = \overline{AD + BC \cdot (\overline{C} + D)}$

Section 3.7   K-map simplification

36. Using a K-map, obtain the simplified sum-of-products equations for Tables 3.26 through 3.29.

37. Using a K-map, obtain the simplified product-of-sums equations for Tables 3.26 through 3.29.

38. Simplify each of the logic expressions given in Problem 28 into sum-of-products form by using the K-map technique.

39. Simplify each of the logic expressions given in Problem 28 into product-of-sums form by using the K-map technique.

40. Simplify each of the logic expressions given in Problem 29 into product-of-sums form by using the K-map technique.

41. Simplify each of the logic expressions given in Problem 29 into sum-of-products form by using the K-map technique.

42. Convert the following logic equations to their equivalent simplified product-of-sums form by using a K-map:
    a. $Q = \overline{A}\overline{B} + AB$
    b. $X = \overline{Q} + PR$
    c. $T = \overline{X}\overline{Z} + YZ + XZ$
    d. $W = \overline{P}\overline{Q}R + \overline{P}QR + PQ\overline{R} + P\overline{Q}R$
    e. $Z = \overline{A} + \overline{C} + BD + A\overline{B}D$
    f. $A = X + \overline{W}Y\overline{Z} + \overline{W}Y\overline{Z}$

43. Convert the following logic equations to their equivalent sum-of-products form by using a K-map:
    a. $Z = (P + Q)(\overline{P} + \overline{Q})$
    b. $X = (\overline{A} + B + C)(A + \overline{C})(\overline{B} + \overline{C})$
    c. $S = (P + Q)(R + \overline{Q})(\overline{P} + \overline{Q})(\overline{R} + \overline{P})$
    d. $A = (W + X + Y + Z)(\overline{X} + \overline{Z})(\overline{W} + X + \overline{Y} + Z)$
    e. $T = R(P + \overline{S})(\overline{P} + Q + S)$
    f. $J = (A + C) \cdot (\overline{A} + B + C + D)$

Troubleshooting

Section 3.8   IC logic and troubleshooting applications

44. The logic circuit shown in Figure 3.70 has a fault. The recorded outputs of the logic gates G1, G2, G3, and G4 are shown below for all possible combinations of the inputs B and C. Use the analysis procedure to troubleshoot the circuit and isolate the fault.

Figure 3.70

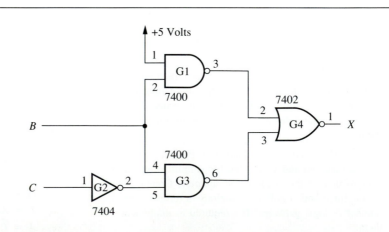

		Outputs			
B	**C**	**G1**	**G2**	**G3**	**G4**
L	L	H	H	H	L
L	H	H	L	H	L
H	L	L	H	L	L
H	H	L	L	H	L

45. The logic circuit shown in Figure 3.71 has a fault. The recorded outputs of the logic gates G1, G2, and G3 are shown below for all possible combinations of the inputs *P*, *Q*, and *R*. Use the analysis procedure to troubleshoot the circuit and isolate the fault.

Figure 3.71

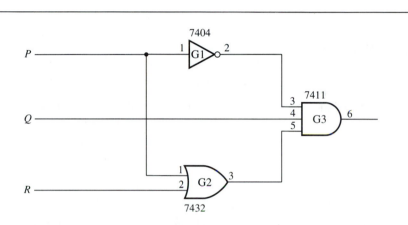

			Outputs		
P	**Q**	**R**	**G1**	**G2**	**G3**
L	L	L	H	L	L
L	L	H	H	L	L
L	H	L	H	L	L
L	H	H	H	L	L
H	L	L	L	H	L
H	L	H	L	H	L
H	H	L	L	H	L
H	H	H	L	H	L

4

APPLICATIONS OF COMBINATIONAL LOGIC

OBJECTIVES

The objectives of this chapter are to:

- Study the basic procedures for designing application circuits from a given specification

- Investigate the design and operation of various types of adder circuits such as half-adders, full-adders, parallel-adders, etc.

- Introduce the operation of the Exclusive-*OR* gate and its application in adder circuits

- Apply the concepts of 1's and 2's complement binary subtraction to the design and operation of subtracter circuits

- Design and analyze various circuits that are used to compare binary numbers

- Study the applications of encoding and decoding binary numbers and the design of encoders and decoders

- Apply the concepts of decoding to code converters such as seven-segment decoders, and introduce the operation of seven-segment LEDs

- Investigate the design and operation of multiplexers and demultiplexers

- Introduce the concepts of tristate logic, tristate logic gates, and their applications in multiplexers and demultiplexers

- Design and analyze various types of circuits to generate and check the parity of binary codes

- Incorporate various application circuits into the design of an arithmetic logic unit

184

One of the most common applications of digital circuits is the calculator. The calculator contains various logic circuits to perform everything from simple arithmetic to complex mathematical operations (Photo courtesy Hewlett-Packard Company).

4.1
Introduction

Logic circuits are used for many applications in the field of digital computers and electronics. Combinational logic circuits are often designed to respond to certain input combinations by producing 1's and 0's at their outputs as required by the application. There are different applications of combinational logic circuits. Many of these circuits are used in digital computers to process binary numbers. For example, circuits such as adders are used to add binary numbers whereas comparators are used to compare binary numbers. This chapter investigates the design and analysis of these application circuits.

The procedures for designing and analyzing application circuits have been studied in Chapter 3. The synthesis procedures will be used to design the various application circuits from a given specification and the analysis procedures will be used to determine the function of various application circuits. Therefore, the procedures that will be used to design and analyze the application circuits covered in this chapter will be identical except for a few variations introduced for certain applications to improve efficiency.

This chapter also introduces another gate that is used frequently in adders, comparators, and parity circuits—the Exclusive-*OR* gate. Also introduced in conjunction with the study of multiplexers and demultiplexers is the concept of tristate logic and logic gates with tristate outputs. Even though this chapter concludes the study of combinational logic circuits, many of the concepts and circuits developed in this and previous chapters will be applied to other circuits and concepts covered in later chapters because this material is often considered to be the foundation of the study of digital circuits.

4.2
Application circuit design

The design of a logic circuit for a specific application generally begins with a specification that describes the function of the circuit. For example, consider the following specification: ''Design a logic circuit that will be used to detect three

Figure 4.1

Logic symbol for circuit to be designed.

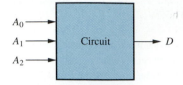

Table 4.1   Truth table for specification

A_2	A_1	A_0	D
0	0	0	0
0	0	1	0
0	1	0	1
0	1	1	0
1	0	0	0
1	0	1	1
1	1	0	1
1	1	1	0

Figure 4.2

K-map for the circuit to be designed.

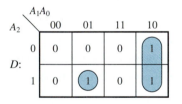

binary numbers. The circuit should produce a logic 1 at its output whenever the binary equivalent of the number 2, 5, or 6 is applied at its input.''

The first step is to draw a *logic symbol* for the circuit that will represent the inputs and outputs of the circuit in the form of a block diagram. Figure 4.1 shows the logic symbol for the circuit to be designed.

The logic symbol for the circuit contains three inputs labeled A_2, A_1, and A_0 because the inputs must be able to accommodate the numbers 2 (010_2), 5 (101_2), and 6 (110_2), and the largest number, 6, is represented by a 3-bit binary number, 110. Also note that the inputs have been labeled with the same variable name A but the subscripts 2, 1, and 0 have been used to identify each bit that makes up the 3-bit binary number A. The subscript 0 will be used to identify the LSB and the highest subscript (2 in this example) will be used to identify the MSB. This scheme of symbolically identifying binary numbers will be used throughout the book. The input to the circuit of Figure 4.1 is the 3-bit binary number A that is made of bits A_2, A_1, and A_0. The output of the circuit is D.

Because the specification usually includes all the information necessary to construct a truth table for the circuit, the next step is to construct a truth table (Table 4.1) that represents the operation of the circuit.

Observe how the input variable A is ordered in Table 4.1 — the LSB (A_0) is placed in the rightmost column and the MSB (A_2) is placed in the leftmost column. Because the circuit is to produce a 1 at its output (D) when the number 010, 101, or 110 is applied to its input, the output variable D in Table 4.1 is a logic 1 for these combinations only.

The logic equation for the circuit can now be obtained by using one of two techniques discussed in Chapter 3 — obtain the unsimplified (sum-of-products or product-of-sums) logic equation and then reduce it by using Boolean algebra, or obtain the simplified equation directly from a K-map. The latter technique will be used for this and most of the other examples to follow. The K-map for the circuit is shown in Figure 4.2.

From the K-map shown in Figure 4.2 the two adjacent 1-cells and the single isolated 1-cell can be combined to produce the following simplified logic equation in sum-of-products form:

$$D = A_2\overline{A}_1 A_0 + A_1\overline{A}_0 \tag{4.1}$$

Figure 4.3

Logic circuit for the specification.

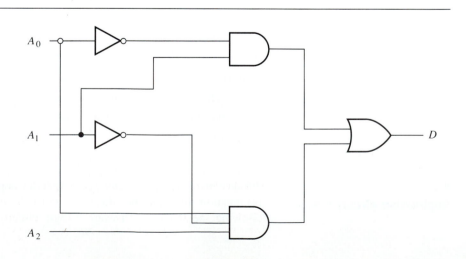

The design can now be completed by constructing the logic circuit for Equation 4.1 as shown in Figure 4.3. Note that the circuit in Figure 4.3 fills in the "box" that represents the logic symbol for the circuit (Fig. 4.1).

Example 4.1

A farmer hires a helping hand who is not very intelligent. The job assigned to the helping hand is to keep the farmer's goat out of the barn while the barn door is open, because the barn contains a pile of corn to be used as cattle feed. Also present in the area is a wolf who considers the goat a delicious treat.

The farmer would like to design a box that has three buttons labeled as follows: "DOOR-OPEN," "GOAT-IN-SIGHT," and "WOLF-IN-SIGHT." The helping hand will be instructed to simply press the buttons according to what he sees at any given time. If the situation is "dangerous," that is, if the conditions are such as to cause harm to the goat or corn, the circuit will activate an alarm to alert the farmer so that proper action can be taken.

Solution

The logic symbol for the circuit to be designed is shown below

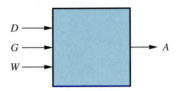

The independent variables (or inputs to the circuit) are as follows:

$$D = \text{DOOR OPEN}$$
$$G = \text{GOAT-IN-SIGHT}$$
$$W = \text{WOLF-IN-SIGHT}$$

Assume that when any of the above inputs is a logic 1, the condition is true. For example, if $G = 1$, this indicates that the goat is in sight, and if $G = 0$, the goat is not in sight.

The output of the circuit is A, which will represent the ALARM to be activated ($A = 1$) if a dangerous condition exists.

The truth table for the circuit is shown below

D	G	W	A
0	0	0	0
0	0	1	0
0	1	0	0
0	1	1	1
1	0	0	0
1	0	1	0
1	1	0	1
1	1	1	1

In the first entry of the truth table, the ALARM is off because neither the goat nor wolf is in sight, and the barn door is not open. In the second entry, the wolf is in sight but the goat is not in sight so there is no cause for alarm. In the third entry, the goat is in sight but the barn door is closed so the corn is safe and there is no cause for setting the alarm. The alarm is turned on in the fourth entry of the truth table because both the goat and the wolf are in sight. In the fifth entry, the barn door is open but the goat is not in sight and therefore the alarm is not activated. In the sixth entry, the wolf is in sight and the barn door is open, but because the wolf has no interest in the corn, there is no cause for alarm. In the seventh entry, the barn door is open and the goat is in sight making the pile of corn vulnerable, and thus the alarm is turned on. In the last entry the alarm is again turned on because the goat and wolf are both in sight and the barn door is open.

The simplified logic equation for the output A can be obtained from the following K-map:

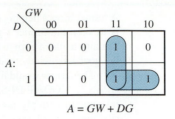

$$A = GW + DG$$

Finally, the logic circuit for the above equation can be implemented as follows:

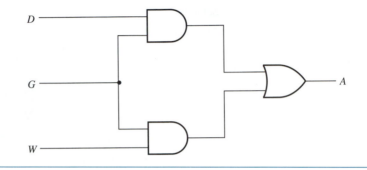

Many application circuits have more than one output. For such circuits, the design procedures remain the same except that each output is designed independently.

For example, consider the design of a logic circuit that will accept all possible combinations of 3-bit binary numbers as input, pass through only the odd combinations, and filter out all even combinations. Thus, if any of the odd numbers, $001, 011,$ $101,$ or 111 are applied to its input, the output will be the same as the input, but if any of the even numbers $000, 010, 100,$ or 110 are applied to its input, the output will be a 000.

The logic symbol for the circuit is shown in Figure 4.4. The variable (s) I represents the input number, and the variable (s) X represents the output number.

The truth table for the specification is shown in Table 4.2. Because Table 4.2 represents the values for the three output variables X_0, X_1, and X_2, it can be treated as three independent truth tables for the purpose of synthesis.

A K-map for each output can now be obtained as shown in Figure 4.5. From Figure 4.5a, the simplified logic equation for X_0 is

$$X_0 = I_0. \tag{4.2}$$

Similarly, from Figure 4.5b

$$X_1 = I_1 I_0 \tag{4.3}$$

and from Figure 4.5c

$$X_2 = I_2 I_0 \tag{4.4}$$

Note that the simplicity of this truth table also enables us to obtain Equations 4.2 through 4.4 by inspection.

The final step is to construct the circuits for Equations 4.2 through 4.4. Even

Figure 4.4

Logic symbol for the "filter" circuit.

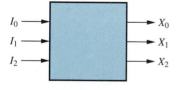

Table 4.2 Truth table for the "filter" circuit

I_2	I_1	I_0	X_2	X_1	X_0
0	0	0	0	0	0
0	0	1	0	0	1
0	1	0	0	0	0
0	1	1	0	1	1
1	0	0	0	0	0
1	0	1	1	0	1
1	1	0	0	0	0
1	1	1	1	1	1

Figure 4.5

K-maps for the "filter" circuit.

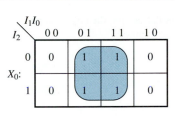

(a)

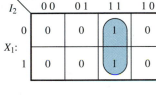

(b)

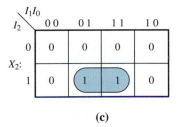

(c)

Figure 4.6

Logic circuit that "filters" out all even numbers.

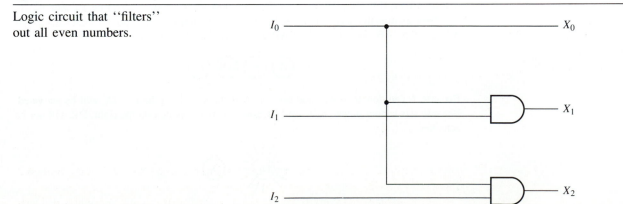

though these equations have been derived independently, to produce three independent circuits, the three circuits function together as one application circuit as shown in Figure 4.6.

Example 4.2

A person can be in any one of the following four blood types: A, B, AB, or O. A person with blood type A can donate blood to another person with blood type A or type AB, but can only receive blood from a person with blood type A or type O. Similarly, a person with blood type B can donate to a person with blood type B or AB, but can only receive blood from a person with blood type B or type O. A person with blood type AB can receive any blood type but can only donate to another person with blood type AB. Type O blood donors can donate blood to anyone but can only receive type O blood.

Design a circuit that will accept any two of the four basic blood types as input and indicate if a blood transfusion is possible, and if possible, identify which of the two types entered can be the donor(s).

Solution

To implement the design, the circuit must have four inputs and two outputs as shown in the following logic symbol:

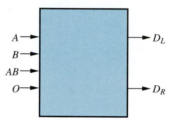

Assume that the four inputs to the circuit will be connected to four switches arranged and labeled as follows:

Also, the two outputs will be connected to two lamp indicators arranged and labeled as follows:

D_L will be activated if the blood type of the leftmost switch activated can be a donor, and D_R will be activated if the blood type of the rightmost switch activated can be a donor.

Now assume that a person would like to know if a transfusion is possible between a type A and a type AB. Switches A and AB will be activated as follows:

Because A can donate to AB, and because A is the switch on the left, D_L will be activated. Because AB cannot donate to A, and because AB is the switch on the right, DR will not be activated:

Therefore, because D_L is activated, the donation can take place from left to right (i.e., A can donate to AB). Because D_R is not activated, the donation cannot take place from right to left (i.e., AB cannot donate to A).

The truth table for the circuit can now be completed as follows. The variable *C* is used instead of *AB* to prevent confusion when writing the equations for each output. A description of each possible transfusion is also included next to each truth table entry.

A	*B*	(AB) *C*	*O*	D_L	D_R	**Description**
0	0	0	0	0	0	No blood types entered
0	0	0	1	1	1	O can donate to O
0	0	1	0	1	1	AB can donate to AB
0	0	1	1	0	1	AB cannot donate to O, O can donate to AB
0	1	0	0	1	1	B can donate to B
0	1	0	1	0	1	B cannot donate to O, O can donate to B
0	1	1	0	1	0	B can donate to AB, AB cannot donate to B
0	1	1	1	0	0	Invalid — more than two types entered
1	0	0	0	1	1	A can donate to A
1	0	0	1	0	1	A cannot donate to O, O can donate to A
1	0	1	0	1	0	A can donate to AB, AB cannot donate to A
1	0	1	1	0	0	Invalid — more than two types entered
1	1	0	0	0	0	A cannot donate to B, B cannot donate to A
1	1	0	1	0	0	Invalid — more than two types entered
1	1	1	0	0	0	Invalid — more than two types entered
1	1	1	1	0	0	Invalid — more than two types entered

In the truth table shown above, observe that if only one blood type is entered (such as in the second entry), both outputs are turned on to indicate that the transfusion is possible both ways. For example, if A is the only input activated, both outputs will be activated to indicate that A can be a donor and a receiver. Also, if more than two inputs are activated (such as in the eighth entry), both outputs are turned off to indicate that no transfusion is possible.

The simplified logic equations can now be obtained for the two outputs using the following K-maps:

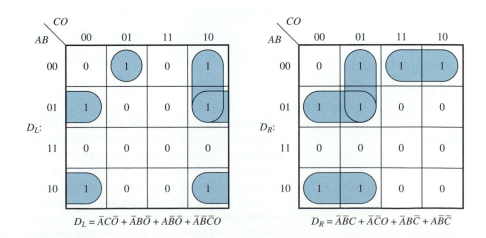

$$D_L = \bar{A}C\bar{O} + \bar{A}B\bar{O} + A\bar{B}\bar{O} + \bar{A}\bar{B}CO$$

$$D_R = \bar{A}\bar{B}C + \bar{A}\bar{C}O + \bar{A}B\bar{C} + A\bar{B}\bar{C}$$

Finally, the logic circuit for the design can be implemented using the logic equations as follows:

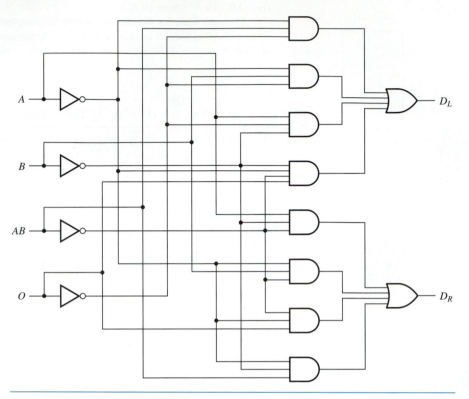

The design procedure outlined in this section will be used to design most of the logic circuit applications discussed in the remainder of this chapter and the rest of the book. This procedure, which can be used to design any logic circuit from a given specification, is referred to as the *conventional approach* to designing application circuits. Even though this may be the most direct and efficient design procedure for most application circuits, this approach can get complicated and does not necessarily produce the most efficient logic circuit.

Review Questions

1. What does a logic symbol represent?
2. Briefly outline the steps involved in designing a circuit using the conventional approach.

4.3
Adders

Adders are logic circuits that are used to perform binary addition. It was stated in Chapter 1 that most digital computers reduce all arithmetic operations to the process of addition, Therefore, adders play an important role in the architecture of a digital computer. This section deals with the design and construction of the building blocks of adders and their applications in adder circuits.

One of the basic elements of binary addition is a circuit known as a *half-adder* whose logic symbol is shown in Figure 4.7. A half-adder is a logic circuit that adds 2 bits (A, B) and produces a single-bit sum (S) and a single-bit carry (C). The process of

Figure 4.7

Logic symbol for a half-adder.

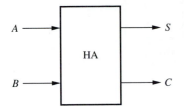

adding all possible combinations of 2 bits was discussed in Chapter 1 and is shown in Table 1.9. The truth table for the half-adder can be constructed as shown in Table 4.3. Note in Table 4.3 that the equation for the output variable C is the equation for the *AND* function and that no simplification is possible for the output variable S. The logic equations for the half-adder circuit are

$$C = AB \tag{4.5}$$

$$S = \overline{A}B + A\overline{B}. \tag{4.6}$$

The logic circuit that implements Equations 4.5 and 4.6 to produce a half-adder is shown in Figure 4.8.

The circuit for Equation 4.6 and consequently the circuit for the half-adder can be simplified even further by replacing it with an *Exclusive-OR* gate, which can only have two inputs and will produce a logic 0 output if and only if the 2 input bits are equal. The truth table for the Exclusive-OR gate is shown in Table 4.4 and its two logic symbols and equivalent *AND-OR* logic circuit are shown in Figure 4.9a, b, and c respectively.

Note in Table 4.4 that the Exclusive-OR gate produces the sum of X and Y (ignoring carries) at its output Z.

The *AND-OR* logic equation for Table 4.4 is

$$Z = \overline{X}Y + X\overline{Y} \tag{4.7}$$

and the logic equation for the Exclusive-OR gate is

$$Z = X \oplus Y \tag{4.8}$$

where the symbol $\oplus$ represents the Exclusive-OR function.

Because Table 4.4 is represented by both Equations 4.7 and 4.8, it can be concluded that they are equivalent.

Note that the values for the variable S in Table 4.3 represent the Exclusive-OR function:

$$S = A \oplus B \tag{4.9}$$

The half-adder circuit shown in Figure 4.8 can be redesigned by replacing the circuit for Equation 4.6 with the circuit for Equation 4.9 and reconstructing the circuit as shown in Figure 4.10.

Table 4.3 Truth table for a half-adder

A	B	C	S
0	0	0	0
0	1	0	1
1	0	0	1
1	1	1	0

Table 4.4 Truth table for an Exclusive-OR gate

X	Y	Z
0	0	0
0	1	1
1	0	1
1	1	0

Figure 4.8

Half-adder circuit.

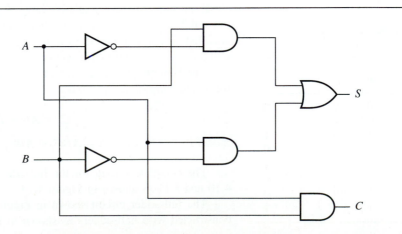

Figure 4.9

Logic symbols for an Exclusive-*OR* gate. (a) Distinctive shape symbol; (b) ANSI / IEEE Std. 91-1984 symbol; (c) equivalent *AND-OR* circuit.

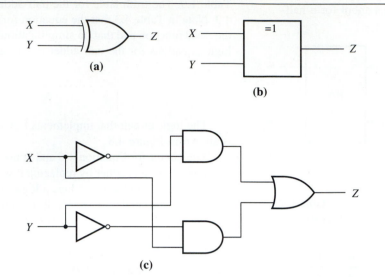

Figure 4.10

Simplified half-adder circuit.

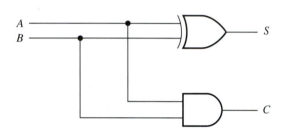

Figure 4.11

Logic symbol for a full-adder.

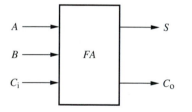

Table 4.5   Truth table for a full-adder

A	B	C_i	C_o	S
0	0	0	0	0
0	0	1	0	1
0	1	0	0	1
0	1	1	1	0
1	0	0	0	1
1	0	1	1	0
1	1	0	1	0
1	1	1	1	1

In Figure 4.10 the original half-adder circuit has been reduced from six gates to two by incorporating the Exclusive-*OR* gate into the circuit.

In performing binary addition of numbers, it is often desirable to be able to add 2 bits at a time; the half-adder accomplishes this task. It is also desirable to be able to add 3 bits at a time; a circuit that does this is known as a *full-adder*. Its logic symbol is shown in Figure 4.11. A full-adder adds 3 input bits (A, B, C_i) and also produces a single-bit sum (S) and a single-bit carry (C_o). The addition of all possible combinations of 3 bits was covered in Chapter 1 and illustrated in Table 1.10. The variable C_i is called the carry-in and the variable C_o is called the carry-out and their functions will be examined later in this section. Table 4.5 shows the truth table for a full-adder.

The K-maps for the output variables S and C_o shown in Figure 4.12 produce the following simplified equations:

$$C_o = AB + BC_i + AC_i \tag{4.10}$$

$$S = \overline{A}\,\overline{B}C_i + \overline{A}B\overline{C_i} + ABC_i + A\overline{B}\,\overline{C_i} \tag{4.11}$$

The complete circuit for the full-adder is finally constructed from Equations 4.10 and 4.11 as shown in Figure 4.13.

The full-adder circuit shown in Figure 4.13 can be greatly simplified if it is constructed with half-adders as shown in Figure 4.14. Observe that the circuit in

Figure 4.12

K-maps for the design of the full-adder.

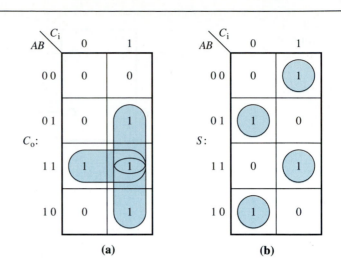

(a)

(b)

Figure 4.13

Full-adder circuit.

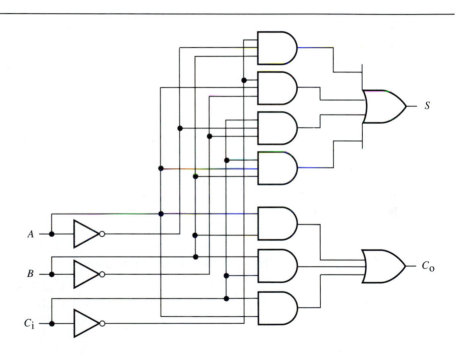

Figure 4.14 requires only five gates (because each half-adder contains two gates), whereas the circuit in Figure 4.13 requires 12 gates.

The circuit in Figure 4.14 adds the 3 input bits A, B, and C_i by first adding the 2 bits, A and B using HA1. The sum of these 2 bits (S_1) is then added to the third input C_i using HA2 to produce the final sum output (S). The final carry output (C_o) is obtained by *OR*ing the carry outputs (C_1 and C_2) of both half-adders, HA1 and HA2. We can prove that

$$C_o = C_1 + C_2 \qquad (4.12)$$

Figure 4.14

Full-adder circuit
constructed from half-
adders.

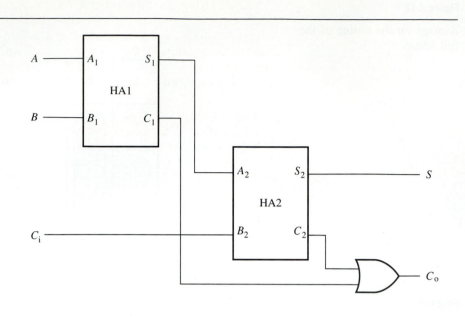

as follows
Because

$$C_1 = AB$$

and

$$C_2 = S_1 C_i$$

Equation 4.12 becomes

$$C_o = AB + S_1 C_i. \tag{4.13}$$

However,

$$S_1 = A \oplus B$$

Therefore, Equation 4.13 becomes

$$
\begin{aligned}
C_o &= AB + (A \oplus B)\, C_i \\
&= AB + (A\overline{B} + \overline{A}B)\, C_i \\
&= AB + A\overline{B}C_i + \overline{A}BC_i \\
&= A\,(B + \overline{B}C_i) + \overline{A}BC_i \\
&= A\,(B + C_i) + \overline{A}BC_i \\
&= AB + AC_i + \overline{A}BC_i \\
&= AB + C_i\,(A + \overline{A}B) \\
&= AB + C_i\,(A + B) \\
&= AB + AC_i + BC_i. \tag{4.14}
\end{aligned}
$$

Note that Equation 4.14 is the same as Equation 4.10.

It is possible to construct other adders that are capable of adding 4 bits or more using the same procedures discussed in this section. However, there are not many applications for such circuits. The main applications of half- and full-adders are in the addition of multibit binary numbers using circuits known as *parallel binary adders*.

Parallel binary adders are classified according to the size of the numbers that are processed (added) by the circuit. One that adds two 2-bit numbers is referred to as a 2-bit parallel binary adder and one that adds two 4-bit numbers is called a 4-bit parallel binary adder. Parallel binary adders can be of any size, and function in the same manner regardless of the size of the numbers processed. For example, consider the 2-bit parallel binary adder whose logic symbol is shown in Figure 4.15. The circuit adds two 2-bit numbers A (A_1A_0) and B (B_1B_0) and produces a 2-bit sum S (S_1S_0) and a carry C. The conventional approach can be used to design this circuit — that is, a truth table can be constructed of all possible combinations of the inputs A_1, A_0, B_1, B_0, and the logic equations and circuits for each output, S_1, S_0, and C determined. However, it is much easier to design the circuit by interconnecting various half- and full-adders to accomplish the addition on a bit-by-bit basis.

Figure 4.15

Logic symbol for a 2-bit parallel binary adder.

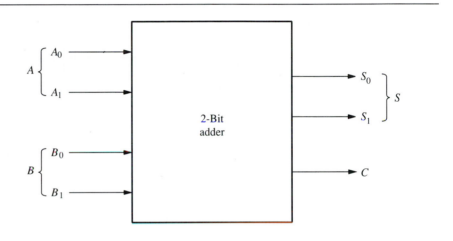

Recall from Chapter 1 that the binary addition of the numbers A and B is accomplished by adding each bit in each number starting with the LSBs and carrying over any overflow from one position to the next up to the MSB. For example:

$$
\begin{array}{ccc}
 & C_0 & \\
 & A_1 & A_0 \\
+ & B_1 & B_0 \\
\hline
C \quad & S_1 & S_0 \\
\end{array}
$$

This addition can be accomplished by using a half-adder

$$
\begin{array}{cc}
 & A_0 \\
+ & B_0 \\
\hline
C_0 \quad & S_0 \\
\end{array}
$$

and a full-adder

$$
\begin{array}{cc}
 & C_0 \\
 & A_1 \\
+ & B_1 \\
\hline
C \quad & S_1 \\
\end{array}
$$

The circuit that implements the 2-bit parallel binary adder is shown in Figure 4.16. Note that the circuit in Figure 4.16 can be very easily extended to add numbers

Figure 4.16

Logic circuit for a 2-bit
parallel binary adder.

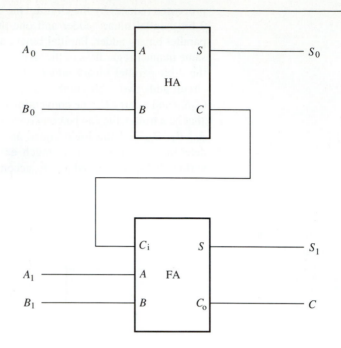

of any size simply by adding more full-adders to the circuit. The following examples
illustrate the concept of expanding the size of a parallel binary adder.

Example 4.3

(a) Design a 4-bit parallel binary adder whose logic symbol is shown in Figure 4.17. Note that
the circuit should incorporate a single-bit carry input (C_i) that will be added to the numbers A
and B to facilitate expansion.

Figure 4.17

Logic symbol for a 4-bit
parallel binary adder.

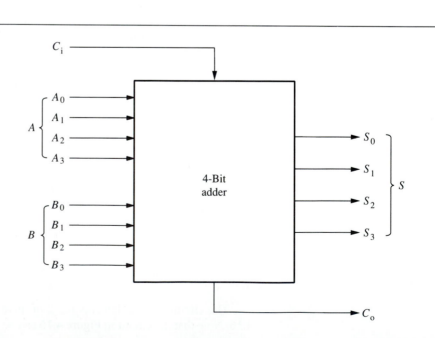

The design of the 4-bit parallel binary adder is shown in Figure 4.18. Note that the only difference between this circuit and the circuit in Figure 4.16 is that two more full-adders have been added to expand the size of each number to 4 bits instead of 2 bits and that the least significant half-adder (in Fig. 4.16) has been replaced by a full-adder to accommodate a carry input.

Figure 4.18

Logic circuit for a 4-bit parallel binary adder.

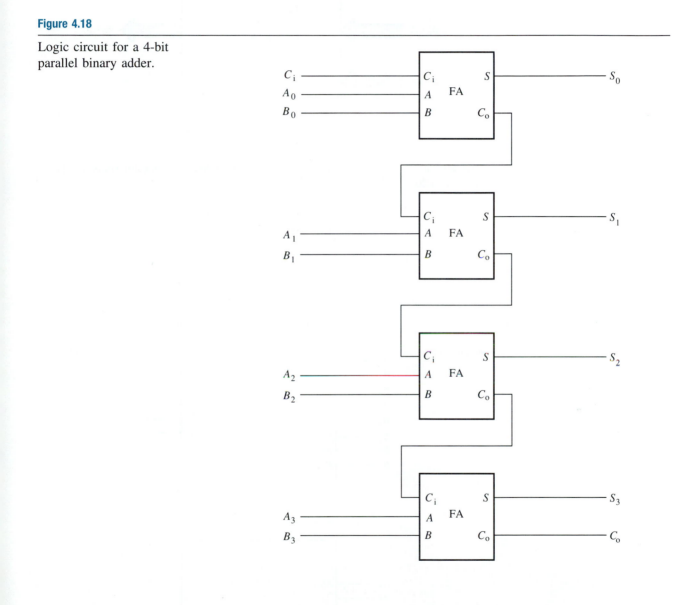

(b) Expand the 4-bit parallel binary adder shown in Figure 4.17 to add two 8-bit numbers as illustrated by the logic symbol shown in Figure 4.19.

The logic symbol shown in Figure 4.19 represents each set of eight input lines as a double arrow, sometimes referred to as a *bus* (more will be said about this in Section 4.11). To add the two 8-bit numbers A and B and obtain an 8-bit sum S, first add the least significant nibbles (4 bits) of each number, and then add the most significant nibbles of both numbers along with the

Figure 4.19

Logic symbol for an 8-bit
parallel binary adder.

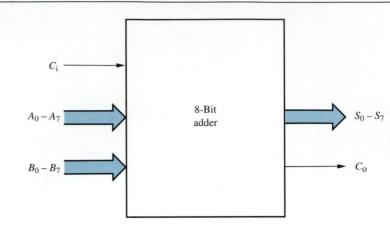

carry generated from the addition of the least significant nibbles. The resulting circuit is shown
in Figure 4.20.

$$
\begin{array}{cccc}
 & & C_i \leftarrow & \\
 & A_7\,A_6\,A_5\,A_4 & & A_3\,A_2\,A_1\,A_0 \\
 & B_7\,B_6\,B_5\,B_4 & & B_3\,B_2\,B_1\,B_0 \\
C_0 & S_7\,S_6\,S_5\,S_4 & C_0 & S_3\,S_2\,S_1\,S_0
\end{array}
$$

Figure 4.20

Logic circuit for an 8-bit
parallel binary adder.

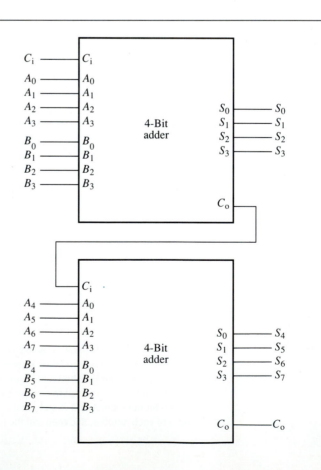

IC logic

The 74xxx series includes two ICs that provide the functions of parallel binary addition with expansion capabilities. The logic symbol and pin configurations of the two ICs are shown in Figure 4.21. The 7482 is a 2-bit parallel binary adder and is similar in function to the circuit designed in Figure 4.16. Σ_1 and Σ_2 are the outputs that represent the sum of the numbers A (A_2A_1) and B (B_2B_1) whereas C_{in} is the carry input and C_2 is the carry output.

Figure 4.21

Logic symbols and pin configurations of (a) the 7482 2-bit parallel adder and (b) the 7483 4-bit parallel adder.

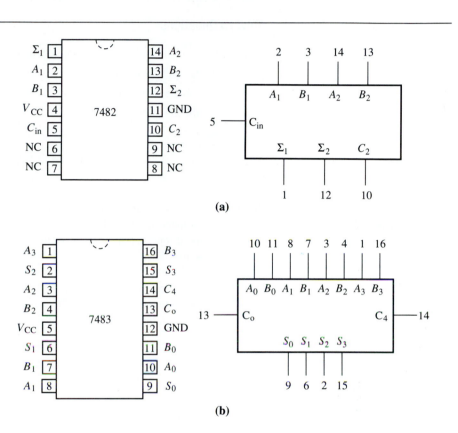

(a)

(b)

Figure 4.22

Pin configuration of the 7486 quad 2-input Exclusive-*OR* gate.

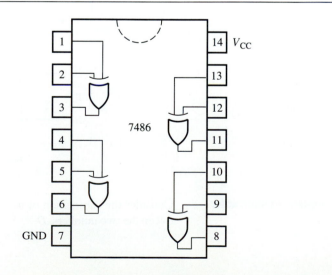

The 7483 is a 4-bit parallel binary adder and is similar in function to the circuit designed in Figure 4.18. The 4-bit inputs A_3–A_0 and B_3–B_0 are added together with a carry input C_0 to produce a 4-bit sum, S_3–S_0, and a carry output, C_4.

The 7486 is an IC that packages four 2-input Exclusive-*OR* gates as shown in Figure 4.22. These gates can be used in the construction of the basic half- and full-adders discussed earlier in this section as well as in some of the other application circuits to be discussed in this chapter and in others.

Review Questions

1. What is a half-adder? How many bits can it add?
2. What is a full-adder? How many bits can it add?
3. Describe the function of an Exclusive-*OR* gate.
4. Describe the operation of a parallel binary adder.
5. What is the purpose of the C_i and C_o input and output in a parallel binary adder?

4.4
Subtracters

Subtracter circuits are used to perform binary subtraction on two numbers. Chapter 1 showed that binary subtraction is performed by adding a negative number (in 1's or 2's complement representation) to a positive number. A few simple modifications to the adder circuits discussed in Section 4.3 will allow these circuits to perform binary subtraction. Figure 4.23 shows the logic symbol for a 4-bit subtracter that will accept as input the two 4-bit numbers A ($A_3A_2A_1A_0$) and B ($B_3B_2B_1B_0$) and obtain the difference between the two numbers D ($D_3D_2D_1D_0$) using either the 1's or 2's complement technique. The sign bit, S, identifies whether the result is negative or positive.

Figure 4.23

Logic symbol for a 4-bit subtracter.

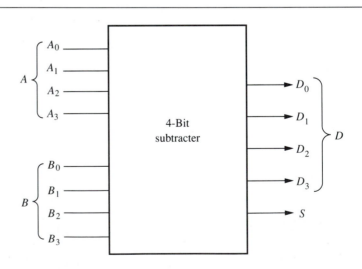

1's complement subtracter

Consider the subtraction of two 4-bit numbers A and B to obtain the difference between the two numbers, D

$$\begin{array}{r} A_3\,A_2\,A_1\,A_0 \\ -\ B_3\,B_2\,B_1\,B_0 \\ \hline D_3\,D_2\,D_1\,D_0 \end{array}$$

The number B is subtracted from the number A to obtain a negative or positive difference, D.

To perform this subtraction using the 1's complement technique, first obtain the 1's complement representation of the negative number B by inverting each bit in the number

$$\overline{B}_3 \, \overline{B}_2 \, \overline{B}_1 \, \overline{B}_0$$

Next, add the 1's complement representation of the negative number, $\overline{B}$, to the positive number, A, to obtain a partial sum, P, and an end-around carry, C.

$$
\begin{array}{r}
A_3 \, A_2 \, A_1 \, A_0 \\
+ \; \overline{B}_3 \, \overline{B}_2 \, \overline{B}_1 \, \overline{B}_0 \\
\hline
C \quad P_3 \, P_2 \, P_1 \, P_0
\end{array}
$$

The end-around carry, C, is then added to the partial sum, P, to obtain the final result D, which is the difference between A and B. The value of C determines whether the result is positive ($C = 1$) or negative ($C = 0$)

$$
\begin{array}{r}
P_3 P_2 P_1 P_0 \\
+ \qquad\qquad C \\
\hline
D_3 D_2 D_1 D_0
\end{array}
$$

The circuit to implement 1's complement subtraction as described above is shown in Figure 4.24. The 1's complement of the number B is first obtained by passing each bit in the number through an inverter. A simple parallel binary adder (identified as Stage 1) is used to add the numbers A and $\overline{B}$ and obtain the partial sum, P. The carry output C of the most significant full-adder in Stage 1 is the sign bit that is added to the LSB of the partial sum, P_0 (Stage 2). The three remaining half-adders in Stage 2 are included to take care of any carry that may be generated during the addition of each set of bits because C must be added to the entire number P rather than just P_0. The outputs D of the half-adders in Stage 2 represent the final result. Note that the carry output of the most significant half-adder in Stage 2 can be ignored because it will always be a logic 0 because the result of any 4-bit subtraction will never require more than 4 bits of representation.

2's complement subtracter

Again, consider the subtraction of two 4-bit numbers A and B to obtain the difference between the two numbers, D

$$
\begin{array}{r}
A_3 \, A_2 \, A_1 \, A_0 \\
- \; B_3 \, B_2 \, B_1 \, B_0 \\
\hline
D_3 \, D_2 \, D_1 \, D_0
\end{array}
$$

As before, the number B is subtracted from the number A to obtain a negative or positive difference, D. To perform this subtraction using the 2's complement technique, first obtain the 2's complement representation of the negative number B by inverting each bit in the number and then adding 1 to the final result

$$
\begin{array}{r}
\overline{B}_3 \, \overline{B}_2 \, \overline{B}_1 \, \overline{B}_0 \\
+ \qquad\qquad\qquad 1
\end{array}
$$

Next, add the 2's complement representation of the negative number to the positive number, A, to obtain the difference D and a final carry, C.

$$
\left.
\begin{array}{r}
A_3 \, A_2 \, A_1 \, A_0 \\
\overline{B}_3 \, \overline{B}_2 \, \overline{B}_1 \, \overline{B}_0 \\
+ \qquad\qquad\qquad 1 \\
\hline
C \; D_3 \, D_2 \, D_1 \, D_0
\end{array}
\right\} \quad \text{2's complement of } B
$$

Figure 4.24

A 4-bit 1's complement
subtracter circuit.

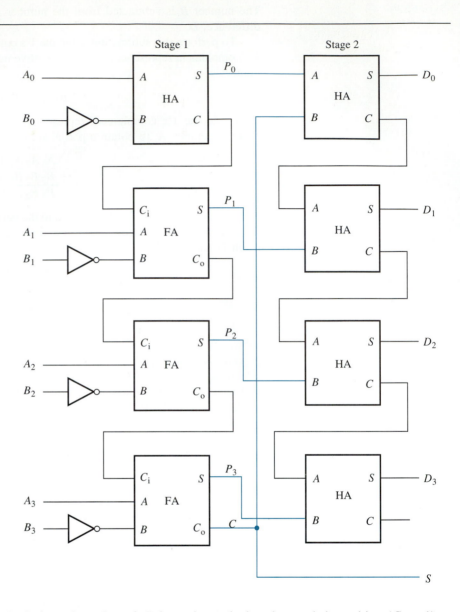

As before, the value of C determines whether the result is positive ($C = 1$) or negative ($C = 0$) and is considered to be the sign bit.

The circuit to implement 2's complement subtraction as described above is shown in Figure 4.25. Note that the 1's complement of the number B is first obtained by passing each bit in the number through an inverter. However, instead of immediately obtaining the 2's complement of the number B by adding 1 to its 1's complement and then adding the 2's complement of B to the number A, the circuit adds the number A, the 1's complement of B, and a logic 1, using a parallel binary adder to effectively produce the same result with a minimum amount of components. The least significant full-adder has its carry input tied to a logic 1 to effectively convert the 1's complement of B into its 2's complement. The difference between the two numbers is produced directly at the output of the adder and the carry output of the most significant full-adder represents the sign bit of the subtracter.

Both the 1's complement and the 2's complement subtracters can be extended to subtract binary numbers of any size by adding more components to accommodate the

Figure 4.25

A 4-bit 2's complement
subtracter circuit.

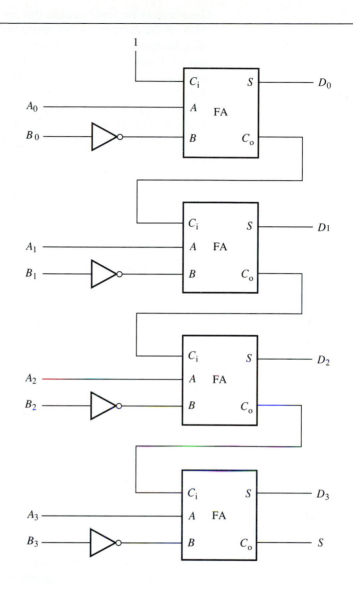

additional bits; the procedure is similar to that discussed in Section 4.3 to expand the
size of parallel binary adders. It is apparent that the design of the 2's complement
subtracter is less complex than the design of the 1's complement subtracter and is the
preferred method of binary subtraction in most digital computers. The conventional
approach can also be used to design subtracters, but the resulting circuits would be
much more complex than the circuits obtained by using parallel binary adders.
Because binary subtracters can be designed using parallel binary adders, there are no
special ICs available that perform binary subtraction.

Review Questions

1. Describe the operation of a 1's complement subtracter.
2. Describe the operation of a 2's complement subtracter.
3. Why is a 2's complement subtracter preferred over a 1's complement
   subtracter?

4.5
Comparators

Comparators are logic circuits that are used to compare two binary numbers. The result of the comparison are single-bit outputs that indicate whether the numbers being compared are equal to each other, or if one number is greater or less than another. Figure 4.26 shows the logic symbol for a comparator (also referred to as a magnitude comparator) that compares two 2-bit binary numbers A (A_1A_0) and B (B_1B_0) and activates one of three outputs, E, G, or L, if A is equal to B, A is greater than B, or A is less than B, respectively. The conventional approach can be used to design the comparator circuit as follows.

The truth table for the 2-bit magnitude comparator is shown in Table 4.6. It includes the decimal equivalents of the numbers A and B for convenience in comparing their values. The outputs of the circuits are *active high,* which means that the state of an output will be a logic 1 (high) when activated; for example, whenever the two numbers are equal, the output E will be (activated to) a logic 1 to indicate equality, otherwise its state will be a logic 0.

Now obtain the simplified logic equations for each of the outputs, E, G, and L from the truth table shown in Table 4.6 by constructing K-maps for each output as shown in Figure 4.27.

$$E = \overline{A}_1\overline{A}_0\overline{B}_1\overline{B}_0 + \overline{A}_1A_0\overline{B}_1B_0 + A_1A_0B_1B_0 + A_1\overline{A}_0B_1\overline{B}_0 \qquad (4.15)$$

$$G = A_1\overline{B}_1 + A_0\overline{B}_1\overline{B}_0 + A_1A_0\overline{B}_0 \qquad (4.16)$$

$$L = \overline{A}_1B_1 + \overline{A}_1\overline{A}_0B_0 + \overline{A}_0B_1B_0 \qquad (4.17)$$

Even though Equation 4.15 cannot be simplified, we can obtain an equivalent circuit that will be less complex by using the Exclusive-*OR* gate to test a pair of bits for equality. The manner in which this is done can be explained as follows.

For the two numbers A and B to be equal, the following conditions must be satisfied:

$$A_0 = B_0$$

and

$$A_1 = B_1$$

Because the output of an Exclusive-*OR* gate will be a logic 0 if and only if both inputs are equal, the following conditions are true:

$$A_0 \oplus B_0 = 0 \qquad (4.18)$$

Figure 4.26

Logic symbol for a 2-bit magnitude comparator.

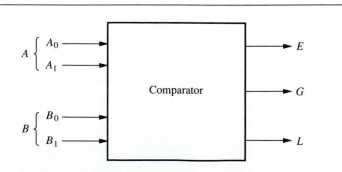

Table 4.6   Truth table for a 2-bit magnitude comparator

A B	A_1	A_0	B_1	B_0	E	G	L
0 = 0	0	0	0	0	1	0	0
0 < 1	0	0	0	1	0	0	1
0 < 2	0	0	1	0	0	0	1
0 < 3	0	0	1	1	0	0	1
1 > 0	0	1	0	0	0	1	0
1 = 1	0	1	0	1	1	0	0
1 < 2	0	1	1	0	0	0	1
1 < 3	0	1	1	1	0	0	1
2 > 0	1	0	0	0	0	1	0
2 > 1	1	0	0	1	0	1	0
2 = 2	1	0	1	0	1	0	0
2 < 3	1	0	1	1	0	0	1
3 > 0	1	1	0	0	0	1	0
3 > 1	1	1	0	1	0	1	0
3 > 2	1	1	1	0	0	1	0
3 = 3	1	1	1	1	1	0	0

Figure 4.27

K-maps for the design of the 2-bit magnitude comparator.

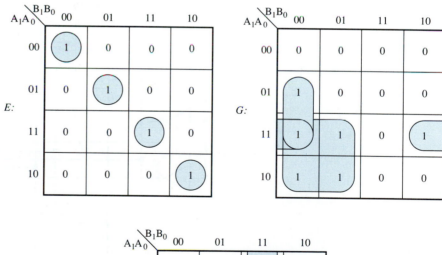

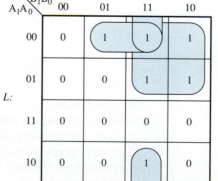

and

$$A_1 \oplus B_1 = 0 \qquad (4.19)$$

if the numbers A and B are equal.

To generate the comparator output E, *NOR* Equations 4.18 and 4.19 so that E is a logic 1 if and only if $A_0 \oplus B_0$ and $A_1 \oplus B_1$ produce logic 0's. The simplified equation for E then becomes

$$E = \overline{(A_0 \oplus B_0) + (A_1 \oplus B_1)}. \qquad (4.20)$$

We can see that the circuit for Equation 4.20 will require fewer gates than the circuit for Equation 4.15.

The complete circuit for the 2-bit magnitude comparator can now be constructed with Equations 4.16, 4.17, and 4.20, as shown in Figure 4.28.

Figure 4.28

The 2-bit magnitude comparator circuit.

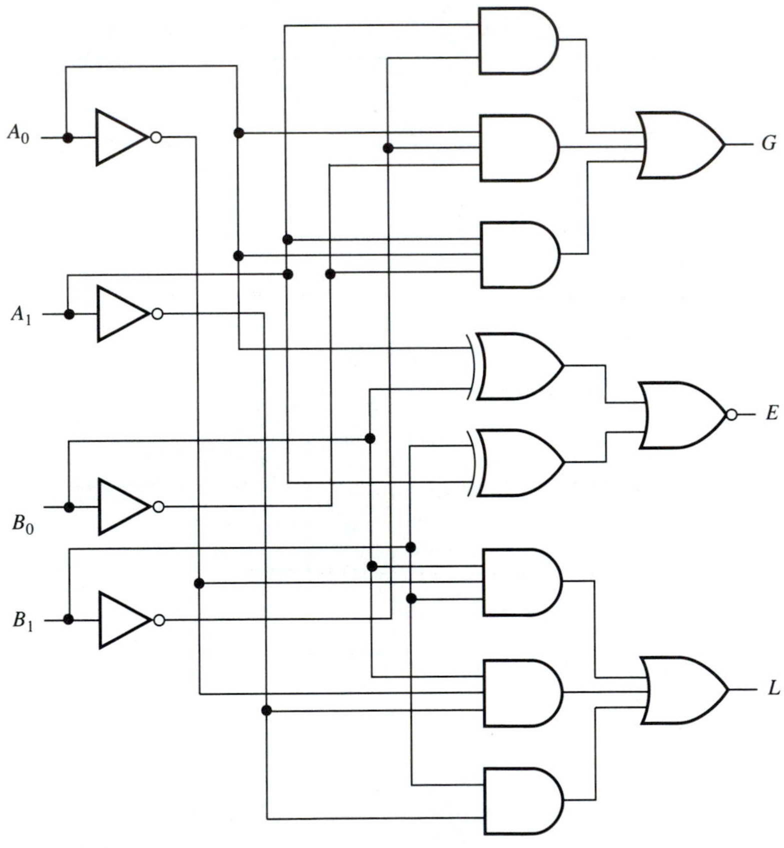

IC logic

The 7485 is similar in operation to the 2-bit magnitude comparator designed in this section, except that the 7485 compares two 4-bit numbers and has the capability to extend the comparison to numbers of any size. The 7485 is called a 4-bit magnitude comparator because of the size of the numbers it compares. The logic symbol and pin configuration of the 7485 are shown in Figure 4.29.

Figure 4.29

Logic symbol and pin configuration of the 7485 4-bit magnitude comparator.

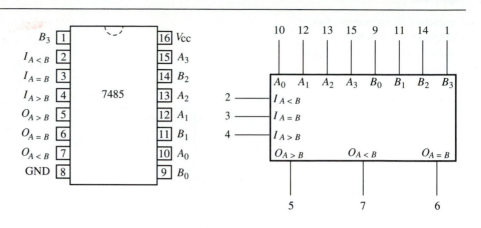

Table 4.7 Truth table for the 7485 magnitude comparator

Inputs	$O_{A>B}$	$O_{A<B}$	$O_{A=B}$
$A > B$	1	0	0
$A < B$	0	1	0
$A = B$	0	0	1

The two 4-bit binary numbers being compared are applied to the A (A_0–A_3) and B (B_0–B_3) inputs of the 7485 and the result of the comparison is indicated by activating (to a logic 1) one of the three outputs, O, as shown in Table 4.7.

The three expansion inputs, $I_{A>B}$, $I_{A<B}$, and $I_{A=B}$, are used to expand the size of the numbers being compared by connecting together several 7485 ICs. If the size of the numbers being compared is less than or equal to 4 bits, that is, if only a single 7485 IC is being used with no expansion, then the $I_{A=B}$ input must be tied to a logic 1; the states of the other two expansion inputs are not significant to the operation of the 7485 under these conditions.

To compare numbers greater than 4 bits, several 7485 ICs can be connected serially. Figure 4.30a shows how two 7485's can be connected together to function as an 8-bit comparator. Note that IC-1 compares the least significant nibble of both numbers and IC-2 compares the most significant nibble. The expansion inputs $I_{A>B}$ and $I_{A<B}$ of IC-1 are tied to a logic 0 state whereas $I_{A=B}$ is tied to a logic 1. The outputs of IC-1 are connected to the expansion inputs of IC-2 so that the final result of the comparison will be based on both parts of each number. The three outputs of IC-2 provide the results of the comparison of the entire 8-bit number. Figure 4.30b shows three 7485s connected to compare two 12-bit numbers. The circuit is similar to the circuit in Figure 4.30a with another 7485 added in series to accommodate the 4 extra bits. Thus, a pair of binary numbers of any size can be compared by using the expansion inputs of the 7485 as shown in the circuits of Figure 4.30.

Figure 4.30

Cascading comparators. (a)
Comparing two 8-bit
numbers; (b) comparing two
12-bit numbers.

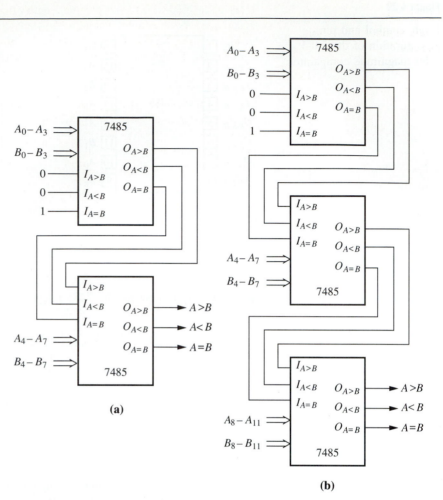

(a)

(b)

Review Questions

1. Describe the operation of a comparator.
2. What is the purpose of the expansion inputs of the 7485 IC?

4.6
Encoders and decoders

Encoders and decoders are digital circuits that are used to generate and detect digital codes. Besides playing an important part in digital communications, these circuits are also used in digital computers to interface many of the various components that make up the digital computer. Applications of encoders and decoders will be seen in this and other chapters in this book.

Encoders

Encoders have several input lines that produce a unique binary code when activated and are generally categorized according to the size of the codes they produce. Typical encoders produce 3-bit (octal) or 4-bit (BCD or hexadecimal) codes. To illustrate the operation and design of a simple encoder, consider the 2-bit encoder whose logic symbol is shown in Figure 4.31. The circuit has four inputs labeled "I_0" through "I_3," and a 2-bit binary output X_1X_0 provides the required code. The input

Figure 4.31

Logic symbol for a 2-bit encoder.

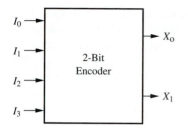

lines are labeled with subscripts that identify the binary code produced when the line is activated. For example, when line I_0 is activated (set to a logic 1), the output code produced is 00, when line I_1 is activated, the output code produced is 01, when line I_2 is activated, the output code produced is 10; and when line I_3 is activated, the output code produced is 11. Therefore, each input line essentially "tells" the encoder to output its unique code.

A 2-bit encoder generally requires 4 (2^2) input lines to produce all possible 2-bit codes. Similarly, a 3-bit encoder will require 8 (2^3) input lines to produce all possible 3-bit codes, and a 4-bit encoder will require 16 (2^4) input lines to produce all possible 4-bit codes. Note, however, that encoders do not have to produce all possible codes and therefore the number of input lines may be less than the maximum for certain applications.

To complete the design of the encoder shown in Figure 4.31, first obtain its truth table as shown in Table 4.8 and then obtain the logic equations for each output. Note that because the inputs to the encoder do not represent a binary number, the order of the variables, I_0 to I_3 is not important in the truth table.

Table 4.8   Truth table for a 2-bit encoder

I_0	I_1	I_2	I_3	X_1	X_0	
0	0	0	0	x	x	
0	0	0	1	1	1	←
0	0	1	0	1	0	←
0	0	1	1	x	x	
0	1	0	0	0	1	←
0	1	0	1	x	x	
0	1	1	0	x	x	
0	1	1	1	x	x	
1	0	0	0	0	0	←
1	0	0	1	x	x	
1	0	1	0	x	x	
1	0	1	1	x	x	
1	1	0	0	x	x	
1	1	0	1	x	x	
1	1	1	0	x	x	
1	1	1	1	x	x	

The function of an encoder is to produce a unique binary code when one of its input lines is activated because (as stated earlier) activating an input line signals the encoder to output the binary code corresponding to that line. Activating more than one line "tells" the encoder to output more than one code on the same set of output lines, which is logically impossible. Therefore, in the design of the encoder we must make the assumption that only one of the input lines will be activated at a given instant and if more than one line is activated we *don't care* what the output will be. These *don't care conditions* are identified by the letter *x* in the truth table shown in Table 4.8. Because an *x* identifies an insignificant logic state, it can be used as a 1 or a 0 when convenient. We also don't care what the output of the encoder is if none of the inputs are activated; therefore, the first entry in the truth table also has *x* states for the outputs. Note that if none of the input lines are activated, ideally no code will be produced at the output (i.e., no logic levels). However, at this time, assume that this is not possible because the output of any logic gate or circuit must be at one of two logic states, 1 or 0. Therefore, when none of the input lines are activated there will be some insignificant code produced at the outputs.

The K-maps for the output variables X_1 and X_0 are shown in Figure 4.32. Because all the logic 1's must be included in the K-map in the largest possible group and with as few groups as possible, in each K-map two of the don't care states have been considered to be logic 1's and the rest of the don't care states to be logic 0's. The presence of don't care states in a K-map can, in most cases, assist greatly in producing simplified equations as long as their use is limited to increasing the size of groups of adjacent 1-cells (or 0-cells for product-of-sums simplification). The logic equations obtained from the K-maps in Figure 4.32 are as follows:

$$X_1 = \bar{I}_0 \bar{I}_1 \qquad\qquad (4.21)$$

$$X_0 = \bar{I}_0 \bar{I}_2 \qquad\qquad (4.22)$$

Figure 4.32

K-maps for the design of the 2-bit encoder.

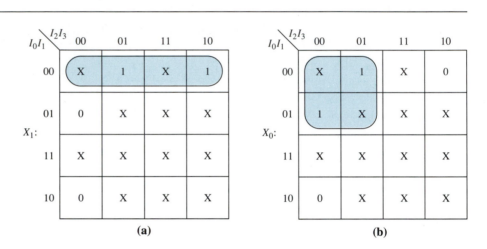

(a)   (b)

The logic circuit for the 2-bit encoder is constructed from Equations 4.21 and 4.22 as shown in Figure 4.33. Note that the outputs of the circuit appear to be independent of the input I_3, which is not connected to anything. This is because don't care outputs were specified for the first entry in the truth table and because of the manner in which the two x cells were combined in each K-map, the output of the circuit when none of the inputs are activated is a 11. Therefore, if I_3 is active or inactive the output code will be a 11. This condition does not pose a problem because the correct code is produced when I_3 is activated even though it is not connected to anything.

An encoder that takes into consideration the possibility of more than one input being activated at the same time is known as a *priority encoder*. A priority encoder therefore resolves all the don't care conditions in Table 4.8 by assigning priorities to each input line. Most priority encoders are high-order data line encoded, which means that if two or more inputs are activated at the same time, the line with the higher code gets priority. For example, if I_1 and I_3 are activated at the same time, the code for I_3 is put out. Table 4.9 is a modification of Table 4.8 to include high-order priority encoding of the inputs.

Observe in Table 4.9 that the first entry in the truth table is still shown with x

Figure 4.33

Logic circuit for a 2-bit encoder.

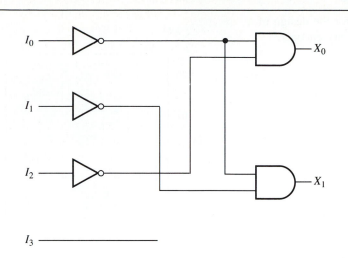

Table 4.9 Truth table for a 2-bit priority encoder

I_0	I_1	I_2	I_3	X_1	X_0
0	0	0	0	x	x
0	0	0	1	1	1
0	0	1	0	1	0
0	0	1	1	1	1
0	1	0	0	0	1
0	1	0	1	1	1
0	1	1	0	1	0
0	1	1	1	1	1
1	0	0	0	0	0
1	0	0	1	1	1
1	0	1	0	1	0
1	0	1	1	1	1
1	1	0	0	0	1
1	1	0	1	1	1
1	1	1	0	1	0
1	1	1	1	1	1

outputs because this entry corresponds to no active input. The K-maps for each of the outputs are shown in Figure 4.34, and the logic equations obtained from the K-maps are

$$X_1 = I_2 + I_3 \tag{4.23}$$

$$X_0 = I_3 + I_1\bar{I}_2. \tag{4.24}$$

The logic circuit for the 2-bit priority encoder shown in Figure 4.35 is obtained from Equations 4.23 and 4.24. Again, note that one of the input lines, I_0, is not connected to anything because in this case if none of the input lines are activated, the output will be a 00.

Encoders are used mainly in keyboards and keypads to produce unique binary codes for each key pressed. For example, an ASCII keyboard on a computer uses an

Figure 4.34

K-maps for the design of a
2-bit priority encoder.

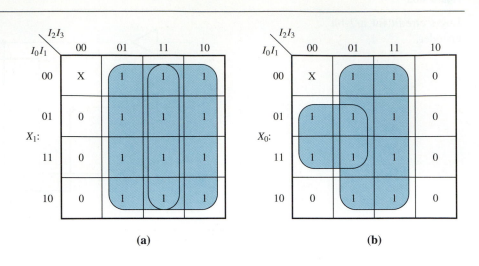

(a)   (b)

Figure 4.35

Logic circuit for a 2-bit
priority encoder.

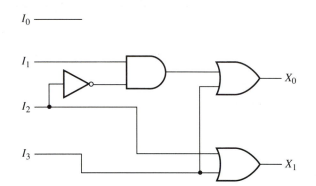

ASCII encoder circuit to generate a 7-bit ASCII code for each of the keys or key combinations pressed. Figure 4.36 shows a typical encoder application with a telephone-style keypad.

The keypad consists of an arrangement of pushbutton switches, each of which outputs a logic 1 on a separate line (labeled "0" through "*") whenever pressed. The 12 switch outputs are connected to a 4-bit encoder that will produce a unique 4-bit code for each of the keys in the keypad as shown in Table 4.10.

Because a 4-bit encoder has the capability to produce 16 codes (2^4) from a set of 16 input lines, 4 of these 16 lines are unused in the keypad circuit because there are only 12 keys. The four unused codes (1100 through 1111) are shown in Table 4.10.

BCD encoders also generate 4-bit codes but do not generate all 16 combinations. They are circuits that output the ten BCD codes (0000 through 1001) and have only ten input lines corresponding to each of these codes. BCD encoders are used in applications where the ten keys of a decimal keypad (such as in a calculator) have to be encoded to BCD.

Decoders

The function of a decoder is the opposite of the function of an encoder. Whereas an encoder *produces* a unique binary code for each of its input lines, a decoder *detects* a unique binary code by activating one of several output lines. Figure 4.37 shows the

Figure 4.36

Encoded keypad circuit.

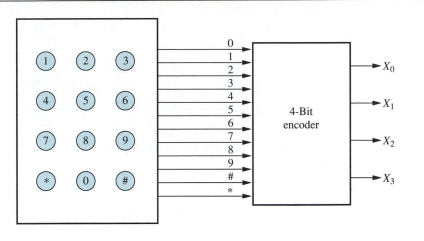

Figure 4.37

Logic symbol for a 1-of-4 decoder.

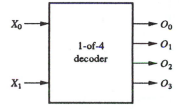

Table 4.10 Key codes for encoded keypad circuit

X_3	X_2	X_1	X_0	Key
0	0	0	0	0
0	0	0	1	1
0	0	1	0	2
0	0	1	1	3
0	1	0	0	4
0	1	0	1	5
0	1	1	0	6
0	1	1	1	7
1	0	0	0	8
1	0	0	1	9
1	0	1	0	#
1	0	1	1	*
1	1	0	0	Not used
1	1	0	1	Not used
1	1	1	0	Not used
1	1	1	1	Not used

logic symbol for a 1-of-4 decoder. The inputs to the circuit, X_1X_0, represent a 2-bit code and the outputs of the circuit, O_0 through O_3, are used to identify the 2-bit code that is applied to the inputs; the subscript used in each output variable identifies the input code that activates it. For example, if the code 00 is applied to the input, output line O_0 is activated; if the code 01 is applied to the input, O_1 is activated; if the code 10 is applied to the input, O_2 is activated; and if the code 11 is applied to the input, O_3 is activated. Thus, at any given instant, only one of four lines will be active when a code is present at the input, hence the term ''1-of-4 decoder.'' Table 4.11 shows the truth table for the 1-of-4 decoder. The logic equations for each one of the outputs of the decoder can be obtained directly from Table 4.11 (because no simplification is

Table 4.11   Truth table for a 1-of-4 decoder

X_1	X_0	O_0	O_1	O_2	O_3
0	0	1	0	0	0
0	1	0	1	0	0
1	0	0	0	1	0
1	1	0	0	0	1

necessary) as follows:

$$O_0 = \overline{X}_1\overline{X}_0 \tag{4.25}$$

$$O_1 = \overline{X}_1 X_0 \tag{4.26}$$

$$O_2 = X_1\overline{X}_0 \tag{4.27}$$

$$O_3 = X_1 X_0 \tag{4.28}$$

Finally, the logic circuit for the decoder can be constructed as shown in Figure 4.38 from Equations 4.25 through 4.28.

Figure 4.38

Logic circuit for a 1-of-4 decoder.

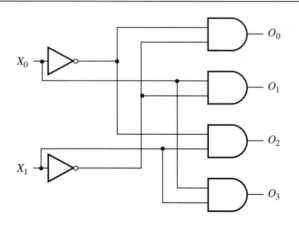

Example 4.4

Design a 1-of-8 decoder whose logic symbol is shown in Figure 4.39.

Solution

The truth table for the 1-of-8 decoder is shown in Table 4.12. The logic equations for O_0 through O_7 are obtained directly from Table 4.12 as follows:

$$O_0 = \overline{X}_2\overline{X}_1\overline{X}_0 \tag{4.29}$$

$$O_1 = \overline{X}_2\overline{X}_1 X_0 \tag{4.30}$$

$$O_2 = \overline{X}_2 X_1\overline{X}_0 \tag{4.31}$$

$$O_3 = \overline{X}_2 X_1 X_0 \tag{4.32}$$

$$O_4 = X_2\overline{X}_1\overline{X}_0 \tag{4.33}$$

$$O_5 = X_2\overline{X}_1 X_0 \tag{4.34}$$

$$O_6 = X_2 X_1\overline{X}_0 \tag{4.35}$$

$$O_7 = X_2 X_1 X_0 \tag{4.36}$$

Figure 4.39

Figure 4.39

Logic symbol for a 1-of-8
decoder.

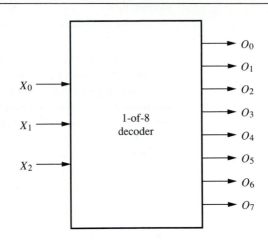

Table 4.12   Truth table for a 1-of-8 decoder

X_2	X_1	X_0	O_0	O_1	O_2	O_3	O_4	O_5	O_6	O_7
0	0	0	1	0	0	0	0	0	0	0
0	0	1	0	1	0	0	0	0	0	0
0	1	0	0	0	1	0	0	0	0	0
0	1	1	0	0	0	1	0	0	0	0
1	0	0	0	0	0	0	1	0	0	0
1	0	1	0	0	0	0	0	1	0	0
1	1	0	0	0	0	0	0	0	1	0
1	1	1	0	0	0	0	0	0	0	1

Figure 4.40

Logic circuit for a 1-of-8
decoder.

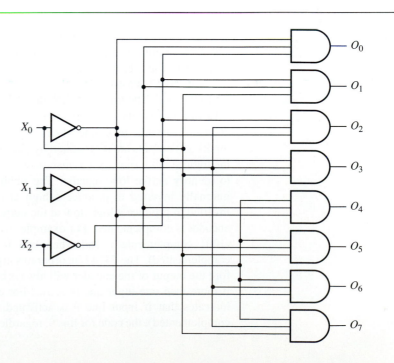

The logic circuit for the 1-of-8 decoder shown in Figure 4.40 is obtained from Equations 4.29 through 4.36.

Encoders and decoders are frequently used in digital communication circuits to transfer information between digital devices. For example, the keys on a keyboard are used to transfer information to a digital computer or circuit by encoding the keys and transmitting the key codes rather than transmitting the signals generated by each individual key. The codes are then decoded at the receiving end into their original form so that the process of encoding and decoding appears transparent to the transfer of information. Figure 4.41 illustrates such a communications scheme where each input line I will indirectly activate the corresponding output line O. For example, activating I_5 will indirectly activate O_5 through the encoding and decoding process. The encoding and decoding circuits make it appear that there is a direct connection between the input (I) and output (O) lines with the advantage of having fewer lines connected between the transmitter and receiver.

Figure 4.41

Digital information transfer using an encoder and decoder.

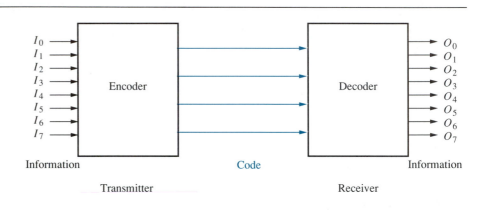

Transmitter   Receiver

IC logic

Priority encoders are available in the form of two ICs — the 74147 and the 74148. The logic symbol and pin configuration of the 74147 and the 74148 are shown in Figure 4.42a and b, respectively.

The 74147 is a 4-bit BCD priority encoder that encodes ten input lines into a 4-bit BCD code. The truth table for the 74147 is shown in Table 4.13. Note in Table 4.13 that the inputs ($\overline{1}$ through $\overline{9}$) are *active low*, meaning that the encoder will encode a particular input line when that line is (activated) at the low state. Active low inputs and outputs are represented by placing a bar over the variable and a bubble in the logic symbol (note: some manufacturers use either the bar or the bubble or both to identify active low signals). The bubble simply indicates that a logic low is internally inverted to produce a high at the input and a logic high is internally inverted to produce a logic low at the output. Also the output code produced by the encoder is complemented. For example, when line $\overline{9}$ is activated the output is 0110 (1001 complemented), and when line $\overline{1}$ is activated the output is 1110 (0001 complemented). The 74147 incorporates highest-order data line encoding and therefore the output of the encoder will always correspond to the highest-order input line if more than one input line is active. For example, in Table 4.13 the second entry indicates that if input line $\overline{9}$ is activated, the output code will be a 0110 (1001 complemented), the code for line $\overline{9}$, regardless of the states of the other input lines (as

Figure 4.42

Logic symbols and pin configurations of (a) the 74147 priority encoder and (b) the 74148 priority encoder.

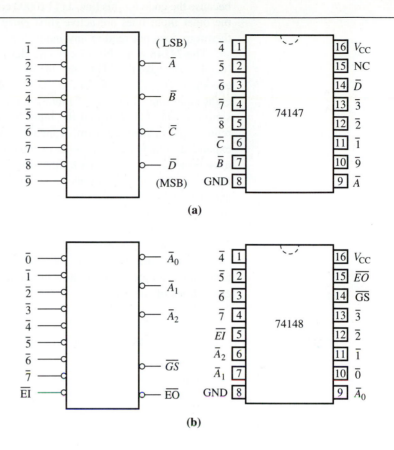

(a)

(b)

indicated by the don't cares — X). Similarly, the fifth entry in Table 4.13 indicates that if line $\overline{6}$ is activated, the output of the encoder will be 1001 (0110 complemented), regardless of the states of lines $\overline{1}$ through $\overline{5}$ because line $\overline{6}$ has a higher priority; however, because lines $\overline{7}$ through $\overline{9}$ have a higher priority than line $\overline{6}$, these lines must be inactive (high state). Also, note in the logic symbol of the 74147 in Figure 4.42a and its truth table in Table 4.13 that input line $\overline{0}$ is missing. This is

Table 4.13 Truth table for the 74147 priority encoder

Inputs									Outputs			
$\overline{1}$	$\overline{2}$	$\overline{3}$	$\overline{4}$	$\overline{5}$	$\overline{6}$	$\overline{7}$	$\overline{8}$	$\overline{9}$	$\overline{D}$	$\overline{G}$	$\overline{B}$	$\overline{A}$
1	1	1	1	1	1	1	1	1	1	1	1	1
X	X	X	X	X	X	X	X	0	0	1	1	0
X	X	X	X	X	X	X	0	1	0	1	1	1
X	X	X	X	X	X	0	1	1	1	0	0	0
X	X	X	X	X	0	1	1	1	1	0	0	1
X	X	X	X	0	1	1	1	1	1	0	1	0
X	X	X	0	1	1	1	1	1	1	0	1	1
X	X	0	1	1	1	1	1	1	1	1	0	0
X	0	1	1	1	1	1	1	1	1	1	0	1
0	1	1	1	1	1	1	1	1	1	1	1	0

because the code for this line, 1111 (0000 complemented), is produced when none of the other input lines are active (first entry in Table 4.13). This implied condition therefore does not require any input line.

The 74148 is a 3-bit priority encoder that encodes eight input lines into a 3-bit (octal) code. Like the 74147, the 74148 incorporates highest-order input line priority encoding. The truth table for the 74148 is shown in Table 4.14. The 74148 encodes eight input lines ($\overline{0}$ through $\overline{7}$) into a 3-bit code at the output lines ($\overline{A_2}\overline{A_1}\overline{A_0}$) and is similar to the 74147 in that the input lines are active low and the output code is represented in inverted form. It includes two enable lines that allow the IC to be expanded to increase the number of input lines encoded. The enable input line ($\overline{EI}$) must be active (low) in order for the 74148 to function and the enable output line $\overline{EO}$ will be at the low state if none of the encoder's input lines ($\overline{0}$ through $\overline{7}$) are active and the $\overline{EI}$ line is active. An example of expanding the 74148 using these lines follows. The $\overline{GS}$ output line is active (low) when the $\overline{EI}$ line is active and when any of the input lines ($\overline{0}$ through $\overline{7}$) are active. This line can therefore be used to indicate if the output of the encoder contains a valid code. For example, if a keypad were connected to the 74148, the closure of any key would activate this line. The following example illustrates how the $\overline{EI}$, $\overline{EO}$, and $\overline{GS}$ lines can be used to expand the capabilities of the 74148.

Table 4.14　Truth table for the 74148 priority encoder

Inputs									Outputs				
$\overline{EI}$	$\overline{0}$	$\overline{1}$	$\overline{2}$	$\overline{3}$	$\overline{4}$	$\overline{5}$	$\overline{6}$	$\overline{7}$	$\overline{A_2}$	$\overline{A_1}$	$\overline{A_0}$	$\overline{GS}$	$\overline{EO}$
1	X	X	X	X	X	X	X	X	1	1	1	1	1
0	1	1	1	1	1	1	1	1	1	1	1	1	0
0	X	X	X	X	X	X	X	0	0	0	0	0	1
0	X	X	X	X	X	X	0	1	0	0	1	0	1
0	X	X	X	X	X	0	1	1	0	1	0	0	1
0	X	X	X	X	0	1	1	1	0	1	1	0	1
0	X	X	X	0	1	1	1	1	1	0	0	0	1
0	X	X	0	1	1	1	1	1	1	0	1	0	1
0	X	0	1	1	1	1	1	1	1	1	0	0	1
0	0	1	1	1	1	1	1	1	1	1	1	0	1

Example 4.5

Figure 4.43b illustrates how two 74148 ICs can be connected together to make up a 4-bit encoder that encodes 16 input lines into a 4-bit code. The logic symbol for such an encoder is shown in Figure 4.43a.

Encoder no. 1 has its $\overline{EI}$ line permanently tied low to enable the encoder at all times. If any of the input lines ($\overline{0}$ through $\overline{7}$) of encoder no. 1 is activated, the enable output line $\overline{EO}$ will be a logic 1 and therefore disable encoder no. 2 because this line is connected to $\overline{EI}$ of encoder no. 2. The outputs of encoder no. 2 ($\overline{A_2}\overline{A_1}\overline{A_0}$) will be high (111) and the outputs of encoder no. 1 will be passed through the *AND* gates (G1, G2, and G3) as the output code ($\overline{A_3}\overline{A_2}\overline{A_1}\overline{A_0}$). The $\overline{GS}$ output of encoder no. 2 represents the MSB of the output code ($\overline{A_3}$) and will be a logic 1 if any of the input lines ($\overline{0}$ through $\overline{7}$) of encoder no. 1 are activated. Thus, encoder no. 1 will produce the codes 1111 through 1000 (0000 through 0111 complemented) at the output lines $\overline{A_3}\overline{A_2}\overline{A_1}\overline{A_0}$ when any of the input lines $\overline{0}$ through $\overline{7}$ are activated. If none of the input lines ($\overline{0}$ through $\overline{7}$) of encoder no. 1 are activated, its $\overline{EO}$ output is low, and therefore encoder no. 2 is enabled. Because the output lines $\overline{A_2}\overline{A_1}\overline{A_0}$ of encoder no. 1 are now at the high state (because

Figure 4.43

(a) Logic symbol for a 4-bit encoder; (b) connecting two 74148's to function as a 4-bit encoder.

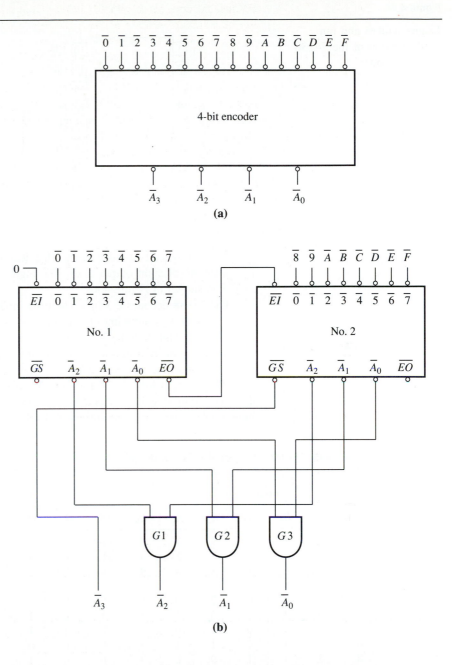

(a)

(b)

its input lines $\overline{0}$ through $\overline{7}$ are inactive), the output lines $\overline{A}_2\overline{A}_1\overline{A}_0$ of encoder no. 2 will be passed through the *AND* gates (G1, G2, and G3) as the output code ($\overline{A}_3\overline{A}_2\overline{A}_1\overline{A}_0$) if any of the input lines ($\overline{8}$ through $\overline{F}$) of encoder no. 2 are activated. The $\overline{GS}$ line of encoder no. 2 is a logic 0 when the encoder's input lines ($\overline{8}$ through $\overline{F}$) are active, and therefore activating any of the input lines of encoder no. 2 will produce the codes 0111 through 0000 (1000 through 1111 complemented) at the output lines $\overline{A}_3\overline{A}_2\overline{A}_1\overline{A}_0$.

Note that this circuit does not incorporate highest-order data input priority encoding between the input lines connected to encoder no. 1 ($\overline{0}$ through $\overline{7}$) and the input lines connected to encoder no. 2 ($\overline{8}$ through $\overline{F}$). For example, if line $\overline{2}$ and line $\overline{A}$ are activated at the same time, the output of the circuit $\overline{A}_3\overline{A}_2\overline{A}_1\overline{A}_0$ will be 1101 (0010 complemented).

Figure 4.44

Logic symbol and pin configuration of the 74138 1-of-8 decoder.

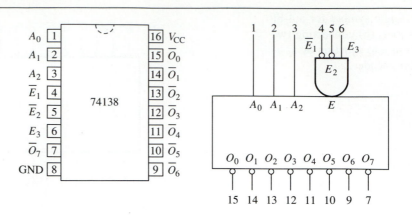

The 74138 is a 1-of-8 decoder IC that decodes a 3-bit binary code at its input and activates one of eight output lines. The logic symbol and pin configuration of the 74138 is shown in Figure 4.44.

The 74138 decodes a 3-bit code at its input lines $A_2A_1A_0$ and activates one of eight (active-low) output lines $\overline{O}_0$ through $\overline{O}_7$. The truth table for the 74138 is shown in Table 4.15. Note that the 74138 has three input enable lines, $\overline{E}_1$, $\overline{E}_2$, and E_3, which must be active in order for the decoder to function. Because $\overline{E}_1$ and $\overline{E}_2$ are active low, these two lines must be tied to a logic 0, whereas the active-high enable input line E_3 must be tied to a logic 1. If any one of these three enable lines are inactive, the decoder's output will be inactive regardless of the states of the other inputs as illustrated in the first three entries of Table 4.15.

Table 4.15   Truth table for the 74138 decoder

Inputs						Outputs							
$\overline{E}_1$	$\overline{E}_2$	E_3	A_2	A_1	A_0	$\overline{O}_0$	$\overline{O}_1$	$\overline{O}_2$	$\overline{O}_3$	$\overline{O}_4$	$\overline{O}_5$	$\overline{O}_6$	$\overline{O}_7$
1	X	X	X	X	X	1	1	1	1	1	1	1	1
X	1	X	X	X	X	1	1	1	1	1	1	1	1
X	X	0	X	X	X	1	1	1	1	1	1	1	1
0	0	1	0	0	0	0	1	1	1	1	1	1	1
0	0	1	0	0	1	1	0	1	1	1	1	1	1
0	0	1	0	1	0	1	1	0	1	1	1	1	1
0	0	1	0	1	1	1	1	1	0	1	1	1	1
0	0	1	1	0	0	1	1	1	1	0	1	1	1
0	0	1	1	0	1	1	1	1	1	1	0	1	1
0	0	1	1	1	0	1	1	1	1	1	1	0	1
0	0	1	1	1	1	1	1	1	1	1	1	1	0

The enable inputs of the 74138 can be used to expand the decoding capabilities of the IC as shown in the following example.

Example 4.6

The 1-of-16 decoder whose logic symbol is shown in Figure 4.45a can be implemented by using two 74138 decoders as shown in Figure 4.45b.

Note in Figure 4.45b that the input lines A_0, A_1, and A_2 of the circuit are connected to the

Figure 4.45

(a) logic symbol of a 1-of-16 decoder; (b) two 74138s connected to function as a 1-of-16 decoder.

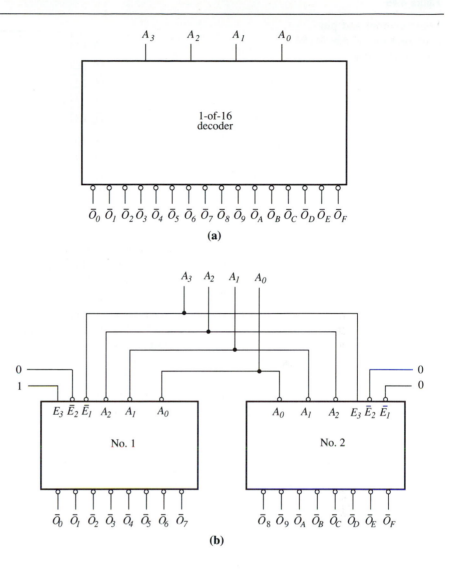

input lines ($A_2A_1A_0$) of both the decoders. However, input line A_3 will either enable decoder no. 1 (when at the logic 0 state) or enable decoder no. 2 (when at the logic 1 state). Therefore, decoder no. 1 will activate its output lines $\overline{O}_0$ through $\overline{O}_7$ whenever the codes 0000 through 0111 appear at the circuit's input lines $A_3A_2A_1A_0$, whereas decoder no. 2 will activate its output lines $\overline{O}_8$ through $\overline{O}_F$ whenever the codes 1000 through 1111 appear at the circuit's input lines $A_3A_2A_1A_0$. Note that the remaining enable lines of decoders no. 1 and 2 are unused and therefore tied active.

The 74154 is a 1-of-16 decoder that functions in a manner similar to the circuit designed in Example 4.6. The logic symbol and pin configuration of the 74154 are shown in Figure 4.46. The 74154 decodes a 4-bit code appearing on the input lines $A_3A_2A_1A_0$ and activates 1 of 16 active-low outputs, $\overline{O}_0$ through $\overline{O}_{15}$. It has two active-low enable inputs, $\overline{E}_0$ and $\overline{E}_1$, that must both be at the low state in order for the IC to function.

The 74139 is a dual 1-of-4 decoder IC that incorporates two independent 1-of-4

Figure 4.46

Logic symbol and pin
configuration of the 74154
1-of-16 decoder.

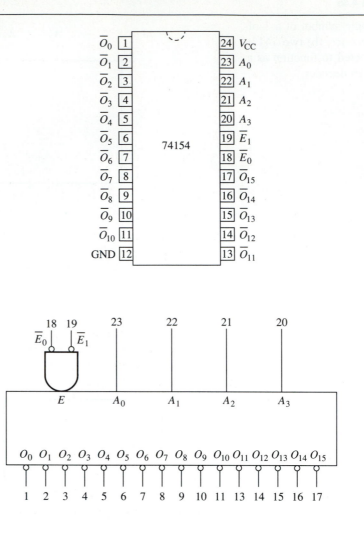

Figure 4.47

Logic symbol and pin configuration of the 74139 dual 1-of-4 decoder.

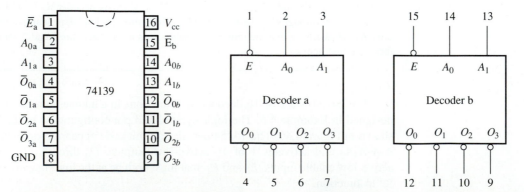

decoders in one package. The logic symbol and pin configuration for the 74139 are shown in Figure 4.47.

Each decoder in the 74139 IC decodes a 2-bit code appearing on the input lines A_1A_0 and activates one of four active-low output lines $\overline{O}_0$ through $\overline{O}_3$. The decoders are similar in operation to the circuit designed in Figure 4.38. Each decoder also has an active-low enable line, $\overline{E}$, which must be tied to the logic 0 state in order to enable the circuit.

Review Questions

1. Describe the operation of an encoder.
2. What is a priority encoder?
3. What is meant by high-order data line encoding?
4. Describe an application of an encoder.
5. Describe the function of a decoder.
6. Why are most decoders known as 1-of-n decoders?

4.7

Seven-segment decoders

The decoders discussed in Section 4.6 detected binary codes at their inputs by activating a unique output line whenever a particular binary combination appeared at the input. The output line that was activated could then be used to identify the binary code present at the input of the decoder. There are many digital applications where it is necessary to decode a binary code directly into a form that represents that code. For example, a typical application may require that a binary number be decoded into a form that can be displayed on a seven-segment light-emitting diode (LED). This section deals with decoders that are used to drive seven-segment LEDs and are actually known as code converters because they effectively convert one code into another.

A seven-segment LED consists of a set of seven bar-shaped LEDs arranged in the shape of the number 8 as shown in Figure 4.48a. The seven segments are labeled *a* through *g* for identification purposes. There are two common configurations available, the common cathode (Fig. 4.48c) and the common anode (Fig. 4.48b). Note in Figure 4.48c that all the cathodes of the seven-segment LED are tied together at one point (cc) that is grounded (logic 0). Because a LED must be forward-biased (i.e., current must pass from anode to cathode) in order for it to emit light, a logic 1 on any of the segment connections *a* through *g* will illuminate the respective segment. In Figure 4.48b because all the anodes of the seven-segment LED are tied together at one point (ca), that is tied to a positive voltage V^+, a logic 0 on any of the segment connections *a* through *g* will forward bias the respective diode and illuminate the segment. Thus, common cathode seven-segment LEDs require a logic 1 to illuminate a segment whereas common anode seven-segment LEDs require a logic 0 to illuminate a segment. There are certain design considerations necessary to supply or limit the flow of current through the LED segments, but these will be examined in a later chapter.

We can now investigate the binary codes necessary to display all the hexadecimal digits (0 through F) on a common cathode seven-segment LED. Because the hexadecimal digit set includes all the binary, octal, and decimal digits, these codes can be used in any application requiring a display in any one of the four numbering systems. The display format for the hexadecimal digits is shown in Figure 4.49. Note

Figure 4.48

Seven-segment LEDs. (a) Configuration; (b) common anode; (c) common cathode.

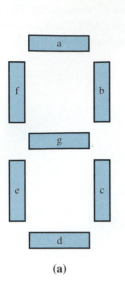
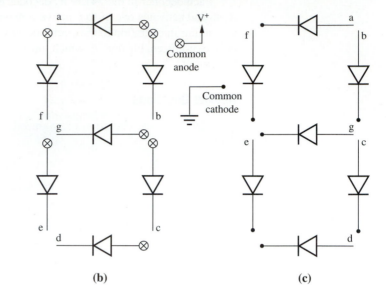

(a)   (b)   (c)

that the digits B and D must be displayed as lowercase letters *b* and *d*, respectively. The pictorial representation of each digit is then translated into the binary codes required to activate the appropriate segments for each digit as shown in Table 4.16. For example, to display the digit 5 on the seven-segment LED, segments *a*, *c*, *d*, *f*, and *g* must be illuminated. In Table 4.16 these segments have been set to logic 1's for the digit 5. Note that the codes shown in Table 4.16 are common cathode codes and must be complemented for common anode seven-segment LEDs.

   A BCD to seven-segment decoder decodes (actually converts) a 4-bit BCD code that represents the decimal digits 0 through 9 into the (first ten) seven-segment codes listed in Table 4.16 so that when the output of the decoder is connected to a seven-segment LED, the appropriate segments will be activated to display the decimal digit corresponding to each one of the ten BCD codes. A hexadecimal to seven-segment decoder decodes (or converts) all possible combinations of 4 bits into the seven-segment codes to display the hexadecimal digits that represent the 4-bit codes. Because a hexadecimal to seven-segment decoder has the capability to function as a BCD to seven-segment decoder, the design of the former circuit will be investigated. The truth table for a hexadecimal to seven-segment decoder is shown in Table 4.17 and its logic symbol is shown in Figure 4.50.

Figure 4.49

Seven-segment display of hexadecimal digits.

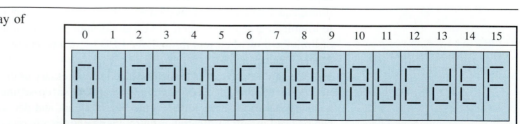

| 0 | 1 | 2 | 3 | 4 | 5 | 6 | 7 | 8 | 9 | 10 | 11 | 12 | 13 | 14 | 15 |

Table 4.16 Common cathode seven-segment codes

Digit	a	b	c	d	e	f	g
0	1	1	1	1	1	1	0
1	0	1	1	0	0	0	0
2	1	1	0	1	1	0	1
3	1	1	1	1	0	0	1
4	0	1	1	0	0	1	1
5	1	0	1	1	0	1	1
6	1	0	1	1	1	1	1
7	1	1	1	0	0	0	0
8	1	1	1	1	1	1	1
9	1	1	1	0	0	1	1
A	1	1	1	0	1	1	1
B	0	0	1	1	1	1	1
C	1	0	0	1	1	1	0
D	0	1	1	1	1	0	1
E	1	0	0	1	1	1	1
F	1	0	0	0	1	1	1

From Table 4.17, the simplified logic equations for each one of the outputs, a through g, are obtained through the use of the K-maps shown in Figure 4.51.

$$a = A_2A_1 + \overline{A}_3A_1 + A_3\overline{A}_0 + \overline{A}_2\overline{A}_0 + \overline{A}_3A_2A_0 + A_3\overline{A}_2\overline{A}_1 \tag{4.37}$$

$$b = \overline{A}_3\overline{A}_2 + A_3\overline{A}_1A_0 + \overline{A}_2\overline{A}_0 + \overline{A}_3\overline{A}_1\overline{A}_0 + \overline{A}_3A_1A_0 \tag{4.38}$$

$$c = A_3\overline{A}_2 + \overline{A}_3A_2 + \overline{A}_1A_0 + \overline{A}_3\overline{A}_1 + \overline{A}_3A_0 \tag{4.39}$$

$$d = A_3\overline{A}_1\overline{A}_0 + A_2\overline{A}_1A_0 + A_2A_1\overline{A}_0 + \overline{A}_2A_1A_0 + \overline{A}_3\overline{A}_2\overline{A}_0 \tag{4.40}$$

$$e = \overline{A}_2\overline{A}_0 + A_3A_2 + A_1\overline{A}_0 + A_3A_1 \tag{4.41}$$

$$f = \overline{A}_1\overline{A}_0 + A_3\overline{A}_2 + A_2\overline{A}_0 + A_3A_1 + \overline{A}_3A_2\overline{A}_1 \tag{4.42}$$

$$g = A_3\overline{A}_2 + A_3A_0 + \overline{A}_2A_1 + A_1\overline{A}_0 + \overline{A}_3A_2\overline{A}_1 \tag{4.43}$$

The logic circuit for the hexadecimal to seven-segment decoder is obtained by implementing Equations 4.37 through 4.43 as shown in Figure 4.52.

It is apparent that seven-segment decoders are actually code converters for they convert binary codes from one form to another. The circuits discussed in this section are not limited only to seven-segment LED displays but can be modified or redesigned for more complex LED designs, including dot-matrix LEDs and printheads for printers, *cathode-ray tube* (CRT) character generation, and *liquid crystal displays* (LCDs).

IC logic

The 9368 seven-segment decoder / driver / latch is an IC that incorporates a hexadecimal to seven-segment decoder and is designed to drive common cathode LEDs. The logic symbol and pin configuration of the 9368 are shown in Figure 4.53.

The 9368 is similar in operation to the circuit designed in Figure 4.52 except that it incorporates a few additional features. The 9368 accepts as input a 4-bit binary code ($A_3A_2A_1A_0$) and decodes it into a seven-segment code (a through g) to display

Table 4.17 Truth table for a hexadecimal to seven-segment decoder

Digit	Hex code $A_3A_2A_1A_0$	Seven-segment code						
		a	b	c	d	e	f	g
0	0 0 0 0	1	1	1	1	1	1	0
1	0 0 0 1	0	1	1	0	0	0	0
2	0 0 1 0	1	1	0	1	1	0	1
3	0 0 1 1	1	1	1	1	0	0	1
4	0 1 0 0	0	1	1	0	0	1	1
5	0 1 0 1	1	0	1	1	0	1	1
6	0 1 1 0	1	0	1	1	1	1	1
7	0 1 1 1	1	1	1	0	0	0	0
8	1 0 0 0	1	1	1	1	1	1	1
9	1 0 0 1	1	1	1	0	0	1	1
A	1 0 1 0	1	1	1	0	1	1	1
B	1 0 1 1	0	0	1	1	1	1	1
C	1 1 0 0	1	0	0	1	1	1	0
D	1 1 0 1	0	1	1	1	1	0	1
E	1 1 1 0	1	0	0	1	1	1	1
F	1 1 1 1	1	0	0	0	1	1	1

Figure 4.50

Logic symbol for a hexadecimal to seven-segment decoder.

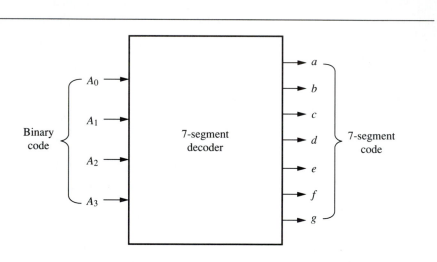

the hexadecimal digit corresponding to the input binary number. The 9368 has a *ripple blanking input* ($\overline{RBI}$) and a *ripple blanking output* ($\overline{RBO}$) that are used for leading zero suppression in circuits containing several 9368 ICs. *Leading zero suppression* is a technique that is used to blank out leading zeros in LED displays; for example, the number 0645 is displayed as 645 with leading zero suppression. If the binary number 0000 is applied to the input of the 9368 and $\overline{RBI}$ is at the low state, the LED is blanked by deactivating all segments and the $\overline{RBO}$ output is set to a logic 0. If $\overline{RBI}$ is at the high state when the binary number 0000 is applied to the input of the 9368, the number 0 is displayed on the LEDs and the $\overline{RBO}$ output is set to a logic 1. Thus, $\overline{RBI}$ and $\overline{RBO}$ can be connected such that leading zeros are blanked out on a multidigit display containing several 9368 ICs. To incorporate leading zero suppression in multidigit displays, the $\overline{RBI}$ input of the MSD is tied low so that the MSD is

Figure 4.51

K-maps for the design of the
seven-segment decoder.

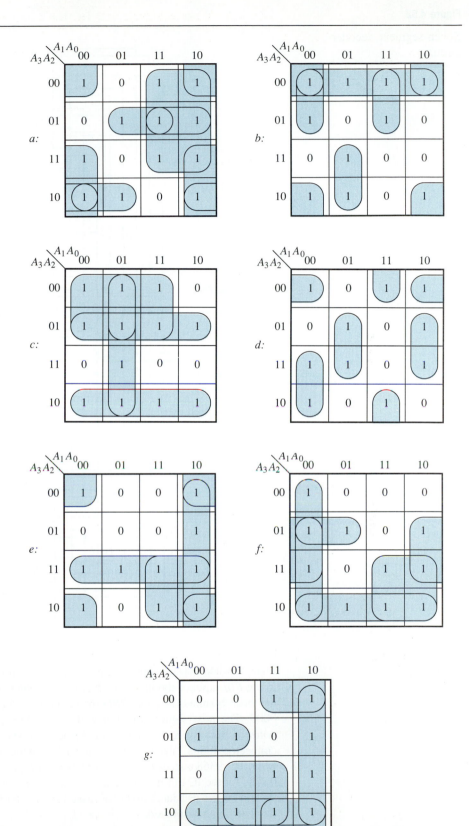

Figure 4.52

Seven-segment decoder
circuit.

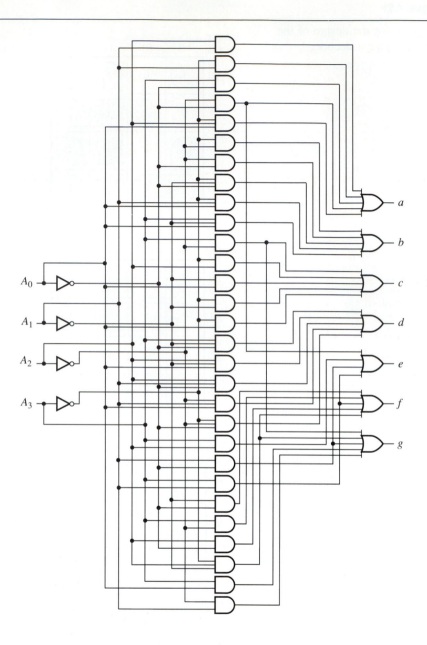

blanked when 0000 is applied to the 9368. The $\overline{RBO}$ output of the MSD is connected to the $\overline{RBI}$ input of the next significant digit, which will only be blanked if the MSD has been blanked (because $\overline{RBO}$ of the MSD = 0) and the number at the input is 0000. The next significant digit will not be blanked if the MSD is displaying a nonzero because $\overline{RBO}$ will be at the logic 1 state. The $\overline{RBI}$ and $\overline{RBO}$ lines are connected in this ''daisy chain'' manner for all the other digits in sequence up to the LSD. The 9368 also has an active-low latch enable input, $\overline{E}_L$, that must be at the logic 0 state for the IC to function. This input controls the operation of an internal storage system known as a latch (to be discussed in Chapter 5) that is used to store the binary number being displayed.

Figure 4.53

Logic symbol and pin
configuration of the 9368
seven-segment decoder /
driver / latch.

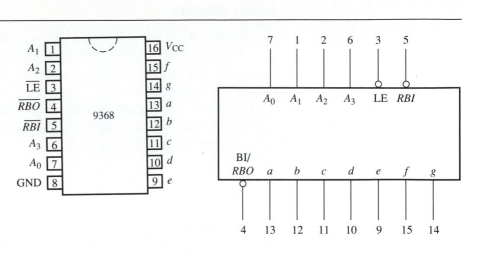

Figure 4.54

Logic symbol and pin
configuration of the 7448
BCD to seven-segment
decoder.

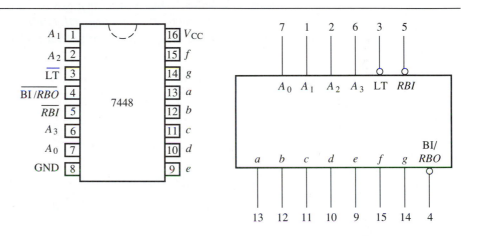

The 7448 BCD to seven-segment decoder IC (Fig. 4.54) is similar in operation to the 9368 except that the 7448 decodes a 4-bit BCD code in the range 0000 to 1001 into the seven-segment codes required to display the decimal digits 0 through 9, respectively. If an attempt is made to decode the codes 1010 to 1111, the resulting seven-segment codes produce unintelligible displays on the seven-segment LED connected to the 7448. Instead of a latch enable line, $\overline{E}_L$, the 7448 has an active-low lamp test, $\overline{L}_T$, line that is used to test the segments of the seven-segment LED connected to the 7448. When this line is low, all seven segments are activated and any defective segments can be identified. The $\overline{RBI}$ and $\overline{RBO}$ lines function in the same manner as the 9368 and are used for ripple blanking and leading zero suppression.

The 9368 and 7448 are interchangeable, that is, their pin configurations are identical. Because the 9368 can display all the decimal digits, it can substitute for the 7448 but not vice versa because the 7448 cannot display the hexadecimal digits A through F. There are also identical ICs available that are designed with inverted outputs (*a* through *g*) to allow them to interface to common anode – type LEDs.

Review Questions

1. What is the difference between a standard decoder (such as the ones discussed in Section 4.6) and a seven-segment decoder?

2. What is the difference between a common cathode and a common anode seven-segment LED?

3. Why is a seven-segment decoder also known as a code converter?

4. What is the purpose of the ripple blanking inputs and outputs in a seven-segment decoder IC?

4.8

Multiplexers and tristate logic

In many digital applications, particularly in digital computers, it is often necessary to route or channel data in the form of binary numbers from different circuits for processing. The circuit that provides this routing or multiplexing of data is known as a data selector or *multiplexer*. A multiplexer can be thought of as a multiposition switch as shown in Figure 4.55 that can be used to connect one of several input channels (*A* through *F*) to a common *OUTPUT*. Each channel can represent a binary number that can be a single bit or several bits in size.

Figure 4.55

Conceptual multiplexer circuit.

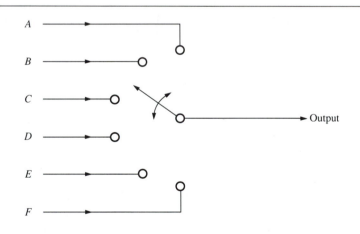

The circuit shown in Figure 4.55 provides data selection by mechanically moving the switch to the desired channel. A digital multiplexer provides data selection by means of a binary code that identifies a particular channel. Figure 4.56 shows the logic symbol for a two-channel multiplexer that multiplexes two input channels, *A* and *B*, to the output, *OUT*. The input channels (and consequently, the output) are 1 bit wide. The multiplexer has a *SELECT* line that is used to select the channel being multiplexed. Assume that if *SELECT* is a logic 1, then channel *A* is multiplexed (connected) to the output, and if *SELECT* is a logic 0 then channel *B* is multiplexed (connected) to the output. Using this information, a truth table is constructed for the multiplexer as shown in Table 4.18.

From Table 4.18 it can be seen that when *SELECT* = 0, the output, *OUT*, of the multiplexer is equal to the logic states of channel *B*; it appears as if the output *OUT* is

Figure 4.56

Logic symbol for a two-channel 1-bit multiplexer.

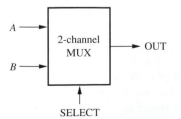

Table 4.18   Truth table for a two-channel 1-bit multiplexer

SELECT	A	B	OUT
0	0	0	0
0	0	1	1
0	1	0	0
0	1	1	1
1	0	0	0
1	0	1	0
1	1	0	1
1	1	1	1

Figure 4.57

K-map for the design of the two-channel 1-bit multiplexer.

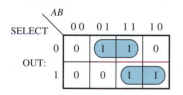

Figure 4.58

Logic circuit for the two-channel 1-bit multiplexer.

directly connected to *B*. Similarly, when *SELECT* = 1, the output *OUT* is equal to the logic states of channel *A* making it seem as if *OUT* is directly connected to *A*.

The conventional approach is used to obtain the logic equations and logic circuit to implement the truth table for the multiplexer shown in Figure 4.56. The K-map for the output of the multiplexer is shown in Figure 4.57 and its logic equation is as follows:

$$OUT = \overline{SELECT} \cdot B + SELECT \cdot A \qquad (4.44)$$

The logic circuit for Equation 4.44 shown in Figure 4.58 functions as a two-channel, 1-bit multiplexer. Figure 4.58 shows that the logic levels of *A* and *B* are passed through the *AND* gates by applying a logic 1 on the second input. For example, when *SELECT* = 1, because one of the inputs of the *AND* gate G1 is a logic 1, the output will be equal to the logic state of *A* (identity law). However, when *SELECT* = 1, one of the inputs of the *AND* gate G2 is a logic 0 and therefore the output will be a logic 0 (dominance law). Because one of the inputs of the *OR* gate is a logic 0, the output will be equal to the state of the other input (identity law) — in this case, the state of *A*. Similarly when *SELECT* = 0 gate G1 effectively blocks *A* and passes *B* by applying a logic 0 and a logic 1 at the second inputs of gates G1 and G2, respectively. Thus, we can multiplex as many single-bit channels as desired simply by using a two-input *AND* gate for each channel and applying a logic 1 to one of its inputs to pass the logic level of the channel through the gate and by applying a logic 0 to one of its inputs to prevent the logic level of the channel from passing through the gate. The outputs of each of these *AND* gates can then be *OR*ed together to obtain the multiplexed output. The following example illustrates the design of a four-channel, 1-bit multiplexer.

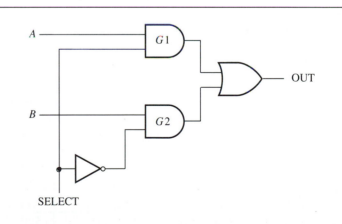

Example 4.7

The logic symbol for a four-channel multiplexer is shown in Figure 4.59. The circuit multiplexes one of four input channels (*A* through *D*) to the output, *X*. Because one of four channels is being multiplexed, the circuit requires two select lines ($S_1 S_0$) so that each 2-bit combination applied to the select lines will select a particular channel to be multiplexed as shown in Table 4.19.

Note in Table 4.19 that if $S_1 S_0 = 00$, then the logic state of the output *X* is equal to channel *A*; if $S_1 S_0 = 01$, then the logic state of the output *X* is equal to channel *B*, etc. Instead of using the conventional approach to design the circuit to implement the multiplexer shown in

Figure 4.59

Logic symbol for a four-channel 1-bit multiplexer.

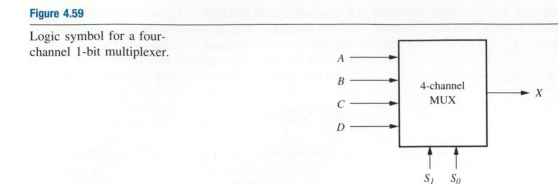

Table 4.19   Truth table for a four-channel 1-bit multiplexer

S_1	S_0	X
0	0	A
0	1	B
1	0	C
1	1	D

Figure 4.59, we can apply the operation of the circuit in Figure 4.58 along with our knowledge of decoders (from Section 4.6) to obtain the circuit shown in Figure 4.60.

The circuit in Figure 4.60 is similar to the circuit in Figure 4.58 as far as the arrangement of the *AND* gates and the *OR* gate is concerned. That is, as stated earlier we have used *AND* gates for each input channel and the outputs of the *AND* gates are *OR*ed together to provide the final multiplexed output, *X*. However, to enable each *AND* gate (to pass the logic level of the channel connected to it) by applying a logic 1 at its other input, we have used a 1-of-4 decoder. The decoder accepts one of four possible binary codes at its inputs $S_1 S_0$ and activates (logic 1) a single output, which is used to enable the respective *AND* gate and the appropriate channel.

Figure 4.60

Logic circuit for a four-channel 1-bit multiplexer.

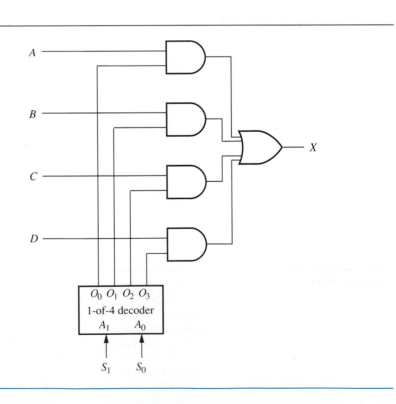

Table 4.20 Truth table for a two-channel 2-bit multiplexer

SELECT	X_1X_0
0	B_1B_0
1	A_1A_0

Many digital applications require larger binary numbers to be routed to different circuits using multiplexers. Multiplexers can be designed to handle channels larger than 1 bit by combining the single-bit multiplexers investigated so far. For example, Figure 4.61 shows the logic symbol for a two-channel, 2-bit multiplexer that multiplexes two channels A and B to the output, *OUT*. However, note that channels A and B are actually 2-bit numbers, A_1A_0 and B_1B_0, respectively, and the output is a 2-bit output, X_1X_0. The truth table for the multiplexer is shown in Table 4.20. Instead of using the conventional approach to implement the two-channel, 2-bit multiplexer, the circuit is implemented by using two 2-channel, 1-bit multiplexers (from Fig. 4.56) as shown in Figure 4.62. Note in Figure 4.62 that each multiplexer is used to multiplex a single bit that makes up the binary number and therefore the circuit can be extended to multiplex two numbers of any size. Similarly, the number of channels can be increased by using single-bit multiplexers with the desired number of channels.

Figure 4.61

Logic symbol for a two-channel 2-bit multiplexer.

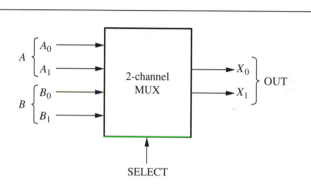

Figure 4.62

Logic circuit for a two-channel 2-bit multiplexer.

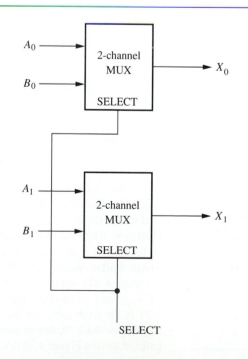

Tristate logic

At this stage it may be appropriate to introduce the concept of tristate logic and tristate logic gates. They can greatly simplify the design of multiplexers and are also used extensively in other digital computer circuits that will be examined in later chapters.

The logic gates examined so far have all had *two-stated* outputs, that is, the output of a logic gate is either a logic 1 (5 volts) or a logic 0 (ground potential, 0 volts) and must exist in one of these two states. A logic gate with *three-stated* outputs, also referred to as a *tri-state* logic gate, can have its output exist in one of three states — a logic 1 (5 volts), a logic 0 (ground potential), or no logic level (no voltage). This third state is often referred to as a *high-impedance* state and can be explained as follows. If the output of a two-stated logic gate is viewed as a two-position switch that provides either a logic 1 or a logic 0 by connecting the output to 5 volts or ground potential, respectively, as shown in Figure 4.63a, then the output of a tristated logic gate can be viewed as a three-position switch (Fig. 4.63b) that provides a logic 1 and a logic 0 as before but includes a third position that is neither a logic 1 nor a logic 0 but represents no voltage. Observe that the third position of the switch in Figure 4.63b is not connected to any logic state but is floating (or open), and therefore the impedance or resistance of the output in this state is infinite or extremely high. This state is therefore referred to as the high-impedance, or Z, state.

Figure 4.63

Conceptual circuits for two-state and tristate outputs.

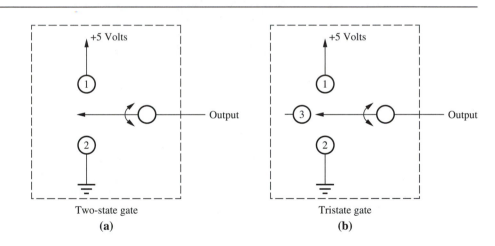

Two-state gate	Tristate gate
(a)	**(b)**

Table 4.21 Truth table for a tristate buffer

E	A	B
0	0	Z
0	1	Z
1	0	0
1	1	1

Tristate logic is often incorporated into the outputs of many of the basic logic gates. However, one of the more common gates that is available with a tristate output is the buffer whose logic symbol is shown in Figure 4.64a. The buffer has an additional input labeled "*E*" that is used to enable or disable the output of the buffer. When the output of the buffer is enabled ($E = 1$) as shown in Figure 4.64b, the buffer functions normally and the states of the output B are equal to the input A. However, when the buffer is disabled ($E = 0$) as shown in Figure 4.64c, the output of the buffer is in a high-impedance state (Z) and independent of the input A. The tristate buffer can therefore be viewed as a switch that is opened or closed as shown in Figure 4.64b and c, respectively. When closed, the output is connected to the input ($B = A$) and when open, the output is floating at a high-impedance state (Z). Table 4.21 is the truth table for the tristate buffer shown in Figure 4.64a.

Figure 4.65 illustrates some of the other basic gates with tristate outputs. All the gates shown in Figure 4.65 operate normally (i.e., according to their functions) when

Figure 4.64

The tristate buffer. (a) Logic symbol; (b) output enabled; (c) output disabled.

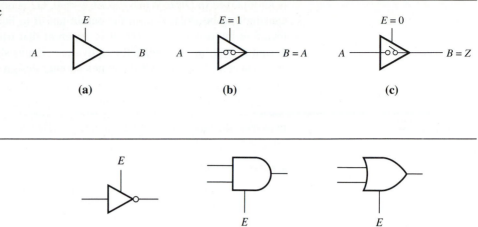

(a)   (b)   (c)

Figure 4.65

Other types of gates with tristate outputs.

the output is enabled ($E = 1$). However, when the output is disabled ($E = 0$) it is effectively cut off from the gate and is in a high-impedance state.

The tristate gates investigated so far have active-high enable lines (E). However, there are many gates that have active-low enable lines ($\overline{E}$) that must be tied to a logic 0 in order to enable the output.

When two or more outputs of logic gates are connected together there exists a *conflict of logic levels*, especially when the outputs are at different logic states. This condition cannot be defined logically because there are no precedence rules for logic states. For example, if two outputs are connected together and the two outputs are at different logic states, it cannot be assumed that a logic 0 state will override a logic 1 state or vice versa. Also, from an electronics perspective this means that the 5 volts (logic 1) is effectively shorted to ground (logic 0). Even though the resulting state would be a logic 0, the connection could damage the internal electronic circuitry of the gates (there are exceptions to this rule that will be covered in Chapter 12) and should be avoided. Keeping this in mind, the multiplexer in Figure 4.56 is redesigned using tristate buffers as shown in Figure 4.66.

Observe in Figure 4.66 that the A and B channels have been connected through a set of tristate buffers. When *SELECT* = 1, the buffer G1 is enabled and G2 is disabled. The output of G2 is therefore at a high-impedance state effectively cutting

Figure 4.66

A two-channel 1-bit multiplexer implemented with tristate logic.

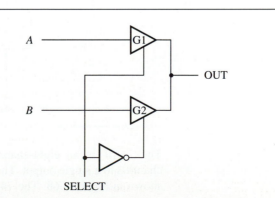

off *B* from the output *OUT*, whereas the output of G1 is equal to *A*. Therefore, input *A* is multiplexed to *OUT*. When *SELECT* = 0, G2 is now enabled and G1 is disabled, causing *A* to be isolated from the output and *B* to be connected to the output; *B* is therefore multiplexed to *OUT*. It is apparent that tristate buffers can simplify the design of a multiplexer when the equivalent circuits shown in Figures 4.66 and 4.58 are compared. Example 4.8 illustrates another design of a multiplexer using tristate logic.

Example 4.8

Implement the four-channel, 1-bit multiplexer whose logic symbol is shown in Figure 4.59 using tristate buffers.

Solution

The logic circuit for the multiplexer is shown in Figure 4.67. The design is similar to the circuit in Figure 4.66 except that the 1-of-4 decoder is used to enable one of the four tristate buffers to multiplex the input to the output.

Figure 4.67

A four-channel 1-bit multiplexer implemented with tristate logic.

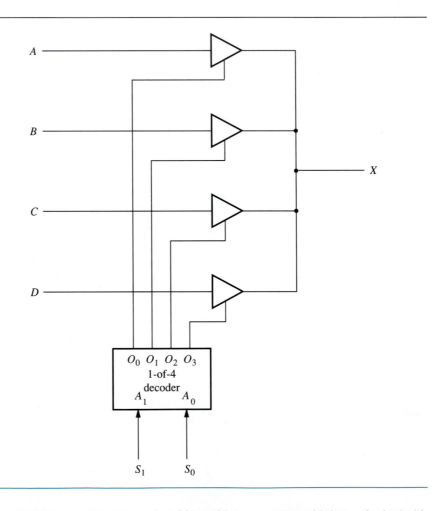

IC logic

The 74151 is an eight-channel, 1-bit multiplexer that multiplexes 8 single-bit channels to a single output. The logic symbol and pin configuration of the 74151 are shown in Figure 4.68. The 74151 has three select lines, $S_2 S_1 S_0$, that select one of

Figure 4.68

Logic symbol and pin configuration of the 74151 eight-input multiplexer.

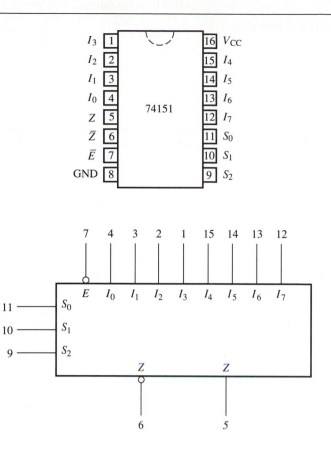

eight channels to be multiplexed. Each one of the input lines I can be multiplexed to the output Z when the binary code for a particular line is applied to the select lines $S_2S_1S_0$. The subscript that identifies each input line also identifies the binary code that selects it. For example, to multiplex I_3 to Z, the code at the select lines must be 011. Note that the 74151 also provides the output Z in complementary form, $\overline{Z}$, as a convenience. The 74151 also has an active-low enable line, $\overline{E}$, that must be tied low for the multiplexer to function; if $\overline{E} = 1$ the output of the multiplexer Z is held permanently low. The enable line $\overline{E}$ is used to expand the 74151 (i.e., increase the number of input channels) as shown in Figure 4.69.

The circuit in Figure 4.69 illustrates the use of the 74151's enable line $\overline{E}$ to interconnect two ICs to double the number of input channels multiplexed to 16. The 16 inputs to be multiplexed (I_0 through I_{15}) are evenly broken up between 74151 no. 1 (I_0 through I_7) and 74151 no. 2 (I_8 through I_{15}). Because the circuit has a total of 16 channels it requires four select lines, $S_3S_2S_1S_0$, to select 1 of 16 channels. The three select lines, $S_2S_1S_0$, select the basic eight channels of each IC, whereas the most significant select line, S_3, selects either IC no. 1 or IC no. 2 when at the 0 and 1 states, respectively, because this line is connected to the enable lines $\overline{E}$ of IC no. 1 and IC no. 2. Thus, when S_3 is a logic 0, IC no. 1 will be enabled and the eight combinations of $S_2S_1S_0$ will select the first eight inputs (I_0 through I_7). Also, when S_3 is a logic 0, the tristate buffer G1 is enabled (G2 is disabled) and the output, Z, of IC no. 1 is passed to *OUT*. When S_3 is a logic 1, IC no. 2 is enabled and the eight combinations of $S_2S_1S_0$ will select the next eight inputs (I_8 through I_{15}). Also, when S_3 is a logic 1,

Figure 4.69

Sixteen-channel multiplexer constructed from two 74151s.

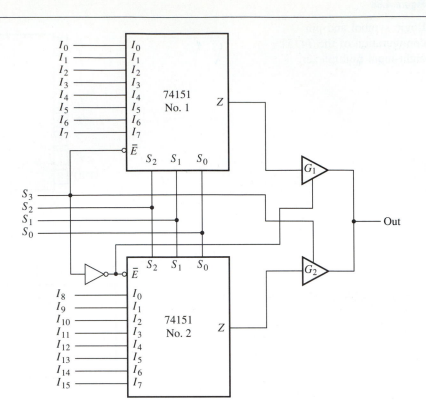

the tristate buffer G2 is enabled (G1 is disabled) and the output, Z, of IC no. 2 is passed to *OUT*.

The 74150 IC is a 16-input multiplexer that effectively provides the same function as the circuit in Figure 4.69. The logic symbol and pin configuration of the 74150 are shown in Figure 4.70. Like the 74151, the 74150 can be expanded to multiplex any number of input channels by using its enable line $\overline{E}$.

The 74153 is a dual 4-input multiplexer IC that contains two 4-channel multiplexers in one package. The logic symbol and pin configuration of the 74153 are shown in Figure 4.71. The two 4-channel multiplexers of the 74153 multiplex four input lines I_0 through I_3 to the output Z. The letters a and b identify the two multiplexers that make up the 74153. The two multiplexers function independently and are enabled separately by the active-low enable lines, $\overline{E}$. However, only one set of select lines S_1S_0 selects the input channels to be multiplexed. For example, when $S_1S_0 = 01$, I_{1a} and I_{1b} will be multiplexed to outputs Z_a and Z_b, respectively. The 74153 can therefore be used as a four-channel, 2-bit multiplexer as well.

The 74157 quad 2-input multiplexer is similar to the 74153 except that the 74157 contains four separate two-channel multiplexers. The logic symbol and pin configuration for the 74157 are shown in Figure 4.72. The four channels of the 74157 are identified by the letters a, b, c, and d. Like the 74153, a single select line is used to select one of two input channels of all four multiplexers together. For example, if $S = 1$ then $Z_a = I_{1a}$ and $Z_b = I_{1b}$ and $Z_c = I_{1c}$ and $Z_d = I_{1d}$. Thus, the 74157 can also be used as a two-channel, 4-bit multiplexer. Unlike the 74153, the 74157 only has a single active-low enable line that enables the entire IC.

The tristate buffers introduced in this section are available in the form of two

Figure 4.70

Logic symbol and pin
configuration of the 74150
16-input multiplexer.

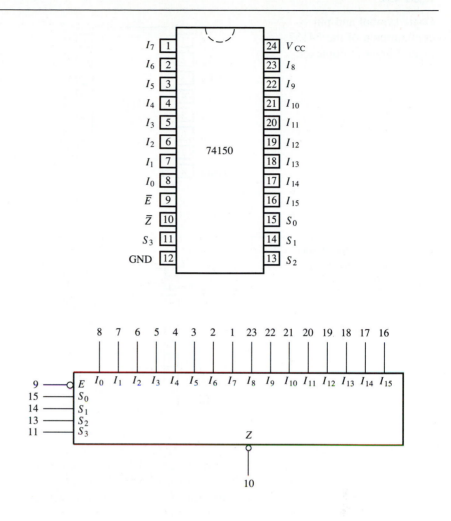

Figure 4.71

Logic symbol and pin configuration of the 75153 dual 4-input multiplexer.

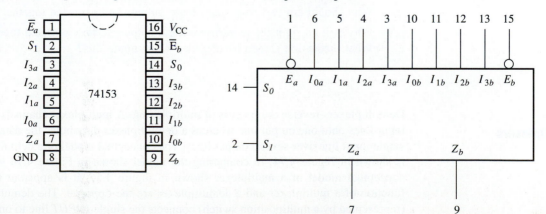

Figure 4.72

Logic symbol and pin configuration of the 74157 quad 2-input multiplexer.

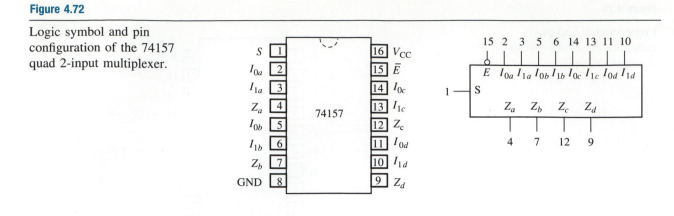

Figure 4.73

Pin configurations of the 74125 and 74126 quad buffers.

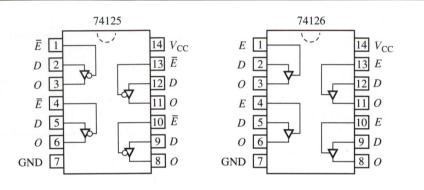

ICs — the 74125 and 74126. Their pin configurations are shown in Figure 4.73. The 74126 contains four tristate buffers with active-high enable lines and the 74125 contains four tristate buffers with active-low enable lines.

Review Questions

1. Describe the operation of a multiplexer.
2. State a possible application of a multiplexer.
3. Explain tristate logic and the high-impedance state.
4. Why should not two (two-state) logic outputs be connected together?
5. Why is it all right for two tristate logic outputs to be connected together? What precaution should be observed when doing this?

4.9

Demultiplexers

Demultiplexers reverse the process of multiplexing. A multiplexer channels several input lines onto one output line whereas a demultiplexer distributes the data from a single input line over several output lines. The conceptual model of a demultiplexer is shown in Figure 4.74. In comparing the model shown in Figure 4.74 with the conceptual model of a multiplexer shown in Figure 4.55, it is apparent that the function of a multiplexer and a demultiplexer are the opposite. The demultiplexer (represented by a multiposition switch) connects the single *INPUT* line to one of six

Figure 4.74

Conceptual demultiplexer circuit.

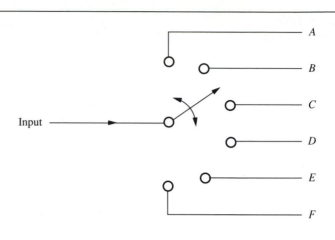

Figure 4.75

Logic symbol for a two-channel 1-bit demultiplexer.

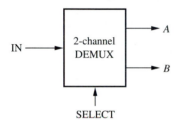

Table 4.22 Truth table for a two-channel demultiplexer

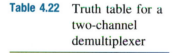

SELECT	IN	A	B
0	0	X	0
0	1	X	1
1	0	0	X
1	1	1	X

output lines, *A* through *F*, by changing the position of the switch. Of course, a logic circuit would not make use of a mechanical switch to select the output channel but would use logic gates to provide the demultiplexing function.

Figure 4.75 shows the logic symbol for a two-channel single-bit demultiplexer. Note that the demultiplexer shown in Figure 4.75 reverses the function of the multiplexer shown in Figure 4.56. The demultiplexer takes the logic state of the input *IN* and channels it to the output *A* if *SELECT* = 1, or channels it to the output *B* if *SELECT* = 0. The truth table for the demultiplexer is shown in Table 4.22. Note in Table 4.22 that when *SELECT* = 0 the state of *B* is equal to the state of *IN* and we don't care about the state of *A*. Similarly, when *SELECT* = 1, the state of *A* is equal to the state of *IN* and we don't care about the state of *B*. The simplified logic equations for the outputs *A* and *B* are obtained directly from Table 4.22 by assuming that the don't care conditions are zeros

$$A = SELECT \cdot IN \tag{4.45}$$

$$B = \overline{SELECT} \cdot IN \tag{4.46}$$

The logic circuit for the demultiplexer can be constructed by implementing Equations 4.45 and 4.46 as shown in Figure 4.76. Again, note that a logic 1 appearing at one of the inputs of each *AND* gate (connected to *SELECT* or $\overline{SELECT}$) in Figure

Figure 4.76

Logic circuit for a two-channel 1-bit demultiplexer.

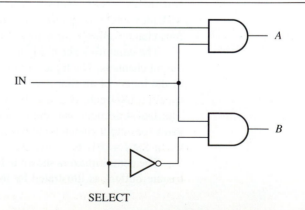

Figure 4.77

A four-channel 1-bit
demultiplexer. (a) Logic
symbol; (b) logic circuit.

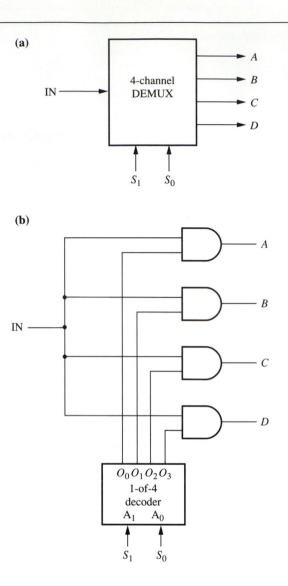

4.76 is used to pass the logic level at the second input (*IN*) to the output of the *AND* gate (identity law). Using this concept, a four-channel demultiplexer can be designed without going through the conventional approach. The logic symbol and circuit for a four-channel, single-bit demultiplexer are shown in Figure 4.77.

The demultiplexer in Figure 4.77 requires two select lines to select one of four output channels. The logic circuit for the demultiplexer works by applying *IN* to one of the inputs of all four *AND* gates. However, *IN* will be passed to the output of a specific *AND* gate if and only if the second input of the *AND* gate is a logic 1. The 1-of-4 decoder uses the 2-bit code at its input, $S_1 S_0$, to select one of four *AND* gates (or output channels) to pass the input *IN*. Thus, when $S_1 S_0 = 00$, $A = IN$ and when $S_1 S_0 = 01$, $B = IN$, etc.

The demultiplexers shown in Figure 4.76 and 4.77 can also be constructed using tristate buffers as illustrated by the following example.

Example 4.9

The two-channel demultiplexer whose logic symbol is shown in Figure 4.75 is implemented using tristate buffers as shown in Figure 4.78.

The circuit in Figure 4.78 uses two tristate buffers (G1 and G2) to connect the input line *IN* to either one of the outputs *A* or *B*. When *SELECT* = 1, G1 is enabled (G2 is disabled because it has an active-low enable line) and *A* = *IN*. When *SELECT* = 0, G2 is enabled while G1 is disabled and *B* = *IN*.

Figure 4.78

Logic circuit for a two-channel 1-bit demultiplexer using tristate logic.

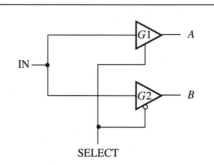

Figure 4.79 illustrates the use of tristate buffers to implement the four-channel demultiplexer whose logic symbol is shown in Figure 4.77a. Again, the tristate buffers are used to connect one of four outputs, *A* through *D*, to the input, *IN*. The select lines $S_1 S_0$ when decoded by the 1-of-4 decoder determine which one of the four tristate buffers are enabled.

Figure 4.79

Logic circuit for a four-channel 1-bit demultiplexer using tristate logic.

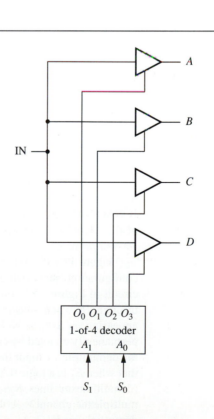

Just as several multiplexers could be connected together to increase the size of the channels being multiplexed, so could several demultiplexers to increase the size of the channels being demultiplexed. Figure 4.80a shows the logic symbol for a two-channel demultiplexer that demultiplexes a 2-bit input channel $I_1 I_0$ to one of two output channels $A_1 A_0$ and $B_1 B_0$. The logic circuit for the demultiplexer is shown in Figure 4.80b and is constructed by using two 2-channel 1-bit demultiplexers (Fig. 4.75), one for I_0 and another for I_1. Because the *SELECT* lines of both the demultiplexers are connected together, when *SELECT* = 0, $I_1 I_0$ will be connected to $B_1 B_0$ and when *SELECT* = 1, $I_1 I_0$ will be connected to $A_1 A_0$.

Figure 4.80

A two-channel 2-bit demultiplexer. (a) Logic symbol; (b) logic circuit constructed from two 2-channel 1-bit demultiplexers.

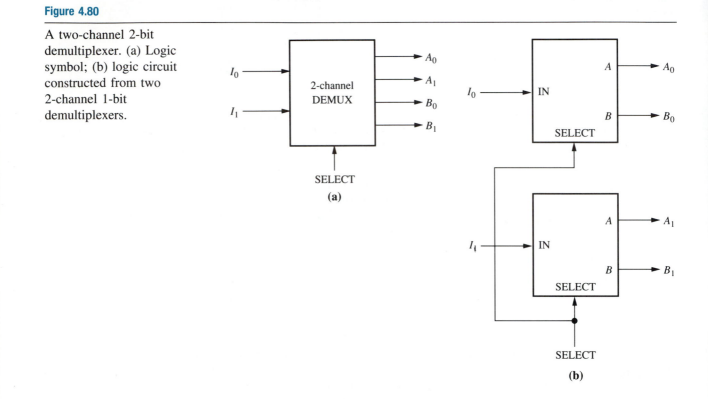

IC logic

The 74xxx series does not include any ICs that are dedicated to the demultiplexing function because many *decoders* can be configured for demultiplexing. Hence, many of the decoder ICs available in the 74xxx series are often referred to as *decoders/demultiplexers*.

Figure 4.81 illustrates the use of the 74138 1-of-8 decoder (logic symbol and pin configuration shown in Fig. 4.44) as an eight-channel 1-bit demultiplexer. The circuit in Figure 4.81 connects the input line *IN* to one of eight output channels, *A* through *H*, when selected by one of eight possible binary combinations (codes) appearing on $S_2 S_1 S_0$. Note that the two enable lines of the decoder, $\overline{E_2}$ and E_3, are permanently enabled by connection to ground and +5 volts, respectively. However, the demultiplexer input line *IN* is connected to the third active-low enable line $\overline{E_1}$, so that when $\overline{E_1}$ is a logic 0, the 74138 is enabled, and depending on the code at $A_2 A_1 A_0$ (demultiplexer lines $S_2 S_1 S_0$) one of the eight output lines $\overline{O_0}$ through $\overline{O_7}$ (demultiplexer channels *A* through *H*) is activated (logic 0). When the input of the demultiplexer *IN* $(\overline{E_1})$ is a logic 1, the 74138 is disabled and all its outputs $\overline{O_0}$ through

Figure 4.81

Figure 4.81

74138 configured as an eight-channel 1-bit demultiplexer.

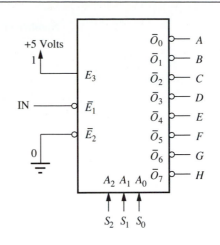

$\bar{O}_7$ (demultiplexer lines *A* through *H*) are inactive at the logic 1 state. However, because a logic 1 appears at all the output channels, it appears as if the input *IN* has been directed to the appropriate output channel, because *IN* is at the logic 1 state.

Review Questions

1. Describe the operation of a demultiplexer.
2. State a possible application of a demultiplexer.
3. Why are decoders also known as demultiplexers?

4.10
Parity circuits

In Chapter 1, the use of binary codes and their applications in digital communication systems were introduced. Recall that many digital communication systems transmit binary codes with a parity bit for error checking. The type of parity used can be even or odd. In odd parity systems all binary codes are transferred with an odd number of 1-bits. Therefore, if the number of 1-bits in the code is an odd number, the parity bit is set to a logic 0 to keep the total number of 1-bits transferred an odd number. If the number of 1-bits in the code is an even number, the parity bit is set to a logic 1 to make the total number of 1-bits transferred an odd number. In even parity systems all binary codes are transferred with an even number of 1-bits. If the number of 1-bits in the code is an even number, the parity bit is set to a logic 0 to keep the total number of 1-bits transferred an even number. If the number of 1-bits in the code is an odd number, the parity bit is set to a logic 1 to make the total number of 1-bits transferred an even number. This section deals with an application circuit that is used to generate the parity bit for a binary code and also to determine the parity of a binary code.

Figure 4.82

Logic symbol for a 4-bit parity generator.

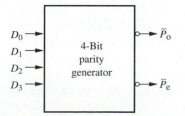

Figure 4.82 shows the logic symbol for a 4-bit *parity generator* circuit. The circuit accepts as input a 4-bit number $D_3D_2D_1D_0$ and activates one of two active-low outputs $\bar{P}_o$ or $\bar{P}_e$, depending on the parity of the code applied to the inputs. The truth table for the parity generator circuit of Figure 4.82 is shown in Table 4.23.

The reason for active-low outputs, $\bar{P}_e$ and $\bar{P}_o$, is that they represent the parity bit for even and odd parity generation, respectively, and must be set to a logic 0 if the

Table 4.23   Truth table for a 4-bit parity generator

Number of 1-bits	D_3	D_2	D_1	D_0	$\overline{P}_e$	$\overline{P}_o$
0	0	0	0	0	0	1
1	0	0	0	1	1	0
1	0	0	1	0	1	0
2	0	0	1	1	0	1
1	0	1	0	0	1	0
2	0	1	0	1	0	1
2	0	1	1	0	0	1
3	0	1	1	1	1	0
1	1	0	0	0	1	0
2	1	0	0	1	0	1
2	1	0	1	0	0	1
3	1	0	1	1	1	0
2	1	1	0	0	0	1
3	1	1	0	1	1	0
3	1	1	1	0	1	0
4	1	1	1	1	0	1

total number of bits (data bits and parity bit) is even or odd, respectively. For example, the number 0111 with even parity is

$$\begin{array}{cc} 1 & 0\,1\,1\,1 \\ \overline{P}_e \end{array} .$$

The number 1100 with even parity is

$$\begin{array}{cc} 0 & 1\,1\,0\,0 \\ \overline{P}_e \end{array} .$$

The number 0111 with odd parity is

$$\begin{array}{cc} 0 & 0\,1\,1\,1 \\ \overline{P}_o \end{array} .$$

The number 1100 with odd parity is

$$\begin{array}{cc} 1 & 1\,1\,0\,0 \\ \overline{P}_o \end{array} .$$

Figure 4.83 shows the K-map obtained from Table 4.23 for the output $\overline{P}_e$. Because no cells are adjacent no simplification is possible. The logic equation for $\overline{P}_e$ from the K-map is as follows:

$$\overline{P}_e = \overline{D}_3\overline{D}_2\overline{D}_1 D_0 + \overline{D}_3\overline{D}_2 D_1\overline{D}_0 + \overline{D}_3 D_2\overline{D}_1\overline{D}_0 + \overline{D}_3 D_2 D_1 D_0$$
$$+ D_3 D_2\overline{D}_1 D_0 + D_3 D_2 D_1\overline{D}_0 + D_3\overline{D}_2\overline{D}_1\overline{D}_0 + D_3\overline{D}_2 D_1 D_0 \quad (4.47)$$

Using the distributive law on Equation 4.47 we have

$$\overline{P}_e = \overline{D}_3\overline{D}_2(\overline{D}_1 D_0 + D_1\overline{D}_0) + \overline{D}_3 D_2(\overline{D}_1\overline{D}_0 + D_1 D_0)$$
$$+ D_3 D_2(\overline{D}_1 D_0 + D_1\overline{D}_0) + D_3\overline{D}_2(\overline{D}_1\overline{D}_0 + D_1 D_0) \quad (4.48)$$

Figure 4.83

K-map for the design of the 4-bit parity generator.

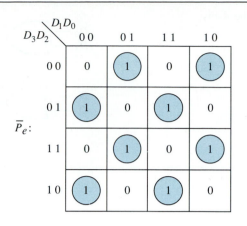

In Equation 4.48 we see that

$$\overline{D_1}D_0 + D_1\overline{D_0} = D_1 \oplus D_0 \qquad \text{(Exclusive-}OR\text{ function)}$$

and

$$\overline{D_1}\,\overline{D_0} + D_1 D_0 = \overline{D_1 \oplus D_0} \qquad \text{(this proof is left as an exercise)}$$

Therefore, Equation 4.48 now becomes

$$\overline{P}_{\rm e} = \overline{D_3}\,\overline{D_2}(D_1 \oplus D_0) + \overline{D_3}D_2(\overline{D_1 \oplus D_0}) + D_3 D_2(D_1 \oplus D_0) + D_3\overline{D_2}(\overline{D_1 \oplus D_0})$$

Using the distributive law again

$$\overline{P}_{\rm e} = (D_1 \oplus D_0)(\overline{D_3}\,\overline{D_2} + D_3 D_2) + (\overline{D_1 \oplus D_0})(\overline{D_3}D_2 + D_3\overline{D_2})$$

Now applying the Exclusive-*OR* function

$$\overline{P}_{\rm e} = (D_1 \oplus D_0)(\overline{D_3 \oplus D_2}) + (\overline{D_1 \oplus D_0})(D_3 \oplus D_2) \qquad (4.49)$$

Note that Equation 4.49 is in the *AND-OR* form that is suitable for conversion to Exclusive-*OR* form

$$\overline{P}_{\rm e} = (D_1 \oplus D_0) \oplus (D_3 \oplus D_2) \qquad (4.50)$$

Because $\overline{P}_{\rm o}$ is the complement of $\overline{P}_{\rm e}$, the logic equation for $\overline{P}_{\rm o}$ is

$$\overline{P}_{\rm o} = \overline{\overline{P}_{\rm e}} = P_{\rm e} \qquad (4.51)$$

The logic circuit for the 4-bit parity generator shown in Figure 4.84 is implemented from Equations 4.50 and 4.51.

In Figure 4.84 the circuit effectively adds the four input bits D_0 through D_3 because each Exclusive-*OR* gate is capable of adding 2 bits (without a carry). It can also be observed (from Table 4.23) that the sum of all the input bits will be a logic 0 if there are an even number of 1-bits and a logic 1 if there are an odd number of 1-bits. This information can be used to design a circuit that can be used to check the parity of a 4-bit code.

A *parity checking* circuit must be able to accept as input the data bits (code) as well as the parity bit, and determine if the parity of all the input bits is even or odd. Figure 4.85 shows the logic symbol for a 4-bit parity generator / checker. The circuit has a 4-bit input $D_3 D_2 D_1 D_0$ for the code, and a set of inputs *OI* and *EI* for the parity

Figure 4.84

Logic circuit for the 4-bit parity generator.

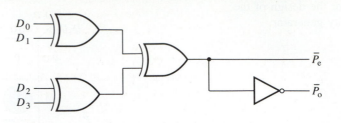

Figure 4.85

Logic symbol for a 4-bit parity generator / checker.

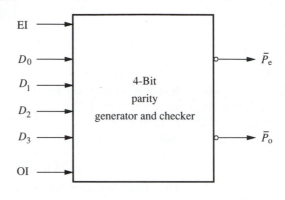

bit of the 4-bit code. If even parity checking is being performed by the circuit, the parity bit of the code is applied to *EI* (EVEN INPUT) but if odd parity checking is being performed by the circuit, the parity bit of the code is applied to *OI* (ODD INPUT). The active-low outputs $\overline{P}_e$ and $\overline{P}_o$ as before indicate whether the input code has even or odd parity. Note that the circuit can operate as a parity generator as well as a checker; the manner in which this is done will be seen shortly.

Figure 4.86 shows the logic circuit for the 4-bit parity generator / checker. The portion of the circuit that is made up of gates G1, G2, and G3 is similar to the circuit shown in Figure 4.84 and simply sums (without carries) the input code (bits D_0 through D_3). Thus, the output of this subcircuit S will be a logic 0 if the parity of $D_3D_2D_1D_0$ is even and a logic 1 if the parity of $D_3D_2D_1D_0$ is odd.

For even parity checking the output of the gate G4 will be a logic 0 (indicating even parity) if S is a logic 0 and the parity bit (*EI*) is a logic 0. If, however, S is a logic 1, indicating that $D_3D_2D_1D_0$ has odd parity but the parity bit (*EI*) is a logic 1 to force

Figure 4.86

Logic circuit for a 4-bit parity generator / checker.

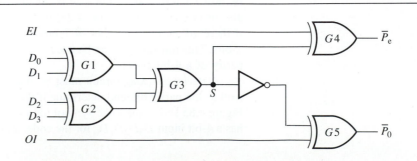

it even, then the output of gate G4 will be a logic 0. All other conditions at the input of G4 will produce a logic 1 out, indicating a parity error.

For odd parity checking the output of gate G5 will be a logic 0 (indicating odd parity) if S is a logic 1 and the parity bit (OI) is a logic 0. If, however, S is a logic 0 indicating that $D_3D_2D_1D_0$ has even parity but the parity bit (OI) is a logic 1 to force it odd, then the output of gate G5 will be a logic 0. All other conditions at the input of G5 will produce a logic 1 out, indicating a parity error.

Thus, for even parity checking, a logic 0 at $\overline{P}_e$ indicates that there is no parity error whereas a logic 1 at $\overline{P}_e$ identifies a parity error. Similarly, for odd parity checking, a logic 0 at $\overline{P}_o$ indicates that there is no parity error. A logic 1 at $\overline{P}_o$ identifies a parity error.

The 4-bit parity generator/checker circuit shown in Figure 4.86 can be used to generate the parity bit for the 4-bit input code $D_3D_2D_1D_0$ by tying EI to a logic 0 (even parity generation) or by tying OI to a logic 0 (odd parity generation). In Figure 4.86 if EI is a logic 0 then $\overline{P}_e$ will be a logic 0 when S is a logic 0, and $\overline{P}_e$ will be a logic 1 when S is a logic 1. However, S will be a logic 0 (and therefore $\overline{P}_e$ will be a logic 0) if the code $D_3D_2D_1D_0$ has even parity. Similarly, if OI is a logic 0 then $\overline{P}_o$ will be a logic 0 when S is a logic 1, and $\overline{P}_o$ will be a logic 1 when S is a logic 0. However, S will be a logic 1 (and therefore $\overline{P}_o$ will be a logic 0) if the code $D_3D_2D_1D_0$ has odd parity. Thus, $\overline{P}_e$ will be a logic 0 when $D_3D_2D_1D_0$ has even parity, and $\overline{P}_o$ will be a logic 0 when $D_3D_2D_1D_0$ has odd parity. For parity generation, the circuit operates like the circuit in Figure 4.84.

IC logic

The 74180 8-bit parity generator/checker IC is similar in concept to the circuit shown in Figure 4.86 except that it processes an 8-bit number. The logic symbol and pin configuration of the 74180 are shown in Figure 4.87. The operation of the 74180 is slightly different from the circuit in Figure 4.86 as illustrated by its truth table shown in Table 4.24.

For even parity checking, the 8-bit code $I_7I_6I_5I_4I_3I_2I_1I_0$ is applied to the IC and the parity bit is applied to EI. If the parity of all 9 bits is even then the Σ_E output will be a logic 0. For odd parity checking, the parity bit is applied to OI, and if the parity of all 9 bits is odd, then Σ_E will be a logic 0. Thus, the output Σ_E is used to identify the parity for both even and odd parity checking configurations, unlike the circuit in Figure 4.86 that uses two outputs.

Parity generation in the 74180 is also slightly different from the parity generation produced by the circuit in Figure 4.86. For even parity generation, the EI line is tied low and the OI line is tied high. An 8-bit code at $I_7I_6I_5I_4I_3I_2I_1I_0$ with even parity will produce a 0 parity bit at Σ_E and a 1 parity bit at Σ_E if the code has odd parity. For odd parity generation, the EI line is tied high and the OI line is tied low. An 8-bit

Figure 4.87

Logic symbol and pin configuration of the 74180 8-bit parity generator/checker.

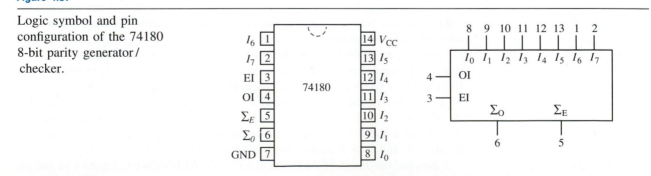

Table 4.24   Truth table for the 74180 parity generator/checker

Inputs			Outputs	
Sum of 1's at I_0 through I_7	EI	OI	Σ_E	Σ_O
EVEN	1	0	1	0
ODD	1	0	0	1
EVEN	0	1	0	1
ODD	0	1	1	0
X	1	1	0	0
X	0	0	1	1

code at $I_7I_6I_5I_4I_3I_2I_1I_0$ with odd parity will produce a 0 parity bit at Σ_E and a 1 parity bit at Σ_E if the code has even parity. Thus, for even or odd parity generation the output Σ_E supplies the required parity bit. Because Σ_O is the complement of Σ_E it can be used if the complement of the parity bit is desired.

Review Questions

1. How can one determine the parity of a binary number?
2. Describe the operation of a parity generator circuit.
3. Describe the operation of a parity checking circuit.

4.11
Arithmetic logic units

This section incorporates several of the application circuits designed in this chapter into a logic circuit known as an *arithmetic logic unit* (ALU). Arithmetic logic units provide an important function in the operation of a digital computer and are designed to provide the computer with the basic arithmetic and logical operations that are applied to binary numbers. Figure 4.88 shows the logic symbol for a typical ALU that operates on two 4-bit numbers. The truth table for the ALU is shown in Table 4.25.

The ALU shown in Figure 4.88 accepts two 4-bit numbers A ($A_3A_2A_1A_0$) and B ($B_3B_2B_1B_0$) and produces a 4-bit result F ($F_3F_2F_1F_0$) as output. The *MODE* input, M, to the ALU selects it for either an arithmetic (0) or a logic (1) operation. The *SELECT* inputs S_1S_0 select one of four arithmetic or logic operations depending on the state of M. Note in the truth table shown in Table 4.25 that the four arithmetic operations produce as a result the sum of A and B, the difference between A and B, the negated value (in 2's complement representation) of B, and the incremented value of A. The outputs C_o and C_s shown in Figure 4.88 provide the carry-outs from addition and subtraction, respectively. The four logic operations are performed on the numbers A and B on a bit-by-bit basis and include the logical sum and product of A and B, the complement of A, and A exclusively ORed with B. For example, if A = 0101 and B = 0110, then the logic product ($A \cdot B$) of A and B would be 0100 if each bit that makes up the numbers A and B is ANDed on a bit-by-bit basis — $A_0 \cdot B_0$, $A_1 \cdot B_1$, $A_2 \cdot B_2$, $A_3 \cdot B_3$. The ALU also incorporates a comparator that compares the data inputs A and B and produces the results of the comparison at the outputs, A = B, $A < B$, and $A > B$ as discussed in Section 4.5.

Because the design of the logic circuit for the ALU shown in Figure 4.88 can get

Figure 4.88

Logic symbol for a 4-bit
ALU.

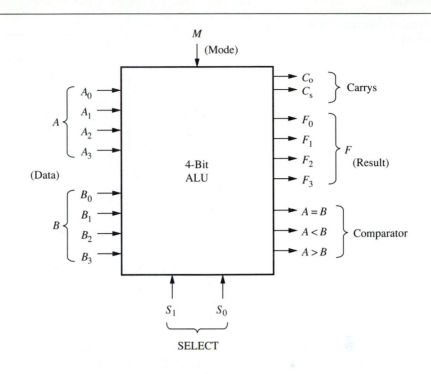

fairly complex, another approach to obtain the circuit will be taken. First, the internal architecture, or layout, of the ALU in terms of various function blocks will be examined. The operation and purpose of each function block will then be investigated. Finally, these function blocks will be implemented using various logic circuits, some of which have already been developed in this chapter.

Figure 4.89 shows the internal architecture of the ALU whose logic symbol is shown in Figure 4.88. The ALU can be viewed as having two main processing units — the arithmetic unit that is responsible for providing the four basic arithmetic functions listed in Table 4.25, and the logic unit that is responsible for providing the four basic logic functions also listed in Table 4.25. The data inputs for the two 4-bit numbers, A and B, are represented in Figure 4.89 as busses for simplicity. Recall that a bus is a collection of lines grouped together to represent a binary number. Thus, the double arrows are busses that represent the data lines, A_0–A_3, B_0–B_3, and F_0–F_3; these busses will be referred to as A, B, and F, respectively.

Table 4.25 Truth table for a 4-bit ALU

Function select		Function	
S_1	S_0	$M = 1$ (logic)	$M = 0$ (arithmetic)
0	0	$A \cdot B$	A plus B
0	1	$A + B$	A minus B
1	0	$A \oplus B$	Minus B (2's complement)
1	1	$\overline{A}$	A plus 1

Figure 4.89

Internal architecture of a
4-bit ALU.

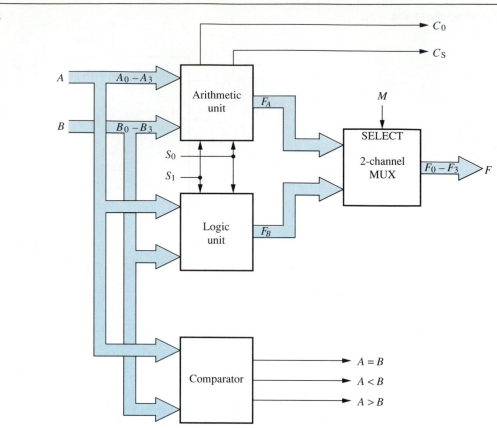

The arithmetic unit accepts as input the numbers A and B and produces the number F_A as output (result). The actual operation performed depends on the combination present at the select inputs $S_1 S_0$ (*see* Table 4.25). The outputs C_o and C_s are the carry outputs from addition and subtraction, respectively.

The logic unit accepts as input the numbers A and B and produces the number F_L as output (result). Again, the actual operation performed depends on the combination present at the select inputs $S_1 S_0$ (*see* Table 4.25).

The internal operation of the arithmetic and logic function-blocks shown in Figure 4.89 will be discussed later.

The 4-bit outputs of the arithmetic unit and logic unit are connected to a two-channel 4-bit multiplexer (discussed in Section 4.8) that will multiplex F_A to the output (F) when M (*MODE*) is at the logic 0 state and F_L to the output (F) when M (*MODE*) is at the logic 1 state. Note in Table 4.25 that the *MODE* (M) input selects between logic and arithmetic functions. The data inputs A and B are also fed into a 4-bit magnitude comparator (discussed in Section 4.5) that produces three outputs $A = B$, $A > B$, and $A < B$ to indicate the results of the comparison.

The internal architecture of the arithmetic unit function-block is shown in Figure 4.90. Each one of the four arithmetic functions listed in Table 4.25 is implemented by a separate circuit. The adder and subtracter function like the adders and sub-tracters discussed in Sections 4.3 and 4.4, respectively. The adder and subtracter

Figure 4.90

Architecture of the
arithmetic unit.

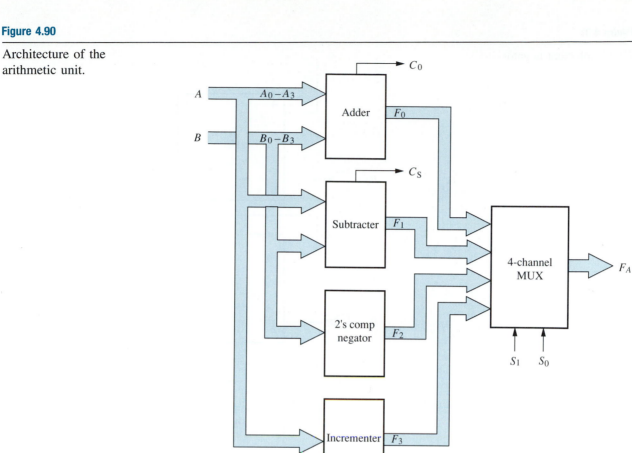

circuits accept as input the 4-bit numbers A and B and produce their results at $F0$ and $F1$, respectively, with carries C_o and C_s, respectively.

The 2's complement negator circuit obtains the 2's complement of the 4-bit number B by first complementing each bit that makes up the number B and then adding 1 to the resulting number using a series of half-adders as shown in Figure 4.91. The output of the circuit, $F2$, is the 2's complement of the number B.

The incrementer circuit adds 1 to the 4-bit number A (increments A) by using a parallel adder circuit similar to the one shown in Figure 4.91. The incrementer circuit is shown in Figure 4.92. Note in Figure 4.92 that the series of half-adders add 1 to the LSB of A and provide for the carries to ripple through the other bits for addition if necessary. The incremented value of A is made available at the output of the circuit, $F3$.

The outputs of the four circuits that make up the arithmetic unit represent the 4-bit results of each function listed in Table 4.25. Because the output of the ALU, F_A, should only supply the arithmetic operation specified by the select lines $S_1 S_0$, a four-channel 4-bit multiplexer is used to multiplex F_0 through F_3 onto the output lines F_A of the ALU.

The internal architecture of the logic unit function-block is shown in Figure 4.93. Like the arithmetic unit, each one of the four logic functions listed in Table

Figure 4.91

2's complement negator
circuit.

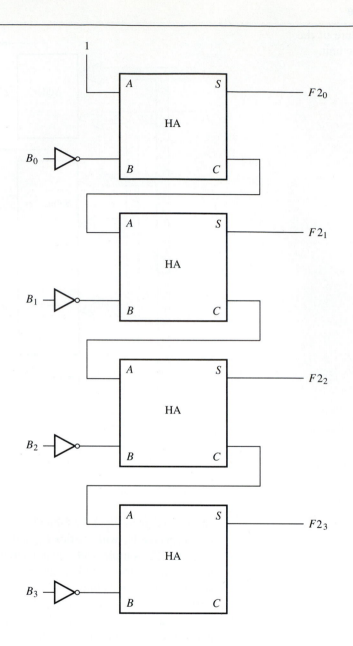

4.25 is implemented by a separate circuit. The 4-bit data inputs to the ALU, A and B,
are applied to the inputs of each circuit that makes up the logic unit. Each circuit is
responsible for performing a particular logical operation on either A or B or both and
providing the result at one of the outputs F_4 through F_7.

It was stated earlier that logical operations are performed on the 4-bit numbers A
and B on a bit-by-bit basis. The circuits shown in Figure 4.94 show the four circuits
that provide the logic unit with its four basic functions and illustrate the manner in
which the two numbers are operated on in a bit-by-bit basis. The circuit in Figure
4.94a *ANDs* the numbers A and B and produces the result at the outputs F_4. Similarly,
the circuit shown in Figure 4.94b *ORs* the numbers A and B and produces the result at

Figure 4.92

Incrementer circuit.

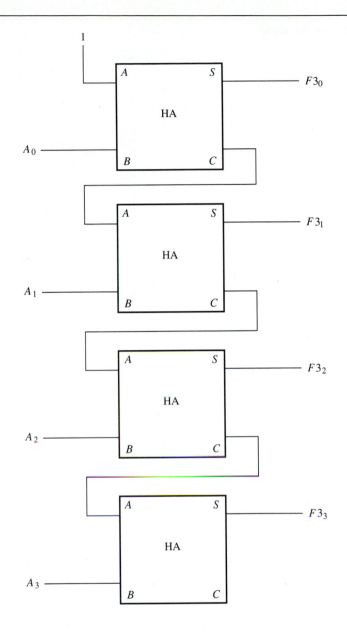

the output F_5, whereas the circuit shown in Figure 4.94c Exclusive-*OR*s the numbers A and B and produces the result at the output F_6. The circuit in Figure 4.94d obtains the 1's complement of the number A at the output F_7 by simply inverting each bit in the number.

Note in Figure 4.93 that the 4-bit outputs of each one of the four circuits shown in Figure 4.94 are connected to a four-channel 4-bit multiplexer that channels the appropriate result to the output of the logic unit F_L, depending on the combination at the select inputs $S_1 S_0$.

The design of the ALU described in this section is a conceptual (but functional) design. An actual ALU could be more efficiently designed by eliminating many of

Figure 4.93

Architecture of the logic unit.

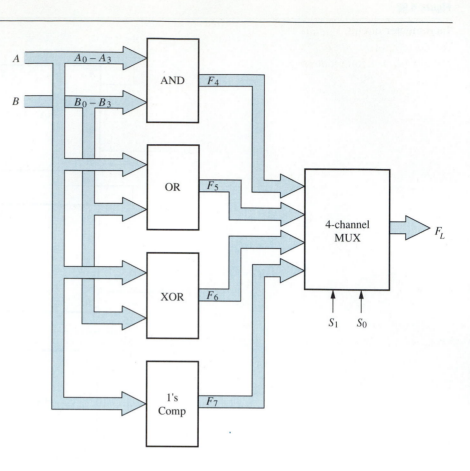

the redundancies inherent in this conceptual model. However, studying the operation of a complex circuit such as an ALU is more easily done by viewing it as a collection of function blocks rather than as a single entity.

IC logic

The 74181 4-bit ALU is an IC that is capable of performing 16 logic operations on two 4-bit input numbers plus a variety of arithmetic operations. The logic symbol and pin configuration of the 74181 are shown in Figure 4.95 and its truth table is shown in Table 4.26.

There are a lot of similarities and differences between the 74181 ALU and the ALU designed in this section (Fig. 4.88). Both ALUs process two 4-bit data inputs A $(A_3A_2A_1A_0)$ and B $(B_3B_2B_1B_0)$ and provide the results at the output F $(F_3F_2F_1F_0)$. The *MODE* select input M selects either the arithmetic unit or the logic unit in both ALUs. However, notice that the 74181 ALU has four function select lines $S_3S_2S_1S_0$ that allow it to (theoretically) select 16 arithmetic and 16 logic functions as listed in Table 4.26. The 74181 also incorporates a comparator that compares the numbers A and B but provides only one output, $A = B$, to indicate whether the two numbers are equal. The 74181's carry outputs P and G are for addition and subtraction operations performed by its arithmetic unit. The carry input, C_n, and carry output, C_{n+4}, can be used to expand the capabilities of the 74181 besides serving other functions. The 74181 ALU also has the capability to process active-low data at the inputs A and B and produce active-low results at the output F; in this mode, however, the functions

Figure 4.94

Logic unit (a) *AND* circuit;
(b) *OR* circuit; (c) *XOR*
circuit; (d) 1's complement
circuit.

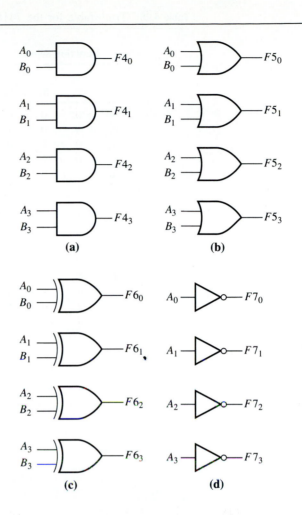

Figure 4.95

Logic symbol and pin
configuration of the 74181
4-bit arithmetic logic unit.

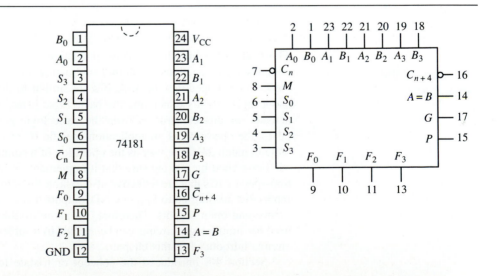

Table 4.26   Truth table for the 74181 4-bit ALU

Function select				Function	
				Logic	Arithmetic
S_3	S_2	S_1	S_0	$M = 1$	$M = 0$
0	0	0	0	$\overline{A}$	A
0	0	0	1	$\overline{A + B}$	$A + B$
0	0	1	0	$\overline{A}B$	$A + \overline{B}$
0	0	1	1	0	Minus 1
0	1	0	0	$\overline{AB}$	$A + A\overline{B}$
0	1	0	1	$\overline{B}$	$(A + B)$ plus $A\overline{B}$
0	1	1	0	$A \oplus B$	A minus B minus 1
0	1	1	1	$A\overline{B}$	AB minus 1
1	0	0	0	$\overline{A} + B$	A plus AB
1	0	0	1	$\overline{A \oplus B}$	A plus B
1	0	1	0	B	$(A + \overline{B})$ plus AB
1	0	1	1	AB	AB minus 1
1	1	0	0	1	A plus A
1	1	0	1	$A + \overline{B}$	$(A + B)$ plus A
1	1	1	0	$A + B$	$(A + \overline{B})$ plus A
1	1	1	1	A	A minus 1

assigned to each one of the codes appearing at S are in a different order and are not as shown in Table 4.26.

Review Questions

1. Describe the operation of an ALU.
2. Which ALU line selects between an arithmetic or logic operation?
3. How are the actual arithmetic or logic functions selected?
4. What purpose does the multiplexer shown in Figure 4.89 serve?
5. What purpose do the multiplexers shown in Figures 4.90 and 4.93 serve?

4.12
Troubleshooting techniques

This chapter has introduced many application-specific ICs that implement the more complex logic functions of adding, subtracting, encoding, decoding, etc. The troubleshooting procedures for these ICs are similar to the procedures used to troubleshoot logic circuits containing the basic logic gates. This is because these ICs are actually circuits constructed from the basic logic gates. For troubleshooting purposes, the operation of an application-specific IC can be verified by using its truth table in much the same way as the operation of a simple logic gate or a logic circuit can be verified by making sure that it implements its truth table correctly. Application-specific ICs that are defective also exhibit the same symptoms as logic gates — inputs that are shorted to V_{CC} or GND, outputs that are shorted to V_{CC} or GND, open inputs, and open outputs. Therefore, the same troubleshooting procedures that are used for simple logic circuits can be applied to troubleshoot the more complex logic circuits introduced in this chapter.

Section 4.8 introduced the concept of tristate logic and the logic gate with tristate outputs. When a logic circuit that contains gates with tristate outputs is troubleshooted, remember that a gate whose output is in a high-impedance state will exhibit the symptoms of an open output. For example, in Figure 4.96a, the 74126 tristate buffer gate is enabled by connecting the enable input to a logic high (H). The logic probe connected to the output indicates a logic high state because the input of the gate is connected to a logic high. If the input is connected low (L) as shown in Figure 4.96b, then the logic probe indicates a low state at the output. However, if the enable input of the 74126 gate is tied low (to disable the gate) as shown in Figure 4.96c and d, then the output of the gate is in a high-impedance state (open circuited) and therefore the logic probe indicates no logic level. Note that if a gate with a two-stated output produced this condition the gate would be considered defective due to an open output.

Figure 4.96

Identifying the outputs of a tristate logic gate.

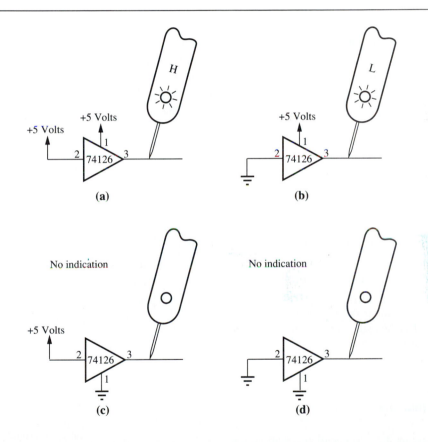

Review Questions

1. Why do the inputs and outputs of application-specific ICs exhibit the same faults as the simple logic gates?

2. Why is it possible to apply the troubleshooting procedures for simple logic circuits to complex circuits that contain application-specific ICs?

3. How does the logic probe indicate that the output of a logic gate is in a high-impedance state?

Summary

- The logic symbol for a circuit represents the circuit as a block and identifies its inputs and outputs.

- Adders are logic circuits that are used to add binary numbers.

- A half-adder adds 2 bits and produces a single-bit sum and carry.

- A full-adder adds 2 bits plus a carry input and produces a single-bit sum and carry-out.

- An Exclusive-*OR* gate can have only two inputs.

- The Exclusive-*OR* gate produces a logic 1 at its output if both inputs are not equal.

- The output of an Exclusive-*OR* gate is the sum (ignoring carries) of its two inputs.

- A parallel binary adder adds multibit binary numbers.

- Subtracter circuits use parallel binary adders and can be of two types — 1's complement and 2's complement.

- A 1's complement subtracter subtracts two numbers by using the 1's complement binary subtraction technique and requires two addition steps.

- A 2's complement subtracter subtracts two numbers by using the 2's complement binary subtraction technique and requires one addition step.

- A comparator is a logic circuit that compares the magnitude of two binary numbers.

- Encoders are logic circuits that produce a unique binary code when one of their input lines is activated.

- A priority encoder allows more than one of its inputs to be activated by assigning priorities to each input.

- Most priority encoders use high-order data line encoding that causes the encoder to output the code of the line that has the highest code if more than one input line is activated at a time.

- A decoder detects a unique binary code by activating one of several output lines to identify the code.

- Most decoders are known as 1-of-n decoders because they will only activate one of several (n) output lines in response to an input code.

- Seven-segment decoders convert a binary code into a code suitable for activating the segments of a seven-segment LED.

- A ripple blanking circuit is used to blank out leading zeros in a seven-segment LED display system.

- Multiplexers channel binary data from one of several input channels onto a single output channel.

- Tristate logic gates have outputs that can exist in one of three states — 1, 0, or Z (high impedance).

- A demultiplexer reverses the process of multiplexing.

- Decoders can be configured to provide demultiplexing and are also known as decoders / demultiplexers.

- Parity generators are designed to produce a parity bit for an input binary number.

- The parity of a binary number can be obtained by adding all the bits in the number (ignoring carries) and examining the single-bit sum for a 0 (even number of 1-bits) or a 1 (odd number of 1-bits).

- Parity checkers determine if a number has even or odd parity.

- The parity generating and checking functions can usually be combined into one circuit called a parity generator / checker.

- Arithmetic logic units are logic circuits that provide the basic arithmetic and logic functions for operation on two binary numbers.

Problems

Section 4.2   Application circuit design

1. Design a logic circuit that will accept all possible combinations of 3-bit binary numbers as input, pass through only the even combinations, and filter out all the odd combinations. For example, if any of the even numbers 000, 010, 100, or 110 are applied to its input, the output will be the same as the input, but if any of the odd numbers 001, 011, 101, or 111 are applied to its input, the output will be 111.

2. The pH scale is often used to determine the acidity or alkalinity of water and is given by a number in the range 0 through 14. A pH of 7 is neutral while a pH less than 7 is acidic and a pH greater than 7 is alkaline. A logic circuit is to be designed to monitor and control the pH level of a swimming pool. The pH level of the water is applied to the input of the circuit as a binary number and the circuit is to maintain the pH level between 6 and 7 by activating a valve to release acid if the pH is greater than 7 or by activating another valve to release base if the pH is less than 6. The circuit should also have four LED indicators that are activated under the following conditions:

a. If the pH is between 6 and 7, a green LED should be illuminated to indicate the optimum setting.
b. If the pH is greater than 7, a blue LED should be illuminated to indicate alkalinity.
c. If the pH is less than 7, a yellow LED should be illuminated to indicate acidity.
d. If the pH is greater than 9 or if the pH is less than 5, a red LED should be illuminated as a warning.

Section 4.3   Adders

3. Prove that the following equation is true:

$$\overline{X \oplus Y} = X \cdot Y + \overline{X} \cdot \overline{Y}$$

4. Design a 2-bit parallel binary adder that will add two 2-bit numbers, A_1A_0 and B_1B_0, and produce a sum, S_1S_0, and a carry-out C_o. Use the conventional approach. Compare this circuit with the circuit shown in Figure 4.16 in terms of the number of gates required.

5. Design a parallel binary adder that will add two 16-bit numbers. Use only 4-bit adders in your design.

6. Construct a schematic diagram of a 4-bit adder using 7482 ICs.

7. Construct a schematic diagram of an 8-bit adder using 7483 ICs.

Section 4.4   Subtracters

8. Design a 2-bit subtracter that will subtract the 2-bit number A_1A_0 from B_1B_0 and produce the difference D_1D_0 and a sign S. Use the conventional approach.

9. Design an 8-bit 2's complement subtracter using a series of parallel binary adders.

10. Design an 8-bit 1's complement subtracter using a series of parallel binary adders.

Section 4.5   Comparators

11. Design a comparator that will compare all numbers from 0 to 7 with the number 5. If the number is less than 5, then the output of the circuit will be a logic 1, otherwise the output of the circuit will be a logic 0. Use the conventional approach.

12. Design a 4-bit magnitude comparator by using two 2-bit magnitude comparators whose logic symbol is shown in Figure 4.26.

Section 4.6   Encoders and decoders

13. The encoder circuit shown in Figure 4.33 was designed with no priorities. However, when more than one line is activated, the output will be at some logic state that will represent some code. Therefore, there will be some priorities built into the circuit by default. Analyze the circuit and determine the priorities of the input lines; that is, obtain the complete truth table for the circuit through analysis.

14. Design a BCD encoder (no priorities) that will encode ten (active-high) input lines into the BCD code.

15. Design a BCD encoder with low-order data line encoding that will encode ten input lines into the BCD code. The inputs and outputs of the encoder should be active high.

16. Design a 1-of-8 decoder with active-low outputs.

17. Design a 1-of-8 decoder with active low outputs and an active low enable line. If the enable line is a logic 1, the outputs of the decoder will be high.

18. Construct a 1-of-16 decoder using two 1-of-8 decoders. Assume that the 1-of-8 decoders have an active low enable line.

19. Determine the priorities of all 16 input lines of the encoder circuit shown in Figure 4.43.

Section 4.7   Seven-segment decoders

20. Design a BCD to a seven-segment decoder to drive a common anode seven-segment LED.

21. Incorporate a $\overline{RBI}$ and a $\overline{RBO}$ into the circuit designed in Problem 20.

22. The *Gray code* is a 4-bit code that has its combinations organized such that only 1 bit changes from one code to the next in sequence

```
0 0 0 0
0 0 0 1
0 0 1 1
0 0 1 0
0 1 1 0
0 1 1 1
0 1 0 1
0 1 0 0
1 1 0 0
1 1 0 1
1 1 1 1
1 1 1 0
1 0 1 0
1 0 1 1
1 0 0 1
1 0 0 0
```

Design a code conversion circuit that will convert the Gray code to the 4-bit binary code.

23. Design a code conversion circuit that will convert the 4-bit binary code to the Gray code described in Problem 22.

24. The excess-3 code is obtained by adding 3 (0011) to each 4-bit BCD code as shown in the following table. When the 1's complement of the excess-3 code of a decimal digit is obtained, the excess-3 code of the 9's complement of the decimal digit is obtained. For example, the excess-3 code for 5 is 1000; the 1's complement of 1000 is 0111, which is the excess-3 code for 4. 4 is the 9's complement of 5.

Decimal	BCD	Excess-3
0	0000	0011
1	0001	0100
2	0010	0101
3	0011	0110
4	0100	0111
5	0101	1000
6	0110	1001
7	0111	1010
8	1000	1011
9	1001	1100

Design a code conversion circuit that will convert the BCD code to the excess-3 code.

25. Design a code conversion circuit that will convert the excess-3 code described in Problem 24 to the BCD code.

Section 4.8   Multiplexers and tristate logic

26. Design a three-channel multiplexer that will channel three 2-bit input numbers onto a single 2-bit output channel. Use regular (two-state) logic in your design.

27. Using tristate logic, design a three-channel multiplexer that will channel three 2-bit input numbers onto a single 2-bit output channel.

28. Analyze the circuit shown in Figure 4.97 by constructing a truth table for the circuit. Use the letter Z to identify all high-impedance states.

29. Design a two-channel 4-bit multiplexer using regular (two-state) logic.

30. Using tristate logic gates, design a two-channel 4-bit multiplexer.

31. A single-bit multichannel multiplexer can very easily be used to implement a truth table by using the select lines as the independent variables and the output line as the dependent variable. The input channel lines are set to the values of the dependent variable so that these values are multiplexed to the output as each combination is applied to the select inputs.

Make the multiplexer shown in Figure 4.98 emulate the function of an Exclusive-*OR* gate by appropriately connecting the input channels to logic 1's or 0's. Assume that a 00 on the select lines selects channel A, a 01 on the select lines selects channel B, etc.

Figure 4.98

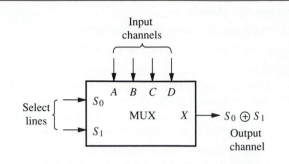

32. Using the information provided in Problem 31, make the 74150 16-input multiplexer implement the following truth table:

A	B	C	D	X
0	0	0	0	1
0	0	0	1	0
0	0	1	0	0
0	0	1	1	1
0	1	0	0	0
0	1	0	1	1
0	1	1	0	1
0	1	1	1	0
1	0	0	0	0
1	0	0	1	1
1	0	1	0	1
1	0	1	1	0
1	1	0	0	1
1	1	0	1	0
1	1	1	0	0
1	1	1	1	1

Figure 4.97

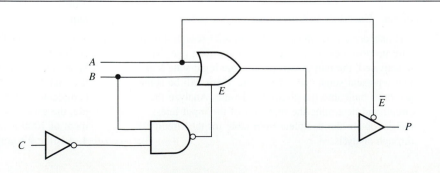

Compare the circuit obtained with the circuit that would be obtained if the logic equation for the truth table were implemented. Compare in terms of the number of ICs required.

33. If the multiplexer shown in Figure 4.98 has its input channels configured as follows, $A = 1$, $B = 1$, $C = 0$, $D = 1$, construct a truth table for the output X for all possible combinations of the inputs $S_1 S_0$.

Section 4.9 Demultiplexers

34. Design a two-channel 4-bit demultiplexer using two-channel 1-bit demultiplexers whose logic symbol is shown in Figure 4.75.

35. Design a four-channel 2-bit demultiplexer using a four-channel 1-bit demultiplexer whose logic symbol is shown in Figure 4.77.

36. Configure the 74154 decoder IC (*see* Fig. 4.46) to function as a 16-channel 1-bit demultiplexer.

Section 4.10 Parity circuits

37. Design an 8-bit parity generator circuit that will accept an 8-bit input number and indicate whether the number has even or odd parity by activating a single output that will be a logic 1 if the parity is even or a logic 0 if the parity is odd.

38. Expand the parity generator/checker circuit in Figure 4.86 to check an 8-bit input number.

Section 4.11 Arithmetic logic units

39. If the following two numbers were applied to the 74181 ALU:

$$A = 1010 \quad B = 0110$$

what would the output of the circuit F be for the following conditions:

a. $M = 1$, $S = 0001$
b. $M = 0$, $S = 1001$
c. $M = 1$, $S = 0111$
d. $M = 0$, $S = 1111$

40. Design a 4-bit ALU to implement the following truth table:

S_2	S_1	S_0	Function
0	0	0	$\overline{A \cdot B}$
0	0	1	$\overline{A + B}$
0	1	0	$\overline{A \oplus B}$
0	1	1	$\overline{B}$
1	0	0	A plus A
1	0	1	A minus 1
1	1	0	B minus A
1	1	1	Minus A (2's complement)

Troubleshooting

Section 4.12 Troubleshooting techniques

41. The circuit shown in Figure 4.43b uses two 74148 encoders to encode 16 inputs into a 4-bit active-low output code. When the circuit is set up and tested, the input lines

Figure 4.99

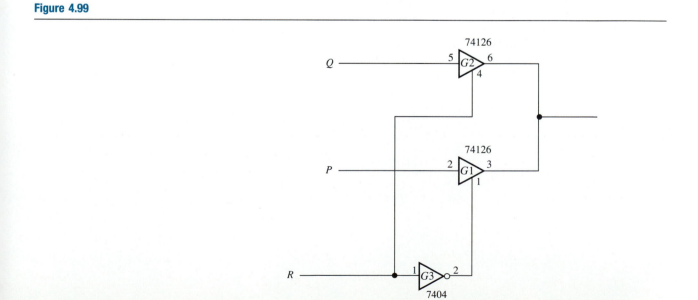

0 through 7 produce the correct codes 1111 through 1000. However, when the input lines 8 through F are activated the output of the circuit appears to be stuck at 1111. Troubleshoot the circuit and suggest possible faults that could produce this situation.

42. The logic circuit shown in Figure 4.99 has a fault. The recorded outputs of the logic gates G1, G2, and G3 are shown below for all possible combinations of the inputs *P*, *Q*, and *R*. Use the analysis procedure to troubleshoot the circuit and isolate the fault.

			Outputs		
P	*Q*	*R*	G1	G2	G3
L	L	L	L	L	H
L	L	H	Z	Z	L
L	H	L	L	L	H
L	H	H	Z	Z	L
H	L	L	H	H	H
H	L	H	Z	Z	L
H	H	L	H	H	H
H	H	H	Z	Z	L

Note that the outputs of G1 and G2 were connected together as shown in Figure 4.99 when the above data were recorded.

5

INTRODUCTION TO SEQUENTIAL LOGIC

OBJECTIVES

The objectives of this chapter are to:

- Study the design and operation of a simple ungated and gated latch

- Introduce the timing diagram technique of analyzing sequential logic circuits

- Introduce the basic operation of a flip-flop as compared to a gated latch

- Study the operation of the *S-R* flip-flop and develop its state transition table and equation

- Study the operation of the *T* flip-flop and develop its state transition table and equation

- Study the operation of the *D* flip-flop and develop its state transition table and equation

- Investigate the various ICs that implement the *D* flip-flop in hardware

- Study the operation of the *J-K* flip-flop and develop its state transition table and equation

- Investigate the various ICs that implement the *J-K* flip-flop in hardware

- Examine the characteristics and operation of master-slave flip-flops

- Examine the versatility of the *D* and *J-K* flip-flops by designing circuits to convert from one type to another

- Study the design and operation of the monostable and astable multivibrator

- Examine the switching characteristics of flip-flops and interpret data sheet parameters

- Investigate the procedures to test and troubleshoot flip-flop circuits and identify the typical faults that could exist in a sequential logic circuit

The basic unit of storage in early computers were the magnetic toroids shown in the photograph. Today semiconductor logic circuits have completely eliminated the need for these devices in digital computers (Courtesy IBM).

5.1
Introduction

The logic circuits studied so far are known as Combinational logic circuits because their output(s) depend only on binary combinations applied to the input(s). This chapter introduces *sequential logic circuits*. Like combinational logic circuits, they have output(s) that could depend on various combinations applied to the input(s), but the outputs also depend on the past history of the state of the outputs. Therefore, a memory element must be incorporated because these circuits must remember the past states of their outputs to produce an appropriate output. One of the characteristics of sequential logic circuits is feedback — a technique by which an output logic level is treated like another input to the circuit. Circuits that incorporate feedback are sequential logic circuits, those that do not are combinational logic circuits.

Sequential logic circuits play an important role in the operation of a digital computer. Besides serving as the primary means of storage, these circuits are also used for timing and counting. This chapter introduces the basic unit of a sequential logic circuit — the *flip-flop*. Chapters 6 and 7 will then investigate the various applications of flip-flops and sequential logic circuits. Because the design and operation of the flip-flop is based on the basic logic gates and concepts covered in earlier chapters, a thorough understanding of this elementary material is vital to the study of sequential logic circuits.

5.2
The latch

A *latch* is a sequential logic circuit that can exist in one of two possible states — *SET* or *RESET*. When a latch is *SET,* its output is at the logic 1 state and when the latch is *RESET*, its output is at the logic 0 state. The term *CLEAR* is also used to represent a *RESET* condition. Figure 5.1a shows the logic symbol of a *SET-RESET* (*S-R*) latch. The latch has two inputs labeled "*SET* (*S*)" and "*RESET* (*R*)" and two outputs labeled "*Q*" and "$\overline{Q}$" that are always complements of each other. The state of the latch is always described in terms of the Q output, that is, if $Q = 1$ then the latch is said to be *SET* and if $Q = 0$ the latch is said to be *RESET*. The $\overline{Q}$ output is made available as a convenience only.

Figure 5.1

The *S-R NOR* latch. (a) Logic symbol; (b) logic circuit.

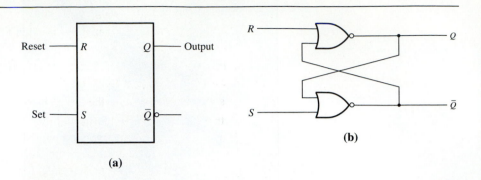

(a)

(b)

To set the *S-R* latch, activate the *S* input by applying a logic 1 to it. Only one of the two inputs must be active at a time and the *R* input must be (inactive) at a logic 0. Similarly, to reset the *S-R* latch, activate the *R* input by applying a logic 1 to it while keeping the *S* input inactive at the logic 0 state. Figure 5.2 illustrates the *timing diagram* for the *S-R* latch. A timing diagram describes the operation of a sequential logic circuit by graphically showing the output states of the circuits as the inputs change. In the timing diagram of Figure 5.2, the state of the latch has been traced for several possible (arbitrarily picked) states of the inputs *S* and *R*. Each time *S* or *R* changes state, it must be determined if the output state of the latch has changed. The intervals during which *S* and *R* remain constant are labeled in Figure 5.2 with the numbers "1" through "8" for reference purposes in the following analysis.

1. Both *S* and *R* are inactive, and the initial state of the output of the latch (*Q*) is *RESET*. Therefore, the latch remains *RESET*. This is a *no-change* condition.

2. *S* is active and commands the latch to *SET*. The output of the latch is therefore *SET*.

Figure 5.2

Timing diagram for the *S-R NOR* latch.

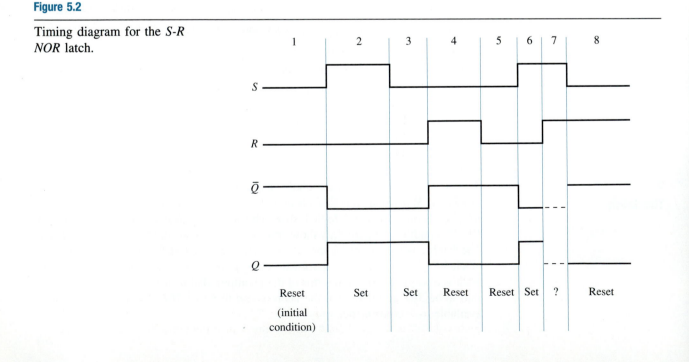

3. *S* is inactive now (while *R* remains inactive) but the latch remains *SET* because this is a no-change condition.

4. *R* is now active and commands the latch to *RESET*. The output of the latch is therefore *RESET*.

5. *R* is inactive now (while *S* remains inactive) but the latch remains *RESET* because this is a no-change condition.

6. *S* is activated again and the latch changes to the *SET* state.

7. Both *S* and *R* are now active and this causes *both* outputs to go to the logic 0 state. This effectively commands the latch to *SET* and *RESET* at the same time. Because this is impossible, the state of the latch is unknown. This is an *undefined* or *illegal* state.

8. *S* is now inactive while *R* is still active, thus commanding the latch to *RESET*.

In the timing diagram shown in Figure 5.2, the $\overline{Q}$ output is the complement of the *Q* output. Also, the *S* and *R* inputs function as command inputs that instruct the latch to either *SET* or *RESET* its output. The memory element characteristic of a sequential logic circuit is apparent in intervals 3 and 5 of the timing diagram. In interval 2 the latch was *SET* by activating the *S* input. However, when the *S* input was returned to its inactive state in interval 3, the latch remained *SET*. That is, the output was latched or held at the *SET* state even though the stimulus that produced it was removed. By momentarily activating the *S* input, a logic 1 has been *stored* at the output of the latch. Similarly, in interval 4 the *R* input was activated to *RESET* the latch. In interval 5, the latch remained *RESET* even though *R* returned to the inactive state. By momentarily activating the *R* input, a logic 0 has been stored at the output of the latch. To understand how the latch accomplishes its function, its logic circuit must be analyzed.

The logic circuit for the *S-R* latch is shown in Figure 5.1b. It is made up of two *NOR* gates and is sometimes referred to as a *NOR* latch. The circuit incorporates the feedback from output to input that is typical of most sequential logic circuits. To analyze the *NOR* latch circuit, assume some initial state for the latch (for example, the same initial conditions specified in interval 1 of the timing diagram in Fig. 5.2) and trace through each of the following intervals to verify its validity.

The circuits in Figure 5.3a through 5.3h show the trace of the logic levels for each of eight intervals of the timing diagram in Figure 5.2. Recall that if any input of a *NOR* gate is a logic 1 the output is a logic 0, and if both inputs of the *NOR* gate are logic 0's the output is a logic 1. Figure 5.3a shows the logic circuit of the *S-R* latch during interval 1 (initial conditions). Because the latch is *RESET* ($Q=0$) and both *R* and *S* are inactive ($R=0$, $S=0$), the output of gate *G2* ($\overline{Q}$) is a logic 1, which is fed back to the input of gate *G1*, producing a logic 0 at its output (*Q*) and keeping the latch *RESET*. During interval 2, the *S* line changes to a logic 1, immediately causing the output of *G2* to change to a logic 0 and consequently the output of *G1* to change to a logic 1. The latch is now *SET* as shown in Figure 5.3b. During interval 3 the *S* input changes back to a logic 0, but the output of *G2* does not change because its other input is still held at the logic 1 state by *Q*. The latch remains *SET* as shown in Figure 5.3c. During interval 4 the *R* input changes to a logic 1 state. This causes the output of *G1* to change to a logic 0 and the output of *G2* to a logic 1. The latch is now *RESET* as shown in Figure 5.3d. During interval 5 the *R* input returns to the logic 0 state, but the output of *G1* does not change because its other input is held high by $\overline{Q}$. The latch remains *RESET* as shown in Figure 5.3e. During interval 6 the *S* input again changes to a logic 1 state, which causes the output of *G2* to change to a logic 0

Figure 5.3

Trace of the various states of the *NOR* latch.

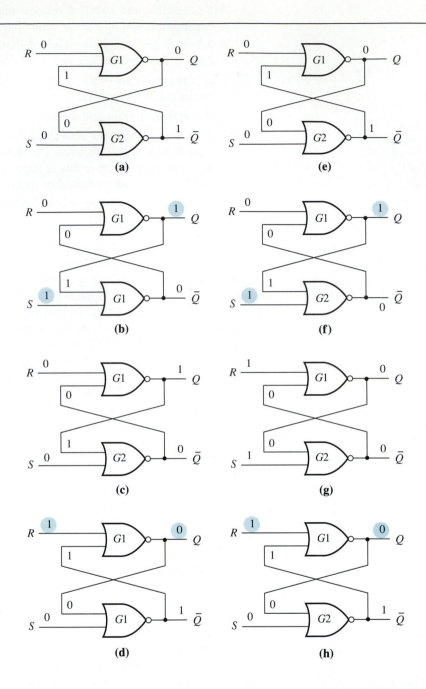

and the output of $G1$ to change to a logic 1, thus setting the latch as shown in Figure 5.3f. During interval 7 while S is active at the logic 1 state, the input R now changes to a logic 1, which causes the output of $G1$ to change to a logic 0. However, the output of $G2$ remains at a logic 0 state because S is a logic 1. Both the outputs of the latch are at the logic 0 state as shown in Figure 5.3g. This is an invalid condition because the outputs of the latch are defined to be complements of each other. During interval 8 the input S returns to the logic 0 state, while R remains active at the logic 1 state. This causes the output of $G2$ to change to a logic 1 and the output of $G1$ to change to a logic 0, thus resetting the latch.

An *S-R* latch can also be designed with active-low inputs. The operation of such a latch is identical to the active-high input *S-R* latch shown in Figure 5.1, except that a logic 0 on the *S* and *R* inputs is required to set and reset the latch, respectively. The logic symbol and circuit for the active-low input *S-R* latch is shown in Figure 5.4a and b, respectively. This latch is sometimes referred to as a *NAND* latch because of its *NAND* gates.

Figure 5.4

The *S-R NAND* latch. (a) Logic symbol; (b) logic circuit.

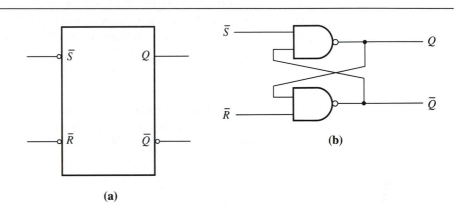

The timing diagram for the *S-R NAND* latch is shown in Figure 5.5. The state of the latch for various (arbitrarily picked) values of $\overline{S}$ and $\overline{R}$ has been analyzed. The timing diagram is divided into five time intervals during which $\overline{S}$ and $\overline{R}$ do not change.

Figure 5.5

Timing diagram for the *NAND* latch.

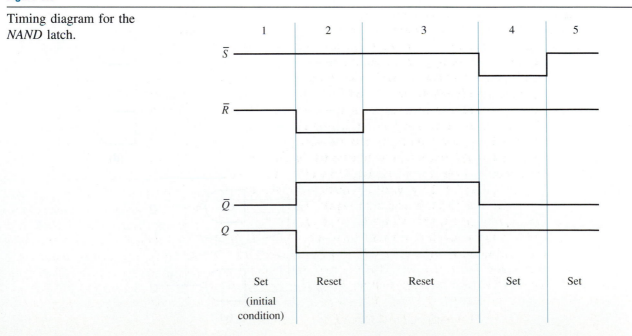

1. During this interval it is assumed that the latch is currently *SET* ($Q = 1$) and that the $\overline{S}$ and $\overline{R}$ inputs are inactive at the logic 1 state. This is the initial condition of the latch.

2. $\overline{R}$ is now active (0) while $\overline{S}$ is inactive (1), and therefore the latch is *RESET* and the output Q changes to the logic 0 state.

3. $\overline{R}$ returns to the inactive state (1) but the latch remains *RESET* ($Q = 0$) because $\overline{S}$ is also inactive (1). This is a no-change condition.

4. $\overline{S}$ is now active (0) while $\overline{R}$ is inactive (1), and therefore the latch is now *SET* and the output Q changes to the logic 1 state.

5. $\overline{S}$ returns to the inactive state (1) but the latch remains *SET* ($Q = 1$) because $\overline{R}$ is also inactive (1). This is also a no-change condition.

To understand the operation of the *NAND* latch, analyze its logic circuit shown in Figure 5.4b for each of the five time intervals of the timing diagram shown in Figure 5.5.

The circuits in Figure 5.6 show the trace of the logic levels for each of the five intervals of the timing diagram in Figure 5.5. Recall that if any input of a *NAND* gate is a logic 0 the output is a logic 1, and if both inputs of the *NAND* gate are logic 1's

Figure 5.6

Trace of the various states of the *NAND* latch.

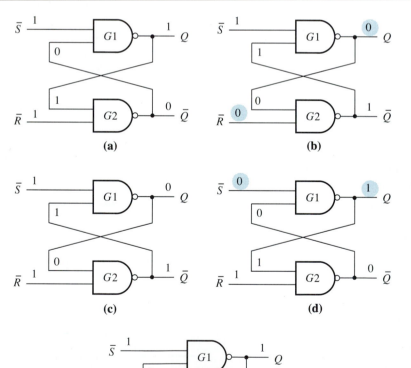

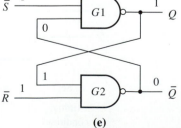

the output is a logic 0. Figure 5.6a shows the logic circuit of the *S-R* latch during interval 1 (initial conditions). Because the latch is *SET* ($Q = 1$) and both $\overline{R}$ and $\overline{S}$ are inactive ($\overline{R} = 1, \overline{S} = 1$), the output of gate *G2* ($\overline{Q}$) is a logic 0, which is fed back to the input of gate *G1*, producing a logic 1 at its output (Q) and keeping the latch *SET*. During interval 2, the $\overline{R}$ line changes to a logic 0, immediately causing the output of *G2* to change to a logic 1 and consequently the output of *G1* to change to a logic 0. The latch is now *RESET* as shown in Figure 5.6b. During interval 3 the $\overline{R}$ input changes back to a logic 1, but the output of *G2* does not change because its other input is still held at the logic 0 state by Q. The latch remains *RESET* as shown in Figure 5.6c. During interval 4 the $\overline{S}$ input changes to a logic 0 state. This causes the output of *G1* to change to a logic 1 and the output of *G2* to a logic 0. The latch is now *SET* as shown in Figure 5.6d. During interval 5 the $\overline{S}$ input returns to the logic 1 state, but the output of *G1* does not change because its other input is held low by $\overline{Q}$. The latch remains *SET* as shown in Figure 5.6e.

The *S-R* latches discussed so far could be set or reset by activating the *S* and *R* inputs, respectively. Activating either input caused the latch to change states immediately. A *gated S-R* latch has another input called the *gate* that determines when the latch will change states. The logic symbol and circuit for the gated *S-R* latch is shown in Figure 5.7. The *S* and *R* inputs function as described earlier and are used to *SET* and *RESET* the latch. However, the latch will not change states until a logic 1 is applied to the gate input *G*. The gate *G* effectively blocks (0) or passes (1) the *S* and *R* inputs by means of the two *NAND* gates *G1* and *G2* as shown in Figure 5.7b. If *G* = 0, then the outputs of *G1* and *G2* are 1's, regardless of the states of *S* and *R* (dominance law). Because *G3* and *G4* make up an active-low *NAND* latch, the logic 1's on the inputs of *G3* and *G4* make the latch inactive, and therefore there is no

The gated *S-R (NAND)* latch. (a) Logic symbol; (b) logic circuit.

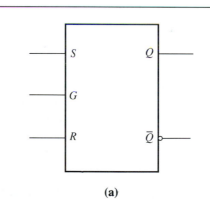

(a)

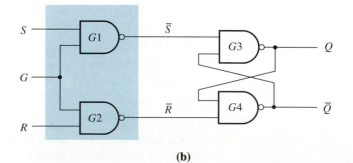

(b)

change in state. If $G = 1$, the outputs of $G1$ and $G2$ are equal to $\overline{S}$ and $\overline{R}$, respectively (identity law). Therefore, the *NAND* latch changes states as dictated by the logic levels applied to S and R. The G input thus functions like an enable line that either allows S and R to control the state of the latch or disables the influence of S and R on the latch. The timing diagram for the gated *S-R* latch is shown in Figure 5.8 and has nine time intervals identified for the following analysis.

Figure 5.8

Timing diagram for the gated *S-R* latch.

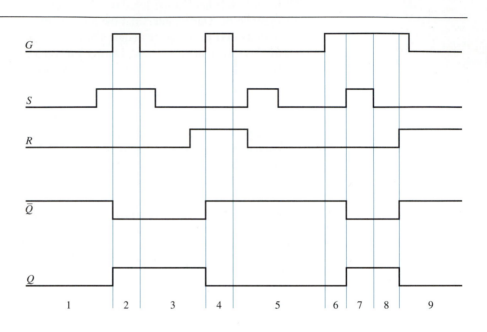

1. The latch is initially *RESET* but will not change states even though S changes to a logic 1 during this interval because G is a logic 0.
2. G is now a logic 1 and the latch inputs S and R are enabled. Because S is a logic 1 during this interval, the latch is now *SET*.
3. G is now a logic 0 and the latch inputs are disabled. However, because the previous state is latched, there is no change and the latch remains *SET*. S and R change during this interval but it has no effect on the state of the latch because $G = 0$.
4. G is now a logic 1 while R is active thus causing the latch to *RESET*.
5. G is now a logic 0 and any change in R and S does not affect the state of the latch. The latch remains *RESET*.
6. G is now a logic 1 but S and R are inactive (no-change command) and the latch remains *RESET*.
7. S is now active while $G = 1$, which causes the latch to *SET*.
8. S and R are inactive (no change) while $G = 1$ and the latch remains *SET*.
9. R is now active while $G = 1$, which causes the latch to *RESET* and remain at that state.

In a gated latch, the S and R inputs function as command or control inputs and determine if the latch will be *SET* or *RESET*, respectively, only when the gate is enabled (1).

The *S-R NOR* latch of Figure 5.1b can also be converted to a gated *S-R NOR* latch as shown in Figure 5.9. Again, the two *AND* gates will either pass the logic states of *R* and *S* to the *NOR* latch if $G = 1$ (identity law) or block the logic states of *R* and *S* to the *NOR* latch if $G = 0$ (dominance law). The operation of the circuit in Figure 5.9 is identical to the operation of the circuit in Figure 5.7b.

Figure 5.9

Logic circuit for the gated *S-R (NOR)* latch.

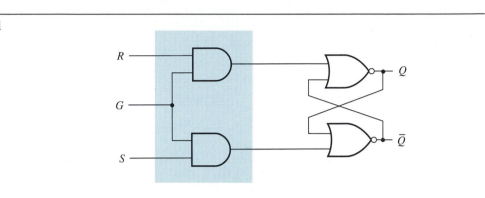

An application of a latch

One of the most common applications of an ungated latch is in a circuit known as a *debounce circuit*, which is used to eliminate the effects of contact bounce that occurs when the contacts of a switch open or close. For example, consider the simple switch circuit shown in Figure 5.10a. The switch in the circuit is a momentary pushbutton switch whose contacts are normally open. The output of the circuit is at the high state (logic 1) when the switch is open because the output is at a 5 volt potential. When the pushbutton switch is depressed momentarily, the contacts close, the output of the circuit is at ground potential (logic 0), the contacts then open, and the output of the switch is back to the logic 1 state. The voltage pulse generated by momentarily depressing the switch is shown in Figure 5.10b. The circuit can therefore be used to produce a pulse. Figure 5.10b indicates an ideal condition, that is, the pulse shown is what we would like the circuit to produce. However, the contacts of the switch are mechanical and when the poles of the switch strike the contact, the contact actually vibrates or bounces. This causes intermittent contact to be made for a short duration of time, the output voltage oscillating between a logic 0 and a logic 1 until the contact comes to rest. The effect of this contact bounce produces a pulse shown in Figure 5.10c. The circuit does not generate a "clean" pulse, which can often be unacceptable to many types of digital circuits. Therefore, a solution is needed to eliminate the effect of contact bounce and to produce a pulse that is ideal. The circuit in Figure 5.11a eliminates the effect of contact bounce and produces a clean pulse like the one shown in Figure 5.10b. An active-low *NAND* latch is used along with a momentary SPDT switch. The switch is normally at position 1. When the switch is depressed, it goes to position 2, and when released goes back to position 1. Because the switch is at position 1, the $\overline{S}$ input to the latch has a ground potential (logic 0) applied to it while the $\overline{R}$ input is at the 5.volt level (logic 1). This causes the output of the latch to be *SET* as shown in Figure 5.11b. When the switch is depressed, the contact moves to position 2 and applies a logic 0 on the $\overline{R}$ input; the $\overline{S}$ input is automatically pulled up to 5 volts. The latch now *RESETS*. If there is any contact bounce, it will produce oscillations on the $\overline{R}$ input of the latch and continuously try to reset the latch. This will have no effect on the latch because it is already reset. When the contact returns to position 1, the latch will be set but again there will be contact bounce that will produce oscillations on the $\overline{S}$ input of the latch, attempting to repetitively set it.

Figure 5.10

The pulse generated by a
switch. (a) Switch circuit;
(b) ideal output assuming no
contact bounce; (c) actual
output with contact bounce.

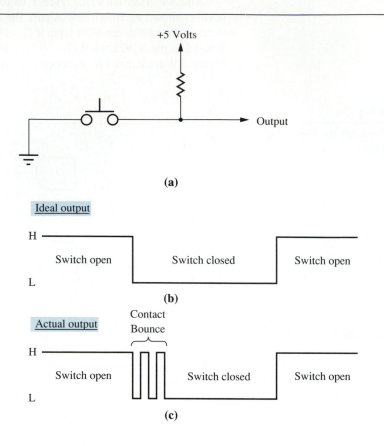

(a)

Ideal output

(b)

Actual output

(c)

Figure 5.11

(a) The debounce circuit; (b)
output waveform.

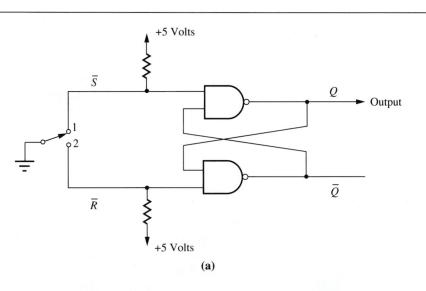

(a)

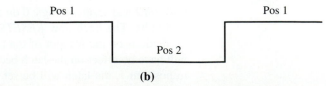

(b)

Because the latch is already set, it will remain at that state. The resulting pulse produced and shown in Figure 5.11b is an exact reproduction of the ideal pulse.

Review Questions

1. Define the terms "*SET*," "*RESET*," and "*CLEAR*."
2. Describe the operation of a latch.
3. What is the purpose of a timing diagram?
4. Describe the operation of a gated latch.

5.3
The flip-flop

A flip-flop is a sequential logic circuit that functions like a gated latch. The only difference between a flip-flop and a latch is the manner in which the control or command inputs (such as S and R) are enabled. In a gated S-R latch, the gate input G enables the S and R control inputs to affect the state of the latch. In other words, it is the gate (G) that causes the latch to actually change states. This is done by applying a logic level (logic 1 for the previously discussed circuits) at the G input to cause the latch to change states. A flip-flop has a *clock input C* that functions like the gate input to a latch but requires a transition for the flip-flop to change states as dictated by the settings of the control inputs.

A *transition* is a change in state from one logic level to another. A *positive transition* is a change from a logic 0 to a logic 1. A *negative transition* is a change from a logic 1 to a logic 0. Examples of negative and positive transitions are shown in Figure 5.12. The waveforms shown in Figure 5.12 are known as *clock pulses* and are often used to *trigger* or change the state of a flip-flop. Figure 5.12a illustrates the positive transitions (identified by the arrows) of two types of clock pulses. Because the positive edges of the pulses appear to rise, they are also referred to as *rising edges*. Figure 5.12b illustrates the negative transitions (identified by the arrows) of two types of clock pulses. Because the negative edges of the pulses appear to fall, they are also referred to as *falling edges*. Flip-flops are designed to change states on either transition. A flip-flop that changes states on a negative transition of the clock is called a *negative edge-triggered* flip-flop, whereas a flip-flop that changes states on a positive transition of the clock is called a *positive edge-triggered* flip-flop. The logic circuits for flip-flops are similar to those for gated latches except that a circuit to

Figure 5.12

Clock transitions used to trigger a flip-flop. (a) Positive edge; (b) negative edge.

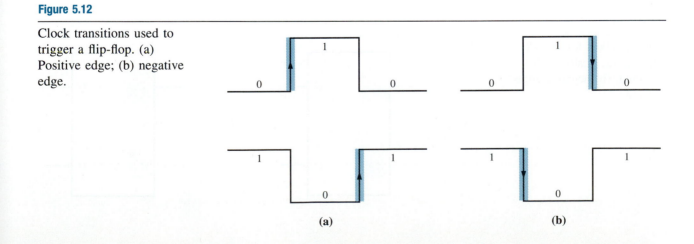

(a)   (b)

detect a transition on the clock input is added to the circuit. Because of the difference in the manner in which flip-flops and latches change states, the timing diagrams may be quite different. For example, in Figure 5.8 if the *G* input of the *S-R* latch was positive edge triggered, the state of the latch during intervals 6, through 9 would be *RESET*, because the positive transition of *G* occurs during the beginning of interval 6 when *S* and *R* are logic 0's (no change) and the current state of the latch is *RESET*.

Flip-flops are the basic logic elements of sequential logic circuits. The remainder of this chapter is dedicated to the study of the various types of flip-flops available and their operational characteristics. Four different types will be introduced — the *S-R*, the *T*, *D*, and *J-K* flip-flops. The *S-R* and *T* flip-flops are conceptual, that is, there are no commercial implementations (in the form of 74xxx ICs); they serve as models to illustrate the basic concepts of flip-flop operation. The *D* and *J-K* flip-flops are available commercially and can emulate the functions of the *S-R* and *T* flip-flops as will be seen in Sections 5.6 and 5.7. The specific applications of flip-flops in various types of sequential circuits will be examined in Chapters 6 and 7.

Review Questions

1. How does a flip-flop differ from a latch?
2. What does the term "trigger" mean with reference to the operation of flip-flops?
3. What are the different ways in which a flip-flop can be triggered?

5.4
The *S-R* flip-flop

The *S-R* flip-flop functions like the *S-R* latch. The *S* and *R* control inputs command the latch to *SET* or *RESET* when the clock *C* makes a transition. Figure 5.13 shows the logic symbols for positive and negative edge-triggered *S-R* flip-flops. The dynamic input indicator shown as a triangle at the clock input identifies the device as being edge triggered. The bubble at the input of the clock (Fig. 5.13) indicates a negative edge-triggered flip-flop whereas no bubble indicates a positive edge-triggered flip-flop (Fig. 5.13a). The bubble symbol at the clock input identifies inversion, meaning that a negative edge-triggered flip-flop is essentially a positive edge-triggered flip-flop with its clock pulse inverted. This concept can be used to convert one type of flip-flop to another. For example, in Figure 5.14 a negative edge-triggered flip-flop has been converted to a positive edge-triggered flip-flop by

Figure 5.13

Logic symbols for an *S-R* flip-flop. (a) Positive edge-triggered; (b) negative edge-triggered.

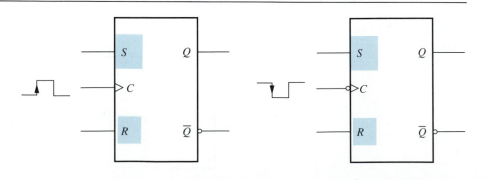

Figure 5.14

Converting a negative edge-triggered flip-flop to a positive edge-triggered flip-flop.

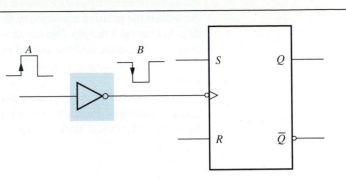

Figure 5.15

Converting a positive edge-triggered flip-flop to a negative edge-triggered flip-flop.

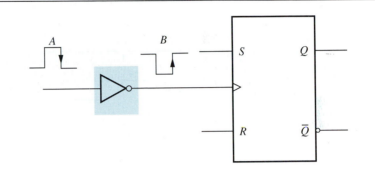

inverting the clock. The rising edge of clock pulse *A* is converted to a falling edge in clock pulse *B* triggering the flip-flop. Similarly, in Figure 5.15 a positive edge-triggered flip-flop is converted to a negative edge-triggered flip-flop by converting the falling edge of input pulse *A* to a rising edge in pulse *B* that triggers the flip-flop.

Figure 5.16 shows the timing diagram for a positive edge-triggered *S-R* flip-flop. The clock input is fed by a continuous train of periodic pulses, and each positive

Figure 5.16

Timing diagram for a positive edge-triggered flip-flop.

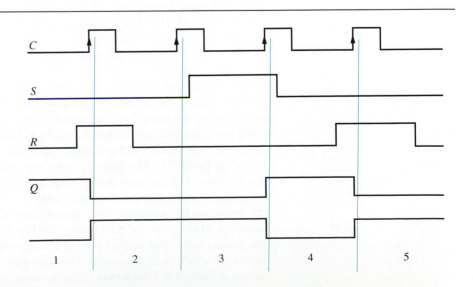

transition of a clock pulse causes the flip-flop to change states. The state of R and S at the instant the positive transition of the clock takes place dictates the state of the flip-flop. In interval 3 the flip-flop remains *RESET* because S changed from a 0 to a 1 after the positive transition of the second clock pulse took place. During interval 4 the flip-flop is *SET*, because S is a logic 1 when the third clock pulse makes a positive transition.

To illustrate the effects of changing the edge at which a flip-flop triggers, the timing diagram of Figure 5.16 has been analyzed for a negative edge-triggered *S-R* flip-flop. The revised timing diagram is shown in Figure 5.17.

Figure 5.17

Timing diagram for a negative edge-triggered flip-flop.

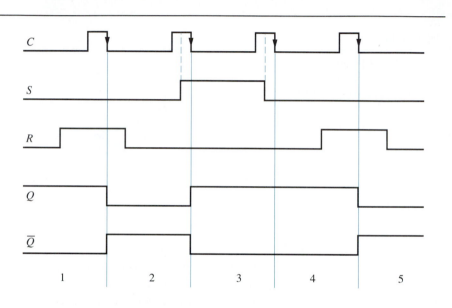

Figure 5.17 shows that each negative transition of the clock pulse causes the flip-flop to change states. The state of R and S at the instant the negative transition of the clock takes place dictates the state of the flip-flop. In interval 3 the flip-flop is *SET* because S changed from a 0 to a 1 before the negative transition of the second clock pulse took place. During interval 4 the flip-flop is still *SET* because S is a logic 0 (and R is also a logic 0) when the third clock pulse makes a negative transition. The resulting output Q in Figure 5.17 is considerably different from the output Q in Figure 5.16, even though the same clock and inputs to S and R were used.

Just as truth tables were used to describe the operation of the basic gates and combinational logic circuits, *state transition tables* are used to describe the operation of flip-flops and sequential logic circuits. They identify the *next state* of a flip-flop given the present conditions, the next state being the state of the flip-flop after the clock pulse makes a transition and triggers it. Table 5.1 shows the summarized state transition table for an *S-R* flip-flop. The next state of the *S-R* flip-flop in Table 5.1 is identified by Q^{n+1}. The letter Q represents the output of the flip-flop, and the superscript $n+1$ indicates that it is the output of the flip-flop after the clock pulse triggers it. A complete state transition table for a flip-flop is often more appropriate for describing its operation. The complete state transition table, or simply the state transition table, shows the next state of the flip-flop Q^{n+1} as a function of S, R, and the *present state*. The present state of the flip-flop, Q^n, is the state of the flip-flop before the clock pulse triggers it. The state transition table for the *S-R* flip-flop is shown in Table 5.2. Like a truth table, the state transition table of Table 5.2 lists all

Table 5.1   Summarized state transition table for an *S-R* flip-flop

S	R	Q^{n+1}	
0	0	Q^n	No change
0	1	0	*RESET*
1	0	1	*SET*
1	1	x	Invalid

Table 5.2   State transition table for an *S-R* flip-flop

S	R	Q^n	Q^{n+1}	
0	0	0		No change
0	0	1		
0	1	0		Reset
0	1	1		
1	0	0		Set
1	0	1		
1	1	0		Invalid
1	1	1		

possible combinations of *S, R,* and Q^n. The value of Q^{n+1} can then be determined for each combination from the summarized state transition table. For example, when *S* = 0 and *R* = 0 there is no change in states when the clock pulse triggers the flip-flop. In other words, the next state of the flip-flop, Q^{n+1}, is equal to the present state, Q^n. When *S* = 0 and *R* = 1, the next state of the flip-flop is a logic 0 (*RESET*) regardless of the present state. Similarly, when *S* = 1 and *R* = 0, the next state of the flip-flop is a logic 1 (*SET*) regardless of the present state. The *S-R* flip-flop does not support the condition when *S* = 1 and *R* = 1, and therefore we don't care what the next state of the flip-flop is under these conditions since they should never be used for an *S-R* flip-flop.

From Table 5.2 a K-map can be obtained for the output variable Q^{n+1} as shown in Figure 5.18 and the following logic equation:

$$Q^{n+1} = S + \overline{R}Q^n \qquad (5.1)$$

Equation 5.1 is called the *state transition equation* for the *S-R* flip-flop and its application in the analysis of sequential logic circuits will be examined in the next chapter. This equation can be used to determine the next state of the *S-R* flip-flop given the logic levels of *S, R,* and its present state, Q^n.

Review Questions

1.   What does the dynamic input indicator identify?
2.   What symbol identifies the flip-flop as being negative edge-triggered?
3.   Describe the operation of the *S-R* flip-flop.
4.   Why is *R* = 1, *S* = 1 an invalid condition for the *S-R* flip-flop?
5.   What is the purpose of a state transition table?

Figure 5.18

K-map for the *S-R* flip-flop.

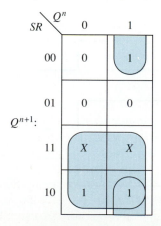

5.5

The *T* flip-flop

The *T* or *toggle* flip-flop has no control inputs like the *S-R* flip-flop but only contains a clock input labeled "*T*" as shown in Figure 5.19. Every time the clock pulse triggers the flip-flop it *toggles,* that is, it changes to the complement of the present state. For example, if the present state is a logic 1, when the clock pulse makes a transition the next state will be a logic 0, and if the present state is a logic 0 the next state will be a logic 1 when the clock pulse makes a transition.

Figure 5.20 show the timing diagrams for positive edge-triggered and negative edge-triggered *T* flip-flops. The outputs of both types of flip-flops are identical. Also, the *T* flip-flop effectively divides the input clock frequency at the *T* input by 2. The following example illustrates how several *T* flip-flops can be connected in series to reduce the frequency of a clock pulse. More examples of frequency division will be seen in Chapter 6.

Figure 5.19

Logic symbols for a *T* flip-flop. (a) Negative edge-triggered; (b) positive edge-triggered.

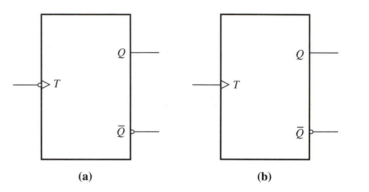

(a)   **(b)**

Figure 5.20

Timing diagrams for a *T* flip-flop. (a) Positive edge-triggered; (b) negative edge-triggered.

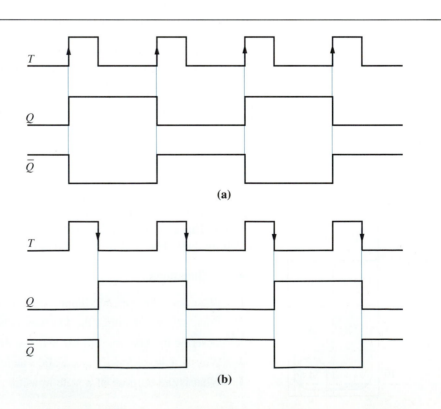

(a)

(b)

Example 5.1

Construct a timing diagram for the circuit shown in Figure 5.21 showing the outputs Q_A, Q_B, and Q_C as a function of nine *CLOCK* pulses. Also determine the frequency of the pulses at each of the outputs if the clock frequency is 800 Hz.

Figure 5.21

Frequency division using T flip-flops.

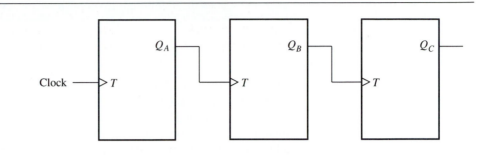

Solution

The timing diagram for Figure 5.21 is shown in Figure 5.22, assuming that Q_A, Q_B, and Q_C start off at the logic 0 states. In the timing diagram of Figure 5.22, Q_A divides the *CLOCK* frequency (800 Hz) by 2 to produce a frequency of 400 Hz. Q_B divides Q_A by 2 to produce a frequency of 200 Hz, and Q_C divides Q_B by 2 to produce a frequency to 100 Hz. Therefore

$$Q_A = CLOCK \div 2$$
$$Q_B = CLOCK \div 4$$
$$Q_C = CLOCK \div 8$$

Figure 5.22

Timing diagram for Figure 5.21.

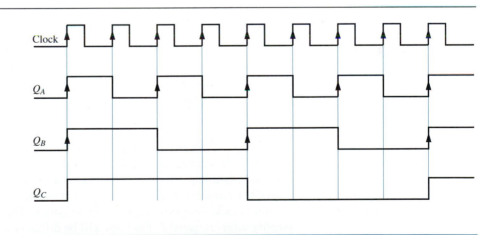

Table 5.3 State transition table for a T flip-flop

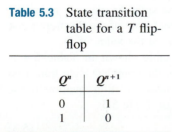

Q^n	Q^{n+1}
0	1
1	0

The state transition table for the T flip-flop is shown in Table 5.3. Because there are no control inputs, the only input variable in the table is the present state Q^n. We can now obtain the state transition equation for the T flip-flop directly from Table 5.3

$$Q^{n+1} = \overline{Q^n} \tag{5.2}$$

Equation 5.2 states that the next state of a T flip-flop is equal to the complement of the present state — a toggle condition.

Review Questions

1. Describe the operation of the *T* flip-flop.
2. How does the operation of the *T* flip-flop compare with the operation of the *S-R* flip-flop?
3. What does the word *toggle* mean?
4. How does the *T* flip-flop accomplish frequency division?

5.6
The *D* flip-flop

The *D* or *data*-type flip-flop has a single control input *D* and a positive edge-triggered or negative-edge triggered clock input as shown in Figure 5.23. The logic level at the *D* input determines the next state of the flip-flop. For example, if *D* is a logic 1 before the clock pulse makes a transition the next state of the flip-flop will be a logic 1, and if *D* is a logic 0 before the clock pulse makes a transition, the next state of the flip-flop will be a logic 0. The *D* flip-flop can be used to store 1 bit of data, and hence it is known as a *D* or data flip-flop.

Figure 5.23

Logic symbols for a *D* flip-flop. (a) Positive edge-triggered; (b) negative edge-triggered.

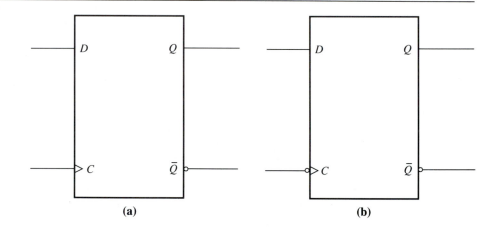

(a)   (b)

Figure 5.24 show the timing diagram for positive and negative edge-triggered *D* flip-flops. The outputs of the two flip-flops are different even though *D* and *C* are the same in both timing diagrams. This is because in Figure 5.24a *D* is a logic 1 when the third clock pulse makes a positive transition and in Figure 5.24b, *D* is a logic 0 when the third clock pulse makes a negative transition. Thus, the outputs of negative and positive edge-triggered *D* flip-flops will be different only if *D* changes between the positive and negative transitions of a clock pulse. The summarized state transition table for the *D* flip-flop is shown in Table 5.4. The next state of the flip-flop is not dependent on the present state but is only dependent on the value of *D*.

Even though the summarized state transition table shown in Table 5.4 is adequate to describe the operation of the *D* flip-flop, the complete state transition table can also be obtained as shown in Table 5.5. The state transition equation for the *D* flip-flop can be obtained directly from Table 5.4 or Table 5.5

$$Q^{n+1} = D \qquad (5.3)$$

Table 5.4 Summarized state transition table for a *D* flip-flop

D	Q^{n+1}
0	0
1	1

Figure 5.24

Timing diagrams for a *D* flip-flop. (a) Positive edge-triggered; (b) negative edge-triggered.

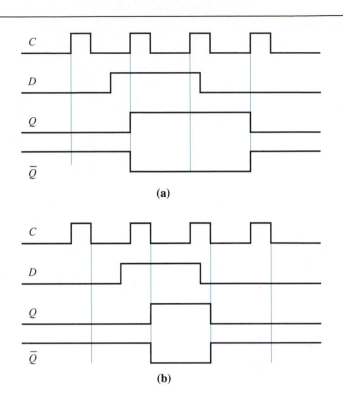

(a)

(b)

Table 5.5 State transition table for a *D* flip-flop

D	Q^n	Q^{n+1}
0	0	0
0	1	0
1	0	1
1	1	1

Table 5.6 Modified state transition table for a *D* flip-flop

Q^n	Q^{n+1}	D
0	0	0
0	1	1
1	0	0
1	1	1

Equation 5.3 states that the next state of the *D* flip-flop will be equal to the present state of *D*.

When the *D* flip-flop is used in the synthesis or design of sequential logic circuits (*see* Chapter 6), it is often more convenient to view the state transition table from a different perspective. The state transition table in Table 5.5 tells us what the next state of the *D* flip-flop would be, given its present state and the value of *D*. From Table 5.5 it can also be determined what *D* has to be for the flip-flop to change from one state to another. This information is shown in Table 5.6, which lists all possible transitions that a flip-flop can make in going from one state to another. These transitions are listed in the first two columns Q^n and Q^{n+1}. The third column lists the values of *D* required for the flip-flop to make the transition. For example, for the output of a *D* flip-flop to change from a 0 to a 1, *D* must be a 1. Similarly, for a *D* flip-flop to change from a 1 to a 0, *D* must be a 0. For the *D* flip-flop to "change" from a 0 to a 0 (no change), *D* must be a 0. Similarly, for the *D* flip-flop to "change" from a 1 to a 1 (no change), *D* must be a 1. For two of the entries in the table, even though no change (or transition) actually takes place, those two conditions are still considered to be transitions.

IC logic

The 7474 dual *D* flip-flop contains two independent positive edge-triggered *D*-type flip-flops in one IC package. The logic symbol and pin configuration for the IC are shown in Figure 5.25. Each flip-flop has the standard *D* and *C* (labeled "*CP*") inputs for the data and clock, respectively, and the standard *Q* and $\overline{Q}$ outputs. However, each flip-flop also has two additional inputs labeled $\overline{S}_D$ (*SET*) and $\overline{C}_D$ (*CLEAR*). The *D* input to a *D* flip-flop is known as a *synchronous* input because it is synchronized with the transition of the clock pulse. This means that if the logic level of *D* changes, the flip-flop will not change states unless the clock pulse triggers the flip-flop. The $\overline{S}_D$

Figure 5.25

Logic symbol and pin
configuration of the 7474
dual *D*-type positive edge-
triggered flip-flop.

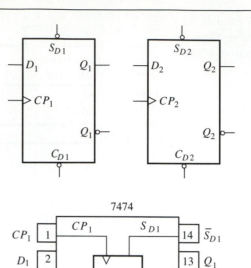

and $\overline{C}_D$ inputs to the 7474 *D* flip-flops are known as *asynchronous* inputs because
they are used to directly *SET* or *CLEAR* (*RESET*) the output of the flip-flop and do
not depend on a clock pulse transition. Therefore, their effect is not synchronized
with the clock — it is, asynchronous. Both $\overline{S}_D$ and $\overline{C}_D$ are active-low inputs and a
logic 0 on $\overline{S}_D$ will *SET* the flip-flop and a logic 0 on $\overline{C}_D$ will *CLEAR* the flip-flop
regardless of the states of *D* or *CP*. Obviously, both asynchronous inputs cannot be
connected to a logic 0 at the same time. If these inputs are not used in the *D* flip-flops
they must be connected to the inactive logic 1 state. Because the $\overline{C}_D$ and $\overline{S}_D$ inputs to
the flip-flop can directly *CLEAR* or *SET* the flip-flop, respectively, they are often
referred to as the *direct set* and *direct clear* inputs.

Review Questions

1. Describe the operation of the *D* flip-flop.

2. Compare the *SET* and *RESET* operations of the *D* flip-flop with the *SET*
   and *RESET* operations of the *S-R* flip-flop.

3. What are asynchronous inputs? How do they differ from synchronous inputs?

5.7
The *J-K* flip-flop

The *J-K* flip-flop emulates the operation of the *S-R* flip-flop, that is, the *J* input
corresponds to *SET* and the *K* input corresponds to *RESET*. There is no significance
to the letters *J* and *K* used to identify the control inputs of the flip-flop. There is one

difference between the *J-K* and *S-R* flip-flops. In an *S-R* flip-flop, if $S = 1$ and $R = 1$, the condition is defined as invalid and the next state of the flip-flop is undefined. However, in a *J-K* flip-flop, if $J = 1$ and $K = 1$, the next state of the flip-flop is equal to the complement of its present state; in other words, the flip-flop toggles. The logic symbols for a negative edge-triggered and a positive edge-triggered *J-K* flip-flop are shown in Figure 5.26. Figure 5.27 shows the timing diagram for a positive edge-triggered *J-K* flip-flop. During the first transition of the clock pulse, $J = 1$ and $K = 0$ and the output Q of the flip-flop is *SET* during interval 1. At the second transition of the clock pulse if $J = 1$ and $K = 1$ and the present output of the flip-flop is a logic 1, the flip-flop toggles and the next state of the flip-flop is *RESET* (logic 0) during interval 2. At the third transition of the clock pulse, $J = 0$ and $K = 1$, which tells the flip-flop to *RESET;* however, the flip-flop is already *RESET,* so there is no change during interval 3. At the fourth transition of the clock, $J = 0$ and $K = 0$, indicating a no-change condition, and the flip-flop remains *RESET* during interval 4. At the fifth transition of the clock, $J = 1$ and $K = 0$, and the flip-flop is *SET* during interval 5. At

Figure 5.26

Logic symbols for a *J-K* flip-flop. (a) Negative edge-triggered; (b) positive edge-triggered.

(a)

(b)

Figure 5.27

Timing diagram for a positive edge-triggered *J-K* flip-flop.

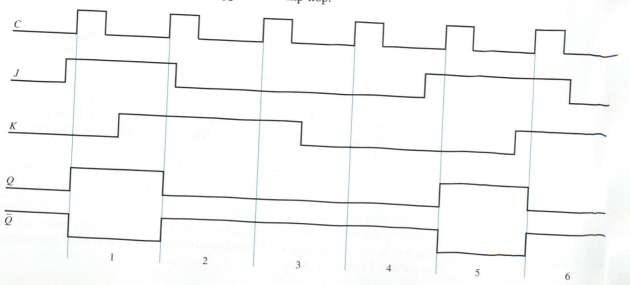

Table 5.7   Summarized state transition table for a *J-K* flip-flop

J	K	Q^{n+1}	
0	0	Q^n	No change
0	1	0	RESET
1	0	1	SET
1	1	$\overline{Q^n}$	Toggle

the sixth transition of the clock $J = 1$ and $K = 1$ (toggle) and the next state of the flip-flop is the complement of the present state; the flip-flop is *RESET* during interval 6.

The summarized state transition table for the *J-K* flip-flop is shown in Table 5.7. The next state of the *J-K* flip-flop is dependent on *J, K,* and the present state Q^n. To thoroughly describe the operation of the *J-K* flip-flop and obtain its state transition equation, construct the complete state transition table as shown in Table 5.8. Using Table 5.8, the state transition equation for the *J-K* flip-flop can be obtained from the K-map shown in Figure 5.28

$$Q^{n+1} = J\overline{Q^n} + \overline{K}Q^n \tag{5.4}$$

Equation 5.4 indicates that the next state of the *J-K* flip-flop depends on *J, K,* and the present state Q^n of the flip-flop.

Figure 5.28

K-map for the *J-K* flip-flop.

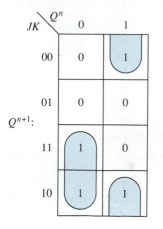

Q^{n+1}:

Table 5.8   State transition table for a *J-K* flip-flop

	J	K	Q^n	Q^{n+1}	
0	0	0	0	0	No change
1	0	0	1	1	
2	0	1	0	0	RESET
3	0	1	1	0	
4	1	0	0	1	SET
5	1	0	1	1	
6	1	1	0	1	Toggle
7	1	1	1	0	

It was stated in Section 5.6 that it is sometimes convenient to reorganize the state transition table to allow determination of what the control input(s) should be for a particular transition to take place at the output of a flip-flop. The eight entries have been numbered for reference purposes in Table 5.8 (0 through 7) and entries 3 and 7 show the output of the *J-K* flip-flop making a transition from a 1 (Q^n) to a 0 (Q^{n+1}). For these two entries *K* must be a logic 1 but *J* can be either a 0 or a 1. In other words, for the output of a *J-K* flip-flop to change from a 1 to a 0, *K* must be a 1 and we don't care what *J* is

Q^n		Q^{n+1}	J	K
1	$\longrightarrow$	0	x	1

Similarly, with reference to entries 4 and 6, the *J-K* flip-flop changes from a 0 (Q^n) to a 1 (Q^{n+1}). For these two entries *J* must be a logic 1 but *K* can be either a 0 or a 1. In other words, for the output of a *J-K* flip-flop to change from a 0 to a 1, *J* must be a 1 and we don't care what *K* is

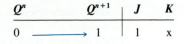

There are two possibilities for no change to occur in the output of the *J-K* flip-flop. Entries 0 and 2 indicate that the flip-flop changes from a 0 to a 0. For these two entries *J* must be a logic 0 but *K* can be either a 0 or a 1. In other words, for the output of a *J-K* flip-flop to remain constant at a 0, *J* must be a 0 and we don't care what *K* is

Q^n		Q^{n+1}	J	K
0	⟶	0	0	x

Similarly, entries 1 and 5 indicate that the flip-flop changes from a 1 to a 1. For these two entries *K* must be a logic 0 but *J* can be either a 0 or a 1. In other words, for the output of a *J-K* flip-flop to remain constant at a 1, *K* must be a 0 and we don't care what *J* is

Q^n		Q^{n+1}	J	K
1	⟶	1	x	0

From this analysis a modified state transition table can be constructed for the *J-K* flip-flop as shown in Table 5.9.

Chapter 2 examined a simple circuit (Fig. 2.32) to control a garage door opener system and discovered some of the limitations of combinational logic. The following example addresses those limitations and solves some of the potential problems by including sequential logic. The example also serves to illustrate a realistic application of a circuit that makes use of the *J-K* flip-flop.

Table 5.9   Modified state transition table for a *J-K* flip-flop

Q^n	Q^{n+1}	J	K
0	0	0	x
0	1	1	x
1	0	x	1
1	1	x	0

Example 5.2

The following circuit is similar to the one designed in Example 2.13 (Chapter 2) and is used to control a garage door opener system. This circuit makes use of two momentary (normally open) pushbutton switches, S1 and S2 to detect the position of the garage door — fully up or down. Assume that switches S1 and S2 are mounted at opposite ends of the garage door track (or rail) such that when the door is fully down, S1 is momentarily closed, and when the door is fully up, S2 is momentarily closed. Switch S3 is also a momentary normally open pushbutton switch that is used to start/stop and reverse the direction of the garage door.

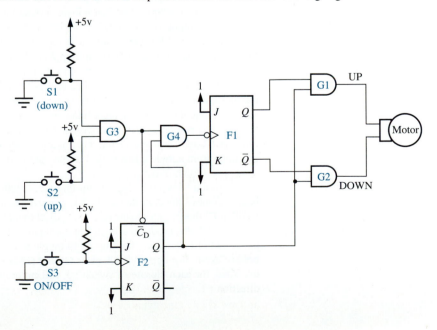

The system operates as follows. Assuming that the garage door is in the *UP* position, when switch S3 is pressed, the garage door motor is started and the door is moved in the *DOWN* direction. If S3 is pressed again (before the door is completely down) the garage door stops moving. If S3 is pressed again, the door reverses direction and begins to move up. When the door is fully up, switch S2 will stop the motor; the next time S3 is pressed, the door will move down. When the door is fully down, switch S1 will stop the motor; the next time S3 is pressed the door will move up.

The outputs of *AND* gates $G1$ and $G2$ control the direction and state of the motor as follows:

$G2$ output	$G1$ output	Motor rotation
0	0	Off
0	1	Door moves *UP*
1	0	Door moves *DOWN*
1	1	Fault

Unlike the circuit in Figure 2.32 this circuit does not require the motor's control circuit to have any latching mechanism. To move the door up, the output of gate $G1$ must be held at the logic 1 state, and to move the door down the output of gate $G2$ must be held at the logic 1 state. If both outputs are logic 0s, the motor immediately stops. Flip-flop $F1$ is a *J-K* flip-flop configured to toggle by tying J and K to a logic 1 state. The Q and $\overline{Q}$ outputs of flip-flop $F1$ are connected to one of the inputs of gates $G1$ and $G2$, respectively. The other inputs of $G1$ and $G2$ are connected together to the output of flip-flop $F2$ that is also a *J-K* flip-flop configured to toggle. If the output of flip-flop $F2$ is set, gates $G1$ and $G2$ are enabled. If gates $G1$ and $G2$ are enabled, the outputs of $G1$ and $G2$ will be equal to the Q and $\overline{Q}$ outputs of flip-flop $F1$, respectively. The outputs of $G1$ and $G2$ will always be complementary when enabled. If gates $G1$ and $G2$ are not enabled, their outputs will be logic 0's. Thus, flip-flop $F1$ is used to control the direction of the motor — up when $F1$ is set, down when reset, and flip-flop $F2$ is used to start and stop the motor — start when set, stop when reset.

To understand the operation of the circuit, assume that the garage door is moving in the down direction (output of $G1 = 0$, output of $G2 = 1$). This means that flip-flop $F1$ is reset, and $F2$ is set. When the door is fully down, switch S1 is closed, applying a logic 0 at the input of gate $G3$. The output of gate $G3$ goes to a logic 0 state and clears flip-flop $F2$ (recall that the direct clear input, $\overline{C}_D$ of the flip-flop will clear the flip-flop without requiring a clock pulse). The output of flip-flop $F2$ is then a logic 0 (reset) and is used to disable gates $G1$ and $G2$. Because the outputs of $G1$ and $G2$ are logic 0's, the motor immediately stops. At the same time when the output of G3 goes to a logic 0, the output of $G4$ makes a negative transition from a logic 1 to a logic 0 and triggers flip-flop $F1$. Flip-flop $F1$'s previous state was reset, and it therefore toggles to a set state. Attempt to start the motor by pressing switch S3. When S3 is pressed, a negative transition is applied to flip-flop $F2$. Flip-flop $F2$, which was reset, is now toggled to a set state. The output of flip-flop $F2$ enables gates $G1$ and $G2$. But because $F1$ was set, the output of $G1$ is now a logic 1 and the motor moves the door up. When the door reaches the fully up position, switch S2 is closed. The output of $G2$ again goes to a logic 0 that clears flip-flop $F2$ and stops the motor. The logic 0 at the output of $G3$ also produces a negative transition from a logic 0 to a logic 1 at the output of $G4$ causing flip-flop $F1$ to toggle to the reset state (down direction).

If switch S3 is pressed again, flip-flop $F2$ is again set, enabling $G1$ and $G2$, and the motor begins to move the door down. If S3 is pressed again (before the door is completely down), flip-flop $F2$ then toggles to the reset state, disabling $G1$ and $G2$, and the motor is stopped. The output of $F2$ also produces a logic 0 at the input of $G4$; this produces a negative transition at the output of $G4$ causing flip-flop $F1$ to toggle to the set state (up direction). If switch S3 is pressed again, flip-flop $F2$ is set, gates $G1$ and $G2$ are enabled, and the motor moves the door up. Thus, for each closure of switch S3, the motor either is stopped or starts in the reverse direction.

The only potential problem in this design is the effects of contact bounce produced by switches S1, S2, and S3. However, this can easily be solved by including a debounce circuit such as the one shown in Figure 5.11.

IC logic

The 74112 dual *J-K* flip-flop contains two independent negative edge-triggered *J-K*-type flip-flops in one IC package. The logic symbol and pin configuration for the IC are shown in Figure 5.29. Like the 7474 *D* flip-flop discussed in the previous section, the 74112 flip-flop has the regular synchronous inputs *J* and *K* as well as the asynchronous active-low inputs $\overline{S}_D$ and $\overline{C}_D$ for each flip-flop. A logic 0 on $\overline{S}_D$ will *SET* the output of the flip-flop and a logic 0 at $\overline{C}_D$ will *CLEAR* (reset) the flip-flop without requiring a clock transition. If the asynchronous inputs are not being used, they should be connected to the inactive logic 1 state. The *CP* input accepts the clock pulses that work in conjunction with the *J* and *K* inputs.

Figure 5.29

Logic symbol and pin configuration of the 74112 dual *J-K* negative edge-triggered flip-flop.

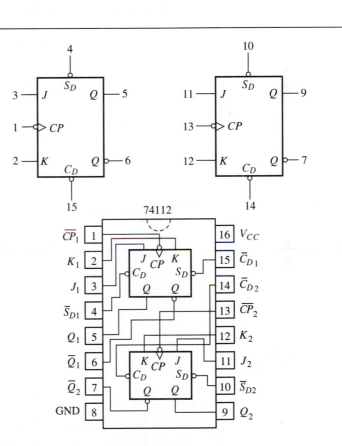

Review Questions

1. Describe the operation of the *J-K* flip-flop.
2. How does the *J-K* flip-flop emulate the operation of the *S-R* flip-flop?
3. How is the *J-K* flip-flop set up to toggle?
4. How is the *J-K* flip-flop set up to *SET* and *RESET*?

5.8

Master-slave flip-flops

The flip-flops discussed so far changed states on either a positive transition or negative transition of a clock pulse and were known as edge-triggered flip-flops. All that was required to trigger the flip-flop was the edge of a clock pulse, and not a complete clock pulse. There are many applications that require flip-flops to change states on receiving a complete clock pulse, that is, a clock pulse that makes both a negative and a positive transition. A flip-flop that changes states on receiving a complete pulse is known as a *master-slave* or *pulse-triggered* flip-flop.

The most common type of master-slave flip-flop available is the *J-K* flip-flop although the master-slave design can also be applied to the *D*, *S-R*, and *T* flip-flops. Because the design of the master-slave *S-R* flip-flop is the simplest to understand, it will be used as a model. Figure 5.30 shows the logic symbol for a master-slave *S-R* flip-flop. The symbol does not have a dynamic input indicator because the flip-flop is not edge-triggered. Instead it has a postponed output indicator (⌐) at each output because the output of the flip-flop does not change until the entire pulse (both transitions) has appeared at the clock input.

The master-slave *S-R* flip-flop actually consists of two edge-triggered *S-R* flip-flops connected in series as shown in Figure 5.31. The *S*, *R*, and *C* inputs to the master-slave flip-flops are applied to the *S*, *R*, and *C* inputs of a positive edge-

Figure 5.30

Logic symbol for a master-slave *S-R* flip-flop.

Figure 5.31

Equivalent circuit for a master-slave *S-R* flip-flop.

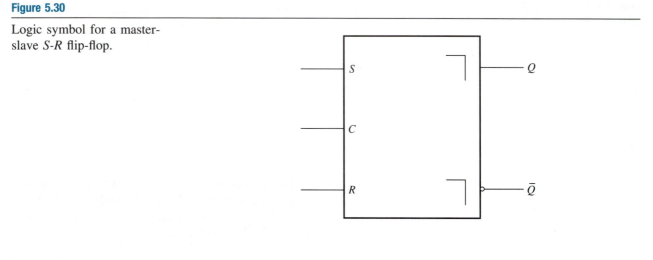

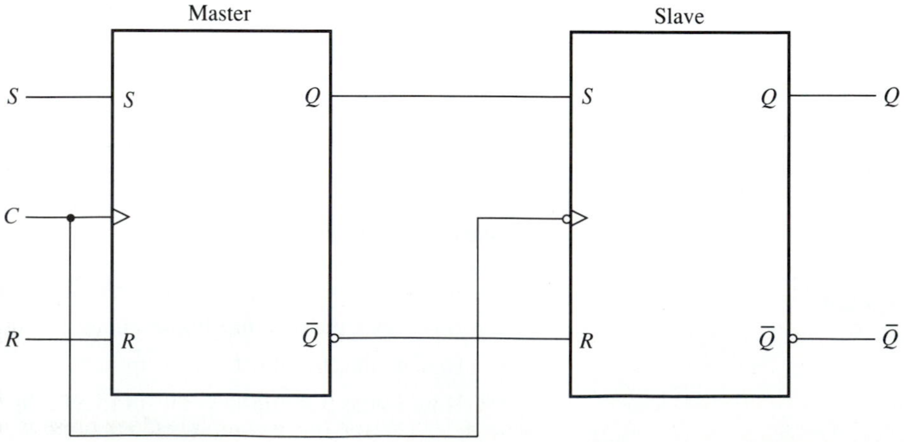

triggered *S-R* flip-flop known as the *master,* and the Q and $\overline{Q}$ outputs of the master-slave *S-R* flip-flop are obtained from a negative edge-triggered flip-flop known as the *slave.*

To set the flip-flop, the S input is set to a logic 1 and the R input to a logic 0 and a pulse is applied to the clock C as shown in Figure 5.32. Because the clocks of the master and the slave flip-flops are connected together, both flip-flops receive the pulse. However, because the master flip-flop is positive edge-triggered, it will change to the *SET* state on the rising transition of the pulse and the logic levels of S and R will be latched to Q and $\overline{Q}$, respectively. But Q and $\overline{Q}$ of the master flip-flop are connected to S and R of the slave flip-flop and the states of S and R of the master flip-flop have been effectively transferred to S and R of the slave flip-flop on the rising transition of the clock pulse. On the falling edge of the clock pulse, the negative edge-triggered slave flip-flop is now *SET* and its outputs (which are the outputs of the master-slave flip-flop) are at the same state as the outputs of the master. Therefore, the master flip-flop only responds to the rising edge and the slave flip-flop only responds to the falling edge of the clock pulse.

Figure 5.32

Pulse used to trigger a master-slave flip-flop.

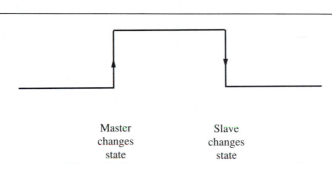

Master
changes
state

Slave
changes
state

Similarly, to *RESET* the flip-flop, the S input is *SET* to a logic 0 and the R input to a logic 1. On the positive edge of the clock pulse, the master flip-flop changes states and its Q and $\overline{Q}$ outputs are 0 and 1, respectively. Because the slave flip-flop's S and R inputs are at the same logic states as the master flip-flop's Q and $\overline{Q}$ outputs, respectively, on the negative transition of the clock pulse, the slave flip-flop is *RESET*.

For a no-change condition, $R = 0$ and $S = 0$, and the output of the master flip-flop does not change on the positive transition of the clock pulse. The output of the master is the same as the output of the slave. Therefore, on the negative transition if the slave is in the reset state it is *RESET* again, and if the slave is in the set state it is *SET* again, producing no change at the output of the slave.

Because the output of a master-slave flip-flop changes only after the entire pulse appears at its clock, its timing diagram is exactly the same as the timing diagram of a negative edge-triggered flip-flop and therefore in many applications is interchangeable with such flip-flops.

IC logic

The 7476 dual *J-K* flip-flop contains two independent master-slave *J-K* flip-flops in one IC package. The logic symbol and pin configuration for the IC are shown in Figure 5.33.

The 7476 *J-K* flip-flops function like the 74112 *J-K* flip-flops (discussed in the Section 5.7) except that a complete clock pulse is required to make the outputs of the

Figure 5.33

Logic symbol and pin
configuration of the 7476
dual *J-K* master-slave flip-
flop.

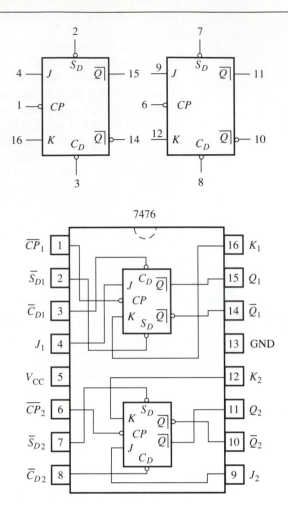

7476

flip-flops change states. Notice that each flip-flop has the clock pulse *CP*, a set of
active-low synchronous inputs *J* and *K*, and a set of asynchronous inputs $\overline{S}_D$ and $\overline{C}_D$.
The bubble shown at *CP* is sometimes used to indicate that the outputs of the flip-flop
will change states on the negative edge of the clock pulse.

Review Questions

1. How does one identify a master-slave flip-flop?
2. How does the operation of a master-slave flip-flop differ from the opera-
    tion of an edge-triggered flip-flop?
3. State another name for a master-slave flip-flop.

5.9

Flip-flop conversions

It was stated earlier that the only commercially available flip-flops are the *D* and *J-K*
types. This is because these two flip-flops can very easily emulate the functions of
the *S-R* and *T* flip-flops and can also be configured to emulate the functions of each

other. Because adapting one flip-flop to function as another can be very helpful in the design of digital circuits, this section deals with the configuration of the D and J-K types of flip-flops to emulate the functions of other types of flip-flops.

J-K flip-flop configurations

Figure 5.34

J-K flip-flop configured as an S-R flip-flop.

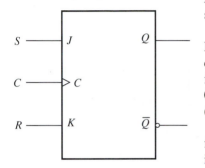

Because the J-K flip-flop functions like the S-R flip-flop for the first three binary combinations of J and K, the J input can be defined to be the S and the K inputs to the R input as shown in Figure 5.34. The condition where $S = 1$ and $R = 1$ is invalid for an S-R flip-flop and will not be used in the circuit shown in Figure 5.34.

To make the J-K flip-flop function like a T-type flip-flop the J and K input have to be connected to a logic 1 as shown in Figure 5.35. Recall that when $J = 1$ and $K = 1$, the next state of the J-K flip-flop is equal to the complement of the present state and the flip-flop toggles.

The circuit to make the J-K flip-flop function like a D flip-flop is shown in Figure 5.36. The input of the D flip-flop is directly connected to J and is indirectly connected to K via an inverter. If D is a logic 1, then $J = 1$ and $K = 0$ and the J-K flip-flop is *SET* and its output is a logic 1 (equal to the state of D). When D is a logic 0, then $J = 0$ and $K = 1$ and the J-K flip-flop is *RESET* and its output is a logic 0 (equal to the state of D).

The versatility in configuring the J-K flip-flop to function like the three other flip-flops is quite apparent. This is due to the capability of the J-K flip-flop to incorporate the functions of the S-R, T, and D flip-flops by changing the logic levels at the J and K control inputs. The D flip-flop can also be configured to function like the three other flip-flops, but it requires a considerable amount of additional circuitry.

D flip-flop configurations

Figure 5.35

J-K flip-flop configured as a T flip-flop.

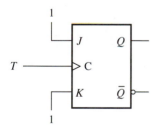

To configure the D flip-flop to function as an S-R flip-flop, first examine the transition equations for the D and S-R flip-flops given in Equations 5.3 and 5.1, respectively, and rewritten here as follows:

For the D flip-flop

$$Q^{n+1} = D$$

and for the S-R flip-flop

$$Q^{n+1} = S + \overline{R}Q^n$$

Because the two transition equations represent the next states of the D and S-R flip-flops and because we would like to make the next state of the D flip-flop the same as the next state of the S-R flip-flop, the two equations must be equated as follows:

$$D = Q^{n+1} = S + \overline{R}Q^n$$

Therefore,

$$D = S + \overline{R}Q^n. \tag{5.5}$$

Figure 5.36

J-K flip-flop configured as a D flip-flop.

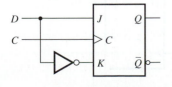

The circuit to implement Equation 5.5 to function like an S-R flip-flop is shown in Figure 5.37.

The circuit to configure the D flip-flop to function like a T flip-flop is shown in Figure 5.38. The D input is connected to the $\overline{Q}$ output so that each time the clock pulse makes a transition, the logic level at D and Q toggles. For example, if Q is initially a logic 1, $\overline{Q}$ is a logic 0 and D is a logic 0. When the clock pulse triggers the flip-flop, the Q output changes to a logic 0 (the state of D) and $\overline{Q}$ (and D) change to a logic 1. On the next transition of the clock pulse Q is back to a logic 1 and the process

Figure 5.37

D flip-flop configured as an
S-R flip-flop.

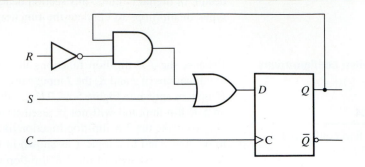

repeats itself. Mathematically, using the state transition equations of the *D* and *T* flip-flops (Eqs. 5.3 and 5.2, respectively)

$$Q^{n+1} = D$$

and

$$Q^{n+1} = \overline{Q}^n$$

the two must be equated to make the next state of the *D* flip-flop the same as the next state of the *T* flip-flop

$$D = Q^{n+1} = \overline{Q}^n$$

Therefore,

$$D = \overline{Q}^n. \tag{5.6}$$

Figure 5.38

D flip-flop configured as a *T*
flip-flop.

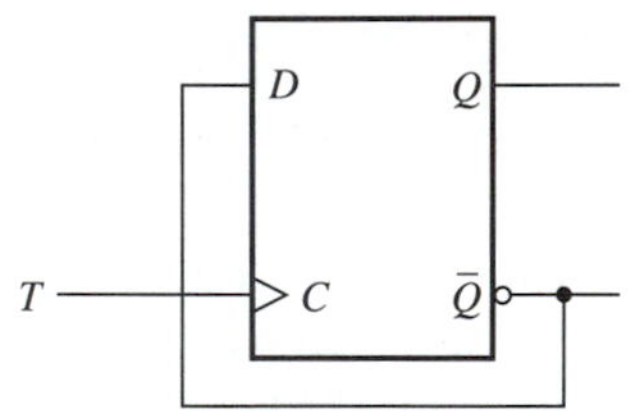

The circuit shown in Figure 5.38 is obtained from Equation 5.6. To configure the *D* flip-flop to function as a *J-K* flip-flop, the procedure is similar to the one used to configure the *D* flip-flop to function like an *S-R* flip-flop.

First examine the transition equations for the *D* and *J-K* flip-flops given in Equations 5.3 and 5.4, respectively, and rewritten here as follows.
For the *D* flip-flop

$$Q^{n+1} = D$$

and for the *J-K* flip-flop

$$Q^{n+1} = J\overline{Q}^n + \overline{K}Q^n$$

Figure 5.39

D flip-flop configured as a
J-K flip-flop.

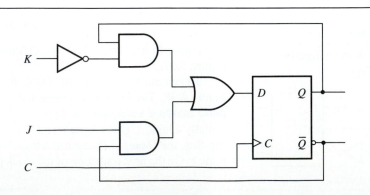

Because the two transition equations represent the next states of the D and J-K flip-flops and because we would like to make the next state of the D flip-flop the same as the next state of the J-K flip-flop, the two equations must be equated as follows:

$$D = Q^{n+1} = J\overline{Q}^n + \overline{K}Q^n$$

Therefore,

$$D = J\overline{Q}^n + \overline{K}Q^n. \tag{5.7}$$

The circuit to implement Equation 5.7 to function like a J-K flip-flop is shown in Figure 5.39.

Review Questions

1. Discuss the versatility of the J-K flip-flop in functioning like an S-R and T flip-flop.

2. Discuss the versatility of the D flip-flop in functioning like an S-R and T flip-flop.

3. Why is it important for the J-K and D flip-flops to be able to function like the T and S-R flip-flops?

Figure 5.40

Manufacturer's data sheet showing the switching characteristics of the 7474 D-type flip-flop.

Switching Characteristics at $V_{CC} = 5$ volts and $T_A = 25°C$

Parameter	From (input) to (output)	Symbol	$C_L = 15$ pF, $R_L = 2K$		Units
			Minimum	*Maximum*	
Maximum clock frequency		f_{MAX}	25		MHz
Propagation delay time LOW to HIGH level output	Clock to Q or $\overline{Q}$	t_{PLH}		25	ns
Propagation delay time HIGH to LOW level output	Clock to Q or $\overline{Q}$	t_{PHL}		30	ns
Propagation delay time LOW to HIGH level output	Preset to Q	t_{PLH}		25	ns
Propagation delay time HIGH to LOW level output	Preset to $\overline{Q}$	t_{PHL}		30	ns
Propagation delay time LOW to HIGH level output	Clear to $\overline{Q}$	t_{PLH}		25	ns
Propagation delay time HIGH to LOW level output	Clear to Q	t_{PHL}		30	ns
Setup time HIGH	D to Clock	t_s (H)	20		ns
Hold time HIGH	D to Clock	t_h (H)	5		ns
Setup time LOW	D to Clock	t_s (L)	20		ns
Hold time LOW	D to Clock	t_h (L)	5		ns
Clock pulse width		t_w (H) t_w (L)	30 37		ns
Clear or preset pulse width (LOW)		t_w (L)	30		ns

5.10
**Switching characteristics
of flip-flops**

Flip-flops exhibit the same input and output electrical characteristics as the basic logic gates. The same is true for loading rules and noise margins. These parameters were discussed in Chapter 2 and apply to all the flip-flops in the 74xxx TTL series of ICs. However, it is important to be able to understand some of the other characteristics that are typical of flip-flops, specifically, the switching characteristics.

IC data books often include the switching or alternating current (AC) characteristics of flip-flops along with the other electrical and logic information about the IC. Figure 5.40 shows the switching characteristics of a 7474 *D*-type flip-flop obtained from the manufacturer's data book. Most of these parameters relate to the timing of the flip-flop.

Maximum clock frequency

The data sheet includes the maximum frequency of the clock pulses that are applied at the clock input, f_{MAX}. If this frequency is exceeded, the flip-flop will not respond properly to the clock pulses and its operation will be impaired.

Propagation delays

Note in the data sheet that the flip-flop has several propagation delays associated with its operation. These parameters have the same symbol, t_P, but describe delays between different inputs and outputs of the flip-flop.

The *propagation delay time* between the clock pulse and the output of the flip-flop is measured from the time the clock pulse makes a transition to the time the Q or $\overline{Q}$ output changes states. Two values are specified for this propagation delay — t_{PLH}, the propagation delay when the output changes from a *LOW* to a *HIGH,* as shown in

Figure 5.41

Propagation delays between the clock pulse and the outputs of a flip-flop. (a) *LOW* to *HIGH*; (b) *HIGH* to *LOW*.

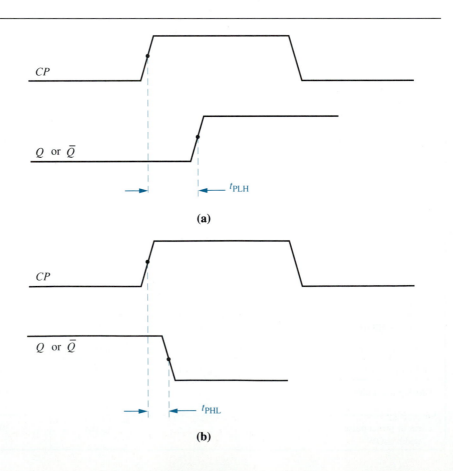

Figure 5.41a, and t_{PHL}, the propagation delay when the output changes from a *HIGH* to a *LOW* as shown in Figure 5.41b. Propagation delays are measured from the midpoint of each transition. The waveforms in Figure 5.41 assume a positive edge-triggered flip-flop.

The other propagation delays given in the data sheet are the delay times between the direct set (preset) and direct clear (reset) asynchronous inputs of the flip-flop and the Q or $\bar{Q}$ outputs. The delays are shown in Figure 5.42. t_{PLH} is the propagation delay from the time the direct set ($\bar{S}$) or direct clear ($\bar{C}$) input is activated (to a *LOW*) to the time the (Q or $\bar{Q}$) output changes state from a *LOW* to a *HIGH*. t_{PHL} is the propagation delay from the time the $\bar{S}$ or $\bar{C}$ input is activated to the time the (Q or $\bar{Q}$) output changes state from a *HIGH* to a *LOW*.

Figure 5.42

Propagation delays between the asynchronous control inputs and the outputs of a flip-flop.

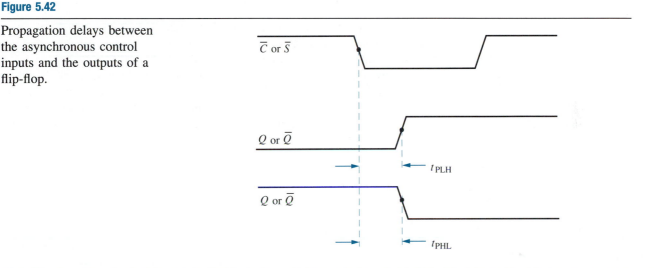

Setup time

The setup time of a flip-flop is the minimum amount of time before the clock makes its transition that the control input(s) (D in this example) have to be stable. $t_s(H)$ is the time interval between the application of a *HIGH* at the D input and the transition of the clock pulse as shown in Figure 5.43a. $t_s(L)$ is the time interval between the application of a *LOW* at the D input and the transition of the clock pulse as shown in Figure 5.43b. Again, both waveforms for the clock pulse (CP) shown in Figure 5.43 assume a positive edge-triggered clock.

Hold time

The hold time of a flip-flop is the minimum amount of time after the clock makes its transition that the control input(s) (D in this example) have to be stable. $t_h(H)$ is the time interval between the transition of the clock pulse and a change in D from the *HIGH* state as shown in Figure 5.44a. $t_h(L)$ is the time interval between the transition of the clock pulse and a change in D from the *LOW* state as shown in Figure 5.44b. Again, both waveforms for the clock pulse (CP) shown in Figure 5.44 assume a positive edge-triggered clock.

Pulse widths

The data sheet also includes the minimum pulse widths required at the clock pulse (CP) and asynchronous inputs ($\bar{S}$ and $\bar{C}$) for reliable operation. For asymmetrical

Figure 5.43

Setup times for a flip-flop.
(a) *D* goes *HIGH*; (b) *D*
goes *LOW*.

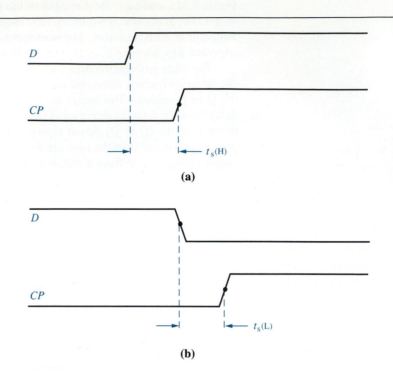

(a)

(b)

Figure 5.44

Hold times for a flip-flop.
(a) *D* goes *LOW*; (b) *D* goes
HIGH.

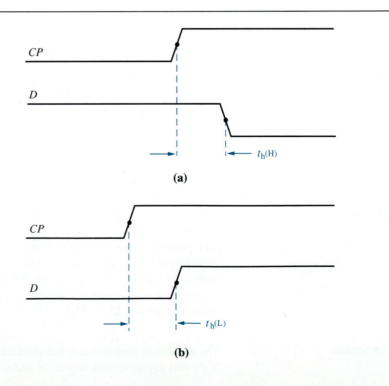

(a)

(b)

clock pulses applied to the *CP* input of the flip-flop, $t_w(H)$ is the minimum width of the *HIGH* cycle and $t_w(L)$ is the minimum width of the *LOW* cycle as shown in Figure 5.45a. For an active-low pulse applied to the direct set ($\overline{S}$) and direct clear ($\overline{C}$) asynchronous inputs, the minimum width of the pulse is $t_w(L)$ as shown in Figure 5.45b.

Figure 5.45

Pulse width measurements.
(a) Clock input; (b)
asynchronous control inputs.

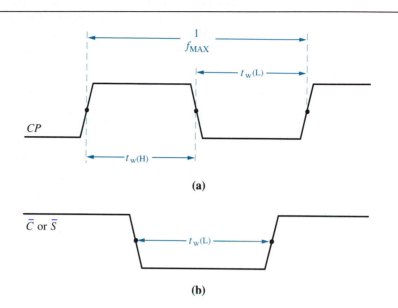

(a)

(b)

Rise and fall times

The transitions of the clock pulses applied to the clock inputs of flip-flops usually require an extremely fast transition from the *HIGH* to *LOW* or *LOW* to *HIGH* state. Typically the transition has to be around 6 to 13 ns. This means that the rise and fall times of clock pulses must be extremely small to properly trigger the flip-flop. If the rise and fall times of the pulse(s) applied to the clock input of the flip-flop are too long, the flip-flop often produces undesirable oscillations as it passes through the undefined area between a *HIGH* and a *LOW*. Most pulse generator circuits such as astable multivibrators (to be examined later) do produce pulses with quick rise and fall times, but due to the effect of capacitance in a circuit the pulses could be distorted by longer rise and fall times.

Review Questions

1. How are the different types of propagation delays measured for a flip-flop.
2. What is the setup time for a flip-flop?
3. What is the hold time for a flip-flop?
4. What effect does a slow rise or fall time generally have on the clock input of a flip-flop?

5.11
The Schmitt trigger

There are several 74xxx IC gates that have special types of inputs known as *Schmitt trigger* inputs. Gates with these inputs are designed to respond reliably to slow transitions. For example, Figure 5.46 shows the logic symbol and timing diagram for a 7414 hex Schmitt trigger inverter. The symbol (⊓) inside the logic symbol for the inverter identifies a gate with a Schmitt trigger input. Assume that a pulse with a very slow rise and fall time is applied to input A of the inverter shown in Figure 5.46. The output of the inverter changes from a logic high to a logic low when the input reaches the threshold voltage V_T^+ (typically around 1.7 volts) and changes from a logic low to a logic high when the input voltage reaches the threshold voltage V_T^- (typically around 0.6 volts). Once the threshold voltage has been detected at the input and the output changes states, the output will remain at that state even if the input voltage fluctuates slightly. Therefore, a pulse with a slow transition can be converted to a pulse with a fast transition using a gate with a Schmitt trigger input.

Figure 5.46

A Schmitt trigger inverter.
(a) Logic symbol; (b) timing diagram.

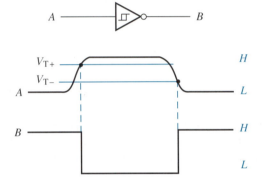

To realize the need for a gate with a Schmitt trigger input and understand how it works, consider the circuit shown in Figure 5.47a. The circuit represents a logic gate (*NAND* gate number 1) transmitting logic signals (such as pulses) to several other logic gates (*NAND* gates 2, 3, 4, . . ., n) over a transmission path. Circuit conditions and other factors can degrade the shape of the pulse during transmission. This can result in an increase of positive and negative edge transition times (rise time and fall time) and / or a marked decrease in the logic 1 voltage level. The capacitance shown in the circuit is dotted in to represent a parasitic capacitance due to loading (connecting an output of a circuit to one or more inputs and / or the capacitance due to circuit lead length). It is this capacitance that degrades the rise and fall times of the pulse. Figure 5.47b represents an ideal pulse that would be produced by *NAND* gate number 1 if loading conditions do not degrade its output. This ideal input pulse has rise and fall times of zero and a logic 1 level of 3.6 volts (a well-shaped pulse with a logic 1 voltage level well above the 2-volt minimum for an acceptable logic 1). Figure 5.47c, which represents the output pulse after it has been degraded by the transmission path, shows that the rise and fall times have increased due to the parasitic capacitance. Also, the logic 1 voltage has been decreased to a level that at certain times falls below the minimum voltage level of 2 volts required for a logic 1. The logic 1 voltage level decrease is due to an overload of inputs (the inputs of *NAND* gates 2, 3, 4, . . ., n) connected to the output of *NAND* gate number 1.

To reshape the waveform of Figure 5.47c to one that is more ideal, there are two options. The loading can be reduced, which is often not possible, or a waveform

Figure 5.47

Pulse degradation.

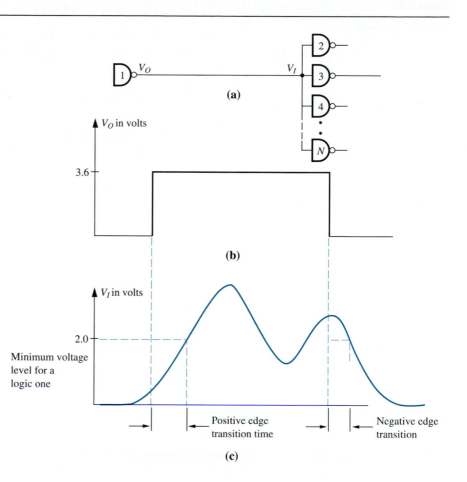

(a)

(b)

(c)

reshaping circuit at the receiving end of the transmission path can be used. The most practical choice is the reshaping circuit. A circuit with a *hysteresis trigger* characteristic (whose explanation follows) will correct the waveshape problems of the pulse shown in Figure 5.47c.

Figure 5.48a shows the logic symbol of a buffer / driver that has a hysteresis trigger characteristic. The symbol (⊓) in the center of the logic symbol represents hysteresis and is inserted in the logic symbol to indicate that the logic device has this trigger characteristic. To understand what a trigger with hysteresis means, refer to the hysteresis curve of Figure 5.48b. First, the two ends of the hysteresis curve have been opened so that the travel direction of the curve may be more easily traced. The ends of a hysteresis curve are actually lines without any openings as illustrated in Figure 5.49 (Fig. 5.49 has been rotated 90 degrees from the orientation of Figure 5.48).

Figure 5.48b shows that when V_I is equal to zero volts the output voltage V_O is a logic 0. As the input voltage is increased the hysteresis curve is traversed from left to right, as indicated by the arrowheads superimposed on the curve. While increasing the input voltage the output voltage V_O remains a logic 0 until the *positive-going threshold voltage* V_T^+ is reached. When the input voltage is increased greater than V_T^+, the output voltage goes from a logic 0 to a logic 1. It can be said that the input voltage level V_T^+ is the input trigger voltage level necessary to drive the output voltage V_O from a logic 0 to a logic 1. The hysteresis curve of Figure 5.48b also

Figure 5.48

Hysteresis curve.

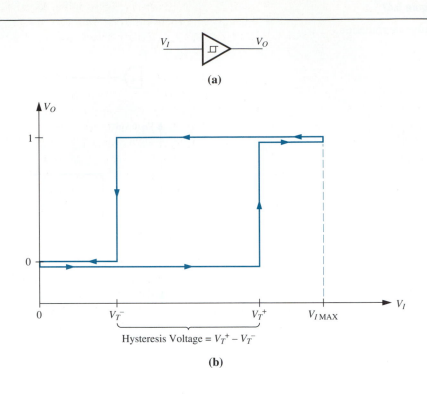

(a)

Hysteresis Voltage $= V_T{}^+ - V_T{}^-$

(b)

illustrates that as the input voltage is increased beyond the trigger voltage $V_T{}^+$ the output remains in the logic 1 state. Suppose that the input voltage was increased to some maximum voltage $V_{I\ MAX}$ and then decreased. As indicated by the arrow directions superimposed on the hysteresis curve, the input voltage would continue to traverse the hysteresis curve (from left to right) until it reaches $V_{I\ MAX}$. Then V_I would turn around on the hysteresis curve (as the input voltage is decreased) and begin to traverse from right to left. This decrease in input voltage does not affect the output voltage until the *negative-going threshold voltage $V_T{}^-$* is reached. When the input voltage is decreased below voltage level $V_T{}^-$, the output voltage makes the transition from the logic 1 level to the logic 0 level. The negative-going threshold voltage $V_T{}^-$ is the input trigger voltage level that triggers the output voltage from a logic 1 to a logic 0. Decreasing input voltage V_I below trigger voltage level $V_T{}^-$ does not affect the output voltage logic level; it remains a logic 0.

In summary, it can be concluded from the analysis of Figure 5.48b that any circuit with an input trigger with hysteresis characteristic has two trigger points — one for triggering the output from a logic 0 to a logic 1 known as the positive-going threshold voltage ($V_T{}^+$) and the other, the negative-going threshold voltage, which triggers the output from a logic 1 to a logic 0 ($V_T{}^-$). An input trigger with this characteristic is known as a Schmitt trigger, and it is this trigger that provides the desired waveshaping properties.

To visualize how a Schmitt trigger works, refer to Figure 5.49. Figure 5.49 has three graphs, a hysteresis curve like that of Figure 5.48b, but rotated 90 degrees so that the V_I axis is now vertical. The second graph is the input voltage V_I that shares the same vertical axis V_I with the hysteresis curve. This graph is also the graph of the input voltage (V_I) applied to the input of the buffer/driver shown in Figure 5.48a. This input voltage is to represent a distorted waveform, such as that shown in Figure 5.47c. The third graph is the output voltage V_O that is produced by the output of the buffer / driver shown in Figure 5.48a. As can be seen in Figure 5.49, the input voltage is very similar to the waveform distorted by noise that was shown in Figure 2.44 (Chapter 2). The following are some important observations about the voltage V_I shown in Figure 5.49:

1. It has very long rise and fall times. Some logic circuits will not trigger properly on such a long positive or negative edge.

2. It intersects the logic 1 voltage level V_{IH} four times (t_4, t_6, t_{10}, and t_{12}). A desired input voltage waveform would remain greater than the logic 1 minimum voltage level V_{IH} once that voltage level had been crossed (at t_4) and remain so during the duration of the pulse. Without a Schmitt trigger, the output voltage would change from a logic 0 to a logic 1 at time t_4, then from a logic 1 to a logic 0 at time t_6. At time t_{10} it would be triggered back from a logic 0 to a logic 1, and at time t_{12} the output would again be triggered back to a logic 0.

To see what happens with Schmitt triggering refer to the waveform of input voltage V_I shown in Figure 5.49. Trace through the waveform of V_I (beginning at time t_0 and ending at time t_{14}) and see what effect the hysteresis curve to the left of the input voltage will have on the output voltage V_O (shown in the plot above V_I). On the plot of V_I at time t_0 there is 0 volts. On the hysteresis curve to the left, this time is identified as event 0. The hysteresis curve shows that for a voltage less than the trigger level $V_T{}^+$ (1.5 volts) the output voltage V_O is a logic 0. The graph of the

Figure 5.49

The effects of hysteresis triggering on the output logic levels of a driver.

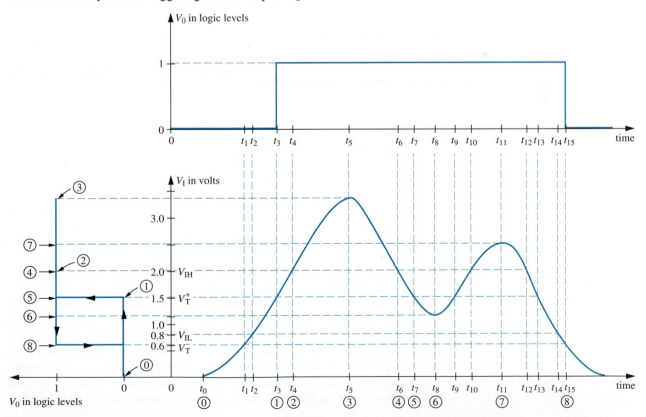

output voltage shows it to be a logic 0. Moving to time t_1 on the graph of V_I, the input voltage has reached the trigger magnitude of V_T^- volts. Referring to the hysteresis curve, increasing the input voltage to V_T^- volts does not cause the buffer/driver circuit output to change logic levels. Therefore, the output voltage remains in the logic 0 state. At time t_3 the input voltage increases to a magnitude just a little greater than the trigger voltage level V_T^+. This event is identified as 1 on the hysteresis curve. The hysteresis curve indicates that for this magnitude of input voltage the output V_O makes a transition from a logic 0 to a logic 1. This is confirmed by viewing the plot of V_O. At time t_4 the input voltage has increased to 2.0 volts (V_{IH}); this is identified as event 2 on the hysteresis curve. The hysteresis curve shows that the output remains in the logic 1 state. The input voltage continues to increase until it reaches its maximum voltage (3.7 volts) at time t_5. This is identified as event 3 on the hysteresis curve. At event 3, the hysteresis curve shows that the output is a logic 1. Therefore, the output voltage remains in the logic 1 state as shown in the plot of output voltage V_O. Now the input voltage decreases to a level of V_{IH} (2.0 volts) at time t_6. This is marked as event 4 on the hysteresis curve. Again, the hysteresis curve shows that the output voltage is to remain in the logic 1 state. To reiterate, if this buffer/driver did not possess Schmitt triggering, the output would have changed to a logic 0 at this time. Continuing, at time t_7 the input voltage has decreased to V_T^+ volts (event 5 on the hysteresis curve), however, the output voltage remains at the logic 1 state. The input voltage decreases to approximately 1.3 volts at time t_8 (event 6) and as the hysteresis curve shows, the output V_O remains at a logic 1. The input voltage increases to a voltage of 2.5 volts at time t_{11} (marked as event 7 on the hysteresis curve). As can be seen from the hysteresis curve, this means that at event 6 the input voltage reversed its direction of travel on the hysteresis curve and traveled back up the curve to the point marked as event 7; during this time the output remained in the logic 1 state. Also notice that at times t_9 and t_{10} the input voltage passed through voltage levels V_T^+ and V_{IH} but this did not affect the output logic level, keeping it at a logic 1 state. After time t_{11} the input voltage began to decrease again. At times t_{12}, t_{13}, and t_{14} the input traversed V_{IH}, V_T^+, and V_{IL}, respectively, none of which affected the logic 1 level maintained by the output. In fact, all that occurred was that at t_{11} (event 7) the direction of travel on the hysteresis curve was again reversed to the down direction. At time t_{15} the input voltage decreased below trigger voltage level V_T^- (event 8 on the hysteresis curve) and at that point the output voltage went from a logic 1 to a logic 0.

By comparing the input voltage V_I with the output voltage V_O of Figure 5.49, it can be seen that the waveform was reshaped to a pulse with fast rise and fall times and has a constant output voltage level during the interval between t_3 and t_{15}. This reshaping is due to the characteristics of a trigger with hysteresis. For once triggered in the logic 1 state the output cannot be driven to the logic 0 state unless the triggering voltage falls below the V_T^- trigger level.

Any logic gate can have a Schmitt trigger. An example discussed earlier is the 7414 hex Schmitt trigger inverter, illustrated in Figure 5.46. Logic gates with more than one input can also employ Schmitt triggering, however, for such gates it is the voltage level of all the inputs that must be considered when determining the positive-going and negative-going threshold voltages. To illustrate the operation of logic gates with more than one Schmitt triggered input, trace through the truth table of a two-input *AND* gate. Begin with both inputs at the logic 0 state (00). As is the case for any *AND* gate with all of its inputs in the logic 0 state, the output would be a logic 0. Now imagine that the logic state of one of the inputs goes to a logic 1 (01 or 10). Because only one of the two inputs traversed the hysteresis curve to a magnitude above the positive-going threshold voltage level the output remains in the logic 0 state. However, when the remaining input is forced to a logic 1 (the logic states of the inputs are now 11) it intersects the positive-going threshold voltage $V_T{}^+$. At this time the output is triggered from a logic 0 to a logic 1. If either of the two inputs (or both) are then forced back to the logic 0 state (00, 01, or 10) the hysteresis curve of the *AND* gate is traversed through the negative-going threshold voltage $V_T{}^-$. In passing through $V_T{}^-$ the output goes from a logic 1 to a logic 0. The output will remain in the logic 0 state until, as seen before, one input is a logic 1 and the other serves as the trigger voltage by being forced to a logic 1. As this trigger voltage traverses the hysteresis curve through the positive-going threshold voltage the output is triggered from a logic 0 to a logic 1.

The circuit shown in Figure 5.50 illustrates an application of a Schmitt trigger. Section 5.2 showed how a latch could be used to eliminate the effects of contact bounce produced by an SPDT switch (Fig. 5.11). The circuit shown in Figure 5.50 accomplishes the same thing but debounces a SPST switch. With the switch open the capacitor charges to V_{CC} volts ($+5$ volts) through the resistor R. The rate of charge and discharge of the capacitor is determined by the RC time constant ($R \times C$), and requires $5 \times R \times C$ seconds (50 ms for this design) to fully charge to V_{CC}. As the capacitor charges to a magnitude greater than $V_T{}^+$ it triggers the 7414 to produce a logic 0 at its output. The output remains in that logic state until the switch is closed. When the switch is closed the capacitor is shorted to ground and quickly discharges (the *RC* time constant is now zero) to zero volts. As the capacitor discharges, the threshold voltage $V_T{}^-$ is crossed and the output of the 7414 is triggered to a logic 1. Of course, when the switch is closed the contacts bounce (as was explained in Section 5.2). Because the *RC* time constant is so small when the switch is closed, the capacitor discharges to zero volts (or at least below the threshold voltage $V_T{}^-$) on the first contact closure of the switch. As the contacts bounce, it open circuits the short to ground and the capacitor again begins to charge. Because this open circuit is due to contact bounce, it is only momentary. As a result of the relatively long *RC* time constant for the switch open condition, the capacitor charges very little, remaining well below the trigger threshold voltage $V_T{}^+$, and the output remains in the logic 1 state. As the contacts continue to bounce the same events are repeated over and over again, keeping the output at a logic 1 state. If one is concerned about too much surge current when the switch is first closed, a small resistance (around 100 ohms) may be placed in the path to ground to limit the discharge current.

Figure 5.50

Switch debounce circuit using a 7414.

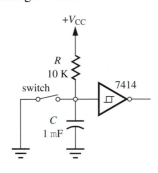

Review Questions

1. How can a gate with a Schmitt trigger input quicken the rise and fall times of a pulse?

2. Explain the significance of the Schmitt trigger's hysteresis curve with respect to the reshaping of an input voltage's rise and fall times.

3. Explain how a Schmitt trigger reshapes waveforms with distorted logic levels.

5.12
The monostable multivibrator

The flip-flop is often called a *bistable multivibrator* because it can exist in one of two (bi) possible stable states — *SET* or *RESET* — and it is capable of switching (vibrating) between these two states and remaining permanently in either one. There are two other types of multivibrators that are used in digital circuits — the *monostable multivibrator* and the *astable multivibrator*.

Figure 5.51

Logic symbol for a
monostable multivibrator
(one-shot).

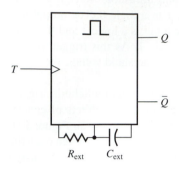

Monostable multivibrators have outputs Q and $\overline{Q}$, just like bistable multivibrators (flip-flops), as is illustrated in Figure 5.51. The difference between the outputs of a bistable multivibrator and a monostable multivibrator is the stability of their logic states. Whereas bistable multivibrators have two stable states, monostable multivibrators have just one, which is usually the *RESET* state ($Q = 0$). When activated or triggered, a monostable multivibrator switches from its stable state to an unstable state and remains there for a set period of time, after which it returns to its stable state. Having only a single stable state is the reason why they are known as monostable (one stable state) multivibrators. The logic symbol for the monostable multivibrator is shown in Figure 5.51.

As illustrated in the timing diagram of Figure 5.52, applying a trigger pulse to input T of a monostable multivibrator will cause the outputs to be temporarily driven out of the stable *RESET* state and into the unstable *SET* state ($Q = 1$). The trigger input of a monostable multivibrator is edge sensitive and, depending on the specific monostable multivibrator, may be either positive or negative edge-triggered. Figure 5.52 shows that for a positive edge-triggered monostable multivibrator (the middle timing diagram), output Q is driven to the unstable logic 1 state at time t_0, t_2, and t_6. For negative edge-triggering (the bottom timing diagram) this occurs at times t_1, t_3, and t_7. In both timing diagrams, the output Q was not affected at t_4 for positive edge-trigger, or at t_5 for negative edge-trigger. This is because the output of this type of monostable multivibrator is not affected by input T while in the unstable state. Monostable multivibrators with this triggering characteristic are known as *nonretriggerable* monostable multivibrators. Those that are affected by input T, regardless of the logic state of the output are known as *retriggerable* monostable multivibrators.

Figure 5.52

Timing diagram for a one-
shot.

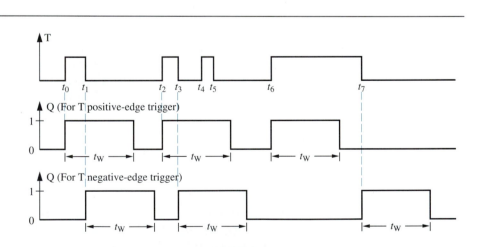

The duration of the unstable state is identified as t_W in Figure 5.52. Its length is determined by external components R_{ext} and C_{ext}. After the time t_W has elapsed the output Q automatically returns to the stable state and remains there until triggered again. To the casual observer the unstable state appears to be stable during interval t_W. For this reason the unstable state is also known as the *quasistable state*.

The monostable multivibrator is also known by other names, such as *one-shot* because "one shot" of the input T is needed to trigger the output to its quasistable state and *pulse stretcher* because the duration of the input pulse at T appears to have been ''stretched'' to duration t_W at the output (*see* timing diagram in Fig. 5.52).

The 74121 monostable multivibrator is a nonretriggerable one-shot. The pin configuration, logic symbol, and example circuit configuration for the 74121 are shown in Figure 5.53. It is important to point out that the pin configuration shows that there is an internal resistor identified as R_{int} (approximately 2 K ohm in value). As shall be seen later, this internal resistor, instead of an external resistor R_{ext}, may be used to determine the output pulse duration t_W.

Figure 5.53

Logic symbol, pin configuration, and triggering combinations of the 74121 monstable multivibrator.

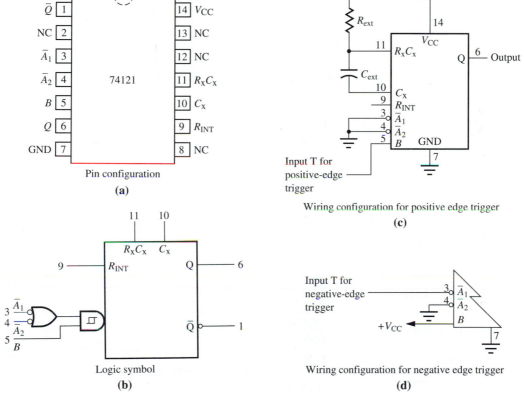

Pin configuration

(a)

Logic symbol

(b)

Wiring configuration for positive edge trigger

(c)

Wiring configuration for negative edge trigger

(d)

The logic symbol of Figure 5.53b shows that inside the 74121 is a two-input Schmitt-triggered *AND* gate. It is the output of this *AND* gate that provides the trigger input T (see Fig. 5.51) to the multivibrator. With two inputs there are two paths from which the trigger source can originate — either from the output of the *OR* gate (inputs $\overline{A}_1$ or $\overline{A}_2$), or input B. Recall from earlier discussion that Schmitt triggering will provide waveshaping for the input trigger pulse T. Because triggering of the 74121 is edge sensitive, slow rise and fall times (positive and negative slopes, respectively) are of special concern. Table 5.10 shows the recommended operating conditions for the 74121 obtained from the manufacturer's data sheets. In Table 5.10, if the parameter dv/dt (the slope of the triggering edge) is examined, the 74121 is recommended for minimum rise and fall times of 1 volt/s for input B and a

Table 5.10   Recommended operating conditions for the 74121

Parameter			74			Unit
			Min	Nom	Max	
V_{CC}	Supply voltage		4.75	5.0	5.25	V
I_{IK}	Input clamp current				-12	mA
I_{OH}	*HIGH*-level output current				-400	μA
I_{OL}	*LOW*-level output current				16	mA
dv/dt	Rate of rise or fall of input pulse	B input	1			V / s
		$\overline{A}_1, \overline{A}_2$ inputs	1			V / μs
T_A	Operating free-air temperature		0		70	°C

Table 5.11   DC Electrical characteristics of the 74121

Parameter		Test conditions[1]		74121			Unit
				Min	Typ[2]	Max	
V_{T+}	Positive-going threshold at $\overline{A}$ and B	V_{CC} = MIN				2.0	V
V_{T-}	Negative-going threshold at $\overline{A}$ and B	V_{CC} = MIN		0.8			V
V_{OH}	*HIGH*-level output voltage	V_{CC} = MIN, V_{IH} = MIN, V_{IL} = MAX, I_{OH} = MAX		2.4	3.4		V
V_{OL}	*LOW*-level output voltage	V_{CC} = MIN, V_{IH} = MIN, V_{IL} = MAX, I_{OL} = MAX			0.2	0.4	V
V_{IK}	Input clamp voltage	V_{CC} = MIN, $I_I = I_{IK}$				-1.5	V
I_I	Input current at maximum input voltage	V_{CC} = MAX, V_I = 5.5 V				1.0	mA
I_{IH}	*HIGH*-level input current	V_{CC} = MAX, V_I = 2.4 V	$\overline{A}_1, \overline{A}_2$ inputs			40	μA
			B input			80	μA
I_{IL}	*LOW*-level input current	V_{CC} = MAX, V_I = 0.4 V	$\overline{A}_1, \overline{A}_2$ inputs			-1.6	mA
			B input			-3.2	mA
I_{OS}	Short circuit output current[3]	V_{CC} = MAX		-18		-55	mA
I_{CC}	Supply current (total)	V_{CC} = MAX	Quiescent		13	25	mA
			Triggered		23	40	mA

[1]For conditions shown as MIN or MAX, use the appropriate value specified under recommended operating conditions for the applicable type.
[2]All typical values are at V_{CC} = 5 V, T_A = 25°C.
[3]I_{OS} is tested with V_{OUT} = $+0.5$ V and $V_{CC} = V_{CC}$ MAX $+0.5$ V. Not more than one output should be shorted at a time and duration of the short circuit should not exceed 1 second.

minimum of 1 volt/μs for inputs $\overline{A}_1$ and $\overline{A}_2$. Input B is used for positive edge-triggering with a positive edge (slope) of no less than 1 volt/s (which is quite slow), whereas inputs $\overline{A}_1$ and $\overline{A}_2$ are for negative edge-triggering with the trigger pulse falling edge being a minimum of 1 volt/μs. Table 5.11 shows the 74121's DC electrical characteristics obtained from the manufacturer's data sheets. Table 5.11 shows that the Schmitt trigger threshold voltages are 2.0 volts maximum for V_T^+ and a minimum of 0.8 volts for V_T^-. For positive edge-triggering, the input trigger pulse will trigger the one-shot at a maximum of 2.0 volts and for negative edge-triggering it occurs at a minimum of 0.8 volts.

Let us examine the triggering possibilities of the 74121 in greater depth. To reiterate, Figure 5.53b shows that there are two paths from which trigger source T may originate; from either the *OR* gate inputs $\overline{A}_1$ or $\overline{A}_2$, or input B. This of course provides the user with the choice of positive or negative edge-triggering. Because inputs $\overline{A}_1$ and $\overline{A}_2$ are inverted they would be used for the trigger source input when negative edge-triggering is desired and input B would be used when positive edge-triggering is to be used. Either input $\overline{A}_1$ or $\overline{A}_2$ can serve as the input for the trigger source or they may be connected to act as a single input. When either input $\overline{A}_1$ or $\overline{A}_2$ singly serves as the trigger input T, the other unused input must be at the logic 0 level, and input B must be at a logic 1. The circuit of Figure 5.53d is setup for negative edge-triggering using input $\overline{A}_1$ as the trigger source input T. The circuit could have been wired so that $\overline{A}_2$ served as input T, with $\overline{A}_1$ grounded. In either case, input B would have to be at the logic 1 level when the triggering pulse appeared. For positive edge-triggering input B serves as the trigger input T and inputs $\overline{A}_1$ and $\overline{A}_2$ are forced to a logic 0. Figure 5.53c illustrates the circuit configuration for positive edge-triggering.

The function table of the 74121 obtained from the manufacturer's data sheets is shown in Table 5.12. This function table is much like a truth table except that in addition to listing logic 0's (L) and 1's (H) it also has positive and negative trigger edges. The trigger edges are represented by vertical arrows. The arrow points up ($\uparrow$) to represent a positive edge and points down ($\downarrow$) for a negative edge.

Table 5.12 shows that for the input logic states of the first four rows, the outputs of the 74121 one-shot are in the stable *RESET* state ($Q = 0$ and $\overline{Q} = 1$). At no time

Table 5.12   74121 Function table

Inputs			Outputs	
$\overline{A}_1$	$\overline{A}_2$	B	Q	$\overline{Q}$
L	X	H	L	H
X	L	H	L	H
X	X	L	L	H
H	H	X	L	H
H	↓	H	⊓	⊔
↓	H	H	⊓	⊔
↓	↓	H	⊓	⊔
L	X	↑	⊓	⊔
X	L	↑	⊓	⊔

Abbreviations: H, *HIGH* voltage level; L, *LOW* voltage level; X, Don't care; ↑, *LOW*-to-*HIGH* transition; ↓, *HIGH*-to-*LOW* transition.

are they driven to the quasistable state, represented by the symbol of a pulse ($\sqcap$ or $\sqcup$), for the conditions shown in the first four rows. This is because there is no edge-triggering at any of the inputs. The fifth row indicates that input $\overline{A}_2$ receives a negative edge-trigger, which causes the outputs to be triggered to the quasistable state, as is indicated by the pulses in the output columns. Negative edge-triggering occurs for rows six and seven. Positive edge-triggering occurs when input B receives a positive edge-trigger and inputs $\overline{A}_1$ or $\overline{A}_2$ are a logic 0, as is the case for rows eight and nine.

Comparison of Figures 5.51 and 5.53c shows that the external components R_{ext} and C_{ext} of Figure 5.51 are connected to pins 10 and 11 of the 74121. The capacitor C_{ext} is connected between pins 10 and 11 and resistor R_{ext} is connected from pin 11 to the voltage source V_{cc}. The duration t_W of the output is established according to the following relationship:

$$t_W = 0.69\,R_{ext}\,C_{ext} \tag{5.8}$$

The manufacturer's data sheets for the 74121 include a graph of output pulse width versus timing resistor value (*see* Fig. 5.54). Rather than using Equation 5.8, Figure 5.54 could be used to graphically determine the component values for a desired duration t_W.

Figure 5.54

74121 output pulse width v/s timing resistor value.

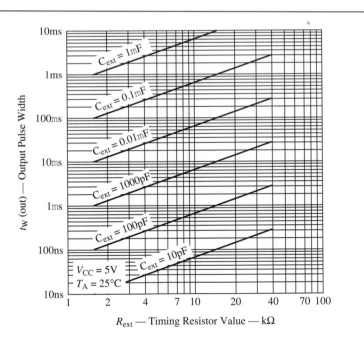

Example 5.3

From Figure 5.54 select a value for R_{ext} to produce a pulse duration of 650 μs when the value of C_{ext} is 0.1 μF. Use Equation 5.8 to verify if the answer is correct.

Solution

From Figure 5.54 locate the curve for $C_{ext} = 0.1$ μF. On the vertical axis locate the horizontal line for 650 μs (as best possible). At the intersection of the 650 μs horizontal line and the curve for C_{ext}, drop a vertical line to the R_{ext} axis, which is 10 K. This is the value of R_{ext} for a duration of 650 μs. To verify this using Equation 5.8, substitute $R_{ext} = 10$ K and $C_{ext} = 0.1$ μF into Equation 5.8 as follows:

$$t_W = 0.69 \times 10^4 \times 10^{-7} = 690 \ \mu s$$

Considering the fact that values taken from a graph are approximations, then indeed $R_{ext} = 10$ K is within range. When implementing the design, if closer tolerances are required for t_W, then an adjustable resistor may be used for R_{ext}.

In most design applications the designer knows only the desired pulse duration t_W and must determine R_{ext} and C_{ext}. Figure 5.54 will assist us with choosing values for both components. Examining Figure 5.54 shows that each C_{ext} curve has a range of t_w over which it can be used. That is, the curve for $C_{ext} = 1 \ \mu$F is valid for the range of 1 to 10 ms for t_W. For pulse durations from 100 μs to 3 ms the manufacturer recommends that a C_{ext} value of 0.1 μF be used.

Example 5.4

Using a 74121 one-shot, stretch a 0.5 μs pulse to a 20 μs pulse. Use the positive edge of the 0.5 μs pulse to trigger the one-shot.

Solution

Because positive edge-triggering is being used, the circuit wiring of Figure 5.53c will be used. Figure 5.54 shows that $t_W = 20 \ \mu$s falls within the range of the $C_{ext} = 0.01 \ \mu$F curve. Dropping vertically from the intersection of the horizontal line for $t_W = 20 \ \mu$s and the $C_{ext} = 0.01 \ \mu$F curve to the R_{ext} axis a value of 3 K is indicated for R_{ext}. These values can be verified using Equation 5.8.

$$t_W = 0.69 \times 10^{-8} \times 3 \times 10^3 = 20.7 \ \mu s$$

Considering the accuracy of obtaining values from a graph, they are within acceptable tolerances.

As stated earlier, the 74121 has an internal resistor R_{int} that can be used instead of external resistor R_{ext} and the value of this resistor is approximately 2 K ohms. The wiring configuration of Figure 5.55 illustrates this application. Note that pin 9 (R_{int}) is tied to the power supply V_{CC}. For variety, negative edge-triggering is used this time. For this circuit configuration Equation 5.8 is modified to:

$$t_W = 0.69 \, R_{int} \, C_{ext} \tag{5.9}$$

Figure 5.55

74121 wiring configuration using internal resistor R_{INT}.

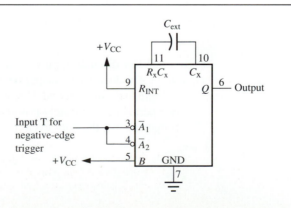

Example 5.5

What value of C_{ext} is required for the circuit of Figure 5.55 to produce an output pulse width of 70 ns?

Solution

Solving Equation 5.9 for parameter C_{ext} and assuming R_{int} is 2 K, we have:

$$C_{ext} = \frac{t_W}{0.69 \times R_{int}} = \frac{70 \times 10^{-9}}{0.69 \times 2 \times 10^3} = 50.7 \text{ pF}$$

To verify the assumed value for R_{int} refer to the AC electrical characteristics of the 74121 obtained from the manufacturer's data sheets and shown in Table 5.13. Table 5.13 shows that a C_{ext} of 80 pF (R_{int} to V_{CC}) will result in a t_W pulse width within a range of 70 ns minimum to a maximum of 150 ns. This says that (using Eq. 5.9 to solve for R_{int}) R_{int} has an actual value that falls within a range from a minimum of 1.268 K to a maximum of 2.717 K. Of course, to make t_W a precise value one would adjust the value of C_{ext} accordingly.

Also note from the AC electrical characteristics in Table 5.13 that t_W minimum (minimum output pulse width) is 20 to 50 ns with $C_{ext} = 0$ pF (no external capacitance) and pin 9 tied to V_{CC}. The necessary RC time constant for t_W is provided by the inherent stray capacitance of the IC.

Table 5.13   AC Electrical characteristics of the 74121

| | | | 74 | | |
| | | | $C_L = 15$ pF, $R_L = 400$ ohms | | |
Parameter		Test conditions	Min	Max	Unit
t_{PLH} t_{PHL}	Propagation delay $\overline{A}$ input to Q and $\overline{Q}$ output	Waveform 1 $C_{ext} = 80$ pF, R_{int} to V_{CC}		70 80	ns
t_{PLH} t_{PHL}	Propagation delay B input to Q and $\overline{Q}$ output	Waveform 2 $C_{ext} = 80$ pF, R_{int} to V_{CC}		55 65	ns
t_W	Minimum output pulse width	$C_{ext} = 0$ pF, R_{int} to V_{CC}	20	50	ns
t_W	Output pulse width	$C_{ext} = 80$ pF, R_{int} to V_{CC}	70	150	ns
		$C_{ext} = 100$ pF, $R_{ext} = 10$ K ohms	600	800	ns
		$C_{ext} = 1$ μF, $R_{ext} = 10$ K ohms	6.0	8.0	ms

When discussing pulse duration, another quantity must be introduced — *duty cycle*. The duty cycle of a periodic pulse is the percentage of time the pulse is present (a logic 1) during a period. Using the conventional symbol T to represent the time for one period (unfortunately it is also the same symbol used to symbolize the one-shot trigger input), the duty cycle may be expressed mathematically as:

$$\text{duty cycle} = \frac{t_W}{T} \times 100 \qquad (5.10)$$

Example 5.6

Trigger pulses of 1 μs duration are applied to pin 5 (positive edge-trigger input B) of a 74121 configured, as illustrated in Figure 5.53c, every 100 μs. The external components have the following values:

$$R_{ext} = 4.7 \text{ K and } C_{ext} = 0.01 \ \mu\text{F}$$

What is the duty cycle of the output at pin 6 of the 74121?

Solution

Use Equation 5.10 to calculate pulse duration t_W of the output. The period T is the same as that of the source of the trigger input, that is, $T = 100 \ \mu$s.

$$t_W = 0.69 \ R_{ext} C_{ext} = 0.69 \times 4.7 \times 10^3 \times 0.01 \times 10^{-6} = 32.43 \ \mu\text{s}$$

Substituting the values for t_W and T into Equation 5.10, we find that

$$\text{duty cycle} = \frac{32.43 \times 10^{-6}}{100 \times 10^{-6}} \times 100 = 32.43\%.$$

This says that for 32.43% of the period the output is a logic 1.

Table 5.14 shows the AC setup requirements of the 74121 as listed in the manufacturer's data sheets. It shows that there are two maximum output duty cycle limits given — 67% for $R_{ext} = 2$ K and 90% for $R_{ext} = 40$ K. For a resistance range from 2 to 40 K for external resistor R_{ext}, the duty cycle can range from 67% to 90%. The reason for this limit is the recovery time needed for the output to return from its quasistable state before it can again be retriggered to its quasistable state.

Table 5.14   AC Setup requirements for the 74121

	Parameter	Test conditions	Min	Max	Unit
			\ \ \ \ \ 74		
t_W	Minimum input pulse width to trigger	Waveforms 1 and 2	50		ns
R_{ext}	External timing resistor range		1.4	40	K ohms
C_{ext}	External timing capacitance range		0	1000	μF
	Output duty cycle	$R_{ext} = 2$ K ohms		67	%
		$R_{ext} = R_{ext}(\text{Max})$		90	%

Example 5.7

For the component values and period of Example 5.5:

$$R_{ext} = 4.7 \text{ K} \quad C_{ext} = 0.01 \ \mu\text{F} \quad T = 100 \ \mu\text{s}$$

What is the maximum allowable pulse duration of the quasistable state?

Solution

From Equation 5.10, solve for the pulse duration t_W for a maximum duty cycle of 67%. Since R_{ext} is 4.7 K and is relatively close to the 2 K specified for the lower limit duty cycle of 67%, solve Equation 5.10 for t_W, substituting 0.67 for DC:

$$t_W = DC \times T = 0.67 \times 100 \times 10^{-6} = 67 \ \mu s$$

This states that for a period of 100 μs with R_{ext} equal to 4.7 K the maximum pulse duration is 67 μs.

Retriggerable and resettable one-shot

The 74122 is a retriggerable, resettable monostable multivibrator. To understand what the terms "retriggerable" and "resettable" imply, study the timing diagram of Figure 5.56c. Before studying the timing diagram, examine the logic symbol of the 74122 in Figure 5.56b to become familiar with its operation. The 74122 is very much like the 74121 except that input B now has two inputs (B_1 and B_2), rather than one.

Figure 5.56

Pin configuration, logic symbol, and timing diagram for the 74122 retriggerable, resettable multivibrator.

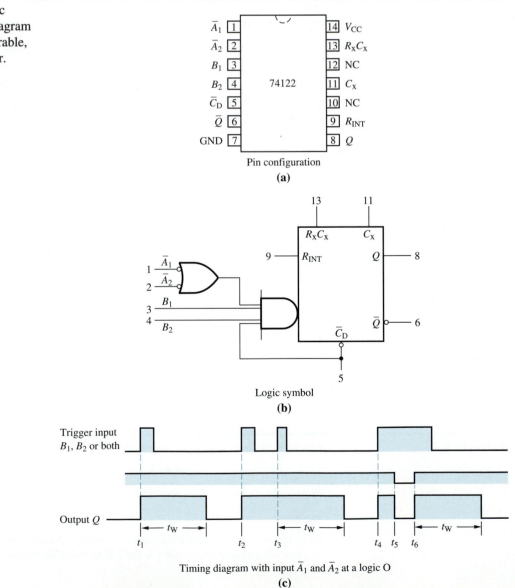

Pin configuration
(a)

Logic symbol
(b)

Timing diagram with input $\bar{A}_1$ and $\bar{A}_2$ at a logic O
(c)

The 74122 is positive edge-triggered by the output of the *AND* gate. Any one of the four inputs can provide the positive edge as long as the other inputs are at the logic 1 state. There is also an active low input ($\overline{C_D}$) to clear (reset the output to its stable state) the output. The $\overline{C_D}$ input also connects to the *AND* gate, whose output is the trigger source. Thus, when the clear signal input to pin $\overline{C_D}$ goes from a logic 0 to a logic 1, creating a positive edge, it will trigger the 74122 to the quasistable state assuming that the other inputs to the *AND* gate are a logic 1.

The timing diagram of Figure 5.56c illustrates that at time t_1, the trigger input (B_1 or B_2 or both) provides a positive edge-trigger, and $\overline{C_D}$ is in the inactive state (logic 1), resulting in the output being triggered to its unstable state (logic 1) for t_W seconds. After sufficient recovery time has passed, a second trigger pulse is applied at time t_2. The output is again triggered to the unstable state. However, before t_W seconds have elapsed, and the output returns to its stable state, another pulse is applied to the B input (t_3). For a nonretriggerable one-shot this pulse would be ignored, but, as we can see, it does affect the duration of the logic 1 state at the output. That is, at time t_3 the interval t_W begins again. Simply stated, the output has been retriggered. Continuing with our study of the timing diagram and its interpretation, at time t_4 another positive edge-trigger is applied to the input. The output is triggered to the unstable state but before t_W seconds have elapsed a clear pulse is applied to pin $\overline{C_D}$. When $\overline{C_D}$ goes low the output is reset to the stable logic 0 state. However, as soon as $\overline{C_D}$ goes inactive (high) this creates a positive edge (t_6) that is applied to the input of the *AND* gate. The output is then triggered and driven to its unstable state for a duration of t_W.

Table 5.15 shows the function table for the 74122. When the clear input is a logic 0 (L), as shown in the first row, the output is reset, as previously discussed. The only other parts of the table that are not self-explanatory are the last two rows. For these two rows, when inputs A and B have the proper logic levels to enable their respective inputs, and a positive edge-trigger appears at input $\overline{C_D}$, this will trigger the 74122. This confirms what has already been discussed with reference to the logic symbol and timing diagram of the 74122.

Table 5.15   Function table for the 74122

	Inputs				Outputs	
Clear	A_1	A_2	B_1	B_2	Q	$\overline{Q}$
L	X	X	X	X	L	H
X	H	H	X	X	L	H
X	X	X	L	X	L	H
X	X	X	X	L	L	H
X	L	X	H	H	L	H
H	L	X	↑	H	⊓	⊔
H	L	X	H	↑	⊓	⊔
H	X	L	H	H	L	H
H	X	L	↑	H	⊓	⊔
H	X	L	H	↑	⊓	⊔
H	H	↓	H	H	⊓	⊔
H	↓	↓	H	H	⊓	⊔
H	↓	H	H	H	⊓	⊔
↑	L	X	H	H	⊓	⊔
↑	X	L	H	H	⊓	⊔

To summarize, the 74122 is a retriggerable, resettable monostable multivibrator. The trigger inputs to the 74122 work in the same manner as the trigger inputs to the 74121 except that the 74122 has two B inputs, B_1 and B_2, instead of a single B input like the 74121. Because B_1 and B_2 are ANDed together, they must both be high if one of the A lines is being used to trigger the one-shot. The 74122 also provides a direct clear input ($\overline{C_D}$) that can be used to reset the one-shot to its stable state after it has been triggered into the quasistable state. The duration of the output pulse is established by the following relationship for the 74122:

$$t_W = 0.32\, R_{ext}\, C_{ext} \left[1 + \frac{0.7}{R_{ext}} \right]$$

The 74123 is also a retriggerable one-shot but contains two independent one-shots in one IC package as shown in Figure 5.57. Like the 74122, the 74123 can also be directly cleared while it is in a quasistable state by applying a low to the $\overline{C_D}$ input. However, each one-shot has only two trigger inputs — $\overline{A}$ and B. The negative edge of a pulse applied to the $\overline{A}$ input (while B is high) will trigger the one-shot, or the positive edge of a pulse applied to the B input (while $\overline{A}$ is low) can also be used to trigger the one-shot. The output pulse duration can be individually set for each one-shot by attaching the appropriate RC circuit.

Logic symbol and pin configuration of the 74123 dual retriggerable, resettable multivibrator.

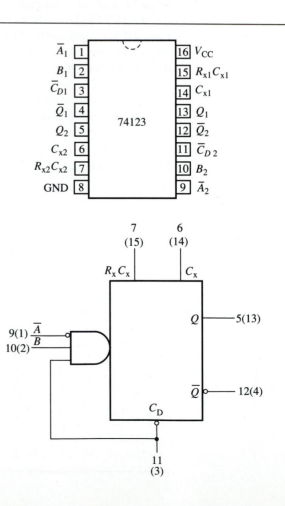

Most of the applications of one-shots are in timing applications and applications that require extremely short-duration pulses to be detected or stretched-out.

Review Questions

1. Why is the flip-flop known as a bistable multivibrator?
2. Describe the operation of a monostable multivibrator.
3. What is the difference between a nonretriggerable and retriggerable one-shot?
4. What is the purpose of the one-shot's $\overline{C}_D$ line?

5.13
The astable multivibrator

Two types of multivibrators have been discussed—the bistable multivibrator (flip-flop), which has two stable states, and the monostable multivibrator (one-shot), which has one stable state. There is a third type of multivibrator that has no stable states—the *astable multivibrator*. Figure 5.58 illustrates the timing for the output logic states of an astable multivibrator. The timing diagram shows that the output is in the logic 1 state for t_H seconds and then automatically switches to the logic 0 state for t_L seconds. This cyclic operation of automatically switching from one logic state to the other (oscillation) continues repeatedly as long as the astable multivibrator has power applied to it. The free running operation of the astable multivibrator gives it another name—the *free-running multivibrator*. Regardless of what it is known as, it is an oscillator with a *pulse repetition frequency* (PRF) of

$$\frac{1}{T} = \frac{1}{t_H + t_L}.$$

Figure 5.58

Timing diagram for an astable multivibrator.

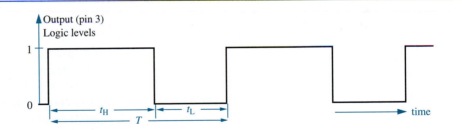

Applications for astable multivibrators are varied, but they most often serve as digital clocks that provide periodic clock pulses for timing purposes.

There are various astable multivibrator circuits. The circuit shown in Figure 5.59 is an astable multivibrator implemented with two 74123's. Another circuit shown in Figure 5.60a uses a 555 Timer IC configured as an astable multivibrator.

Figure 5.59 illustrates how two one-shots can be connected to produce a free-running multivibrator. Because the 74123 contains two one-shots, the circuit could easily be set up using one IC. The operation of the circuit can be explained with reference to the timing diagram shown in Figure 5.58 as follows.

Assuming that multivibrator M1 in Figure 5.59 is triggered on power-up, after a period t_1 when M1 returns to its stable state, it will trigger M2 on the negative edge

Figure 5.59

Two one-shots (74123) connected to function as a free-running (astable) multivibrator.

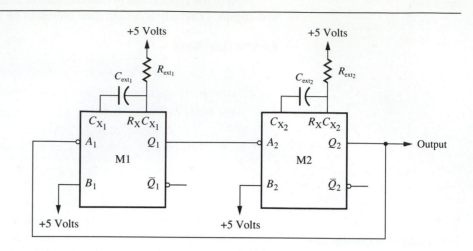

of the pulse. M2 will go to the quasistable state for a period t_2. When M2 returns to the stable state it will retrigger M1 on the negative edge, and the process continues repeatedly.

The duty cycle of the output stream of pulses can be varied by changing the values of R_{ext} and C_{ext} for each one-shot. If the values for these components are equal, then the output pulse will have a 50% duty cycle (the high and low durations will be equal). Assuming that the circuit is set up for a 50% duty cycle, the frequency (f) of the output pulse stream will be

$$\frac{1}{t_1 + t_2}.$$

The 555 Timer is a versatile IC with many applications. However, it will be used for only one application — the astable multivibrator shown in Figure 5.60a. Shown are its architecture (the area shaded in blue) as well as the required wiring configuration of the external components R_A, R_B, C, and the 0.01 μF capacitor. The DC power supply V_{CC} can be any magnitude within the range 4.5 to 18 volts. There is a voltage divider made up of three 5 K resistors (this is the reason why it is known as the "555") labeled "R," which create reference voltages for comparators 1 and 2. At the " $-$ " input of comparator number 1 is the reference voltage $\frac{2}{3} V_{CC}$ volts. All voltages applied to the " $+$ " input of comparator number 1 will be compared to this reference voltage. Similarly, all voltages applied to the " $-$ " input of comparator number 2 will be compared to a reference voltage of $\frac{1}{3} V_{CC}$. As will be seen, it is these comparisons that determine the state of the outputs at pins 3 and 7. The 555 output at pin 3 is the Q output of a flip-flop, and its output will always be a logic level. The discharge output (pin 7) is the collector of the transistor labeled "D". Its state (ON or OFF) is determined by the $\overline{Q}$ output of the flip-flop. When the flip-flop is set, the transistor is off and a high impedance exists between the collector and ground. However, when the flip-flop is reset the transistor is turned on and a low resistance path exists between the collector and ground. The 0.01 μF capacitor connected to pin 5, the control input, is to filter out any reference voltage fluctuations and thus provide constant reference voltages.

To understand how the 555 Timer functions as an astable multivibrator, first analyze its architecture and establish the cause-effect relationship between the input control voltage v_c at the threshold and trigger inputs (pins 6 and 2, respectively), and

Figure 5.60

The 555 Timer astable
multivibrator.

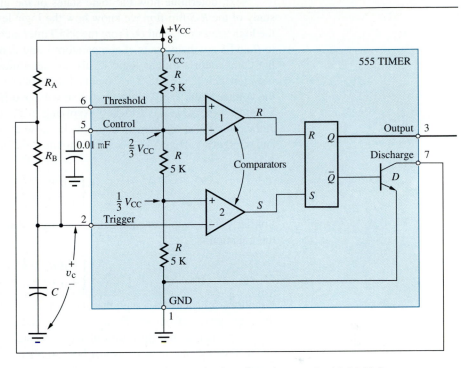

555 Timer IC architecture and circuit configuration as an Astable Multivibrator

(a)

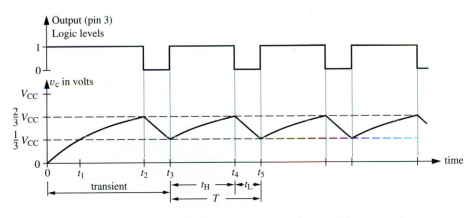

Timing-diagram coordinating 555 Timer input control voltages and the output voltage

(b)

the outputs at pins 3 and 7. This relationship can best be determined by beginning at
the outputs and retracing to the inputs, establishing cause-effect relationships along
the way. Using this method of analysis, Figure 5.60a shows that output pins 3 and 7
are controlled by the logic state of the *R-S* flip-flop. When the flip-flop is set, the
output at pin 3 is a logic 1 and transistor *D* at output pin 7 is off (a logic 0 at the base-
emitter junction of the transistor is not sufficient voltage to forward bias it). When
the flip-flop is reset, pin 3 is a logic 0 and the base-emitter junction of transistor *D* is
biased on, which in turn, turns transistor *D* on and effectively shorts its collector to
ground. We now understand how output pins 3 and 7 are controlled by the logic
states of the flip-flop.

Next, determine how the logic states of the flip-flop are controlled. From our study of the *R-S* flip-flop we know how the logic levels at inputs *R* and *S* determine the logic states of Q and $\overline{Q}$. From the 555 Timer architecture, the logic levels applied to *R* and *S* are the outputs of comparators 1 and 2, respectively. Tracing backwards, the comparator's output logic levels are determined by the difference in voltage between their "+" and "−" inputs. If the voltage at the "+" input is greater than the voltage at the "−" input, a positive voltage difference, the comparator output is a logic 1, otherwise it is a logic 0. Applying this reasoning to comparator number 1, and expressing it mathematically, it is because v_c is the voltage at the "+" input and reference voltage $\frac{2}{3}V_{CC}$ is applied to the "−" input, the logic state of the output *R* is

$$R = \begin{cases} 0 \text{ if } v_c < \frac{2}{3}V_{CC} \\ 1 \text{ if } v_c > \frac{2}{3}V_{CC}. \end{cases} \tag{5.11}$$

Similarly, for comparator number 2, the logic state of its output *S* is determined by the equation:

$$S = \begin{cases} 0 \text{ if } v_c > \frac{1}{3}V_{CC} \\ 1 \text{ if } v_c < \frac{1}{3}V_{CC}. \end{cases} \tag{5.12}$$

Equations 5.11 and 5.12 state that there are two reference voltage levels for which the logic levels of outputs *R* and *S* are dependent—$\frac{2}{3}V_{CC}$ and $\frac{1}{3}V_{CC}$.

Now analyze the operation of the 555 astable multivibrator shown in Figure 5.60a. Assume that power has just been applied to the circuit. This instant of time corresponds to $t = 0$ on the graph shown in Figure 5.60b, and at this instant, the capacitor (*C*) begins to charge through the circuit path of R_A and R_B. Voltage v_c increases toward the reference voltage $\frac{1}{3}V_{CC}$. During the time interval between 0 and t_1 ($0 < t < t_1$), Equations 5.11 and 5.12 will be used to determine the logic states at output pins 3 and 7, via comparator outputs *R* and *S*. During this interval the voltage across the capacitor, v_c is less than $\frac{1}{3}V_{CC}$ volts. Referring to Equations 5.11 and 5.12, if v_c is less than $\frac{1}{3}V_{CC}$, then $R = 0$ and $S = 1$. This sets the flip-flop and causes the logic level at pin 3 to be a logic 1, as shown in Figure 5.60b, and turns transistor *D* off (a logic 0 on the base turns the transistor off). With the transistor off (collector to emitter is an open circuit) the charge path from V_{CC} through R_A and R_B is maintained and the capacitor continues to charge. During the interval between t_1 and t_2 ($t_1 < t < t_2$), voltage v_c is greater than $\frac{1}{3}V_{CC}$ but less than $\frac{2}{3}V_{CC}$. Applying this voltage of v_c to Equations 5.11 and 5.12, $R = 0$ and $S = 0$, which is a no-change condition for the flip-flop and it therefore remains set. Output pin 3 remains a logic 1 and transistor *D* remains off, as indicated in Figure 5.60b. At time t_2 the capacitor charges just slightly greater than $\frac{2}{3}V_{CC}$. Referring to Equations 5.11 and 5.12 if v_c is greater than $\frac{2}{3}V_{CC}$, $R = 1$ and $S = 0$. This condition resets the flip-flop, which drives output pin 3 to a logic 0 and turns transistor *D* on (a logic 1 on the base turns *D* on). With transistor *D* on, its collector-to-emitter junction is (ideally) a short circuit, thus short circuiting the junction of resistor R_A and R_B to ground. As a result, capacitor *C* no longer has a power source from which it can draw current to continue its charging and therefore begins to discharge. Figure 5.60b indicates that the capacitor continues to discharge during the interval between t_2 and t_3 ($t_2 < t < t_3$). As previously determined, whenever v_c is between $\frac{1}{3}V_{CC}$ and $\frac{2}{3}V_{CC}$, as is the case during the interval between t_2 and t_3 ($t_2 < t < t_3$), $R = 0$ and $S = 0$. When $R = 0$ and $S = 0$, this is a no-change condition for the flip-flop, which remains in the reset state. The capacitor continues to discharge to a magnitude slightly less than $\frac{1}{3}V_{CC}$ volts at t_3. Because v_c is less than $\frac{1}{3}V_{CC}$, it can be concluded from Equations 5.11 and 5.12 that at time t_3, the comparator's outputs *R* and *S* are driven to $R = 0$ and $S = 1$. At t_3 the flip-flop is set. This drives pin 3 back to a logic 1 and turns transistor D off. With transistor D off, a charge path for capacitor C, via R_A and R_B, is again established and the capacitor begins to charge. This cyclic process is repeated continually, producing a periodic pulse with a duration of t_H and period T at output pin 3.

The design equations for determining t_H and t_L are as follows:

$$t_H = 0.69\,(R_A + R_B)\,C \qquad (5.13)$$

$$t_L = 0.69\,R_B\,C \qquad (5.14)$$

The duty cycle is

$$\frac{t_H}{T} \times 100 = \frac{t_H}{t_H + t_L} \times 100. \qquad (5.15)$$

Equation 5.13 mathematically states that the RC time constant of the charging circuit (when transistor D is off and the capacitor C charges) is via the two series resistors R_A and R_B and capacitor C; this is consistent with our analysis of the timing diagram of Figure 5.60b. Equation 5.14 states that the discharge circuit for capacitor C is via R_B and has an RC time constant proportional to the product of R_B and C. Equations 5.13 and 5.14 also state that t_H is always larger than t_L, because the quantity $R_A + R_B$ is always larger than R_B.

When Equations 5.13 and 5.14 are substituted into Equation 5.15 and simplified the resulting equation is

$$DC = \frac{R_A + R_B}{R_A + 2\,R_B}. \qquad (5.16)$$

Solving this equation for the ratio of resistors R_A and R_B will enable us to arrive at the duty cycle range over which the 555 astable multivibrator may operate. Solve for the ratio R_A/R_B

$$\frac{R_A}{R_B} = \frac{2\,DC - 1}{1 - DC} \qquad (5.17)$$

Considering the ratio of two resistors, which must be positive, the quantity $2\,DC - 1$ and $1 - DC$ must be positive. Express these inequalities mathematically

$$2\,DC - 1 > 0$$

From this inequality we algebraically solve for DC and find that

$$DC > 0.5 \qquad \text{(the duty cycle must be greater than 50\%)}$$

From the inequality

$$1 - DC > 0$$

we find that

$$DC < 1 \qquad \text{(must be less than 100\%)}$$

It can be concluded that the duty cycle range of operation is between 50% and 100%:

$$50\% < DC < 100\% \qquad (5.18)$$

This agrees with our previous reasoning, from Equations 5.13 and 5.14, and our analysis of the circuit.

In practice, one should not use a resistor with a value less than 500 ohms for R_A. The reason for this restriction is that when transistor D is on, R_A is essentially grounded through the collector-emitter path of the transistor, thus drawing a current equal to V_{CC}/R_A amps through that path. If V_{CC} is the maximum of 18 volts, then 36 mA would be drawn; any current greater than this is excessive for the collector-emitter junction.

Example 5.8

A 555 Timer configured as an astable multivibrator has a 67% duty cycle with a pulse duration of 150 μs.

(a) What is the ratio between R_A and R_B?

(b) For a 0.1 μF capacitor (C), what are the values of R_A and R_B?

(c) What is the PRF?

Solution

(a) From Equation 5.17,

$$\frac{R_A}{R_B} = \frac{2\,DC - 1}{1 - DC} = \frac{2 \times (0.67) - 1}{1 - 0.67} = 1.03$$

(b) Solving for two unknowns, R_A and R_B requires two independent equations. Use Equation 5.13 and the answer from part (a):

$$R_A = 1.03 \times R_B$$

Substituting this equation into Equation 5.13

$$t_H = 0.69 \times [(1.03 \times R_B) + R_B] \times C = 1.4 \times R_B \times C$$

Solving for R_B,

$$R_B = \frac{t_H}{1.4 \times C} = \frac{150 \times 10^{-6}}{1.4 \times (0.1 \times 10^{-6})} = 1071 \text{ ohms}$$

or 1 K ohms using standard component values.

To determine the value for R_A, use the answer to part (a).
Hence,

$$R_A = 1.03\,R_B = 1.03 \times 1000 = 1030 \text{ ohms}$$

or 1 K ohms using standard component values.

As a check, use 1 K ohms for R_A and R_B in Equation 5.13 to see how close t_H will be to our design specifications.

$$t_H = 0.69 \times (2 \times 10^3) \times 0.1 \times 10^{-6} = 138 \text{ } \mu s$$

which is approximately an 8% difference from the 150 μ specified. If greater precision is required, precision components can be used, or potentiometers can be used instead of fixed value resistors.

Also check the duty cycle. Using Equation 5.16,

$$DC = \frac{R_A + R_B}{R_A + 2\,R_B} = \frac{2}{3} = 0.666 \text{ or } 67\%.$$

(c) To determine the PRF, refer to Equation 5.15. Because the values for t_H and DC are known, solve Equation 5.15 for t_L.

$$t_L = \frac{t_H(1 - DC)}{DC} = \frac{150 \times (1 - 0.67) \times 10^{-6}}{0.67} = 73.88 \text{ } \mu s$$

and the period "T" is:

$$T = t_H + t_L = (150 + 73.88) \times 10^{-6} = 223.88 \text{ } \mu sec$$

For a period of 223.88 μs, the PRF is

$$\frac{1}{T} = \frac{1}{223.88 \times 10^{-6}} = 4.47 \times 10^3 \text{ pulses per second.}$$

Review Questions

1. What is the difference between the monostable and the astable multivibrator?
2. What feature of the 74123 allows it to be configured as an astable multivibrator?
3. Why is it not possible for one to achieve a perfect 50% duty cycle with the 555 Timer configured as an astable multivibrator?

5.14
Troubleshooting sequential logic

The logic probe and pulser can be used to test flip-flops and other multivibrators for faults. Figure 5.61 illustrates how a D flip-flop can be tested for proper operation using these two test instruments. The D input is tied high and the pulser is used to inject a pulse into the clock input of the D flip-flop. The logic probe can then be used to examine the Q and $\overline{Q}$ outputs and verify that they are in the high and low state, respectively. Similarly, the D flip-flop can be tested with a low applied to the D input.

Figure 5.61

Testing a D flip-flop for proper operation.

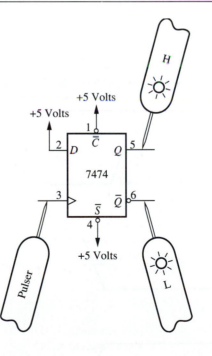

Unconditioned asynchronous inputs are one of the most common problems with flip-flop circuits. If the direct set ($\overline{S}$) and clear ($\overline{C}$) inputs are not being used, they should be tied inactive at the *HIGH* state. If these inputs are left floating, they are susceptible to picking up noise and could cause the flip-flop to set and reset erratically.

Besides the faults that are common in combinational logic circuits (shorted inputs, shorted outputs, floating inputs, etc.), sequential logic circuits (i.e., circuits containing flip-flops) can be affected by faults caused by timing problems. These problems are generally due to the effects of propagation delays at high operating frequencies although many of the faults due to timing problems could also be attributed to improper setup and hold times for flip-flop circuits.

Figure 5.62

Propagation delay at high frequency.

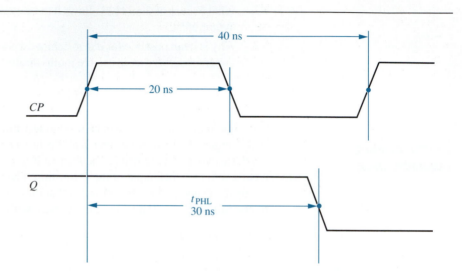

Figure 5.63

Effects of propagation delays.

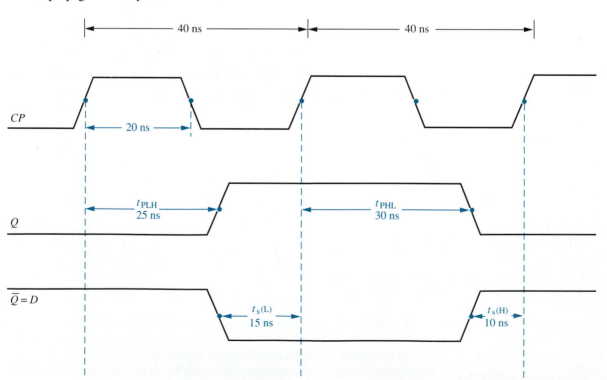

Propagation delays in sequential logic circuits become exaggerated at high frequencies. For example, consider a standard 7474 *D*-type flip-flop being clocked at its maximum frequency of 25 MHz (see data sheet in Figure 5.40). The period of the clock at 25 MHz would be 40 ns. The propagation delay t_{PHL} from the clock to the output would be more than half the period of the clock as shown in Figure 5.62. This could lead to problems in many circuits.

To illustrate the effects of propagation delays at high frequencies consider the timing diagram of the *D* flip-flop that has been configured as a *T* flip-flop as shown in Figure 5.38. Again, assume that the *D* flip-flop is being clocked at its maximum clock frequency of 25 MHz. Now the output *Q* of the flip-flop will toggle from a *LOW* to a *HIGH* 25 ns (t_{PLH}) after the clock makes its transition and will toggle from a *HIGH* to a *LOW* 30 ns (t_{PHL}) after the clock makes the next transition. The timing diagram is shown in Figure 5.63. The $\overline{Q}$ output of the flip-flop will have the same delays. Because the $\overline{Q}$ output is connected to the *D* input of the flip-flop, the setup time $t_s(L)$ will be only 15 ns and the setup time $t_s(H)$ will be only 10 ns. The data sheet in Figure 5.40 specifies minimum setup times of 20 ns for both $t_s(H)$ and $t_s(L)$. This would cause unreliable operation of the toggle flip-flop.

The only solution to this type of a problem is to use a *D*-type flip-flop with a smaller propagation delay or use a *J-K* flip-flop configured as a toggle flip-flop. More examples of such faults and their troubleshooting procedures will be examined in Chapter 6.

Review Questions

1. Why is it important to condition the direct set and clear inputs of a flip-flop?
2. When do propagation delays generally affect the operation of a sequential logic circuit?

Summary

- Combinational logic circuits are circuits whose outputs depend only on binary combinations applied to the inputs.

- Sequential logic circuits are circuits whose outputs depend on binary combinations applied to the inputs as well as on the past history of the outputs.

- A latch is a sequential logic circuit that can exist in either the set or reset states.

- A set state indicates that the output of a sequential logic circuit is a logic 1.

- A reset or clear state indicates that the output of a sequential logic circuit is a logic 0.

- A timing diagram describes the operation of a sequential logic circuit by graphically showing the relationship between the inputs and outputs.

- A gated latch has a gate input that determines when the latch will change states.

- A flip-flop is a gated latch that changes states on the transition of a clock pulse applied to the gate.

- The gate input of a flip-flop is known as the clock input.

- Flip-flops can be made to change states (trigger) by a negative or positive transition of a clock pulse.

- The negative transition of the clock pulse is the falling edge or the transition from a logic 1 to a logic 0.

- The positive transition of the clock pulse is the rising edge or the transition from a logic 0 to a logic 1.

- A negative edge-triggered flip-flop changes states on the negative edge of a clock pulse.

- A positive edge-triggered flip-flop changes states on the positive edge of a clock pulse.

- A negative edge-triggered flip-flop is identified by a bubble at its clock input whereas a positive edge-triggered flip-flop does not have a bubble at its clock input.

- A state transition table describes the operation of a flip-flop by identifying its next state given its present state and the logic levels at its control inputs.

- The *T* or toggle flip-flop has no control inputs but only has a single clock input.

- The next state of a T flip-flop is always equal to the complement of its present state.

- The *D* or data flip-flop has a single control input that determines the next state of the flip-flop.

- The next state of a D flip-flop is equal to the value of D before the clock pulse triggers the flip-flop.

- Flip-flops can have synchronous control inputs that cause the flip-flop to change states only when the clock pulse makes its transition, and asynchronous control inputs that cause the flip-flop to change states without requiring a clock pulse transition.

- The asynchronous inputs to a flip-flop are generally used to directly set or clear the flip-flop and are known as the direct set and clear inputs.

- The *J-K* flip-flop is similar in operation to the S-R flip-flop except that it does not have an invalid combination for the two control inputs.

- The *J* input of the *J-K* flip-flop sets the flip-flop while the K input resets the flip-flop.

- A master-slave flip-flop is also known as a pulse-triggered flip-flop.

- For a master-slave flip-flop to change states, a complete pulse must appear at its clock input.

- The propagation delay time of a flip-flop is usually measured from the time the clock triggers the flip-flop to the time the output changes state.

- The propagation delay time can also be measured from the time the direct set or clear inputs are activated to the time the output changes state.

- The setup time of the flip-flop is the minimum amount of time that the control input of a flip-flop is to remain stable before the clock makes its transition.

- The hold time of the flip-flop is the minimum amount of time that the control input of the flip-flop is to remain stable after the clock makes its transition.

- A gate with a Schmitt trigger input can respond to a slow transition without producing undesirable oscillations at the output.

- The flip-flop is also known as a bistable multivibrator because it can exist in one of two possible stable states.

- A multivibrator that can exist in only one stable state is known as a monostable multivibrator or one-shot.

- A multivibrator that does not have any stable states is known as a free-running or astable multivibrator.

- Propagation delays generally affect the operation of sequential logic circuits at high operating frequencies.

Figure 5.64

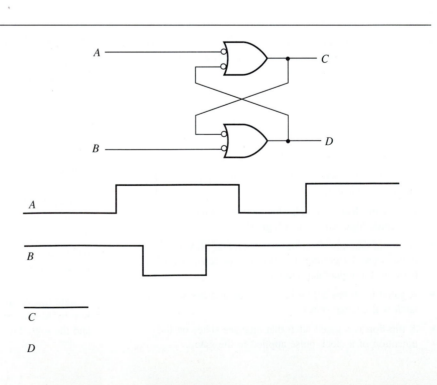

Problems

Section 5.2   The latch

1. Complete the timing diagram for the circuit shown in Figure 5.64.

2. Complete the timing diagram for the circuit shown in Figure 5.65.

3. Identify the Q and $\overline{Q}$ outputs of the circuit in Figure 5.64 and the inputs to *SET* and *RESET* the latch.

4. Identify the Q and $\overline{Q}$ outputs of the circuit in Figure 5.65 and the inputs to *SET* and *RESET* the latch.

Section 5.4   The *S-R* flip-flop

5. Complete the timing diagram shown in Figure 5.66 for a positive edge-triggered *S-R* flip-flop.

6. Complete the timing diagram shown in Figure 5.66 for a negative edge-triggered *S-R* flip-flop.

Figure 5.65

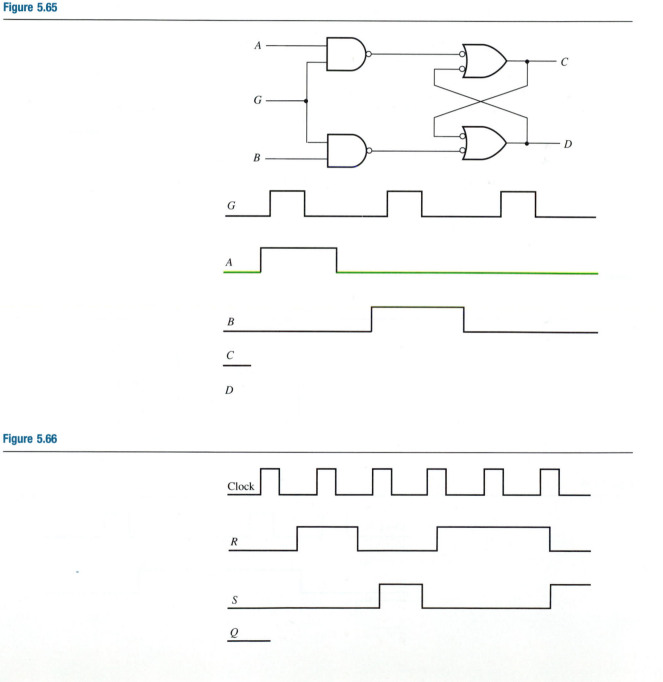

Figure 5.66

7. Complete the timing diagram shown in Figure 5.67 for a negative edge-triggered *S-R* flip-flop.

8. Complete the timing diagram shown in Figure 5.67 for a positive edge-triggered *S-R* flip-flop.

Section 5.5   The *T* flip-flop

9. Draw a timing diagram for the circuit shown in Figure 5.68 illustrating the outputs *X* and *Y* as a function of the clock.

10. Design a circuit using *T* flip-flops to accept a clock pulse having a frequency of 224 Hz as input and producing a clock pulse having a frequency of 7 Hz as output.

Section 5.6   The *D* flip-flop

11. Complete the timing diagram shown in Figure 5.69 for a positive edge-triggered *D* flip-flop.

12. Complete the timing diagram shown in Figure 5.69 for a negative edge-triggered *D* flip-flop.

Figure 5.67

Clock

R

S

$\overline{Q}$

Figure 5.68

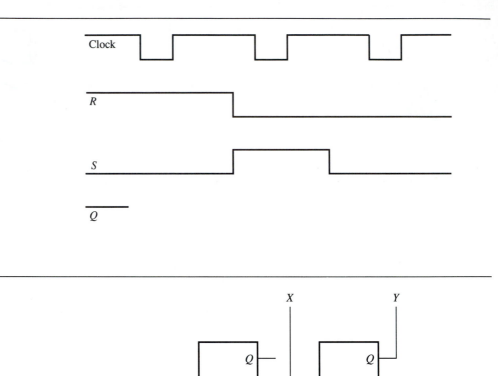

Figure 5.69

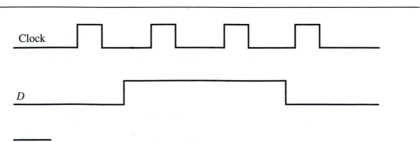

Clock

D

$\overline{Q}$

13. Complete the timing diagram shown in Figure 5.70 for a negative edge-triggered *D* flip-flop.

14. Complete the timing diagram shown in Figure 5.70 for a positive edge-triggered *D* flip-flop.

Section 5.7   The *J-K* flip-flop

15. Complete the timing diagram shown in Figure 5.71 for a positive edge-triggered *J-K* flip-flop.

16. Complete the timing diagram shown in Figure 5.71 for a negative edge-triggered *J-K* flip-flop.

17. Complete the timing diagram shown in Figure 5.72 for a negative edge-triggered *J-K* flip-flop.

18. Complete the timing diagram shown in Figure 5.72 for a positive edge-triggered *J-K* flip-flop.

Figure 5.70

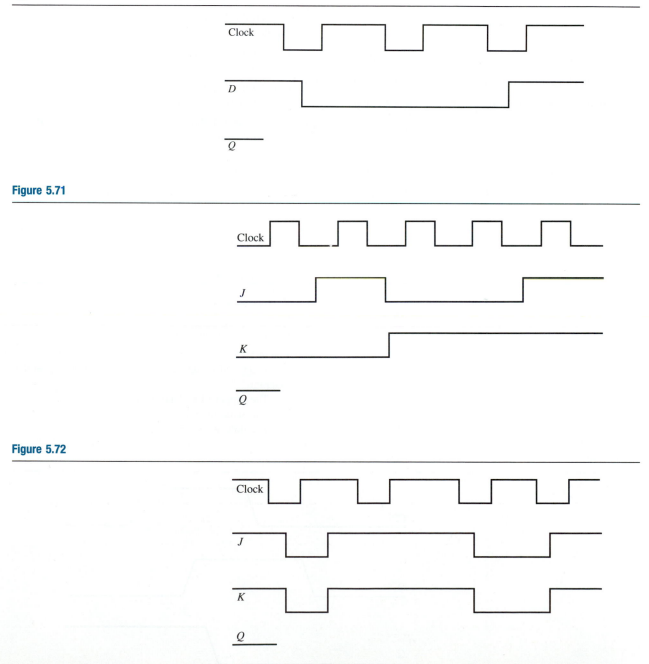

Figure 5.71

Figure 5.72

Section 5.8   Master-slave flip-flops

19. Complete the timing diagram shown in Figure 5.66 for a master-slave *S-R* flip-flop.

20. Complete the timing diagram shown in Figure 5.67 for a master-slave *S-R* flip-flop.

21. Complete the timing diagram shown in Figure 5.69 for a master-slave *D* flip-flop.

22. Complete the timing diagram shown in Figure 5.70 for a master-slave *D* flip-flop.

23. Complete the timing diagram shown in Figure 5.71 for a master-slave *J-K* flip-flop.

24. Complete the timing diagram shown in Figure 5.72 for a master-slave *J-K* flip-flop.

25. Draw a timing diagram for a master-slave *T* flip-flop.

Section 5.9   Flip-flop conversions

26. Configure the *S-R* flip-flop to function like a *J-K* flip-flop.

27. Configure the *S-R* flip-flop to function like a *D* flip-flop.

28. Configure the *S-R* flip-flop to function like a *T* flip-flop.

29. Analyze the circuit shown in Figure 5.37 by constructing a state transition table showing Q^{n+1} as a function of S, R, Q^n, and D.

30. Analyze the circuit shown in Figure 5.39 by constructing a state transition table showing Q^{n+1} as a function of J, K, Q^n, $\overline{Q}^n$, and D.

Figure 5.73

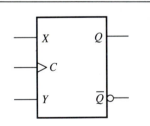

31. The operation of the *X-Y* flip-flop whose logic symbol is shown in Figure 5.73 is described by the following summarized state transition table.

X	Y	Q^{n+1}
0	0	1
0	1	Q^n
1	0	$\overline{Q}^n$
1	1	0

Construct a complete state transition table and obtain a state transition equation for the *X-Y* flip-flop.

32. Convert the *X.Y* flip-flop from Problem 31 to a *D*-type flip-flop.

33. Convert the *X-Y* flip-flop from Problem 31 to a *T*-type flip-flop.

34. Convert the *X-Y* flip-flop from Problem 31 to a *J-K*-type flip-flop.

35. Convert the *X-Y* flip-flop from Problem 31 to a *S-R*-type flip-flop.

36. The transition equation for the *L-M* flip-flop is

$$Q^{n+1} = \overline{L}Q^n + ML$$

Construct a transition table for the flip-flop.

Section 5.10   Switching characteristics of flip-flops

37. Figure 5.74 shows a timing diagram of a positive edge-triggered *D* flip-flop. Identify the following on the diagram:
    a. The propagation delay t_{PHL}
    b. The setup time $t_s(H)$
    c. The hold time $t_h(H)$

38. Figure 5.75 shows a timing diagram of a negative edge-triggered *D* flip-flop. Identify the following on the diagram:
    a. The propagation delay t_{PHL}
    b. The setup time $t_s(L)$
    c. The hold time $t_h(L)$

Figure 5.74

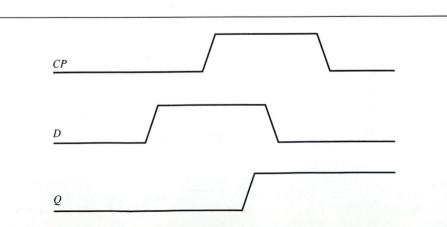

Figure 5.75

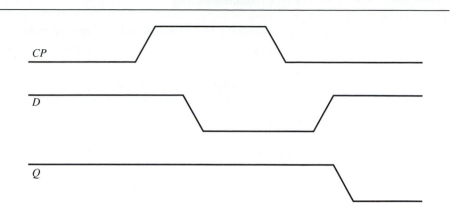

Figure 5.76

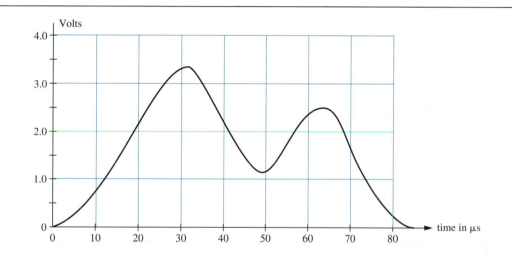

Section 5.11   The Schmitt trigger

39. The waveform shown in Figure 5.76 is applied to the input of a 7414 Schmitt trigger inverter. Make a scaled drawing of the output voltage. For the amplitude of the 7414's output voltage (listed as V_{OH} in the data sheets) assume the typical value of 3.4 volts. Also assume typical values for the threshold voltage levels, V^+ and V^-.

40. A sinusoidal waveform with a peak voltage of 2.5 volts is applied to the input of a Schmitt trigger buffer. The waveform is offset by a $+2.5$ volt DC voltage from the X axis thus producing a sine wave that oscillates between 0 and $+5$ volts. Sketch the output waveform produced by the buffer. Assume that V^+ and V^- are approximately 1.5 and 0.5 volts, respectively.

41. When a Schmitt trigger inverter is used to debounce an SPST switch, how is the input voltage to the 7414 prevented from oscillating above threshold voltage V^+ as the contacts bounce?

Section 5.12   The monostable multivibrator

42. Design a 74121 one-shot circuit to produce an active-high pulse having a duration of 1 ms. Pick any appropriate values of R_{ext} and C_{ext}.

43. Design a 74122 one-shot circuit to produce an active-low pulse having a duration of 0.5 ms. Pick any appropriate values of R_{ext} and C_{ext}.

Section 5.13   The astable multivibrator

44. Determine the values of R_{ext} and C_{ext} in Figure 5.59 to obtain a free-running multivibrator having a symmetric square-wave output with a frequency of 1000 Hz.

45. Determine the values of R_A, R_B, and C for the 555 Timer circuit in Figure 5.60 to produce a free-running multivibrator having a symmetric square-wave output with a frequency of 1000 Hz.

Troubleshooting

Section 5.14   Troubleshooting sequential logic

46. Construct a circuit to test a *J-K* flip-flop. Discuss the procedure that could be used to verify its operation.

47. Referring to the timing diagram of Figure 5.63, calculate a safe frequency to operate the flip-flop while maintaining the setup times at the minimum of 20 ns.

6

COUNTERS

OBJECTIVES

The objectives of this chapter are to:

- Study the design and operation of asynchronous binary counters and their analysis using timing diagrams

- Introduce synchronous counters and their analysis using timing diagrams

- Learn the design procedures (synthesis) used to design various types of synchronous counters

- Develop a formal technique for analyzing synchronous binary counters

- Introduce an application of counters by designing a 24-hour time-of-day digital clock

- Study the application of counters for frequency division

- Investigate the operation and applications of the various IC implementations of counters

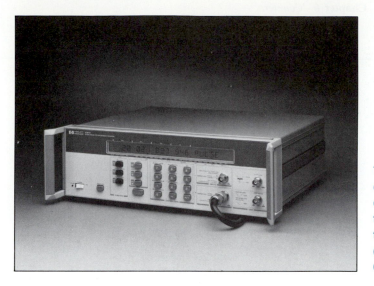

The dynamic operation of digital circuits often requires the use of a digital frequency counter (shown in the photograph) for testing and troubleshooting procedures (Photo courtesy Hewlett-Packard Company).

6.1
Introduction

The basic flip-flop introduced in Chapter 5 has many applications in sequential logic circuits, one of which is in the design of counters. A *counter* is a circuit that produces a numeric count each time an input clock pulse makes a transition. The general logic symbol for a typical counter is shown in Figure 6.1. The clock input to the counter is used to change the state of the counter in much the same way as the clock input to a flip-flop is used to change the state of the flip-flop; the difference is that a counter has several outputs lines (labeled "X" in Figure 6.1) that can exist in several states (called *counts*) and a flip-flop has only one output (not including $\overline{Q}$) that can exist in one of two states. The maximum number of states that a counter can have depends on the relationship 2^n where n is the number of outputs. A 3-bit counter will have 8 possible states and a 4-bit counter will have 16. During each state, the output of the counter has a different value that identifies the state of the counter. The maximum number of states that a counter counts through is known as its *modulus*. The number of outputs dictate the size of the counter (2-bit, 3-bit, 4-bit, etc.), and the actual sequence in which the counter counts (BCD, binary, etc.) at these outputs depends on its design. A counter may have one or more control inputs (labeled "C" in Fig. 6.1) that could be used to modify the counting sequence of the counter.

This chapter studies the operation, design, and analysis of the various types of counters used in digital circuit applications. The counters introduced in this chapter are classified into two broad categories—*synchronous* and *asynchronous*. Asynchronous counters are characterized by very simple logic circuits as long as the count they produce is simple. For more complex counting sequences synchronous counters can be easily designed using the techniques of synthesis covered in Section 6.4. The analysis of these counters is also included in this chapter. The timing diagram is a versatile tool for analyzing all types of counters, but for synchronous counters the analysis techniques discussed in Section 6.5 is more direct. Sections 6.6 and 6.7 in-

Figure 6.1

General logic symbol for a counter.

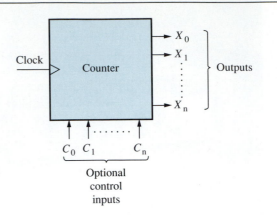

clude the complete design of a counter application — a time-of-day digital clock — and the application of counters in frequency division. Finally, Section 6.8 investigates the operation and applications of various counters implemented in the form of ICs.

6.2
Asynchronous counters

An asynchronous counter is made up of a set of T flip-flops connected in series as shown in Figure 6.2. The use of the T flip-flop is symbolic and could be replaced with the J-K or D implementations designed in Figures 5.35 and 5.38, respectively. The counter has three outputs labeled "X_2", "X_1", "X_0" and a *CLOCK* input that is used to cycle the counter through its various states. It is therefore a 3-bit counter and will have 2^3 or 8 possible states. The inputs and outputs of each flip-flop are labeled with the subscripts 0, 1, and 2 to identify the significance of each bit. The *CLOCK* input is connected to flip-flop $F0$ that will toggle each time a positive transition of the clock pulse takes place. The output of $F0$ is connected to the T input of $F1$ and serves as the clock for $F1$; $F1$ will therefore toggle each time the output of $F0$ makes a positive transition. Similarly, the output of $F1$ is the clock for $F2$ and therefore $F2$ will toggle each time the output of $F1$ makes a positive transition. It is apparent that all flip-flops will not change states at the same time, hence the term *"asynchronous,"* which

Figure 6.2

A 3-bit asynchronous down counter.

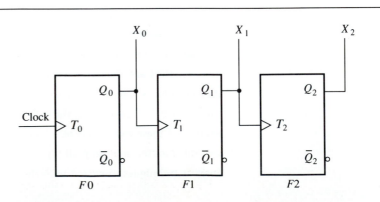

defines events that do not occur at the same time. The asynchronous counter is also called a *ripple counter* because the clock pulse appears to ripple through each flip-flop connected in series.

Figure 6.3 shows the timing diagram for the 3-bit ripple counter of Figure 6.2. The timing diagram shows the outputs of the counter (which are the outputs of the flip-flops) as a function of the *CLOCK* input of the counter. The counter is initially at a state at which its output is 000 (state 0). As each clock pulse of *CLOCK* makes a positive transition, X_0 toggles. Similarly, for each positive transition of X_0, X_1 toggles, and for each positive transition of X_1, X_2 toggles. If the binary code present at the output of the counter is determined for each transition of the main clock pulse *CLOCK,* the sequence is

$$000, 111, 110, 101, 100, 011, 010, 001, 000, 111, \ldots.$$

These outputs are referred to as *states*, and are listed, respectively, as decimal equivalents as follows:

$$0, 7, 6, 5, 4, 3, 2, 1, 0, 7, \ldots$$

From the timing diagram shown in Figure 6.3 the following observations can be made:

1. A *down counter* counts down from 7 to 0.

2. A *binary counter* counts through all possible combinations of 3 bits.

3. The counter functions continuously. Once it *cycles* through all possible states, it *recycles* through those same states in the same sequence.

Figure 6.3

Timing diagram for a 3-bit asynchronous down counter.

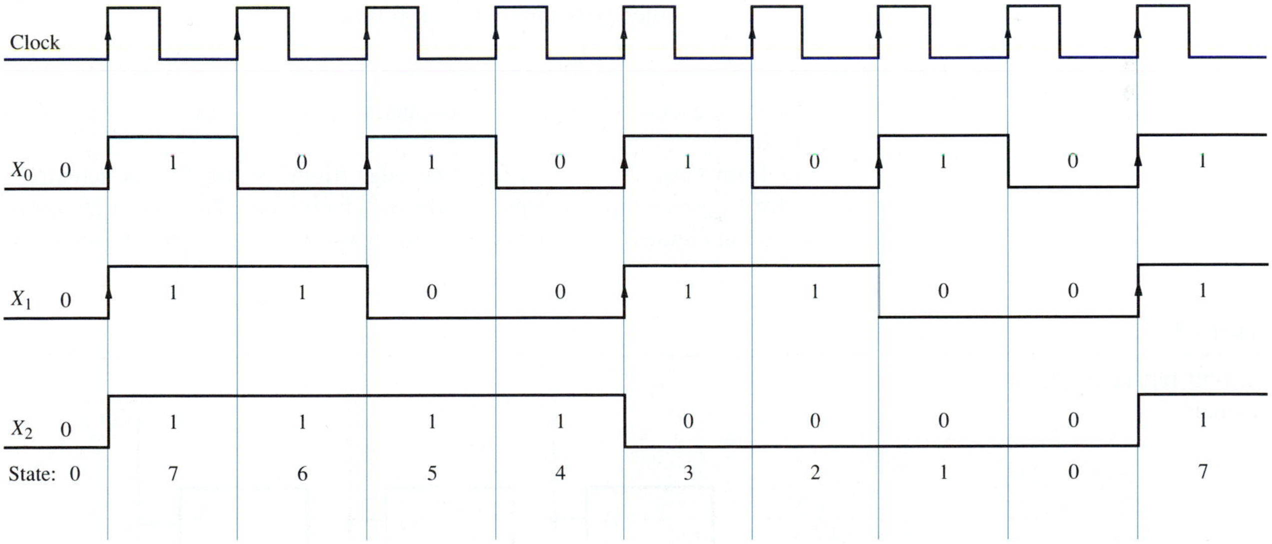

One of the advantages of ripple counters is that they can easily be extended to include more states simply by adding more flip-flops in series. For example, Figure 6.4 shows a 4-bit binary ripple down counter that includes an additional flip-flop $F3$

Figure 6.4

A 4-bit binary ripple down
counter.

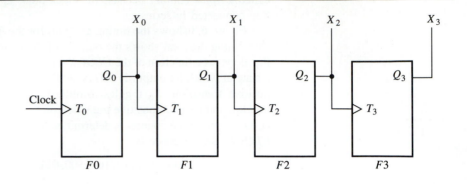

that will toggle each time the output of $F2$ (X_2) makes a positive transition. The counter will have 16 states 0 to 15 and count down in the sequence

$$0000, 1111, 1110, 1101, 1100, 1011, 1010, \ldots, 0000.$$

A 3-bit asynchronous ripple up counter can be constructed by modifying the down counter circuit in Figure 6.2 to contain negative edge-triggered instead of positive edge-triggered flip-flops. The resulting circuit is shown in Figure 6.5. Observe that there is no difference between the circuits of Figures 6.2 and 6.5 other than the type of flip-flops (negative or positive edge-triggered) used. The operation of the up counter can be verified by analyzing the circuit and obtaining a timing diagram as shown in Figure 6.6. The timing diagram analysis in Figure 6.6 is conducted in much the same way as the analysis for the down counter, except that the output of flip-flop $F0$ (X_0) toggles on the negative edge of the clock pulse. Similarly, the output of $F1$ (X_1) toggles on the negative edge of X_0, and the output of $F2$ (X_2) toggles on the negative edge of X_1. For each transition of the *CLOCK,* the sequence of states that the counter goes through is as follows:

$$0, 1, 2, 3, 4, 5, 6, 7, 0, 1, \ldots$$

The count sequence is an up count because it increases from 0 to 7 and then recycles to 0.

Recall from Chapter 5 that a negative edge-triggered flip-flop is actually a positive edge-triggered flip-flop with an inverted clock input. To convert the down counter circuit in Figure 6.2 to an up counter, the positive edge-triggered flip-flops do

Figure 6.5

A 3-bit binary ripple up
counter.

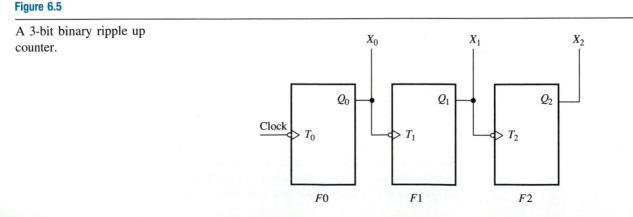

Figure 6.6

Timing diagram for a 3-bit binary ripple up counter.

Clock
X_0
0
1
0
1
0
1
0
1
0
1
X_1
0
0
1
1
0
0
1
1
0
0
X_2
0
0
0
0
1
1
1
1
0
0
State:
0
1
2
3
4
5
6
7
0
1

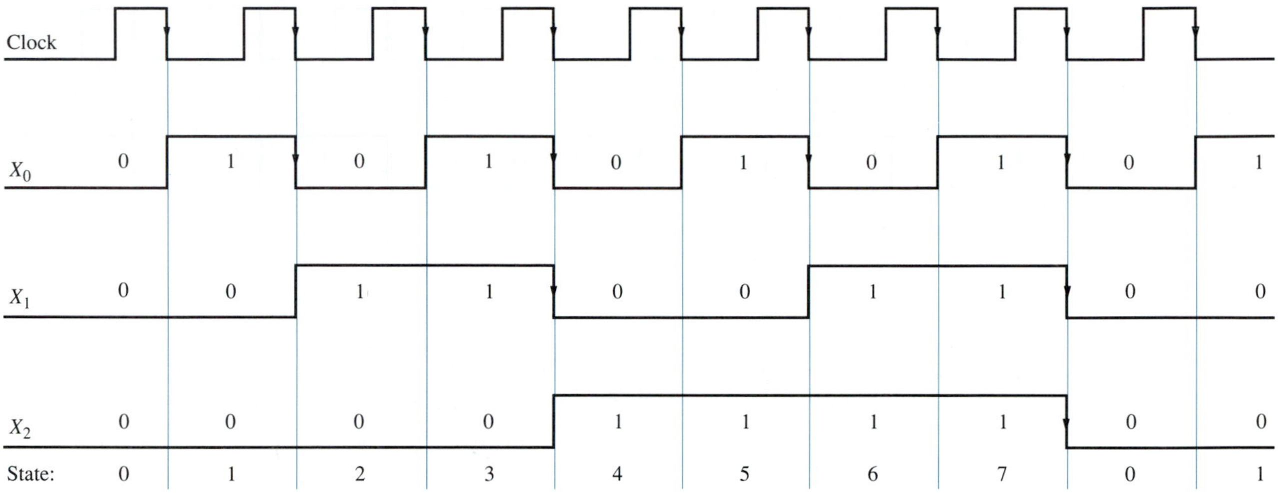

not have to be replaced with negative edge-triggered flip-flops (as shown in Fig. 6.5). Instead simply invert the clock pulses entering flip-flops $F1$ and $F2$. The type of flip-flop used for $F0$ is not significant because it is triggered by the main *CLOCK*. Instead of adding inverters to the circuit, use the $\overline{Q}$ outputs, effectively producing the same results as shown in Figure 6.7.

Figure 6.7

A 3-bit binary ripple up counter using positive edge-triggered flip-flops.

X_0
X_1
X_2
Q_0
Q_1
Q_2
Clock
T_0
T_1
T_2
$\overline{Q}_0$
$\overline{Q}_1$
$\overline{Q}_2$
$F0$
$F1$
$F2$

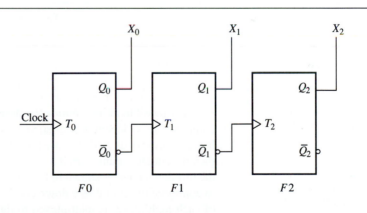

The timing diagram for the up counter in Figure 6.7 is shown in Figure 6.8. The analysis for this counter is more complex because each flip-flop in series is triggered not by the output of the previous flip-flop but by the complement of its output. In Figure 6.8 because $F0$ is directly triggered by the *CLOCK* it changes states on the positive transition of each clock pulse. $F1$ is triggered on the positive edge of $\overline{Q}_0$, the complement of X_0 as shown in the timing diagram; X_1 toggles on the positive edge of $\overline{Q}_0$, which is actually the negative edge of X_0. Similarly, X_2 toggles on the positive edge of $\overline{Q}_1$, which is the negative edge of X_1. The resulting timing diagram is similar to the one shown in Figure 6.6.

Figure 6.8

Timing diagram for Figure 6.7.

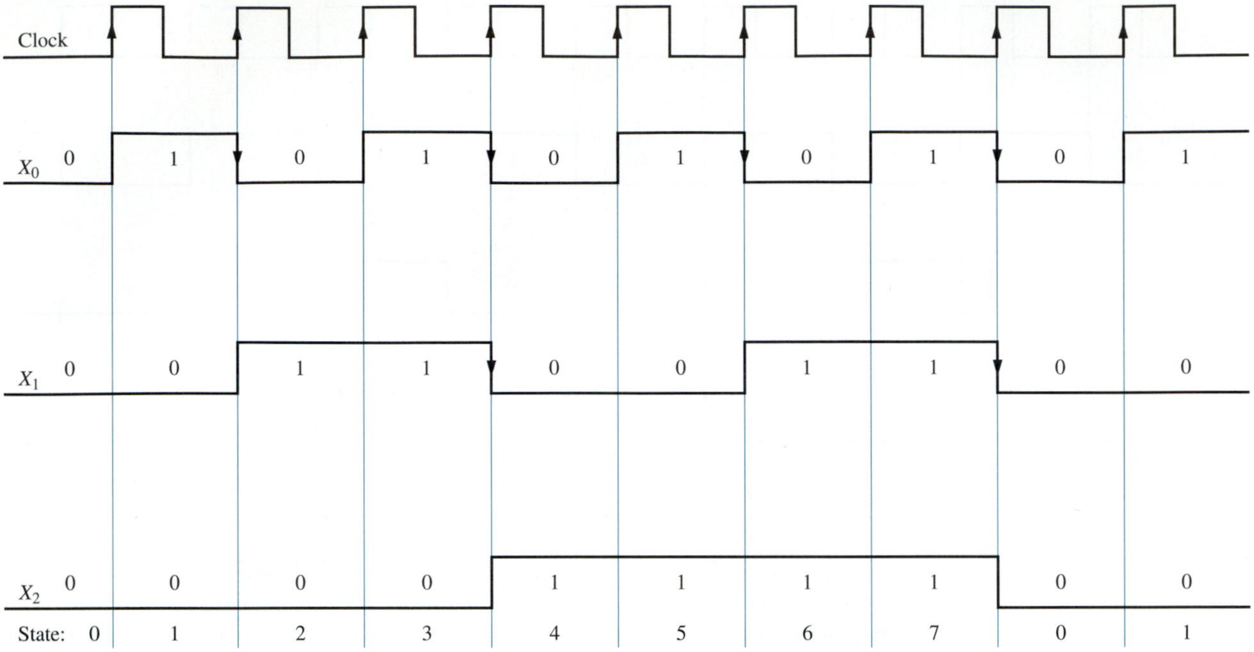

An up counter can be made to count down and a down counter can be made to count up by inverting the outputs of the counter. For example, the counting sequence of a 2-bit up counter with and without inverted outputs would be as follows:

		X_1	X_0	$\overline{X}_1$	$\overline{X}_0$		
up count	0	0	0	1	1	3	down count
	1	0	1	1	0	2	
	2	1	0	0	1	1	
	3	1	1	0	0	0	

It is possible to incorporate the design of Figures 6.2 and 6.7 into an up/down counter. The logic symbol for such a counter is shown in Figure 6.9a. The counter has a control input, S, that is used to select the desired counting sequence. If $S = 0$, then the counter counts *UP*, if $S = 1$ the counter counts *DOWN*. The logic circuit for the up/down counter shown in Figure 6.9b uses two-channel single-bit multiplexers to configure the circuit as a down counter or an up counter. When $S = 0$, channel B of each multiplexer is multiplexed to the output; this effectively "connects" the $\overline{Q}$ output of each flip-flop to the clock of the next flip-flop in series, configuring the circuit like the up counter shown in Figure 6.7. When $S = 1$, channel A of each multiplexer is multiplexed to the output; this effectively connects the Q output of each flip-flop to the clock of the next flip-flop in series, configuring the circuit like the down counter shown in Figure 6.2.

The manner in which the flip-flops are connected and the types of flip-flops used (negative or positive edge-triggered) determine the counting sequence of an asynchronous counter. The following example illustrates the analysis of a 4-bit asynchronous counter that counts in a disordered sequence.

Figure 6.9

An asynchronous up/down
counter. (a) Logic symbol;
(b) circuit.

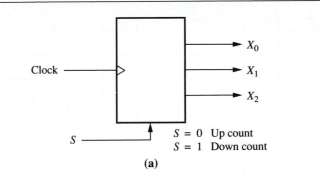

(a)

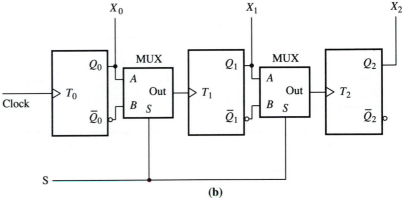

(b)

Example 6.1

Analyze the 4-bit ripple counter shown in Figure 6.10 by obtaining a timing diagram for the outputs X_3, X_2, X_1, X_0. Because the counter will have 16 possible states, the timing diagrams should be analyzed with at least 17 clock pulses to verify that the counter recycles. Assume that the counter starts off at state 13 (1101).

Figure 6.10

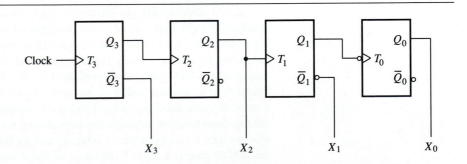

The timing diagram for Figure 6.10 is shown in Figure 6.11. X_3 toggles on each positive transition of the main *CLOCK*. X_2 toggles on each positive transition of Q_3 (the complement of X_3). X_1 toggles on each positive transition of X_2. X_0 toggles on each negative transition of Q_1

Figure 6.11

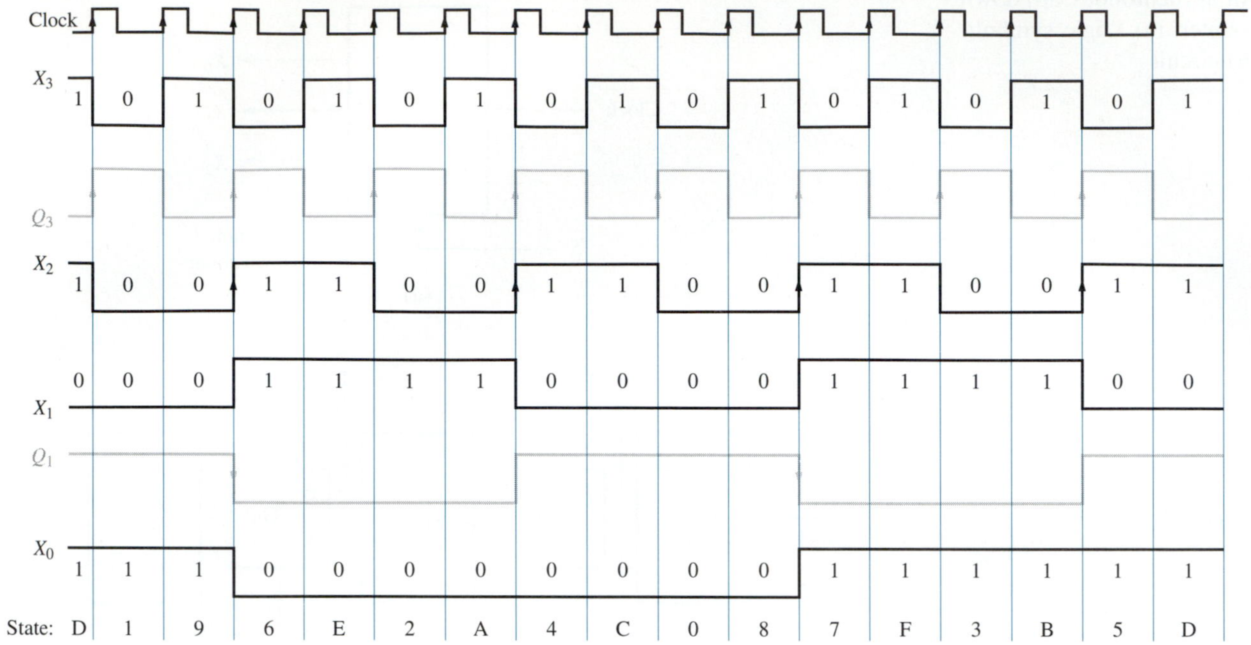

Clock

X_3 1 0 1 0 1 0 1 0 1 0 1 0 1 0 1 0 1

Q_3

X_2 1 0 0 1 1 0 0 1 1 0 0 1 1 0 0 1 1

X_1 0 0 0 1 1 1 1 0 0 0 0 1 1 1 1 0 0

Q_1

X_0 1 1 1 0 0 0 0 0 0 0 0 1 1 1 1 1 1

State: D 1 9 6 E 2 A 4 C 0 8 7 F 3 B 5 D

(the complement of X_1). The sequence of states through which the counter cycles is also shown in the timing diagram.

A counter does not necessarily have to cycle through all possible states. For example, to implement a BCD counter that cycles through ten states, 0 through 9, use a 4-bit counter (to accommodate the numbers 1000 and 1001) that has a maximum of 16 possible states. Therefore, a *BCD counter* is a 4-bit counter implemented with a modulus of ten (uses only 10 of the 16 possible states). Because a BCD counter has ten counts, it is often referred to as a *decade* counter.

To design a BCD ripple counter, first construct a 4-bit binary counter (up or down) and then modify the circuit to automatically reset to 0000 after the count reaches 1001 (9). Figure 6.12 shows the logic circuit for a BCD up counter. *J-K* flip-flops have been used in this circuit with the *K* inputs to each flip-flop permanently connected to a logic 1 state. The *J* inputs of all the flip-flops are connected to the output of a *NAND* gate, which will be a logic 1 for all counts from 0000 to 1000 (states 0 to 8) because either of its inputs X_0 or X_3 will be a logic 0 during these counts. Because the *J* inputs (and also the *K* inputs) to all the flip-flops are logic 1's, the flip-flops function as toggle flip-flops during states 0 through 8. When the counter reaches state 9 and its output is 1001, X_0 and X_3 are both logic 1's, and the output of the *NAND* gate is a logic 0. Because all the *J* inputs are now logic 0's (and all *K* inputs are logic 1's), on the next transition of the clock pulse all the flip-flops reset, the output of the counter is 0000 (state 0), and the counter recycles. Figure 6.13 shows the timing diagram for the BCD counter.

The BCD counter just described uses counter decoding, which uses a combinational logic circuit (or gate) to monitor the output of the counter. When the output of

Figure 6.12

An asynchronous BCD up counter.

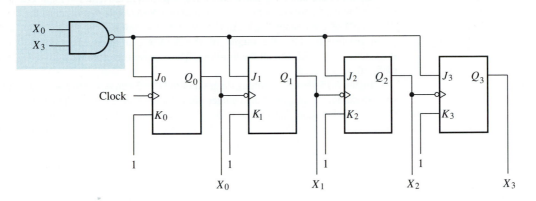

Figure 6.13

Timing diagram for the BCD counter.

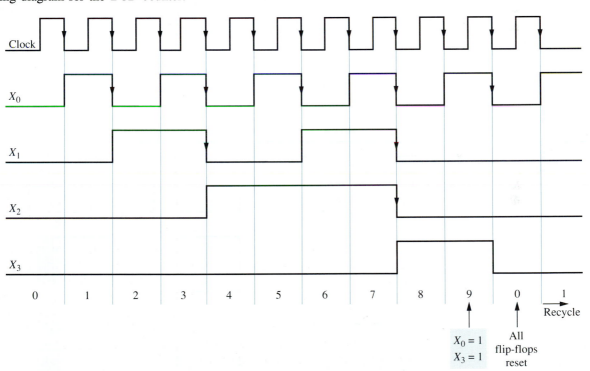

the counter reaches a certain value, the combinational logic circuit resets the counter. The direct set and clear (asynchronous) inputs of the flip-flops can also be used in conjunction with counter decoding. The following examples illustrate the design of a counter that uses the asynchronous inputs of the flip-flops to change the counting sequence.

Example 6.2

Figure 6.14 shows a 3-bit asynchronous counter that counts from 0 to 5 (000 to 101) and then recycles to 0 (000). *J-K* flip-flops configured as *T* flip-flops have been used. Because direct *SET* inputs, $\overline{S}$, are not being used, they have been tied inactive at the logic 1 state. The direct *CLEAR* inputs, $\overline{C}$, are connected to the output of a *NAND* gate that is called the count decoder. The output of the counter X_2, X_1, X_0 is applied to the input of the *NAND* gate, and when the count reaches 6 (110) the output of the *NAND* gate produces a logic 0 that (almost instantaneously) *CLEARS* all the flip-flops. The counter therefore recycles to 000 after a count of 5 (101).

Figure 6.15 shows the timing diagram of the 0 to 5 counter. After a count of 5 (101), for a short duration, the output of the counter does change to 6 (110) but the count decoder detects the 110 combination during this state and resets the counter. However, a spurious output, sometimes called a *glitch*, is produced on the X_1 output. This could be a problem in some applications as will be seen in Section 6.9.

Figure 6.14

Counter implemented with a count decoder.

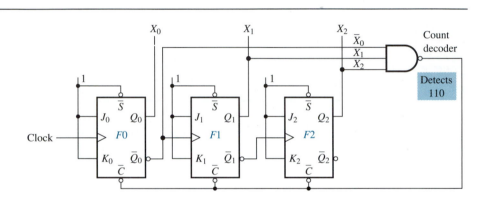

Figure 6.15

Timing diagram for Figure 6.14.

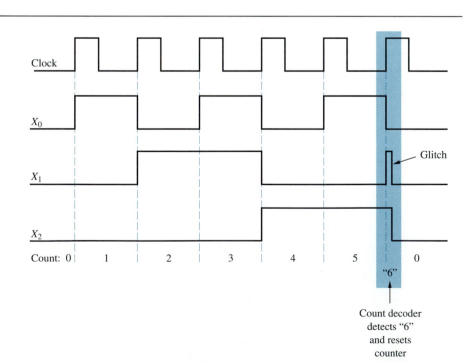

The use of counter decoding does not have to be limited to simply resetting the counter to the 0 state. We can also use counter decoding to recycle the counter to any other state as shown in the following example.

Example 6.3

Figure 6.16 shows a 3-bit asynchronous counter that counts from 3 to 7 (011 to 111) and then recycles to 3 (011). *D* flip-flops configured as *T* flip-flops have been used although *J-K* flip-flops configured as *T* flip-flops could also have been used.

The count decoder will produce a logic 0 when the counter reaches a count of 000. The output of the count decoder is used to *SET* flip-flops *F0* and *F1*, and to *RESET* flip-flop *F2*; this will preset the counter to 011. When the counter reaches a count of 7 (111) on the next count, 0 (000), the count decoder will cause the outputs of the flip-flops to recycle to 3 (011). The timing diagram of the counter is shown in Figure 6.17. Again, because the counter does momentarily recycle to 000 after a count of 111, glitches are produced at the X_0 and X_1 outputs.

Figure 6.16

Counter implemented with a count decoder.

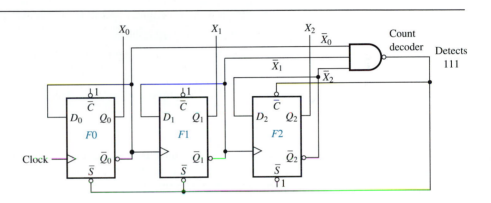

Figure 6.17

Timing diagram for Figure 6.16.

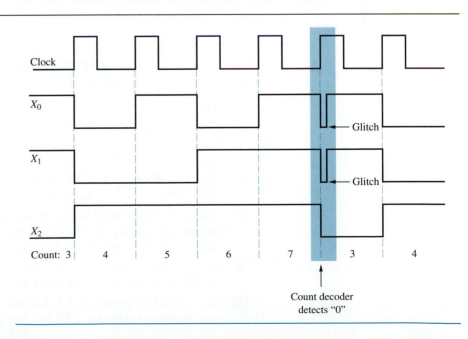

Asynchronous counters are fairly simple logic circuits that can be analyzed easily through the use of timing diagrams. However, the design of asynchronous counters with unusual count sequences is difficult because there is no formal procedure and the design must be conducted on a trial-and-error basis. Asynchronous counters are limited to T flip-flops or D and J-K flip-flops configured as T flip-flops. The next few sections deal with the design and analysis of synchronous counters. Unlike asynchronous counters, there is a formal procedure that can be used to design practically any type of count sequence for a synchronous counter from a given specification. The same procedure can be used in reverse to analyze synchronous counters.

Review Questions

1. Why is the term "asynchronous" used to describe asynchronous counters?

2. Why are asynchronous counters known as ripple counters?

3. What does the term "modulus" mean?

4. What is an up counter and a down counter?

5. What is a binary counter?

6. What determines the counting sequence of an asynchronous counter?

7. What is a BCD counter? What other terms are used to identify it?

8. What is the purpose of counter decoding? How is it done?

6.3
Synchronous counters

A synchronous counter is made up of a set of flip-flops that are triggered simultaneously (in parallel) by the main clock of the counter. Because all the flip-flops change states when the clock makes its transition, the term "synchronous" is used. Unlike asynchronous counters, all the flip-flops used in the counters are not always T flip-flops (or configured as T flip-flops). Furthermore, changing the type of flip-flops used in a synchronous counter (negative or positive edge-triggered) does not affect the counting sequence as long as all flip-flops are of the same type.

Figure 6.18 shows the logic circuit and timing diagram for a 2-bit synchronous binary up counter. Figure 6.18a shows that the clocks of both flip-flops are connected together. Also, flip-flop $F0$ is configured as a toggle flip-flop, but flip-flop $F1$ is not. The timing diagram for the counter is shown in Figure 6.18b for five clock pulses; the intervals between clock pulses have been numbered 0 through 6 for identification purposes. To analyze the circuit and construct the timing diagram shown in Figure 6.18b, first complete the timing for the output X_0 because $F0$ will toggle each time a positive transition of the clock pulse takes place. To trace the output X_1, determine the logic levels at J_1 and K_1 before each clock pulse makes a transition and causes flip-flop $F1$ to change states. The following analysis traces the logic level of X_1 during each interval.

(1)   This is the initial condition $X_1 = 0$ and $X_0 = 0$. Because X_0 is connected to J_1 and K_1, and X_0 is a 0 ($X_0 = J_1 = K_1 = 0$), when the next transition of the clock takes place, $F1$ will not change states.

(2) $X_0 = 1$ and X_1 remains at a 0. Again, $J_1 = K_1 = 1$ (because $X_0 = 1$). When the next transition takes place, $F1$ will toggle to a 1.

(3) $X_0 = 0$ and X_1 toggles to a 1. Because $X_0 = J_1 = K_1 = 0$, when the next transition takes place, $F1$ will not change states.

Figure 6.18

A 2-bit synchronous binary up counter. (a) Logic circuit; (b) timing diagram.

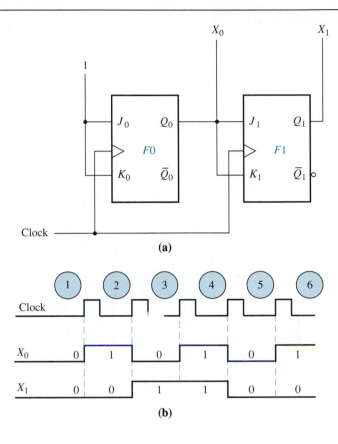

(a)

(b)

④ $X_0 = 1$ and X_1 remains at a 1. Because $X_0 = J_1 = K_1 = 1$, when the next transition takes place, $F1$ will toggle to a 0.

⑤ $X_0 = 0$ and X_1 toggles to a 0. Because $X_0 = J_1 = K_1 = 0$, when the next transition takes place, $F1$ will not change states.

⑥ $X_0 = 1$ and X_1 remains at a 0. Because $X_0 = J_1 = K_1 = 1$, when the next transition takes place, $F1$ will toggle to a 1.

The logic circuits and procedure used to analyze synchronous counters get more complex as the size of the counter increases. Synchronous counters typically contain many of the basic logic gates and have few toggle flip-flops. The control inputs of the flip-flops are dynamically configured as the counter cycles through its states, adding to the complexity of analysis.

Figure 6.19 shows the logic circuit for a 3-bit synchronous binary down counter. Its function is exactly the same as the asynchronous counter shown in Figure 6.2 but its operation is quite different. The main clock is connected in parallel to all three flip-flops, $F0$, $F1$, and $F2$. Only flip-flop $F0$ is a toggle flip-flop. The outputs of the counter are the outputs of the flip-flops and are labeled "X_2", "X_1", and "X_0".

Figure 6.20 shows the timing diagram for the 3-bit synchronous counter. The timing diagram analysis of synchronous counters is somewhat different from the analysis of asynchronous counters because all three flip-flops change states together and their states must be analyzed together, before and after the clock pulse makes a transition. The timing diagram is divided into ten intervals, labeled "1" through "10," for the following analysis.

Figure 6.19

A 3-bit synchronous binary down counter.

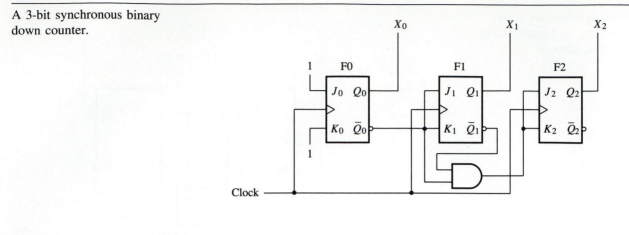

Figure 6.20

Timing diagram for the 3-bit synchronous binary down counter.

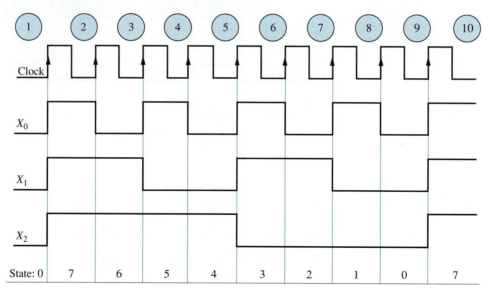

① Before the clock makes a positive transition all flip-flop outputs (X_2, X_1, X_0) are in the logic 0 state (this is the initial condition that has been assumed). This means that $J_1 = K_1 = \overline{Q}_0 = 1$ and $J_2 = K_2 = \overline{Q}_0 \cdot \overline{Q}_1 = 1$. Also, $J_0 = K_0 = 1$ permanently. All three flip-flops are setup to toggle. As a result, on the positive transition of the clock, the outputs of all three flip-flops synchronously change state: flip-flops $F0$, $F1$, and $F2$ change from a logic 0 to a logic 1, driving the circuit into interval 2. This new state is identified in Figure 6.20 as state 7.

② Before the positive transition of the clock pulse, $X_0 = 1$, $X_1 = 1$, and $X_2 = 1$. During this time $J_0 = K_0 = 1$, $J_1 = K_1 = \overline{Q}_0 = 0$, and $J_2 = K_2 = \overline{Q}_0 \cdot \overline{Q}_1 = 0$. Therefore, flip-flop $F0$ is setup to toggle while flip-flops $F1$ and $F2$ are setup for no-change. When the clock makes a transition, the output of flip-flop $F0$ changes from a 1 to a 0 while the outputs of flip-flops $F1$ and $F2$ remain unchanged at the logic 1 state (interval 3). The counter is now in state 6.

(3) Before the positive transition of the clock pulse, $X_0 = 0$, $X_1 = 1$, and $X_2 = 1$. During this time $J_0 = K_0 = 1$, $J_1 = K_1 = \overline{Q}_0 = 1$, and $J_2 = K_2 = \overline{Q}_0 \cdot \overline{Q}_1 = 0$. Therefore, flip-flops $F0$ and $F1$ are setup to toggle while flip-flop $F2$ is setup for no-change. When the clock makes a transition, the output of flip-flop $F0$ changes from a 0 to a 1, the output of flip-flop $F1$ changes from a 1 to a 0, and the output of flip-flop $F2$ remains unchanged at the logic 1 state (interval 4). The counter is now in state 5.

(4) Before the positive transition of the clock pulse, $X_0 = 1$, $X_1 = 0$, and $X_2 = 1$. During this time $J_0 = K_0 = 1$, $J_1 = K_1 = \overline{Q}_0 = 0$, and $J_2 = K_2 = \overline{Q}_0 \cdot \overline{Q}_1 = 0$. Therefore, flip-flop $F0$ is setup to toggle while flip-flops $F1$ and $F2$ are setup for no-change. When the clock makes a transition, the output of flip-flop $F0$ changes from a 1 to a 0, the output of flip-flop $F1$ remains unchanged at a logic 0 state, and the output of flip-flop $F2$ remains unchanged at the logic 1 state (interval 5). The counter is now in state 4.

(5) Before the positive transition of the clock pulse, $X_0 = 0$, $X_1 = 0$, and $X_2 = 1$. During this time $J_0 = K_0 = 1$, $J_1 = K_1 = \overline{Q}_0 = 1$, and $J_2 = K_2 = \overline{Q}_0 \cdot \overline{Q}_1 = 1$. All three flip-flops are setup to toggle. When the clock makes a transition, the outputs of flip-flops $F0$ and $F1$ change from a 0 to a 1 while the output of flip-flop $F2$ changes from a 1 to a 0 (interval 6). The counter is now in state 3.

(6) Before the positive transition of the clock pulse, $X_0 = 1$, $X_1 = 1$, and $X_2 = 0$. During this time $J_0 = K_0 = 1$, $J_1 = K_1 = \overline{Q}_0 = 0$, and $J_2 = K_2 = \overline{Q}_0 \cdot \overline{Q}_1 = 0$. Flip-flop $F0$ is setup to toggle while flip-flops $F1$ and $F2$ are setup for no-change. When the clock makes a transition, the output of flip-flop $F0$ changes from a 1 to a 0, the output of flip-flop $F1$ remains unchanged at a logic 1 state, and the output of flip-flop $F2$ remains unchanged at the logic 0 state (interval 7). The counter is now in state 2.

(7) Before the positive transition of the clock pulse, $X_0 = 0$, $X_1 = 1$, and $X_2 = 0$. During this time $J_0 = K_0 = 1$, $J_1 = K_1 = \overline{Q}_0 = 1$, and $J_2 = K_2 = \overline{Q}_0 \cdot \overline{Q}_1 = 0$. Flip-flops $F0$ and $F1$ are setup to toggle while flip-flop $F2$ is setup for no-change. When the clock makes a transition, the output of flip-flop $F0$ changes from a 0 to a 1, the output of flip-flop $F1$ changes from a 1 to a 0, and the output of flip-flop $F2$ remains unchanged at the logic 0 state (interval 8). The counter is now in state 1.

(8) Before the positive transition of the clock pulse, $X_0 = 1$, $X_1 = 0$, and $X_2 = 0$. During this time $J_0 = K_0 = 1$, $J_1 = K_1 = \overline{Q}_0 = 0$, and $J_2 = K_2 = \overline{Q}_0 \cdot \overline{Q}_1 = 0$. Flip-flop $F0$ is setup to toggle while flip-flops $F1$ and $F2$ are setup for no-change. When the clock makes a transition, the output of flip-flop $F0$ changes from a 1 to a 0, while the outputs of flip-flops $F1$ and $F2$ remain unchanged at the logic 0 state (interval 9). The counter is now in state 0.

(9) The counter is now in State 0. The outputs of the counter are the same as during interval 1 (initial conditions): $X_0 = 0$, $X_1 = 0$, and $X_2 = 0$. The counter has cycled through its eight states and on the next positive

transition of the clock pulse recycles the counting sequence (interval 10).

To determine the outputs of the counter during each timing interval, determine the states of J and K for each flip-flop before the clock pulse triggers. Then, based on the setting of J and K, determine what the next state of the flip-flop is going to be; that is, the state of the flip-flop after the clock pulse triggers, or the state of the flip-flop during the interval.

Recall from the discussion of asynchronous counters that the count of the counter could be reversed by using the opposite type of flip-flops. For example, an asynchronous down counter could be converted to an asynchronous up counter by using negative edge-triggered instead of positive edge-triggered T flip-flops. This is not true for synchronous counters because they are not sensitive to the edge at which the flip-flops trigger. The design of a 3-bit synchronous up counter as shown in Figure 6.21 is considerably different from the 3-bit synchronous down counter in Figure 6.19.

Figure 6.21

A 3-bit synchronous binary up counter.

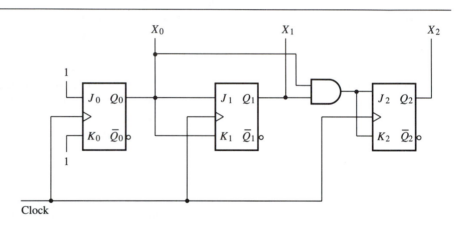

Again observe that the 3-bit synchronous up counter in Figure 6.21 has a common clock for all its flip-flops and that the output of the counter is taken from the output of the flip-flops. This is true for all synchronous counters. The timing diagram for the up counter is shown in Figure 6.22. Again, for the purpose of the analysis that follows, we have identified the ten intervals, labeled 1 through 10, in the timing diagram.

① Before the clock makes a positive transition, all flip-flop outputs (X_2, X_1, X_0) are in the logic 0 state (this is the initial condition that has been assumed). This means that $J_1 = K_1 = Q_0 = 0$ and $J_2 = K_2 = Q_0 \cdot Q_1 = 0$. Also $J_0 = K_0 = 1$ permanently. Flip-flop $F0$ is setup to toggle while flip-flops $F1$ and $F2$ are setup for no-change. As a result, on the positive transition of the clock, the outputs of all three flip-flops synchronously change state: flip-flop $F0$ changes from a logic 0 to a logic 1, while flip-flops $F1$ and $F2$ remain the same (logic 0) driving the circuit into interval 2. This new state is identified in Figure 6.22 as state 1.

② Before the positive transition of the clock pulse, $X_0 = 1$, $X_1 = 0$, and $X_2 = 0$. During this time $J_0 = K_0 = 1$, $J_1 = K_1 = Q_0 = 1$, and $J_2 = K_2 = Q_0 \cdot Q_1 = 0$. Flip-flops $F0$ and $F1$ are setup to toggle while

Figure 6.22

Timing diagram for the 3-bit synchronous binary up counter.

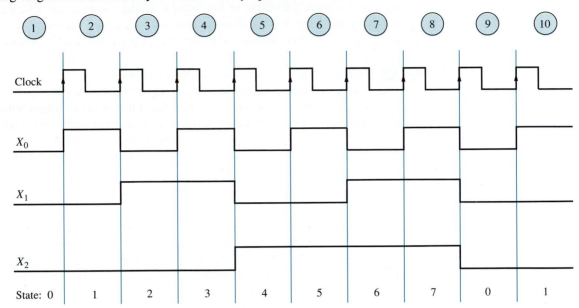

flip-flop $F2$ is setup for no-change. When the clock makes a transition, the output of flip-flop $F0$ changes from a 1 to a 0, the output of flip-flop $F1$ changes from a 0 to a 1, and the output of flip-flop $F2$ remains unchanged at the logic 0 state (interval 3). The counter is now in state 2.

③ Before the positive transition of the clock pulse, $X_0 = 0$, $X_1 = 1$, and $X_2 = 0$. During this time $J_0 = K_0 = 1$, $J_1 = K_1 = Q_0 = 0$, and $J_2 = K_2 = Q_0 \cdot Q_1 = 0$. Flip-flop $F0$ is setup to toggle while flip-flops $F1$ and $F2$ are setup for no-change. When the clock makes a transition, the output of flip-flop $F0$ changes from a 0 to a 1, the output of flip-flop $F1$ remains unchanged at a logic 1 state, and the output of flip-flop $F2$ remains unchanged at the logic 0 state (interval 4). The counter is now in state 3.

④ Before the positive transition of the clock pulse, $X_0 = 1$, $X_1 = 1$, and $X_2 = 0$. During this time $J_0 = K_0 = 1$, $J_1 = K_1 = Q_0 = 1$, and $J_2 = K_2 = Q_0 \cdot Q_1 = 1$. All three flip-flops are setup to toggle. When the clock makes a transition, the outputs of flip-flops $F0$ and $F1$ change from a 1 to a 0 while the output of flip-flop $F2$ changes from a 0 to a 1 (interval 5). The counter is now in state 4.

⑤ Before the positive transition of the clock pulse, $X_0 = 0$, $X_1 = 0$, and $X_2 = 1$. During this time $J_0 = K_0 = 1$, $J_1 = K_1 = Q_0 = 0$, and $J_2 = K_2 = Q_0 \cdot Q_1 = 0$. Flip-flop $F0$ is setup to toggle while flip-flops $F1$ and $F2$ are setup for no-change. When the clock makes a transition, the output of flip-flop $F0$ changes from a 0 to a 1, the output of flip-flop $F1$ remains unchanged at a logic 0 state, and the output of flip-flop $F2$ remains unchanged at the logic 1 state (interval 6). The counter is now in state 5.

⑥ Before the positive transition of the clock pulse, $X_0 = 1$, $X_1 = 0$, and $X_2 = 1$. During this time $J_0 = K_0 = 1$, $J_1 = K_1 = Q_0 = 1$, and $J_2 = K_2 = Q_0 \cdot Q_1 = 0$. Flip-flops $F0$ and $F1$ are setup to toggle while flip-flop $F2$ is setup for no-change. When the clock makes a transition, the output of flip-flop $F0$ changes from a 1 to a 0, the output of flip-flop $F1$ changes from a 0 to a 1, and the output of flip-flop $F2$ remains unchanged at the logic 1 state (interval 7). The counter is now in state 6.

⑦ Before the positive transition of the clock pulse, $X_0 = 0$, $X_1 = 1$, and $X_2 = 1$. During this time $J_0 = K_0 = 1$, $J_1 = K_1 = Q_0 = 0$, and $J_2 = K_2 = Q_0 \cdot Q_1 = 0$. Flip-flop $F0$ is setup to toggle while flip-flops $F1$ and $F2$ are setup for no-change. When the clock makes a transition, the output of flip-flop $F0$ changes from a 0 to a 1, while the outputs of flip-flops $F1$ and $F2$ remain unchanged at a logic 1 state (interval 8). The counter is now in state 7.

⑧ Before the positive transition of the clock pulse, $X_0 = 1$, $X_1 = 1$, and $X_2 = 1$. During this time $J_0 = K_0 = 1$, $J_1 = K_1 = Q_0 = 1$, and $J_2 = K_2 = Q_0 \cdot Q_1 = 1$. All three flip-flops are setup to toggle. When the clock makes a transition, the outputs of all three flip-flops change from a 1 to a 0 (interval 9). The counter is now in state 0.

⑨ The counter is now in State 0. The outputs of the counter are the same as during Interval 1 (initial conditions): $X_0 = 0$, $X_1 = 0$, and $X_2 = 0$. The counter has cycled through its eight states and on the next positive transition of the clock pulse recycles the counting sequence (interval 10).

From the analysis of the two synchronous counter circuits it is apparent that the timing diagram procedure is involved and complex when compared to the timing diagram analysis of asynchronous counters. The two synchronous counters analyzed so far were designed using a formal procedure that will be introduced in the next section. This formal procedure can also be applied (in reverse) to the analysis of synchronous counters (to be examined in Section 6.5) and is considerably less complex than the timing diagram approach.

Review Questions

1. Why is the term "synchronous" used to describe synchronous counters?
2. Compare the construction of a synchronous counter with that of an asynchronous counter.
3. Compare the counting sequence of a synchronous counter with that of an asynchronous counter.

6.4
Synthesis of synchronous counters

The synthesis, or design, of synchronous counters involves a formal six-step procedure. Using this procedure any type of counter can be designed from a given specification. For example, consider the following procedure to design a 2-bit synchronous binary up counter using D flip-flops.

Step 1: state diagram

The state diagram is a graphic representation of the states of the counter and the connections among its states. The state diagram for the 2-bit binary up counter is shown in Figure 6.23. Each state of the counter is identified by a circle and the

Figure 6.23

State diagram for a
synchronous 2-bit binary up
counter.

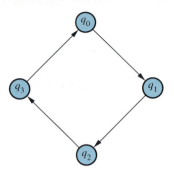

variable q_n where n is the number of the state and corresponds to the output of the counter when in that state. There are four states shown in Figure 6.23 because the size of the counter is 2 bits. Lowercase q is used to identify the state of the counter (uppercase Q is used to identify the state or output of a flip-flop). The connection between the states is shown by arrows and depends on the counting sequence of the counter. For example, for an up counter, if the present state is q_3, the next state is q_0, if the present state is q_0 the next state is q_1, etc. A transition from one state to another occurs when a clock pulse triggers the counter. The state diagram provides almost the same information as a timing diagram because it identifies the sequence in which the counter counts.

Step 2: state table

The state table is a translation of the state diagram into tabular form. The state table for the 2-bit binary up counter is shown in Table 6.1. The first column lists the present state of the counter q^n. The second column lists the next state of the counter, q^{n+1}, that is, the state of the counter after the clock pulse makes a transition. Note the use of the n and $n + 1$ notation to identify the present and next states, respectively; this notation was used in the state transition tables of flip-flops to identify their present and next states.

Table 6.1   State table for a
2-bit binary up
counter

q^n	q^{n+1}
q_0	q_1
q_1	q_2
q_2	q_3
q_3	q_0

To construct a state table, first list all possible present states in column 1 in order starting with state q_0. Refer to the state diagram and determine what the next state of the counter will be when the clock pulse triggers the counter. For example, if the present state is q_0, the next state is q_1 as shown in the first entry in the state table. Similarly, if the present state is q_1, the next state is q_2, if the present state is q_2, the next state is q_3, etc.

Step 3: transition table

The transition table is a translation of the state table and relates each state of the counter to the individual outputs of each flip-flop. The circuit will have two flip-flops that will be referred to as flip-flop $F0$ (LSB) and flip-flop $F1$ (MSB). The output of flip-flop $F0$ is Q_0 and the output of flip-flop $F1$ is Q_1. The transition table for the 2-bit binary up counter is shown in Table 6.2. In constructing the transition table each state of the counter (from the state table) was converted to a 2-bit binary number that represents the outputs of the two flip-flops that make up the counter. For example, when the counter is in state q_0 the outputs of the flip-flops (which are the outputs of the counter) are 00. Therefore, the outputs of the individual flip-flops are

$$Q_1 = 0, \quad Q_0 = 0.$$

Table 6.2 Transition table for a 2-bit binary up counter

$(Q_1\ Q_0)^n$	$(Q_1\ Q_0)^{n+1}$
0 0	0 1
0 1	1 0
1 0	1 1
1 1	0 0

Similarly, when the counter is in state q_1 the outputs of the flip-flops are 01. Therefore, the outputs of the individual flip-flops are

$$Q_1 = 0, \quad Q_0 = 1.$$

etc.

The present state of the outputs of the flip-flops is identified in the transition table by the heading

$$(Q_1 Q_0)^n$$

for convenience and could also be written as

$$Q_1{}^n Q_0{}^n.$$

The next state of the outputs of the flip-flops is identified in the transition table by the heading

$$(Q_1 Q_0)^{n+1}$$

for convenience and could also be written as

$$Q_1{}^{n+1} Q_0{}^{n+1}.$$

The transition table shows us the state transition of each flip-flop as the counter changes from one state to the next. For example, if the present and next states of flip-flop $F0$ (whose output is Q_0) are extracted from Table 6.2, the following will be obtained:

$Q_0{}^n$	$\longrightarrow$	$Q_0{}^{n+1}$
0	$\longrightarrow$	1
1	$\longrightarrow$	0
0	$\longrightarrow$	1
1	$\longrightarrow$	0

Flip-flop $F0$ toggles. Similarly, we can determine how flip-flop $F1$ changes states as the counter counts through its sequence by extracting the $Q_1{}^n$ and $Q_1{}^{n+1}$ columns from Table 6.2

$Q_1{}^n$	$\longrightarrow$	$Q_1{}^{n+1}$
0	$\longrightarrow$	0
0	$\longrightarrow$	1
1	$\longrightarrow$	1
1	$\longrightarrow$	0

Step 4: excitation maps

The previous three steps in the synthesis were not dependent on the type of flip-flop to be used in the counter circuit. Before this step is started, determine the type of flip-flop, *J-K* or *D*, that the counter is to have. Because the specification for this example requires *D* flip-flops, proceed with that assumption.

Excitation maps are used to determine what the control inputs of each flip-flop (*J* and *K*, or *D*) have to be for the flip-flop to make the transitions specified in the transition table. The excitation maps for the *D* input of each flip-flop are shown in Figure 6.24. They are essentially K-maps and are used in the same manner.

The excitation map for flip-flop $F0$ is shown in Figure 6.24a. It represents the control input D_0 of the flip-flop, and therefore the value in each cell is the value of D_0

Figure 6.24

Excitation maps for the
synchronous 2-bit binary up
counter using D flip-flops.

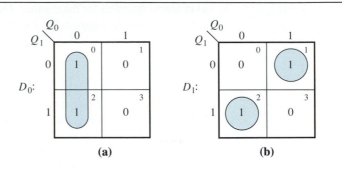

(a)   (b)

needed for the flip-flop to change from its present state to its next state. To determine
the values of D_0 required for the flip-flop to change from one state to another, refer to
the modified transition table for the D flip-flop shown in Table 5.6 and reproduced
here as follows:

Q^n	Q^{n+1}	D
0	0	0
0	1	1
1	0	0
1	1	1

Recall that the table shown above shows what D has to be for a particular transition
to take place.

Because each cell in Figure 6.24a is mapped by $Q_1 Q_0$, the cell number corresponds to the present state of the counter. For example, in cell no. 0, because the
present state of the counter is 0 ($Q_1 = 0$, $Q_0 = 0$) and, according to the transition
table (refer to extracted Q_0^n and Q_0^{n+1} columns), the flip-flop must change from a 0
to a 1 (first entry in the transition table), D_0 must be a 1 (see reproduction of Table
5.6). Similarly, in cell no. 1, because the present state of the counter is 1 ($Q_1 = 0$, Q_0
= 1) and the flip-flop must change from a 1 to a 0 (second entry in transition table),
D_0 must be a 0. In cell no. 2, because the present state of the counter is 2 ($Q_1 = 1$, Q_0
= 0) and the flip-flop must change from a 0 to a 1 (third entry in transition table), D_0
must be a 1. Finally, in cell no. 3, because the present state of the counter is 3 ($Q_1 =$
1, $Q_0 = 1$) and the flip-flop must change from a 1 to a 0 (last entry in transition
table), D_0 must be a 0.

The excitation map for flip-flop no. 1 can be filled in in the same manner as
shown in Figure 6.24b. Refer to the transition table shown in Table 6.2 (or the
extracted Q_1^n and Q_1^{n+1} columns shown earlier) and the reproduction of Table 5.6.

Cell no. 0: The present state of the counter is 0 and the flip-flop changes
from a 0 to a 0. For this to occur D_1 must be a 0.

Cell no. 1: The present state of the counter is 1 and the flip-flop changes
from a 0 to a 1. For this to occur D_1 must be a 1.

Cell no. 2: The present state of the counter is 2 and the flip-flop changes
from a 1 to a 1. For this to occur D_1 must be a 1.

Cell no. 3: The present state of the counter is 3 and the flip-flop changes
from a 1 to a 0. For this to occur D_1 must be a 0.

Step 5: excitation equations

The simplified logic equations obtained from the excitation maps are known as the excitation equations for the counter. From the excitation maps of the 2-bit binary up counter shown in Figure 6.24, the excitation equations are as follows:

$$D_0 = \overline{Q_0} \tag{6.1}$$
$$D_1 = Q_1\overline{Q_0} + \overline{Q_1}Q_0$$
$$\therefore \quad D_1 = Q_1 \oplus Q_0 \tag{6.2}$$

Observe that the excitation equations identify the connections that are to be made between the outputs (Q_0 and Q_1) and the inputs of the flip-flops (D_0 and D_1).

Step 6: logic circuit

The logic circuit for the counter is implemented from the excitation equations. For the 2-bit binary up counter, Equations 6.1 and 6.2 produce the circuit for the counter shown in Figure 6.25. The clocks of both flip-flops are connected to a common *CLOCK* that is used to cycle the counter through each of its states. The type of flip-flop, negative or positive edge-triggered, is not significant to the operation of the circuit. The outputs of the counter (X_1,X_0) are the outputs of the flip-flops (Q_1Q_0) and flip-flop $F0$ will function as a T flip-flop.

Figure 6.25

Logic circuit for the synchronous 2-bit binary up counter using D flip-flops.

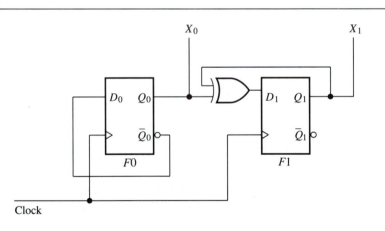

The counter designed in Figure 6.25 can easily be redesigned to use *J-K* flip-flops by modifying steps 4 and 5 of the synthesis as shown in the following example.

Example 6.4

To redesign the 2-bit binary up counter shown in Figure 6.25 to use *J-K* instead of *D* flip-flops, obtain the excitation maps for the two *J-K* flip-flops in step 4 of the synthesis:

Step 4: excitation maps
Because each *J-K* flip-flop has two control inputs, *J* and *K*, there must be excitation maps for each of the inputs of each flip-flop as shown in Figure 6.26. Referring to the modified transition table for a *J-K* flip-flop shown in Table 5.9 and reproduced here as follows:

Q^n	Q^{n+1}	J	K
0	0	0	x
0	1	1	x
1	0	x	1
1	1	x	0

Figure 6.26

Excitation maps for the synchronous 2-bit binary up counter using *J-K* flip-flops.

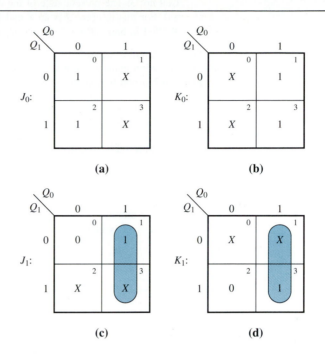

(a)   (b)

(c)   (d)

each set of cells for J_0 and K_0 can be filled in (shown in the maps of Figure 6.26a and b, respectively) as follows:

Cell no. 0: The present state of the counter is 0 and flip-flop *F0* changes from a 0 to a 1 (refer to the transition table shown in Table 6.2 or the extracted Q_0^n and Q_0^{n+1} columns). For this to occur *J* must be a 1 and *K* is an *X*. Cell no. 0 for the J_0 map has a 1 filled in and cell no. 0 for the K_0 map has an *X* filled in.

Cell no. 1: The present state of the counter is 1 and the flip-flop changes from a 1 to a 0. For this to occur *J* is an *X* and *K* must be a 1. Cell no. 1 for the J_0 map has an *X* filled in and cell no. 1 for the K_0 map has a 1 filled in.

Cell no. 2: The present state of the counter is 2 and the flip-flop changes from a 0 to a 1. For this to occur *J* must be a 1 and *K* is an *X*. Cell no. 2 for the J_0 map has a 1 filled in and cell no. 2 for the K_0 map has an *X* filled in.

Cell no. 3: The present state of the counter is 3 and the flip-flop changes from a 1 to a 0. For this to occur *J* is an *X* and *K* must be a 1. Cell no. 3 for the J_0 map has an *X* filled in and cell no. 3 for the K_0 map has a 1 filled in.

Similarly, each set of cells for J_1 and K_1 can be filled in (shown in the maps of Figure 6.26c and d, respectively) as follows:

Cell no. 0: The present state of the counter is 0 and flip-flop no. 1 changes from a 0 to a 0 (refer to Table 6.2 or the extracted Q_1^n and Q_1^{n+1} columns). For this to occur *J* must be a 0 and *K* is an *X*. Cell no. 0 for the J_1 map has a 0 filled in and cell no. 0 for the K_1 map has an *X* filled in.

Cell no. 1: The present state of the counter is 1 and the flip-flop changes from a 0 to a 1. For this to occur *J* must be a 1 and *K* is an *X*. Cell no. 1 for the J_1 map has a 1 filled in and cell no. 1 for the K_1 map has an *X* filled in.

Cell no. 2: The present state of the counter is 2 and the flip-flop changes from a 1 to a 1. For this to occur *J* is an *X* and *K* must be a 0. Cell no. 2 for the J_1 map has an *X* filled in and cell no. 2 for the K_1 map has a 0 filled in.

Cell no. 3: The present state of the counter is 3 and the flip-flop changes from a 1 to a 0. For this to occur J is an X and K must be a 1. Cell no. 3 for the J_1 map has an X filled in and cell no. 3 for the K_1 map has a 1 filled in.

Step 5: excitation equations

The excitation equations obtained from the maps shown in Figure 6.26 are as follows: for Figure 6.26a and b, assuming that the don't cares are 1's,

$$J_0 = 1 \tag{6.3}$$

$$K_0 = 1 \tag{6.4}$$

and for Figure 6.26c and d,

$$J_1 = Q_0 \tag{6.5}$$

$$K_1 = Q_0 \tag{6.6}$$

Step 6: logic circuit

The logic circuit is implemented from Equations 6.3 through 6.6 as shown in Figure 6.27.

Figure 6.27

Logic circuit for the synchronous 2-bit binary up counter using *J-K* flip-flops.

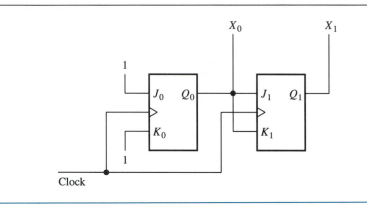

It is apparent that the 2-bit up counter implemented with *J-K* flip-flops as shown in Figure 6.27 has a simpler logic circuit than the 2-bit up counter implemented with *D* flip-flops as shown in Figure 6.25 because of the additional Exclusive-*OR* gate. Generally, counters implemented with *J-K* flip-flops are considerably less complex than counters implemented with *D* flip-flops; this is due to the versatility and flexibility of the *J-K* flip-flop as discussed in Chapter 5. The only steps in the synthesis procedure that are dependent on the type of flip-flop to be used involve the excitation maps and equations. If, however, the counting sequence is to be modified, the synthesis must be started from the beginning. The six steps that make up the synthesis procedure can be applied to the design of any counter. The following example shows the steps involved in the design of a 2-bit down counter.

Example 6.5

Design a 2-bit synchronous binary down counter using *J-K* flip-flops.

Solution

Step 1: state diagram (Fig. 6.28).

Figure 6.28

State diagram for a 2-bit synchronous binary down counter.

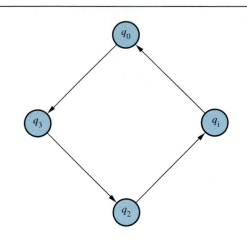

Step 2: state table

q^n	q^{n+1}
q_0	q_3
q_1	q_0
q_2	q_1
q_3	q_2

Step 3: transition table

$(Q_1\ Q_0)^n$	$(Q_1\ Q_0)^{n+1}$
0 0	1 1
0 1	0 0
1 0	0 1
1 1	1 0

Step 4: excitation maps (Fig. 6.29)

Figure 6.29

Excitation maps for the 2-bit synchronous binary down counter using *J-K* flip-flops.

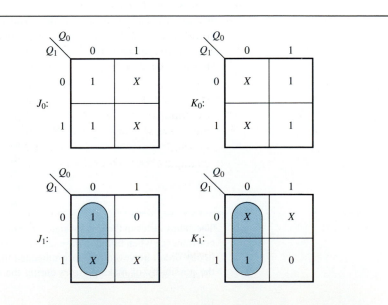

Step 5: excitation equations

$$J_0 = 1$$
$$K_0 = 1$$
$$J_1 = \overline{Q_0}$$
$$K_1 = \overline{Q_0}$$

Step 6: logic circuit (Fig. 6.30)

Figure 6.30

The 2-bit synchronous binary down counter.

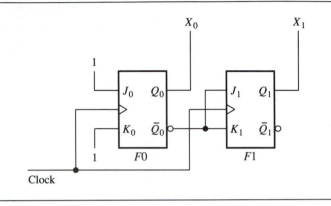

Clock

The examples discussed so far have involved 2-bit binary counters that count in an ordered sequence. There are many applications that require counters that are larger in size and count in some predetermined sequence. The following example illustrates the use of the synthesis procedure to design a 3-bit counter that counts in a predetermined sequence.

Example 6.6

Design a synchronous counter using D flip-flops that counts in the following sequence:

$$6, 3, 5, 0, 2, 6, 3, 5, 0, 2, 6, \ldots$$

Solution

Step 1: state diagram
Because the largest number in the count is 6, the size of the counter must be 3 bits to accommodate this number in its binary representation. A 3-bit counter has eight possible states. However, only five of the eight states (0, 2, 3, 5, and 6) will be used by the counter as shown in the state diagram of Figure 6.31.

Step 2: state table

q^n	q^{n+1}
q_0	q_2
q_2	q_6
q_3	q_5
q_5	q_0
q_6	q_3

The state entries in the present state (q^n) column are listed in increasing order. This is done for convenience when filling in the excitation maps later. The present state entries can be arranged in any order, however, it is recommended that they be arranged in increasing order to reduce the possibility of making errors during the construction of the excitation maps.

Figure 6.31

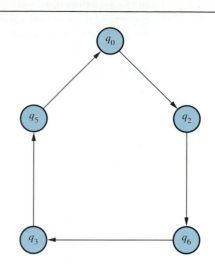

Step 3: transition table

$(Q_2\,Q_1\,Q_0)^n$	$(Q_2\,Q_1\,Q_0)^{n+1}$
0 0 0	0 1 0
0 1 0	1 1 0
0 1 1	1 0 1
1 0 1	0 0 0
1 1 0	0 1 1

Step 4: excitation maps (Fig. 6.32)

Figure 6.32

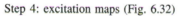

(a)

(b)

(c)

In the excitation maps only cells no. 0, 2, 3, 5, and 6 have been filled in because these correspond to the present states. States 1, 4, and 7 have x (don't cares), because the counter is not designed to use these states and we don't care what the value of D is when the counter is in any of these states.

Step 5: excitation equations

$$D_0 = \underline{Q_1 Q_0} + Q_2 Q_1$$
$$D_1 = \overline{Q_0}$$
$$D_2 = \overline{Q_2} Q_1$$

Step 6: logic circuit (Fig. 6.33)

Figure 6.33

A synchronous counter that counts in a predetermined sequence.

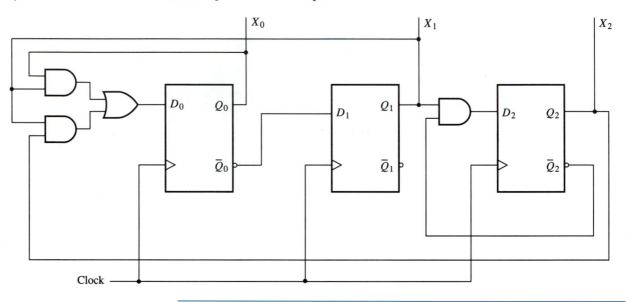

To illustrate the design of a 4-bit synchronous counter consider the implementation of a synchronous BCD counter similar to the one designed in Section 6.2 by using the synthesis procedure. Because the BCD code is a 4-bit code, the counter must be 4 bits in size but is modulus 10 and uses only the first 10 of the 16 possible states. The following example illustrates the design of the BCD up counter.

Example 6.7

Design a synchronous BCD up counter using *J-K* flip-flops.

Solution

Step 1: state diagram (Fig. 6.34)

Figure 6.34

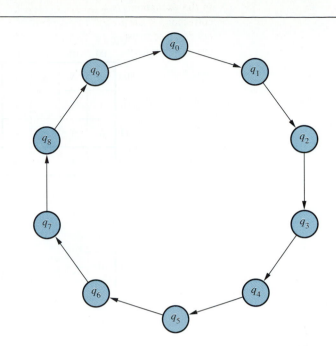

Step 2: state table

q^n	q^{n+1}
q_0	q_1
q_1	q_2
q_2	q_3
q_3	q_4
q_4	q_5
q_5	q_6
q_6	q_7
q_7	q_8
q_8	q_9
q_9	q_0

Step 3: transition table

$(Q_3\ Q_2\ Q_1\ Q_0)^n$	$(Q_3\ Q_2\ Q_1\ Q_0)^{n+1}$
0 0 0 0	0 0 0 1
0 0 0 1	0 0 1 0
0 0 1 0	0 0 1 1
0 0 1 1	0 1 0 0
0 1 0 0	0 1 0 1
0 1 0 1	0 1 1 0
0 1 1 0	0 1 1 1
0 1 1 1	1 0 0 0
1 0 0 0	1 0 0 1
1 0 0 1	0 0 0 0

Step 4: excitation maps (Fig. 6.35)

Figure 6.35

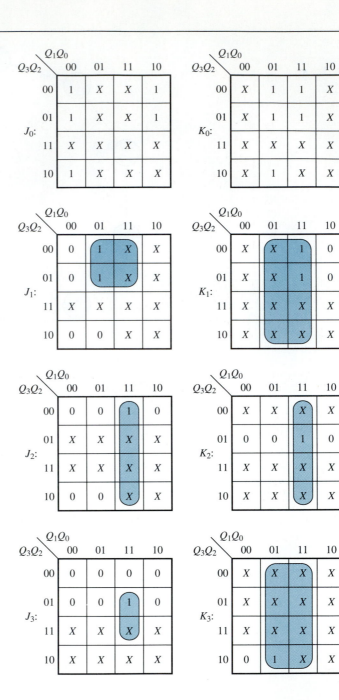

Step 5: excitation equations

$$J_0 = 1$$
$$K_0 = 1$$
$$J_1 = Q_0\overline{Q}_3$$
$$K_1 = Q_0$$
$$J_2 = Q_1Q_0$$
$$K_2 = Q_1Q_0$$
$$J_3 = Q_2Q_1Q_0$$
$$K_3 = Q_0$$

Figure 6.36

A synchronous BCD up counter.

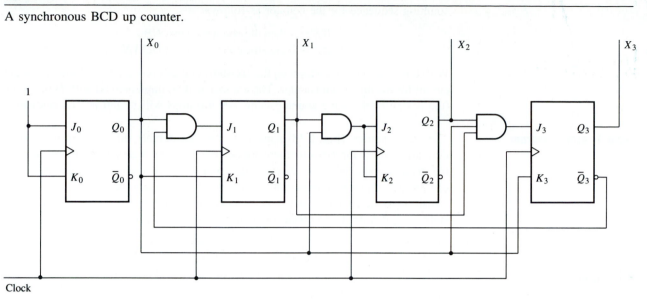

Step 6: logic circuit. (Fig. 6.36).

The counters discussed so far were designed with a dedicated count. That is, each counter was designed for a specific counting sequence that could not be changed unless the entire circuit was redesigned. Section 6.1 stated that a counter could have one or more control inputs that could be used to select the counting sequence of the counter. For example, consider a 2-bit synchronous binary counter that will count up or down on command. This is known as an up/down counter and its logic symbol is shown in Figure 6.37.

Figure 6.37

Logic symbol for a 2-bit up/ down counter.

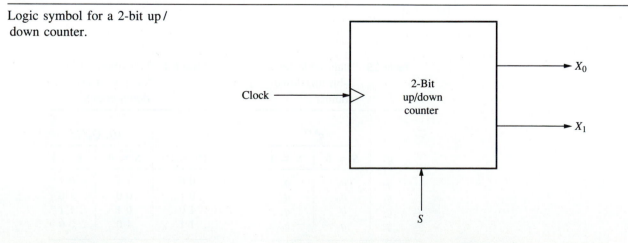

The counter has an additional input S (select) that will be used to select the counting sequence for the counter as follows:

If $S = 1$ then the counter counts *UP*
If $S = 0$ then the counter counts *DOWN*

With this information, go through the six-step synthesis procedure to design the logic circuit for the up / down counter. The counter will be implemented with D flip-flops (the design of the same counter using *J-K* flip-flops will be left as an exercise).

Step 1: state diagram

The state diagram (Fig. 6.38) appears to be considerably different from the state diagrams seen so far. Because there are two possible next states for each present state based on the value of S (1 or 0), each state symbol has two arrows emanating from it, one representing a transition to the next state if $S = 1$, and the other representing a transition to the next state if $S = 0$. For example, if the present state is q_3, the next state is q_0 if $S = 1$ (up count) or q_2 if $S = 0$ (down count).

Figure 6.38

State diagram for the 2-bit up / down counter.

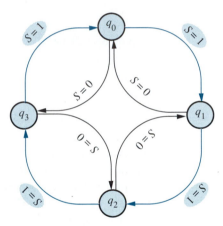

Step 2: state table

Because the up / down counter has two possible next states based on the value of S, the next state column of the state table will have two possible entries as shown in Table 6.3.

Table 6.3   State table for a 2-bit up / down counter

q^n	q^{n+1}	
	$S = 0$	$S = 1$
q_0	q_3	q_1
q_1	q_0	q_2
q_2	q_1	q_3
q_3	q_2	q_0

Table 6.4   Transition table for a 2-bit up / down counter

$(Q_1\,Q_0)^n$	$(Q_1\,Q_0)^{n+1}$	
	$S = 0$	$S = 1$
0 0	1 1	0 1
0 1	0 0	1 0
1 0	0 1	1 1
1 1	1 0	0 0

Step 3: transition table

Because the transition table is simply a translation of the state table to represent the outputs of the flip-flops, the transition table therefore has the same format as the state table as shown in Table 6.4.

Step 4: excitation maps

The excitation maps for each D flip-flop are shown in Figure 6.39. Each cell lists the value of D required for the flip-flop to make the transition to the next state, given the present state and the value of S. For example, consider the excitation map for D_0 (flip-flop $F0$) shown in Figure 6.39a.

> Cell 0: The present state is 0 ($Q_1Q_0 = 00$) and $S = 0$. From the transition table, the next state is 3 ($Q_1Q_0 = 11$) and flip-flop $F0$ changes from a 0 to a 1. Therefore, D_0 has to be a 1.
>
> Cell 2: The present state is 1 ($Q_1Q_0 = 01$) and $S = 0$. From the transition table, the next state is 0 ($Q_1Q_0 = 00$) and flip-flop $F0$ changes from a 1 to a 0. Therefore, D_0 has to be a 0.
>
> Cell 4: The present state is 2 ($Q_1Q_0 = 10$) and $S = 0$. From the transition table, the next state is 1 ($Q_1Q_0 = 01$) and flip-flop $F0$ changes from a 0 to a 1. Therefore, D_0 has to be a 1.
>
> Cell 6: The present state is 3 ($Q_1Q_0 = 11$) and $S = 0$. From the transition table, the next state is 2 ($Q_1Q_0 = 10$) and flip-flop $F0$ changes from a 1 to a 0. Therefore, D_0 has to be a 0.
>
> Cell 1: The present state is 0 ($Q_1Q_0 = 00$) and $S = 1$. From the transition table, the next state is 1 ($Q_1Q_0 = 01$) and flip-flop $F0$ changes from a 0 to a 1. Therefore, D_0 has to be a 1.
>
> Cell 3: The present state is 1 ($Q_1Q_0 = 01$) and $S = 1$. From the transition table, the next state is 2 ($Q_1Q_0 = 10$) and flip-flop $F0$ changes from a 1 to a 0. Therefore, D_0 has to be a 0.
>
> Cell 5: The present state is 2 ($Q_1Q_0 = 10$) and $S = 1$. From the transition table, the next state is 3 ($Q_1Q_0 = 11$) and flip-flop $F0$ changes from a 0 to a 1. Therefore, D_0 has to be a 1.
>
> Cell 7: The present state is 3 ($Q_1Q_0 = 11$) and $S = 1$. From the transition table, the next state is 0 ($Q_1Q_0 = 00$) and flip-flop $F0$ changes from a 1 to a 0. Therefore, D_0 has to be a 0.

Similarly, each cell for D_1 (flip-flop no. 1) can be filled in as shown in Figure 6.39b.

Figure 6.39

Excitation maps for the 2-bit up/down counter.

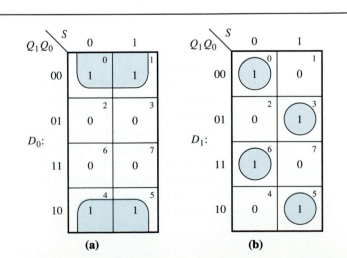

(a)   (b)

Step 5: excitation equations

The excitation equations for D_0 and D_1 can now be obtained from the excitation maps shown in Figure 6.39 as follows:

$$D_0 = \overline{Q_0} \tag{6.7}$$

$$D_1 = \overline{Q_1}\,\overline{Q_0}\,\overline{S} + \overline{Q_1}Q_0 S + Q_1 Q_0 \overline{S} + Q_1\overline{Q_0}S \tag{6.8}$$
$$D_1 = \overline{Q_1}\,\overline{Q_0}\,\overline{S} + Q_1 Q_0 \overline{S} + \overline{Q_1}Q_0 S + Q_1\overline{Q_0}S \qquad \text{(commutative)}$$
$$D_1 = \overline{S}\,(\overline{Q_1}\,\overline{Q_0} + Q_1 Q_0) + S\,(\overline{Q_1}Q_0 + Q_1\overline{Q_0}) \qquad \text{(distributive)}$$
$$D_1 = \overline{S}\,\overline{(Q_1 \oplus Q_0)} + S\,(Q_1 \oplus Q_0) \qquad \text{(Exclusive-}OR\text{)}$$
$$D_1 = \overline{S} \oplus (Q_1 \oplus Q_0) \qquad \text{(Exclusive-}OR\text{)}$$

Step 6: logic circuit

The logic circuit for the 2-bit up/down counter is implemented from Equations 6.7 and 6.8 and shown in Figure 6.40. Flip-flop $F0$ is a toggle flip-flop and is not affected by S because the LSB in a binary count always toggles regardless of whether the count is up or down.

Figure 6.40

Logic circuit for the 2-bit synchronous up/down counter.

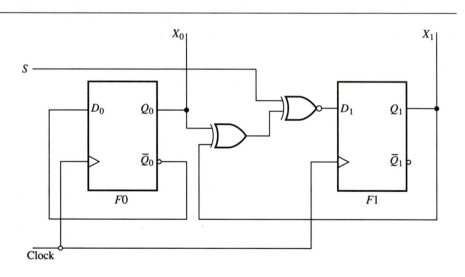

It is obvious that the logic circuits of the synchronous counters designed so far appear to be more complex in many instances than their asynchronous counterparts. This is because the synthesis procedure is generalized for the design of any type of counter and does not account for the various shortcuts that can be taken to make the circuit less complex. However, the synthesis procedure is flexible because it is applicable to any type or any size of counter. There is no formal procedure such as this for the design of asynchronous counters. Another advantage of the synthesis procedures discussed in this section is the ability to work through the steps backward and analyze a given sequential logic circuit without having to go through the tedious process of constructing a timing diagram. This procedure will be the subject of Section 6.5.

Review Questions

1. Briefly outline the six-step procedure for the synthesis of synchronous counters.
2. What is the purpose of the state diagram?
3. What is the purpose of the state table?
4. What is the purpose of the transition table?
5. What is the purpose of the excitation maps?
6. What is the purpose of the excitation equations?

6.5
Analysis of synchronous counters

Section 6.2 introduced the timing diagram procedure for analyzing asynchronous counters and Section 6.3 applied the timing diagram to the analysis of synchronous counters. Timing diagrams can be a very efficient way to analyze a synchronous counter circuit as long as the circuit is relatively simple. The more complex the circuit, the more difficult it is to accurately plot a timing diagram that describes its operation. Because synchronous counters tend to be more complex than asynchronous counters, the timing diagram is not a very suitable means for analyzing synchronous counters. This section introduces a formal procedure of analysis that essentially reverses the six-step synthesis procedure discussed in the previous section.

The analysis of a synchronous counter involves all the steps required to obtain a state diagram from a given circuit. To illustrate the procedure for analyzing a synchronous counter, consider the analysis of the counter shown in Figure 6.25. Assume that the function of this circuit is unknown because the purpose of analysis is to determine what the circuit does. The following steps are used in the analysis of the circuit.

Step 1: excitation equations

The excitation equations for the circuit are obtained by determining the logic equations for D_0 and D_1 from Figure 6.25 as follows:

$$D_0 = \overline{Q_0} \tag{6.9}$$

$$D_1 = Q_1 \oplus Q_0 \tag{6.10}$$

Step 2: transition equations

The transition equations for each flip-flop allow us to determine the next state of each flip-flop, given the present conditions. Recall from Chapter 5 (Section 5.6) that the state transition equation for a D flip-flop is

$$Q^{n+1} = D. \tag{6.11}$$

Therefore, for flip-flop $F0$

$$Q_0^{n+1} = D_0. \tag{6.12}$$

Substitute the excitation equation (Eq. 6.9) into Equation 6.12

$$Q_0^{n+1} = \overline{Q_0} \tag{6.13}$$

Similarly, for flip-flop $F1$

$$Q_1{}^{n+1} = D_1 \qquad\qquad (6.14)$$

Substitute the excitation equation (Eq. 6.10) into Equation 6.14

$$Q_1{}^{n+1} = Q_1 \oplus Q_0 \qquad\qquad (6.15)$$

Equations 6.13 and 6.15 are the transition equations for flip-flops $F0$ and $F1$, respectively. Equation 6.13 states that the next state of flip-flop $F0$ will be equal to the complement of its present state (toggle). Equation 6.15 states that the next state of flip-flop $F1$ will be equal to the present state of flip-flop $F1$ exclusively-ORed with the present state of flip-flop $F0$. Because the right-hand side of a transition equation always has variables that represent the present state, we choose to drop the superscript n for variables on the right-hand side.

Step 3: transition table

From the transition equations the transition table can be constructed for the counter because it is possible to determine the next state of each flip-flop given the present state. The transition table is constructed by first listing all possible combinations of present states

$$(Q_1 \; Q_0)^n$$

0	0
0	1
1	0
1	1

and then determining the next state of each flip-flop from its transition equation as shown in Table 6.5. For example, if the present state of flip-flop $F0$ is a 1, the next state is

$$Q_0{}^{n+1} = \overline{Q_0}$$
$$\therefore \quad Q_0{}^{n+1} = \overline{1} = 0$$

and if the present state of flip-flop $F0$ is a 0, the next state is

$$Q_0{}^{n+1} = \overline{Q_0}$$
$$\therefore \quad Q_0{}^{n+1} = \overline{0} = 1$$

The $Q_0{}^{n+1}$ column in Table 6.5 is obtained by complementing the $Q_0{}^n$ column. Similarly, for flip-flop $F1$ because the next state of this flip-flop is dependent on the present state of both flip-flops $F0$ and $F1$. If the present state of flip-flop $F0$ is 0 and the present state of flip-flop $F1$ is a 0, the next state of flip-flop $F1$ is

$$Q_1{}^{n+1} = Q_1 \oplus Q_0$$
$$Q_1{}^{n+1} = 0 \oplus 0 = 0$$

The $Q_1{}^{n+1}$ column of Table 6.5 can be filled in by simply Exclusively-ORing the $Q_0{}^n$ and $Q_1{}^n$ columns.

Step 4: state table

Because the state table is simply a translation of the transition table and relates the state of the flip-flops to the states of the counter, construct the state table shown in Table 6.6 by converting each binary combination in Table 6.5 to its appropriate state representation.

Step 5: state diagram

The state table shown in Table 6.6 can be represented graphically in the form of a state diagram. To construct the state diagram, first draw the symbols that represent all possible states of the counter. Then, referring to the state table make the appropriate

Figure 6.41

State diagram for the counter in Figure 6.25.

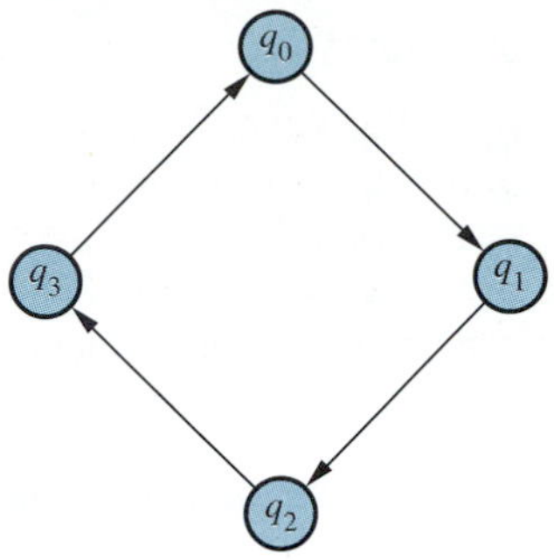

Table 6.5 Transition table for circuit in Figure 6.25

$(Q_1 \, Q_0)^n$	$(Q_1 \, Q_0)^{n+1}$
0 0	0 1
0 1	1 0
1 0	1 1
1 1	0 0

Table 6.6 State table for circuit in Figure 6.25

q^n	q^{n+1}
q_0	q_1
q_1	q_2
q_2	q_3
q_3	q_0

connections between the states as shown in Figure 6.41. The state diagram of Figure 6.41 is the same as the state diagram in Figure 6.23, verifying the analysis. The same procedure can be applied to circuits that contain *J-K* flip-flops. For example, consider the analysis of the circuit shown in Figure 6.30.

Step 1: excitation equations

The excitation equations for the circuit are obtained by determining the logic equations for J_0, K_0, J_1, and K_1 as follows:

$$J_0 = K_0 = 1 \tag{6.16}$$

$$J_1 = K_1 = \overline{Q}_0 \tag{6.17}$$

Step 2: transition equations

Recall from Chapter 5 (Section 5.7) that the state transition equation for a *J-K* flip-flop is

$$Q^{n+1} = J\overline{Q}^n + \overline{K}Q^n.$$

Therefore, for flip-flop *F*0,

$$Q_0^{n+1} = J_0\overline{Q}_0^n + \overline{K}_0 Q_0^n \tag{6.18}$$

Substitute the excitation equation (Eq. 6.16) into Equation 6.18 (and drop the superscript *n* on the right-hand side)

$$Q_0^{n+1} = 1 \cdot \overline{Q}_0 + \overline{1} \cdot Q_0$$
$$Q_0^{n+1} = \overline{Q}_0 + 0 \cdot Q_0$$
$$Q_0^{n+1} = \overline{Q}_0 \tag{6.19}$$

Similarly, for flip-flop *F*1

$$Q_1^{n+1} = J_1\overline{Q}_1^n + \overline{K}_1 Q_1^n \tag{6.20}$$

Substitute the excitation equation (Eq. 6.17) into Equation 6.20 (and drop the superscript *n*)

$$Q_1^{n+1} = \overline{Q}_0\overline{Q}_1 + \overline{\overline{Q}_0}Q_1$$
$$Q_1^{n+1} = \overline{Q}_0\overline{Q}_1 + Q_0 Q_1$$
$$Q_1^{n+1} = \overline{Q}_0 \oplus Q_1 \tag{6.21}$$

Equations 6.19 and 6.21 are the transition equations for flip-flops *F*0 and *F*1, respectively.

Step 3: transition table

From the transition equations construct the transition table for the counter as was done before. The next state of flip-flop *F*0 will toggle, and the next state of flip-flop *F*1 will be equal to the complement of the present state of flip-flop *F*0 Exclusively-ORed with the present state of flip-flop *F*1. The resulting transition table is shown in Table 6.7.

Table 6.7	Transition table for circuit in Figure 6.30

$(Q_1 Q_0)^n$	$(Q_1 Q_0)^{n+1}$
0 0	1 1
0 1	0 0
1 0	0 1
1 1	1 0

Table 6.8	State table for circuit in Figure 6.30

q^n	q^{n+1}
q_0	q_3
q_1	q_0
q_2	q_1
q_3	q_2

Step 4: state table

The state table shown in Table 6.8 is constructed by translating the outputs of the flip-flops in the transition table to the states of the counter.

Step 5: state diagram

The state table shown in Table 6.8 can be represented graphically in the form of a state diagram as shown in Figure 6.42. The state diagram of Figure 6.42 is the same as the state diagram in Figure 6.28, verifying the analysis.

    One of the advantages of this procedure for analyzing synchronous counters is that it provides a thorough description of the circuit being analyzed. For example, in the previous section the 3-bit counter designed in Figure 6.33 counted through five of the eight possible states in the following sequence:

$$0, 2, 6, 3, 5, 0, 2, \ldots$$

Figure 6.42

State diagram for the counter in Figure 6.30.

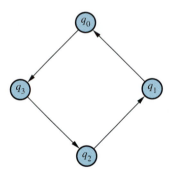

Because the remaining states were not defined in the synthesis (hence the don't cares in the excitation maps) what would happen if the counter started counting in one of the states 1, 4, or 7? To determine this using the timing diagram approach three timing diagrams would have to be drawn, each one starting off in one of the three undefined states. The timing diagrams would then indicate what the next state(s) would be. This is a tedious process. Furthermore, because the objective of analysis is to determine the function of the circuit being analyzed. If this were a circuit whose function was unknown, we would not know at which state to begin the timing diagram analysis and would have to use a trial-and-error approach. Using the formal five-step approach to analysis, one can obtain a complete state diagram that includes the main counting sequence of the counter as well as the connections among the undefined states. To illustrate this fact consider the analysis of the counter shown in Figure 6.33 in the following example.

Example 6.8

Analyze the logic circuit shown in Figure 6.33 by obtaining its state diagram.

Solution

Step 1: excitation equations

$$D_0 = Q_1 Q_0 + Q_2 Q_1$$
$$D_1 = \overline{Q_0}$$
$$D_2 = \overline{Q_2} Q_1$$

Step 2: transition equations

$$Q_0^{n+1} = D_0$$
$$= Q_1 Q_0 + Q_2 Q_1$$
$$Q_1^{n+1} = D_1$$
$$= \overline{Q_0}$$
$$Q_2^{n+1} = D_2$$
$$= \overline{Q_2} Q_1$$

Step 3: transition table

$(Q_2 \, Q_1 \, Q_0)^n$	$(Q_2 \, Q_1 \, Q_0)^{n+1}$
0 0 0	0 1 0
0 0 1	0 0 0
0 1 0	1 1 0
0 1 1	1 0 1
1 0 0	0 1 0
1 0 1	0 0 0
1 1 0	0 1 1
1 1 1	0 0 1

Step 4: state table

q^n	q^{n+1}
q_0	q_2
q_1	q_0
q_2	q_6
q_3	q_5
q_4	q_2
q_5	q_0
q_6	q_3
q_7	q_1

Step 5: state diagram. (Fig. 6.43).

Figure 6.43

State diagram for the
counter in Figure 6.33.

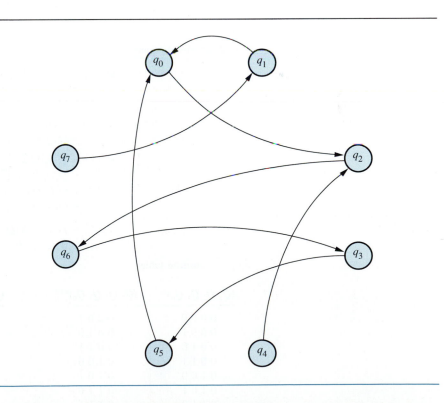

Even though the state diagram obtained in Example 6.8 (Fig. 6.43) does not resemble
the original state diagram (Fig. 6.31) from which the circuit in Figure 6.33 was

designed, by closer inspection of Figure 6.43 the count sequence 0, 2, 6, 3, 5, 0, . . . can be traced. From Figure 6.43 it can also be determined what the next state of the counter will be if it starts off in states 1, 4, or 7. For example, if the counter begins at state 1, the next state is 0, and the count cycles through the main counting sequence. Similarly, if the counter begins at state 4, the next state is 2, and count cycles through the main counting sequence. If, however, the counter begins at state 7, the next state is 1, which connects to the main counting sequence. Observe in Example 6.8 that all possible present states must be listed in the transition table because the counting sequence of the counter to be analyzed is usually not known.

The BCD counter designed in Example 6.7 is another example of a counter that when analyzed will produce a state diagram that is considerably different from the original state diagram from which the counter was designed. This is because the BCD counter uses 10 of the 16 possible states, and the state diagram obtained through analysis will also show the connections between the six undefined states besides the BCD counting sequence. The following example illustrates the analysis of the synchronous BCD counter designed in Figure 6.36.

Example 6.9

Analyze the circuit shown in Figure 6.36 by obtaining a state diagram.

Solution

Step 1: excitation equations

$$J_0 = K_0 = 1$$
$$J_1 = Q_0\overline{Q}_3$$
$$K_1 = Q_0$$
$$J_2 = K_2 = Q_1Q_0$$
$$J_3 = Q_2Q_1Q_0$$
$$K_3 = Q_0$$

Step 2: transition equations

$$Q_0^{n+1} = J_0\overline{Q}_0 + \overline{K}_0Q_0$$
$$= 1 \cdot \overline{Q}_0 + 0 \cdot Q_0$$
$$= \overline{Q}_0$$

$$Q_1^{n+1} = J_1\overline{Q}_1 + \overline{K}_1Q_1$$
$$= Q_0\overline{Q}_3\overline{Q}_1 + \overline{Q}_0Q_1$$

$$Q_2^{n+1} = J_2\overline{Q}_2 + \overline{K}_2Q_2$$
$$= Q_1Q_0\overline{Q}_2 + \overline{Q_1Q_0}\,Q_2$$
$$= Q_1Q_0\overline{Q}_2 + (\overline{Q}_1 + \overline{Q}_0)Q_2$$
$$= Q_1Q_0\overline{Q}_2 + \overline{Q}_1Q_2 + \overline{Q}_0Q_2$$

$$Q_3^{n+1} = J_3\overline{Q}_3 + \overline{K}_3Q_3$$
$$= Q_2Q_1Q_0\overline{Q}_3 + \overline{Q}_0Q_3$$

Step 3: transition table

$(Q_3\,Q_2\,Q_1\,Q_0)^n$	$(Q_3\,Q_2\,Q_1\,Q_0)^{n+1}$	$(Q_3\,Q_2\,Q_1\,Q_0)^n$	$(Q_3\,Q_2\,Q_1\,Q_0)^{n+1}$
0 0 0 0	0 0 0 1	1 0 0 0	1 0 0 1
0 0 0 1	0 0 1 0	1 0 0 1	0 0 0 0
0 0 1 0	0 0 1 1	1 0 1 0	1 0 1 1
0 0 1 1	0 1 0 0	1 0 1 1	0 1 0 0
0 1 0 0	0 1 0 1	1 1 0 0	1 1 0 1
0 1 0 1	0 1 1 0	1 1 0 1	0 1 0 0
0 1 1 0	0 1 1 1	1 1 1 0	1 1 1 1
0 1 1 1	1 0 0 0	1 1 1 1	0 0 0 0

Step 4: state table

q^n	q^{n+1}
q_0	q_1
q_1	q_2
q_2	q_3
q_3	q_4
q_4	q_5
q_5	q_6
q_6	q_7
q_7	q_8

q^n	q^{n+1}
q_8	q_9
q_9	q_0
q_A	q_B
q_B	q_4
q_C	q_D
q_D	q_4
q_E	q_F
q_F	q_0

Step 5: state diagram. (Fig. 6.44)

Figure 6.44

State diagram for the
counter in Figure 6.36.

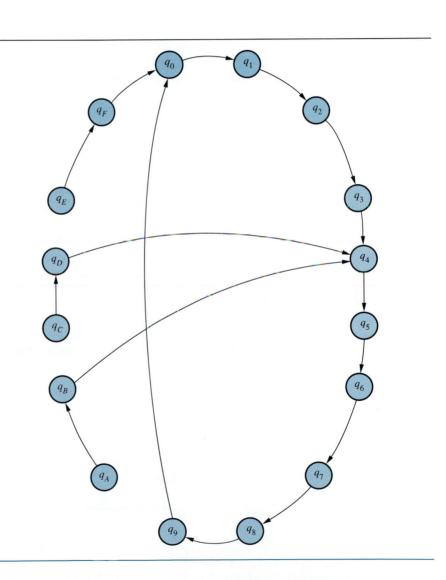

The state diagram in Figure 6.44 illustrates a BCD counting sequence that verifies
the operation of the circuit as a BCD counter. The undefined states *A, B, C, D, E,* and

F are all connected to the main BCD counting sequence. For example, if the counter starts off in state *E*, the next states are *F* and then 0. If the counter starts in state *C*, the next states are *D*, and then 4. Similarly, if the counter starts in state *A*, the next states are *B* and then 4.

It is apparent that the analysis technique discussed in this section has some advantages and disadvantages when compared with the timing diagram approach to analysis. Depending on the type of circuit, the formal approach to analysis can be the quickest path to a solution. However, some circuits are simple enough to be analyzed through a timing diagram. The formal approach is restricted to synchronous counters only whereas the timing diagram can be used to analyze both synchronous and asynchronous counters. Finally, the formal approach to analysis is the most complete and direct approach because the state diagram obtained accounts for all possible states of the circuit.

Review Questions

1. Briefly outline the steps involved in the formal analysis of synchronous counters.
2. What are transition equations?
3. Compare the formal analysis procedure with the timing diagram procedure of analysis.

6.6
An application of counters

One of the many digital circuit applications that makes extensive use of counters is a time-of-day clock. This section applies many of the concepts of flip-flops and counters covered in previous sections to the design of a 24-hour time-of-day clock. Figure 6.45 shows the logic symbol for a time-of-day clock circuit. The circuit has a single clock input, *CLOCK,* which is used to keep the basic time of the counters, and outputs to drive three sets of seven-segment LED displays that will be used to display the hours, minutes, and seconds counts. For each (negative) transition of *CLOCK,* the seconds count will increment once from 00 to 59. Each time the seconds recycles to 00, the minutes count will increment once from 00 to 59. When the minutes count recycles to 00, the hours count will increment once from 00 to 23. The clock circuit can therefore be used to count hours, minutes, and seconds of real time.

Figure 6.45

Logic symbol for a time-of-day clock.

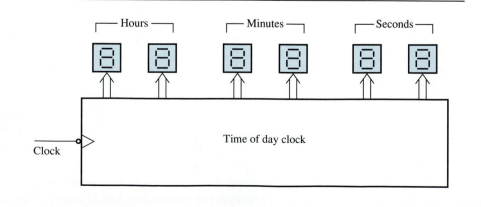

The logic circuit for the 24-hour time-of-day clock is shown in Figure 6.46. The clock circuit consists of a series of counters that are used to count through the units of time — seconds, minutes, and hours. Because the hours, minutes, and seconds are displayed as two decimal digits, there are two seven-segment LEDs for each. Each unit of time (hours, minutes, and seconds) is maintained by a set of counters whose outputs are connected to seven-segment decoders to drive the LEDs. The operation of the circuit can be explained with reference to Figure 6.46 as follows.

Figure 6.46

Logic circuit for a time-of-day clock.

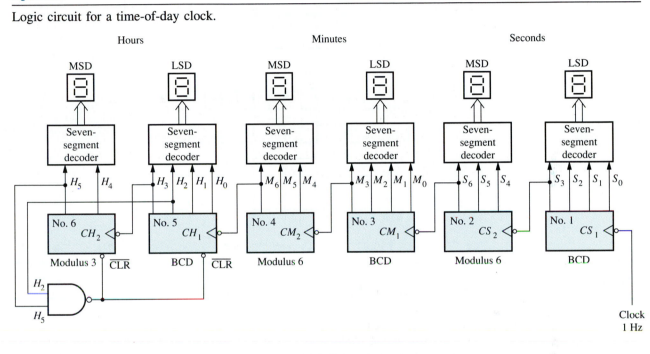

Seconds counter

The seconds counter consists of two counters — one for each of the digits in the seconds count. The seconds count cycles from 00 to 59, and therefore a BCD counter is used to cycle through the 0 to 9 count of the LSD of the seconds count, and a modulus-6 counter is used to cycle through the 0 to 5 count of the MSD of the seconds count.

A periodic clock pulse with a frequency of 1 Hz provides the basic clock pulses for the BCD counter. Because the rate at which the clock pulses trigger the counter's clock (CS_1) is 1 pulse per second, the counter will count from 0 to 9 in 1-second increments. The timing diagram for the outputs of the BCD counter ($S_3S_2S_1S_0$) with reference to its clock input (CS_1) is shown in Figure 6.47. The output of the BCD counter, S_3, makes a negative transition when the counter recycles to 0.

The S_3 output of the BCD counter makes a negative transition every 10 seconds and is used as a clock for the modulus-6 counter (CS_2). The modulus-6 counter maintains the MSD count for seconds and only requires six states, 0 through 5. When S_3 makes a negative transition every 10 seconds, because S_3 is connected to CS_2, the clock input of the modulus-6 counter, the counter will increment as shown in Figure 6.47. The outputs of the modulus-6 counter, $S_6S_5S_4$ will cycle through the counts 0 through 5.

Figure 6.47

Timing diagram for seconds count.

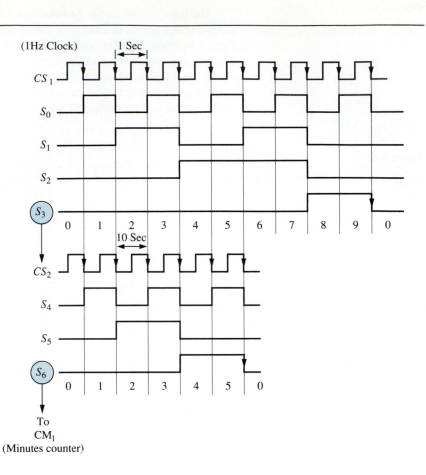

The timing diagram of Figure 6.47 shows that the output of the modulus-6 counter, S_6, makes a negative transition when it recycles to 0. Because each time interval between CS_2 clock pulses is 10 seconds, this negative transition will occur every 60 seconds (1 minute).

Minutes counter

Because the minutes count is exactly the same as the seconds count, the two counters that provide the minutes count are identical to the two counters that provide the seconds count. The minutes counter consists of two counters, one for each of the digits in the minutes count. Because the minutes count cycles from 00 to 59, a BCD counter is used to cycle through the 0 to 9 count of the LSD of the minutes count. A modulus-6 counter is used to cycle through the 0 to 5 count of the MSD of the minutes count.

The clock input of the BCD counter, CM_1, is connected to S_6. Recall that S_6 will make a negative transition once every minute and the BCD counter will change states every minute. The BCD counter will therefore count from 0 to 9 in 1-minute increments. The timing diagram for the outputs of the BCD counter ($M_3 M_2 M_1 M_0$) with reference to its clock input (CM_1) is shown in Figure 6.48. The output of the BCD counter, M_3, makes a negative transition when the counter recycles to 0.

The M_3 output of the BCD counter makes a negative transition every 10 minutes and is used as a clock for the modulus-6 counter. The modulus-6 counter maintains

Figure 6.48

Timing diagram for minutes count.

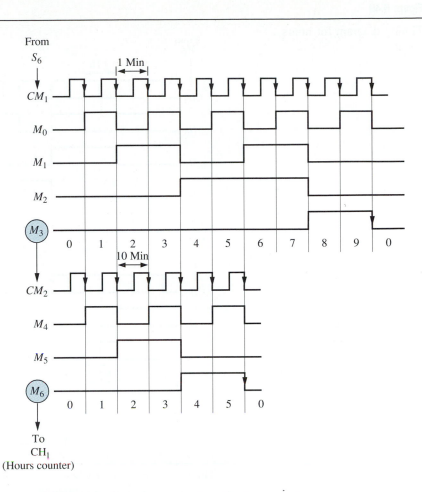

the MSD count for minutes and only requires six states, 0 through 5. When M_3 makes a negative transition every 10 minutes, because M_3 is connected to CM_2 (the clock input of the modulus-6 counter), the counter will increment as shown in Figure 6.48. The outputs of the modulus-6 counter, $M_6M_5M_4$, will cycle through the counts 0 through 5.

The timing diagram of Figure 6.48 shows that the output of the modulus-6 counter, M_6, makes a negative transition when it recycles to 0. Because each time interval between CM_2 clock pulses is 10 minutes, this negative transition will occur every 60 minutes (1 hour).

Hours counter

The hours counter consists of two counters, one for each of the digits in the hours count. Because the hours count cycles from 00 to 23, a BCD counter is used to cycle through the 0 to 9 count of the LSD of the hours count, and a modulus-3 counter is used to cycle through the 0 to 2 count of the MSD of the hours count.

The clock input of the BCD counter, CH_1, is connected to M_6. Recall that M_6 will make a negative transition once every hour and therefore the BCD counter will change states every hour, that is, it will count from 0 to 9 in 1-hour increments. The timing diagram for the outputs of the BCD counter ($H_3H_2H_1H_0$) with reference to its

Figure 6.49

Timing diagram for hours count.

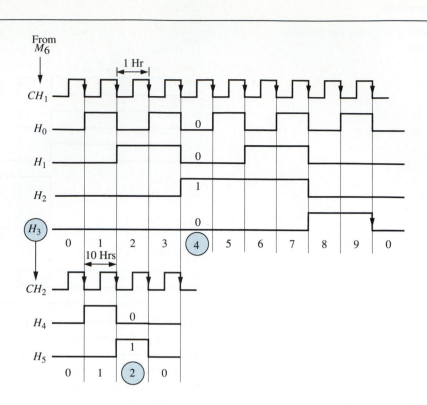

clock input (CH_1) is shown in Figure 6.49. The output of the BCD counter, H_3, makes a negative transition when the counter recycles to 0.

The H_3 output of the BCD counter makes a negative transition every 10 hours and is used as a clock for the modulus-3 counter. The modulus-3 counter maintains the MSD count for hours and therefore only requires three states, 0 through 2. When H_3 makes a negative transition every 10 hours, because H_3 is connected to CH_2 (the clock input of the modulus-3 counter), the counter will increment as shown in Figure 6.49. The outputs of the modulus-3 counter, H_5H_4, will cycle through the counts 0 through 2.

The timing diagram of Figure 6.49 shows that the output of the modulus-3 counter, H_5, makes a negative transition when it recycles to 0. Because each time interval between CH_2 clock pulses is 10 hours, theoretically this negative transition will occur every 30 hours. However, to recycle the counter after the hours count reaches 23 the counter is reset when the hours count reaches 24. Note in the timing diagram of Figure 6.49 that when the hours count reaches 24

$$H_5H_4 = 10 \quad H_3H_2H_1H_0 = 0100$$

and this is the only time when output $H_5 = 1$ and $H_2 = 1$. Therefore, when the two lines are at the logic 1 state both counters must be reset. The *NAND* gate shown in Figure 6.46 functions as a count decoder and will produce a logic 0 only when H_5 and H_2 are logic 1's. This active-low signal is used to directly reset (clear) the flip-flops that make up each counter through their direct $\overline{CLR}$ inputs. This scheme allows the hours counter to recycle after a count of 23.

Review Questions

1. What type of counters could be used to implement the time-of-day clock — synchronous, asynchronous, or any type?
2. Could the entire time-of-day clock circuit be considered a synchronous or asynchronous (ripple) counter?
3. Why are the seven-segment decoders used in the circuit?

6.7
Frequency division

It was stated in Chapter 5 that a T flip-flop could be used to divide its input clock frequency by 2. If the output of the T flip-flop was connected to the clock input of another T flip-flop, the original clock frequency would then be divided by 4. Similarly, if a third T flip-flop was added in series the frequency division would be 8, etc. Therefore, each flip-flop in series divides its input clock by 2 so that the total frequency division is

$$2 \times 2 \times 2 = 8.$$

In general it can be stated that the frequency division is 2^n where n is the number of flip-flops in the counter. The asynchronous counters discussed in Section 6.2 can be used as frequency division circuits because they are made up of T flip-flops connected in series. For example, Figure 6.3 shows that the timing diagram for the 3-bit asynchronous down counter shown in Figure 6.2 divides the clock by 2 at output X_0, by 4 at output X_1, and by 8 at output X_2. Because the timing diagram is the same for the asynchronous and synchronous counters with the same counting sequence, the frequency division can also be accomplished by synchronous counters.

Because sequential binary counters can also be used for frequency division, they are also known as *divide-by-n* counters, where n is the modulus of the counter. For example, a 3-bit binary counter will have eight possible states (modulus 8) and its most significant output will divide the clock frequency by 8. The 3-bit binary counter (modulus-8 counter) could also be referred to as a divide-by-8 counter. Similarly, a 4-bit binary counter would provide a maximum frequency division of 16 at its most significant output, whereas a 4-bit BCD counter (modulus-10 counter) would provide a maximum frequency division of 10 at its most significant output. The general frequency divisors for the outputs of binary counters is shown in Figure 6.50. The divisor by which the clock frequency is divided is a power of 2

$$2^{p+1}$$

where p is the bit position of the counter output. This relationship can also be used to determine the maximum frequency division that could be accomplished by an n-bit binary counter

$$2^n$$

where n is the size of the counter.

Counters can be *cascaded* to increase the frequency division. For example, Figure 6.51 shows how a divide-by-4 counter and a divide-by-8 counter could be connected to produce a divide-by-32 counter. In general, to determine the total frequency division of two cascaded counters, multiply the divisors of each counter.

Figure 6.50

Divided outputs for binary counters.

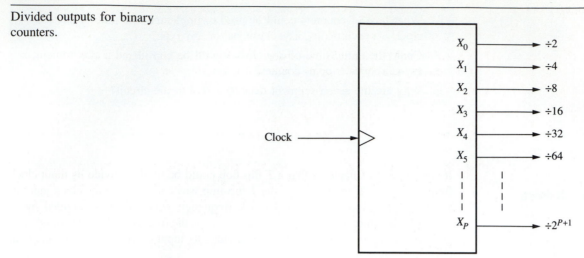

Figure 6.51

Divide-by-32 counter.

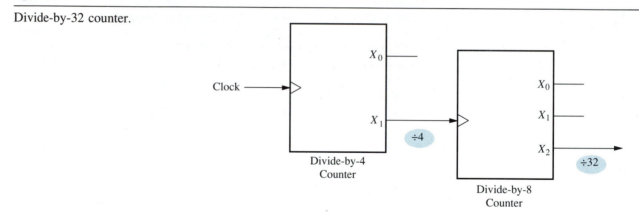

Figure 6.52

Divide-by-160 counter.

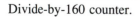

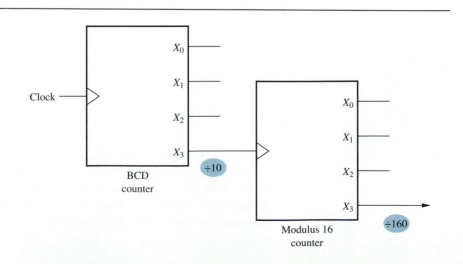

Example 6.10

Design a circuit to divide a periodic clock frequency by a factor of 160.

Solution

To accomplish the frequency division, cascade a divide-by-16 counter with a divide-by-10 counter (BCD counter) to obtain a total frequency division of

$$16 \times 10 = 160.$$

The circuit to do this is shown in Figure 6.52.

An example of frequency division was seen in the previous section for the design of the time-of-day clock. Figure 6.46 shows that the circuit effectively takes a 1 pulse per second (pps) input clock (at CS_1) and uses it to trigger a seconds counter. The seconds counter effectively divides the 1 pps clock first by 10 at S_3 and then by 60 at S_6 to produce a clock frequency of $1/60$ pps or 1 pulse per minute (ppm). Similarly the 1-ppm clock triggers the minutes counters that also divide the 1-ppm clock by 60 and produces a clock frequency of $1/60$ ppm or 1 pulse per hour (pph) at M_6. This clock is then used to trigger the hours counter.

Review Questions

1. Why are binary counters known as divide-by-n counters?
2. Is frequency division limited to asynchronous counters? Why?
3. How can frequency division be increased?
4. What determines the maximum frequency division that can occur in a counter?

6.8

IC logic

The 74xxx series of ICs has several prepackaged counters available in the form of single ICs that can be used for a variety of counter applications. Both synchronous and asynchronous counters are available, many of which have additional control inputs for initializing the state of the counter and/or arrangements for reconfiguring the count.

The logic symbol and pin configuration of the 7493 divide-by-16 counter is shown in Figure 6.53. The 7493 is a four-stage (4-bit) modulus-16 binary ripple

Figure 6.53

Logic symbol and pin configuration of the 7493 divide-by-16 counter.

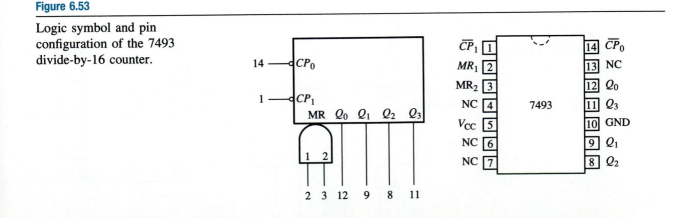

counter. The 4-bit output of the counter is made available at Q_3, Q_2, Q_1, Q_0. Internally the 7493 consists of a divide-by-2 counter (single flip-flop) with clock input CP_0 and output Q_0, and a divide-by-8 counter with clock input CP_1 and outputs Q_2, Q_1, Q_0. The 7493 can therefore operate as two independent counters. By connecting the output of the divide-by-2 counter Q_0 with the clock input of the divide-by-8 counter CP_1, the 7493 can be configured as a divide-by-16 counter with a 4-bit output Q_3, Q_2, Q_1, Q_0 and a clock input CP_0. The master reset lines MR_1 and MR_2 are used to asynchronously reset the counter when logic 1's are applied to both inputs.

The 7492 divide-by-12 counter whose logic symbol and pin configuration are shown in Figure 6.54 is similar in operation to the 7493 except that it is a modulus-12 counter with 4-bit outputs Q_3, Q_2, Q_1, Q_0. Internally the 7492 is designed as two independent divide-by-2 and divide-by-6 counters that, like the 7493, can be connected to produce a divide-by-12 count. CP_0 is the clock for the divide-by-2 counter with output Q_0, and CP_1 is the clock for the divide-by-6 counter with outputs Q_2, Q_1, Q_0. Like the 7493, the 7492's master reset inputs MR_1 and MR_2 can be used to reset the counter outputs when both are at the logic 1 state.

Figure 6.54

Logic symbol and pin configuration of the 7492 divide-by-12 counter.

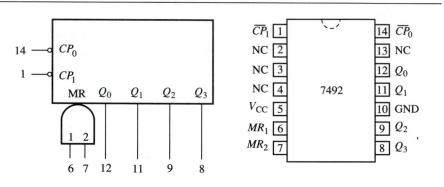

The 7490 decade counter whose logic symbol and pin configuration are shown in Figure 6.55 is also similar in operation to the 7492 and 7493, except that it is a modulus-10 BCD counter with 4-bit outputs Q_3, Q_2, Q_1, Q_0. Internally the 7490 is designed as two independent divide-by-2 and divide-by-5 counters that, like the 7492 and 7493, can be connected to produce a divide-by-10 count. CP_0 is the clock

Figure 6.55

Logic symbol and pin configuration of the 7490 decade counter.

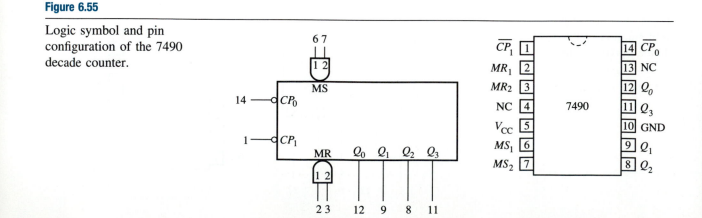

for the divide-by-2 counter with output Q_0, and CP_1 is the clock for the divide-by-5 counter with outputs Q_2, Q_1, Q_0. Like the 7492 and 7493, the 7490's master reset inputs MR_1 and MR_2 can be used to reset the counter outputs when both are at the logic 1 state. The 7490 includes two additional master set inputs, MS_1 and MS_2, that are used to asynchronously set the outputs of the counter to the maximum count, 1001 (9) when both inputs are at the logic 1 state.

The 74177 presettable binary counter is similar in operation to the 7493. It is a 4-bit modulus-16 ripple counter that is internally designed with a divide-by-2 and a divide-by-8 section. The logic symbol and pin configuration for the 74177 are shown in Figure 6.56. Unlike the 7493, however, the outputs of the 74177 can be asynchronously preset to any 4-bit binary combination so that on the next clock pulse the counter begins at a preset state. The data inputs P_3, P_2, P_1, P_0 are used to enter the binary number that is to be preset in the counter and the parallel load ($\overline{PL}$) input is used to load (or latch) the data at the outputs Q_3, Q_2, Q_1, Q_0. Only a single active-low master reset line $\overline{MR}$ is used to reset the counter.

Figure 6.56

Logic symbol and pin configuration of the 74177 presettable binary counter.

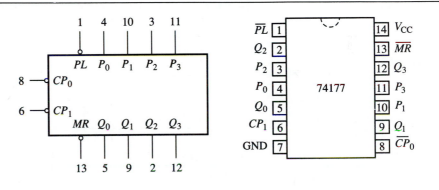

The 74176 presettable decade counter is similar in operation to the 7490. It is a 4-bit modulus-10 ripple counter that is internally designed with a divide-by-2 and a divide-by-5 section. The logic symbol and pin configuration for the 74176 are shown in Figure 6.57. Like the 74177, the outputs of the 74176 can be asynchronously preset to any 4-bit binary combination so that on the next clock pulse the counter begins at a preset state. The data inputs P_3, P_2, P_1, P_0 are used to enter the binary number that is to be preset in the counter and the $\overline{PL}$ input is used to load (or latch)

Figure 6.57

Logic symbol and pin configuration of the 74176 presettable decade counter.

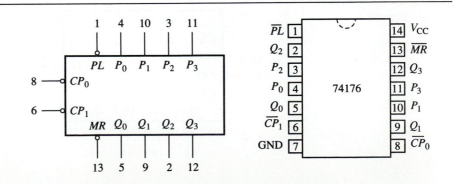

the data at the outputs Q_3,Q_2,Q_1,Q_0. Only a single active-low master reset line $\overline{MR}$ is used to reset the counter.

The 74176 and 74177 are asynchronous counters. For applications that require synchronous counters, the 74160 and 74161 provide the same functions, respectively, using synchronous counters instead.

The 74169 is a synchronous bidirectional modulus-16 binary counter that has data inputs for presetting the counter to any state, as well as a control input to count either *UP* or *DOWN*. The logic symbol and pin configuration for the 74169 are shown in Figure 6.58. The 74169 contains a single modulus-16 binary counter with outputs Q_3,Q_2,Q_1,Q_0. The preset data inputs P_3,P_2,P_1,P_0 are used to initialize the output of the counter to a predetermined state when the parallel enable ($\overline{PE}$) input is low and the clock (*CP*) makes a positive transition. The *UP/DOWN* ($U/\overline{D}$) input causes the counter to count *UP* (logic 1) or *DOWN* (logic 0). For counting to occur, both count enable parallel ($\overline{CEP}$) and count enable trickle ($\overline{CET}$) must be at the logic 0 states. The terminal count ($\overline{TC}$) output of the counter will be a logic 0 when the counter reaches 0000 in the *DOWN* mode or when the counter reaches 1111 in the *UP* mode.

The 74168 is a synchronous bidirectional BCD decade counter that, like the 74169, has data inputs for presetting the counter to any state, as well as a control

Figure 6.58

Logic symbol and pin configuration of the 74169 synchronous bidirectional modulus-16 binary counter.

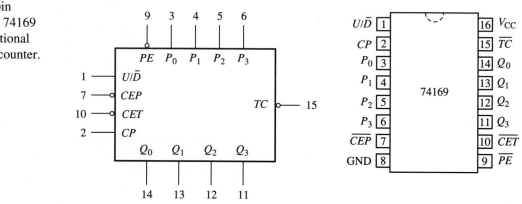

Figure 6.59

Logic symbol and pin configuration of the 74168 synchronous bidirectional decade counter.

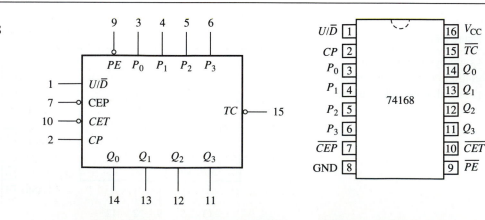

input to make the counter count either *UP* or *DOWN*. The logic symbol and pin configuration for the 74168 are shown in Figure 6.59. The 74168 contains a single modulus-10 BCD counter with outputs Q_3, Q_2, Q_1, Q_0. Like the 74169, the preset data inputs P_3, P_2, P_1, P_0 are used to initialize the output of the counter to a predetermined state when $\overline{PE}$ is low and the clock (*CP*) makes a positive transition. The $U/\overline{D}$ input causes the counter to count *UP* (logic 1) or *DOWN* (logic 0). For counting to occur, both $\overline{CEP}$ and $\overline{CET}$ must be at the logic 0 states. The $\overline{TC}$ output of the counter will be a logic 0 when the counter reaches 0000 in the *DOWN* mode or when the counter reaches 1001 in the *UP* mode.

Review Questions

1. Compare the functions of the 7493, 7492, and 7490 counters.

2. Compare the functions of the 74177 and 74176 counters.

3. Compare the functions of the 74169 and 74168 counters.

6.9
Troubleshooting counter circuits

Most problems and faults occur in counter circuits due to propagation delays. The effects of propagation delays are more common in asynchronous counters and at high frequencies. The effects of high-frequency operation on sequential logic circuits were seen in Chapter 5 where the propagation delays were more pronounced at higher frequencies and did not comply with the recommended setup and hold times of the flip-flop. The effects of propagation delay on counters usually produce incorrect states in the counter.

For example, consider the 3-bit asynchronous binary down counter shown in Figure 6.2. The ideal timing diagram for the counter is shown in Figure 6.3. However, if the circuit is operated at very high clock rates, the actual timing diagram showing propagation delays is quite different as shown in Figure 6.60.

Figure 6.60

Timing diagram of a counter showing propagation delays.

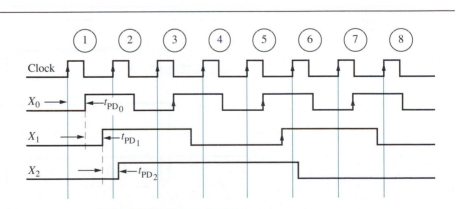

The output X_0 (output of flip-flop *F0*) is delayed from the main clock transition by the propagation delay labeled "t_{PD_0}." The output X_1 (output of flip-flop *F1*) is delayed from the transition of X_0 by the propagation delay labeled "t_{PD_1}." Finally, the output X_2 (output of flip-flop *F2*) is delayed from the transition of X_1 by the propagation delay labeled "t_{PD_2}." Therefore, the output of the last flip-flop in series (*F2*) is delayed from the transition of the main clock by three propagation delays!

Figure 6.60 shows that the propagation delays cause the outputs of the flip-flops to change in the middle or close to the end of each interval, thereby producing an unstable counter state for each transition of the main clock pulse. Also, there are many intervals (interval 1 for example) where the propagation delay causes the output X_2 to change after the next clock pulse has made its transition. This leads to an incorrect counting sequence for interval 1.

One of the possible causes for a fault in the counting sequence of a counter could be due to the flip-flop's propagation delay time or the operating clock frequency of the counter. If a counter is to be operated at high clock frequencies, then flip-flops with very short propagation delays should be used. Counters that display wrong

Figure 6.61

Detecting glitches in the output of a counter. (a) Circuit; (b) timing diagram.

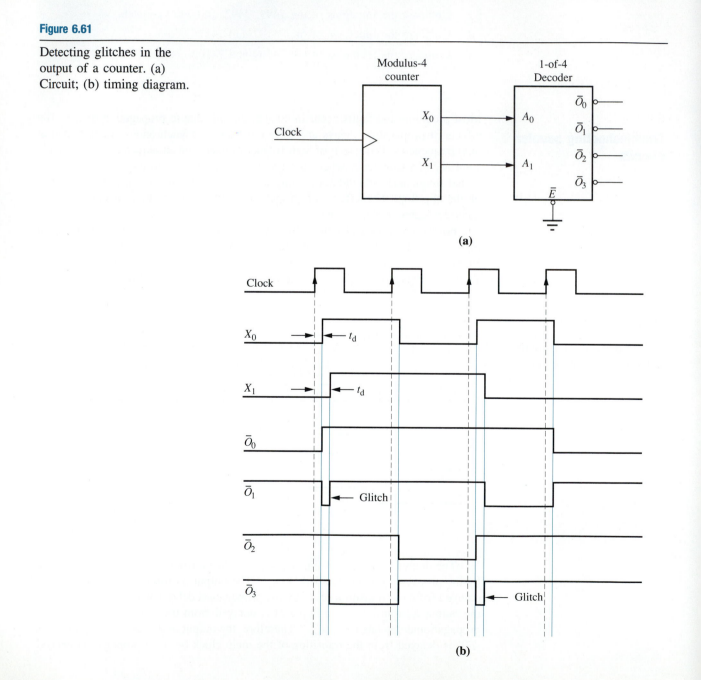

counts or missing counts should be checked for such problems by temporarily reducing the operating frequency of the circuit.

Synchronous counters are not as susceptible to the effects of propagation delay as asynchronous counters because the flip-flops are triggered simultaneously and the propagation delay effects are not cumulative as in asynchronous counters. Therefore, synchronous counters, rather than asynchronous counters, should be used for high-frequency applications.

Another common fault that could cause problems in counter circuits is the effect of glitches in asynchronous counters that use counter decoding to recycle the count. If the counter is being cascaded with other counters, the glitch could cause false

Figure 6.62

Modification of Figure 6.61 to eliminate glitches. (a) Circuit; (b) timing diagram.

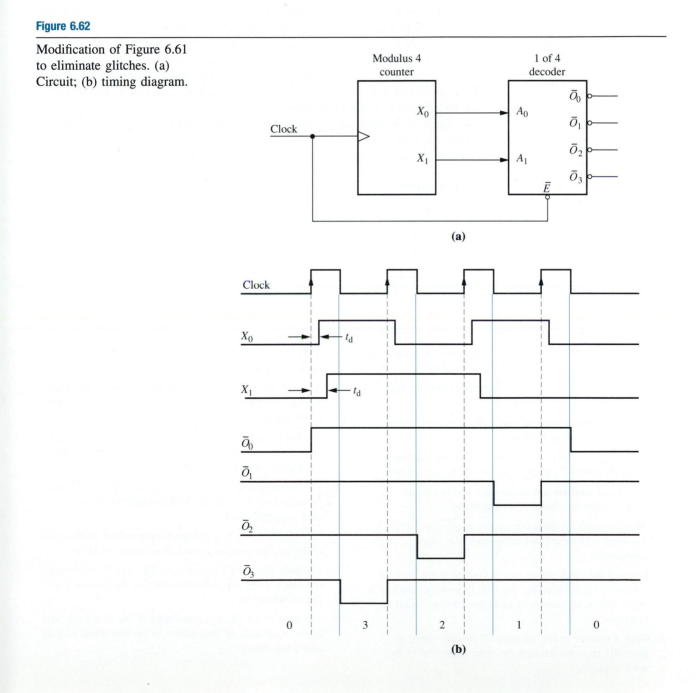

triggering of the other counters in the circuit. These glitches could also be caused by the propagation delays that exist in asynchronous counters. To detect these glitches use a circuit similar to the one shown in Figure 6.61a, that is, an asynchronous modulus-4 counter whose outputs are connected to a 1-of-4 decoder. The decoder will activate one of its four outputs ($\overline{O}_0$ through $\overline{O}_3$) when it detects a particular count. Of course, a larger-size counter could also be used with a larger-size decoder. This type of circuit is used in many applications where it is necessary to determine if the counter is in one or more desired states.

The timing diagram of the circuit shown in Figure 6.61b shows how the small propagation delays generated by the output of each flip-flop in the counter can cause the decoder to falsely activate its output and produce glitches. The glitches on $\overline{O}_1$ and $\overline{O}_3$ are produced due to the cumulative effect of the propagation delays of each flip-flop. If these propagation delays do not exceed the high pulse width of the clock then we can use the low level of the clock to strobe the enable line of the decoder after the glitches disappear. The modifications to the circuit and the resulting timing diagram are shown in Figure 6.62. Now that the decoder is enabled only when the clock goes *LOW,* its output will not be affected by the propagation delays of X_0 and X_1 during the *HIGH* period of the clock pulse.

Again, if the problem of glitches in a circuit cannot be corrected by any other means, the only solution is to replace the asynchronous counters with synchronous counters that are less susceptible to propagation delays.

Review Questions

1. What type of counter could produce incorrect counts at high frequencies?
2. What causes a counter to malfunction at high frequencies?
3. How is a glitch in the output of a counter produced?

Summary

- A counter is a circuit that produces a numeric count each time an input clock pulse makes a transition.
- A counter has several output lines that can exist in several states called counts.
- The maximum number of states that a counter counts through is called its modulus.
- The maximum number of states of a counter depends on the relationship 2^n, where n is the number of outputs.
- All digital counters can be divided into two basic types — asynchronous and synchronous.
- Asynchronous counters are usually constructed with T flip-flops with their clock connected in a series arrangement.
- Because the clock pulses that trigger each flip-flop are generated by the previous flip-flop in series to produce a ripple effect, asynchronous counters are also called ripple counters.
- Once a counter cycles through all possible states it generally recycles through the same states and in the same sequence.

- A *BCD* counter counts from 0 to 9 and is known as a modulus-10 or decade counter.
- A synchronous counter is made up of a set of flip-flops that are triggered simultaneously (in parallel) by a single *CLOCK* pulse.
- The term synchronous implies that all the flip-flops in the counter change states at the same time when the main clock pulse makes its transition.
- The counting sequence of a synchronous counter is not dependent on the type of flip-flop, negative or positive edge-triggered, used.
- The state diagram is a graphic representation of the states of the counter and the connections among its states.
- The state table is a translation of the state diagram and represents in tabular form the states of the counter and their connections.
- The transition table is a translation of the state table and relates each state of the counter to the individual outputs of the flip-flops.

- The excitation maps are used to determine what the control inputs of each flip-flop (J and K, or D) have to be for the flip-flop to make the transitions specified in the transition table.

- The excitation equations are obtained from the excitation maps and identify the connections for the control inputs of the counter's flip-flops.

- The transition equations allow us to determine what the next state of a flip-flop will be, given the present conditions.

- Because sequential binary counters can also be used for frequency division, they are also known as divide-by-n counters where n is the modulus of the counter.

- When two counters are cascaded to divide an input clock, the total frequency division obtained is the product of the divisors of each counter.

- The 74xxx series has both synchronous and asynchronous counters, many of which have additional control inputs for initializing the state of the counter and/or arrangements for reconfiguring the count.

- Asynchronous counters are more susceptible to the effects of propagation delay than synchronous counters.

- The effects of propagation delay at high frequencies could cause a counter to malfunction.

Problems

Section 6.2   Asynchronous counters

1. Draw the logic circuit for a 4-bit asynchronous binary up counter using J-K flip-flops.

2. Draw the logic circuit for a 4-bit asynchronous binary down counter using D flip-flops.

3. Analyze the circuit shown in Figure 6.63 by obtaining its timing diagram. Assume that the initial state of the counter is 000.

4. Analyze the circuit shown in Figure 6.64 by obtaining its timing diagram. Assume that the initial state of the counter is 0000.

Figure 6.63

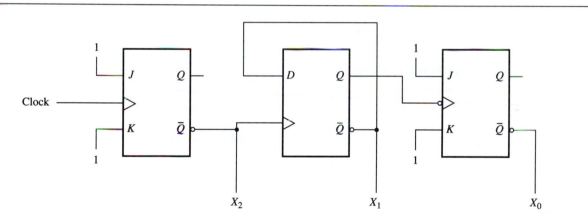

Figure 6.64

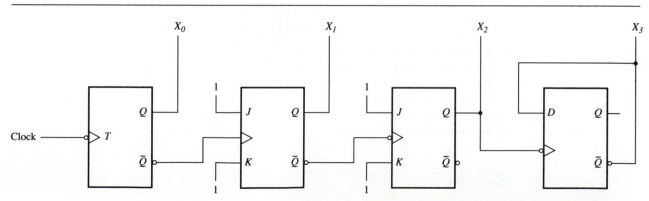

5. Design a modulus-6 ripple up counter using positive edge-triggered T flip-flops.

6. Design an asynchronous count-down timer to count down from 7 to 0 and stop when the count reaches 0 (it should not recycle).

Section 6.3   Synchronous counters

7. Analyze the circuit shown in Figure 6.25 by obtaining its timing diagram. Assume that the initial state of the counter is 00.

8. Analyze the circuit shown in Figure 6.27 by obtaining its timing diagram. Assume that the initial state of the counter is 00.

9. Analyze the circuit shown in Figure 6.30 by obtaining its timing diagram. Assume that the initial state of the counter is 00.

10. Analyze the circuit shown in Figure 6.33 by obtaining its timing diagram. Start the timing diagram at state 1.

11. Analyze the circuit shown in Figure 6.33 by obtaining its timing diagram. Start the timing diagram at state 0.

12. Analyze the circuit shown in Figure 6.33 by obtaining its timing diagram. Start the timing diagram at state 4.

13. Analyze the circuit shown in Figure 6.33 by obtaining its timing diagram. Start the timing diagram at state 7.

14. Compare the results obtained from Problems 10 through 13 with the state diagram of Figure 6.43.

Section 6.4   Synthesis of synchronous counters

15. Design a 3-bit synchronous binary up counter using J-K flip-flops.

16. Design a 3-bit synchronous binary up counter using D flip-flops.

17. Design a 3-bit synchronous binary down counter using J-K flip-flops.

18. Design a 3-bit synchronous binary down counter using D flip-flops.

19. Design a synchronous counter using D flip-flops to count in the sequence

$$9, F, 3, 0, 7, 9, F, \ldots$$

20. Design a synchronous counter using J-K flip-flops to count in the sequence

$$3, 7, 1, 0, 6, 5, 3, \ldots$$

21. Design a 2-bit synchronous up/down counter that will count up or down on command. If the control input M is a logic 1 the counter should count down, and if the control input M is a logic 0 the counter should count up. Use J-K flip-flops in the design.

22. Design a synchronous counter that will count in the sequence: 2, 1, 3, 0, if control input $A = 1$, and in the sequence 0, 1, 3, 0, if control input $A = 0$. Use D flip-flops in the design.

23. Design a synchronous count-down counter that will count down from 7 to 0 and stop when the count reaches 0. Use D flip-flops in the design.

Section 6.5   Analysis of synchronous counters

24. Obtain a state diagram for the circuit shown in Figure 6.65.

25. Obtain a state diagram for the circuit shown in Figure 6.66.

26. Determine what the next state(s) of the counter designed in Problem 20 would be if the counter were to start off in states 2 and 4. Construct a state diagram showing *all* possible state transitions.

Figure 6.65

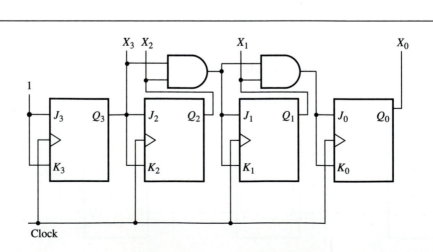

Figure 6.66

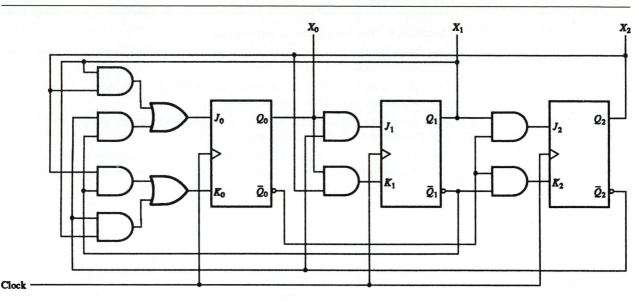

Figure 6.67

27. Determine what the next state(s) of the counter designed in Problem 19 would be if the counter were to start off in states 1, 2, 4, 5, 6, 8, A, B, C, D, E. Construct a state diagram showing *all* possible state transitions.

28. Analyze the circuit obtained for Problem 22 by obtaining its state diagram.

29. Obtain a state diagram for the circuit shown in Figure 7.26.

30. Obtain a state diagram for the circuit shown in Figure 7.28.

Section 6.6 An application of counters

31. Modify the 24-hour time-of-day clock in Figure 6.46 to function as a 12-hour time-of-day clock.

Section 6.7 Frequency division

32. Design a divide-by-40 counter circuit using a BCD counter and a 4-bit binary counter.

33. Determine the frequency of the *OUTPUT CLOCK* in Figure 6.67 if the *INPUT CLOCK* has a frequency of 8 kHz.

34. Determine the frequency of the *OUTPUT CLOCK* in Figure 6.67 if the *INPUT CLOCK* has a frequency of 100 Hz and the *OUTPUT CLOCK* is taken from output X_3 of the modulus-16 counter.

35. Design a divide-by-32 counter using two modulus-16 counters.

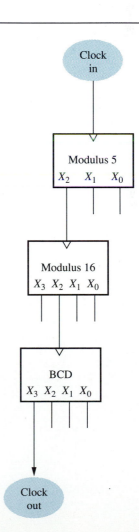

Troubleshooting

Section 6.9   Troubleshooting counter circuits

36. Draw a timing diagram of the counter shown in Figure 6.5 showing the propagation delays produced by each flip-flop.

37. Draw a timing diagram of the counter shown in Figure 6.14 showing the propagation delays produced by each flip-flop.

38. Discuss the timing diagrams produced in Problems 36 and 37 in terms of high-frequency operation of the two counters.

39. The circuit shown in Figure 6.68 produces spurious outputs when in operation. Troubleshoot the circuit by constructing a timing diagram and determining where the glitches occur in the outputs. Suggest possible remedies.

Figure 6.68

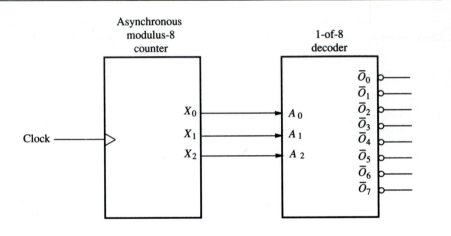

40. Refer to the counter circuit shown in Figure 6.21. How would the circuit behave if one of the following faults existed in the circuit:
    a. The output of the *AND* gate was shorted to *GND*.
    b. The output of the *AND* gate was shorted to V_{CC}.

7

REGISTERS

OBJECTIVES

The objectives of this chapter are to:

- Introduce the concepts of parallel data storage

- Study the operation of parallel registers and the manner in which data is loaded and retrieved

- Introduce the concepts of serial data storage

- Study the design and operation of shift registers and their timing characteristics

- Examine some simple applications of shift registers

- Introduce the concepts of register arrays (register files) and their applications

- Introduce the elementary concepts of memory circuits such as addressing, reading, and writing

- Apply addressing concepts to the expansion of register arrays

- Investigate the characteristics and operation of the various ICs that implement serial and parallel registers and register files

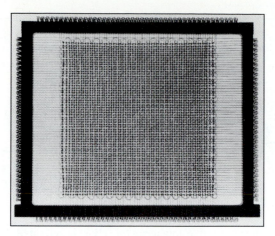

A computer must have the capability to store information in the form of binary numbers. The photograph shows a vintage magnetic memory plane that was used for this purpose. Thousands of these circuits can now be integrated on a single chip (Courtesy IBM).

7.1
Introduction

The storage of binary numbers is an important function of a digital computer and is also essential to several other types of digital circuit applications. Registers are logic circuits that are capable of storing single binary numbers of varying size whereas memory systems (to be examined in Chapter 8) can be viewed as arrays of registers that are capable of holding several binary numbers of varying size. Flip-flops can store single bits, registers can store single binary numbers (several bits), and memories can store several binary numbers.

This chapter deals with the operation and design of various types of registers. Two broad categories of registers are covered here—parallel and serial registers. The two types of registers differ only in the manner in which the binary data (binary numbers) is loaded-in and retrieved from the register. Both are used to store binary data, but serial registers have other applications besides storage as will be seen in this chapter.

The concepts of memory architecture and design are also introduced in conjunction with the study of register arrays (register files). A more detailed study of memories is found in Chapters 8 and 9. The chapter concludes by examining the various hardware implementations of registers in the form of ICs.

7.2
Parallel registers

Parallel registers are used to store binary numbers of different sizes. The size of a parallel register is always equal to the number of bits in the largest binary number that it is capable of storing. For example, a 4-bit register is capable of storing all 4-bit numbers in the range 0000 through 1111. The word ''parallel'' is used to describe these registers because the bits that make up the number are loaded and retrieved into and from the register in *parallel*. That is, they are loaded and retrieved together, simultaneously. This is in contrast to the manner in which serial registers load and/or retrieve data—1 bit at a time, or *serially*.

Figure 7.1 shows the logic symbol for a 4-bit parallel register. The register has four data inputs, $I_3I_2I_1I_0$ that allow a binary number to be stored into the register

Figure 7.1

Logic symbol for a 4-bit
parallel register.

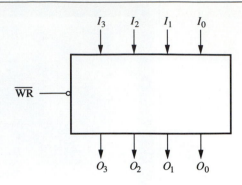

when the active-low *write line* ($\overline{WR}$) is activated (by a logic 0). The data stored in the
register is available at all times at the output lines $O_3 O_2 O_1 O_0$.

The logic circuit for the 4-bit parallel register is shown in Figure 7.2. The circuit
uses a series of D flip-flops to store each bit that makes up the binary number.
Because all the bits of the number are stored into the flip-flops at one time, the clock
inputs to all the flip-flops are connected to make up the $\overline{WR}$ line. Because the flip-
flops are negative edge-triggered, the flip-flops will trigger when the $\overline{WR}$ line
changes to the logic 0 state. The input lines $I_3 I_2 I_1 I_0$ are connected to the D inputs of
the flip-flops to allow each flip-flop to latch a single bit of the input binary number.
The Q outputs of the flip-flops hold the stored bits and are connected to the register
outputs $O_3 O_2 O_1 O_0$.

Figure 7.2

Logic circuit for a 4-bit
parallel register.

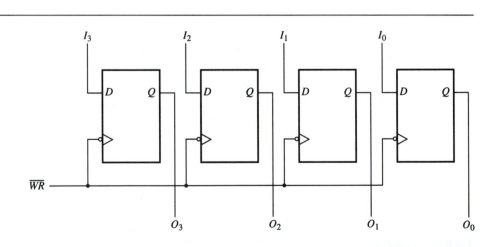

A parallel register such as the one shown in Figure 7.2 extends the concept of
storage of a single bit using a D flip-flop to many bits (a binary number) using a
series of D flip-flops. The size of the register can be expanded by adding more flip-
flops to accommodate larger numbers. For example, Figure 7.3 shows the logic
symbol and circuit for an 8-bit parallel register.

There are many applications that have the data outputs of several registers
connected on a common bus. In such a scheme, the outputs of all the registers must
be isolated from the bus and should only put their data on the bus when selected.
Figure 7.4a shows the logic symbol of a 4-bit parallel register with an active-low

Figure 7.3

An 8-bit parallel register. (a) Logic symbol; (b) circuit.

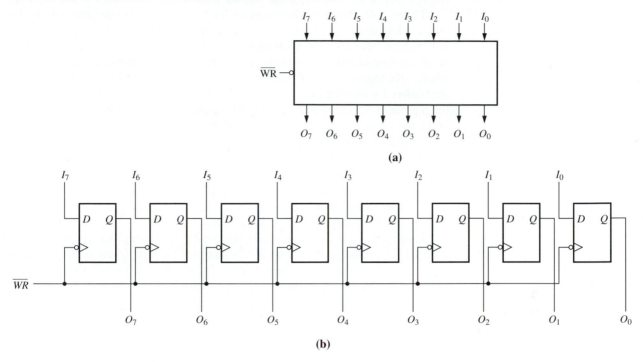

(a)

(b)

Figure 7.4

A 4-bit parallel register with tri-state outputs. (a) Logic symbol; (b) circuit.

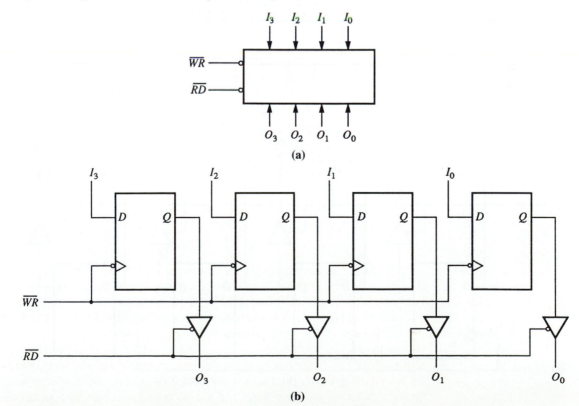

(a)

(b)

read line ($\overline{RD}$). When the $\overline{RD}$ line is inactive (logic 1 state), the outputs of the register $O_3O_2O_1O_0$ are in a high-impedance state (Z). However, when the $\overline{RD}$ line is active (logic 0 state), the outputs of the register contain the data stored in the flip-flops.

Figure 7.4b shows the logic circuit for the 4-bit parallel register with tristate outputs. The circuit is similar to the circuit shown in Figure 7.2 except that tristate buffers have been added to the outputs of the flip-flops. The active-low enable lines of all the tristate buffers are connected so that the $\overline{RD}$ line enables all the buffers when at the logic 0 state and disables all the buffers when at the logic 1 state. When the buffers are enabled ($\overline{RD} = 0$), the outputs of the flip-flops, Q, are connected to the outputs of the register $O_3O_2O_1O_0$. However, when the buffers are disabled ($\overline{RD} = 1$), the outputs of the register are in a high-impedance state and the outputs of the flip-flops are effectively cut off from the outputs of the register.

The parallel registers examined so far have had a separate set of input and output lines to load and retrieve data from the register, respectively. Because at any given instant of time the register is either being read from or written to, but never both, it is more convenient to have a common set of lines to load and retrieve data. Such a register is said to have *bidirectional* data lines.

Figure 7.5a shows the logic symbol for a 4-bit parallel register with bidirectional data lines. To write (store) a 4-bit binary number, the logic levels that represent the number are placed on the data lines $D_3D_2D_1D_0$ and the $\overline{WR}$ line is activated. To read (retrieve) a 4-bit binary number from the register, the $\overline{RD}$ line is activated and the register puts out the data on the same set of lines $D_3D_2D_1D_0$.

Figure 7.5

A 4-bit parallel register with bidirectional data lines. (a) Logic symbol; (b) circuit.

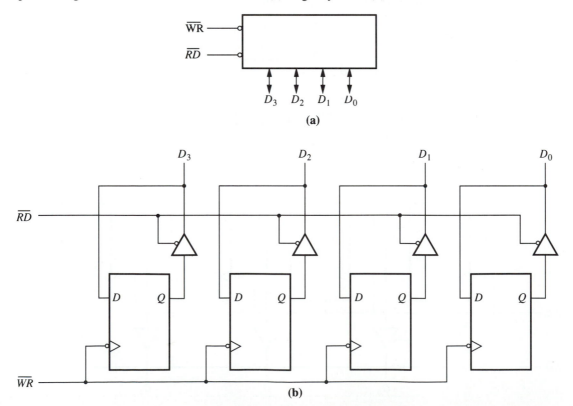

Figure 7.5b shows the logic circuit for the 4-bit parallel register with bidirectional data lines. The circuit is similar to Figure 7.4b except that the input and output lines have been connected to produce a set of bidirectional data lines. At any given instant either the $\overline{RD}$ or the $\overline{WR}$ line may be activated (but never both). When data is to be stored into the register, the $\overline{RD}$ line is inactive, and the tristate buffers isolate the outputs of the flip-flops from the data lines $D_3D_2D_1D_0$. The data on the D lines can then be latched into the flip-flops by activating the $\overline{WR}$ line. When data is to be retrieved, the $\overline{RD}$ line is activated, which allows the outputs of the flip-flops to be connected to the data lines $D_3D_2D_1D_0$. The data retrieved from the outputs of the flip-flops is also fed back to the inputs. However, because the $\overline{WR}$ line is inactive, it does not have any effect on the circuit.

Example 7.1

Figure 7.6 shows how the binary number 1011 is stored and retrieved using the parallel register designed in Figure 7.5.

In Figure 7.6a, because the $\overline{RD}$ line is inactive, the outputs of the tristate gates are open, isolating the outputs from the data lines. The number 1011 is applied to the D inputs and the $\overline{WR}$ line is pulsed to a logic 0. This latches the number 1011 into the flip-flops for storage.

In Figure 7.6b, the $\overline{RD}$ line has been activated while keeping the $\overline{WR}$ line inactive. This enables the tristate gates, thereby connecting the outputs of the flip-flops to the data lines. The data stored in the flip-flops 1011 is therefore retrieved and made available on the data lines D.

Figure 7.6

(a) Writing data into the register; (b) reading data from the register.

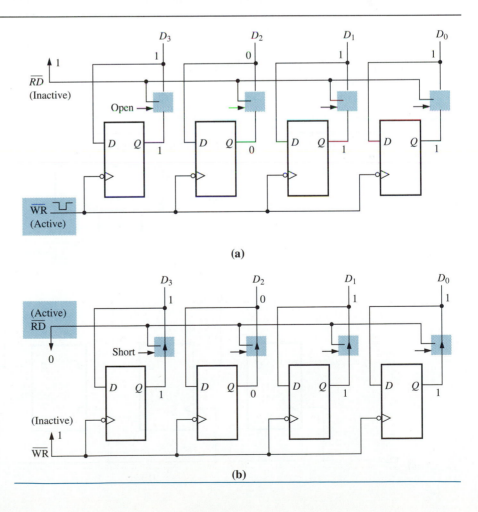

Parallel registers, or *parallel-load* registers, are logic circuits that are used to hold (or store) single binary numbers. All the bits that make up the binary number are loaded and retrieved simultaneously. The process of storing a binary number is known as *writing* data and the process of retrieving a previously stored binary number is called *reading*. Not all registers use a parallel load/retrieve design. The next section examines a type of register in which data may be loaded into or retrieved from a register 1 bit at a time.

Review Questions

1. What is the difference between a parallel register and a serial register?
2. What is the purpose of the $\overline{WR}$ line of a register?
3. What is the purpose of the $\overline{RD}$ line of a register?
4. Define the terms "reading" and "writing."
5. What are bidirectional data lines?

7.3
Serial registers

The parallel registers examined in the previous section are sometimes referred to as parallel-in parallel-out registers because data is loaded into and retrieved from those registers in a parallel format. Serial registers load data serially, retrieve data serially, or both. When data is loaded serially into a register, each bit of the binary number is loaded individually — that is, sequentially, one bit at a time. Similarly, when data is retrieved serially from a register, each bit of the binary number stored in the register is retrieved individually.

Figure 7.7

Serial-in parallel-out shift register. (a) Logic symbol; (b) circuit.

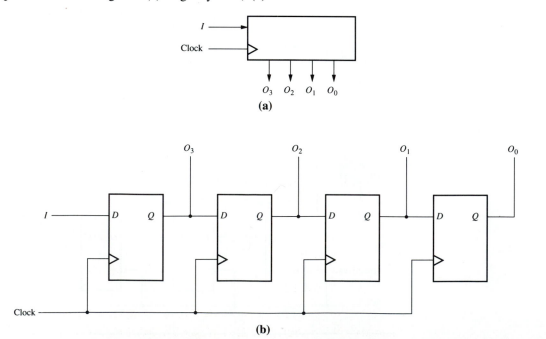

Serial-in parallel-out registers

Figure 7.7a shows the logic symbol for a 4-bit serial-in parallel-out register. It has a single input line I over which the data is loaded, and a 4-bit output $O_3O_2O_1O_0$ from which the stored data can be retrieved in parallel. The clock input is used to load into the register each bit of the binary number to be stored. A 4-bit binary number is loaded into the register by applying each bit that makes up the number to the input line I. The bit is then loaded into the register by applying a clock pulse. The remaining bits are written in similarly. It therefore takes four clock pulses to load a 4-bit binary number into the register. The logic circuit for the register is shown in Figure 7.7b. To understand the manner in which the register functions, consider the loading of the 4-bit binary number 1010. Because there are two 0 bits and two 1 bits in this number, identify each bit by a subscript so that the movement of a bit through the circuit can be viewed

$$1_B0_B1_A0_A$$

The LSB of the number is therefore 0_A and the MSB of the number is 1_B.

Figure 7.8a through 7.8e shows a trace of the binary number for each pulse applied to the clock input. Each step in the trace can be explained as follows.

Clock ①: In Figure 7.8a before the first clock triggers the register, the LSB of the binary number, 0_A, is applied to the I input. The outputs of the flip-flops, which are also the outputs of the register $O_3O_2O_1O_0$, are currently set to some logic levels that we are not concerned about and have identified by the don't care states X_A, X_B, X_C, and X_D. When the clock makes its transition, the output of each flip-flop will be equal to the logic level of its input before the clock pulse makes its transition. The result of the first clock transition is shown in Figure 7.8b. All the bits have shifted right one position so that the output O_3 is now equal to bit 0_A and bit X_D has been discarded.

Clock ②: Figure 7.8b shows the state of the register before the second clock pulse makes a transition. The second significant bit, 1_A, of the binary number has been applied to the I line. When the clock pulse makes a transition, all the bits shift to the right, and the new state of the register is shown in Figure 7.8c. X_C has been discarded and $O_2 = 0_A$, $O_3 = 1_A$.

Clock ③: Figure 7.8c shows the state of the register before the third clock pulse makes a transition. The third significant bit, 0_B, of the binary number has been applied to the I line. When the clock pulse makes a transition, all the bits shift to the right, and the new state of the register is shown in Figure 7.8d. X_B has been discarded and $O_1 = 0_A$, $O_2 = 1_A$, and $O_3 = 0_B$.

Clock ④: Figure 7.8d shows the state of the register before the fourth (and last) clock pulse makes a transition. The MSB, 1_B, of the binary number has been applied to the I line. When the clock pulse makes a transition, all the bits shift to the right and the new state of the register is shown in Figure 7.8e. X_A has now been discarded and $O_0 = 0_A$, $O_1 = 1_A$, $O_2 = 0_B$, and $O_3 = 1_B$.

At the end of four clock pulses, the entire 4-bit binary number 1010 has been shifted in the register because $O_3O_2O_1O_0 = 1010$. The previous number $X_AX_BX_CX_D$ that was stored in the register has been completely discarded. Because of the manner in which the data bits are shifted into (and out of) a serial register, serial registers are often known as *shift registers*.

Figure 7.9 shows the complete timing diagram for the shift register when the number 1010 is shifted in. We have assumed that the don't care states X_A, X_B, X_C,

Figure 7.8

Operation of the shift
register.

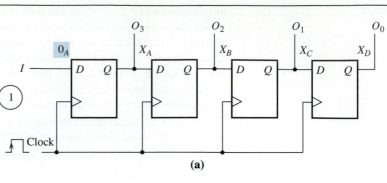

(a)

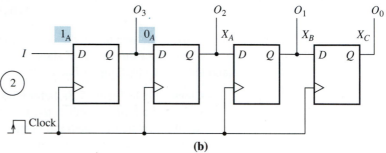

(b)

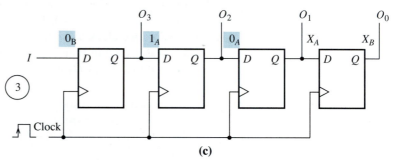

(c)

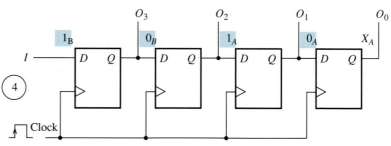

(d)

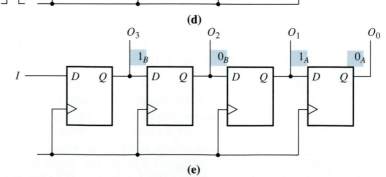

(e)

Figure 7.9

Timing diagram for the shift
register.

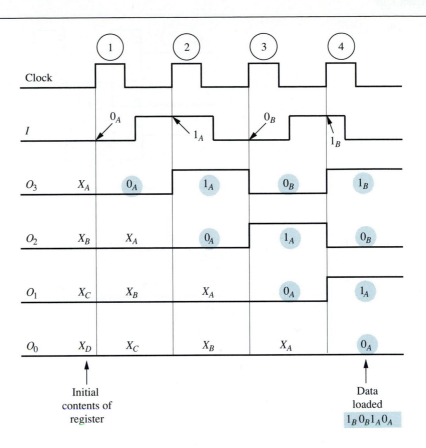

and X_D are zeros. Notice how each bit applied to the *I* input shifts right for each
transition of the clock pulse and at the end of the fourth clock pulse, the outputs of the
register $O_3O_2O_1O_0$ contain the complete number 1010. The following example
illustrates the timing diagram analysis of the shift register when loading another 4-bit
binary number.

Example 7.2

Construct a timing diagram for the shift register shown in Figure 7.7b when the 4-bit binary
number 0110 is shifted in. Assume that the initial state of the register is 1111.

Solution

Identify each bit in the number 0110 as follows:

$$0_A 1_A 1_B 0_B$$

The timing diagram for the shift register is shown in Figure 7.10.

Shift registers can be expanded to store any size numbers by adding more flip-
flops in series. For example, Figure 7.11 shows the logic symbol and circuit for an 8-
bit shift register. The 8-bit shift register will function like the 4-bit shift register but
will require eight clock pulses to completely load a number.

Figure 7.10

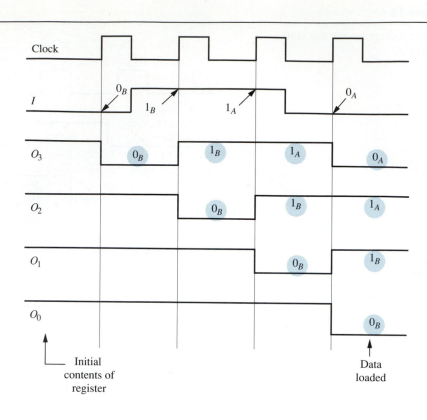

Figure 7.11

Eight-bit shift register.

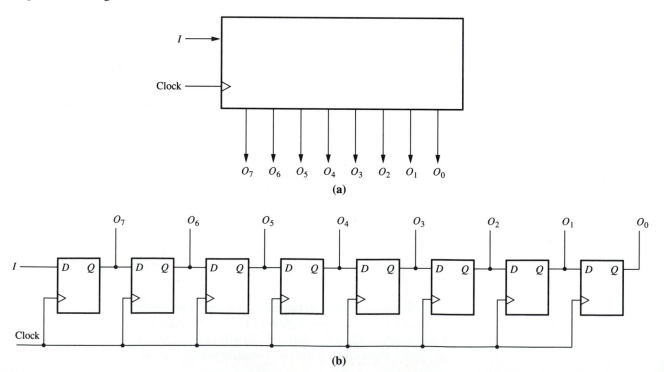

Serial-in serial-out registers

The serial-in serial-out register, which is another type of shift register, requires that data be loaded in and retrieved serially. The logic symbol and circuit for the shift register are shown in Figure 7.12. Figure 7.12a shows that the register has a single input I for loading the data serially, and a single output O for serially retrieving previously stored data. The clock input is used for both loading and retrieving data one bit at a time. The logic circuit for the shift register is a simple modification of the serial-in parallel-out shift register shown in Figure 7.7b. Because the data will not be retrieved in parallel, the individual outputs of the flip-flops are not accessed and are not made available. Instead, the serial output of the shift register is the output of the last flip-flop in series. The loading of serial data is the same as for a serial-in parallel-out register. Recall that as data bits are shifted in on each clock transition, the existing bits in the register are discarded (shifted out) at the output of the last flip-flop. To retrieve existing data in the register, the clock is triggered four times and the data bits appear at output O, LSB first (if the data was loaded-in LSB first).

Figure 7.12

Serial-in serial-out shift register.

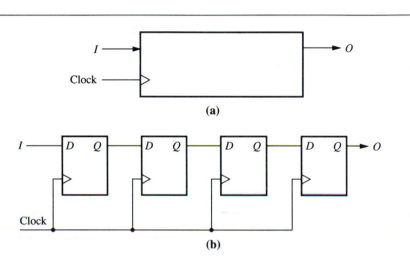

(a)

(b)

Example 7.3

Construct a timing diagram for the shift register shown in Figure 7.12 showing the storage and retrieval of the number 1011.

Figure 7.13a shows the storage of the number 1011 into the serial register of Figure 7.12, MSB first. Each bit is placed on the I line starting with the MSB. The bit is shifted into the register on the positive edge of the clock. After four clock pulses the entire number 1011 has been shifted into the register. The rightmost flip-flop in Figure 7.12 holds the MSB because it was shifted-in first.

Figure 7.13b shows the retrieval of the data 1011 that was previously stored in the register. On the first positive transition of the clock pulse, the MSB appears on the O line, because the MSB was stored in the rightmost flip-flop. The other bits follow on each successive transition of the clock pulse, ending with the LSB. After four clock pulses the entire number 1011 has been retrieved serially.

Parallel-in serial-out registers

The third type of shift register to be studied is known as a parallel-in serial-out shift register because it loads the data in parallel, and retrieves the data serially. The logic symbol of the shift register is shown in Figure 7.14a. The data to be loaded into the

Figure 7.13

Timing diagram for the
serial-in serial-out shift
register.

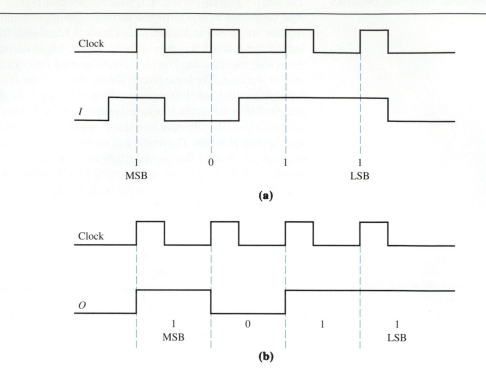

Figure 7.14

Parallel-in serial-out shift
register. (a) Logic symbol;
(b) circuit.

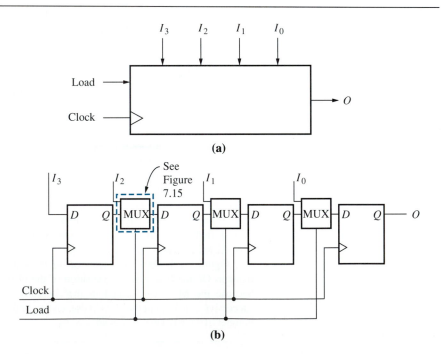

register is placed on the input lines $I_3I_2I_1I_0$ and the *LOAD* line is activated by applying a logic 1 to it; this enables the parallel loading of data. A clock pulse is then applied to the clock input to actually load the data into the register. To retrieve data from the register serially, the *LOAD* line is made inactive by connecting it to a logic 0 (this disables the parallel loading of data) and four clock pulses are applied to the clock input to obtain the stored data serially at the output line *O*, LSB first.

The logic circuit for the parallel-in serial-out shift register is shown in Figure 7.14b. The *D* input of each flip-flop (except the first one) must be connected to one of two sources — the parallel input line I_n (where *n* is the bit number 0, 1, 2, or 3) or the output of the previous flip-flop in series *Q*. To load data into each flip-flop the data to be loaded in parallel, $I_3I_2I_1I_0$, has to be applied to the *D* inputs of the flip-flops and the flip-flops are then triggered by a clock pulse. During loading, the *Q* outputs of the flip-flops must be disconnected from the *D* inputs of the flip-flops in series. During serial retrieval of data, the opposite must occur. The inputs I_3, I_2, I_1, and I_0 must be disconnected from the *D* inputs of the flip-flops and the *Q* outputs of each flip-flop connected to the *D* input of the flip-flop in series so that the bits can be shifted through the circuit. To accomplish this task use a two-channel single-bit multiplexer (such as the one designed in Chapter 4, Figs. 4.58 or 4.66). An expanded view of the multiplexer is shown in Figure 7.15. When the *LOAD* line is a logic 1 the *SELECT* input is a logic 1 and channel *A* is multiplexed to *OUT*; the I_n input is connected to *D* and the *Q* input is cut off from *D*. When the *LOAD* line is a logic 0 the *SELECT* input is a logic 0 and channel *B* is multiplexed to *OUT*; the *Q* input is connected to *D* and the I_n input is cut off from *D*. No multiplexer is required for the first flip-flop because the only input to it is I_3.

Multiplexer used in Figure 7.14.

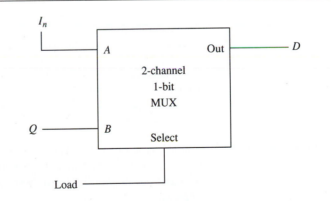

Figure 7.16a shows the configuration of the circuit when *LOAD* = 1. The circuit resembles and functions as a parallel-load register. Once the data has been loaded in, it can be retrieved by setting *LOAD* to a logic 0. The circuit in Figure 7.16b shows the configuration of the circuit when *LOAD* = 0. The circuit now resembles and functions as a serial-out shift register. Also, when *LOAD* = 0, the parallel-in serial-out shift register can also function as a serial-in serial-out shift register with serial input at I_3 and serial output at *O*.

Example 7.4

Figure 7.17a shows the timing diagram to store the number 1011 into the register shown in Figure 7.14. The *LOAD* line is set to a logic 1 and the number 1011 is applied to the *I* inputs of

Figure 7.16

Operation of the parallel-in serial-out register. (a) Loading data into the register; (b) shifting data out.

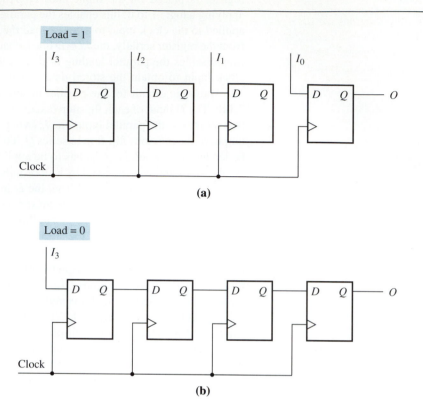

the register. Because the data is loaded in parallel, only a single positive transition of the clock pulse is required to load the data into the register.

Figure 7.17b shows the timing diagram to retrieve the number 1011 that was previously stored in the register. The *LOAD* line is now set to a logic 0, and on each positive transition of the clock pulse the number 1011 is shifted out of the *O* line of the register, LSB first.

Bidirectional shift registers

The last type of shift register to be examined is the bidirectional shift register. It is very similar to a serial-in serial-out shift register except that the data can be shifted into and out of the register in both directions, and it has parallel outputs for data retrieval.

Figure 7.18 shows the logic symbol for a 4-bit bidirectional shift register. The register has a *serial input line* (*SRI*) to shift data into the register in the right direction (i.e., to the right), and a *serial output line* (*SRO*) to shift data out in the right direction. It also has another serial input, *SLI,* and another output line, *SLO,* to shift data into and out of the register in the left direction (i.e., to the left), respectively. The direction in which data is shifted (right or left) depends on the $R/\overline{L}$ (direction) line. If $R/\overline{L} = 1$ then data is shifted in and out in the right direction, and if $R/\overline{L} = 0$ then data is shifted in and out in the left direction. The *CLOCK* input is used to shift the bits in and out of the register. The content of the 4-bit register can be retrieved in parallel from the output lines $O_3O_2O_1O_0$.

The logic circuit for the bidirectional shift register is shown in Figure 7.18b. A

Figure 7.17

Timing diagram for the parallel-in serial-out register. (a) Loading data into the register; (b) shifting data out.

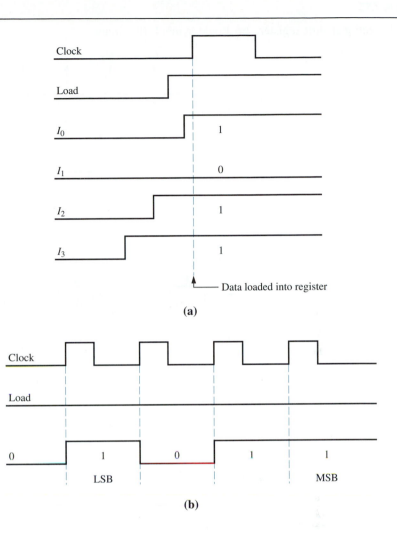

(a)

(b)

series of two-channel single-bit multiplexers (similar to the one shown in Figure 7.15) have been used to configure the "flow" of data to the right, that is, from *SRI* to flip-flops *F*3, *F*2, *F*1, and *F*0, to *SRO* if $R/\overline{L} = 1$, or to the left, that is, from *SLI* to flip-flops *F*0, *F*1, *F*2, and *F*3, to *SLO* if $R/\overline{L} = 0$. The parallel outputs $O_3 O_2 O_1 O_0$ are obtained from the outputs of the flip-flops, and the *CLOCK* input synchronously loads each bit into the flip-flops.

Figure 7.19a shows the configuration of the circuit when the $R/\overline{L}$ line is set to a logic 1—right shift. Because the $R/\overline{L}$ line is connected to the select lines of the multiplexers, this effectively multiplexes channel *A* and connects *SRI* to the input of *F*3, the output of *F*3 to the input of *F*2, the output of *F*2 to the input of *F*1, the output of *F*1 to the input of *F*0, and the output of *F*0 to *SRO*. This accomplishes a right shift.

Figure 7.19b shows the configuration of the circuit when the $R/\overline{L}$ line is set to a logic 0—left shift. The multiplexers now multiplex the other input channel (*B*) so that this effectively connects *SLI* to the input of *F*0, the output of *F*0 to the input of *F*1, the output of *F*1 to the input of *F*2, the output of *F*2 to the input of *F*3, and the output of *F*3 to *SLO*. This accomplishes a left shift. The order of the flip-flops drawn in Figure 7.19b has been rearranged for convenience.

Figure 7.18

A bidirectional shift register. (a) Logic symbol; (b) circuit.

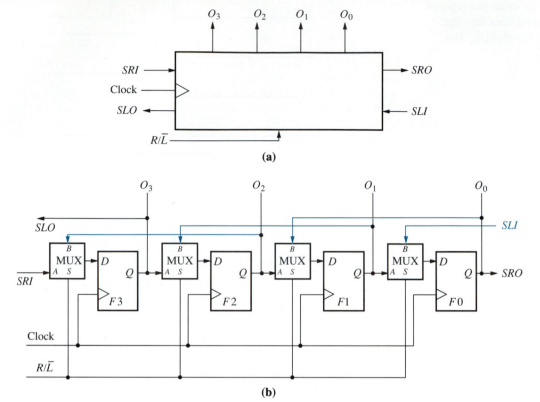

Figure 7.19

Operation of the bidirectional shift register. (a) Right shift; (b) left shift.

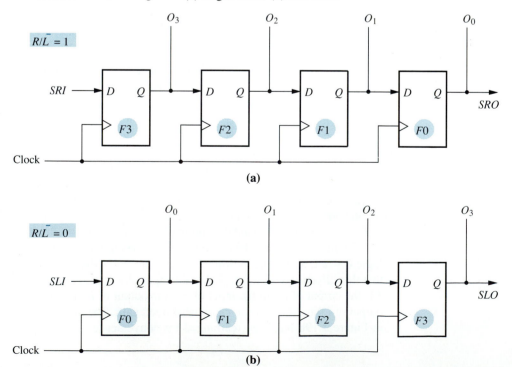

Example 7.5

Assuming that the number 1011 has been shifted into the bidirectional shift register shown in Figure 7.18, draw a timing diagram of the parallel outputs when the $R/\overline{L}$ line is set to a logic 1, and when the $R/\overline{L}$ line is set to a logic 0.

Solution

The timing diagram of the parallel outputs $O_3O_2O_1O_0$ is shown in Figure 7.20a when $R/\overline{L} = 1$. Because *SRI* line is assumed to be tied to a logic 0, after four clock pulses the content of the register is 0000.

Figure 7.20b shows the timing diagram of the parallel outputs $O_3O_2O_1O_0$ when $R/\overline{L} = 0$. Again, because it has been assumed that the *SLI* line is tied to a logic 0, the content of the register is 0000 after four clock pulses.

Figure 7.20

Timing diagram for the bidirectional shift register.
(a) Right shift; (b) left shift.

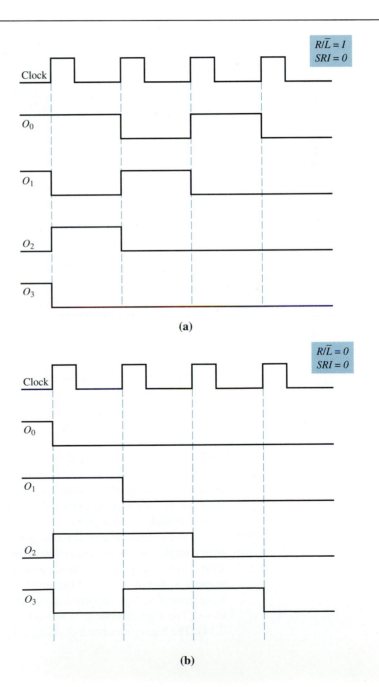

From the two timing diagrams shown in Figure 7.20 it is apparent that the bit pattern that appears on the output lines is different during the right and left shifts.

Both serial and parallel registers accomplish the same objective — storage of binary data. The manner in which data is loaded and/or retrieved is different in these two types of registers. It is apparent that parallel registers are faster than serial registers because data is loaded and retrieved on the transition of a single clock pulse for a parallel register but requires several clock pulses (the number depends on the size of the register) for a serial register. On the other hand, one advantage of serial registers is that the number of lines required to load and/or retrieve data is greatly reduced. The type of register used depends on the application.

Review Questions

1. Why are serial registers known as shift registers?
2. Explain the operation of the four different types of shift registers.
3. What advantage does a parallel register have over a serial register?
4. What advantage does a serial register have over a parallel register?

7.4
Shift register applications

Shift registers have several other applications besides data storage. This section investigates a few of the more common applications of shift registers.

Serial data communications

One of the more important applications of shift registers is in digital data communications systems. One aspect of such a system is to transfer binary data from one place to another. The source and destination of the transfer could often be isolated by fairly long distances. Binary numbers can be transferred from one place (the source) to another (the destination) in one of two forms — parallel or serial. For parallel data transfers the entire binary number is transferred at one instant. Each bit in the number has a separate line dedicated to it and the number of interconnecting data lines would be equal to the number of bits in the number being transferred. Figure 7.21 shows a simple parallel data communications system using two parallel registers. Assume that a 4-bit number $X_3X_2X_1X_0$ is to be transferred from the source A to the destination B. There are four interconnecting lines between A and B for the data, and one line for a common clock. The data is applied to the parallel register at A and the *CLOCK* line is pulsed. The *WR* line of the parallel register at A requires a positive transition to latch the data at its inputs, and the data is transmitted to the destination at the positive transition of the clock pulse. Because the same clock activates the $\overline{WR}$ line of the parallel register at B, the data transmitted by A is latched by this register on the negative transition of the clock pulse. The data $X_3X_2X_1X_0$ is transmitted from A to B on one complete clock pulse.

In serial data communications systems, the source and destination transfer data over a single line. A binary number is transmitted serially over this line using a set of shift registers. Figure 7.22 shows a simplified circuit for transmitting a 4-bit binary number $X_3X_2X_1X_0$ serially from source A to destination B. This circuit has only two interconnecting lines between A and B — one for the serial data and one common clock. The shift register at A is a parallel-to-serial shift register (as shown in Fig. 7.14). The binary number $X_3X_2X_1X_0$ is loaded into the register when the *LOAD* line

Figure 7.21

Figure 7.21

A parallel data communications circuit.

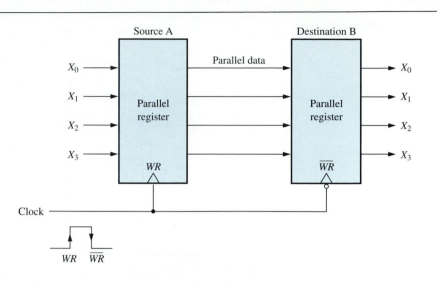

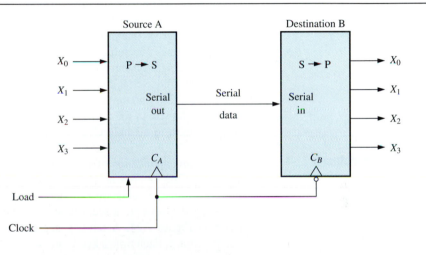

Figure 7.22

A serial data communications circuit.

is activated and four clock pulses are required to shift the data out onto the serial data line. The serial data is shifted out on the positive transition of the clock pulse as shown in the timing diagram of Figure 7.23, which illustrates the transfer of the binary number 1001. The same serial data line connects to a serial-to-parallel shift register (as shown in Fig. 7.7) at destination B. This register is activated by the same clock that shifts the data out of the shift register at A, but the data on the serial data line is shifted in at B on the negative transition of the clock pulse. Shifting the data in on the negative transition allows the clock C_B to latch the data on the serial data line at approximately its midpoint, allowing for minor variations in the timing of the two circuits. If the serial data in was sampled on the positive edge of the clock, any small difference in timing could lead to the wrong logic level(s) being shifted in. This type of serial data communications is known as *synchronous serial data communications* because a common clock is used to synchronize the transmitter A and the receiver B. The circuit in Figure 7.22 transmits the data $X_3X_2X_1X_0$ from A to B on four complete clock pulses. The objective of the two circuits shown in Figures 7.21 and 7.22 was to

Figure 7.23

Timing diagram for serial
data transfer.

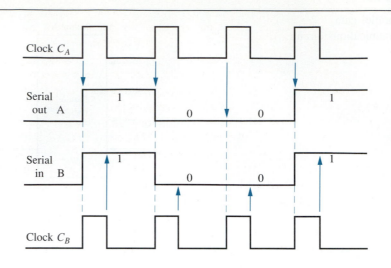

transfer a 4-bit number $X_3X_2X_1X_0$ from the source A to the destination B. The manner
in which this was done was quite different, although the objective was met in both
circuits. It is apparent that the advantage of serial over parallel data transfers is in the
smaller number of interconnecting lines. A parallel data transfer system requires
several data lines depending on the size of the numbers being transferred whereas a
serial data transfer system requires one data line regardless of the size of the numbers
being transferred. Serial data transfers are slow. Each bit must be shifted separately
over a single line, and the time required to completely transfer a number serially will
depend on the size of the number. In contrast, parallel data transfers can transmit an
entire number on a single clock pulse.

**Combination detector
circuit**

An interesting application of a shift register is in a commonly used circuit to detect a
numerical code entered at a keypad. For example, a door could be unlocked on
entering a four-digit code on a numeric keypad, or a six-digit code could be entered
on a keypad to disable a security system. The block diagram for such a circuit is
shown in Figure 7.24. A ten-key decimal keypad is connected to a combination
detector circuit. Each time a key is pressed on the keypad, the keypad outputs a pulse
(assume an active-high pulse) on the line corresponding to that key. For example, if
the "8" key is pressed, a pulse appears on the line labeled "8," if the "4" key is

Figure 7.24

Block diagram for a
combination detector circuit.

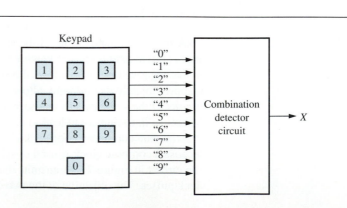

pressed, a pulse appears on the line labeled "4," etc. The combination detector circuit monitors the pulses on these ten lines, and if the right keys are pressed in the right sequence, the circuit activates its output line X to a logic 1 state. This output line could then be used to activate a device that unlocks a door, disables an alarm, or activates or deactivates any other device depending on the application.

Consider the design of a combination detector circuit to detect the combination "6284." When this combination is entered at the keypad, the circuit must set X to a logic 1 and to a logic 0 if any other combination is entered. The circuit to implement this specification is shown in Figure 7.25. The circuit consists of a 4-bit shift register made up of D flip-flops with direct (asynchronous) active-low $\overline{CLR}$ inputs. Recall from Chapter 5 that a logic 0 on the $\overline{CLR}$ input will directly reset the flip-flop without requiring a clock pulse. The first flip-flop in series, $F1$, has its D input connected to a logic 1. The objective is to shift this logic 1 to the output of the last flip-flop in series, $F4$. The output of $F4$ is X, the output of the combination detector circuit. To shift the logic 1 from the input of flip-flop $F1$ to the output of $F4$, flip-flops $F1$, $F2$, $F3$, and $F4$ must be clocked in sequence. When the "6" key is pressed, $F1$ is triggered and the logic 1 is latched to its output. When the "2" key is pressed, $F2$ is triggered and the logic 1 (from $F1$) is latched to its output. When the "8" key is pressed, $F3$ is triggered and the logic 1 (from $F2$) is latched to its output. Finally, when the "4" key is pressed, $F4$ is triggered and the logic 1 (from $F3$) is latched to its output, X. If any other key besides "6," "2," "8," or "4" is pressed, the *NOR* gate produces an active-low pulse that clears all the flip-flops; the entire four-digit combination must then be reentered. If the keys "6," "2," "8," and "4" are not pressed in the right sequence, the logic 1 will never be shifted to X.

Figure 7.25

Logic circuit for a combination detector circuit.

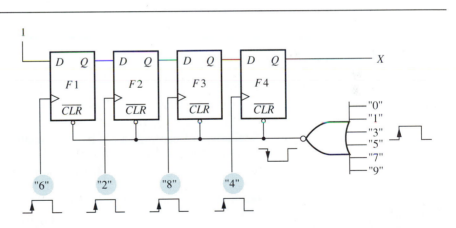

Shift register counters

Another application of shift registers is in special-purpose shift register counters. Two of the more common types are the *ring* and *Johnson* counters.

The logic circuit for a 3-bit ring counter is shown in Figure 7.26. The circuit consists of three D flip-flops configured as shift registers with the output of the last flip-flop $F0$ connected to the input of the first flip-flop $F2$. The outputs of the counter are $X_2X_1X_0$. When the counter is started, the $\overline{CLEAR}$ line is used to directly (asynchronously) clear all the flip-flops so that the initial state is 000. Flip-flop $F2$ is then asynchronously set by activating the preset ($\overline{PRE}$) line. Once the clock pulses are started, the logic 1 at the input of $F1$ is shifted to the next flip-flop, and the next, etc., until it is fed back to the input of $F2$ to recycle the count. The logic 1 appears to

Figure 7.26

A 3-bit ring counter.

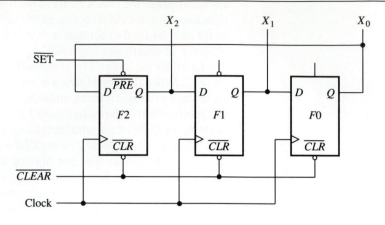

move (or rotate) through the counter in a ring or circular pattern, and hence the term ''ring counter.''

The timing diagram for the counter is shown in Figure 7.27. Note that the logic 1 at X_2 is shifted to X_1 and then to X_0, and then back again to X_2. The count for the 3-bit ring counter follows the sequence

$$4, 2, 1, 4, 2, 1, 4, \ldots$$

or, in terms of the counter's outputs

$$
\begin{array}{ccc}
1 & 0 & 0 \\
0 & 1 & 0 \\
0 & 0 & 1 \\
1 & 0 & 0 \\
0 & 1 & 0 \\
0 & 0 & 1 \\
1 & 0 & 0 \\
& \vdots &
\end{array}
$$

Similarly, one can expect a 4-bit ring counter to follow the sequence:

$$8, 4, 2, 1, 8, 4, 2, 1, 8, \ldots$$

or, in terms of the counter's outputs

$$
\begin{array}{cccc}
1 & 0 & 0 & 0 \\
0 & 1 & 0 & 0 \\
0 & 0 & 1 & 0 \\
0 & 0 & 0 & 1 \\
1 & 0 & 0 & 0 \\
0 & 1 & 0 & 0 \\
0 & 0 & 1 & 0 \\
& & \vdots &
\end{array}
$$

The Johnson counter is basically a modification of the ring counter shown in Figure 7.26. It is constructed by connecting the complement of the output of the last

Figure 7.27

Figure 7.27

Timing diagram for a 3-bit ring counter.

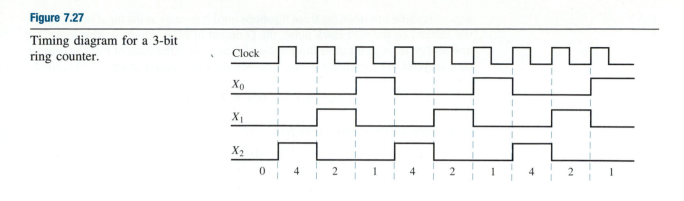

Figure 7.28

A 3-bit Johnson counter.

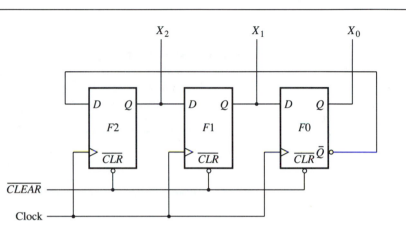

flip-flop in series ($F0$) to the input of the first flip-flop in series ($F2$) as shown in Figure 7.28. This design produces a count sequence that appears to fill up the outputs with 1-bits, and then fill up the outputs with 0-bits, continuously repeating the cycle. The timing diagram for the Johnson counter is shown in Figure 7.29 and illustrates this effect.

Before the clock pulses to the counter are started, the counter is first cleared by activating its $\overline{CLEAR}$ line. The state of the counter's outputs is 000 as shown in Figure 7.29. Because the output of $F0$ is a logic 0, a logic 1 is fed back to $F2$ and this

Figure 7.29

Timing diagram for a 3-bit Johnson counter.

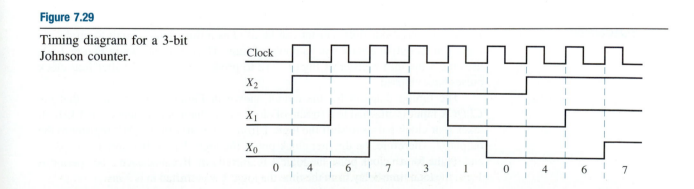

logic 1 is shifted through the three flip-flops until it appears at the input of $F0$. When this occurs, on the next clock pulse, the $\overline{Q}$ output of $F0$ is a logic 0. This logic 0 is now shifted through the three flip-flops until it appears at the input of $F0$. The counter then recycles. The resulting count sequence for the Johnson counter is therefore

$$0, 4, 6, 7, 3, 1, 0, 4, 6, 7, 3, 1, 0, \ldots$$

or, in terms of the counter's outputs

$$
\begin{array}{ccc}
0 & 0 & 0 \\
1 & 0 & 0 \\
1 & 1 & 0 \\
1 & 1 & 1 \\
0 & 1 & 1 \\
0 & 0 & 1 \\
0 & 0 & 0 \\
1 & 0 & 0 \\
1 & 1 & 0 \\
1 & 1 & 1 \\
& \vdots &
\end{array}
$$

Similarly, one can expect a 4-bit Johnson counter to follow the sequence

$$0, 8, C, E, F, 7, 3, 1, 0, 8, C, \ldots$$

or, in terms of the counter's outputs

$$
\begin{array}{cccc}
0 & 0 & 0 & 0 \\
1 & 0 & 0 & 0 \\
1 & 1 & 0 & 0 \\
1 & 1 & 1 & 0 \\
1 & 1 & 1 & 1 \\
0 & 1 & 1 & 1 \\
0 & 0 & 1 & 1 \\
0 & 0 & 0 & 1 \\
0 & 0 & 0 & 0 \\
1 & 0 & 0 & 0 \\
1 & 1 & 0 & 0 \\
& & \vdots &
\end{array}
$$

Time delay circuit

A serial-in serial-out shift register can be used as a time delay circuit to produce a logic 1 at its output after a fixed amount of time. The circuit shown in Figure 7.30a uses the 4-bit shift register of Figure 7.12 to produce a time delay after four clock pulses have elapsed.

The timing diagram for the circuit shown in Figure 7.30b assumes that the *CLOCK* input is attached to a stream of clock pulses having a frequency of 1 kHz. It takes four clock pulses to shift the logic 1 from the input *I* of the shift register to the output *O*. Therefore, on the first clock pulse, the logic 1 is shifted into the register, and on the fourth clock pulse the logic 1 is shifted out. Because each clock period is 1 ms, the total time delay from the time the logic 1 was shifted in is 3 ms.

Figure 7.30

A time delay circuit.

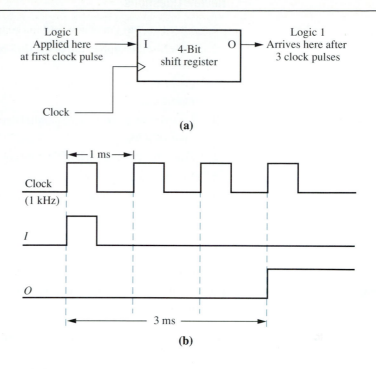

(a)

(b)

The amount of time delay that a serial-in serial-out shift register can produce can be modified by increasing or decreasing the size of the register. The amount of delay produced by a shift register (in clock pulses) will always be equal to one less than its size. For example, a 4-bit shift register will produce a delay of three clock pulses, an 8-bit shift register will produce a delay of seven clock pulses, etc. The amount of time delay produced by this type of circuit can also be increased by cascading shift registers. For example, the circuit shown in Figure 7.31 uses two 4-bit shift registers to produce a delay of seven clock pulses, because the two cascaded 4-bit shift registers function as a single 8-bit shift register. Again, if it is assumed that the clock frequency is 1 kHz, the total time delay will be 7 ms.

This section has investigated a few of the applications of shift registers. It is apparent that shift registers are used in very diverse applications ranging from storage circuits to counters. Parallel registers on the other hand are best suited for storage applications since speed is of primary importance in such applications. The next section examines such an application of parallel registers.

Figure 7.31

Using two shift registers to increase the time delay.

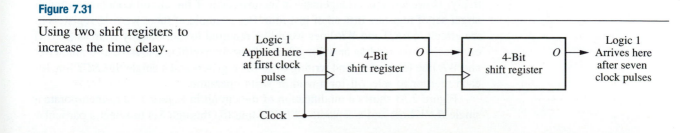

Review Questions

1. State two ways in which data can be transferred from one point to another.
2. Describe the application of the shift register in a data communications system.
3. Describe an advantage of serial data communications over parallel data communications.
4. Describe an advantage of parallel data communications over serial data communications.
5. Describe the operation of the ring counter.
6. Describe the operation of the Johnson counter.
7. What type of shift register can be used to produce a time delay?

7.5
Register arrays

A register array or register file is a collection of registers incorporated into a single package. A register file is similar in concept and operation to a memory system. This section deals with the operation and design of register arrays and also introduces the elementary concepts of storage of binary numbers and memory systems. The concepts developed in this section will then be applied in Chapter 9 to the operation and design of larger and more practical memory circuits.

To illustrate the design of a register array, consider the set of four 4-bit parallel registers with bidirectional data lines shown in Figure 7.32. These registers are the same as the register shown in Figure 7.5 and explained in Section 7.2, except that active-high *RD* and *WR* lines have been used. The bidirectional data lines of all the registers are connected by means of a data bus, $D_3D_2D_1D_0$. Each register is numbered (from 0 to 3) for identification purposes. If a binary number is applied to the data bus, the number is applied to the data lines of all the registers. However, to write (store) the data into register 2, activate the *WR* line of register 2. Similarly, to write the data into register 0, activate the *WR* line of register 0. To read the data stored in a particular register activate the *RD* line of that register. The contents of the selected register will then be loaded onto the data bus from which it can be retrieved. Because the data outputs of all the registers are tristated and because the *RD* line enables the tristate gates at the outputs of each register, this ensures that only one register's *RD* line is enabled at a time to prevent a situation known as *bus contention* from occurring. This occurs when more than one register puts its stored data on the data bus. This generally causes a conflict in logic levels and the presence of invalid data on the bus. For example, if an attempt is made to read from register 1 and register 2 at the same time, and if register 1 has a 0000 stored in it and register 2 has a 1111 stored in it, what number will the data bus contain? Logically this is an undefined condition. Data from a particular register can be stored or retrieved by activating its *WR* or *RD* line, respectively. However, this arrangement is inconvenient if the circuit contained a much larger set of registers instead of just four. For example, if there were 32 registers in the circuit, 64 *RD* and *WR* lines would be required to individually address (select) each register for reading or writing. It would be more efficient to have a common *RD* and *WR* line for reading and writing to all the registers and a single *SELECT* line for each register to select it for a read or write operation.

Figure 7.33 shows a modification of the circuit in Figure 7.32 to incorporate a single $R/\overline{W}$ line, and a separate *SELECT* line (S_0 through S_3) to select a particular

Figure 7.32

Figure 7.32

The building blocks of a
register array.

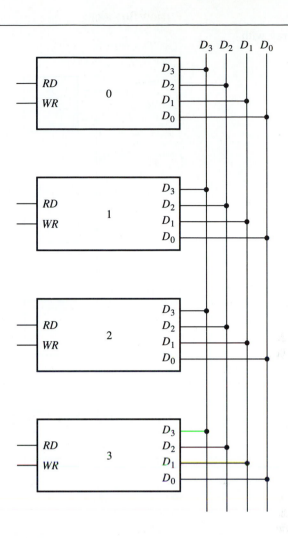

register for either a read or write operation. The data lines have been drawn in the
form of a bus. Because at a given instant we will either be reading from or writing to a
particular register, the functions of the *RD* and *WR* lines can be combined into a single line called $R/\overline{W}$ that will determine whether we are reading from a selected register ($R/\overline{W} = 1$) or writing to a particular register ($R/\overline{W} = 0$). For example, to read
from register 0, the S_0 line is activated to a logic 1, and the $R/\overline{W}$ line is set to a logic 1.
To write to register 0, the S_0 line is activated to a logic 1, and the $R/\overline{W}$ line is set to a
logic 0. Similarly, to read from register 1, the S_1 line is activated to a logic 1, and the
$R/\overline{W}$ line is set to a logic 1. To write to register 1, the S_1 line is activated to a logic 1,
and the $R/\overline{W}$ line is set to a logic 0. A combinational logic circuit (identified by the
box with the ?) is needed that will accept as input, the select line for a particular register S_n (where *n* is the register number) and the $R/\overline{W}$ line, and activate the *RD* and *WR*
lines of the selected register appropriately for reading, writing, or no operation as
shown in Table 7.1.

   In Table 7.1, the first two combinations of $R/\overline{W}$ and S_n will not activate the *RD*
or *WR* line of the register *n* because the register is not selected ($S_n = 0$); *RD* and *WR*
are both logic 0's for the first two combinations. For the third combination $S_n = 1$,

Figure 7.33

The register array with select lines.

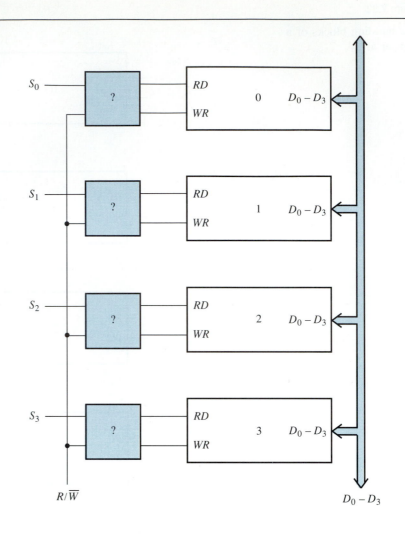

Table 7.1   Truth table for *RD/WR* circuit

Operation	S_n	$R/\overline{W}$	RD	WR
No operation	0	0	0	0
No operation	0	1	0	0
Write	1	0	0	1
Read	1	1	1	0

and the register *n* is selected for a write operation because $R/\overline{W} = 0$. Therefore *WR* = 1 while *RD* is inactive at the 0 state. For the fourth combination $S_n = 1$, the register *n* is selected for a read operation because $R/\overline{W} = 1$. Therefore $RD = 1$ and *WR* is inactive at the 0 state. From the truth table shown in Table 7.1, obtain the logic circuit for the "? boxes" shown in Figure 7.33 by first obtaining the logic equations for *RD* and *WR* as follows

$$RD = R/\overline{W} \cdot S_n \qquad\qquad (7.1)$$

$$WR = \overline{R/\overline{\overline{W}} \cdot S} \tag{7.2}$$

The logic circuit to implement Equations 7.1 and 7.2 is shown in Figure 7.34. It is incorporated into each of the "? boxes" shown in Figure 7.33.

The circuit in Figure 7.33 now only uses a single select line, S_n, to select a particular register, n, for either a read or write operation that is controlled by a common $R/\overline{W}$ line. However, this circuit is still not practical for a large register set. The number of lines required to select each register can be reduced even further to produce a more efficient design that is practical for any number of registers.

Figure 7.34

The read / write and select control circuit for each register.

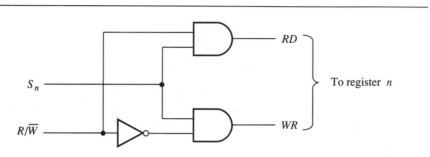

At any given instant, only one of the four registers shown in Figure 7.33 is activated for a read or write operation. Only one of the four select lines S_0 through S_3 will be active at a time. Instead of selecting each register individually through the select lines, each register can be selected by means of a 2-bit binary code using a 1-of-4 decoder. Recall that a 1-of-4 decoder will accept a 2-bit binary code as input and activate one of four outputs to identify a particular code. The modified circuit is shown in Figure 7.35.

In Figure 7.35 the "? boxes" have been replaced with the circuit in Figure 7.34. The select lines are connected to a 1-of-4 decoder such that each 2-bit code applied to its inputs A_1A_0 will select a particular register as shown in Table 7.2.

Because the input lines to the decoder A_1A_0 are used to address or select a register for a read or write operation, they are known as *address lines*. The binary codes that are applied to these lines to select a particular register are known as *addresses*.

The active-low enable input of the decoder $\overline{E}$ allows the decoder to function normally when at the 0 state and disables the decoder when at the 1 state. When the decoder is disabled all its outputs are inactive. If $\overline{E}$ is a logic 1 none of the registers can be selected and no data can be stored or retrieved from any register.

The circuit in Figure 7.35 is a *register array* also known as a 4 × 4 *register file* because it can store four binary numbers each being 4 bits (wide) in size. The logic symbol for the circuit is shown in Figure 7.36.

A register file can be designed for a variety of sizes and configurations. It is usually described in terms of the number of registers (R) and the size of each number stored (S). The product of the number of registers and the size of each register is the total capacity (in bits) of the register file. By convention, the number of registers (R) is always stated first and then the size (S) of each register — R × S. For example, Figure 7.37a shows the logic symbol for an 8 × 4 register file that is capable of storing eight 4-bit numbers for a total capacity of 32 bits. There are three address

Figure 7.35

The complete 4×4 register array (file) circuit.

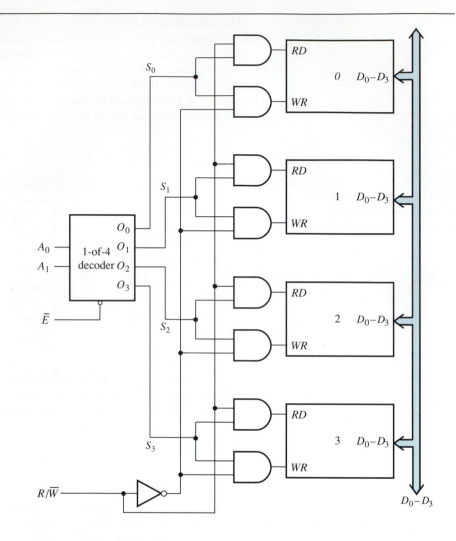

Table 7.2 Addressing table for Figure 7.35

Register	$\overline{E}$	A_1	A_0	S_0	S_1	S_2	S_3
Register 0	0	0	0	1	0	0	0
Register 1	0	0	1	0	1	0	0
Register 2	0	1	0	0	0	1	0
Register 3	0	1	1	0	0	0	1
None	1	x	x	0	0	0	0

lines $A_2A_1A_0$ to address 2^3, or 8, registers and four data lines $D_3D_2D_1D_0$ to store and retrieve 4-bit numbers into and from each register. Figure 7.37b shows the logic symbol for a 4×8 register file that has the same storage capacity (32 bits) as the register file shown in Figure 7.37a. However, the configuration of the register file in Figure 7.37b is much different because it is organized as an array of four registers,

Figure 7.36

Logic symbol for the 4 × 4 register file.

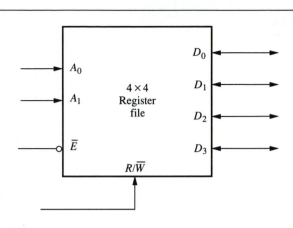

Figure 7.37

Logic symbols for (a) an 8 × 4 register file; (b) a 4 × 8 register file.

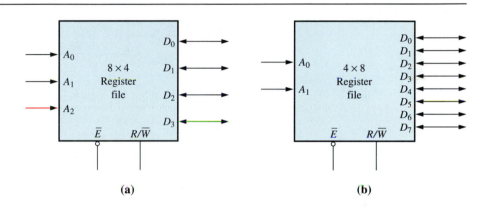

(a)   (b)

each capable of storing an 8-bit number. Address lines A_1A_0 select one of four registers and data lines $D_7D_6D_5D_4D_3D_2D_1D_0$ allow 8-bit numbers to be stored and retrieved into and from each selected register.

Register files can easily be expanded to store more numbers or larger numbers. *Vertical expansion* involves increasing the number of registers in the file and *horizontal expansion* involves increasing the size of each register. For example, Figure 7.38 implements the 8 × 4 register file whose logic symbol is shown in Figure 7.37a by using two 4 × 4 register files.

The register file shown in Figure 7.38 will have a total of eight registers, four of which (0, 1, 2, and 3) will be located in register file no. 0 and the remaining four (4, 5, 6, and 7) in register file no. 1. The data lines $D_3D_2D_1D_0$ of both register files are connected to form a common data bus to store and retrieve data to and from all eight registers. The $R/\overline{W}$ line controls the type of operation (read or write) performed on a selected register. Address lines A_1 and A_0 select the basic eight registers in each register file whereas address line A_2 selects either register file no. 0 (when at the logic 0 state) or register file no. 1 (when at the logic 1 state) as shown in Table 7.3.

Because the two register files in Figure 7.38 are treated as one register file circuit with a total of eight registers, registers 0 through 3 of register file no. 1 are often referred to as registers 4, 5, 6, and 7. Note that the register numbers always correspond to the decimal or hexadecimal equivalents of the addresses.

Figure 7.38

8 × 4 register file
implemented with two
4 × 4 register files.

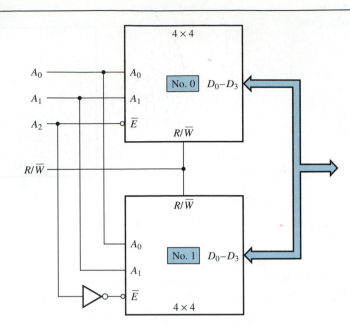

Table 7.3   Addressing table for Figure 7.38

	A_1	A_1	A_0	Register number
Register file no. 0	0	0	0	0
	0	0	1	1
	0	1	0	2
	0	1	1	3
Register file no. 1	1	0	0	0 (4)
	1	0	1	1 (5)
	1	1	0	2 (6)
	1	1	1	3 (7)

Example 7.6

Figure 7.39 shows a 16 × 4 register file constructed by using two of the 8 × 4 register files shown in Figure 7.37a. The addressing for the register file is shown in Table 7.4.

In Figure 7.39 that address lines $A_2A_1A_0$ select the eight registers within each register file. However, when address line A_3 is a logic 0, register file no. 0 is enabled, and when address line A_3 is a logic 1, register file no. 1 is enabled. The registers in register file no. 0 have the addresses 0 through 7 and the registers in register file no. 1 have the addresses 8 through F. As before, the data lines connect to both register files but only the selected (addressed) register in the selected (addressed) register file will be accessed for storage or retrieval of data. The $R/\overline{W}$ line establishes the direction of data flow into or out of the addressed register.

Example 7.7

Figure 7.40 shows a 16 × 4 register file constructed by using four of the 4 × 4 register files shown in Figure 7.36. A 1-of-4 decoder is used to select one of four register files as the most significant address lines A_3A_2 cycle through all possible combinations of 2 bits as shown in Table 7.5.

Figure 7.39

16×4 register file
implemented with two
8×4 register files.

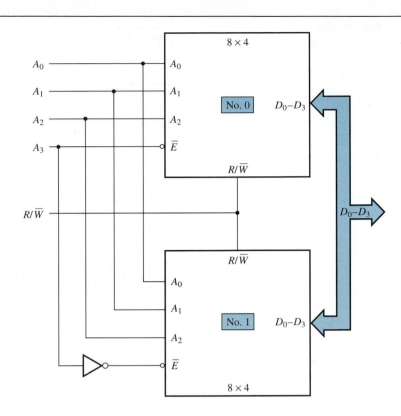

Table 7.4 Addressing table for Figure 7.39

	A_3	A_2	A_1	A_0	Register
Register file no. 0	0	0	0	0	0
	0	0	0	1	1
	0	0	1	0	2
	0	0	1	1	3
	0	1	0	0	4
	0	1	0	1	5
	0	1	1	0	6
	0	1	1	1	7
Register file no. 1	1	0	0	0	8
	1	0	0	1	9
	1	0	1	0	A
	1	0	1	1	B
	1	1	0	0	C
	1	1	0	1	D
	1	1	1	0	E
	1	1	1	1	F

Figure 7.40

16 × 4 register file
implemented with four
4 × 4 register files.

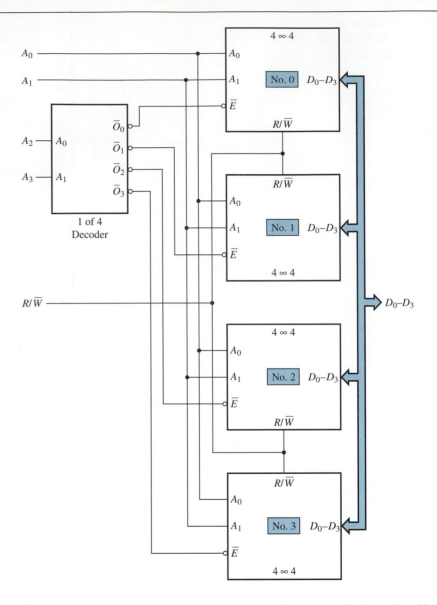

The addressing for the register file is shown in Table 7.5. Figure 7.40 shows that address lines A_1A_0 select the four registers within each selected register file. Address lines A_3A_2 are decoded and select one of the four register files as follows:

A_3A_2	Register file
0 0	0
0 1	1
1 0	2
1 1	3

The registers in register file no. 0 have the addresses 0 through 3, the registers in register file no. 1 have the addresses 4 through 7, the registers in register file no. 2 have the addresses 8 through B, and the registers in register file no. 3 have the addresses C through F. As before, the data lines (D_0 to D_3) connect to all four register files, but only the selected (addressed) register

Table 7.5   Addressing table for Figure 7.40

	A_3	A_2	A_1	A_0	Register
Register file no. 0	0	0	0	0	0
	0	0	0	1	1
	0	0	1	0	2
	0	0	1	1	3
Register file no. 1	0	1	0	0	4
	0	1	0	1	5
	0	1	1	0	6
	0	1	1	1	7
Register file no. 2	1	0	0	0	8
	1	0	0	1	9
	1	0	1	0	A
	1	0	1	1	B
Register file no. 3	1	1	0	0	C
	1	1	0	1	D
	1	1	1	0	E
	1	1	1	1	F

in the selected (addressed) register file will be accessed for storage or retrieval of data. The $R/\overline{W}$ line establishes the direction of data flow into or out of the addressed register.

Figure 7.41

4×8 register file implemented with two 4×4 register files.

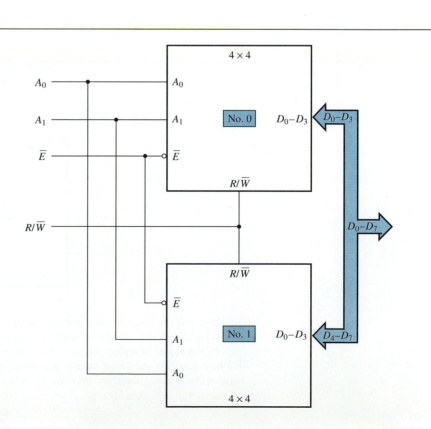

The design procedures for vertical expansion have been examined. The size of the numbers stored in each register has remained the same, but several register files have been combined to increase the total number of registers available. For horizontal expansion, the number of address lines remains the same because the number of registers is not being expanded, but the number of data lines increases to accommodate larger numbers. For example, Figure 7.41 implements the 4×8 register file whose logic symbol is shown in Figure 7.37b by using two 4×4 register files.

The address lines of both register files in Figure 7.41 are connected so that an address selects a register from both register files. However, the 8-bit number $D_7 D_6 D_5 D_4 D_3 D_2 D_1 D_0$ is split between register file no. 0 ($D_3 D_2 D_1 D_0$) and register file no. 1 ($D_7 D_6 D_5 D_4$). When a particular address appears on the address lines, half of the 8-bit number is stored in the addressed register in register file no. 0 and the other half is stored in the addressed register in register file no. 1. Therefore, by using this technique a register file system can be designed to store a binary number of any size as illustrated by the following examples.

Example 7.8

Figure 7.42 shows a 4×16 register file constructed by using two of the 4×8 register files shown in Figure 7.37b.

Figure 7.42

4×16 register file implemented with two 4×8 register files.

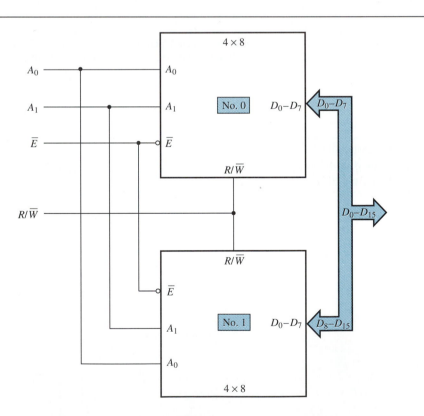

Note that the 16-bit number to be stored in the register file is split into two halves. The least significant 8 bits ($D_0 - D_7$) is stored in register file no. 0 and the most significant 8 bits ($D_8 - D_{15}$) is stored in register file no. 1. The same address is applied to both register files so that both halves of the 16-bit number are stored (or retrieved) at the same location in each register file. Because there are only two address lines there are four storage registers.

Example 7.9

Figure 7.43 shows a 4 × 16 register file constructed by using four of the 4 × 4 register files shown in Figure 7.36.

4 × 16 register file implemented with four 4 × 4 register files.

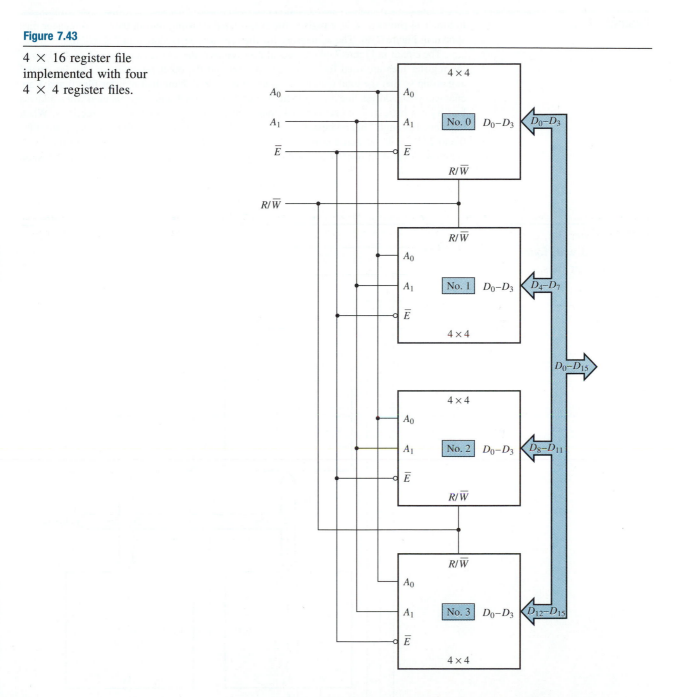

In this circuit, the 16-bit number is split into four parts, each part being stored in a 4 × 4 register file. D_0-D_3 is stored in register file no. 0, D_4-D_7 is stored in register file no. 1, D_8-D_{11} is stored in register file no. 2, and $D_{12}-D_{15}$ is stored in register file no. 3. Again, the same locations in all four register files are addressed in order to store the 16-bit number, and therefore address lines A_1A_0 select each of the four locations available for storage.

Expansion can also be done both horizontally and vertically as illustrated by the following example.

Example 7.10

Figure 7.44 shows an 8×8 register file constructed by using four of the 4×4 register files shown in Figure 7.36. The addressing for the register file is shown in Table 7.6.

The circuit in Figure 7.44 consists of four register files of the same size (4×4). Two of the register files are used for vertical expansion and the other two are used for horizontal expansion so that the total configuration becomes 8×8. Note that register files 0 and 2 are enabled when address line A_2 is a logic 0 and register files 1 and 3 are enabled when address line A_2 is a logic 1. Each set contains four registers to give a total of eight registers. When registers 0 through 3 are addressed, the 8-bit number to be stored is split between register files 0 and 2: D_0 to D_3 is stored in register file no. 0 and D_4 to D_7 is stored in register file no. 2. When registers 4 through 7 are addressed, the 8-bit number to be stored is split between

Figure 7.44

8×8 register file implemented with four 4×4 register files.

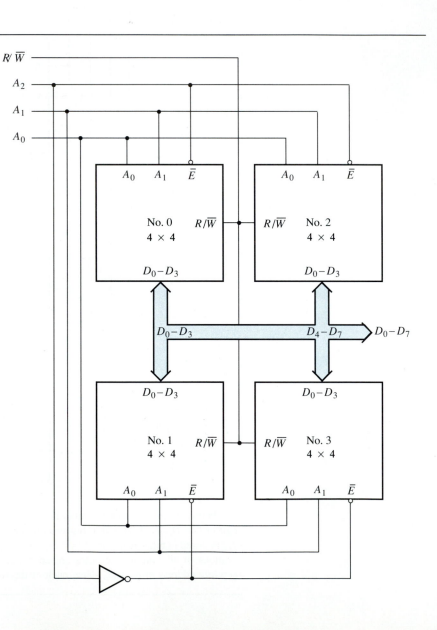

Table 7.6 Addressing table for Figure 7.44

	A_2	A_1	A_0	Register
Register files no. 0 and 2	0	0	0	0
	0	0	1	1
	0	1	0	2
	0	1	1	3
Register files no. 1 and 3	1	0	0	4
	1	0	1	5
	1	1	0	6
	1	1	1	7

register files 1 and 3: D_0 to D_3 is stored in register file no. 1 and D_4 to D_7 is stored in register file no. 3. The $R/\overline{W}$ lines establish the direction of data flow either into or out of an addressed register and are connected to a common $R/\overline{W}$ line for the entire system.

The concepts of storage of binary numbers developed in this section are important in the design and analysis of memory systems. From an external point of view, memory systems are identical to register files, and all the concepts of addressing and expansion can be applied to memory design. The only difference between memory circuits and register files is architectural, that is, the manner in which each bit is stored and retrieved. Also, memory systems are generally used as the primary storage circuits for digital computers and tend to be much larger in size. Chapter 9 investigates such memory systems and expands on the design techniques introduced here to include the design of larger memory systems used in digital computers.

Review Questions

1. What is a register file?
2. How does one specify the configuration of a register file?
3. What is bus contention? How is it prevented?
4. What is the purpose of the address lines?
5. What is an address?
6. How is the size of a register file specified?
7. What is vertical expansion?
8. What is horizontal expansion?

7.6
IC logic

There are several ICs that implement the registers discussed in this section. Some ICs integrate the capabilities of both serial and parallel registers in one package while others provide only the basic capabilities of serial or parallel registers. This section examines a few of the more common ICs available that implement serial and parallel registers as well as a register file.

The 74379 quad parallel register is a simple 4-bit parallel register that is similar in operation to the circuit examined in Figure 7.1. The logic circuit and pin configuration of the 74379 are shown in Figure 7.45. The 4-bit binary number to be stored is applied to the data inputs $D_3D_2D_1D_0$, and a pulse is applied to the write input, CP. On the positive edge of the clock pulse, the data is stored in the register. The stored data is available at all times at the output of the register $Q_3Q_2Q_1Q_0$ or in complementary form at $\overline{Q}_3\overline{Q}_2\overline{Q}_1\overline{Q}_0$. The 74379 also has an active-low enable line $\overline{E}$ that when at a logic 1 state, will disable the register; that is, when $\overline{E} = 1$ no data can be loaded into the register, but when $\overline{E}=0$, the register functions normally. The 74379 does not have tristated outputs (like the register in Fig. 7.4) and there is no read line necessary to retrieve the stored data.

Figure 7.45

Logic symbol and pin configuration of the 74379 quad parallel register.

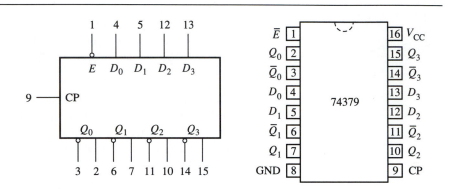

The 74173 4-bit D-type parallel register functions like the circuit shown in Figure 7.4. Its logic symbol and pin configuration are shown in Figure 7.46. Like the 74379, the 74173 has a 4-bit data input $D_3D_2D_1D_0$ to load the data into the register when a pulse on the write line, CP, makes a positive transition. However, the outputs of the register $O_3O_2O_1O_0$ are tristated. To read data stored in the register, the active-low output enable lines $\overline{OE}_1$ and $\overline{OE}_2$ must both be at the logic 0 state. If any one of the output enable lines is inactive at the logic 1 state, the output of the register will be

Figure 7.46

Logic symbol and pin configuration of the 74173 4-bit D-type register.

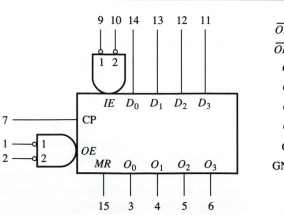
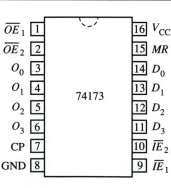

(disabled) in a high-impedance state. The 74173 also has an asynchronous active-high master reset input MR that can be used to reset the register ($MR = 1$) independent of the clock input, CP. The active-low input enable lines $\overline{IE}_1$ and $\overline{IE}_2$ when active (at the logic 0 state) allow the data applied to $D_2D_2D_1D_0$ to be stored into the register. However, if any one of the input enable lines is inactive (at the logic 1 state), the data that was previously stored in the register will not be altered when an attempt is made to write new data into the register by activating CP.

Figure 7.47 shows the logic symbol and pin configuration of the 74178 4-bit shift register. The 74178 can function as a parallel-in parallel-out register, a parallel-in serial-out register, a serial-in parallel-out register, and a serial-in serial-out register.

Figure 7.47

Logic symbol and pin configuration of the 74178 4-bit shift register.

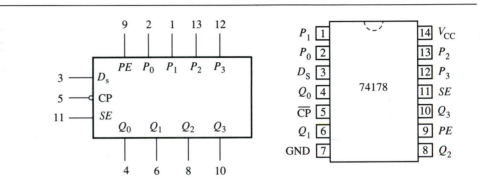

For parallel-in parallel-out operation, the data to be stored into the register is applied to the inputs $P_3P_2P_1P_0$, and the active-high parallel enable input PE is set to a logic 1. The active-high serial enable input SE is set inactive at a logic 0. A clock pulse is applied to the $\overline{CP}$ input, and on the negative edge of the pulse the data at $P_3P_2P_1P_0$ is loaded into the register and is always available at the parallel register outputs $Q_3Q_2Q_1Q_0$.

For serial-in parallel-out operation, the serial data is applied to the serial data input line D_s, MSB first. The SE input is now activated at the logic 1 state and the PE input is made inactive at the logic 0 state. The clock pulses are then applied to $\overline{CP}$, and on the negative transition of each clock pulse each bit is shifted into the register. After four clock pulses the entire 4-bit number is loaded into the register and is available at the parallel register outputs $Q_3Q_2Q_1Q_0$.

For serial-in serial-out operation, the serial data is loaded into the register as described for serial-in parallel-out operations. To retrieve the data from the register serially, the Q_3 output functions as a serial output, and on each negative transition of the clock pulse applied to $\overline{CP}$ the data is shifted out, MSB first.

For parallel-in serial-out operation, the parallel data is loaded into the register as described for parallel-in parallel-out operations. To retrieve the data in the register serially, SE is activated while PE is deactivated. As before, the Q_3 output functions as a serial output, and on each negative transition of the clock pulse applied to $\overline{CP}$ the data is shifted out, MSB first.

The 74194 is a 4-bit bidirectional universal shift register that incorporates the functions of all the different types of registers discussed in this chapter. It is capable of parallel-in parallel-out, serial-in parallel-out, parallel-in serial-out, serial-in serial-

Figure 7.48

Logic symbol and pin configuration of the 74194 4-bit bidirectional universal shift register.

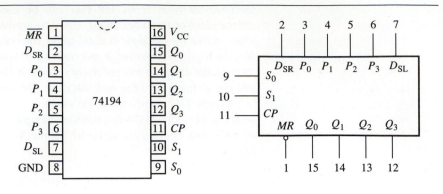

out, and bidirectional left-right shift functions. The logic symbol and pin configuration of the 74194 are shown in Figure 7.48.

The 74194 has four parallel data inputs, $P_3P_2P_1P_0$, and four parallel data outputs, $Q_3Q_2Q_1Q_0$, that are used to store and retrieve data in parallel format. The Q_0 and Q_3 outputs can also be used to retrieve data serially. The master reset input $\overline{MR}$ is an active-low input that can be used to reset the contents of the register while the CP input is used to shift data in or out of the register on the positive edge of the clock pulse. The D_{SR} and D_{SL} inputs are used to shift data into the register. Because the 74194 is a bidirectional shift register, data is applied to the D_{SR} input when the shift direction is to the right and data is applied to the D_{SL} input when the shift direction is to the left. The two mode-control inputs S_1S_0 configure the 74194 to operate in one of the four following modes:

S_1	S_0	Function
L	L	Hold (do nothing)
L	H	Shift Right
H	L	Shift Left
H	H	Parallel Load

Figure 7.49

Logic symbol and pin configuration of the 74670 4 × 4 register file.

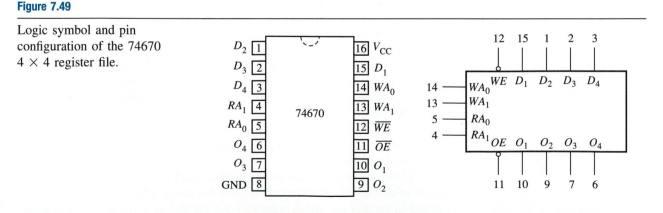

The 74670 is a 4 × 4 register file whose function is similar to the circuit shown in Figure 7.36. Its logic symbol and pin configuration are shown in Figure 7.49. The 74670 register file has separate data inputs and data outputs, unlike the circuit of Figure 7.36. The data to be stored is applied to the data input lines $D_4 D_3 D_2 D_1$, and the active-low write enable input $\overline{WE}$ is activated by applying a logic 0 to it. To retrieve data stored in a register, the active-low output enable line $\overline{OE}$ is activated by applying a logic 0 to it. This enables the tristated output lines $O_4 O_3 O_2 O_1$ and makes the data available at these outputs.

The D and O lines can be connected to produce a set of bidirectional data lines. To address one of four internal 4-bit registers the address must be applied to $WA_1 WA_0$ or $RA_1 RA_0$ depending on whether data is being written or read, respectively. These two sets of lines can be combined as a single set of address lines $A_1 A_0$ to address the register file registers for read and write operations.

Review Questions

1. Compare and contrast the operation of the 74379 with the circuits shown in Figures 7.2, 7.4, and 7.5.

2. Compare and contrast the operation of the 74173 with the circuits shown in Figures 7.2, 7.4, and 7.5.

3. Compare and contrast the operation of the 74178 with the circuits shown in Figures 7.7, 7.12, and 7.14.

4. Compare and contrast the operation of the 74194 with the circuit shown in Figure 7.18.

5. Compare and contrast the operation of the 74670 with the operation of the circuit shown in Figure 7.35.

7.7
Troubleshooting register circuits

Register circuits exhibit the same electrical faults as combinational logic circuits because their building blocks are flip-flops, which present the same faults as the basic logic gates. Therefore, register circuits that do not function properly can be tested for common problems such as shorted or open inputs or outputs using the basic test instruments — the logic probe, logic pulser, and current tracer. For example, consider the following troubleshooting case study.

Example 7.11

The circuit in Figure 7.50 shows two 74178 ICs connected to produce an 8-bit Johnson counter. On testing the circuit we find that following a reset the output of the counter produces the following sequences:

$$
\begin{array}{l}
00000000 \\
10000000 \\
11000000 \\
11100000 \\
11110000 \\
11110000 \\
11110000 \\
\quad \vdots
\end{array}
$$

Figure 7.50

Register circuit containing a
fault.

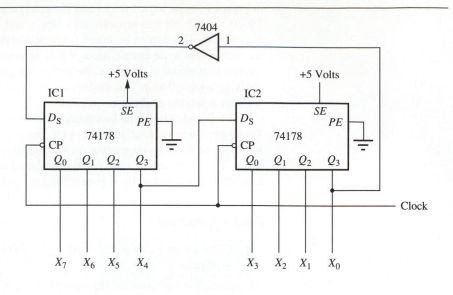

When a count of 1111000 is reached the counter stays in that state continuously until reset. A
functional counter would produce the following sequence after a count of 11110000:

$$11110000$$
$$11111000$$
$$11111100$$
$$11111110$$
$$11111111$$
$$01111111$$
$$\vdots$$

From observations of the counting sequence it can concluded that the fault lies in the
connection between IC1 and IC2. The Q_3 output of IC1 appears to be functioning properly
because the output of the counter, X_4, produces the correct levels during the valid counts. The
fault must be in the D_s input of IC2. D_s is probably internally open circuited, because if it were
shorted to GND or V_{CC}, X_4 would be held at a constant logic level.

Registers, like counters, are also made up of flip-flops and are susceptible to
faults at high operating frequencies, due to propagation delays. However, because
the flip-flops of shift registers are synchronously clocked, they are not as affected by
propagation delays as asynchronous counters. Most of the timing problems that
occur in shift register circuits are primarily due to faults in the design of the circuit
rather than faults due to propagation delays. For example, consider the timing
diagram of the serial data communications system shown in Figure 7.23. Recall that
the data was shifted out of the source on the positive edge of the clock and shifted
into the destination on the negative edge of the clock. This was done to prevent any
timing problems that could exist due to various factors such as delays due to line
capacitance and setup times that are too short. Assume that the two clock inputs C_A
and C_B shifted data in and out on the same edge of the clock pulse. Now if the rise
times of the clocks C_A and C_B were different, this could cause the two clocks to be
out of synchronization, which could lead to an error in data transmission. Also the
setup time of the clock C_B would be too short because the serial data changes at

almost the same time as the clock makes its transition. The remedy, of course, is to trigger the data out and the data in at different edges of the clock pulse as shown in Figure 7.23.

Many of the troubleshooting procedures developed in earlier chapters can be applied to isolate faults in register circuits. There are more advanced techniques of troubleshooting sequential logic circuits using more complex and expensive test equipment; applications of these techniques and equipment will be seen in Chapters 8 and 9 because they are designed to troubleshoot more complex circuits. For the simple circuits and their applications examined so far, most faults can be isolated and corrected with the basic test instruments such as the logic probe, pulser, and current tracer, and through the use of analytical reasoning.

Review Questions

1. In what ways are the faults found in register circuits similar to those found in other logic circuits?

2. Why are register circuits not as susceptible to the effects of propagation delay as asynchronous counters?

Summary

- Registers are logic circuits that are capable of storing single binary numbers of varying sizes.

- Parallel registers or parallel-load registers simultaneously load and retrieve all the bits that make up a binary number.

- Serial registers load and/or retrieve a binary number 1 bit at a time.

- The write line of a register is used to store (load), or write, data into the register.

- The read line of a register is used to retrieve, or read, data from the register.

- A register that stores and retrieves data over a common set of lines is said to have bidirectional data lines.

- Parallel registers are known as parallel-in parallel-out registers because data is stored and retrieved in parallel.

- Serial registers are often called shift registers because the bits that make up a binary number are shifted into or out of the register 1 bit at a time.

- Serial-in parallel-out registers load data into the register serially but supply the data out in a parallel format.

- Serial-in serial-out registers load data into the register serially and supply the data out of the register serially.

- Parallel-in serial-out registers load data into the register in parallel but supply the data out of the register serially.

- Bidirectional shift registers can shift data into and out of a register in two directions — left and right.

- Parallel registers are characterized by fast loading and retrieving of data whereas serial registers are much slower.

- Serial registers require fewer lines to load and/or retrieve data whereas parallel registers require several lines to accomplish the same task.

- Shift registers are also used in special purpose counters known as shift register counters.

- Two of the more common types of shift register counters are the ring counter and the Johnson counter.

- A register array or register file is a collection of registers incorporated into a single package.

- The configuration of a register file is specified as $R \times S$ where R is the number of registers and S is the size of each register.

- The total storage capacity of a register file (in bits) is the product of the number of registers (R) and the size of each register (S).

- A register file is similar in concept and operation to a memory system.

- Bus contention is a conflict of logic levels that occurs when more than one register loads its data on a bus.

- Address lines are a set of lines that are used to select a particular register for reading or writing.

- An address is a binary number that uniquely identifies a particular register.

- Register files can easily be expanded to store larger or more numbers.

- Vertical expansion involves increasing the number of registers in a register file.

- Horizontal expansion involves increasing the size of each register in a register file.

- Register circuits exhibit the same faults as other sequential and combinational logic circuits.

- Registers are not as susceptible to the effect of propagation delays at high frequencies as asynchronous counters because the flip-flops are synchronously clocked.

Problems

Section 7.2   Parallel registers

1. Design an 8-bit parallel register using D-type flip-flops with bidirectional data lines and active-high RD and WR lines.

2. Implement the circuit in Problem 1 using J-K flip-flops instead of D flip-flops.

3. Design a 4-bit parallel register with bidirectional data lines and a single R/W line that will allow data to be loaded into the register when $R/\overline{W} = 0$ and data to be retrieved from the register when $R/\overline{W} = 1$.

4. Discuss what would happen if the $\overline{RD}$ and $\overline{WR}$ lines of Figure 7.5b were activated at the same time.

Section 7.3   Serial registers

5. Construct a timing diagram for the circuit shown in Figure 7.7b when the binary number 1100 is shifted in. Assume that the initial state of the register is 1010.

6. Construct a timing diagram for the circuit shown in Figure 7.7b when the binary number 0101 is shifted in. Assume that the initial state of the register is 1011.

7. Construct a timing diagram for the circuit shown in Figure 7.12b that shows the number 1010 shifted into the register and then shifted out.

8. Construct a timing diagram for the circuit shown in Figure 7.12b that shows the number 0101 shifted into the register and then shifted out.

9. Construct a timing diagram to *LOAD* the number 1010 into the register shown in Figure 7.14 and then retrieve the number from the register.

10. Construct a timing diagram to *LOAD* the number 0101 into the register shown in Figure 7.14 and then retrieve the number from the register.

11. Assuming that the number 0101 has been shifted into the bi-directional shift register shown in Figure 7.18, draw a timing diagram of the parallel outputs when the $R/\overline{L}$ line is set to a logic 1, and when the $R/\overline{L}$ line is set to a logic 0. Assume that SLI and SRI are both logic 0's.

12. Assuming that the number 1010 has been shifted into the bidirectional shift register shown in Figure 7.18,

draw a timing diagram of the parallel outputs when the $R/\overline{L}$ line is set to a logic 1, and when the $R/\overline{L}$ line is set to a logic 0. Assume that SLI and SRI are both logic 0's.

13. Design a 4-bit shift register that is capable of loading and retrieving data in all possible formats — parallel-in parallel-out, serial-in serial-out, serial-in parallel-out, and parallel-in serial-out. Draw a logic symbol for this circuit and show the complete logic circuit for the shift register.

Section 7.4   Shift register applications

14. Construct a timing diagram for the circuit shown in Figure 7.22 when the binary number 1100 is transmitted from A to B.

15. Construct a timing diagram for the circuit shown in Figure 7.22 when the binary number 0101 is transmitted from A to B.

16. Expand the circuit in Figure 7.22 to transfer 8-bit numbers from A to B.

17. Construct a timing diagram for the circuit in Problem 16 when the binary number 10110010 is transmitted from A to B.

18. Modify the circuit in Figure 7.25 to detect the code "3916."

19. Expand the circuit in Figure 7.25 to detect the 5-bit combination "30217."

20. Design a 6-bit ring counter using J-K flip-flops.

21. Design a 6-bit Johnson counter using J-K flip-flops.

22. Draw a timing diagram for the circuit designed in Problem 20. Assume that the initial state of the counter is 0.

23. Draw a timing diagram for the circuit designed in Problem 21. Assume that the initial state of the counter is 0.

24. Determine the amount of delay that would occur in the circuit shown in Figure 7.30 if the clock had a frequency of 2 MHz.

25. Determine the amount of delay that would occur in the circuit shown in Figure 7.30 if the clock had a frequency of 10 MHz.

26. Design a time delay circuit that would produce a delay for exactly 15 clock periods using the registers discussed in this chapter.

27. Design a time delay circuit that would produce a delay for exactly 11 clock periods using the registers discussed in this chapter.

Section 7.5   Register arrays

28. Design the logic circuit for an 8×4 register file that has eight 4-bit registers. The circuit should be similar to the one shown in Figure 7.35.

29. Design the logic circuit for a 4×8 register file that has four 8-bit registers. The circuit should be similar to the one shown in Figure 7.35.

30. Design an expanded 16×4 register file system that has sixteen 4-bit registers using only 4×4 register files.

31. Design an expanded 16×8 register file system that has sixteen 8-bit registers using only 4×4 register files.

32. Design an expanded 64×8 register file system that has 64 8-bit registers using only 16×4 register files.

Section 7.6   IC logic

33. Expand the 74178 4-bit shift register to accommodate 8 bits.

34. Design a 16-bit ring counter using 74178 ICs.

35. Design a 16-bit Johnson counter using 74178 ICs.

36. Construct the serial data communications circuit shown in Figure 7.22 using 74194 ICs.

37. Construct an 8-bit bidirectional shift register using two 74194 ICs.

38. Implement the 8×4 register file shown in Figure 7.37a using only 74670 ICs and any other logic gates that may be required.

39. Implement the 4×8 register file shown in Figure 7.37b using only 74670 ICs and any other logic gates that may be required.

Troubleshooting

Section 7.7   Troubleshooting register circuits

40. The 8-bit Johnson counter shown in Figure 7.50 produces the following count sequence when tested:

$$
\begin{array}{l}
00000000 \\
10000000 \\
11000000 \\
11100000 \\
11110000 \\
11111000 \\
11111100 \\
11111110 \\
11111111 \\
11111111 \\
11111111 \\
\vdots
\end{array}
$$

Discuss the possible fault(s) that could exist in the circuit.

Figure 7.51

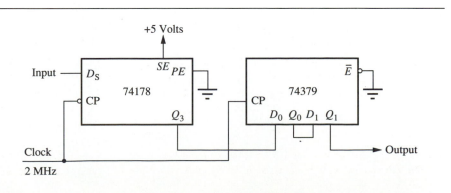

41. The time delay circuit shown in Figure 7.51 uses a 74178 and a 74379 IC to produce a delay of five clock pulses, or 2.5 μs. When the circuit is tested the output pulse appears after only 2.25 μs. Troubleshoot the circuit and determine the possible fault. Suggest a solution for correcting the fault.

42. The ring counter shown in Figure 7.26 is to be operated at the maximum possible frequency that the flip-flops can handle. Determine if the circuit will work properly at that frequency by constructing a timing diagram showing the propagation delays. Use the flip-flop AC characteristics given in Chapter 5.

MEMORY CONCEPTS AND DESIGN

OBJECTIVES

The objectives of this chapter are to:

• Explain the operational concepts of various types of memories

• Introduce the concept of memory organization

• Explore methods of accessing memory locations and I/O devices

• Explain the method of read / write control when accessing a memory location or I/O device

• Study the hazards of operating on a memory location

• Categorize memory devices according to characteristics that are useful when implementing a memory design

• Explain fundamental concepts in the design of a memory system

• Develop procedures to troubleshoot faults in memory or I/O devices

Magnetic tape is often used in large digital computers that have to backup large volumes of information in certain types of nonvolatile memory (Sperry Corporation).

8.1
Introduction

Computers are the most commonly used digital systems, which is why the authors have chosen to include some rather extensive computer-related material. An integral part of any computer system are the memories within that system. The purpose of this chapter is to identify some of those memories and to develop related concepts and design procedures such as:

1. Memory classifications according to usage.
2. Internal organization of various IC memories.
3. Characteristics and classifications of various types of IC memories.
4. Addressing memory.
5. Providing control and timing for read/write operations.
6. Develop design procedures for a computer main memory.

Before continuing with the main topics of this chapter, pause for a moment and think about what the word "computer" means. Included in the classification are those special-purpose computers that do not necessarily interface with a human as a personal computer (PC) does, but rather with another machine. This class of computer is often described as an embedded controller. Embedded controllers appear in applications such as automobile emissions control, household appliances, robots, and in a host of other machine control applications. In modern control technology it would be difficult to not find an embedded controller somewhere in the control process.

Our introduction to computer memories will begin by identifying those that are found in computer systems. Take note that the names given to these memories are indicative of their usage.

The first type of memory to be examined is the memory within which computer programs and data are stored. A computer program is a sequence of instructions written by a programmer (the human master) that tells the computer the precise sequence of operations (add, subtract, *AND*, *OR*, etc.) to be executed. Like humans recalling a sequence of instructions to perform a task, the computer program must first be stored (memorized) in the computer's memory. To execute the program, the computer refers to this memory and fetches and executes the first instruction of the program. It then returns to memory, fetches, and executes the next instruction. This cycle of fetching and executing instructions from memory continues until all the instructions of the program have been executed and the desired task has been achieved.

In addition to storing programs in memory it is also necessary to store and retrieve data. While there are a variety of memories within a computer, the type of memory from which the computer fetches and executes program instructions, as well as stores and retrieves data associated with the program being executed, is known as the computer's *main memory*. The design of main memory will be a focal point of this chapter and Chapter 9.

Because of the finite size of main memory, computers have a limited capacity to store programs and data. Programs and data not currently being operated upon by the computer can be, and often are, stored in an extended memory that is separate from the main memory. This memory has a large capacity for mass storage of programs and data. The devices that provide *mass storage memory* are known as *mass storage devices* and include floppy and hard disks, magnetic tape, *magnetic bubble memories* (MBM), and CD *read only memory* (CD-ROM).

Another type of computer memory is the *queue*. A small capacity memory is provided within some computers to store instructions in a queue ("queue" means waiting in line to be served) while waiting to be executed. The use of a queue memory can speed up the process of fetching and executing instructions by enabling instructions to be prefetched (i.e., instructions can be fetched in advance of execution and stored in the queue while waiting to be executed). Queue memories will be examined in detail when microprocessors are covered in Chapter 10.

Some computers have an internal memory within which partial calculations and other temporary data can be stored until needed, much like the internal memory of a calculator. This type memory is known as a *scratch pad* memory.

Peripheral devices such as printers and monitors also have memory within which data can be stored. The data in main memory that is to be printed or displayed by the peripheral device can be transferred from main memory to the peripherals' memory. This frees up the computer and its main memory space by enabling peripherals to operate fairly independent of the computer and its main memory. A memory of this type is known as a *buffer memory*.

This brief introduction described a variety of memories, serving a different need. The internal format of main memory will now be examined. Many of the concepts developed are also applicable to other types of memories.

As previously noted, all computers are an integration of digital logic circuits. Because all digital circuits work with binary logic levels (1's and 0's) so do computers. This means that all computer instructions and data stored in the main memory are in a binary format. Main memory is composed of memory locations that have a fixed number of bits. The number of bits per location are multiples of 8 (8, 16, and 32 bits), with the usual being 8 bits or 1 byte. For instance, if a program is 1000 bytes and operates on 200 bytes of data, then the minimum memory capacity required of main memory is 1200 memory locations with each location occupying 1 byte.

8.2
Computer architecture

To understand the material presented in this and later chapters, it is necessary to continually reexamine the parts of a computer and their operations. For now, the part of most interest is memory.

Figure 8.1 is an architectural representation of a computer, a symbolic representation that has been divided into functions, represented by function-blocks. The interactions among the function-blocks are indicated by connecting lines with arrowheads indicating the flow of operations.

Figure 8.1

Computer architecture.

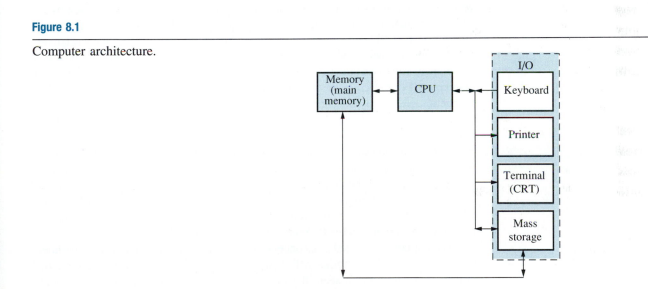

There are three function blocks shown in Figure 8.1.

1. The *central processing unit* (CPU) that provides the intelligence and processing capability to the system. As Chapter 10 will show, the CPU is implemented with a *microprocessor unit* (MPU).

2. Main memory, as already discussed, is where programs to be executed and data to be processed by the CPU are stored.

3. *Input and output* (I/O) devices describe all system peripheral devices in a generic fashion. I/O devices, which serve as communication links to the outside world for the CPU, include such peripherals as the system printer, keyboard, monitor, and its mass storage system (floppy and hard disk drives).

Figure 8.1 shows that the connecting line between the CPU and main memory is bidirectional, because the CPU must both retrieve (read) from and store (write) to this memory. Mass storage also has a bidirectional connection between it and main memory. This is to enable mass storage memory to load directly into main memory a program or data or both to be executed by the CPU, as well as to enable main memory to free up some of its space by placing programs and program results (data) that are not currently being executed into mass storage for later reference. Recall that for a program to be executed or for data to be accessible to the CPU, it must be stored in main memory. Some mass storage devices such as floppy disks also provide portability of programs from one computer to another.

Main memory is essential to the computer system because the CPU works from that memory. Mass storage memory is convenient, but, from a conceptual point of view, is not essential for basic operation of the system. However, as every user knows, mass storage soon becomes an essential item from a practical viewpoint because the main memory runs out of space as more and bigger programs and data are loaded in main memory.

Memory technology may be utilized as either a main or mass storage memory depending on the speed of operation and the memory capacity. The CPU operates at relatively fast speeds (microseconds to nanoseconds). Because it gets its instructions and data from main memory, main memory must operate at similar speeds to keep up with the CPU. To meet or exceed these timing requirements, as established by the CPU, main memory is implemented with semiconductor IC memory devices that are faster than the CPU. Unfortunately, semiconductor memory devices have relatively small memory capacity, which makes them unsuitable for mass storage applications. In contrast, for mass storage, storage capacity and portability are more important than speed. For these reasons mass storage devices use high-capacity media such as magnetic tape, floppy and hard disks, and magnetic bubble memories (MBM). At present, these mass storage devices have relatively slow operating speeds (milliseconds).

The main memory of Figure 8.1 is composed of one or more semiconductor memory devices. To avoid any prerequisite study of electronics, semiconductor memory devices will be discussed in this chapter from a systems point of view and the study of their internal electronic design will be deferred until Chapter 12. Readers who prefer to know the electronic principles before they study semiconductor memory devices may refer to Sections 12.1 to 12.3 of Chapter 12.

Having better clarified our concept of the parts of a computer, via the architecture of Figure 8.1, memories will now be categorized according to their internal organization. As will be seen, the internal organization of a memory often determines its application.

Review Questions

1. What are the three essential function blocks of a computer?
2. Relative to the CPU, what is the difference between main and mass storage memory?
3. Why is there a bidirectional link between the main memory of Figure 8.1 and mass storage?
4. Why is main memory considered an essential part of a computer system whereas mass storage is not?
5. Why are main memories implemented with semiconductor devices?

8.3
Types of memory organizations

There are five categories that describe one or more characteristics or abilities of a memory:

1. The organization of the memory and memory media.
2. The manner of accessing memory locations within the memory.
3. The ability to alter the contents of memory.
4. The ability of memory to retain its contents (volatility) when power is removed.
5. The ability of memory to retain its contents even with power applied to the memory device, without continual refreshing of the contents.

These characteristics are used by the system designer to determine whether or not a specific memory device can be used to implement a memory design. This section will investigate the organization of memories.

Memory internal organization

A memory is made up of many locations, each of which is made up of a fixed number of memory cells similar to register arrays examined in Chapter 7. Each cell can store a binary bit (1 or 0). The number of memory cells and their arrangement are collectively known as the organization of that memory. Some memories are organized such that they can store only 1 bit at each memory location, whereas others can store as many as 8 bits at each location; the size of the location depends on the number of memory cells per location. The more common memory cell organizations for semiconductor memories are 1, 4, or 8 bits per location. For example, a memory organized such that it has 256 locations and each location contains four memory cells is known as a 256×4 memory (256 by 4 bits), or a 1024-bit memory. Note that the ''$\times$'' is treated as a dimensional indicator (by) rather than an arithmetic operator (multiply), even though the product of 256 and 4 is 1024, which is the total number of bits for that memory. If each of the 256 memory locations contained eight memory cells, it would be a 256×8 (2048-bit) memory.

Figure 8.2 illustrates two simplified (only four locations) memory organizations, representing a 4×1-bit and a 4×4-bit memory. Each row of memory cells represents a memory location, and the number of memory cells per memory location represents the number of storage bits per location. Other examples of this type of memory organization are 1024×4 (4096 bits organized such that there are 1024

Figure 8.2

Examples of memory organization.

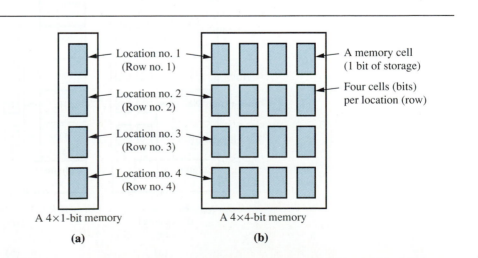

Location no. 1 (Row no. 1)

Location no. 2 (Row no. 2)

Location no. 3 (Row no. 3)

Location no. 4 (Row no. 4)

A memory cell (1 bit of storage)

Four cells (bits) per location (row)

A 4×1-bit memory

A 4×4-bit memory

(a)

(b)

locations and 4 bits of storage per location), 2048 × 8 (16,384 bits), or 2048 × 1 (2048 bits).

Expanding on the concepts of memory organization, the organizations of memory cells can be placed into one of four categories based on the dimensional characteristics of the memory.

1. One-dimensional array
2. Two-dimensional array
3. Three-dimensional array
4. Dynamic one-dimensional array

One-dimensional memory arrays

One-dimensional memory arrays are memory cell organizations in which each memory location is represented as a row, each row having a fixed number of memory cells, such as the organizations illustrated in Figure 8.2.

Two-dimensional memory arrays

Two-dimensional memory arrays represent memory cell organization as a two-dimensional x-y coordinate system, such as those illustrated in Figure 8.3. Figure 8.3a shows eight memory cells arranged in an 8 × 1-bit array (8 locations with 1 bit of storage per location), whereas Figure 8.3b shows a 16 × 1-bit organization. For the convenience of memory cell identification, each cell in the arrays of Figure 8.3 has been labeled with an identifying hexadecimal number: 0 through 7 for Figure 8.3a and 0 through F for Figure 8.3b. The location of any memory cell within either array of Figure 8.3 can be specified by stating the row (x coordinate) and the column (y coordinate) within which it appears. That is, memory cell 3 of Figure 8.3a is in row 01 and column 1.

Figure 8.3 shows that the rows and columns of the arrays are identified with binary numbers; these are specified in the index tables at the bottom and to the left of the array and are labeled "Row (x) no." and "Column (y) no". The binary value created by combining the row and column binary numbers within which the cell is located, where the LSBs of that combination are formed by the column and the MSB

Figure 8.3

Two-dimensional memory cell organizations.

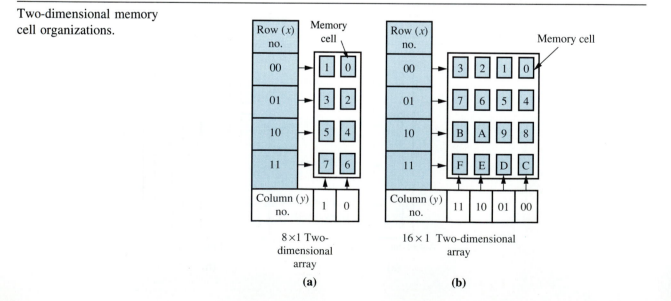

(a)   8 × 1 Two-dimensional array

(b)   16 × 1 Two-dimensional array

positions are formed by the row, is numerically equivalent to the hexadecimal number used to identify the cell. For example, using this *x-y* coordinate technique, cell no. 6 resides at location 110 (row 11 and column 0) in Figure 8.3a and at location 0110 (row 01 and column 10) in Figure 8.3b.

Example 8.1

From Figure 8.3a and 8.3b, develop a table that specifies the location of each memory cell in binary, using the *x-y* coordinate system.

Solution

The example above shows that a cell's location in Figure 8.3 is specified by combining the row and column binary values with the *y*-coordinate (column) forming the LSBs and the *x*-coordinate (row) forming the MSBs. The resulting binary values are listed in the "location" column in Table 8.1. The column labeled "Cell" contains the hexadecimal equivalent of the binary number in the "location" column. That is, for Figure 8.3a, row 00 and column 0 locate cell 0, and when row and column binary numbers are combined the location number 000 is formed. When row 00 and column 00 of Figure 8.3b are combined, location number 0000 is formed, which is the binary location of memory cell no. 0. Likewise, 001 is the location of cell 1 in Figure 8.3a and 0001 is the location of cell 1 in Figure 8.3b.

Table 8.1   Memory location and memory cell per location for Figure 8.3a and 8.3b

Figure 8.3a

Row	Column	Location	Cell
00	0	000	0
00	1	001	1
01	0	010	2
01	1	011	3
10	0	100	4
10	1	101	5
11	0	110	6
11	1	111	7

Figure 8.3b

00	00	0000	0
00	01	0001	1
00	10	0010	2
00	11	0011	3
01	00	0100	4
01	01	0101	5
01	10	0110	6
01	11	0111	7
10	00	1000	8
10	01	1001	9
10	10	1010	A
10	11	1011	B
11	00	1100	C
11	01	1101	D
11	10	1110	E
11	11	1111	F

Table 8.1 is basically a truth table in which the column labeled "Location" has all possible binary combinations present, 8 combinations for Figure 8.3a and 16 for Figure 8.3b, and (as previously stated) the cell identification number given in the column labeled "Cell" is the hexadecimal equivalent of the location number.

It is important to understand the x-y coordinate system of identifying the location of a cell.

Three-dimensional memory arrays

Figure 8.4 illustrates the concept of a three-dimensional memory cell array. As before, a small number of cells (64) are being used to minimize the complexity of the illustration. As can be seen in Figure 8.4a, this organization has duplicate two-dimensional arrays that are stacked in a third dimension, which is called the z-coordinate. These stacked two-dimensional arrays are identified as planes (plane 0, plane 1, etc.), and the number of planes determines the number of memory cells for each memory location. The bit size of each memory location for the memory of Figure 8.4a is four because there are four planes of memory.

Figure 8.4

Three-dimensional memory cell organization.

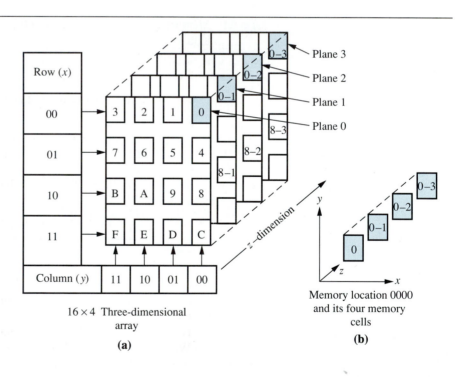

16 × 4 Three-dimensional array

(a)

Memory location 0000 and its four memory cells

(b)

Memory locations of a three-dimensional memory are identified with the same x-y coordinate scheme as a two-dimensional array except that rather than identifying one memory cell in plane 0, identically positioned memory cells in the other planes are also identified. For example, memory location 0000 of Figure 8.4a contains four memory cells, cell 0 of plane 0, cell 0-1 of plane 1, cell 0-2 of plane 2, and cell 0-3 of plane 3, as also illustrated in Figure 8.4b. Likewise, memory location 1000 consists of memory cells 8, 8-1, 8-2, and 8-3. It can be concluded that the x-y coordinates determine the number of memory locations, which is 16 (0 through F) for the array of

Figure 8.4a, and the number of planes determines the storage bit size of each location. The memory of Figure 8.4 is a 16 × 4 memory. If one wished to devise a memory similar to that of Figure 8.4a but with eight memory cells per location (16 × 8-bit memory), then four more planes of memory cells would have to be added to the memory of Figure 8.4a.

In reality, when a three-dimensional array is manufactured using semiconductor technology, which uses a two-dimensional IC wafer, each plane of memory cells is integrated on the same IC wafer. The concept of a three-dimensional array, as illustrated in Figure 8.4a, is a convenient way to visualize how multiple cells can reside at the same memory location.

Example 8.2

In regard to the three-dimensional array of Figure 8.4a
1. What memory cells are at memory locations
   (a) 0111   (b) 1001   (c) 1111?

2. At what memory location are memory cells
   (a) 2, 2-1, 2-2, and 2-3   (b) A, A-1, A-2, and A-3?

Solution

The solution for part (1) is to decode the two LSBs of the memory location to obtain the column and the two MSBs to identify the row.

(a) Memory location 0111 selects column 11 and row 01, which, when cross-referenced using an x-y coordinate system, selects memory cell 7 of plane 0, cell 7-1 of plane 1, cell 7-2 of plane 2, and 7-3 of plane 3.

(b) 1001 selects column 01 and row 10, which selects memory cells 9, 9-1, 9-2, and 9-3.

(c) 1111 selects column 11 and row 11, which selects memory cells F, F-1, F-2, and F-3.

The solution for part (2) is the reverse of the solution for part (1)—the creation of a binary number that is equivalent to the hexadecimal memory location number.

(a) Memory cells 2, 2-1, 2-2, and 2-3 reside at memory location 2H(hexadecimal). The binary equivalent of 2H is 0010_2; hence 0010 is the binary location number for memory cells 2, 2-1, 2-2, and 2-3.

(b) Memory cells A, A-1, A-2, and A-3 reside at memory location AH. Because AH = 1010_2, 1010 is the correct memory location number.

A memory cell array can be configured in many different combinations, so long as both the number of rows and columns is a power of two (2, 4, 8, 16, etc.). The eight memory cell array of Figure 8.3a could have been configured as two rows and four columns. With this organization the row numbers (x) would be 0 and 1 and the column numbers (y) would be 00, 01, 10, and 11 for determining the cell location number. The 16 memory cells of Figure 8.3b could have been organized with two rows and eight columns or with eight rows and two columns. For the former organization the row numbers (x) would be 0 and 1 and the column numbers (y) would be 000, 001, 010, 011, 100, 101, . . . , 111 for cell location identification. Take note of binary patterns of the row and column numbers.

Dynamic one-dimensional memory arrays

These are one-dimensional arrays in which the array is mobile (dynamic). Either these memory cells are on a movable memory medium, such as a magnetic tape, or the array itself is movable.

Review Questions

1. What is a memory cell?
2. How are the memory cells of an IC semiconductor memory organized?
3. How are the number of memory locations and the bit size of each location within a memory device specified?
4. What are the four types of dimensional arrays and how do they differ?
5. What is the essence of an *x-y* coordinate addressing scheme?
6. How is the third dimension formed in a three-dimensional array?
7. What characteristic is associated with dynamic one-dimensional arrays?

8.4
Memory addressing and access time

Each memory location must be capable of being accessed so that binary bits representing either instructions or data may be stored or retrieved. Imagine each memory location as analogous to a person's residence; like a residence, each memory location has a unique address. Accessing a memory location is done by supplying the address of the location to the semiconductor memory. Once given the address the memory IC decodes it and locates the addressed location. Just as was done in Example 8.2.

The parameter that indicates the speed with which a location within a memory device can be accessed is known as *access time*. Access time is the elapsed time between application of a memory location address to a semiconductor memory and when the contents of that memory location are available at the output of the memory IC.

There are conceptually four techniques for accessing memory locations: *linear accessing*, *random accessing*, *first-in first-out* (FIFO), and *last-in first-out* (LIFO).

Linear accessing of memory

The linear accessing technique is used to access dynamic one-dimensional cell arrays. The access time for linear accessing varies with how far the contents of the addressed location must travel before being output. If memory location 100 is being accessed and location 101 is to be accessed next, the access time is much shorter than if location 500 were the next location to be accessed.

Random accessing of memory

The random access technique accesses memory locations using the *x-y* coordinate system, such as the two- and three-dimensional arrays of Figures 8.3 and 8.4. The access time for this technique is constant, that is, any memory location within the array has the same access time regardless of the previous location being accessed. Access time for this technique is not location dependent. To understand why this is so, recall that to locate a memory cell in a two- or three-dimensional array, the row and column of the memory location are identified by the memory location address. The memory device decodes that address and then selects the addressed row and column. Within any memory IC the same amount of time is required to decode any row address. The same is true for any column address. As a result, the access time of any combination of row and column is the same, regardless of the location. Because the access time of any memory location chosen at random is the same, a memory device with this characteristic is known as a *random access memory* (RAM). All modern main memories are composed of semiconductor RAM memory devices. As will be shown, the term "RAM" is often misused.

Example 8.3

If 400 ns is required to select a column and row for a given memory location of Figures 8.3 and 8.4, what is the access time of those memory devices?

Solution

Because a memory location address simultaneously selects column and row for either array, then 400 ns is the access time for any memory location in those arrays. Hence, the access time of either memory device is 400 ns.

FIFO and LIFO accessed memories

"FIFO" and "LIFO" are acronyms for the method of accessing a certain type of memory. The data within a FIFO is accessed in a first-in first-out manner, and the data stored within a LIFO is accessed in a last-in first-out manner. The first datum written into a FIFO memory is the first that can be read from it, whereas the last datum written into a LIFO is the first that can be read from it.

Conceptually there are similarities in the accessing techniques of FIFO and LIFO memories.

FIFO accessed memories

A 4 × 4 FIFO accessed memory is illustrated in Figure 8.5a. Figure 8.5a shows that the FIFO is a dynamic one-dimensional array (i.e., the data is mobile) that has been implemented with shift registers such as those studied in Chapter 7. Memory cells are the flip-flops of the shift registers. The arrows indicate the direction of the data shift between registers (cells). For simplicity, the clock and control logic are not shown. Figure 8.5a shows that there is access to the FIFO only at those registers that make up location no. 1 and no. 4. Location no. 1, known as the top of the FIFO, is used to input the datum, and location no. 4, known as the bottom of the FIFO, is used for outputing the datum. The rectangles identified as input and output data represent a datum being written into the FIFO and a datum being read from it. The implementation of these rectangles is most often accomplished with registers.

Some FIFOs are designed so that a datum written into an empty FIFO (no data have been previously stored) via the top of the FIFO memory location is automatically shifted by control circuitry (not shown in Figure 8.5a) to the bottom of the FIFO, as indicated in Figure 8.5b. Figure 8.5b indicates that the datum written into the FIFO (1101) was written into the top of the FIFO and then automatically shifted to the bottom of the FIFO. This action fills the bottom location and makes that datum available to be read at the output (location no. 4); hence the datum is first-in first-out. The next datum written into the FIFO (0011) is automatically shifted from the top of the FIFO to the next available memory location (an empty one), which is the location preceding the bottom location, that is, location no. 3, as depicted in Figure 8.5c. As data are written into the FIFO this process continues until all locations of the FIFO are filled, as illustrated in Figure 8.5d if 1010 and 1110 were written into the FIFO.

Another method of filling a FIFO is to shift the entered datum one location at a time. The datum written into this type of FIFO memory will push (shift) all subsequent data toward the bottom by one location. For a FIFO with four locations this would require that the four data entries be made before the first entry is available to be read at the bottom of the FIFO memory. For example, if the FIFO of Figure 8.5d had been filled using this technique, then the first datum written into the FIFO (1101) would have been written into the top of the FIFO but not automatically shifted to the bottom of the FIFO. Hence, location no. 1 has 1101 written into it and the remaining locations are empty. Next, 0011 is written into the top of the FIFO (location no. 1), which shifts 1101 from location no. 1 to location no. 2. Then location no. 1 contains 0011, location no. 2 contains 1101, and location no. 3 and no.

Figure 8.5

Accessing a FIFO memory.

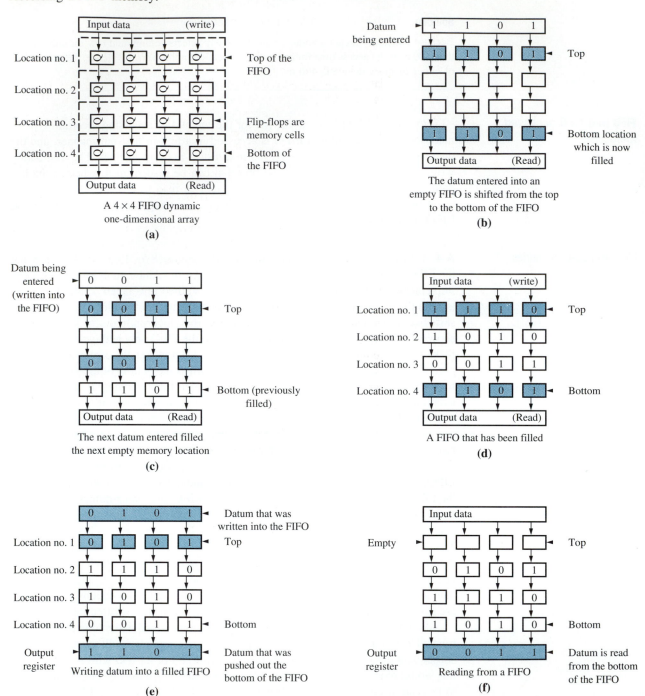

A 4 × 4 FIFO dynamic
one-dimensional array

(a)

The datum entered into an
empty FIFO is shifted from the top
to the bottom of the FIFO

(b)

The next datum entered filled
the next empty memory location

(c)

A FIFO that has been filled

(d)

Writing datum into a filled FIFO

(e)

Reading from a FIFO

(f)

4 are empty. The next datum, 1010, is written into the top of the FIFO (location no.
1), which shifts the two previous data down by one location. As a result, location no.
1 contains 1010 (the new datum written into the FIFO), location no. 2 contains 0011
(shifted from location no. 1), location no. 3 contains 1101 (shifted from location no.
2), and location no. 4 is empty. Finally, 1110 is written into the top of the FIFO and

all previous data are shifted down by one location, which results in the FIFO of Figure 8.5d.

The application dictates which FIFO is used; regardless of the technique, the most important point is that the first datum into the FIFO is the first datum that can be read from it.

Because the data are mobile in a FIFO memory and its organization is one dimensional, it is classified as a dynamic one-dimensional organization.

When a datum is written into an already filled FIFO memory, care must be taken to avoid losing data. This can be seen by comparing the previously filled FIFO of Figure 8.5d to the FIFO memory of Figure 8.5e. Figure 8.5e represents the FIFO memory of Figure 8.5d after 0101 has been written into it. Figure 8.5e shows that the datum at location no. 1 in Figure 8.5d has been shifted to location no. 2 and location no. 1 contains the newly entered datum. Likewise, the datum at location no. 2 in Figure 8.5d has been shifted to location no. 3 in Figure 8.5e. This shifting of data continues in the FIFO memory until the datum that was at the bottom of the FIFO (location no. 4) is shifted out of the FIFO, which means that this datum is lost if it is not read and saved by another device. It is important that a user of a FIFO-type memory realize that improper use of a FIFO (entering more data than there are locations) results in lost data unless that data are read and processed as they are pushed out the bottom of the FIFO.

Reading data from a FIFO memory shifts the datum at the bottom of the FIFO memory into the output register. The data in the other locations are then shifted down by one location, which empties the top location and makes it available to have a new datum written into it. This action is illustrated in Figure 8.5f, which assumes that the filled FIFO memory of Figure 8.5e is being read. The datum in location no. 4 of Figure 8.5e has been read (shifted to the output register), and the data at the other locations in Figure 8.5e have been shifted down one location in Figure 8.5f. The top location has been emptied as a result of reading this FIFO memory.

Example 8.4

Suppose that 1100 were written into the FIFO memory of Figure 8.5d. Beginning with memory location no. 1, give the resulting data in each location.

Solution

1100 is written into location no. 1, which is the top of the FIFO memory, and all previous data are shifted by one location. Hence, after 1100 has been written into the FIFO memory of Figure 8.5d, which is a filled FIFO memory, the content of location no. 1 is 1100, location no. 2 contains 1110, location no. 3 contains 1010, and location no. 4 contains 0011. The datum that was previously in location no. 4 (1101) was pushed out of the FIFO memory.

LIFO accessed memories

In contrast to a FIFO accessed memory is a *last-in first-out* (LIFO) accessed memory, that is, the last datum written in memory is the first that can be read from it. To illustrate the LIFO accessing technique, a representative simple dynamic one-dimensional 4 × 2-bit array is used, as shown in Figure 8.6. Figure 8.6a indicates that memory location no. 1 is the only accessible location in the array and will be used to input and output the datum to and from this LIFO memory. The bidirectional arrows between the shift register flip-flops indicate that data can be shifted in either direction, "down" or "up," where down is toward the bottom location and up is toward the top location. A datum is written (pushed) into the top location of a LIFO memory and all subsequent data are shifted one location down, as in a FIFO accessed memory. The datum read from a LIFO is shifted out the top location and all subsequent data are shifted one location up.

Figure 8.6

Accessing a LIFO memory.

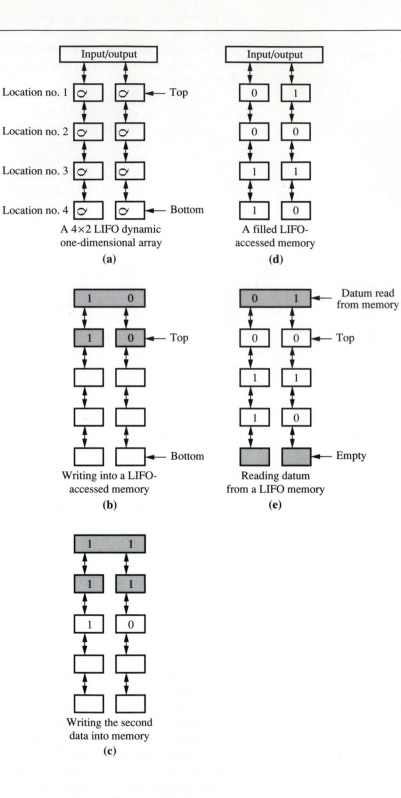

Location no. 1   ← Top

Location no. 2

Location no. 3

Location no. 4   ← Bottom

A 4×2 LIFO dynamic
one-dimensional array

(a)

Input/output

A filled LIFO-
accessed memory

(d)

← Top

← Bottom

Writing into a LIFO-
accessed memory

(b)

Datum read
from memory

← Top

← Empty

Reading datum
from a LIFO memory

(e)

Writing the second
data into memory

(c)

Figure 8.6b contains an example of writing data into a LIFO-type memory. The datum being written into memory is 10, and as illustrated, it is stored in the top location. Figure 8.6c illustrates that the second datum (11) written into memory is stored at location no. 1 and the previously stored 10 has been shifted to location no. 2. Data are written into memory in this manner until the array is filled, which is illustrated in Figure 8.6d. So far there have been no differences in operation between a LIFO and a FIFO, excluding those FIFOs that automatically shift the datum being entered from the top to the bottom location. The difference occurs when a datum is being read from a LIFO accessed memory.

When a datum is read from a LIFO memory, such as the memory shown in Figure 8.6d, the top location is the one that is read. The datum stored at that location is shifted out of memory, indicated in Figure 8.6e by showing that the previous contents of location no. 1 of Figure 8.6d are now shifted into the input / output register of Figure 8.6e. All subsequent data stored are shifted up one location when a read occurs, which can be seen by comparing the shifted data of Figure 8.6e with the original in Figure 8.6d. The last datum entered is the first that can be read from a LIFO memory, and also that reading data from the top location of a LIFO memory empties the bottom location, as indicated in Figure 8.6e.

As with a FIFO memory, if more data are written into a LIFO memory than there are memory locations, the excess data are shifted out the bottom location and lost, which requires the user or designer to keep track of the number of locations available and the amount of data to be stored. The jargon used to describe this situation is the statement that "the excess data was pushed (shifted) through the bottom (of memory) and lost in the bit-bucket," where the bit-bucket is imagined as a place that all bits to be discarded are "thrown."

Example 8.5

Determine the data in the LIFO of Figure 8.6d if 10 is first pushed on the LIFO and then 11 is pushed on the LIFO.

Solution

Pushing 10 on the LIFO results in data 10, 01, 00, and 11 being stored, with 10 (stored at the bottom location) being pushed out the bottom and into the bit-bucket. Pushing 11 on the LIFO next would result in data 11, 10, 01, and 00 in the LIFO, with 11 being pushed into the bit-bucket.

As will be seen in the study of microprocessors, memories that use FIFO and LIFO techniques of accessing data are utilized to implement microprocessor designs and applications. We shall also discover that FIFO-type memories are well suited for some I/O devices. The FIFO serves as temporary storage of data that are to be processed by the I/O device, such as when main memory sends data to a printer.

Review Questions

1. What is access time?
2. What is meant by RAM?
3. What is linear accessing?
4. How are FIFO and LIFO accessing similar and how do they differ?
5. How can data be lost in a FIFO and LIFO accessed memory?

8.5
Alteration of memory contents: read / write and ROM memory devices

In some memory devices, memory locations can be written to and read from by a function-block(s) within the system of Figure 8.1. A memory device with this characteristic is known as a *read / write memory* (R / W memory). When the contents of a memory cannot be altered by a function-block(s) within the system, that memory is known as a *read only memory* ROM.

The bits are written into a ROM by a manufacturing process or the action of an apparatus (a ROM programmer) that is not part of the system illustrated in Figure 8.1. The process of storing the bit patterns in each memory location of a ROM is known as programming the ROM. Because no function-block within the system of Figure 8.1 can write data into a ROM, reading its contents is the only operation that can be performed by the system — hence the term "read only memory."

There is an important fact to be remembered about R / W semiconductor memory devices. In the beginning of the evolution of the computer, main memories were often magnetic core memories, which were the only R / W memories that could be accessed as a RAM. For that reason these R / W memories were referred to as RAMs; hence the acronym RAM came to be synonymous with read / write memory. This is a misuse of the term "RAM," which actually indicates the accessing technique, whereas R / W indicates the alterability of memory contents. Nonetheless, reference to a RAM usually means R / W memory. We will follow convention and also use the acronym RAM to indicate an R / W memory, except when it would be clearer in meaning to use R / W.

Review Questions

1. What is an R / W memory?
2. What is a ROM memory?
3. When would one use an R / W memory rather than a ROM?
4. Is it correct to refer to a memory as being a RAM when an R / W memory is meant? Why?
5. In industry is the acronym RAM misused, and if so when?

8.6
ROM classifications

There are four types of semiconductor ROMs, each requiring a different method to program it. The method of programming a ROM is directly related to two essential considerations in the design of a memory system — the time required to program the ROM and the cost of the programmed ROM. In deciding which type of ROM to use, the designer must consider the cost of both the ROM itself and its programming, as well as the time required for it to be programmed, which can cause delays in project construction. The four types of ROMs are:

1. Mask ROMs (MROM).
2. Fuse link programmable ROMs (PROM).
3. Ultraviolet light erasable ROMs (EPROM), also termed UV-EPROM.
4. Electrically erasable ROMs (EEPROM).

Each has both positive and negative characteristics for a given application; the designer must weigh one characteristic against the others to make the optimal choice.

It should be noted that all modern ROM semiconductors are accessed as RAM, that is, all memory locations have the same access time even when locations are accessed at random.

MROM

A mask ROM (MROM) is programmed at the time of manufacture using a series of photo masks. The data stored is permanent and cannot be changed. The process can take weeks, and the cost is very high for just a few ROMs; however, for large quantities, such as in high-volume production, the process becomes very inexpensive. MROMs are suited to high-volume applications that have sufficient turnaround time to include the manufacturing process. This excludes MROMs from a research and development environment as well as low-volume production applications.

PROM

Programmable ROMs (PROMs) are manufactured so that they can be programmed by the user by means of a special apparatus known as a PROM programmer. When they are manufactured each memory cell has a logic 0 programmed into it. The PROM programmer supplies to those memory cells that are to be programmed in the logic 1 state a sufficiently large current to blow a fuse in the cell, resulting in an unalterable logic 1 state. Once programmed by the user, the content of PROM is permanent. The cost of a PROM is relatively low. The greatest disadvantages of PROMs are (1) the inability to alter its contents and (2) the time required to program a large number of PROMs. They are suited for applications with a medium-volume production. The trade-off between PROMs and MROMs is the time involved to program a PROM versus the high cost and long turnaround time for manufacturing a small number of MROMs. PROMs are not suited to research and development environments where programs are frequently changed.

EPROM (UV-EPROM)

Exposing the IC wafer of an erasable PROM (EPROM) to ultraviolet (UV) light for approximately 15 minutes erases the content of each memory cell (actually, it changes all memory cells to a logic 1). Once the contents of the EPROM have been erased, it can be reprogrammed using an EPROM programmer. EPROMs are easily recognizable by their quartz window that exposes the IC wafer. Once the EPROM has been programmed, this window is covered with opaque tape to block out any UV light.

EPROMs are higher in cost than PROMs, but because their contents can be erased and reprogrammed this cost is offset in environments in which the contents of ROM are frequently changed, such as in research and development. EPROMs are also well suited for very low-volume production, such as when one or two items are being produced.

EEPROM

Electrically erasable PROMs (EEPROMs) are the only ROMs that can be programmed by a function-block within the computer system (usually the CPU). The CPU can write data into an EEPROM, which erases the old data by overwriting it with the new data. One might question why a memory device with this characteristic is not classified as an R/W memory rather than a ROM. The reason is the time required for the CPU to write to an EEPROM memory location. It is relatively long (on the order of a millisecond); as a result, if an EEPROM were used to implement an R/W memory device in main memory, it would slow down the system considerably. Therefore, EEPROMs should be limited to applications that require very few write operations, such as back-up memory for power-sensitive R/W memory devices (*see* Section 8.7). For a back-up memory application, the contents of main memory are copied into an EEPROM or EEPROMs (depending on the number required for back-up) when an impending power failure is detected by the system.

EEPROMs are manufactured with two planes of memory within the IC: One plane is a typical R/W memory and the other duplicates the first plane in organization but is an EEPROM. Data can be copied from one plane to the other via command signals. For instance, when the CPU detects an impending power failure, it signals (via a pulse output to the EEPROM) that the memory contents in the R/W memory plane are to be written into the EEPROM plane. As a result, the data stored in the R/W memory plane will not be lost owing to a power failure. After power has been restored the process can be reversed, that is, the contents of the EEPROM can be copied into the R/W plane.

An EEPROM can also be viewed as an R/W memory with its own internal mass storage. Can EEPROMs therefore be used as mass storage devices? The answer is theoretically yes, but for most applications no, because the number of memory locations available in an EEPROM is relatively small (remember that there are two memories in one IC) and an EEPROM is not very portable (one would have to remove the chip from the system, vs. removing a disk or tape). However, *programmable logic controllers* (PLCs) do use them as a means to transport programs.

Because applications of EEPROMs are limited to read-mostly operations rather than R/W operations, they are sometimes referred to as *read-mostly memories*.

Review Questions

1. How many types of ROMs are there?
2. What are the advantages and disadvantages of each type of ROM relative to their possible applications?
3. Why might one imagine an EEPROM as two memories in one?

8.7
Memory volatility

Another way of categorizing memories is by their memory content retention when power is removed from the memory device. If the memory contents are not lost or altered owing to the removal and reapplication of power to the memory device, then the memory device is known as a *nonvolatile memory*. A memory that does not retain its contents when power is removed is known as a *volatile memory*. Because the contents of ROMs are permanent, they are nonvolatile memories. In contrast, the majority of semiconductor R/W memory devices are volatile. Of course, this is why R/W memory devices within a main memory must unload their content into mass storage, or an EEPROM, if that content is to be saved when power is removed from the system. It should be noted that some computers have a battery back-up to retain the contents of their volatile memory.

Another name for an EEPROM is *nonvolatile RAM*. The reason for this name is that it would normally function as an R/W (RAM) that can upon command transfer its data to a nonvolatile memory plane.

Mass storage memory devices (disk, magnetic tape, and bubbles) are also nonvolatile memories. However, they are not referred to as nonvolatile RAMs; rather, that term is reserved for semiconductor nonvolatile R/W memories.

Review Questions

1. What are the four types of ROMs?
2. Under what conditions would each of the four types be used?
3. Why are ROMs also RAMs?

8.8
Memory cell data retention: static and dynamic memories

The final category of memory is a characteristic associated with semiconductor R / W memory. It is characterized by the ability, or inability, of a memory cell to retain the stored logic level even with power applied. This characteristic is not to be confused with volatility — note the key words ''with the power applied.'' If a semiconductor memory cell can retain its logic level as long as power is applied to the semiconductor memory device, it is known as a *static memory cell* and the semiconductor memory device is known as a *static R / W memory device*, or *static RAM*. However, if the logic level of a memory cell is lost over a period of time (usually less than 3 ms), even with power applied, and therefore must be continually refreshed (before 3 ms has elapsed) to retain the logic level, then the cell is known as a *dynamic memory cell.* A memory device made up of these type of cells is known as a *dynamic R / W memory* or *dynamic RAM* (DRAM). The important thing to remember is that once data is written into a static RAM it will retain (remember) that data so long as power is applied to the semiconductor memory device, or until the data is changed, whereas a dynamic RAM will not retain its data beyond approximately 3 ms and therefore requires that the data be continually refreshed. Insofar as serving as a memory, there is no difference between a static and a dynamic memory device. The difference is the required refreshing of the data within the dynamic memory.

The interval of time when refreshing of the data occurs is known as a *refresh cycle*, and refreshing is usually performed and controlled by a refresh cycle controller (DRAM controller). Refresh cycle controllers are available as ICs and are discussed in Chapter 9.

The reason for the existence of dynamic memory devices is that they contain a larger number of memory cells than static memory devices. Of course, the disadvantage is the additional refresh cycle circuitry that is required.

The technology of a dynamic memory device basically involves the parasitic capacitance of a *field effect transistor* (FET) as the memory cell, greatly reducing the number of components needed to construct a memory cell. To write a logic 1 into a dynamic memory cell, the parasitic capacitor is charged to the logic 1 level; however, because parasitic capacitance has a large leakage current, the logic 1 level voltage quickly discharges (in approximately 3 ms), which is the reason for the required refresh cycle.

All memory devices discussed henceforth are assumed to be static unless stated otherwise.

Review Questions

1. What is the difference between static and dynamic memories?
2. Why are DRAMs sometimes preferred over static RAMs?
3. Why is it necessary to have a refresh cycle for DRAMs but not for static RAMs?

8.9
Addressing and read / write operation control

To read or write from or to memory or I/O, two events are required and must occur in the following order: (1) Locating the memory location or I/O device. (2) Controlling when the read or write operation is to take place. The first of these two events occurs as a result of addressing the memory location or I/O device. The second occurs as a result of applying the proper control signal (a read or write control signal) to the memory addressed or I/O device at the proper time. The control signal controls the actual read or write operation being performed by enabling the data output pin for read operations or serving as a strobe signal for write operations.

For example, if the CPU is to read a location in main memory, it must begin the process by applying the address of the memory location to main memory. Logic circuitry within main memory decodes that address and selects the appropriate memory location, as was done for addressing register arrays in Chapter 7. Once the addressed memory location has been selected, the CPU signals that memory location when to output its contents so that the CPU can read it. To provide the timing the CPU furnishes a read control signal (a write control signal when writing to memory) at the appropriate time after it has sent the address to memory.

The following sections will discuss how addressing and timing are implemented for various memory and I/O devices.

Review Questions

1. How is memory or I/O accessed?
2. Once memory or I/O has been addressed, how does the CPU control the read or write operation?

8.10

Semiconductor memory architecture

The sections covered so far have investigated memories in a rather general manner. It is now appropriate to become more specific. This section will cover semiconductor memory device architecture and its application to main memory will be covered in the next section.

As previously stated, to perform a read or write operation at a memory location, two events must take place. First, the address of the memory location to be accessed is sent by the CPU to memory. Second, the operation (read or write) that is to be performed must be initiated at the proper time.

Addressing semiconductor memories

From our study of two- and three-dimensional arrays we know that the address of a memory location within an array is the binary combination of the row and column. To access a memory location within a semiconductor memory, its address must be applied to the memory device and decoded by that device so that the appropriate row and column of the array are activated. The address decoder is an integral part of the memory device. The 8×2-bit three-dimensional memory device of Figure 8.7a will be used to illustrate this concept.

There are two decoders in Figure 8.7a, one to decode the two MSBs of the binary address (A_1 and A_2) and the other to decode the LSB (A_0) of the address. Decoding address bits A_1 and A_2 selects the row ($R0$, $R1$, etc.) by driving the appropriate row line to a logic high and decoding address bit A_0 selects the column ($C0$ or $C1$). *No cell can be read or written to electronically* (to be explained in Chapter 12) *without both its row and column select lines being active (high in this case)*. From the truth table of Figure 8.7b, address 000 will select row $R0$ and column $C0$. Using an *x-y* coordinate system, $R0$ and $C0$ will select memory location 0H (as before, an H is used to indicate base 16). Location 0H contains two memory cells, 0 from plane 0 (shaded) and 0-1 from plane 1, as indicated in the truth table. Address 001 selects row $R0$ and column $C1$, which in turn selects memory location 1H. Memory location 1H contains memory cells 1 and 1-1, which can be verified from the truth table of Figure 8.7b. Using this same scheme, the remainder of the truth

Figure 8.7

Address decoding of an
8 × 2 three-dimensional
array.

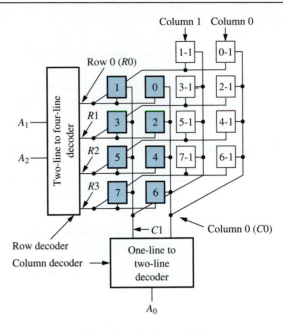

Address decoding for an 8 × 2 array

(a)

Address $A_2A_1A_0$	Row (A_2A_1)	Column (A_0)	Memory location in hex	Memory cells at location
0 0 0	R0	C0	0	0, 0-1
0 0 1	R0	C1	1	1, 1-1
0 1 0	R1	C0	2	2, 2-1
0 1 1	R1	C1	3	3, 3-1
1 0 0	R2	C0	4	4, 4-1
1 0 1	R2	C1	5	5, 5-1
1 1 0	R3	C0	6	6, 6-1
1 1 1	R3	C1	7	7, 7-1

Truth Table for Figure 8-7(a)

(b)

table is self-explanatory. It is important to realize that both the row and the column of
the memory cell must be selected (activated) before any read/write operation can
take place within that cell.

There is an equation that can be used to calculate the number of memory
locations within a memory device:

$$N = \text{number of binary combinations} = 2^n \qquad (8.1)$$

N is the number of binary combinations for n independent variables (n inputs). For
this application, N is the number of memory locations for n address pins. The actual
binary addresses of those locations are the N binary combinations possible. Applying
this above statement to the memory array of Figure 8.7a, $n = 3$ because there are

three independent variables A_0, A_1, and A_2. Setting $n = 3$ in Equation 8.1 we find that

$$N = 2^3 = 8.$$

There are eight memory locations and the actual addresses of these locations are the eight binary combinations possible. To verify this refer to the truth table of Figure 8.7b and find that indeed under the address column there are eight consecutive binary numbers, beginning with 000, the first address, and ends with the last address 111.

Example 8.6

From the truth table of Figure 8.7b and from what has been previously stated, the hexadecimal memory location number and the binary address are equivalent values.
1. What is the binary address for memory location
   (a) 2H   (b) 5H   (c) 7H?

2. What is the hexadecimal address for binary address
   (a) 001   (b) 101   (c) 110?

Solution

We convert hexadecimal values to binary values.
1. (a) 2H $= 010_2$ (b) 5H $= 101_2$ (c) 7H $= 111_2$

Now convert binary values to hexadecimal values.

2. (a) $001_2 = $ 1H   (b) $101_2 = $ 5H   (c) $110_2 = $ 6H

Verify these addresses and memory location values using the truth table of Figure 8.7b.

System bus architecture and buffering of the data bus

Now that the concepts of addressing a memory location within a semiconductor memory device via its address are understood, the concepts of controlling the read/write operation to be performed on that addressed location will be covered next. Read/write operations must be synchronized with the CPU for it is in control of performing the read/write operation. This synchronization is provided by read and write control signals that are generated and output by the CPU.

To develop the concepts associated with the control of read/write operations and broaden those addressing concepts already learned, reexamine the computer architecture of Figure 8.1. Figure 8.8a, is a modified version of Figure 8.1 in which:

1. The connecting lines between function-blocks have been expanded into busses, resulting in a three-bus architecture (address, data, and control busses)

2. An I/O selector is added, which is the I/O address decoder.

3. Mass storage is omitted because it has little to do with the concepts being developed here

Because Figure 8.8a has three busses, the term "bus" needs definition and explanation. A bus is a line (a wire or other low-resistance medium) or lines used to electrically connect two or more devices in parallel. For example, the three devices of Figure 8.9 have four connections so that the three devices are in parallel. These four lines form a 4-bit bus.

If any two or more of the devices of Figure 8.9 were to use the bus at the same time by outputting data on it, there would be competition among the devices for use of the data bus, or more simply stated, bus contention would exist. Bus contention

Figure 8.8

(a) A simplified computer architecture with a three-bus structure; (b) timing for a memory read operation; (c) timing for a memory write operation.

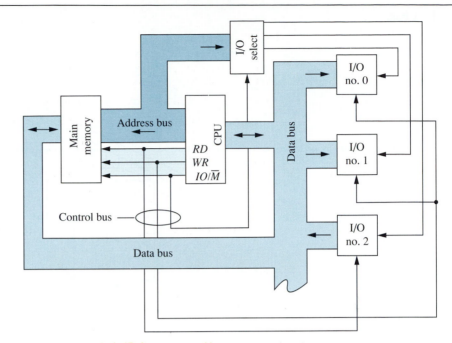

A simified computer architecture with a three-bus structure.

(a)

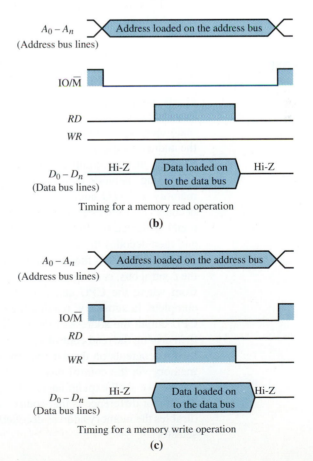

Timing for a memory read operation

(b)

Timing for a memory write operation

(c)

Figure 8.9

A 4-bit bus.

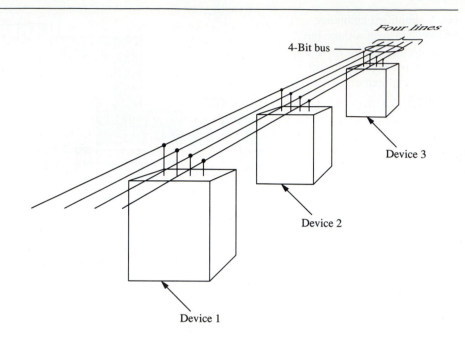

must be avoided. As stated in an earlier chapter, bus contention results in unintelligible data on a bus while it is in the state of contention (imagine the result if each device contending for the bus loaded a different binary bit pattern on it at the same time — a conflict would exist).

Read/write control

In the process of read/write operations, the data bus is used for transmitting data to and receiving data from different function-blocks within a system. The data bus of Figure 8.8a connects the CPU, main memory, and I/O devices. Consider the sequence of operations that occurs when the CPU fetches an instruction or data from main memory, as shown in Figure 8.8b. The CPU begins the process by outputting the address of the memory location containing the instruction or data on the address bus and driving control signal $IO/\overline{M}$ to the appropriate logic level, a low for addressing memory or a high for addressing I/O. $IO/\overline{M}$ differentiates between memory and I/O addresses by enabling either main memory or the I/O address decoder (I/O selector) of Figure 8.8a. When $IO/\overline{M}$ is driven low, the semiconductor memory device of main memory (which contains that memory location) is enabled and then decodes the address on the address bus and selects the addressed location within it. The CPU then sends a read control signal (RD) to the memory device via the control bus, causing the addressed location to transfer its contents to the data bus, from where the CPU can read it. The reading of that memory location is now complete. In summary, reading the contents of a memory location requires that the CPU output the address of the location on the address bus, driving $IO/\overline{M}$ low, and then driving the read control signal RD high. RD will cause the addressed location to load its contents on the data bus from where the CPU reads it. RD is transmitted to memory via the control bus.

As we shall explain later, reading an I/O device is similar to reading memory. The major difference is that control signal $IO/\overline{M}$ is driven high by the CPU, which disables the memory address decoders and enables the I/O address decoder (the I/O selector).

For the CPU to write to the main memory of Figure 8.8a, the CPU again begins the operation by outputting the address of the memory location on the address bus and driving control signal $IO/\overline{M}$ low as illustrated in Figure 8.8c. The CPU then outputs the data to be stored on the data bus. The CPU next generates the write (WR) control signal and transmits it to main memory via the control bus. This causes the addressed semiconductor memory device to latch that data from the data bus and store it in the addressed memory location.

Just as a point of interest, for the CPU to write to I/O, the process is similar to writing to memory except that the control signal $IO/\overline{M}$ is driven high by the CPU. As previously stated, with $IO/\overline{M}$ high main memory is disabled and the I/O address decoder is enabled, so that the address on the address bus selects an I/O device.

Figure 8.8a shows that the data bus is shared among many devices (main memory, CPU, and three I/O devices). Because it is shared, the potential for bus contention exists. To prevent bus contention, all devices that have outputs connected to the data bus must have their outputs buffered with tristate logic. With buffered outputs, only the addressed memory location or I/O device has its buffers on (enabled). This electronically connects those outputs, and only those outputs, to the data bus. The other output buffers are disabled (high-impedance [Hi-Z]), thus electronically isolating those outputs from the data bus. The enabling and disabling of a device's output buffers are done with control signals ($IO/\overline{M}$, RD, and WR).

Having reviewed the fundamental concepts of a data bus and bus contention, the following will show how R/W semiconductor memory devices achieve buffering from the data bus as well as the control of read/write operations. Figure 8.10 is a

Figure 8.10

Read/write control logic of an R/W memory location.

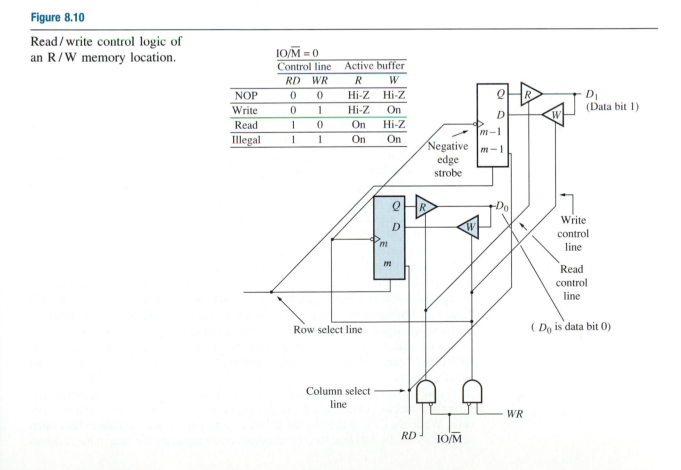

representative example of a 2-bit memory location that has memory cell m in plane 0 and memory cell m-1 in plane 1. It will be used to illustrate the concepts of buffering and the control of read/write operations. Figure 8.10 uses D-type flip-flops to implement the memory cells. There are a variety of semiconductor memory cells that could be used, but from a conceptual point of view, the D-type flip-flop can be used as a general representation. Consider the input D as the means to write data into the memory cell and the output Q as the means to read the data stored in the memory cell. As a convenience the D inputs are not drawn in the conventional location in Figure 8.10. Take note of the row and column select lines. These are similar to the row and select lines of Figure 8.7. Recall that for a cell to be read or written to both its row and column select lines must be active.

The memory location of Figure 8.10 is one representative location from an $N \times$ 2-bit memory, where N is the number of memory locations within the array. For example, if the memory location of Figure 8.10 is to represent a memory location from the array of Figure 8.7a, then $N = 8$ and m is a number from 0 to 7. Specifically, if the memory location of Figure 8.10 is to represent memory location 010 of Figure 8.7a, then $m = 2$. However, if it is to represent memory location 101 in Figure 8.7a, then $m = 5$.

Figure 8.10 has two tristate buffers for each memory cell labeled ''R'' and ''W.'' The R buffers are for read operations and the W buffers are for write operations. The R buffers provide buffering between data bus lines D_0 and D_1 and the Q output of the memory cell flip-flops. The W buffers buffer these same data bus lines from the memory cells flip-flops D inputs.

To understand how the state of these buffers is controlled, refer to the AND gates of Figure 8.10 and the accompanying truth table. As previously discussed, when the CPU is addressing a memory location it outputs the address on the address bus, which activates the row and column lines of the addressed cell, and also the CPU drives the control signal $IO/\overline{M}$ low. The requirement for $IO/\overline{M}$ to be low is stated at the top of the truth table. The logic circuit in Figure 8.10 shows that control signal $IO/\overline{M}$ being low enables both AND gates. If the operation being performed by the CPU is a read, the operation indicated in the third row of the truth table, the CPU activates the read control signal RD by driving it to a logic 1. As shown in Figure 8.8b, the CPU drives RD to a logic 1 after the address has been placed on the address bus and keeps it a logic 1 for a sufficient time necessary for it to complete the read process. Coincidental with driving RD to a logic 1 the CPU will also drive the write control signal WR inactive (low) during this period. The read AND gate of Figure 8.10, to the left, outputs a logic 1 while RD is high and the write AND gate, to the right, outputs a logic low. These AND gate output logic levels will turn on the R buffers and put the W buffers in the Hi-Z state. While the R buffers are on, the logic states of the memory cells (outputs Q) are loaded on to the data bus from where the CPU can read (latch) them. This completes a read operation.

If the CPU were doing a write operation, the conditions shown in the second row of the truth table in Figure 8.10, the CPU would drive RD low (inactive) and WR high, as is illustrated in Figure 8.8c. The timing for driving these control signals to the logic levels indicated would be the same as it was for the read operation, after it loaded the address of the memory location on the address bus. With RD low the R buffer AND gate will output a logic 0, putting the R buffers in the Hi-Z state and isolating the memory cells Q outputs from the data bus. And with WR being active (high) during this period the W buffers are on. With the W buffers on, whatever data is on the data bus lines D_0 and D_1 is also being input to the D inputs of the memory cells. While the CPU is holding the W buffers on, it also loads the data to be written to memory on the data bus, thereby simultaneously inputting the data to the D inputs of the memory cells. The CPU then drives the *WR* control signal inactive (from a logic 1 to a logic 0) before removing the data being written to memory from the data bus. The transition of *WR* from a logic 1 to a logic 0 supplies the needed negative edge trigger for the memory cell flip-flops to latch the data that is on the data bus, which completes the write operation.

It is of interest to take note of the last row of the truth table of Figure 8.10. It states that for both *RD* and *WR* to be a logic 1 at the same time is an illegal condition because it is impossible to execute a read and write at the same time. If an microprocessor (CPU) would output *RD* and *WR* a logic 1 at the same time the CPU has failed and needs to be replaced.

We now understand how read / write operations are controlled by the CPU via its control signals — $IO/\overline{M}$, *RD*, and *WR*.

In summary, the read (*R*) buffers of Figure 8.10 buffer the outputs of the memory cells (*Q* of the flip-flops) from the data bus lines D_0 and D_1. The state (on or off) of these buffers is controlled by the signal *RD*. From a systems point of view, the *RD* control signal turns on the *R* buffer(s) of the addressed memory device and turns off all other memory device *R* buffers, thus preventing bus contention. While the write (*W*) buffers of Figure 8.10 are not related to bus contention (recall that only the outputs of multiple devices connected to a common bus can cause bus contention), they do reduce loading problems. A memory that is constructed with memory devices with buffered inputs reduces the loading of the data bus because a data bus line sees the input loading of only a single *W* buffer rather than the *D* inputs of all the memory cells.

To expand the number of memory locations of Figure 8.10 into a three dimensional 8 × 2 R/W (RAM), as illustrated in Figure 8.11, duplicate the basic 2-bit single memory location of Figure 8.10 into two planes of eight identical locations. The result is the same memory of Figure 8.7a. Memory cells 0 to 7 are in plane 0 and cells 0-1 to 7-1 are in plane 1. Memory cells 0 and 0-1 form memory location 0, cells 1 and 1-1 form location 1, etc. Data is input and output via the I/O pins I/O_0 and I/O_1, connected to the data bus lines D_0 and D_1, respectively. The memory locations of Figure 8.11 are accessed in exactly the same manner as the 8 × 2 array of Figure 8.7a. Each memory cell has a row and column line going to it, and both must be selected before the cell can be accessed for either a read or write operation (the electronic principles are explained in Chapter 12). To simplify Figure 8.11, the row select lines (row decoder outputs R_0, R_1, R_2, and R_3) and the column select lines (C_0 and C_1) are not shown connected to the memory cells.

The memory cell *Q* outputs of each memory plane in Figure 8.11 are connected to a single R buffer. However, only the addressed cell from each plane will actually output its *Q*'s logic level to the R buffer for its plane. The write and read control hardware are structured similarly. There is one W buffer for each plane of memory cells. The output of the W buffer appears at the *D* input of each memory cell in that plane. However, only the addressed memory cell is active and is capable of latching the data at its *D* input. The actual latching of the data occurs when the control signal *WR* forms a negative edge by going inactive (from a logic 1 to a logic 0).

To see why Figure 8.11 was chosen to represent the architecture of semiconductor R/W memories, briefly refer to Figure 9.4 of Chapter 9. Figure 9.4 is a manufacturer's data sheet for the architecture of a 2148 semiconductor memory device. Comparing Figures 8.11 and 9.4, on the data pins (I/O_1 to I/O_4) of Figure 9.4 are four *R* and *W* buffers, but they are not labeled. The *R* buffers input data from the I/O pins to the input data control function-block. The *W* buffers output data from that function-block to the I/O pins. Also not shown are the four planes of memory cells that are contained within the function-block identified as the memory array.

Figure 8.11

An 8 × 2 R/W memory device.

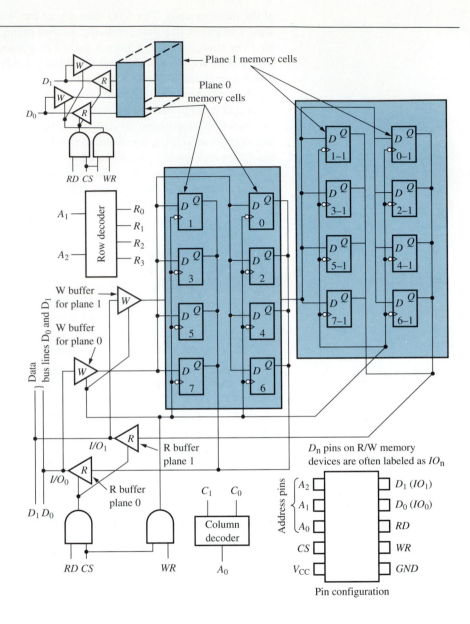

There are four planes of memory cells because a plane of memory cells exists for each bit within a memory location, and, because this memory has 4 bits per location, there must be four planes of memory cells. The memory array function-block of Figure 9.4 has 64 rows and 64 columns. To easily verify that there are 64 rows use Equation 8.1; for six row address pins (A_4 to A_9) there are 64 binary combinations and hence 64 rows. There are four column address pins (A_0 through A_3) and hence 16 columns (2^4). However, because there are four planes of memory cells and each plane has 16 columns, there are a total of 64 columns. It is hoped that this comparison has justified our conceptual representation of semiconductor memories, because industry has a similar representation.

The input *CS* pin (chip select) of Figure 8.11 enables and disables the semiconductor memory device. The logic gates of Figure 8.11 shows that if the level of input *CS* is a logic 0, then both *AND* gates are disabled (outputting logic 0's) and the *R* and

W buffers are in the Hi-Z state. As a consequence, the memory device is electronically isolated from the data bus lines D_0 and D_1. However, if input CS is active (a logic 1), then the memory cells of the addressed memory location can be read (if RD = 1) or written into (if $WR = 1$). Compare the *AND* gates of Figures 8.11 and 9.4 to verify their similarities in function. There are no conceptual differences between the memory IC of Figure 8.11 and the 2148 of Figure 9.4, however there are some differences in implementation. The most obvious is that the 2148 uses the logic 0 level as the active level for chip select $\overline{CS}$. Also there is a single pin to implement read and write operations. Pin $\overline{WE}$ functions as both the read and write pin. When low the 2148 performs a write operation and when high it performs a read on the addressed location.

The R/W operations of Figure 8.11 are basically the same as those of the memory location illustrated in Figure 8.10. If $CS = 1$, $RD = 1$, and $WR = 0$ (a read operation), the R buffer of each plane of Figure 8.11 is on, and the output of each addressed memory cell (recall that only an addressed cell with both row and column lines active can be accessed) in both planes 0 and 1 is electronically connected to data bus lines D_0 and D_1 via their respective R buffers. This loads the data bits of those cells onto the data bus. Once the data is on the data bus it can be read by the CPU.

If a write operation is to be performed then $CS = 1$, $WR = 1$, and $RD = 0$. When input WR is driven high, it activates the W buffers of both planes that electronically connect data bus lines D_0 and D_1 to the D inputs of each addressed memory cell in planes 0 and 1. However, the data lines are loaded (load capacitance and any input current requirements) with the input of the W buffer only for each plane rather than the D inputs of every memory cell (eight in this case) in its plane. Even though the data bit is at the D input to every memory cell within the plane, only one memory cell in each plane can be written into and it is the one being addressed (both row and column lines are active). When WR creates a negative edge by going inactive (1 to 0), the data at the D inputs of the addressed memory location is strobed into those memory cells and the write operation is completed.

Example 8.7

For the stated logic levels of the indicated input pins of Figure 8.11, determine the memory location being accessed and the operation to be performed.

1. $CS = 1$, $A_2 = 0$, $A_1 = 1$, $A_0 = 0$, $RD = 1$ and $WR = 0$

2. $CS = 1$, $A_2 = A_1 = A_0 = 1$, $RD = 0$ and $WR = 1$

3. $CS = 0$, $A_2 = A_1 = A_0 = 1$, $RD = 0$ and $WR = 1$

4. $CS = 1$, $A_2 = A_1 = A_0 = 1$, $RD = WR = 1$

Solution

1. Because $CS = 1$, the memory device is enabled. The address input to the memory device is 2H, which accesses memory location 2H (memory cells 2 and 2-1). $RD = 1$ and $WR = 0$ drive the W buffers into the Hi-Z state and turn the R buffers on, which reads the data bits from memory cells 2 and 2-1 onto data bus lines D_0 and D_1, respectively.

2. The memory device is enabled ($CS = 1$) and the accessed memory location is 7H. Because $RD = 0$ and $WR = 1$, a write operation is to be performed on memory cells 7 and 7-1; hence, the data from data bus lines D_0 and D_1 is written into these cells when the negative edge of WR occurs (WR goes from 1 to 0).

3. Because $CS = 0$, the memory device is disabled (R and W buffers are in the Hi-Z state) and no read or write operation can be performed.

4. The memory device is enabled, but because $RD = WR = 1$, an illogical operation (read and write at the same time) is requested from the device by the CPU. The data will be read from the memory cell and then, on the negative edge of WR, that data will be written back into the cell. If the CPU drove $RD = WR = 1$, then the CPU would be malfunctioning and must be replaced.

If the memory device of Figure 8.11 actually existed, the pin configuration of its dual in-line package (DIP) might appear as shown.

There are applications for which the designer or manufacturer of a memory device is not concerned with loading of the data bus lines for a write operation, such as if the array is small. Under this condition the designer or manufacturer would eliminate the W buffers.

Conceptually a ROM has the same logic circuits for addressing and reading memory locations as the R/W memory device of Figure 8.11. Obviously, for a ROM there is no write control circuitry. A memory location for ROM might appear as illustrated in Figure 8.12. The memory cells are represented as flip-flops without the ability to have their logic levels changed (the D input and strobe input have been eliminated). From previous discussion it is known that the memory location of Figure 8.12 is first accessed via an address that selects the row and column of the memory cells within that memory location. This activates memory cells m and m-1 of that addressed memory location. After the memory cells have been activated, control signal RD (on a ROM this input is often labeled "OE") is applied to the R buffers, which turn them on. When the R buffers are on, the data bits from the memory cells m and $m - 1$ are loaded on data bus lines D_0 and D_1, respectively.

Figure 8.12

Read control logic of a
ROM memory location.

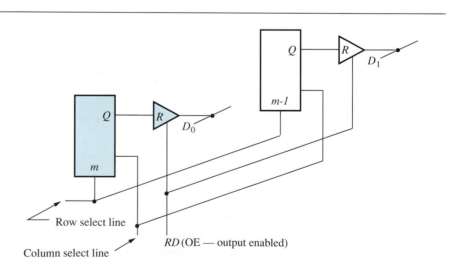

To incorporate the memory location of Figure 8.12 into an $N \times 2$ ROM, use reasoning similar to that used to develop the R/W array of Figure 8.11. In fact, removing the W buffers, the input WR, and the write control *AND* gate would convert it into a ROM.

The memory cell array organization of any memory device can be configured in a variety of ways, as was discussed in Section 8.3. From this discussion and Equation

8.1, the address pins (A_0, A_1, A_2, etc.) can be distributed in any fashion to address row and column, but their sum when applied to Equation 8.1 equals the number of memory locations. If the memory device array of Figure 8.11 were organized such that there were two rows and four columns, then the row decoder would have a single input A_2 and the column decoder would have two inputs, A_1 and A_0. With three address pins there are eight memory locations, according to Equation 8.1. Manufacturers often organize the array of a semiconductor memory device so that the number of row-related address pins is as close as possible to the number of column-related address pins. This minimizes the access time of the semiconductor memory device because the burden of address decoding is evenly distributed between row and column decoders.

Review Questions

1. For all read/write operations two events must occur. Name the events and place them in the order in which they must occur.
2. What is the essence of an *x-y* coordinate accessing scheme?
3. How can one calculate the number of binary combinations for any number of inputs and also determine what those combinations are?
4. What purpose do the three busses of Figure 8.8 serve?
5. Why is it that only the CPU furnishes addresses, as indicated in Figure 8.8?
6. Why is it necessary to buffer all memory cell outputs? While it is not essential, why is it desirable to buffer the inputs of memory cells?
7. Before a memory cell can be either read from or written to, what condition must exist?
8. What purpose does the *CS* input serve for both ROM and R/W memory devices?
9. How are read/write operations performed on R/W memory devices?
10. How are read operations performed on ROMs?
11. How many address pins are there on a 4096 $\times$ 8 semiconductor memory IC?
12. For a five-address pin IC, how many pins are used to address rows and how many to address columns of the memory cell array?

8.11

Introduction to the fundamental concepts of main memory design

This section will focus on the design concepts of main memory systems. As a review, recall that any time the CPU wishes to access main memory, it must first load the address of the memory location to be accessed onto the address bus and then drive control signal $IO/\overline{M}$ to a logic 0. That address input to the semiconductor memory device via its address pins is decoded by the memory IC and as a result locates the desired memory location. Once the memory location has been selected, the CPU must generate the appropriate control signal (*RD* or *WR*) to perform the intended read or write operation.

Addressing techniques

There are two basic types of addressing, *fully decoded* and *linear addressing*. In a fully decoded addressing scheme the address lines of the address bus are decoded. This scheme maximizes the number of addresses available. For instance, if eight

Figure 8.13

Main memory addressing
schemes.

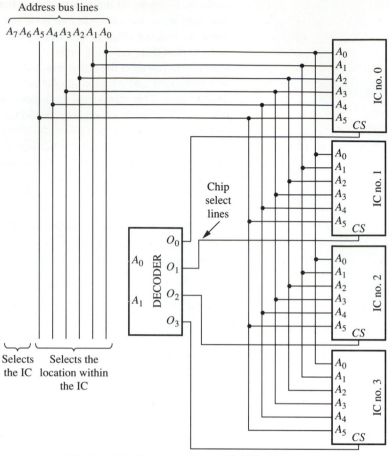

A fully decoded addressing scheme with 256 addresses

(a)

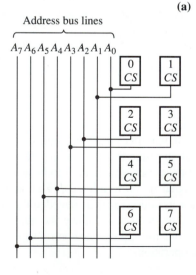

A linear addressing scheme

(b)

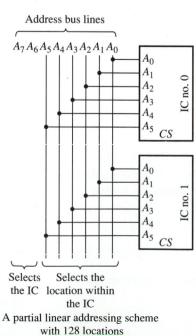

A partial linear addressing scheme
with 128 locations

(c)

address lines were used to address memory locations and all eight were decoded, there would be $2^8 = 256$ memory addresses available, according to Equation (8.1). However, if two of those address bus lines are not decoded, leaving six that are decoded, the number of available addresses is reduced to $2^6 = 64$ (this is an important concept, especially when comparing fully decoded addressing to partial linear addressing).

Figure 8.13a illustrates the concept of a fully decoded addressing scheme. Address bus lines A_6 and A_7 are applied to the 1-of-4 decoder, and the outputs of the decoder are used to select a semiconductor memory IC via the IC's *CS* pin. The remaining address bus lines are decoded by the row and column decoders internal to each IC, and as a result the addressed memory location within the IC is selected via the *x-y* coordinate system previously discussed. Because each address line is decoded, the memory system of Figure 8.13a is a fully decoded addressing scheme with 2^8 (256) memory locations. To verify this, each memory IC has 64 (2^6) locations. Because there are four ICs then there is a total of 256 (4 × 64) locations.

From Figure 8.13a it can be conclude that for a fully decoded addressing scheme, those address lines serving as inputs to the decoder select an IC, whereas the address bus lines serving as inputs to the address pins of the ICs select the specific memory location within the selected IC. Being able to address any location within a memory IC, as well as addressing a specific IC, allows the addressing of any location in the memory system.

Linear addressing refers to a method of addressing where the address bus lines are not decoded, but instead are used to select a semiconductor memory device. Figure 8.13b illustrates the concept of linear addressing. Each address line is connected to the *CS* of an IC. Hence, whenever the connecting address line has a logic 1 on it, the corresponding IC is selected. There is no linear method of selecting the memory location within the selected IC, because, as previously discussed all semiconductor memory devices use a decoding process for internal memory location selection. As a result, linear addressing by itself cannot be used to address main memory locations. However, as will be shown, linear addressing can be used for addressing I/O.

Fully decoded and linear addressing can be combined to produce a usable scheme for addressing main memory, which will be referred to as *partial linear addressing*. A partial linear addressing scheme uses those address bus lines not decoded by the memory ICs to select and enable the memory IC being addressed. Figure 8.13c illustrates a partial linear addressing scheme. Address bus lines A_0 through A_5 are used to select a memory location within an IC, and address lines A_6 and A_7 serve to select the IC that contains the location being addressed. With this addressing scheme there are a total of 128 locations, 64 (2^6) locations per IC and there are two ICs.

Comparing fully decoded addressing to partial linear addressing, it can be seen that a fully decoded scheme yields more addresses for a given number of address lines, whereas partial linear addressing simplifies the hardware (no decoder). The addressing technique used in a design depends on the required memory capacity of the design and the ability to expand its capacity versus the desire to minimize design cost.

Memory organization

In a fully decoded addressing scheme, memory addresses are segmented according to the device within which they reside. As an example of this segmentation, consider the address segmentation of Figure 8.14. It is composed of four IC memory devices (ROM-0, ROM-1, RAM-2, and RAM-3), each having four address pins (A_0 to A_3), resulting in 16 memory locations per memory device ($2^4 = 16$). With four such

Figure 8.14

A fully decoded addressed main memory system.

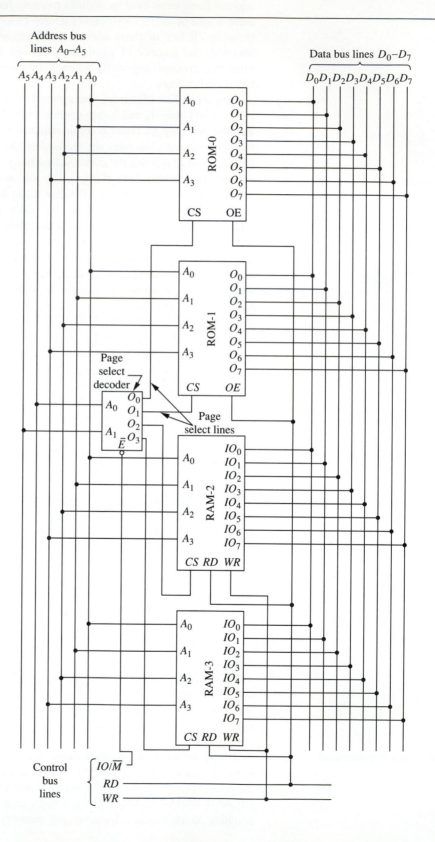

memory devices there are a total of 64 memory locations in this memory system. The number of connections from the outputs of each memory device (O_n for ROM and IO_n for RAM, where RAM is to mean R/W memory) to the data bus lines indicates the bit size of each memory location, which in this case is eight. Therefore, this is a 64×8 memory.

The 64 memory locations of Figure 8.14 can be divided into four equal blocks, one block of memory per IC, with each block of memory locations being defined as a page of memory. For the memory of Figure 8.14 a page of memory is 16 memory locations. It should be mentioned that often the contents of main memory are analogously viewed as made up of many pages, much like a book, and the contents of each IC memory device comprise one or more pages of that book. This is the reason for the notation page *n* of Table 8.2 and page select lines of Figure 8.14.

Table 8.2 Decoding the address bus of Figure 8.14

						A_5	A_4	A_3	A_2	A_1	A_0
Address bus lines											
Address pins of the ROM and RAM ICs								A_3	A_2	A_1	A_0
		Page select decoder		*Output on*		A_1	A_0				
ROM-0 addresses	Page 0	0	0H	$O_0 = 1$		0	0	0	0	0	0
		0	1H			0	0	0	0	0	1
		0	2H			0	0	0	0	1	0
		0	3H			0	0	0	0	1	1
		0 $\vdots$	$\vdots$			$\vdots$	$\vdots$	$\vdots$	$\vdots$	$\vdots$	$\vdots$
		0	FH			0	0	1	1	1	1
ROM-1	Page 1	1 $\vdots$	0H $\vdots$	$O_1 = 1$		0 $\vdots$	1 $\vdots$	0 $\vdots$	0 $\vdots$	0 $\vdots$	0 $\vdots$
		1	FH			0	1	1	1	1	1
RAM-2	Page 2	2 $\vdots$	0H $\vdots$	$O_2 = 1$		1 $\vdots$	0 $\vdots$	0 $\vdots$	0 $\vdots$	0 $\vdots$	0 $\vdots$
		2	FH			1	0	1	1	1	1
RAM-3	Page 3	3 $\vdots$	0H $\vdots$	$O_3 = 1$		1 $\vdots$	1 $\vdots$	0 $\vdots$	0 $\vdots$	0 $\vdots$	0 $\vdots$
		3	FH			1	1	1	1	1	1

Hexadecimal addresses   Page select   Memory location address within the memory device

If we were to determine the hexadecimal address for each of the 16 memory locations within each block of memory or memory device of Figure 8.14, we would find that the addresses would begin at 0 and end with F, as was found for Figure 8.3b (*see* Table 8.1). To assign a unique address to each memory location within the memory system of Figure 8.14, assign page numbers to each memory device, such as page 0 for ROM-0, page 1 for ROM-1, etc., and then prefix all memory locations within a memory device with the page number of that memory device. (*Note*: The suffix "H" is used to identify all hexadecimal numbers.) Memory address 0FH is memory location FH within page 0 (ROM-0), memory address 1FH is memory location FH within page 1 (ROM-1), memory address 2FH is memory location FH within page 2, and memory address 3FH is memory location FH within page 3. These addresses can be verified by referring to Table 8.2.

Table 8.2 can serve as both a wiring table and a truth table for the address bus lines of Figure 8.14. The wiring connections are shown in the first three rows of the rightmost six columns. The truth table is formed by the remaining rows under those six columns. The top row of Table 8.2 has the six address bus lines listed (A_0 through A_5). The next two rows are address pins for the ROMs and RAMs of Figure 8.14 as well as the page select decoder. These IC address pins are connected to address bus lines in their respective columns, as indicated in Table 8.2. The second row of Table 8.2 states that address pins A_0 through A_3 of the ROMs and RAMs are connected to the corresponding address bus lines, whereas the third row of Table 8.2 states that address pins A_0 and A_1 of the decoder of Figure 8.14 are connected to address bus lines A_4 and A_5, respectively.

As previously stated, in addition to indicating address bus line connections to device pins, Table 8.2 also serves as a truth table for the address bus lines of Figure 8.14. That is, all possible binary combinations shown are addresses of memory locations. Of course, it is not necessary to show each combination to understand the concept; therefore, for the sake of convenience, many are not shown and are instead represented by dots. The binary address of each memory location is the binary combination of each of these rows. The corresponding hexadecimal address is shown in the column identified as the hexadecimal addresses. The page number for each block of memory addresses is shown in Table 8.2, as well as the specific IC (ROM-0, ROM-1, RAM-2, and RAM-3) containing those addresses. The page number corresponds to the MSD of the hexadecimal address. Because there are four pages (0 to 3), each page contains 16 addresses (0H to FH) that can be verified by viewing the LSD position of the hexadecimal address. The column labeled "O_n" is the page select column and shows which output of the page select decoder is active for any given address (logic 1 — the other outputs are at a logic 0 state).

For any fully decoded addressing scheme, it is the page number prefix that is used to logically distinguish one page of addresses from another. The page number prefix of a memory address is decoded by the page select decoder, and the resulting active output of that decoder is used to enable the addressed page of memory via the CS input of the memory device. Refer again to the main memory system of Figure 8.14 for an example of this concept. The selection of memory pages is accomplished by connecting address bus lines A_4 and A_5 to inputs A_0 and A_1 of the page select decoder, which decodes the logic levels on address lines A_4 and A_5 and then activates the appropriate output (O_0 through O_3). This output is used to enable the selected memory device by driving its CS pin high. As can be seen from Table 8.2, if address bus lines A_5 and A_4 have logic levels 00 on them, the page select decoder activates its O_0 output, which would select ROM-0 (page 0), whereas if logic level 01 is on these address bus lines, the decoder activates its output O_1 and ROM-1 is selected. Similarly, if logic levels 10 are on address bus lines A_5 and A_4, RAM-2 is selected, and if logic levels 11 are on these address bus lines, RAM-3 is selected.

Figure 8.14 shows that the page select decoder is enabled via its enable pin $\overline{E}$ that is driven by control bus line $IO/\overline{M}$. Recall that the CPU uses $IO/\overline{M}$ to isolate (partition) memory addresses from I/O addresses. If the address output by the CPU is a memory address the CPU drives $IO/\overline{M}$ low and for I/O addresses it drives $IO/\overline{M}$ high. When addressing memory the low on $IO/\overline{M}$ enables the page select decoder and it decodes address lines A_4 and A_5, which in turn, enables the addressed

IC by driving its *CS* pin high. When enabled, the addressed memory IC decodes address bus lines A_0 through A_3, thus selecting the addressed memory location. Once the addressed memory location has been selected (both its row and column are active) the desired read or write operation may be preformed by activating either *RD* or *WR*.

If the CPU is addressing I/O the $IO/\overline{M}$ is high and the page selector is disabled. As a result, all of the page selectors' outputs are low. This pulls the *CS* pins of all memory ICs low, which disables them (their R and W buffers are in the Hi-Z state) and no read or write operation can take place.

Example 8.8

For the bus logic levels indicated below, refer to Figure 8.14 and determine
1. The hexadecimal address of the memory location being accessed.

2. The page select decoder output that is active (high) and the resulting memory device being enabled via its *CS* input.

3. The specific memory location within the selected IC memory device being accessed.

4. The read or write operation being performed.
The bus logic levels are

(a) Address bus lines (A_5 through A_0) = 001001, *RD* = 1, and *WR* = 0.

(b) Address bus lines = 100010, *RD* = 0, and *WR* = 1.

(c) Address bus lines = 011110, *RD* = 0, and *WR* = 1.

(d) Address bus lines = 110101, *RD* = 1, and *WR* = 0.
For each of the above conditions $IO/\overline{M}$ = 0.

Solution

(a) The hexadecimal address for address bus logic levels 001001 is 09H ($001001_2 = 09_{16}$). The two most significant address bus lines are decoded by the page select decoder, and for input logic levels 00 output O_0 is active, which enables ROM-0. The remaining bits (1001) are decoded by ROM-0, which selects location 9_{16}. Because *RD* = 1 and *WR* = 0, a read operation is to be performed at the addressed memory location 09H.

(b) The hexadecimal address is 22H ($100010_2 = 22_{16}$), which activates output O_2 of the page select decoder. This output enables RAM-2 and selects location 2_{16} (0010_2) within RAM-2. Because *WR* = 1, a write operation is performed at location 22H.

(c) The memory location being accessed is 1EH, which activates output O_1 of the page select decoder and, in turn enables ROM-1. Location EH (1110_2) within ROM-1 is being addressed. According to the logic levels of the control signal, a write operation is to take place. This is an illogical and illegal command (it cannot be done), because ROM-1 is a ROM. Therefore, there is an error in either the address or the control signal.

(d) Memory location 35H is being read. Output O_3 is active and enables RAM-3, and location 5H within RAM-3 is read.

Recall that the difference between fully decoded addressing and partial linear addressing is that there is no page select decoder in a partial linear addressing scheme. Rather, an address line from the address bus serves the function of selecting the addressed memory IC. Figure 8.15 is the memory system of Figure 8.14 modified to use partial linear addressing. In comparison to Figure 8.14, the partial linear addressing scheme of Figure 8.15 requires two additional address bus lines (A_6 and A_7) because there must be one address bus line per IC and there are four ICs. Those

Figure 8.15

A partial linear addressed main memory system.

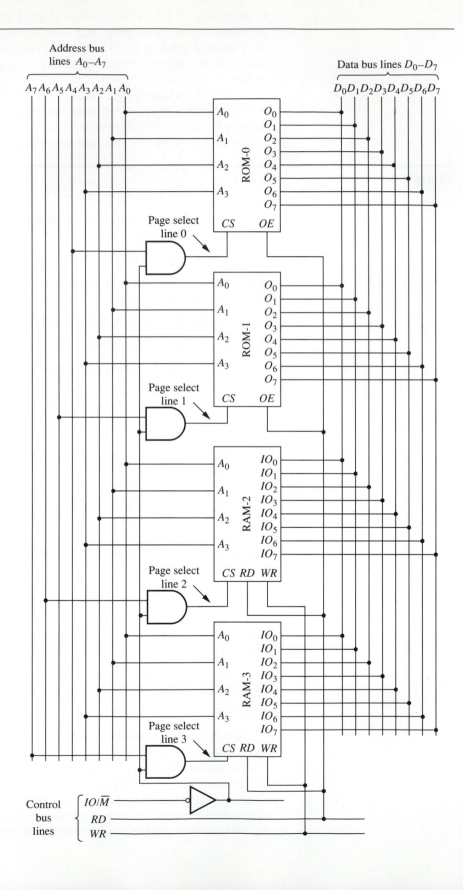

address bus lines that are to serve as IC select lines (A_4 through A_7) are *AND*ed with control signal $IO/\overline{M}$ (inverted) to distinguish between memory and I/O addresses. For the system of Figure 8.15, address bus line A_4 enables ROM-0 (page 0), A_5 selects ROM-1 (page 1), etc.

Table 8.3 provides a convenient reference for the available addresses of Figure 8.15. The top six rows of Table 8.3 indicate which memory device pin is connected to which address bus line. For instance, the *CS* of ROM-0 is connected to address bus line A_4, as indicated by the fifth column from the right. When $A_4 = 1$, ROM-0 is selected. Likewise, the sixth column from the right states that address bus line A_5 is connected to *CS* of ROM-1, and when $A_5 = 1$, ROM-1 is selected. Address bus lines A_0 through A_3 are connected to address pins A_0 through A_3 of the RAMs and ROMs and select any one of 16 memory locations from the IC whose chip select pin is active.

Table 8.3 Decoding the address bus of Figure 8.15

Address bus lines		A_7	A_6	A_5	A_4	A_3	A_2	A_1	A_0
ROMs and RAMs						A_3	A_2	A_1	A_0
ROM-0					*CS*				
ROM-1				*CS*					
RAM-2				*CS*					
RAM-3			*CS*						
ROM-0 addresses (Page 0)	10H ⋮ 1FH	0 ⋮ 0	0 ⋮ 0	0 ⋮ 0	1 ⋮ 1	0 ⋮ 1	0 ⋮ 1	0 ⋮ 1	0 ⋮ 1
ROM-1 addresses (Page 1)	20H ⋮ 2FH	0 ⋮ 0	0 ⋮ 0	1 ⋮ 1	0 ⋮ 0	0 ⋮ 1	0 ⋮ 1	0 ⋮ 1	0 ⋮ 1
RAM-2 addresses (Page 2)	40H ⋮ 4FH	0 ⋮ 0	1 ⋮ 1	0 ⋮ 0	0 ⋮ 0	0 ⋮ 1	0 ⋮ 1	0 ⋮ 1	0 ⋮ 1
RAM-3 addresses (Page 3)	80H ⋮ 8FH	1 ⋮ 1	0 ⋮ 0	0 ⋮ 0	0 ⋮ 0	0 ⋮ 1	0 ⋮ 1	0 ⋮ 1	0 ⋮ 1

Table 8.3 shows that the biggest disadvantage to a partial linear addressing scheme is the decrease in the number of addresses available for the number of address lines used. It indicates that the first available address is 10H and that addresses 30H through 3FH, 50H through 5FH, 60H through 6FH, and 70H through 7FH are not available, that is, not all addresses possible are available and those available addresses are not continuous. Had those eight address bus lines of Figure 8.15 been fully decoded, there would have been $2^8 = 256$ available addresses rather than the 64 using partial linear addressing techniques. However, for designs requiring small numbers of memory locations, partial linear addressing is suitable, especially because linear addressing simplifies the hardware (no page select decoder) and reduces cost.

Example 8.9

For the addresses given, determine from Figure 8.15 the enabled memory IC and the address of the accessed memory location within that memory device. Assume that $IO/\overline{M}$ is low.
(1) 25H   (2) 4AH   (3) 65H

Solution

1. $25H = 00100101_2$. Because bit 5 is a logic 1, $A_5 = 1$, which via the *AND* gate, enables ROM-1. Bits 3-0 have logic levels 0101, which is equivalent to 5H. Therefore, memory location 5H of ROM-1 is being accessed.

2. $4AH = 01001010_2$. Because bit 6 is a logic 1, memory device RAM-2 is enabled. Memory location AH within RAM-2 is being accessed.

3. $65H = 01100101_2$. Two address lines have logic 1 level (bits 5 and 6), which would enable both ROM-1 and RAM-2. This is illegal (a nonexistent address); hence there is a fault with the address. If $RD = 1$ and $WR = 0$ for this address, bus contention would exist (both ROM-1 and RAM-2 would be contending for the data bus).

It is the responsibility of the memory designer to identify valid addresses using a table similar to Tables 8.2 and 8.3. It is the responsibility of the user to use only valid addresses when writing programs.

Review Questions

1. What is the principle of operation of a fully decoded addressing scheme?
2. What is the principle of operation of a linear addressing scheme?
3. How does partial linear addressing combine linear and fully decoded addressing?
4. How can one determine the memory capacity of a memory system by examining its address and data bus configuration?
5. How are read/write operations executed?
6. Why must read/write operations be synchronized with the CPU?

Isolated I/O and memory-mapped I/O

Note from the computer architecture of Figure 8.1 that if I/O peripheral devices were redefined such that I/O devices were simply "all devices external to the CPU," then main memory would also be considered an I/O. In reality this is the case; however, because of the very special nature and specific role of main memory, it tends to be treated as if it were not an I/O device. Because main memory is in reality an I/O device (although it is never referred to as I/O), the techniques used to access it are similar to those for traditional I/O devices. In previous coverage of accessing memory locations, many of the essentials of accessing other I/O devices were also covered. When the CPU is accessing an I/O device it must first output its address on the address bus and then drive control signal $IO/\overline{M}$ high, enabling the I/O selector so that it can decode that address (if decoding is used). The CPU then generates the appropriate read or write control signal. As you now realize, the process of accessing main memory and I/O is conceptually the same.

A computer system that distinguishes between memory and I/O addresses is known as an *isolated I/O system*. Isolated I/O systems utilize the control signal $IO/\overline{M}$ or a similar control signal to make such distinctions. The computer system of

Figure 8.8 is an isolated I/O system because it utilizes control signal $IO/\overline{M}$. The memory systems of Figures 8.14 and 8.15 are also isolated I/O memory systems, as indicated by the use of the control signal $IO/\overline{M}$.

Those computer systems that do not differentiate between memory and I/O addresses are referred to as *memory-mapped I/O systems.* They treat I/O devices as memory locations and assign addresses to both memory and I/O with no distinction between the two. This implies that the control signal $IO/\overline{M}$ is not required.

Converting the isolated I/O memory systems of Figures 8.14 and 8.15 to a memory-mapped I/O system requires the elimination of the control signal $IO/\overline{M}$. As a result, the system of Figure 8.14 would be modified so that the enable pin of the page select decoder would be tied low (strapped to ground). Eliminating $IO/\overline{M}$ and the *AND* gates of Figure 8.15 would convert that memory to a memory-mapped I/O system once I/O was incorporated into the design (to be discussed in Chapter 10). The number of address bus lines would also be increased to accommodate the additional addresses for I/O.

The computer system designer divides the available addresses into two parts — those assigned to memory and those assigned to I/O, a process known as mapping the address space. The addressing is designed so that there is no overlapping of the memory space and I/O space (i.e., there are no addresses common to both a memory location and an I/O device).

8.12
Fundamental concepts of troubleshooting

Any fault in either a main memory or an I/O system can be detected by first observing the logic levels of the address and control signals and then detecting the resultant operation via the data bus. These events were illustrated in the timing diagram of Figure 8.8b and 8.8c, where:

1. An address is loaded on the address bus lines A_0 through A_n by the CPU. The address activates the *CS* of the addressed IC memory or I/O device, via decoders if an address decoding technique is used or via an address line if linear addressing is used.

2. The control signal $IO/\overline{M}$ is activated by the CPU at approximately the same time that the address is loaded on the address bus (it is driven high for I/O addresses and low for memory addresses). This control signal is held at the appropriate logic for as long as the address is held on the address bus.

3. The CPU must activate the read (*RD*) or write (*WR*) control signal at the proper time — when the CPU is ready to read the datum from the data bus or when it has loaded the datum to be written to the addressed device onto the data bus for a write operation.

When the data bus is not used for the reading or writing data it is in a Hi-Z state, meaning that all output buffers of both memory and I/O are turned off. When the data bus is in use, only the addressed device (memory IC or I/O) has its output buffers turned on, giving it exclusive use of the data bus during that time.

When troubleshooting, keep in mind these events and the timing diagrams of Figure 8.8b and 8c, which serve as a guide to troubleshooting diagnosis. Using the timing diagrams of Figure 8.8b and 8.8c as a guide:

1. Apply and/or verify that there is a valid address on the address bus.

2. Apply and/or verify that $IO/\overline{M}$ is at the proper logic level (if it is an isolated I/O system).

3. Verify that the logic level on the *CS* pin of the addressed device is valid for the given address.

4. Apply and/or verify that the proper read (*RD*) or write (*WR*) control signal is active for the intended operation.

5. Verify that the datum on the data bus is correct and is present at the proper time. For a read operation this means that the datum in the addressed memory location must be known so that it can be verified when it appears on the data bus. For a write operation the datum being written by the CPU must be known and also verified via the data bus.

"Apply and/or verify" means that if the memory system under test is integrated into a system, such as that in Figure 8.8a, the CPU applies the logic levels for the address and control bus signals and it must be verified that they are correct. However, if the system under test has not been integrated into a completed system, these signals must be applied as part of the test environment.

There are two methods of troubleshooting. One involves static testing techniques, whereas the other involves dynamic testing. In static testing, the logic 1's and 0's applied to the logic circuits under test use DC levels (static) to implement those logic levels, or logic levels change so slowly that their transition from one state to the other can be observed with the use of a volt-ohm meter (VOM) or logic probe. Static testing has until now been the major diagnostic tool. In dynamic testing, the logic 1's and 0's input to the logic circuitry under test change logic levels faster than can be observed with static-type test instruments. Dynamic conditions are those experienced in the real world. For example, when a modern CPU is operating, it is outputting a new address on the address bus and reading or writing a datum from or to the data bus every few microseconds, as well as outputting control signals. To detect and observe dynamic logic levels, dynamic test equipment must be used.

Static troubleshooting

To statically troubleshoot memory refer to the troubleshooting guide previously listed. Consider the memory system of Figure 8.14 as the system to be statically tested; let it further be assumed that it has not been integrated into a complete system, meaning there is no CPU or I/O. Because there is no CPU to input addresses and control signals to this memory, the test environment must be configured to do so. The address bus lines A_0 through A_n. ($n = 5$) are being driven by the address static logic level generator. The logic level of each output of this logic level generator, as well as all the other static logic level generators, is determined by the position of the slide switch of Figure 8.16b. If the slide switch is in the "left" position, a logic 0 is output; otherwise, a logic 1 is output. Applying the intended logic levels to control bus lines $IO/\overline{M}$, *RD*, and *WR*, use the control signal static logic generator that has the same structure as the address logic level generator. Once the address and appropriate $IO/\overline{M}$ logic level have been output, the logic level of the *CS* pin of the addressed memory IC may be verified using a logic probe, as indicated in Figure 8.16a, or a logic level detector. If a datum is to be written into memory, the data static logic level generator serves that purpose via the data bus lines D_0 through D_m ($m = 7$). If a datum is being read from memory, the data static logic level detector is used. Each data line is connected to an input of the data static logic level detector, such as the one illustrated in Figure 8.16c. If a logic 1 is present at the input of the detector, the LED is on and a light is emitted; otherwise, the LED is off. Both the data static logic level detector and generator are buffered. These buffers have the opposite logic level for control of their output states (active or Hi-Z). These buffers are used to isolate the read test equipment from the write test equipment and to eliminate bus contention in

Figure 8.16

Static testing.

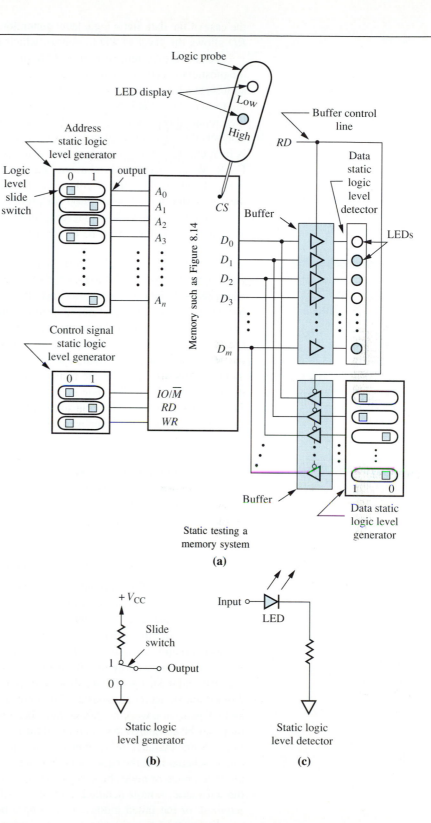

Static testing a
memory system

(a)

Static logic
level generator

(b)

Static logic
level detector

(c)

the case of the data static logic level generator. Connecting the buffer control line to *RD* allows the level of *RD* to control which buffer is on.

To test the system of Figure 8.16a in a thorough and organized manner, the diagnostician needs to create a truth table that shows all binary combinations for the address bus lines and the resulting active output of the page select decoder, similar to the truth table of Table 8.2 for the memory system of Figure 8.14. The diagnostician then inputs those addresses, one at a time, via the test system of Figure 8.16a, and also sets the logic levels of the control bus lines. The *CS* pin of the addressed device is then checked. Last, a read and/or write on each memory location is performed. For R/W memory (RAM) a known binary pattern must first be written into memory before being read out for verification of storage. For ROM, the contents must be known so that a read verification for proper content and operation can be made.

Static testing is limited to testing for faults resulting from an improperly wired system or the failure of an IC or other component.

Dynamic troubleshooting

Dynamic testing is conceptually the same as static testing, the difference being that for a dynamic system the logic levels on the busses and the *CS* lines are rapidly changing — at the rate of microseconds or nanoseconds. Under dynamic conditions, problems can arise that are not present during static testing. These types of faults may be timing problems due to propagation delay times or exceeding the operation speed of some IC or other component in the system. Actual working conditions are the best environment in which to test a system, which means dynamic testing. The memory systems under test are considered to be integral parts of a total system, like the system of Figure 8.8. This means those logic levels that are produced by the CPU — that is, the address, control signal, and data — are rapidly changing. What is needed is test equipment capable of equally fast detection and display of logic levels in a form that can be interpreted. Because logic levels are changing in the microsecond or nanosecond range, the results cannot be displayed at the rate at which they appear (real time). Instead, they must be stored and then presented at a slower rate that will permit scrutiny by the observer.

Test equipment for dynamic logic level detection and display includes:

1. Signature analyzers

2. Logic analyzers

3. *Microprocessor* development systems

Discussion will be limited to signature analyzers and logic analyzers, for their concepts of operation are also basic to microprocessor development systems.

The signature of an electrical point, known as a node, is determined by counting the logic 1's and the order in which they appear in an interval of time between two events (start and stop markers). From that count a four-digit code (0 through 9, A, C, F, H, P, U) is derived by the signature analyzer. This code is statistically unique to that particular string of data flow and can serve as the signature of that node. Obviously, signature analysis is of diagnostic value only when the nodes (bus lines and *CS* pins) of a working system have been previously determined and recorded so they may be compared to signatures of the same nodes of an identical system that has failed. Some manufacturers of electronic equipment print the signature of each node on the schematic of the equipment. When a system fails, the diagnostician can locate the failed node or nodes by comparing signatures of the failed system with those on the schematic. A more detailed analysis, possibly using a logic analyzer, may then be required, or the failed module may simply be replaced.

Figure 8.17 is a photograph of a signature analyzer manufactured by Hewlett-Packard. It has a 100% probability of detecting single-bit errors and a 99.98%

Figure 8.17

Hewlett-Packard Signature
Analyzer, model 5006A
(courtesy of Hewlett-
Packard).

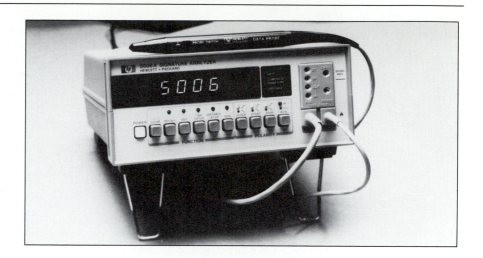

probability of detecting multiple-bit errors. There is no maximum time between its
start and stop limits. Up to 32 signatures can be stored for recall. It operates at a
frequency of 25 MHz.

Logic analyzers perform much the same function as movie cameras. Movie
cameras freeze a moment of time with each frame snapshot. Each frame that is a
snapshot is a static representation of a moment that is stored on film. They create the
illusion of movement by taking a series of frames and playing them back in a
properly timed succession. Logic analyzers take snapshots of the logic levels present
on system busses and *CS* lines whenever commanded to do so by a trigger signal.
After taking the snapshot the logic analyzer stores the logic levels in a FIFO memory
in the same order in which they occurred. Just as importantly, each snapshot or series
of snapshots of logic levels can be displayed at any rate desired, much as a frame of
movie film can be viewed.

A representative architecture of a logic analyzer is presented in Figure 8.18. The
architecture can be modularized into five function-blocks.

1. Input section, which consists of input channels, comparators, and voltage
   level detectors.

2. Logic level storage unit, which is made up of an addressable FIFO and
   its control circuitry.

3. Trigger source unit.

4. Clock source.

5. Display.

The input section inputs data from the nodes being tested. Each system node
under test has an input channel (C_0 through C_N) connected to it. To ensure that only
good clean logic levels are being input to the logic level storage unit (the FIFO
memory), comparators are used to shape up the input waveforms. Also, the voltage
level detector allows the logic analyzer to be adjusted for different logic families that
differ in the voltage levels that constitute a logic 1 or 0 (voltage requirements for
various logic families are discussed in Chapter 12). Logic analyzers are available
with a variety of channels. The least is 16 ($N = 16$) and the most is usually 136 ($N =
136$). For instance, if each node of the memory system of Figure 8.14 were to be

Figure 8.18

The architecture of a logic analyzer.

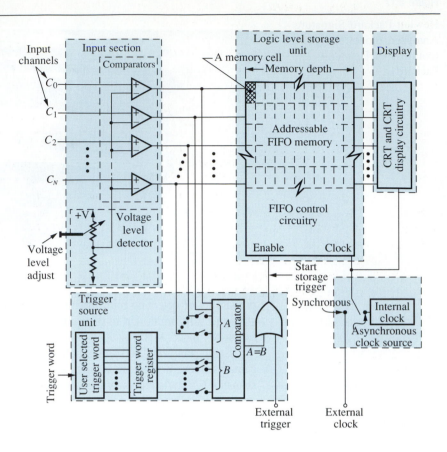

tested, a minimum of 22 channels would be required (six address lines, eight data lines, three control bus lines, four *CS* lines, and one enable pin on the page select decoder). As can be easily seen, the number of input channels can rapidly increase as the system being tested becomes more complex.

As the nodes under test input data to the logic analyzer of Figure 8.18, that data is passed on to the logic level storage unit and also to input *A* of the comparator in the trigger source unit. The data is not stored in the FIFO unless its enable input is high, driven by the start storage trigger output by the trigger source unit. As can be seen by examining the trigger source unit function-block, the logic level of the start storage trigger is determined by the output of the comparator (*A* = *B*) or the external trigger input. If data storage in the FIFO is to be initiated when a certain binary bit pattern appears at the selected input channels (note that not all channel data need be input to the comparator — see the switches), that binary bit pattern, known as the trigger word, is loaded into the trigger word register via a keyboard or panel switches. When the trigger word appears at input *A* of the comparator, its output (*A* = *B*) will go high and the start storage trigger output will be driven high, which enables the FIFO for data storage. When the external trigger of the trigger source unit is used to drive the start storage trigger output high, it is connected to the node within the system that generates the signal with which data storage is to coincide. For example, if data collection and storage are to begin when *RD* goes high, then the external trigger input is connected to the *RD* line (node).

Enabling the FIFO to store data is determined by the logic level of the enable input of the FIFO control circuitry, but it is the clock input of the FIFO control circuitry that actually strobes the data into the FIFO. As Figure 8.18 shows, an external clock can be used to strobe data into the FIFO, or an internal clock can serve as the clock source. When an external clock source is used that is synchronized with the system being tested, the data collected is stored and displayed in synchrony with the tested system. For this reason, external timing of this type is known as the synchronous mode of operation. On the other hand, when an internal clock is used, no relationship exists between data collection, storage, and display and system timing; the internal timing is known as an asynchronous mode of operation.

The FIFO is identified as addressable, which means that even though data is collected and shifted in a FIFO fashion, the data at any location can be addressed. This allows the user of a logic analyzer to define a reference point and to display the collected data before or after this reference point.

The collected data is displayed on a CRT. Because the data to be displayed is in FIFO memory and therefore in binary form, it can be displayed in any format desired, assuming that the logic analyzer has the proper logic circuits. The only requirement is that the data be converted to the desired format before being displayed. For instance, if the address bus lines are being tested, it would be more convenient to display a hexadecimal value rather than a logic level diagram. The stored logic levels could be converted to hexadecimal characters for display on the CRT.

Figure 8.19 illustrates the storing of channel logic levels. Figure 8.19a represents the logic levels being input to channels C_0 through C_3, which could be any four nodes, such as address lines, data lines, or control bus lines. These channel inputs are shaped up and input to the FIFO via the comparators. At time t_1 the logic levels at the input of the FIFO, which is the output of the input section, are strobed into the FIFO, as illustrated in Figure 8.19b. At t_2 the channel datum has changed from 0000 to 0001 and is strobed into the FIFO, which also shifts the first datum by one location, as shown in Figure 8.19c. At t_3 the process is repeated, and the FIFO appears as in Figure 8.19d. Data is sampled and written into the FIFO until it is filled. On command by the user, the FIFO outputs the data to the display in the desired format. If the sampled data is displayed as logic levels with the same timing at which it was collected, the displayed data is shifted to the right by one half clock period, as illustrated in Figure 8.19e. This is why the output is really a logic level diagram rather than a timing diagram. If an accurate timing diagram is required, the logic analyzer must increase the sample time of the input logic levels by increasing its sample rate (increase the clock frequency).

The size of the FIFO is usually between 256 and 1028 words of storage (depth), and the word size (the number of bits stored per location) corresponds to the number of input channels.

Figure 8.20 is a photograph of the Hewlett-Packard Logic Analyzer model 1660A. Channel input connections are made via the five channel modules as shown, where only one module has the individual channel connectors inserted. The 1660A is a 136-channel logic analyzer that operates at 250 MHz. Among other features, the 1660A has six clock inputs and ten clock qualifiers (a clock qualifier serves the same function as a trigger word) that enable the user to determine when the logic analyzer will begin storing the data at the nodes. It has methods to examine the stored data according to patterns or range of samples taken, and the displayed data can be formatted in a variety of ways, such as timing diagrams, state testings in hexadecimal, or inverse assembly language (assembly language will be discussed in Chapter 10). The test system performance can also be displayed as a bar graph.

Both static and dynamic test equipment will be used when troubleshooting memory and I/O designs in the chapters to follow.

Figure 8.19

Filling the FIFO memory of
the logic analyzer of
Figure 8.18.

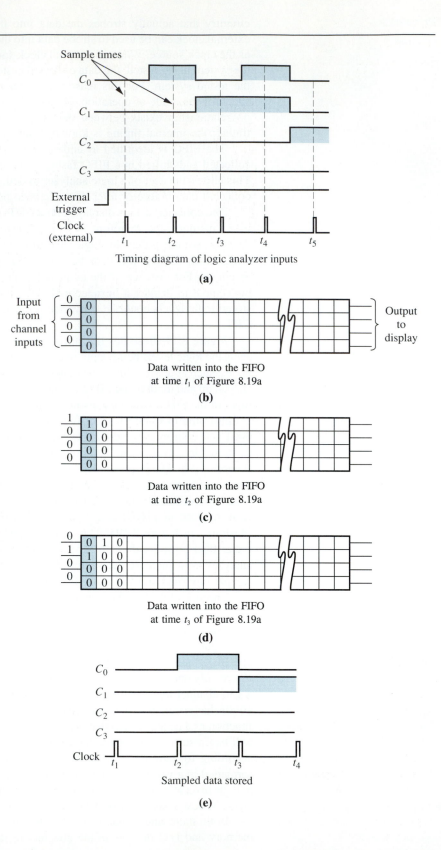

Sample times

C_0

C_1

C_2

C_3

External
trigger

Clock
(external)

t_1 t_2 t_3 t_4 t_5

Timing diagram of logic analyzer inputs

(a)

Input
from
channel
inputs

Output
to
display

Data written into the FIFO
at time t_1 of Figure 8.19a

(b)

Data written into the FIFO
at time t_2 of Figure 8.19a

(c)

Data written into the FIFO
at time t_3 of Figure 8.19a

(d)

C_0

C_1

C_2

C_3

Clock

t_1 t_2 t_3 t_4

Sampled data stored

(e)

Figure 8.20

Hewlett-Packard Logic
Analyzer, model 1660A
(courtesy of Hewlett-
Packard).

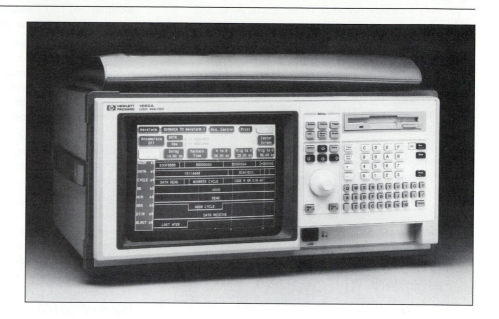

Review Questions

1. When troubleshooting a read or write operation, whether statically or dynamically, what nodes would be tested to make a complete diagnosis of the system and why should those nodes be monitored?

2. What are the five troubleshooting steps to be used as a guide?

3. What is the essential difference between static and dynamic testing?

4. When is static testing the preferred choice?

5. When is dynamic testing necessary?

6. How does a signature analyzer function?

7. How does a logic analyzer function?

8. Under what test conditions can a signature analyzer and logic analyzer be used?

Summary

- Memories can be categorized in a variety of ways. A classification based on CPU access to the data stored has two dominant memories: main memory and mass storage memory.

- Main memory is the working memory of the system and its contents are programs and data that are currently being operated on by the CPU.

- Mass storage memory stores mass amounts of data from program execution or programs that are to be executed at a later time.

- The portability of the mass storage medium allows convenient storage of the medium (tape or disk) as well as transport of programs and data between systems.

- Semiconductor memory devices used in the design of main memory have memory cell organizations that are either two- or three-dimensional arrays. If the width of the memory location is 1 bit, it is a two-dimensional array; if the width is greater than 1 bit, it is a three-dimensional array.

- There are four different techniques of accessing a memory location—dynamic linear, RAM, FIFO, and LIFO.

- The access time for dynamic linear accessing is a variable because it is position dependent.

- The access time for RAM devices is the same for all memory locations within the device.

- The access time for FIFOs and LIFOs is a function of how many shifts must be made to access the desired data.

- A FIFO memory has accessible input and output memory locations, but those locations between the two are usually not accessible (the FIFO of a logic analyzer is an exception).

- A LIFO memory has only one accessible memory location that is defined as the top of those memory locations comprising the LIFO.

- The ability of the CPU to alter the contents of a semiconductor memory device categorizes semiconductor memory devices into one of two types—either a ROM or an R/W memory.

- The CPU can only read the contents of a ROM, whereas it can both read from and write to an R/W memory device. The acronym RAM is erroneously used as a synonym for R/W memory.

- There are many types of ROMs, each having its advantages and disadvantages.

- EEPROMs are read-mostly R/W memories and are sometimes referred to as nonvolatile RAMs.

- A memory device that cannot retain the contents of its memory locations on the removal of power is known as a volatile memory device.

- Memory devices that do retain the contents of their memory locations, even without power, are known as nonvolatile memory devices.

- Static memory devices retain their contents so long as power is applied to the memory device.

- *Dynamic memories* cannot retain their contents without being refreshed.

- The CPU generated signals *RD* and *WR* control read and write operations.

- There are three basic techniques for addressing memory locations and I/O devices—fully decoded addressing, linear, and partial linear addressing.

- The fully decoded addressing technique decodes the address bus. These address bus lines are decoded by the memory device and the page select decoder. The decoder internal to the memory device selects the addressed location within the memory device, and the page select decoder selects the memory I/C.

- Partial linear and linear addressing are different from fully decoded addressing in that there are no select decoders; instead, address bus lines are used to select memory and I/O devices.

- There are two techniques for allocating addresses to memory and I/O—isolated and memory-mapped I/O. Isolated I/O systems partition memory addresses from I/O addresses via the $IO/\overline{M}$ control signal. Memory-mapped I/O systems do not distinguish between memory and I/O addresses.

- Memory-mapped I/O systems require less hardware to implement (no I/O select decoder), but they have fewer addresses available.

- In static troubleshooting, DC voltages are input to the circuit under test, and static test equipment, such as a logic probe, is used to detect the resultant logic levels.

- Static testing is acceptable for troubleshooting catastrophic failures such as IC failure and wiring problems.

- Dynamic troubleshooting tests for catastrophic failures and dynamic-related problems such as timing and coupling due to stray capacitance.

- Signature analysis is a dynamic test but requires that the signatures of the system nodes be known.

- Logic analyzers are also dynamic test equipment. They sample the data on nodes according to a trigger signal and store that data in a FIFO in the order sampled. When the collected data is displayed there are a variety of formats from which the user can choose. The user can compare the displayed data with that of a truth table for diagnostic purposes.

Problems

Section 8.1   Introduction

1. What are program instructions and what purpose do they serve?

2. What are programs and what purpose do they serve?

3. Why must a computer system have a memory?

Section 8.2   Computer architecture

4. What does the term "architecture" mean?

5. How does the function of main memory differ from that of mass storage memory?

6. What are the criteria of main memory and mass storage

memory? As part of your answer, provide a rationale for your statements.

7. Why does the CPU work from main memory rather than mass storage memory?

8. Why is main memory implemented with semiconductor memory devices?

Section 8.3   Types of memory organization

9. What are the five basic categories that can be used to classify memories?

10. For the given memory organizations determine the number of memory locations, memory cells per location, and total bit capacity.
    a. 256 × 4
    b. 256 × 8
    c. 1024 × 4
    d. 1024 × 8
    e. 2048 × 8
    f. 4096 × 1

11. For a 64 × N-bit memory device, which is either a two- or three-dimensional array, determine the possible number of row and column configurations of the array using an x-y coordinate system, as was used in Figures 8.3 and 8.4.

12. Repeat Problem 8.11 for a 32 × N-bit memory device.

13. From the evolving patterns of Problems 11 and 12, develop a table that indicates the combinations of columns and rows for N memory locations.

14. If it is desired to have the number of rows and columns equal, or as nearly equal as possible, for a memory cell organization, how many columns and rows would exist in Problems 11 and 12?

15. Using the same x-y coordinate scheme used to identify the memory locations of Figures 8.2 and 8.4, develop a memory cell organization for a 16 × 1-bit memory device that has eight rows and two columns for memory location identification. How would the addressing be affected if each memory location had 4 bits of storage?

16. For the memory cell array of Problem 15, develop a table similar to Table 8.1 showing the binary address of each memory location and the cells at those locations. Assume this to be a 16 × 2 memory. Compare this table with Table 8.1 and note the results are the same except for the number of cells per location.

Section 8.4   Memory addressing and access time

17. If the access time of a RAM is 2 μs, what is the access time of any location within that RAM? Explain your answer.

18. Suppose the FIFO of Figure 8.5a requires 0.20 μs to shift data from one location to the next. If the FIFO were empty and a datum was written into the FIFO,

how long would it take for the datum to be available at the output location?

19. If the FIFO of Problem 18 had another 4 bits of data written into it, what would be the delay time before this datum appeared in the next available location?

20. Suppose that the LIFO of Figure 8.6 requires 0.2 μs to shift data from one location to the next. How much time is required to read a datum that has had two other data entries pushed on top of it?

21. Suppose that the LIFO of Problem 20 had three 2-bit datum entries pushed on it. How much time is required to push this data down and return the original datum to the top location?

Section 8.5   Alteration of memory contents: R/W and ROM devices

22. When would a ROM rather than an R/W memory be used in an application?

23. Can a ROM also be a RAM? Why?

24. Why must all mass storage devices have read/write capability?

Section 8.6   ROM classifications

25. In a manufacturing environment (large-volume production), what type of ROMs might be used?

26. In a research and development environment, what type of ROMs might be used?

27. What type of ROM might be used by an engineering firm that produces one to three copies of a product?

28. If you were designing a computer that processed data so important that you did not dare lose it as a result of a power failure to the system, what type of ROM could you use as a nonvolatile R/W memory back-up?

Section 8.7   Memory volatility

29. What is volatility?

30. What is a nonvolatile RAM?

Section 8.8   Memory cell data retention: static and dynamic memories

31. When would a dynamic RAM be chosen over a static RAM?

32. What is the disadvantage of using dynamic RAMs?

33. Are both static and dynamic RAMs volatile?

Section 8.9   Addressing and read/write operation control

34. When accessing memory, what two events are required and in what order?

35. How does the CPU control read and write operations?

Section 8.10   Semiconductor memory architecture

36. The number of address pins is listed for four different memory devices. Determine the number of memory locations within each device. Also, would the number of bits per location affect the answers. Assume each of these memory locations to have N bits of storage.
    a. 8
    b. 10
    c. 11
    d. 12

37. For the number of address pins listed in Problem 36, what combination of row and column address pins might you expect? Refer to Problem 14.

38. The number of address pins of three memory devices are listed. In each case, determine the type of row and column address pin decoder, such as specified in Figure 8.7a.
    a. 10
    b. 11
    c. 13

39. For a 12-address pin memory device, how many columns and rows are there and what is the total number of locations?

40. For a memory IC with 13 address pins how are the addresses divided among row and column and specify the number of row addresses, column addresses, and total addresses.

41. Why is it logical that the CPU of a system, such as the computer illustrated in Figure 8.8, be the function-block that supplies addresses (both memory and I/O)?

42. The illustration of Figure 8.8 shows the CPU as the origin of the control signals $IO/\overline{M}$, RD, and WR. Explain why it is logical that the CPU is the source for control signals.

43. To access the memory cells of Figure 8.10 for either a read or a write operation, what is the first event that must occur? Explain why it must occur first.

44. Explain the purpose of the R and W buffers of the memory cells of Figure 8.11.

45. In Figure 8.11 the outputs (Q) of all memory cells within a plane are connected. Why doesn't this cause bus contention on that connecting line?

46. Suppose you were asked to describe the internal design of a memory device that has the pin configuration of Figure 8.21. Using Figure 8.7 as a reference, give the requested description.

47. Using the ROM memory cell of Figure 8.12, design an 8×2 ROM.

Section 8.11   Introduction to the fundamental concepts of main memory design

48. In a fully decoded address scheme, what function does the page select decoder serve?

49. What is meant by the expression "a page of memory"? Express it in mathematical terms using Equation 8.1, if an IC with N address pins is said to contain one page of memory.

50. Define a page of memory to be equal to the least number of memory locations within any memory device used to implement a main memory system. Using this definition, how many memory locations would there be in a page of memory if there were 12 address pins on the system's ROMs and 10 address pins on the system's RAMs?

51. If you designed a main memory system using ROMs and RAMs that had 11 address pins per memory device:
    a. How many memory locations would each memory device contain?
    b. How many memory locations would a page of memory contain if the number of locations within the memory systems ICs defined a page of memory?
    c. For a fully address-decoded 8 K memory (8192 memory locations) how many address bus lines are required?
    d. Using partial linear addressing, how many address bus lines are required for an 8 K memory?
    e. What control signal is used to separate memory addresses from I/O addresses?

52. For part (c) of Problem 51, what address bus lines would serve as inputs to a page select decoder?

53. For part (d) of Problem 51, what address bus lines would be used to activate the CS inputs of the memory devices?

54. Design an 8 K fully decoded main memory that is applicable for either an isolated or memory-mapped I/O

Figure 8.21

The pin configuration of a memory IC.

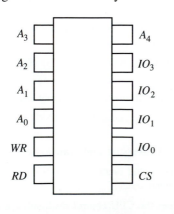

Figure 8.22

Pin configurations for
memory IC's *X* and *Y*.

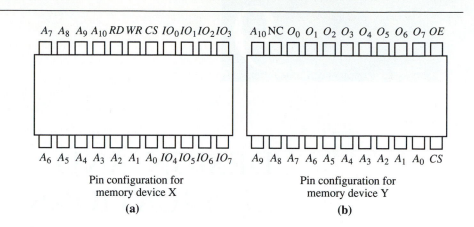

A_7 A_8 A_9 A_{10} RD WR CS IO_0 IO_1 IO_2 IO_3

A_6 A_5 A_4 A_3 A_2 A_1 A_0 IO_4 IO_5 IO_6 IO_7

Pin configuration for
memory device X

(a)

A_{10} NC O_0 O_1 O_2 O_3 O_4 O_5 O_6 O_7 OE

A_9 A_8 A_7 A_6 A_5 A_4 A_3 A_2 A_1 A_0 CS

Pin configuration for
memory device Y

(b)

system using the memory devices of Figure 8.22, such
that there are 2 K of ROM and 6 K of RAM. ROM is
to begin at address zero. Draw a schematic of your design similar to Figure 8.14. Create a table that shows
the connections of component pins to the address bus
lines and also the beginning and ending address of each
page of memory, similar to Table 8.2. Show the necessary modifications required to your design to make it
an isolated I/O system and then do the same for memory-mapped I/O.

55. Repeat Problem 54 but use partial linear addressing.
Show your design for both isolated and memory-
mapped I/O.

56. What is the essential difference between the static
memories of Problems 54 and 55 and similar designs
using dynamic memory devices?

Troubleshooting

Section 8.12 Fundamental concepts of troubleshooting

57. To the design of Figure 8.14 add 5 I/O devices (3 input and 2 output) and then
devise a static test. Make a drawing similar to Figure 8.16. Assume each I/O is
enabled via a chip select (CS) pin.

58. Devise a dynamic test for the design of Problem 57. Create a drawing similar in
concept to Figure 8.16 using a logic analyzer.

59. For the dynamic test of Problem 57 create a timing diagram for reading memory
locations 0000H, 0001H, and 0002H. At location 0000H data 21H is stored, at
0001H data 31H is stored, and at 0002H data 44H is stored. For these three addresses assume that samples were made and stored in a logic analyzer.

60. Can a signature analyzer be used to diagnose a fault with the newly constructed
design of Problem 58?

9

DESIGN AND APPLICATIONS OF SEMICONDUCTOR MEMORIES

OBJECTIVES

The objectives of this chapter are to:

- Introduce various types of memory systems and applications

- Explain the characteristics of operation of some specific semiconductor memory devices

- Explain semiconductor memory timing parameters and how they relate to system requirements

- Apply the fundamental concepts set forth in this and the previous chapter to the design of main memory systems

- Apply troubleshooting techniques to the memory designs developed in this chapter

Technology now allows us to design highly complex digital circuits and fabricate them in a tenth of the space required a few years ago. The photograph shows a complete digital computer circuit on a single printed circuit board
(Courtesy Chips & Technology, Inc.).

9.1
Introduction

Chapter 8 covered fundamental concepts that were concerned primarily with memory cell organization, various accessing techniques, and the control of read / write operations. This chapter will apply that information to various memory applications and to the design of main memory systems.

When designing a semiconductor memory system, choices must be made concerning the specific memory devices to be used in the implementation of the design, which often requires choosing among various technologies, such as *transistor-transistor logic* (TTL), *metal oxide semiconductor* (MOS), and *complementary MOS* (CMOS). In this chapter, the chosen technologies and the reasons for choosing those technologies will be a matter of stated fact. When we study the logic technologies in Chapter 12 we will explain the reasons for those choices.

Because this is a broad-based text on digital fundamentals, rather than one limited to the design of memory systems, a presentation on the design of memory systems must be somewhat selective. The authors believe that the focus of memory design should be on main memory systems because main memory is an essential part of a computer system. In addition, the principles of main memory design also apply to the design of other memory systems. In fact, often the only differences between a main memory and other types of memory systems are the name and its application; the principles of design are the same. To complete the study of memories, take a quick look at some other commonly used semiconductor memories and their applications in computers.

9.2
Applications of semiconductor memories

When referring to a specific type of memory system, that system is classified according to its application. Each application has operational characteristics that determine the type of memory system and the semiconductor memory devices to be used in implementation of that design. For instance, if designing the portion of main memory from which the CPU is fetching and executing a bootstrap program (a program that gives the CPU its beginning instructions), a ROM must be used to store that program.

In Chapter 8 various types of memories were discussed. That list will now be expanded to include five types of memories, all of which are implemented with semiconductor technology:

1. Scratch pad memory
2. Buffer memory
3. Stack memory
4. Queue memory
5. Look-up table memory

The following material will discuss the primary application of each of these five memory types.

Scratch pad memory

Scratch pad memory serves as a memory for the storage of data considered to be temporary. This is analogous to a scratch pad used by a person calculating his or her federal income tax. The intermediate values (temporary data) are calculated and saved on a scratch pad so as to be readily accessible for other calculations. Once these intermediate values have served their purpose the scratch pad data can be discarded. In contrast, recording data on an IRS form, which then becomes a permanent record, is analogous to storing data in main memory or mass storage memory, depending on how permanent the information is meant to be.

There are several assumptions behind the concept of scratch pad memory: (1) The data stored is considered temporary, meaning that it should quickly serve its purpose in the execution of a program so that it can be discarded to make room for other scratch pad data. (2) Scratch pad data is readily available. Data retrieval time from scratch pad memory should be faster than from main memory; otherwise, why not store the temporary data in main memory? (3) The capacity of a scratch pad memory is small (perhaps only eight memory locations) because the amount of intermediate data requiring temporary storage is small.

Scratch pad memories are often implemented with registers and are manufactured as an internal part of the CPU. Registers A, B, C, and D of Figure 9.1 are representative of scratch pad memory within a CPU. The ALU, which is also an integral part of the CPU but is not shown, performs all calculations and requires a scratch pad memory for storage and retrieval of intermediate data. Manufacturing the scratch pad memory as part of the CPU allows the read/write operations required with intermediate calculations to be performed much faster than if the CPU had to go to an external memory, such as main memory (recall that to access main memory the CPU must generate addresses and control signals).

Sometimes programmers set aside a portion of main memory to act as a scratch pad memory. This serves only to partition scratch pad data from other data (from a programming organization point of view) and does not offer the readily available aspect of a true scratch pad memory. A programmer's view of memory can be different from that of a designer.

Figure 9.1

A LIFO stack memory.

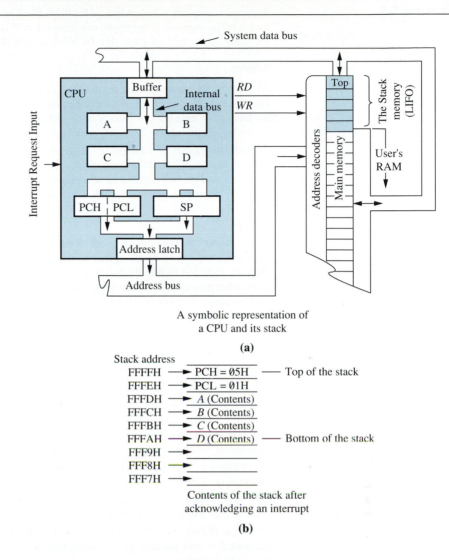

A symbolic representation of
a CPU and its stack

(a)

Stack address

FFFFH	→	PCH = 05H	— Top of the stack
FFFEH	→	PCL = 01H	
FFFDH	→	*A* (Contents)	
FFFCH	→	*B* (Contents)	
FFFBH	→	*C* (Contents)	
FFFAH	→	*D* (Contents)	— Bottom of the stack
FFF9H	→		
FFF8H	→		
FFF7H	→		

Contents of the stack after
acknowledging an interrupt

(b)

Buffer memory

Buffer memories serve as a buffer between a fast operating device, such as a computer or CPU, and a slow device, such as an I/O. An example is a computer sending data to a printer. The computer can send out data much faster than the printer can print it, which would result in an ineffective use of the computer if it had to wait for 1 byte of data to be printed before sending the next byte. A solution is for the printer to have a large-capacity buffer memory to which the computer can write its data at a high rate of speed. The printer can fetch the data to be printed from the buffer memory at its own rate, releasing the computer for other duties during printing.

A FIFO memory such as illustrated in Figure 8.5 can serve as a buffer memory. However, a memory with a larger capacity than that in the figure is required.

The memory device used to implement a buffer memory must have a write cycle time equal to or faster than the CPU. As stated in Chapter 8, shift registers can be used to construct a FIFO-style memory. A memory much like main memory could also be used, with an added controller (to furnish addresses and control signals) that would access its memory locations as a FIFO.

Example 9.1

Suppose a computer were sending a 1K (1024) block of ASCII data to a printer that does not have a buffer memory. If the printer can print 100 ASCII characters per second, how long will the computer be tied up while this 1K block of ASCII data is being printed?

Solution

At a rate of 100 characters per second, it takes $1/100$ second (10 ms) for the printer to print one character. To print 1024 characters therefore takes 10.24 seconds (1024×0.01), which is the amount of time the computer will be tied up.

Example 9.2

To make the printer of Example 9.1 more efficient, add a FIFO buffer memory that has a capacity of at least 1 K. If the computer can transmit an ASCII character to the buffer memory every 5 μs and the write cycle time of the buffer memory is sufficiently fast, how long will the computer be tied up printing this 1K block of ASCII characters?

Solution

The computer can write this 1K block of data in 5.12 ms ($1024 \times 5\ \mu$s). Once the computer has transmitted the 1K block of data to the buffer memory, it can perform other functions. Once the CPU has written the block of data in the buffer memory, the printer can retrieve it from buffer memory (1 byte at a time) and print it at its own pace, which from Example 9.1 is 10.24 seconds. Obviously, 5.12 ms of computer time is more efficient than 10.24 seconds.

Stack memory

A stack memory is a portion of main memory whose purpose is to store data temporarily as the result of an interrupt request to the CPU. To understand the function of a stack, imagine that a CPU is executing a program, which we shall call the main program. At some point in time the CPU is requested by an I/O device to interrupt its execution of the main program to service it. To do so, the CPU must leave the main program and jump to another program, referred to as the I/O service subroutine, which instructs the CPU on how to service the requesting I/O device. Before the CPU jumps to the service routine, it stores (pushes) the address of the next instruction in the main program to be executed, referred to as the return address, on the stack. Also pushed on the stack are CPU register contents, such as the contents of scratch pad memory, which otherwise might be lost as a result of servicing the interrupt. When the CPU has finished executing the I/O subroutine, it pops (reads) the data that was pushed on the stack, which loads the CPU registers with the same data before the interrupt and restores the main program address (the return address) it had at the time of the interrupt. As a result, the next address output by the CPU for an instruction fetch, is the return address of the main program. This pushing and popping of data and return address on and off stack allows the CPU to be interrupted, then jump to a subroutine and execute that subroutine, and then return and resume execution of the main program without loss of data or place of execution within the main program.

Figure 9.1 illustrates the concepts of a stack. Figure 9.1a is a symbolic representation of both the CPU and its stack memory. It shows the stack as a portion of the main memory; as indicated by the bidirectional arrow of the data bus, it is an R/W memory. There are two points of access to this main memory, one via the top of the stack when accessed as a LIFO, which is its normal mode of access; and any location within the stack when accessed as any standard RAM (we mean RAM in its true sense here).

To reiterate, the first observation to be made is that the stack is a portion of main memory RAM. This portion of RAM is user defined, the user decides how much stack space is needed and then reserves that many memory locations, calling them the stack. The stack is accessed as a LIFO, that is, storage and retrieval of data are done through the top of the stack under control of the CPU. The memory spaces immediately following the stack are user RAM. These locations are accessed as RAMs (RAM is meant in its true meaning here) by the address decoders decoding the address on the address bus. User RAM is where programs and data are to be stored.

By observation one can conclude that if the user fails to define the stack large enough, and more bytes are pushed on the stack than there are stack memory locations, then the stack will fill and overflow into user RAM. Overflowing into user RAM will write over programs and/or data being stored in user RAM, therefore losing those bytes. The need to keep these two memory spaces separate, and to provide an orderly and organized means to address them, provides justification for having two different CPU addressing registers to supply these addresses. One to address memory for stack operations and the other for addressing user RAM for fetching program instructions and data.

Also note from Figure 9.1a that there are two busses, one internal to and the other external to the CPU. The internal bus is known as the *internal data bus* and the external bus is known as the *system data bus.* The two are buffered from each other by the buffer shown. Realize that it is over these busses that data and instructions are transmitted and received between memory and the CPU registers (the same can also be said of the CPU and I/O).

To see how the CPU addresses the stack and user RAM, examine Figure 9.1a. All addresses are supplied by the CPU, however, as previously stated, there are dedicated CPU registers for addressing memory when fetching program instructions and another for addressing the stack. The program counter (PC) is for addressing memory (user RAM) for instruction fetches and the *stack pointer* (SP) is for addressing memory locations in the stack. Assume addresses to be 16 bits (2 bytes), then both the PC and SP are 16-bit registers.

The PC is divided into two 8-bit registers, PCH and PCL. PCH is the high-order byte of the address and PCL is the low-order byte. The PC is divided into bytes for storage of its contents on the stack, providing the return address when the CPU service interrupts. When the CPU acknowledges an interrupt request it begins by pushing the contents of the PC (the return address) on the stack. The return address is stored on the stack in two consecutive memory locations, one for the PCL and the other for the PCH. After storing the return address on the stack the CPU then jumps to the service routine for servicing the request, which resides in another portion of memory (user RAM). At the end of that service routine is a return instruction that tells the CPU to pop the return address off the stack and into the PC. With the return address in its PC, the CPU will address the memory location following the memory location being addressed when the interrupt occurred. Beginning with the instruction at this address the CPU will resume fetching and executing instructions. When the CPU stores (pushes) and retrieves (pops) the return address on and off the stack, the SP provides the stack addresses.

As an example of the concepts just discussed, imagine the CPU being interrupted by an I/O device when the PC is 0500H and the SP is FFFFH. This means that the CPU is in the process of fetching an instruction from memory location 0500H and the SP is pointing to the next available memory location (FFFFH) at the top of the stack. When the CPU finishes execution of that instruction, it acknowledges the interrupt request by transferring control (jumps) to the I/O service subroutine. Before the CPU leaves the main program it automatically increments its PC and pushes the contents of the PC which is the return address 0501H, on the stack. It then transfers (jumps) to the subroutine, where the first instructions tell the

CPU to also push the contents of its scratch pad memory (registers A, B, C, and D) on the stack. To accomplish this requires the following steps: (1) The contents of the SP (SP = FFFFH) are latched by the address latch of Figure 9.1a and output on the address bus, which accesses memory location FFFFH (top of the stack). (2) The contents of the PC (PC = 0501H) are pushed on the stack by first writing the contents of PCH (PCH = 05H) into memory location FFFFH, via the internal data bus and system data bus (*see* Fig. 9.1b). (3) The SP is then decremented by one by the CPU (SP = FFFEH) and the content of PCL (PCL = 01H) is pushed on the stack. (4) The CPU jumps to the subroutine, where it is instructed to push the contents of register A on the stack. The SP is decremented again (SP = FFFDH), and the content of register A is pushed on the stack (*see* Fig. 9.1b). This pushing of register contents continues until all scratch pad memory contents are saved on the stack. To pop data off the stack the process is reversed, that is, the stack is read and the SP is incremented by one. The last data in is the first data out, which, as discussed in Chapter 8, is LIFO-type accessing.

Example 9.3

Using Figure 9.1b as a model, pop data from the stack and explain how the stack is accessed as well as where the data is to go.

Solution

When the stack of Figure 9.1b was filled, the content of the SP was FFFAH (the address of the bottom location of the stack). To pop data from the stack, the content of the SP is latched by the address latch, from which it is output on the address bus. This accesses location FFFAH. The CPU then activates control signal *RD*, and the content of location FFFAH (contents of register D) is loaded on the system data bus and also on the internal data bus (the buffer is active), from which it is latched by register D. The SP is incremented (SP = FFFBH) and the content of stack location FFFBH is loaded into register C. The SP is again incremented (SP = FFFCH) and the content of stack location FFFCH is loaded into register B. The SP is incremented (SP = FFFDH) and the content of stack location FFFDH is loaded in register A. The SP is again incremented (SP = FFFEH) and the content of stack location FFFEH, which is 01H, is loaded in PCL. The SP is again incremented (SP = FFFFH) and the content of stack location FFFFH, which is 05H, is loaded in PCH. The stack has been emptied and the CPU is back to its original state at the time of the interrupt.

Figure 9.2

A representative I/O service subroutine format.

PUSH A
PUSH B
PUSH C
PUSH D

The actual
service
subroutine

POP D
POP C
POP B
POP A
RET

It should again be noted that these pushes and pops are under program control, except for pushing the content of the PC on the stack, which is an automatic consequence of an interrupt acknowledge. A representative program to push and pop data on and off a stack is shown in Figure 9.2.

Figure 9.2 shows that the data of scratch pad memory is pushed on the stack (PUSH and POP are actual 8085 and 8086 microprocessor instructions). After the PUSH instructions are carried out, the service routine is executed. After execution of the subroutine, the scratch pad data is POPped off the stack. Last, a return instruction (RET) causes the return address to be popped from the stack and loaded into the PC (recall that the return address is automatically pushed on the stack when the interrupt request is acknowledged).

To reiterate, the stack memory is just a designated portion of main memory. What makes it different from main memory is not the hardware, for it is physically the same as main memory, but rather the addressing sequence executed by the CPU. The CPU addresses this portion of main memory so that it functions as a LIFO-type memory. It could be said that in this case the CPU functions as a LIFO controller for accessing the stack.

Queue memory

A queue memory is internal to the microprocessor (MPU) implementing the CPU and speeds up the instruction fetch and execute cycle, as well as makes more efficient use of the system data bus. To accomplish this, the CPU fetches instructions from main memory and stores them in proper sequence for execution in the queue, which is much like a small-capacity buffer memory (shift register) with FIFO-type accessing. To make better use of the system data bus, the CPU fills its queue at the same time it is executing instructions that do not require use of the system data bus. As a result, when the CPU finishes the execution of one instruction and is ready to fetch and execute the next, it fetches the next instruction from its queue rather than from main memory, which speeds up the instruction-fetching process, because internal fetches are much faster than external ones. This process is known as *pipelining*. Six- and 4-byte queues are common in some MPUs, such as the 8086 family.

Figure 9.3 is a simplified illustration of an MPU queue memory. Instructions are loaded in the queue as they are fetched from main memory. Because the queue is accessed as a FIFO, the first instruction fetched is the first one executed. The CPU fetches an instruction for execution from the queue and loads it into its *instruction register* (IR) where it is decoded for execution. As an instruction is fetched from the queue, the above procedure empties the entry location, allowing the CPU to fetch another instruction for the queue at a convenient time, that is, when the CPU is not using the data bus to execute an instruction.

Figure 9.3

An instruction queue.

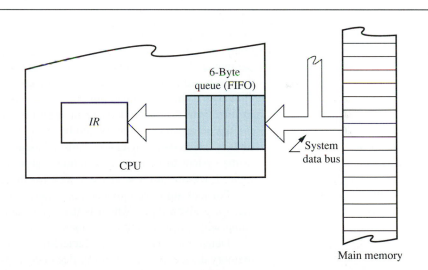

Example 9.4

Of the instructions listed below, which ones could the CPU execute while at the same time fetching another instruction from main memory to fill its queue?

(a) ADD A,B. This instruction adds the contents of the CPU's registers A and B and then stores the sum in register A.

(b) MOV D,C. This instruction moves (copies) the data in register C into register D.

(c) MOV B,DATA_1. This instruction moves the datum at memory location DATA_1, which is a symbolic representation of an address, into register B.

Solution

(a) Executing ADD A,B does not require use of the system data bus because both registers are internal to the CPU. Therefore, the CPU could execute this instruction and at the same time fetch another instruction for its queue.

(b) The CPU could execute MOV D,C and also fetch another instruction for the queue for the same reason as in Problem a.

(c) To execute MOV B,DATA_1, the CPU must use the system data bus to fetch the datum from memory location DATA_1; therefore, it cannot fetch another instruction at the same time.

Although the CPU cannot simultaneously execute and fetch instructions 100% of the time, two times out of three is not a bad average.

Look-up table (translation table)

There are many situations in which data must be converted (translated) from one format to another, as done by the encoders described in Chapter 4. One method of data conversion for a computer is for a program to manipulate and perform calculations on the data in order to transform it into a usable form. Another method is to use a look-up table to translate the data from one form to another, such as converting a person's name into that person's telephone number. For this conversion, the look-up table is the telephone book. Another example is use of a table of sine values to find the sine of 30 degrees.

To manipulate or perform calculations on data via a program requires CPU time, which for some applications may be more time than the application can tolerate. For instance, suppose a spacecraft had an on-board navigational computer. If it took this computer 10 seconds to perform the calculations necessary to convert coordinate parameters into a navigational decision that had to be made in less than 3 seconds, that on-board computer would not be an appropriate device for such an application. However, if this computer could be relieved of having to perform those calculations by having the CPU use a look-up table approach for the answer, which required just 5 μs, the on-board computer would be an appropriate device. Also, there are applications in which data conversions must be made, but there is no "intelligence" (CPU) in the system to perform calculations, as in a system composed of combinational logic. Look-up tables work very well for these types of applications.

For look-up table applications, a device is needed that can be programmed so that for a given input, which is the binary number to be converted, the converted value will appear as output. A memory device is well suited for this application.

Determine some other characteristics of this look-up table so that the type of memory device to be used can be decided. (1) Is there a CPU in this system that can write the look-up table values in the memory device? (2) Must the look-up table be present from a cold start? If the answer to the first question is no, a ROM must be used to implement the look-up table. If the answer to the second question is yes, again a ROM must be used. Only if the answer to the first question is yes and the second question is no might the system designer use a RAM. Even under these conditions a ROM might still be a better choice, for the designer may not want the look-up table taking up RAM memory, which is usually at a premium.

Example 9.5

As an example of a ROM look-up table application, program a ROM so that it will convert (encode) 4-bit binary numerical values into their corresponding binary ASCII values. Recall from Chapter 1 that ASCII is a coding scheme for representing characters on a keyboard.

Table 9.1 ROM look-up table to convert binary to ASCII

ASCII character	Input				Contents								ASCII hex value
	A_3	A_2	A_1	A_0	m_7	m_6	m_5	m_4	m_3	m_2	m_1	m_0	
0	0	0	0	0	0	0	1	1	0	0	0	0	3 0
1	0	0	0	1	0	0	1	1	0	0	0	1	3 1
2	0	0	1	0	0	0	1	1	0	0	1	0	3 2
3	0	0	1	1	0	0	1	1	0	0	1	1	3 3
4	0	1	0	0	0	0	1	1	0	1	0	0	3 4
5	0	1	0	1	0	0	1	1	0	1	0	1	3 5
6	0	1	1	0	0	0	1	1	0	1	1	0	3 6
7	0	1	1	1	0	0	1	1	0	1	1	1	3 7
8	1	0	0	0	0	0	1	1	1	0	0	0	3 8
9	1	0	0	1	0	0	1	1	1	0	0	1	3 9
A	1	0	1	0	0	1	0	0	0	0	0	1	4 1
B	1	0	1	1	0	1	0	0	0	0	1	0	4 2
C	1	1	0	0	0	1	0	0	0	0	1	1	4 3
D	1	1	0	1	0	1	0	0	0	1	0	0	4 4
E	1	1	1	0	0	1	0	0	0	1	0	1	4 5
F	1	1	1	1	0	1	0	0	0	1	1	0	4 6

Solution

Begin by constructing a table that shows the logic states of the independent variables (the inputs) and the desired logic states of the dependent variables (the outputs). The results are given in Table 9.1. The leftmost column of Table 9.1 lists the ASCII characters that are to be translated into their equivalent binary ASCII values. The next four columns comprise the inputs to the ROM, which are address pins (A_3 through A_0). Under those four columns are listed all possible addresses in binary, each address corresponding to the hexadecimal equivalent of the ASCII character given in the leftmost column. The next eight columns are the contents of each memory cell (m_7 through m_0) at the corresponding address. The binary patterns listed beneath them are the codes representing each ASCII character. The rightmost column gives the hexadecimal equivalent of the ASCII binary code. At memory location 0000_2 the ROM will have been programmed with 00110000_2, which is equivalent to hexadecimal value 30.

Hexadecimal 30 is the code for ASCII character 0. At location 0001_2 the binary value 00110001_2 is programmed, etc., until at the last location, 1111_2, the binary value 01000110_2 is programmed, the binary value for ASCII character F.

This look-up table application may be called a binary-to-ASCII encoder.

It is important to realize the relationship between the addresses and the data stored at those addresses because it is often the key to encoder solutions.

Example 9.6

Suppose a 16 × 8-bit ROM is programmed according to Table 9.1. What is the output of this ROM if the input is (a) 5H   (b) 7H (c) AH   (d) EH?

Solution

Referring to Table 9.1 (a) 5H is address 0101, which will output 00110101 (35H) (b) 37H   (c) 41H (d) 45H.

Review Questions

1. What are the five memory applications discussed in this section?

2. What are the essential characteristics of each of the five memory applications?

3. Why is a scratch pad memory often manufactured as an integral part of the CPU?

4. When would a scratch pad memory be a portion of main memory?

5. When a memory is referred to as a buffer memory, what is implied about its application?

6. What type of accessing is best for a buffer memory? Why?

7. What is the major function of a stack memory?

8. The stack memory of a computer system is a designated portion of main memory. How is a stack memory different from a main memory?

9. What function does a queue memory serve, and what type of accessing is used?

10. What is a look-up table and how is it implemented?

9.3

Memory device architecture and timing characteristics

Before designing a memory, choose the semiconductor memory devices to be used in the design. These choices are based on the performance requirements of the memory system, such as memory capacity and organization, speed of operation (access time and write cycle time), and power consumption. Cost is also a major consideration. Selection of a memory device is often based on the technology used in its fabrication. The technologies from which semiconductors are manufactured [TTL, emitter coupled logic (ECL), N-channel metal-oxide semiconductor (NMOS), and complementary MOS (CMOS)] have characteristics that make one better suited than another for a particular application. When we study the electronics of these technologies we will understand why they differ, but such knowledge is not necessary at this time. Table 9.2 lists characteristics of semiconductor technologies.

Summarizing the information of Table 9.2, the bipolar technologies (bipolar transistors are the active devices) TTL and ECL are the fastest, but they have less memory capacity than the others. NMOS and CMOS [metal-oxide-semiconductor

Table 9.2   Characteristics of semiconductor technologies

Technology		Bit capacity	Speed of operation	Power
RAMs	TTL	576	45 ns	1W
	ECL	1–16 K	20 ns	0.7–1W
	NMOS (dynamic)	16–64 K	150 ns	0.2–0.3W
	NMOS (static)	4–16 K	50 ns	0.6–0.8W
	CMOS	16 K	150 ns	0.3W
PROMs	NMOS EPROM	16–64 K	200–450 ns	0.5–0.8W
	CMOS EPROM	16–64 K	250–300 ns	40mW/MHz
	TTL PROM	1–32 K	35–65 ns	0.8–1W

field-effect transistors (MOSFETS) are the active devices] have larger memory capacities but are slower. CMOS devices consume less power.

There is a large selection of semiconductor memory devices. The priority of the three characteristics parameters listed in Table 9.2 depends on the application. The following computer systems are examined to demonstrate the relative importance of these three parameters.

1. Mainframe computers are powerful and fast, with a large memory capacity. The trade-off is between speed and memory capacity. If speed is the primary requirement, ECL technology must be used, but additional ECL memory devices are necessary to provide the desired memory capacity. However, if slower operating speed and fewer memory devices are acceptable, then NMOS technology can be used.

2. Personal computer (PC). A PC trades operating speed for memory capacity, because by their very nature they are to be small enough to fit on a desktop. Also, the CPU of a PC is a microprocessor (to be studied in Chapter 10) that is usually fabricated from one of the MOS technologies, which lessens the need for an exceptionally fast main memory. As a result, main memories of PCs are also fabricated from either NMOS or CMOS. If a PC has a battery back-up power supply, power consumption can be critical.

3. Portable PCs have the same basic requirements as a desktop PC except that power consumption must also be considered. Hence, CMOS should be a primary consideration for implementing main memory.

4. Microcontrollers are microprocessor-based computers that are dedicated to the control of some function or process. Such applications are often referred to as an embedded application of the microcontroller and the microcontroller is referred to as an embedded controller. Because they usually interface with other machines, they generally do not have a keyboard or terminal. Owing to their limited abilities and narrowness of purpose, the programs they execute require relatively little memory capacity. Microcontrollers must often be physically small so that they can be integrated into the system they are controlling, such as a robotic mechanism. As a result of the physical constraints and microprocessor operating speed, a microcontroller memory is most often of the MOS (NMOS or CMOS) technology.

Designing a main memory with any of the technologies listed in Table 9.2 is conceptually the same; it is basically a matter of choosing the technology that meets the system's memory capacity and speed of operation requirements. Cost is also a major factor, especially in a high-production environment.

To become familiar with semiconductor memory devices, we will study some representative architectures and timing characteristics of dominant memory device technologies, which are NMOS and CMOS. The architecture of those memory ICs are conceptually the same as those discussed in Chapter 8. The objective here is to take those concepts and see how they apply to some real semiconductor memory ICs. As for the timing characteristics of these memory devices, we shall learn the symbols used by manufacturers to represent the various timing parameters as well as their significance as they relate to the design and operation of main memory. Therefore, when reading the following material on timing characteristics, keep in mind that these parameters define the timing constraints under which a semiconductor memory IC must operate and these timing constraints must be adhered to when designing a memory system. Data sheets and excerpts from data sheets of the semiconductor memory devices studied are found in Appendix C.

Figure 9.4 shows the architecture and pin configuration and Figures 9.5 and 9.6 show the read/write timing diagrams for the 2148 static RAM (SRAM). The 2148 has ten address pins (A_0 to A_9) and 1024 (2^{10}) memory locations. The four pins labeled I/O_1 to I/O_4, which are similar in function to the I/O pins of Figure 8.11, are for reading and writing data. From the address and data pins it can be reasoned that the 2148 is a 1024 × 4 (4096)-bit memory. The memory array block states that there are 64 rows, which for six address pins (A_4 through A_9) agrees with the equation 2^6, but that same block also states that there are 64 columns, for which there are only four address pins (A_0 through A_3). As pointed out in Chapter 8, the manufacturer has included in this number the columns formed by the z-coordinate

---

Figure 9.4

The architecture and pin configuration of a 2148 NMOS 1024 × 4-bit static RAM. (Courtesy of FUJITSU Microelectronics, Inc.)

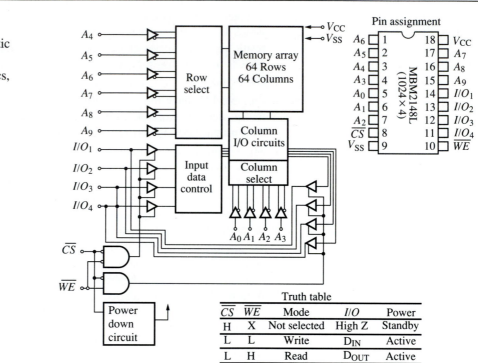

Figure 9.5

Basic read timing diagram for a 2148.

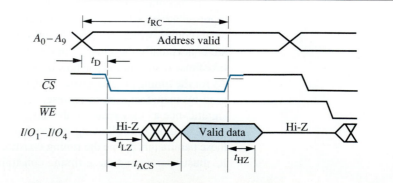

Figure 9.6

Basic write timing diagram for a 2148.

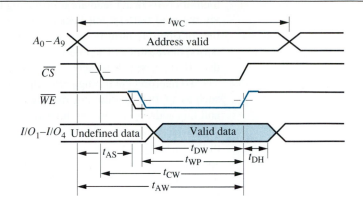

planes (*see* Fig. 8.11). Knowing this, we find that for four address pins and four planes of bits we have $2^4 \times 4 = 64$ columns, which agrees with the manufacturer.

The $\overline{CS}$ pin of Figure 9.4 serves to activate the 2148, just as pin $\overline{CS}$ did in Figure 8.11; however, the active logic level is a low. If $\overline{CS} = 1$, the chip is disabled (not selected) and all read/write buffers are in the Hi-Z state. When disabled the chip is in the standby mode, and the power consummation is reduced by as much as 60%. When pin $\overline{CS}$ of the 2148 is a logic 0, the *AND* gates to which it inputs are enabled. The other input to those *AND* gates, $\overline{WE}$ (write enable), controls whether the operation to be performed at the addressed location is a read or a write.

If $\overline{WE}$ is a logic 0, which is the active logic level for this pin, a write is to be performed; if $\overline{WE}$ is a logic 1, a read is to be done. This can be reasoned from the architecture of Figure 9.4. If $\overline{WE} = 0$ and $\overline{CS} = 0$, the top *AND* gate to which these signals are input will output a logic 1, which activates the data (I/O_n) buffers that allow data to flow from the data pins into the memory cell array via the input data control function block. When $\overline{WE} = 1$ and $\overline{CS} = 0$, the bottom *AND* gate outputs a logic 1, which activates the read buffers.

These concepts should be recognized from Figure 8.11; the application of these concepts aids in understanding the architecture of Figure 9.4.

Figure 9.5 is a basic read timing diagram for many SRAMs but has been personalized, via pin identification, to the 2148. Industry has standardized timing specification parameters in a generic manner, but they become personalized when parameter values for a specific memory device are supplied by the manufacturer in the data specification sheets of that device. These timing specifications usually appear in table form and are labeled "AC characteristics."

The timing diagram of Figure 9.5 shows the logic levels of the address pins A_0 through A_9, the chip select pin $\overline{CS}$, the write enable pin $\overline{WE}$, and the data pins I/O_1 through I/O_4. To interpret the meanings of these symbolic representations, first consider the address pins. When an address is input to a memory device, each address pin has either a logic 0 or a logic 1 applied to it. To simplify and generalize the timing diagram of those multiple address pins, they are represented as a group (A_0 through A_9) with both logic levels present. The crossover, or X, is the point of possible transition from one logic level to the other, such as when a change of address is being applied. When the address is stable (and usable), it is identified as "address valid" for that interval of time. The same symbolic representation applies to the data pins I/O_1 through I/O_4 as to the address pins. Each datum being read is present on those data pins in the interval labeled "valid data." The hash-marked interval just before the valid data interval is a period of transition. That is, the memory cells being read are transferring their respective logic levels to the data pins via the read buffers, but because all four read buffers do not activate at the same instant (there is turn-on time that is inherently different for each buffer), there is a period of transition. The interval labeled ''Hi-Z'' occurs when the I/O buffers are in the off state, which isolates those pins.

To interpret the timing parameters of Figure 9.5, recall that the accessing device, such as a CPU, first outputs the address of the memory location on the address bus, and this address is input to the address pins of the memory device as well as to the page selector decoder. The active page select decoder output then drives the $\overline{CS}$ pin of the memory device to its active logic low level, which enables the chip. There is some delay between input of the address to the page select decoder and output of the page selector. The authors have arbitrarily labeled this delay time ''t_D'' in Figure 9.5. Once the $\overline{CS}$ pin is driven low and the write enable pin is high, the addressed memory location outputs its content at pins I/O_4 through I/O_1. Owing to the time required to turn on the output buffers of the 2148, there is a delay between activation of the $\overline{CS}$ pin and availability of the datum at the I/O pins. This delay time is the access time of the 2148 measured relative to the $\overline{CS}$ pin being enabled. As indicated in Figure 9.5, the *chip select access time* is identified as ''t_{ACS}.'' Because the data pins of the 2148 are connected to the system data bus, the datum being read from the 2148 is also available on the system data bus after t_{ACS} seconds have elapsed. Parameter t_{HZ} indicates the time from deselection of the chip ($\overline{CS} = 1$) until the read buffers actually go into the Hi-Z state. The CPU must read the datum while it is available and stable on the system data bus. Using CPU read control signal $\overline{RD}$ to drive the $\overline{CS}$ pin, this occurs at the positive edge of $\overline{CS}$. The *read cycle time* (t_{RC}) measures the minimum time required to complete a read operation.

The primary parameters of Figure 9.5 are the t_{ACS} and the t_{RC}. These parameters are used as a gauge for the read operation speed of a 2148 RAM.

Note from Figure 9.5 that the address was input to the device before the $\overline{CS}$ pin was driven low. This timing diagram would also be valid if the $\overline{CS}$ pin were driven low at the same instant ($t_D = 0$) the address was input to the address pins, which is the case when partial linear addressing is used. The timing of Figure 9.5 would not be valid if the $\overline{CS}$ pin of the 2148 were driven low before a valid address was applied to the address pins.

The write timing parameters of Figure 9.6 are interpreted in the following manner: The minimum time before the write enable pin can be driven active ($\overline{WE} = 0$), as measured from the time the address is input to the address pins, is t_{AS} (the *address set-up time*). Because timing parameter t_{AS} is a timing requirement of the 2148, this ensures that address decoding will be completed by the 2148 before the write enable is activated. Timing parameter t_{WP} is the minimum pulse width allowable for a write operation to occur. This specifies a minimum requirement that must be met by the device driving the write enable pin ($\overline{WE}$), which is the CPU. Timing parameter t_{CW} specifies a minimum time for the $\overline{CS}$ pin to be active. Timing parameter t_{DW} specifies the minimum time that the datum being written into memory must be present before the write enable pin goes inactive ($\overline{WE} = 1$). Timing parameter t_{DH} is the minimum time that the datum must be held on the data pins after the positive edge of $\overline{WE}$. Parameter t_{AW} is the minimum time that the positive edge of $\overline{WE}$ can appear after the address has been input to the address pins. The *write cycle time* (t_{WC}) is the minimum time required to complete a write operation. Timing parameters t_{DW}, t_{AW}, and t_{WC} are the most critical, for they basically determine the speed at which the memory device can be written to.

There is a situation that is potentially damaging when a write operation is to be done. The condition occurs when the chip select is active before the write enable (as shown in Fig. 9.6), which is the logic state for a memory read. Under this condition the read buffers of the 2148 are active, resulting in output of the datum on the data pins (labeled as undefined data). Also, if during this period the device writing to the 2148 loads the datum being written on the data bus, bus contention exists until the write enable pin goes active and puts the read buffers in the Hi-*Z* state. Whether damage results depends on the current capabilities (to be studied later) of the 2148 and the device writing to it. A way to avoid this problem is to activate the chip select at the same time as or after the write enable, or to make sure that the CPU does not write on the data bus except during interval t_{WP}. The latter criterion will be used in our designs.

The parameters listed in Table 9.3 are representative timing values for one variety of 2148 (there are different versions of the 2148).

Table 9.3 Representative AC parameters of a 2148

Read parameters	Write parameters
$t_{ACS} = 55$ ns max	$t_{DW} = 20$ ns min
$t_{LZ} = 20$ ns min	$t_{WP} = 40$ ns min
$t_{HZ} = 20$ ns max	$t_{AW} = 50$ ns min
$t_{RC} = 55$ ns min	$t_{WC} = 55$ ns min

Table 9.3 shows that the read access time (t_{ACS}) is 55 ns, as is t_{RC}, for a 2148. The write parameters are as follows: t_{DW} is 20 ns, t_{WP} is 40 ns, and t_{WC} is 55 ns. The fastest access time of this variety of 2148 for a read or write operation is 55 ns. The values of t_{RC} and t_{WC} are used as an overall gauge of speed for 2148 RAM, but t_{ACS}, t_{AW}, t_{WP}, and t_{DW} are essential for detailed read/write timing coordination. It is interesting to note that the 2148 timing parameter t_{DH} has a value of zero, as does t_{AS} (not shown in Table 9.3).

It is important that the significance of timing parameters t_{ACS}, t_{DW}, t_{WP}, and t_{AW} be well understood.

Figure 9.7 shows the architecture and pin configuration of an 8128 NMOS SRAM that has a memory capacity of 16,384 (2048 × 8) bits. There are 11 address pins (A_0 through A_{10}), which is equivalent to 2048 memory locations, and eight data pins (I/O_1 through I/O_8), which indicates that each location is 8 bits wide. The 8128 has an input pin with a name different from any pin on the 2148 — the output enable pin $\overline{OE}$. From the architecture it can be reasoned that to activate the read buffers, the output of the three-input *NAND* gate must be high. This requires that the chip enable input ($\overline{CE}$) be low, the write enable ($\overline{WE}$) high, and the output enable pin ($\overline{OE}$) low. The output enable pin $\overline{OE}$ provides more control over read operations than does the 2148. A read for a 2148 is controlled by the chip select and the write enable, whereas a read is controlled by the chip select, write enable, and output enable for a 8128.

Figure 9.8 is a basic read timing diagram for an 8128. Comparing the timing diagrams of Figures 9.5 and 9.8, the addition of the output enable ($\overline{OE}$) is the major difference. The timing parameters t_{RC}, t_{ACE}, and t_{OE} are of most interest, for they are measures of how fast a read operation can be performed. Time parameter t_{ACE}, the chip enable access time, provides the maximum elapsed time between activation of

Figure 9.7

Architecture and pin configuration of an 8128 NMOS 2048 × 8-bit static RAM. (Courtesy of FUJITSU Microelectronics, Inc.)

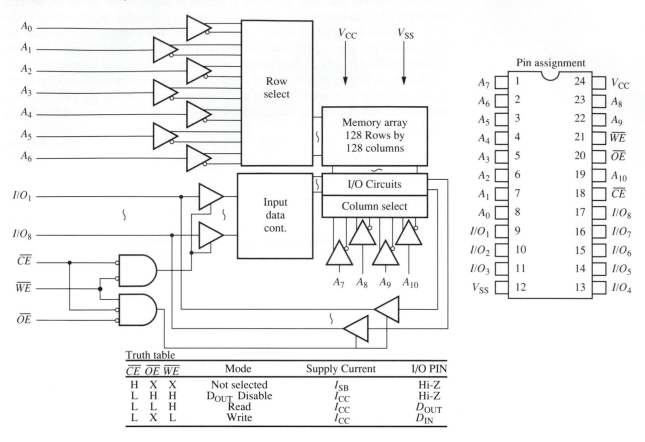

Truth table					
$\overline{CE}$	$\overline{OE}$	$\overline{WE}$	Mode	Supply Current	I/O PIN
H	X	X	Not selected	I_{SB}	Hi-Z
L	H	H	D_{OUT} Disable	I_{CC}	Hi-Z
L	L	H	Read	I_{CC}	D_{OUT}
L	X	L	Write	I_{CC}	D_{IN}

the chip enable and output of valid data (datum) at the I/O pins, assuming that the output enable pin has been activated for at least t_{OE} seconds. Time parameter t_{OE} is the maximum time that the output enable pin $\overline{OE}$ must be low before valid data (datum) is output at the I/O pins, assuming that the chip enable pin $\overline{CE}$ has been active for at least t_{ACE} seconds.

The write timing diagram for the 8128 is basically the same as Figure 9.6. The major difference is the output enable $\overline{OE}$, which for a write operation requires $\overline{OE}$ to be driven high (verify this from the logic of Fig. 9.7). For one type of 8128, some representative AC parameter values are given in Table 9.4.

Figure 9.9 shows the architecture of a CMOS SRAM that has 16,384 bits of storage capacity. Because there are 11 address pins and 8 I/O pins, the memory cell organization is 2048 × 8 bits. It appears to be drawn differently from those previously studied; most logic gates, decoders, etc. are represented as function-blocks rather than discrete logic gates and function-blocks. The former approach entails simplifying the architectural drawing by omitting some details; however, it has the same conceptual meaning.

This memory device has three input pins labeled "$\overline{WE}$," "$\overline{CS}$," and "$\overline{CE}$." By this time you should be able to decipher the function of these inputs. $\overline{WE}$ is the write enable pin and $\overline{CE}$ is a chip enable pin, and both are active low. $\overline{CS}$ is a chip select

Figure 9.8

Basic read timing diagram for an 8128.

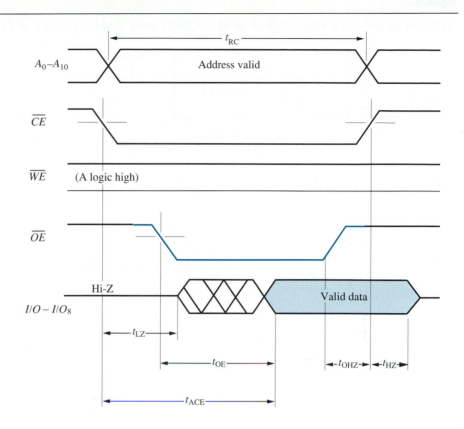

Table 9.4 Representative AC parameters of an 8128

Read parameters	Write parameters
$t_{ACE} = 100$ ns max	$t_{DW} = 40$ ns min
$t_{OE} = 50$ ns max	$t_{WP} = 85$ ns min
$t_{LZ} = 0$ ns min	$t_{AW} = 95$ ns min
$t_{RC} = 100$ ns min	$t_{WC} = 100$ ns min

pin. The $\overline{CS}$ input activates those buffers by which it is directly input as well as those with inputs labeled "CSB" (chip select buffer). Likewise, input $\overline{CE}$ enables all buffers with "CEB" (chip enable buffer) as an input. The I/O pins require $\overline{CE}$ (CEB) and $\overline{WE}$ to be active. Both the row and column access pins buffers are enabled by CEB ($\overline{CE}$), and the write buffers are activated by $\overline{WE}$ and $\overline{CS}$ (CSB). From the architecture it can be reasoned that both the $\overline{CE}$ and $\overline{CS}$ pins must be active to perform a read operation, because CEB ($\overline{CE}$) is an input to the I/O pin buffers and CSB ($\overline{CS}$) is an input to the write buffers, as well as $\overline{WE}$. In fact, there are six different timing diagrams possible for a read or write operation. These are:

1. A $\overline{WE}$-controlled read ($\overline{CE}$ = low, $\overline{CS}$ = low)
2. A $\overline{CE}$-controlled read ($\overline{WE}$ = high, $\overline{CS}$ = low)

Figure 9.9

Architecture and pin configuration of an 8417 CMOS static RAM. (Courtesy of FUJITSU Microelectronics, Inc.)

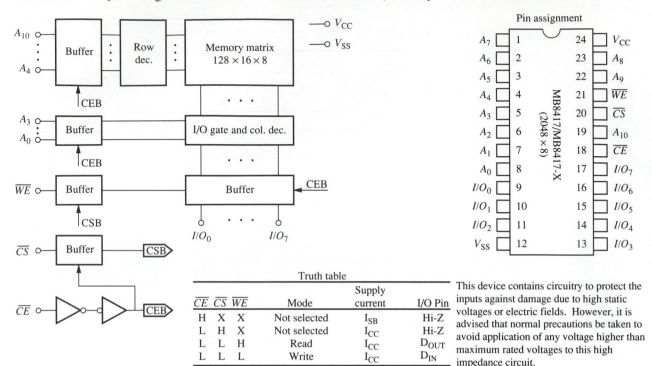

Truth table

$\overline{CE}$	$\overline{CS}$	$\overline{WE}$	Mode	Supply current	I/O Pin
H	X	X	Not selected	I_{SB}	Hi-Z
L	H	X	Not selected	I_{CC}	Hi-Z
L	L	H	Read	I_{CC}	D_{OUT}
L	L	L	Write	I_{CC}	D_{IN}

This device contains circuitry to protect the inputs against damage due to high static voltages or electric fields. However, it is advised that normal precautions be taken to avoid application of any voltage higher than maximum rated voltages to this high impedance circuit.

3. A $\overline{CS}$-controlled read ($\overline{WE}$ = high, $\overline{CE}$ = low)

4. A $\overline{WE}$-controlled write ($\overline{CE}$ = low, $\overline{CS}$ = low)

5. A $\overline{CE}$-controlled write ($\overline{WE}$ = low, $\overline{CS}$ = low)

6. A $\overline{CS}$-controlled write ($\overline{WE}$ = low, $\overline{CE}$ = low)

Each of these diagrams is unique, yet they are conceptually similar. The timing diagram for the read condition in number three and the write condition in number four will be studied. From these two representative timing diagrams the others can be determined.

The 8417 read timing diagram of Figure 9.10a is chip select controlled, which means that $\overline{CS}$ is the controlling pin for a read operation. The architecture shows that $\overline{WE}$, $\overline{CS}$, and $\overline{CE}$ are involved in all read / write operations and that for input pin $\overline{CS}$ to control the read operation, the write enable ($\overline{WE}$) must be high and the chip enable ($\overline{CE}$ is not shown) must be low before the chip select input becomes active. Figure 9.10a shows that the $\overline{CS}$ access time t_{ACS} is measured from the time that $\overline{CS}$ goes low until the datum is output (D_{OUT} valid) at the I/O pins. The $\overline{CS}$ access time is the maximum time required to read a datum from an 8417 using $\overline{CS}$ as the read control pin. Parameter t_{AA} is the address access time, the maximum time required to read the content of a memory location, measured from when the address is applied to the address pins. The $\overline{CS}$ pin must have been active t_{ACS} seconds, and $\overline{WE}$ must be at a logic high level and $\overline{CE}$ at a logic low. Parameter t_{AS} is the address set-up time; for the 8417 the minimum required time is 0 ns.

Figure 9.10b is a timing diagram for a write operation in which the write enable pin ($\overline{WE}$) is the write control pin. For the $\overline{WE}$ to control the write operation, the chip

Figure 9.10

Representative read and
write timing diagrams for an
8417 CMOS RAM.
(Courtesy of FUJITSU
Microelectronics, Inc.)

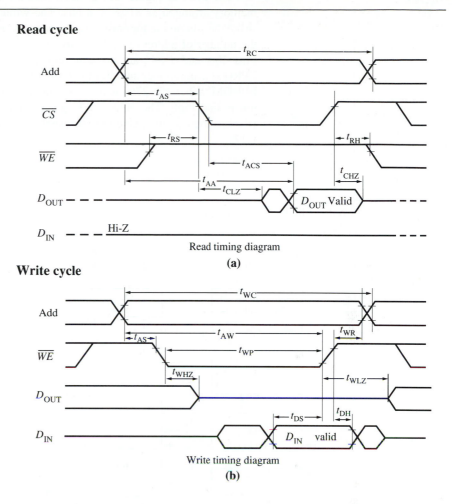

Read cycle

Read timing diagram

(a)

Write cycle

Write timing diagram

(b)

enable pin ($\overline{CE}$) must be active prior to $\overline{WE}$ being active. Figures 9.6 and 9.10b are
very similar, except that the chip enable and select pins of Figure 9.10b are assumed
to be low and are not shown. Parameter t_{DS} is the data set-up time, the minimum time
that the datum must be at the data pins prior to the rising edge of the write enable
signal $\overline{WE}$ (the datum is actually written into the 8417 on the positive edge of signal
$\overline{WE}$). The $\overline{WE}$ pin must be active for a minimum of t_{WP} seconds. AC parameter t_{AS} is
the address set-up time and is 0 ns for the 8417. Table 9.5 is a listing of some 8417
AC parameters.

Table 9.5   Representative AC parameters of an 8417

Read parameters	Write parameters
t_{ACS} = 100 ns max	t_{DS} = 60 ns min
t_{AA} = 200 ns max	t_{WP} = 140 ns min
t_{AS} = 0 ns min	t_{AW} = 160 ns min
t_{RC} = 200 ns min	t_{WC} = 200 ns min

The 2764 of Figure 9.11 is an NMOS UV EPROM device with 65,536 bits of memory arranged as 8192×8 bits. The window in the chip permits UV light to strike the silicon die for erasures. The memory organization should be confirmed via the number of address and data pins (the data pins are output pins labeled "O_n"). Pins V_{PP} and $\overline{PGM}$ are programming pins and are not discussed, because the EPROM programmer supplies the proper voltages and timing for programming the EPROM. When the 2764 is in the memory system for normal operations, the $\overline{PGM}$ pin is tied high. The chip enable pin $\overline{CE}$ enables the chip for a read operation, whereas the output enable pin $\overline{OE}$ enables the read buffers, as illustrated in Figure 8.12.

Figure 9.11

Architecture and pin configuration of a 2764 8192×8-bit NMOS UV EPROM. (Courtesy of FUJITSU Microelectronics, Inc.)

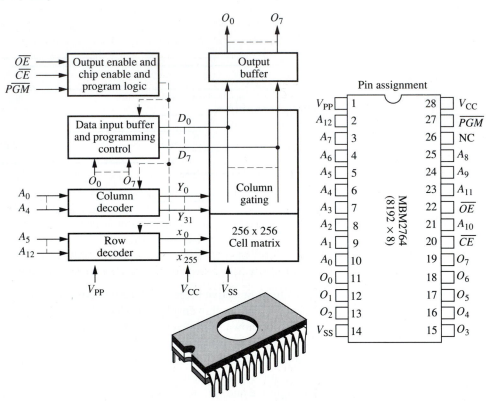

The read timing diagram is illustrated in Figure 9.12. The 2764 chip enable pin $\overline{CE}$ must be low t_{CE} seconds before the datum can be output on the data pins O_0 through O_7. When the output enable pin $\overline{OE}$ is driven low, t_{OE} seconds elapse before the datum is output at the output pins. From the time $\overline{OE}$ is deactivated (0 to 1) the datum may remain at the data pins for t_{DF} seconds. An overall measure of how fast a read operation can be performed is the access time, t_{ACC}, but this measurement requires that the timing specification of $\overline{CE}$ and $\overline{OE}$ be met. t_{ACC} will be used as the primary indicator of how fast the 2764 can be read. For one variety of 2764, $t_{ACC} = 200$ ns(max), $t_{CE} = 200$ ns(max), and $t_{OE} = 70$ ns(max).

Figure 9.12

Timing diagram for a 2764. (Courtesy of FUJITSU Microelectronics, Inc.)

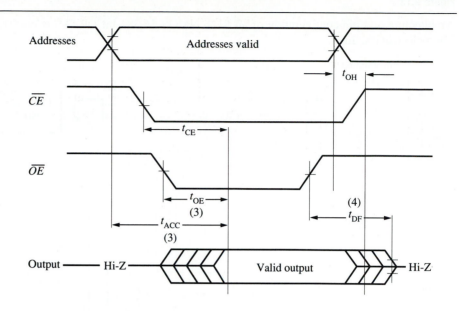

Notes: (3). $\overline{OE}$ may be delayed up to $t_{ACC} - t_{OE}$ after the falling edge of $\overline{CE}$ without impact on t_{ACC}.
(4). t_{DF} is specified from $\overline{OE}$ or $\overline{CE}$, whichever occurs first.

The architecture and pin configuration of Figure 9.13 are for an 8116 NMOS DRAM. The 8116 is a 16,384-bit memory that is organized as 16,384 × 1 bit. To address 16,384 memory locations, 14 address pins (2^{14}) are required, yet the 8116 has only 7 address pins (A_0 through A_6). These seven address pins are multiplexed for row and column addresses as can be seen from the architecture of Figure 9.13. Multiplexing pin function allows the pin size to be minimized. The address pins are input in parallel to both the row and column address buffers, which are latches in this case, and decoders. From the architecture it can be reasoned that the signal input to the row address strobe ($\overline{RAS}$) pin acts as a strobe to latch the address on the address pins into the row address buffer, and the signal input on the column address strobe ($\overline{CAS}$) pin provides the strobe for latching the bit levels of the address pins into the column address buffer. Obviously, the full 14-bit address is divided into two 7-bit parts.

The read and write timing diagram of Figure 9.14 shows that the 7-bit row address must be applied to the address pins first, followed by the 7-bit column address. Both timing diagrams also show the negative edge (1 to 0, where these levels are identified as V_{IHC} and V_{IL}) of $\overline{RAS}$ strobes row address into the "row address buffer and decoders" function-block (via clock generator number 1) and the negative edge of $\overline{CAS}$ strobes column address into the "column address buffer and decoder" (via clock generator number 2). Both the architecture and pin configuration of Figure 9.13 show that there are two data pins, D_{IN} and D_{OUT}. With proper timing of the signals input to $\overline{WE}$ and $\overline{CAS}$, the two data pins can be connected to form one I/O data pin, which is preferred. The timing is oriented to first identifying the type of access to be made (read or write) and requires that $\overline{WE}$ go active before the negative edge of $\overline{CAS}$ occurs.

Figure 9.14 shows that either a read or a write operation begins with the negative edge of $\overline{RAS}$. The row address must be on the address pins for a minimum of t_{ASR}

Figure 9.13

Architecture and pin configuration of an 8116 NMOS 16,384-bit DRAM. (Courtesy of FUJITSU Microelectronics, Inc.)

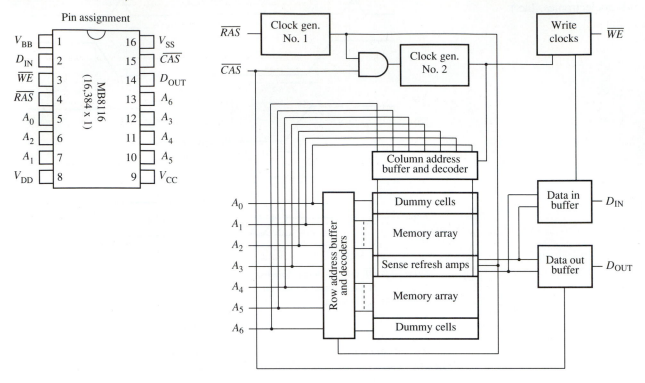

seconds (the row address set-up time) prior to the negative edge of $\overline{RAS}$. For the 8116E, t_{ASR} is 0 ns, meaning that the address may be applied at the same time $\overline{RAS}$ goes low. This row address must be held on the address pins for t_{RAH} seconds (the row address hold time), which for the 8116E is a minimum of 25 ns. After t_{RAH} seconds the address pins function as column address pins. The column address is input to the address pins, and on the negative edge of $\overline{CAS}$ this address will be latched by the column address buffer. From either timing diagram the column address hold time, t_{CAH}, is measured with reference to the negative edge of $\overline{CAS}$, which for the 8116E is 55 ns minimum. This completes latching of the row and column address for either a read or a write operation.

The timing diagram of Figure 9.14a indicates that to perform a read operation the write enable pin $\overline{WE}$ must be a logic high and the $\overline{RAS}$ and $\overline{CAS}$ pins must be low. There are two read access times, one for $\overline{RAS}$ and the other for $\overline{CAS}$. The read row access time, t_{RAC}, is measured from the negative edge of $\overline{RAS}$, and for the 8116E is a maximum of 200 ns. The read column access time, t_{CAC}, is measured from the negative edge of $\overline{CAS}$, and for the 8116E is a maximum of 135 ns. Because both $\overline{RAS}$ and $\overline{CAS}$ are required for a read operation, and t_{RAC} is the longer of the two access times, t_{RAC} is the limiting access time and is considered the access time of the 8116E.

The write timing diagram of Figure 9.14b indicates that, in this case, the write operation is controlled by the negative edge of $\overline{CAS}$. The reasoning behind this statement is based on the timing reference line in the "middle" of "valid data." The write strobe should be timed so that the datum being written into memory is stable on the data pin(s) and not in transition. Because the datum should be stable near the

Figure 9.14

Read and write timing diagrams for an 8116 DRAM. (Courtesy of FUJITSU Microelectronics, Inc.)

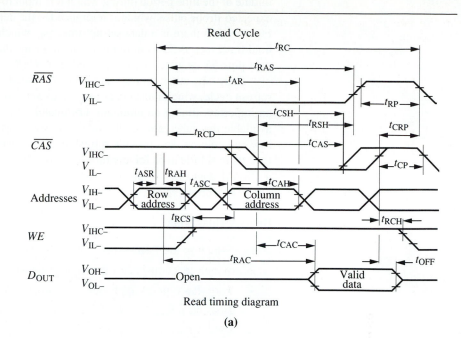

Read timing diagram

(a)

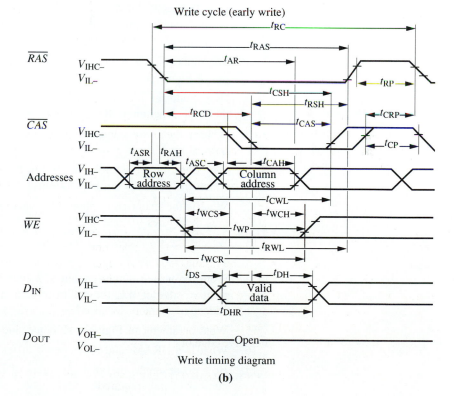

Write timing diagram

(b)

middle of the time period during which it is input to the data pin(s), look there for the expected strobe pulse, which is referenced to the negative edge of $\overline{CAS}$. According to Figure 9.14b there is a data set-up time, t_{DS}, which is 0 ns for the 8116E. The data hold time, t_{DH}, is 55 ns minimum, which means that the datum must be held on the data pins 55 ns after the negative edge of $\overline{CAS}$.

There are other timing diagrams associated with DRAMs. For instance, the refresh cycle, which must occur every 2 to 3 ms to refresh the data in the DRAM, is an important timing requirement. Fortunately, there are chips (DRAM controllers) available that take care of all read/write signal generation, divide the address into row and column, and provide refresh cycles as well. For the intended level of this text, there is little need for any further detailed timing diagrams of a DRAM, because a DRAM controller would be used to provide the proper signals and timing to a DRAM. A brief explanation of a DRAM controller will be given in the next section.

Review Questions

1. Which is the fastest technology — TTL, NMOS, or CMOS?

2. Which technology has the highest density?

3. For what type of applications is power consumption critical?

4. How do microcontroller memories differ from microcomputer memories?

5. What is the essence of the access time of a memory device?

6. How is t_{ACS} of Figure 9.5 similar to t_{ACE} of Figure 9.8?

7. What is the relationship between t_{ACE} and t_{OE} in Figure 9.8, relative to performing a read operation?

8. The input pin CS of Figure 8.11 is similar in function to which input pin of Figure 9.7?

9. How could the inputs RD and WR of Figure 8.11 be modified to serve the same function as $\overline{WE}$ of Figure 9.7?

10. What is meant by stand-by mode, and why does this mode consume less power?

11. Why does the write strobe occur "in the middle" of the period during which the datum is on the data pins?

12. What is the major conceptual difference between the architectural drawings of Figures 9.7 and 9.9?

13. Why is the $\overline{CS}$ access time t_{ACS} used to gauge the access time of the 8417?

14. From the write timing diagram of Figure 9.10b, explain why, on the positive edge of $\overline{WE}$, the datum on the data pin(s) is written into the addressed memory location?

15. What parameter of Figure 9.12 determines the access time of 2764?

16. What is a DRAM, and how does it differ from a static RAM?

17. How do designers and manufacturers of DRAMs economize on the number of pins required?

18. Why are DRAMs attractive for large memory applications?

19. What is the purpose of the inputs $\overline{RAS}$ and $\overline{CAS}$ of Figure 9.14?

20. What parameter of Figure 9.14a determines the access time of the 8116 DRAM?

9.4

AC compatibility

This section applies the concepts of Section 8.5 to the design of main memory systems. In designing a memory system one of many criteria that must be met is operational speed compatibility between the CPU and main memory. Assume that the memory systems being designed are for microprocessor-based systems; that is, an MPU is used to implement the CPU. As a result, an MPU will be implementing those operations attributed to a CPU. Because main memory is supportive to the MPU, the MPU sets the criteria that are to be met by main memory. Figure 9.15 shows two representative timing diagrams, one for an MPU read operation (a) and the other (b) for an MPU write operation. For either operation the MPU first outputs the address of the memory location on its address pins. Recalling those concepts developed from Figure 8.8b and 8.8c and relating them to the timing diagrams of Figure 9.15, the address is applied to both the address pins of the memory device and the page select decoder, via the address bus. The MPU control pins $\overline{RD}$ and $\overline{WR}$ differ from the control signals RD and WR only in their active logic level. Next, the MPU activates the appropriate control signal. If performing a memory read it activates its read control signal $\overline{RD}$ but if it is writing to memory then it activates its write control signal $\overline{WR}$. As stated in the previous chapter, the MPU is controlling the use of the data bus through these two control signals. If the MPU is reading memory then $\overline{RD}$ is used as an input to the semiconductor memory IC causing it to load the data of the addressed memory location on to the data bus, from where the MPU can read (latch) it on the positive edge of $\overline{RD}$. If the MPU is writing to memory then $\overline{WR}$ is used by the MPU to signify that it has written the data to be stored onto the data bus. The

Figure 9.15

A representative timing diagram for a microprocessor.

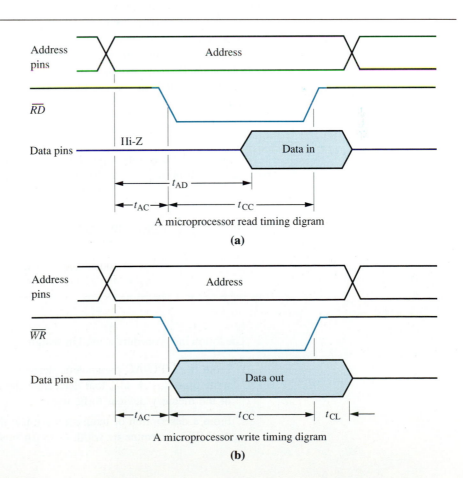

A microprocessor read digram

(a)

A microprocessor write digram

(b)

positive edge of $\overline{WR}$ is used as an input to the semiconductor memory IC to strobe the data on the data bus into the addressed memory location.

To examine the details of MPU timing and the meaning of the timing parameters refer to Figure 9.15. The read timing diagram of Figure 9.15a indicates that the datum being read from a memory location, labeled "Data in," must be at the MPU's data pins on or before t_{AD} seconds. For an MPU and a memory device to be time compatible, the access time t_{ACC} of the memory device must be less than (faster than) the read time t_{AD} of the MPU, where t_{ACC} is meant as a generic symbol. Mathematically stated, this is expressed as

$$t_{AD} > t_{ACC}. \tag{9.1}$$

The positive edge of $\overline{RD}$ serves as a strobe for the MPU to read (latch) the datum from the data bus, which brings the datum internal to the MPU via the MPU buffer of Figure 9.1. The positive edge of $\overline{RD}$ occurs in the middle of "Data in," (ensuring the data is on the data pins and is stable). The latching of the data by the MPU occurs at t_{AC} plus t_{CC} after the address is output by the MPU on the address bus.

The write timing diagram of Figure 9.15b indicates that the MPU outputs on the data bus the datum it is writing to memory t_{AC} seconds after it outputs the address of the memory location on the address bus. The MPU signals that it has written this datum on the data bus by pulling its write control pin $\overline{WR}$ low. The MPU holds $\overline{WR}$ low t_{CL} seconds less than it holds the datum on the data bus, which is t_{CC} plus t_{CL} seconds. Because the positive edge of $\overline{WR}$ occurs while the output datum is stable on the data bus, it can be used as a write strobe pulse to the addressed memory location so as to latch that datum from the data bus.

For the MPU and a memory device to be compatible in timing for a write operation, the memory device must be capable of storing a datum faster than the MPU can write a datum to it. This can be stated mathematically by using write control parameters t_{AC} and t_{CC} of Figure 9.15b and the memory write enable parameter t_{AW}, such as shown for 2148 and 8417 in Figures 9.6 and 9.10b, respectively.

$$t_{AC} + t_{CC} > t_{AW} \tag{9.2}$$

Table 9.6 contains some representative AC parameter values for an MPU. These parameters will be used as criteria for the AC parameters of memory ICs.

Table 9.6   Representative MPU AC parameters

$t_{AD} = 600$ ns max	$t_{AC} = 250$ ns min
$t_{CC} = 450$ ns min	$t_{CL} = 50$ ns min

The following procedure is used in designing a memory system:

1. From the MPU AC parameters, determine the read access time t_{AD} and write time $t_{AC} + t_{CC}$, that determine the minimum read and write times of the memory devices to be used.

2. From a description of main memory use determine its capacity, and from the MPU determine its width (8 or 16 bits). By knowing the length and

purpose of programs and data to be stored the capacity of memory can be estimated, and by knowing the size of the data bus the width of memory can be determined.

3. From the volatility of programs to be stored, determine how much of main memory is to be ROM and how much is to be RAM.

4. From the previous steps the designer knows the speed, capacity, width, and make-up (ROM and RAM) of main memory. With this information, along with power consumption, cost, and availability considerations, the designer can decide on the optimal semiconductor memory devices to be used in implementing a design.

As examples of implementing the first step of the design procedures outlined above, compare the MPU AC parameter values of Table 9.6 with the semiconductor memory device AC parameters of Tables 9.3, 9.4, and 9.5 to determine which of those devices are AC compatible with the MPU. As a first approximation of compatibility, compare the sum of t_{AC} and t_{CC} of the MPU with the read and write cycle time of the memory devices, which must be faster than the read and write time of the MPU.

$$t_{AC} + t_{CC} > \text{read / write cycle time of memory} \qquad (9.3)$$

From Table 9.6 we find for this MPU

$$t_{AC} + t_{CC} = (250 + 450)\text{ns} = 700\,\text{ns}$$

Returning to Figure 9.5, t_{RC} is used to reference the read cycle time of the 2148, and Table 9.3 shows that this parameter is 55 ns. Because 55 ns is less than 700 ns, Equation 9.3 is satisfied and the 2148 meets the first criterion for timing compatibility for a read operation. For a first approximation of write compatibility, Figure 9.6 shows that the 2148 uses t_{WC} to specify the write cycle time; according to Table 9.3, parameter t_{WC} is also 55 ns. Equation 9.3 is satisfied, and the 2148 has passed the first test for write timing compatibility.

Comparing the read / write timing parameters for the 8417, as represented in Figure 9.10, the read and write cycle times are also represented by t_{RC} and t_{WC}. Table 9.4 indicates that for this CMOS device the times are 100 ns for the read and write parameters, which are almost twice as long as for the 2148, but these times still easily satisfy Equation 9.3. On a first approximation basis, the 8128 qualifies as AC compatible with this MPU.

The read / write cycle times for the 8116 DRAM are specified by t_{RC}, which is 375 ns. The 8116 also has sufficient speed to be compatible with this MPU.

Because the 2764 is an EPROM read timing is the only timing of concern. In Figure 9.12 t_{RC} is not a specified AC parameter; therefore, decide which 2764 parameter(s) can be used as a measure of AC compatibility. In the MPU read timing diagram of Figure 9.15a the MPU latches the datum at its data pins on the positive edge of its read signal $\overline{RD}$, $t_{AC} + t_{CC}$. Therefore, the memory device being read — the 2764 in this case — must have the datum being read from it on the data bus, via its data pins, in that amount of time. In fact, Figure 9.15a indicates that the datum must be on the data pins t_{AD} seconds after the MPU outputs the address on the data bus, and the datum must be held on the data bus by the memory device until $t_{AC} + t_{CC}$. If at that point the AC parameter t_{ACC} of the 2764 is less than t_{AD} of the MPU, the two are AC compatible. As a note of interest for later use, in our design we used the read control signal $\overline{RD}$ of the MPU to hold the datum being read from the 2764 on the data bus. As stated earlier, t_{ACC} for the 2716 is 200 ns and as indicated in Table

9.6 t_{AD} is 600 ns; as a first approximation the 2764 and the MPU are AC compatible. Because the detailed analysis of the 2764's timing diagram has already started, it will be convenient to continue and finish the AC compatibility test.

Figure 9.15a shows that the MPU requires that the datum being read be held on the data bus for a minimum of

$$\text{time data held} = (t_{CC} + t_{AC}) - t_{AD} \qquad (9.4)$$

If the MPU read control signal $\overline{RD}$ is used as an input to the output enable pin $\overline{OE}$ of the 2764 (*see* Fig. 9.17), then t_{OE} of Figure 9.12 must be less than the "time data held" of Equation 9.4. Applying the appropriate data of Table 9.6 to Equation 9.4, the "time data held" is equal to 100 ns. It was stated earlier that t_{OE} for the 2764 is 70 ns (maximum), so the 2764 and this MPU are AC compatible and no further AC test need be made. It is important to understand the concepts of AC parameter comparison for the 2764 and MPU, for they will serve as a basis for timing analysis of the other memory devices.

For a more detailed comparison of AC read parameters, determine if the memory devices are quick enough to meet the MPU specification of t_{AD}, as already done for the 2764. By using the MPU read control signal $\overline{RD}$ to activate the chip select pin $\overline{CS}$ of the 2148 (the write enable pin $\overline{WE}$ must be high), it can be estimated from Figure 9.5 that t_{ACS} must be less than t_{AD}. Comparing those values of Tables 9.3 and 9.6 (55 ns and 600 ns, respectively), the 2148 is much faster than the MPU and is therefore AC compatible for a read operation. It is important to understand how Figure 9.5 was analyzed to use MPU control signal $\overline{RD}$ to drive $\overline{CS}$ of the 2148.

It can be concluded from the read timing of Figure 9.8 that the output enable pin $\overline{OE}$ of the 8128 can be used to control a read operation ($\overline{CE}$ could also be used). If $\overline{OE}$ is used, then t_{OE} specifies the read time relative to the negative edge of $\overline{OE}$. This implies that, in a memory design, the MPU read control signal $\overline{RD}$ can be used to drive $\overline{OE}$ of the 8128 (*see* Fig. 9.17). Because t_{OE} is 50 ns and t_{AD} is 600 ns, the 8128 is more than fast enough to be AC compatible with this MPU for read operations.

It can be concluded from Figure 9.10a that the MPU control signal $\overline{RD}$ can be used to drive chip select pin $\overline{CS}$ of the 8417 and, if so, t_{ACS} must be compared to t_{AD}. Table 9.5 shows that t_{ACS} is 100 ns, which is much faster than 600 ns. Hence, the 8417 is AC compatible for read operations.

Now determine write timing compatibility of the memory devices using reasoning similar to that for read operations. Reviewing the MPU write timing diagram of Figure 9.15b, the MPU writes the datum that is to be stored in memory on the data bus, via its data pins, t_{AC} seconds after the MPU outputs the address on the address bus. The MPU holds that datum on the data bus for t_{CC} plus t_{CL} seconds and signals that the datum is on the data bus by driving its write control pin $\overline{WR}$ low during that time. It can be seen that the positive edge of $\overline{WR}$ occurs while the datum is still present (and stable) on the data bus; therefore, control signal $\overline{WR}$ can serve as a strobe pulse. As a result, the write control signal $\overline{WR}$ of the MPU will be used to drive the write enable pin $\overline{WE}$ of the RAM memory devices.

It can be reasoned from the timing diagram of the 2148, as illustrated in Figure 9.6, that if the write control signal of the MPU is used to drive the write enable pin $\overline{WE}$ of the 2148 (*see* Fig. 9.23), the datum must be on the data pins of the 2148, via the data bus, a minimum of t_{DW} seconds before the positive edge of $\overline{WE}$. Of course $\overline{WE}$ has the same timing as $\overline{WR}$ if $\overline{WR}$ and $\overline{WE}$ are connected. Because memory must respond to the write control of the MPU, t_{DW} of the 2148 must be less than t_{CC} of the MPU. Table 9.3 shows that t_{DW} is 20 ns and Table 9.6 shows that t_{CC} is 450 ns, indicating write timing compatibility. Lastly, Figure 9.6 indicates that $\overline{WE}$ must be active for a minimum of t_{WP} seconds. Because the MPU write control signal will be driving the $\overline{WE}$ pin and (from Fig. 9.15b) will be active for t_{CC} seconds (450 ns) and t_{WP} is 40 ns for the 2148, this requirement is also satisfied. Hence, the 2148 is AC compatible with this MPU for both read and write operations.

Likewise, as indicated in Table 9.4 t_{DW} is 40 ns, indicating that the 8128 is also AC compatible for write operations. It can be concluded that the 8128 is AC compatible with this MPU.

By using the MPU control signal $\overline{WR}$ to drive the 8417 write enable pin $\overline{WE}$, it can be reasoned from the 8417 timing diagram that the MPU must write the datum to the 8417 data pins, via the data bus, t_{DS} seconds before the positive edge of $\overline{WE}$ (MPU control signal $\overline{WR}$) occurs. Because the MPU writes the datum on the data bus at the negative edge of $\overline{WR}$ and holds it there for at least t_{CC} seconds, at which time the positive edge of $\overline{WR}$ ($\overline{WE}$) occurs, the 8417 AC parameter t_{DS} must be less than the MPU AC parameter t_{CC}. As indicated in Table 9.5 t_{DS} is 60 ns and as indicated in Table 9.6 t_{CC} is 450 ns, which agrees with the criterion. Also, Figure 9.10b indicates that $\overline{WE}$ must be active for a minimum t_{WP} seconds or 140 ns; because t_{CC} is 450 ns, this criterion is also met.

It is evident that modern memory devices are certainly fast enough to be compatible with an MPU that operates at speeds indicated in Table 9.6. The specifications of Table 9.6 are representative of an NMOS 8-bit MPU.

There are other timing diagrams that can be applied to the sampling of memory devices studied so far, as well as timing diagrams for devices not yet studied. The principal concepts of timing diagrams have been sufficiently explained to enable you to interpret other timing diagrams with little assistance.

To demonstrate the remaining three steps of the design procedure, some design examples will be considered in the next section. Refer to and review the design concepts established in Chapter 8.

Review Questions

1. Why do MPU AC parameters determine the timing criterion for memory?

2. From the read timing diagram of Figure 9.15a, at what time does the MPU actually latch (read) the datum from its data pins?

3. From the timing diagram of Figure 9.15b, when does the MPU write its output datum to its data pins?

4. When the MPU writes data to a device (memory) via the data bus, how does the MPU signal the device that the datum is available on the data bus?

5. Generally speaking, what is the critical timing between the MPU and memory for a read or write operation?

6. For RAMs without an output enable pin $\overline{OE}$ and instead a write enable pin $\overline{WE}$ to control write operations, how can read operations be controlled?

7. For RAMs with both chip enable and output enable pins, such as the 8128, could a read operation be controlled by the chip enable pin rather than the output enable pin, as illustrated in Figure 9.8?

8. Why does the 8116 DRAM require input signals $\overline{RAS}$ and $\overline{CAS}$?

9. Why must a DRAM have a refresh cycle? How often must a DRAM be refreshed?

9.5
Designing static memories

All the memory devices of the previous section are AC compatible with the MPU AC specifications of Table 9.6. Therefore, to have AC compatibility for the design examples of this section assume MPU specifications from Table 9.6 and limit selection of memory devices to those covered in the previous section. To learn the process of memory design consider this example.

Example 9.7

An 8-bit microprocessor–based system requires a fully decoded isolated I/O memory with 5 K of memory for its bootstrap loader and 3.5 K of user memory space, memory in which the user can store programs and data. Note the implication that the bootstrap loader (recall that the bootstrap loader is a program that instructs the MPU from a cold start) is not a user program; rather, it is a permanently installed program that cannot be altered by the MPU.

Solution

From the description of memory, 5 K of ROM is required, because the bootstrap loader must be stored in a nonvolatile memory, and 3.5 K of RAM is specified, because the user memory space must be capable of read and write operations. Because the MPU is 8 bits, the data bus will be 8 bits wide; hence, 8-bit memory locations are required. Having only the 2764 EPROM available, implement the design with a single 2764, resulting in an 8 K $\times$ 8 of ROM rather than 5 K. For the 3.5 K of user memory use the 2148 (1 K $\times$ 4), the 8128 (2 K $\times$ 8), or the 8417 CMOS (2 K $\times$ 8). Using the 2148 would require that two 2148's be paralleled to achieve 8-bit memory locations and to get 3.5 K locations would require four pairs of 2148 memory devices. Better options are available, for example, the 8128 and 8417. Use an NMOS device, which is the 8128. Using the 8128 means that the user memory space is 4 K because two 8128's must be used to get at least 3.5 K of memory.

Now that the memory devices have been identified, assign memory device address pins to system address bus lines, as was done in Table 8.2. Find out the size of the address bus, which is determined by the number of address pins of the MPU. Most 8-bit MPUs have 16 address pins; assume that the address bus is composed of 16 address lines identified as A_0 through A_{15}. Table 9.7 shows these address bus lines as the top row. In the next row list the address pins of the memory device whose addresses must begin at 0000H (on a cold start this MPU goes to address 0000H for the first instruction), which is the 2764. The 2764 has 13 address pins

Table 9.7 Decoding the address bus of Example 9.7

System address bus lines		A_{15}	A_{14}	A_{13}	A_{12}	A_{11}	A_{10}	A_9	A_8	A_7	A_6	A_5	A_4	A_3	A_2	A_1	A_0
2764 Address pins					A_{12}	A_{11}	A_{10}	A_9	A_8	A_7	A_6	A_5	A_4	A_3	A_2	A_1	A_0
8128 Address pins (no. 1 and no. 2)							A_{10}	A_9	A_8	A_7	A_6	A_5	A_4	A_3	A_2	A_1	A_0
74LS138 pins (page selector)		A_2	A_1	A_0													
2764-0 Address range	0000H to	0	0	0	0	0	0	0	0	0	0	0	0	0	0	0	0
	1FFFH	0	0	0	1	1	1	1	1	1	1	1	1	1	1	1	1
8128-1 Address range	2000H to	0	0	1	X	X	0	0	0	0	0	0	0	0	0	0	0
	3FFFH	0	0	1	X	X	1	1	1	1	1	1	1	1	1	1	1
8128-2 Address range	4000H to	0	1	0	X	X	0	0	0	0	0	0	0	0	0	0	0
	5FFFH	0	1	0	X	X	1	1	1	1	1	1	1	1	1	1	1

identified as A_0 through A_{12} and they form the second row of Table 9.7. Align address pins A_0 through A_{12} of the 2764 to address bus lines of the same subscript. As a result, the lower 13 bits on the address bus pinpoint one of 8192 (2^{13}) locations within the 2764 memory device when the 2764 is chip enabled. The columns formed by these pin and bus line alignments in Table 9.7 are interpreted as bus connects when implemented in hardware. The third row of Table 9.7 lists the address pins of the 8128's, which have 11 pins each.

Table 9.7 shows that the number of address pins for the 2764 and the 8128's is not the same, which was not the case for the simplified design of Figure 8.13a. Follow procedures similar to those used to design Figure 8.13a and use address bus lines A_{13} through A_{15} as inputs to a page select decoder. Use a 74LS138 decoder whose logic symbol and truth table are shown in Figure 9.16. The fourth row of Table 9.7 represents the input pins A_0 through A_2 of the 74LS138 that are connected to address bus lines A_{13} through A_{15}. These first four rows serve as a wiring table for connecting device pins to address bus lines.

Figure 9.16

Logic symbol and truth table for a 74LS138.

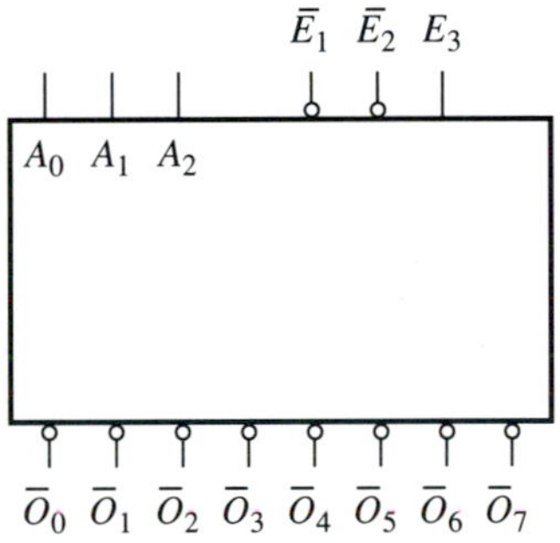

$0 \equiv$ Logic low
$1 \equiv$ Logic high

Inputs		Outputs
$\bar{E}_1 \bar{E}_2 E_3$	$A_2 A_1 A_0$	$\bar{O}_0 \bar{O}_1 \bar{O}_2 \bar{O}_3 \bar{O}_4 \bar{O}_5 \bar{O}_6 \bar{O}_7$
0 0 1	0 0 0	0 1 1 1 1 1 1 1
	0 0 1	1 0 1 1 1 1 1 1
	0 1 0	1 1 0 1 1 1 1 1
	0 1 1	1 1 1 0 1 1 1 1
	1 0 0	1 1 1 1 0 1 1 1
	1 0 1	1 1 1 1 1 0 1 1
	1 1 0	1 1 1 1 1 1 0 1
0 0 1	1 1 1	1 1 1 1 1 1 1 0
1 0 1	X X X	1 1 1 1 1 1 1 1
0 1 1	X X X	1 1 1 1 1 1 1 1
0 0 0	X X X	1 1 1 1 1 1 1 1

Use Table 9.7 to determine the first and last addresses within each memory device because the range of addresses will then be known. To determine the first address, apply 0's to the address pins of the memory device being studied; to determine the last address, apply all 1's to those address pins. The page selector pins A_0 through A_2 have a binary value that corresponds to the assigned page number of the device. Because 2764-0 contains the first page (page 0), the 74LS138 address pins will have 0's applied to them. The beginning address for the 2764, as indicated by the first address row of Table 9.7, has the 2764 address pins equal to 0's, whereas the second address row has logic 1's input to those address pins, representing the last address within the 2764. The 2764 has an address range of 0000H to 1FFFH.

The next range of addresses shown in Table 9.7 is for the 8128-1, where the suffix distinguishes between 8128's. To deselect the 2764 and instead select 8128-1, the logic levels input to address pins A_2 through A_0 of the 74LS138 must change to 001, as is indicated in Table 9.7. Because the 8128-1 has only 11 address pins, rather than 13 as the 2764 has, when the 8128-1 is selected address bus lines A_{11} and A_{12} have no function. As a result, the logic levels on those two address bus lines do not matter. When the logic level of a bus line or pin has no meaning, the logic level is known as a "don't care" and is symbolized by an X. 0's were used for the X's in the beginning address and 1's were used for the X's in the last address of the 8128-1. For this condition, the address range of the 8128-1 is 2000H to 3FFFH. Within this range of addresses there are 8192 memory locations ($3FFF_{16} - 2000_{16} + 1 = 8192_{10}$). However, because the 8128 has only 11 address pins, it has only 2048 memory locations, which is four times fewer than the 8196 addresses allocated to it. This means that every 8128-1 memory location has four addresses. For instance, any one of the following addresses accesses the same location: 2000H, 2800H, 3000H, and 3800H. Verify this by beginning with address

Figure 9.17

Memory design of Example 9.7.

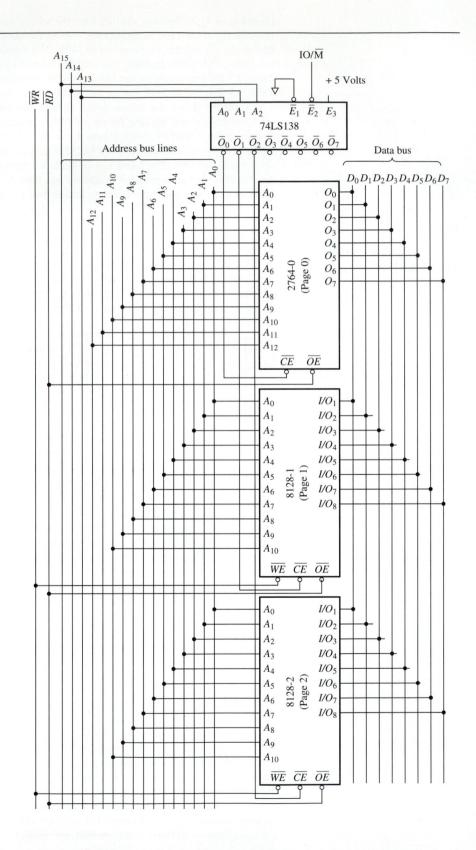

2000H and then substituting all binary combinations for the two don't care (XX) address bus lines A_{11} and A_{12}. The reason there are four addresses for each location is that there are two don't care address bus lines ($2^2 = 4$). The terminology used to describe the condition of multiple addresses for a location is *address foldback*. To have address foldback is to have wasted addresses. Foldback may be acceptable in small memory systems but not in large ones. How to avoid address foldback will be covered later in this chapter.

To select the 8128-2, the logic levels of address bus lines A_{15} through A_{13} must be changed to 010. Repeating the procedures used to determine the address range for the 8128-1, the 8128-2 has an address range of 4000H to 5FFFH. For the reasons already discussed for the 8128-1, the 8128-2 will also have address foldback.

Table 9.7 indicates how to connect the address pins of each chip to the address bus lines of the system. That is, those address pins in the same column are connected to the address bus line in that column. For instance, the A_0 pins of both 8128's and the 2764 are connected to address bus line A_0, and address pin A_0 of the 74LS138 is connected to A_{13} of the address bus. Figure 9.17 illustrates the implementation of the wiring indicated in Table 9.7. The outputs of the page select decoder (74LS138) go to the chip enable ($\overline{CE}$) of the corresponding page number; for example, output $\overline{O}_0$ is connected to the chip enable of page 0. Outputs $\overline{O}_3$ to $\overline{O}_7$ of the 74LS138 are available for memory expansion.

An address locates and selects the memory location to be accessed; however, there must also be proper control of the read or write operation that is to be performed. MPU-generated control signals $\overline{RD}$ and $\overline{WR}$ are used to control memory read and write operations. From our analysis of timing diagrams, we have already reasoned that the MPU write control signal line $\overline{WR}$ will be connected to the write enable pin $\overline{WE}$ of the 8128's to control write operations. To control read operations the MPU read control signal line $\overline{RD}$ will be connected to the output enable $\overline{OE}$ of each memory device as shown in Figure 9.17. MPU control signal $IO/\overline{M}$ provides separation between memory and I/O addresses by enabling the 74LS138 when a logic 0.

From the address ranges of Table 9.7 a memory map can be constructed for the system of Figure 9.17. A memory map is a geometric representation (a rectangle) of assigned memory addresses for a given memory system and is a convenient reference for both hardware and software designers. The hardware designer may refer to it when expanding or otherwise modifying memory, and the programmer may refer to it to obtain available addresses for programming. Figure 9.18 is a memory map for the memory of Figure 9.17. Keep in mind that foldback is present due to the don't cares.

Figure 9.18

Memory map of the memory in Figure 9.17.

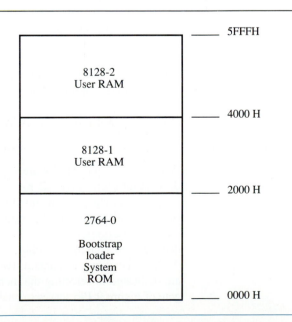

Before designing the next memory, outline the design procedures of Example 9.7.

1. Based on memory specifications, memory allocation is divided into ROM and RAM.

2. By knowing the required memory capacity and width of ROM and RAM, initial choices of memory devices can be made to implement the design.

3. These initial choices are checked for AC compatibility with the MPU.

4. An address bus decoding table (Table 9.7) is constructed, which indicates the wiring of the memory devices and page selector to the address bus as well as supplying memory addresses for a memory map.

5. The MPU must exercise control over all read and write operations, which is accomplished by having the MPU read control signal $\overline{RD}$ drive the out enable pin $\overline{OE}$ of the memory device and the MPU write control signal drive the write enable pin $\overline{WE}$ of the memory device. Of course an output from the page selector drives the chip enable pin $\overline{CE}$ (or chip select) of the memory device.

The address foldback of Example 9.7 is due in part to the different number of address pins for each memory device. Table 9.7 shows that address bus lines A_0 through A_{12} are dedicated to memory location selection within a device, the memory device with the greater number of address pins determining which address bus lines are to be used for this purpose. Address bus lines A_{13} through A_{15} select the memory IC. Because the 2764 has two more address pins than the 8128's, there are 2^2 times as many addresses in the 2764 as there are in either 8128, which is why there were four addresses for every memory location within each 8128. To eliminate address foldback due to a mismatch of the number of address pins, the memory device with the least capacity determines the address bus lines to be used for addressing a memory location within a memory device. Using this addressing scheme, a page of memory is defined as the number of memory locations within the memory device with the least capacity. Hence, a page of memory in Example 9.7 would be 2^{11} (2048) locations, as determined by the number of address pins on an 8128. Those memory devices with a greater capacity are divided into pages of the same size, as defined by the device with the smallest capacity. The number of pages assigned to a memory device will be

$$\text{number of page assignments (NPA)} = 2^m / 2^n \qquad (9.5)$$

where m is equal to the number of address pins of the memory device in question and n is equal to the number of address pins of the memory IC within the system with the least memory capacity. For a memory system composed of 2764's and 8128's, the n is 11 and m is 13 for the 2764. According to Equation 9.5, the NPA for the 2764 is

$$\text{NPA(2764)} = 2^{13} / 2^{11} = 4$$

When 2764's and 8128's are used in the same system, a 2764 is imagined to be divided into four pages (2 K per page).

When any page of memory locations within a memory IC is addressed, the page selector will select that memory device. If a memory IC contains one page of memory locations, one page select line (an output of the page select decoded) will be dedicated to enabling its chip enable, or chip select, pin, as illustrated in Figure 9.19a. If a memory IC contains two pages of memory locations, two page select lines are dedicated to selecting the memory IC, as illustrated in Figure 9.19b. The *AND* gate of Figure 9.19b serves to enable the memory device if either of its two pages is

Figure 9.19

Examples of page select line
assignments to memory
IC's.

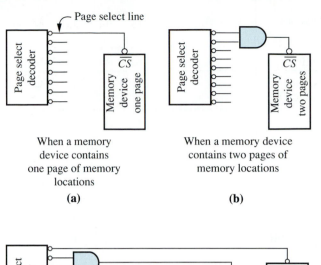

When a memory
device contains
one page of memory
locations

(a)

When a memory device
contains two pages of
memory locations

(b)

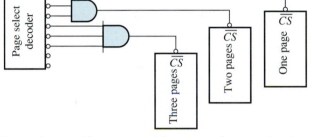

Memory devices with one, two, and three pages of memory locations

(c)

addressed. Figure 9.19c shows three memory ICs, each having a different number of memory pages, but in each case there is a one-to-one correspondence between the number of dedicated page select lines and the number of pages.

Example 9.8

Redesign the memory system of Example 9.7 such that address foldback is eliminated and the addressing is converted to memory-mapped I/O (eliminate $IO/\overline{M}$).

Solution

Begin with the creation of the address bus decoding table (similar to Table 9.7). The difference between the address bus pin assignments for this table and that of Table 9.7 is assignment of the page select decoder address pins A_0 through A_2. Because the page select decoder is used to select pages of memory locations rather than just memory IC, its address pins are used to decode those address bus lines that have the page addresses. These are the address bus lines that are numerically to the left of the most significant address pin of the memory IC with the least number of address pins, which is the memory IC that defines the number of memory locations in a page of memory. As seen from Table 9.7, address bus line A_{11} is the first numerically more significant address bus line to the left of the most significant address pin of the 8128's, which defines a page of memory to be 2^{11} memory locations. The next three address bus lines, A_{11} to A_{13}, are decoded by the page select decoder (74LS138) to select the addressed page of memory, as indicated in Table 9.8.

Table 9.8 Fully decoded address bus of Example 9.8

		A_{15}	A_{14}	A_{13}	A_{12}	A_{11}	A_{10}	A_9	A_8	A_7	A_6	A_5	A_4	A_3	A_2	A_1	A_0
System address bus lines		A_{15}	A_{14}	A_{13}	A_{12}	A_{11}	A_{10}	A_9	A_8	A_7	A_6	A_5	A_4	A_3	A_2	A_1	A_0
2764 Address pins					A_{12}	A_{11}	A_{10}	A_9	A_8	A_7	A_6	A_5	A_4	A_3	A_2	A_1	A_0
8128 Address pins (no.1 and no.2)							A_{10}	A_9	A_8	A_7	A_6	A_5	A_4	A_3	A_2	A_1	A_0
74LS138 Page selector				A_2	A_1	A_0											
Page	Address																
2764-0 Page 0	0000H to 07FFH	0 0	0 0	0 0	0 0	0 0	0 1	0 1	0 1	0 1	0 1	0 1	0 1	0 1	0 1	0 1	0 1
2764-0 Page 1	0800H to 0FFFH	0 0	0 0	0 0	0 0	1 1	0 1	0 1	0 1	0 1	0 1	0 1	0 1	0 1	0 1	0 1	0 1
2764-0 Page 2	1000H 17FFH	0 0	0 0	0 0	1 1	0 0	0 1	0 1	0 1	0 1	0 1	0 1	0 1	0 1	0 1	0 1	0 1
2764-0 Page 3	1800H 1FFFH	0 0	0 0	0 0	1 1	1 1	0 1	0 1	0 1	0 1	0 1	0 1	0 1	0 1	0 1	0 1	0 1
8128-1 Page 4	2000H 27FFH	0 0	0 0	1 1	0 0	0 0	0 1	0 1	0 1	0 1	0 1	0 1	0 1	0 1	0 1	0 1	0 1
8128-2 Page 5	2800H 2FFFH	0 0	0 0	1 1	0 0	1 1	0 1	0 1	0 1	0 1	0 1	0 1	0 1	0 1	0 1	0 1	0 1

Determine the number of memory pages in each memory IC using Equation 9.5. As already determined for this system, $n = 11$, because the 8128 has the least number of address pins. The NPA for the 8128's is

$$NPA(8128) = 2^{11}/2^{11} = 1 \text{ page}$$

and for the 2764

$$NPA(2764) = 2^{13}/2^{11} = 4 \text{ pages.}$$

Each of the two 8128's is assigned one page of addresses and the 2764 is assigned four pages of addresses. For this memory system there is a total of six pages of memory as shown in Table 9.8.

Table 9.8 indicates that the 2764-0 has four pages of memory that are identified with page address 0, 1, 2, and 3. The 8128-1 has one page of memory and that page address is 4. The 8128-2 also has one page of memory, which is assigned page address 5. The numerical suffix after each memory device no longer can be associated with a page number, as in Example 9.7, but rather is used as chip identification. The binary patterns of the page selector in Table 9.8 correspond to the page addresses. Throughout the entire range of memory addresses (0000H to 07FFH) for page 0, the 74LS138 address pins A_2 through A_0 have the binary bit pattern 000, for page 1 (0800H to 0FFFH) the bit pattern of these three address pins is 001, for page 2 it is 010, for page 3 it is 011, etc. It can be stated that for this memory system address bus lines A_{13} through A_{11} are used to address memory pages.

Address bus lines A_{14} and A_{15} are not used for addressing and are not decoded. As a result, they could be considered don't cares. However, this would allow address foldback due

Table 9.9 Truth table for A_{14} and A_{15}

		Output			
A_{15}	A_{14}	*AND*	*NAND*	*OR*	*NOR*
0	0	0	1	0	1
0	1	0	1	1	0
1	0	0	1	1	0
1	1	1	0	1	0

to unused address bus lines and, as in Example 9.7, there would be 2^2 addresses for each location (verify that addresses 0000H, 4000H, 8000H, and C000H address the same location if A_{14} and A_{15} are considered don't cares). To prevent foldback due to unused address bus lines, incorporate those unused lines in the addressing scheme. They are used to enable one of the page selectors' enable pins ($\overline{E}_1$, $\overline{E}_2$, or E_3). If logic levels of A_{14} and A_{15} are 0 for a valid address, as is assumed in Table 9.8, we can determine from Table 9.9 that we must use a *NOR* gate if E_3 is used (E_3 requires a logic 1 input) or an *OR* gate if either $\overline{E}_1$ or $\overline{E}_2$ (they require a logic 0 input) is used to prevent foldback. Let us somewhat arbitrarily choose to enable E_3. Then a two-input *NOR* gate is used to detect when both address bus lines are low, and its output is used to enable the page selector via its E_3 pin, as indicated in Figure 9.20.

Figure 9.20

Memory design of Example 9.8 (no address foldback).

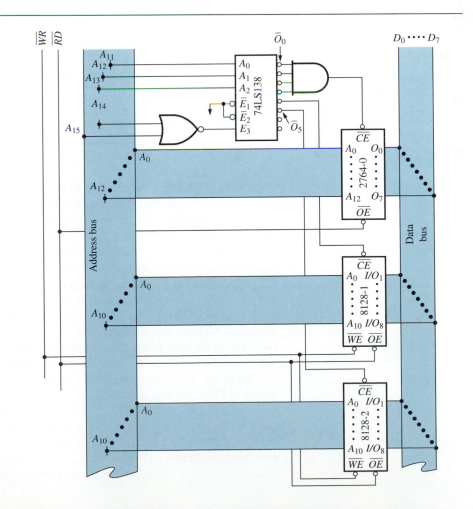

The design of this example is basically the same as that of Example 9.7, except for the connections of the 74LS138 address pins and enable pin E_3. For that reason, to simplify the wiring diagram of this memory design, symbolically abbreviate bus connections and detail just the 74LS138 connections. For detailed bus connections for the memory devices, refer to Figure 9.17. Figure 9.20 illustrates the design. It should also be observed that the MPU control signal $IO/\overline{M}$ is not present due to this memory system being memory mapped I/O.

The memory map for the design of Figure 9.20 can be obtained from the address ranges listed under the Address heading in Table 9.8. This memory map is shown in Figure 9.21.

Figure 9.21

Memory map for the memory of Figure 9.20.

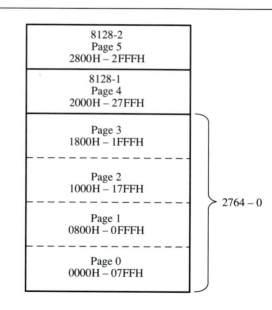

The 74LS138 has eight outputs that limit to eight the number of pages that can be addressed by a single 74LS138. To expand the number of page addresses available, up to eight 74LS138's can be paralleled to yield 64 pages of memory. To parallel 74LS138's, their address pins A_0 through A_2 are connected and then tied to the address bus lines that are used to address memory pages. Other numerically more significant address lines are used to address (enable) the desired page selector. Figure 9.22 shows four 74LS138's that have been paralleled. Address bus lines A_{11} through A_{15} were arbitrarily chosen to represent those address bus lines that are to furnish page addresses. Using five address bus lines means that 32 page addresses are available. Address bus lines A_{14} and A_{15} are used to enable the page selectors according to the logic levels of the truth table shown. Any time a logic 0 is to enable a pin, either $\overline{E}_1$ or $\overline{E}_2$ can be used, otherwise enable pin E_3 is used.

There are many semiconductor memory devices that do not have memory widths of 8 bits, yet they are used in designs with 8-bit widths. A very common width in DRAMs is 1 bit, as for the 8116 (*see* Fig. 9.13). Memory ICs may be paralleled to achieve the desired width. For instance, eight 8116's could be paralleled for each memory location to have an 8-bit width, or a pair of 2148's (Fig. 9.4) could be paralleled to achieve an 8-bit width.

Figure 9.22

Expanded page addressing.

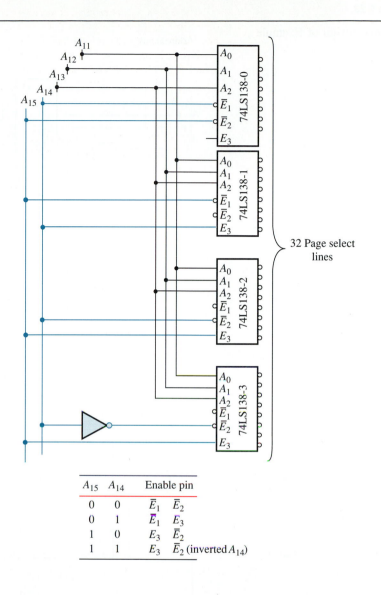

A_{15}	A_{14}	Enable pin	
0	0	$\bar{E}_1$	$\bar{E}_2$
0	1	$\bar{E}_1$	E_3
1	0	E_3	$\bar{E}_2$
1	1	E_3	$\bar{E}_2$ (inverted A_{14})

Example 9.9

Design a fully decoded isolated I/O 9K memory system that does not have address foldback. The system is to have 4 K of ROM and 5 K of RAM. Use a 2732 for the ROM and 2148's for RAM. A 2732 has pin functions similar to a 2764 ($\overline{OE}$, $\overline{CE}$, etc.), but it has 12 address pins.

Solution

Using 2148's requires pairing to achieve 8-bit memory locations, which is indicated in Figure 9.23, where the suffix *A* identifies the lower nibble (D_0 through D_3) and suffix *B* identifies the upper nibble (D_4 through D_7). Table 9.10 shows decoding of the address bus. Address bus lines A_{10} through A_{15} could be used for page addressing (A_9 is the most significant bus line common to all memory devices). To determine the number of memory pages required, calculate the number of memory pages per IC using Equation 9.5. Because the number of address pins of the 2148 is the least, *n* is equal to 10 and the 2148's define the number of memory locations to a page of memory (1024). Thus,

$$\text{NPA (2732)} = 2^{12}/2^{10} = 4 \text{ pages}$$

Figure 9.23

Memory design of Example 9.9.

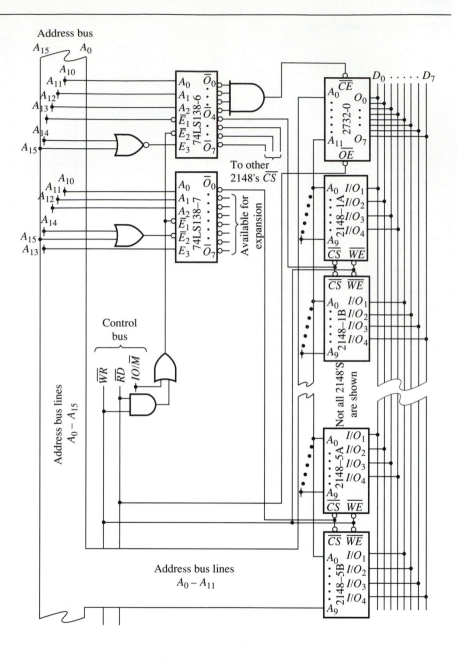

and

$$\text{NPA (2148)} = 1 \text{ page.}$$

Because the system is to have 9 K of memory, and a page is 1 K, nine pages of memory are required. The 2732 has four pages and five 2148 pairs have five pages. To address nine pages of memory, four address bus lines ($2^4 = 16$) are required. Table 9.10 indicates that these four address bus lines are A_{10} through A_{13}, where A_{13} is used to enable (E) the appropriate page selector enable pin. With one address bus line used to enable the appropriate page selector, there can be only two page selectors ($2^1 = 2$). When A_{13} is a logic 0, 74LS138-6 of Figure 9.23 is enabled via its $\overline{E}_1$ pin, and when it is a logic 1, 74LS138-7 is enabled via its E_3 pin. Address bus lines A_{14} and A_{15} are again used to prevent address foldback as indicated in Figure 9.23. If the logic levels of the address lines are to be 0's, a *NOR* gate must be used when

Table 9.10 Fully decoded address bus of Example 9.9

System address bus lines		A_{15}	A_{14}	A_{13}	A_{12}	A_{11}	A_{10}	A_9	A_8	A_7	A_6	A_5	A_4	A_3	A_2	A_1	A_0
2732 Address pins						A_{11}	A_{10}	A_9	A_8	A_7	A_6	A_5	A_4	A_3	A_2	A_1	A_0
2148 Address pins								A_9	A_8	A_7	A_6	A_5	A_4	A_3	A_2	A_1	A_0
74LS138 pins				E	A_2	A_1	A_0										
2732-0 Page 0	0000H to 03FFH	0 0	0 0	0 0	0 0	0 0	0 0	0 1	0 1	0 1	0 1	0 1	0 1	0 1	0 1	0 1	0 1
2732-0 Page 1	0400H to 07FFH	0 0	0 0	0 0	0 0	0 0	1 1	0 1	0 1	0 1	0 1	0 1	0 1	0 1	0 1	0 1	0 1
2732-0 Page 2	0800H to 0BFFH	0 0	0 0	0 0	0 0	1 1	0 0	0 1	0 1	0 1	0 1	0 1	0 1	0 1	0 1	0 1	0 1
2732-0 Page 3	0C00H to 0FFFH	0 0	0 0	0 0	0 0	1 1	1 1	0 1	0 1	0 1	0 1	0 1	0 1	0 1	0 1	0 1	0 1
2148-1 Page 4	1000H to 13FFH	0 0	0 0	0 0	1 1	0 0	0 0	0 1	0 1	0 1	0 1	0 1	0 1	0 1	0 1	0 1	0 1
2148-2 Page 5	1400H to 17FFH	0 0	0 0	0 0	1 1	0 0	1 1	0 1	0 1	0 1	0 1	0 1	0 1	0 1	0 1	0 1	0 1
2148-3 Page 6	1800H to 1BFFH	0 0	0 0	0 0	1 1	1 1	0 0	0 1	0 1	0 1	0 1	0 1	0 1	0 1	0 1	0 1	0 1
2148-4 Page 7	1C00H 1FFFH	0 0	0 0	0 0	1 1	1 1	1 1	0 1	0 1	0 1	0 1	0 1	0 1	0 1	0 1	0 1	0 1
2148-5 Page 8	2000H to 23FFH	0 0	0 0	1 1	0 0	0 0	0 0	0 1	0 1	0 1	0 1	0 1	0 1	0 1	0 1	0 1	0 1

a logic 1 output is required and an *OR* gate when a logic 0 output must occur (*see* Table 9.9). As a point of interest, the binary bit patterns of address bus lines A_{13} through A_{10} in Table 9.10 are a truth table sequence equivalent to the decimal value that identifies the page number.

To control the read and write operations MPU control signals $\overline{RD}$ and $\overline{WR}$ must be used. However, the 2148's do not have an output enable pin with which to control read operations as the 8128's do. To control read operations use the $\overline{RD}$ control signal to enable any page selector that is addressed to select a page of memory from a 2148. In this case, that is both 74LS138's. At first it would seem logical to connect $\overline{RD}$ to either $\overline{E}_1$ or $\overline{E}_2$ of the 74LS138's. However, the 74LS138's could not be enabled for write operations. However, inputting $\overline{RD}$ and $\overline{WR}$ into an *AND* gate and having the output of that gate drive either $\overline{E}_1$ or $\overline{E}_2$ enables the 74LS138 for a read or a write operation. To isolate memory addresses from I/O addresses, the output of this *AND* gate is ORed with control signal $IO/\overline{M}$. This logic circuit is illustrated in Figure 9.23.

Verify that the timing is compatible because the timing will be referenced to the negative edge of $\overline{RD}$ or $\overline{WR}$ via the 2148's input signal $\overline{CS}$. The timing diagrams of Figures 9.5, 9.6, and 9.15 indicate that for the timing to be compatible,

$$t_{\text{ACS}} < t_{\text{AD}}$$

and

$$t_{\text{WP}} < t_{\text{CC}}.$$

Tables 9.3 and 9.6 satisfy these conditions.

A memory map can be constructed easily from Table 9.10.

To convert the fully decoded address memory systems of Figures 9.17, 9.20, and 9.23 to partial linear address memory systems, eliminate the page selectors and enable the memory devices using address bus lines. Using an address bus line to enable a memory device means that it must be dedicated solely to enabling a memory device and cannot be used for any other purpose. This often leads to wasted addresses, in that addresses exist for nonexistent memory locations.

Example 9.10

Redesign the memory system of Figure 9.20 using partial linear isolated I/O addressing instead of fully decoded memory mapped I/O addressing. Construct a memory map and compare it with the memory map for Figure 9.20.

Solution

As in previous designs, begin constructing the address bus by developing an address bus table as is shown in Table 9.11. This table indicates that there are three address bus lines available for selecting (enabling) memory ICs. These address bus lines are A_{13}, A_{14}, and A_{15} and will be used to enable memory IC 2764-0, 8128-1, and 8128-2, as shown in Figure 9.25. To provide isolation between the memory and I/O address spaces, control signal $IO/\overline{M}$ is ORed with these address bus lines. Address bus lines A_{11} and A_{12} are don't cares (X) when an 8128 is addressed. Therefore, foldback exists. The don't cares are 0's in the address ranges of Table 9.11. By comparing Figures 9.20 and 9.25 it is easy to see that the hardware of a partial linear system is simpler. However, comparing the memory maps of Figures 9.21 and 9.24 shows the disadvantages of partial linear addressing—the wasted memory addresses due to foldback and the number of addresses for which no memory exists (0000H to 6000H, for example).

Table 9.11 Partial linear addressing of Example 9.10

System address bus lines		A_{15}	A_{14}	A_{13}	A_{12}	A_{11}	A_{10}	A_9	A_8	A_7	A_6	A_5	A_4	A_3	A_2	A_1	A_0
2764 Address pins					A_{12}	A_{11}	A_{10}	A_9	A_8	A_7	A_6	A_5	A_4	A_3	A_2	A_1	A_0
8128 Address pins							A_{10}	A_9	A_8	A_7	A_6	A_5	A_4	A_3	A_2	A_1	A_0
2764-0	C000H to DFFFH	1 1	1 1	0 0	0 1	0 1	0 1	0 1	0 1	0 1	0 1	0 1	0 1	0 1	0 1	0 1	0 1
8128-1	B000H B7FFH	1 1	0 0	1 1	X X	X X	0 1	0 1	0 1	0 1	0 1	0 1	0 1	0 1	0 1	0 1	0 1
8128-2	6000H 67FFH	0 0	1 1	1 1	X X	X X	0 1	0 1	0 1	0 1	0 1	0 1	0 1	0 1	0 1	0 1	0 1

Figure 9.24

Memory map for the memory of Figure 9.25.

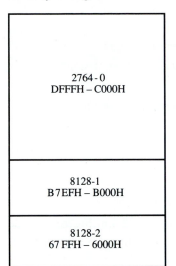

Partial linear addressing is often used in systems that require smaller memory capacities, as these systems do not require utilization of every possible address. The cost and space savings resulting from the reduced hardware are a real benefit. The user would be furnished a memory map to identify which memory addresses exist. When a system designer must make the most of the available memory space, a fully decoded system is the solution.

There are MPUs that require that ROM be placed at the beginning address of memory (0000H) and others that require that ROM be placed at the top end of memory ($A_{15} = 1$). This difference is due to the design of MPUs that causes them to automatically go (vector) to a specific memory location to fetch the first instruction of the bootstrap loader as a result of either a reset or a cold start. Under these conditions, some MPUs are designed to vector to location zero, whereas others vector to the top of memory.

A designer must be able to manipulate the placement of memory pages within the memory map. In partial linear addressing this simply means that the designer must choose the address line and its logic level to enable a memory IC. For instance, to place 2764-0 at memory location zero, address lines A_{15} and A_{14} could be inverted before being input to the chip OR gates which drive the enable pins of Figure 9.25. Under this condition, to enable 2764-0 the logic levels of the address lines are such that $A_{13} = 0$ and $A_{14} = A_{15} = 0$, which places 2764-0 at address zero. To enable 8128-1 requires that $A_{13} = 1$, $A_{14} = 1$, and $A_{15} = 0$, and to enable 8128-2 requires that $A_{13} = 1$, $A_{14} = 0$, and $A_{15} = 1$.

Figure 9.25

Partial linear addressing for Example 9.10.

To manipulate which ICs contain what memory pages within a fully address decoded memory, the designer selects which decoder output is used to select each IC. There are various combinations that exist for the placement of memory pages.

Review Questions

1. What is the rationale of a fully decoded address design?
2. What is the rationale of a partial linear address memory design?
3. How is the page selector(s) enabled?
4. How are the timing of RAM read/write operations controlled by an MPU?
5. How are ROM read operations controlled by an MPU?
6. What is foldback?
7. How can foldback be prevented?
8. What is a memory map, and of what value is it?
9. What is the significance of Table 9.9 to the 74LS138's of Figure 9.23?
10. Explain the function of the *AND* gate being driven by inputs $\overline{WR}$ and $\overline{RD}$ of Figure 9.23?
11. Without going into the details of specific AC parameters, what timing criteria must be met for MPU-initiated read/write operations, relative to memory ICs?
12. What is paging?
13. How is paging implemented?
14. How is expanded page addressing implemented?

9.6
Dynamic memories

The memories designed so far have had SRAMs. However, many memory systems are better suited to DRAMs because of their low power consumption and high-density memory capacity. The disadvantage is their required refresh cycle, which means additional circuitry and timing considerations. A survey approach will be used to study and design dynamic memory systems.

Recall from studying the 8116 DRAM that memory address pins are multi-plexed via address pins A_0 through A_6. These seven pins first serve as seven row address pins and then as seven column address pins, resulting in a 14-bit address (16,384 locations). The timing diagrams of Figure 9.14 indicate that a row ($\overline{RAS}$) and column strobe ($\overline{CAS}$) must be input to the 8116 so that it latches the 7-bit row address applied to its address pins and then the 7-bit column address. To the MPU there is no difference between a SRAM and a DRAM address. As a result, the MPU outputs this address as any standard 16-bit binary number on the system's 16-line address bus, just as it did for memories with SRAMs. It is necessary to take the 16-bit address and divide it in half, a low byte and a high byte, and apply those bytes to 8116 address pins A_0 through A_6. When these address bytes are being input to those address pins, strobe pulses must be generated and applied to 8116 input pins $\overline{RAS}$ and $\overline{CAS}$ at the appropriate time. Rather than waste engineering time designing the logic to do this, as well as engineering the logic to execute a memory refresh cycle, it is better to purchase a chip manufactured for this specific task. Such a device is known as a *refresh* or a *DRAM controller*. Figure 9.26 illustrates some of the essential ad-dressing functions of a DRAM controller. The memory refresh functions, which

Figure 9.26

DRAM controller for a 48 K
dynamic memory.

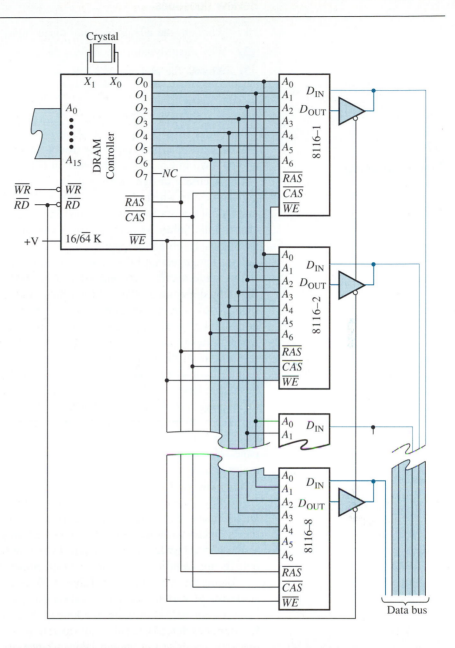

require a better knowledge of MPU technology, have been omitted. The DRAM
controller of Figure 9.26 has 16 address pins that are connected to the system address
bus lines. If the $16/\overline{64}$ K pin is strapped high, the DRAM controller treats the 16-bit
address as a 14-bit address (16 K); if strapped low, all 16 address pins are valid (64
K). The DRAM controller divides the 16-bit address into two 7-bit addresses that are
output at its output pins O_0 through O_6 (O_7 is not connected). The timing signals $\overline{RAS}$
and $\overline{CAS}$ are generated by the DRAM controller and are input to the 8116's so that
the row and column addresses are latched by the 8116's as previously discussed. The
DRAM controller has its own crystal because its refresh cycles are independent of
the MPU's clock. The DRAM has MPU control signals $\overline{RD}$ and $\overline{WR}$ as inputs so that
DRAM read and write operations are synchronized with the MPU.

Review Questions

1. What are the advantages and disadvantages of dynamic memories?
2. Why must dynamic memories be refreshed?
3. What is the essence of a DRAM controller?

9.7
Troubleshooting

The concepts and procedures for troubleshooting memory systems were developed and outlined in Chapter 8. This section will apply those concepts and procedures to testing memory designs with both static and dynamic tests.

To decide whether static or dynamic testing is appropriate, the diagnostician must recall that static testing is limited to diagnosing circuit continuity problems and catastrophic failure of components. Even then the failed design must be such that static logic levels can be applied, such as in the case of a newly constructed memory not yet integrated into the system. All other cases require dynamic testing.

A static testing environment will be devised for the memory of Figure 9.17. Begin by determining the number of test nodes and then classifying them as either input nodes, output nodes, or bidirectional nodes. As illustrated in Figure 8.16, input nodes require static logic generators to input the desired logic levels to the system, whereas the output nodes require static logic level detectors to detect the logic levels resulting from the logic levels applied to the input nodes. The bidirectional nodes require both static logic level generators and detectors. The control and address busses are input nodes while the data bus lines are bi-directional nodes. The outputs of the page selector are output nodes.

The memory system of Figure 9.23 has 19 input nodes composed of 16 address bus lines and 3 control lines. The system has 14 output nodes, if the outputs of the page selector and logic gates are monitored. Lastly, there are eight bidirectional nodes, which are the data bus lines. Figure 9.27 is an illustration of the static test configuration for the memory of Figure 9.23.

Apply static logic level generators to the input nodes as illustrated in Figure 9.27. Connect both static logic level generators and detectors, which are buffered, to the bidirectional nodes (data bus lines) as was done in Figure 8.16. Figure 9.27 shows a read / write switch to control these buffers (to turn them on and off). The outputs are identified in Figure 9.27 as circles with numbers.

Using the truth table of Table 9.10 as a map of the memory space, select addresses of various memory locations and then either read the contents of those locations for ROMs or write in a known data byte and read it back in the case of a RAM. For each read or read / write operation use a logic probe to test the corresponding logic levels at the outputs. For instance, to test location 0510H, put the static logic levels generator slide switches for the address bus lines to output logic levels 0000010100010000. Table 9.10 indicates that page 1 of the 2732-0 is being addressed. Because this is a ROM and can only be read, drive control signal $\overline{RD}$ low via its static logic level generator. $\overline{WR}$ must be driven high and $IO/\overline{M}$ low via their static logic level generators. With the address and control signal logic levels properly set, output number 14 can be checked for its resultant logic level, which must be a logic 0 to enable 2732-0. If this test point is low and the logic level on $\overline{OE}$ is also low, then the content of the memory location (which must be known in advance of the test) should be present on the data bus and detected by the static logic level detectors.

If the correct datum does not appear on the data bus with the proper address and control signals applied, begin to diagnose the problem starting at the point of the

Figure 9.27

Static testing of the memory of Figure 9.23.

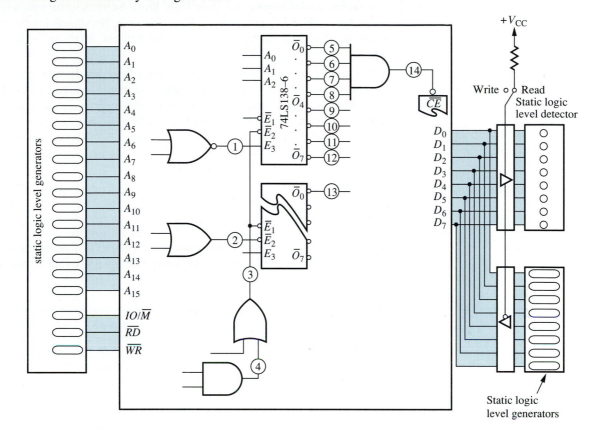

detected fault and working back to its source. Suppose the content of location 0510H of Figure 9.27 is not what is on the data bus, but rather the bus is floating (it is in a Hi-Z state). This means that the 2732-0 is not being enabled. Verify the logic level of output 14 and $\overline{OE}$. The output enable is connected to $\overline{RD}$ and its logic level is found to be low. However, output 14 was tested with a logic probe and it was high. The problem exists with the *AND* gate or output $\overline{O}_0$ of the page select decoder 74LS138-6. If the output of $\overline{O}_0$ was a logic 1, the decoder has failed or its enable circuitry is not functioning properly. To test the enable circuitry use a logic probe to verify the logic levels on outputs 1, 2, and 3, that is, a logic 1 at E_3 and logic 0's at $\overline{E}_1$ and $\overline{E}_2$. If these are correct, the decoder has failed and must be replaced. If one or more of these output nodes has the wrong logic level on it, continue to backtrack through the logic. Suppose that output node 3 had a logic high; this means that either the *OR* or *AND* gate has failed. If output 4 has a logic 0 on it, either the *OR* gate or the *AND* gate has failed.

To summarize and formalize an approach to troubleshooting memory systems or any other digital system, do the following:

1. Configure the test environment by deciding what is to be tested. Once the test objectives are decided, such as reading the content of a memory location, the diagnostician can reason from a truth table(s) and a system schematic(s) what nodes must be under control of the test as well as those to be monitored.

2. Input nodes are driven by logic level generators, which are not part of the system, and output nodes are to be monitored with logic level detectors. Output nodes require that their logic levels be monitored with test equipment such as a logic probe or logic analyzer. By classifying the test nodes, the diagnostician has also determined how the nodes are to be tested.

3. Decide what nodes are to be tested for correct logic levels. The logic levels of the input nodes are checked first, and then the output nodes that respond to those input nodes are tested for their logic levels. A truth table can often be of value by providing a reference showing the responses to various inputs. Once an incorrect response is found, the diagnostician can retrace the fault through the logic circuits until its source is identified.

4. Trace back through the logic looking for the source of the fault. The diagnostic techniques for tracing through logic circuits were discussed in the early chapters, beginning with Chapter 2.

The above four steps apply to both static and dynamic testing. Review the previous analysis of the static test of Figure 9.27 and relate the above four steps to the diagnosis.

Example 9.11

Suppose that for the test configuration of Figure 9.27 it was determined that the datum read from address 1250H was not the datum previously written into that location. All input nodes were tested and had the correct logic levels. All output nodes were tested and these logic levels were detected using a logic probe: 1 = 1, 2 = 0, 3 = 0, 4 = 0, 5 through 8 = 1, 9 = 0, 10 through 12 = 1, 13 = 0, and 14 = 1. What could be causing the fault?

Solution

Before beginning the diagnosis, understand what is implied by the statement, "the datum read . . . was not the datum previously written into that location." Address 1250H is in page 4, as can be determined from Table 9.10. Page 4 is made up of two 2148's, which are RAMs. Because there is no way of knowing for certain the content of a RAM memory location, test it by first writing a known data byte into that location and then reading the content of that location to verify that the written datum was indeed stored. To really be certain of a memory location, AAH should be written into it and then read from it and then the write and read process repeated for the complement of AAH(55H). This ensures that each memory cell can store both logic levels (a cell could have failed such that its logic level is always 1 or 0).

To begin the diagnosis, refer to Table 9.10 and the test configuration of Figure 9.27. Table 9.10 indicates that address 1250H is in page 4 of the memory space and that page 4 is contained in the 2148-1. Figure 9.27 (and Fig. 9.23) show that page 4 is enabled by page select line $\overline{O}_4$ of 74LS138-6, which is output 9. This is the start of the backward trace. By knowing that output 9 enabled (chip select) the 2148-1 pair, we also know that the correct logic level is a low, which was the logic level measured, and that the other outputs of the page selectors must be high. Logic 1's were measured at the other decoder test points except for 13, which was measured as low. This identifies a fault and diagnostic efforts can be focused on 74LS138-7 and from there fault tracing can be continued. To determine if the decoder is at fault, check the logic levels at its enable pins. Test point 2 should be high and 3 low. Test points 2 and 3 were measured to be logic 0's. This leads to the conclusion that the *OR* gate has failed, unless the $\overline{E}_2$ pin of 74LS138-7 has failed by shorting to ground. This can be checked when the *OR* gate is removed.

As stated in Chapter 8, dynamic troubleshooting is conceptually similar to static troubleshooting. The differences are the logic level generators and detectors, that is, the test equipment. Both generators and detectors must be dynamic. If the memory is an integral part of a system, the MPU acts as the logic level generator (it outputs addresses and control signals, and when writing it also outputs the datum). If the memory is isolated from a system (no MPU present), there must be a dynamic logic level generator that generates addresses and control signals. A counter and logic gates could be utilized for this task. In either case, the logic level detectors are implemented with a signature or logic analyzer.

To dynamically test the memory of Figure 9.23 a logic analyzer must be used because no correct signatures are given (none is on the schematic). As outlined by the four steps to be followed for troubleshooting, identify the objective of the test. If it is to verify the content of a memory location, the principal nodes of logic level verification are the data bus lines. In addition to the principal nodes also monitor input nodes such as enable pins and control lines for proper logic levels. The test configuration of Figure 9.27 is valid if the static generators and detectors are removed. There is no replacement required for generators if an MPU is present; otherwise there must be dynamic generators instead of static ones. We are assuming that an MPU is present. To verify the logic levels on the input nodes, an input channel of the logic analyzer must be connected to each input node, which requires 16 such nodes for the address bus and 3 for the control bus. In addition, eight input channels of the logic analyzer are needed to detect the logic levels on the data bus lines. So far we are requiring that 27 nodes be continuously monitored, which requires that the input analyzer have 27 input channels. To monitor test points 1 through 14 of Figure 9.27 requires 14 more channels. If the logic analyzer has a sufficient number of channels, an input channel of the logic analyzer may be connected to each of these nodes, a total of 41 channels. If the logic analyzer does not have enough channels, a channel would have to be shared for the remaining nodes and moved from one node to another for each sample interval. We will assume a 136-channel logic analyzer.

Recall that there must be some event (either a word or a trigger pulse) with which to start the data sampling. Determine what event is to act as the reference point. For Example 9.11 we could use the address 1250H as the trigger word. Under this condition, when address 1250H was output on the address bus, the logic analyzer would sample the nodes and display their logic levels on a CRT. This display is then analyzed for the various logic levels, just as if it were a static test.

Example 9.12

Suppose that when memory location 2000H of the memory system of Figure 9.23 is addressed and read, the data bus is found to be in the Hi-Z state. Using the nodes of Figure 9.27, diagnose the problem.

Solution

A logic analyzer is used to detect the logic levels on the nodes for the various sample times. Figure 9.28 is a representation of the logic analyzer's display, where the channels connected to the address and data bus lines are not displayed individually, as they actually are, but rather are grouped together symbolically. Also for simplicity and convenience, test points 6 through 12 are not shown.

Because address 2000H is the memory location in question, use 2000H as the trigger word to start the logic analyzer reading the logic levels at those nodes connected to its channels and storing that sampled data in its FIFO memory. Recall that, once triggered, the logic analyzer continues to collect data until the logic analyzer buffer memory (the FIFO) is full. The sampled data can then be displayed in a variety of formats. Display the data as a timing diagram, as illustrated in Figure 9.28.

Figure 9.28 indicates that outputting the address 2000H on the address bus triggered the logic analyzer, which then began to collect the nodal data at the rate indicated by the sample

Figure 9.28

The display of a logic
analyzer for Example 9.12.

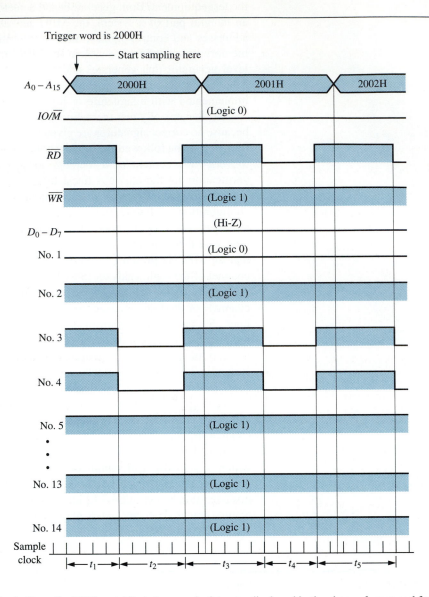

clock. Once the FIFO was filled, the sample data was displayed in the chosen format and from
that display we have Figure 9.28, which can be used much like a truth table. That is, the timing
diagram is divided into intervals of time (t_1, t_2, t_3, etc.), with the boundaries of these intervals
defined by a change in logic state of any of the input nodes that are responsible for a
corresponding change in an output node. The logic changes of $\overline{RD}$ define the time interval
boundaries (nos. 3 and 4 are outputs). Having defined the intervals, analyze the logic levels
within the intervals for correctness.

From the statement of the detected fault (data bus in Hi-Z) the addressed memory IC is
not being selected. From Table 9.10 address 2000H should select and enable page 8 of
memory, which means the 2148-5 should be enabled. From Figures 9.23 and 9.27 this means
that output 13 is to be low for a read (when $\overline{RD}$ is low). Therefore, it is when $\overline{RD}$ is low that is
of most interest. During interval t_2 of Figure 9.28 test point 13 is high, as it is also in t_4. Thus,
the page selector 74LS138-7 is not being enabled or the decoder has failed. This leads to
examination of the logic levels of test nodes 2 and 3, each of which must be low during these
intervals of t_2 and t_4. In both intervals test node 2 is high, whereas test node 3 is correct. Hence,
it is the *OR* gate of test node 2 that has failed.

In our examples trigger words were used to signal the logic analyzer to begin sampling and storage of nodal data. If our diagnosis were to be focused on a read or write operation, control signal $\overline{RD}$ or $\overline{WR}$ could be used to trigger the logic analyzer. Often an MPU has a signal that triggers the beginning of a fetch and execute cycle of an instruction (*ALE* for an 8085 or 8086 and *VMA* for a 6800). These signals can be used to sample data based on the instruction fetch and execute cycle.

Summary

- There are basically seven types of memories:

  1. Main memory

  2. Mass storage memory

  3. Scratch pad memory

  4. Buffer memory

  5. Stack memory

  6. Queue memory

  7. Look-up table memory

- Scratch pad memory stores intermediate and other temporary-type data. It should be easily accessed so as to minimize the CPU's data retrieval time, which often means that it is internal to the MPU.

- Buffer memories serve as a buffer between two devices, such as a computer and a printer. A buffer memory can be used to free up the computer from the printer.

- Stack memory also contains temporary data, such as MPU register contents and the return address as the result of an interrupt, but it differs from scratch pad memory in the manner of access and its physical location. Stack memories are usually a portion of main memory and are accessed in a LIFO fashion.

- Queue memory is internal to an MPU and holds instructions that were prefetched from main memory. As the MPU is ready to fetch an instruction, it does the fetching from its queue rather than from main memory, which speeds up the instruction fetch time and does not tie up the system data bus for an instruction fetch.

- Look-up table memories have been programmed with values that translate an input (address) to a desired output, which is the content of the addressed memory location.

- Manufacturers furnish data sheets for their semiconductor memory devices, and among other information these data sheets contain an architectural drawing and timing diagram. A memory designer can get a feel for how the memory device functions from its architecture, and this also helps the designer determine its organization. Timing diagrams indicate timing specifications of the memory device.

- Timing specifications for the MPU dictate timing parameters that must be met by memory devices

(including I/O devices). The MPU requires that the datum being read from memory be on the data bus in t_{AD} seconds from the time it outputs an address and be held there until the positive edge of $\overline{RD}$. For write operations the MPU outputs the datum it is writing to memory t_{AC} seconds from the time it outputs the address on the address bus. It holds that datum on the data bus for t_{CC} seconds.

- All memory devices must be checked for AC compatibility by determining if their read / write response time is faster than that of the MPU.

- In designing a memory, the designer chooses memory devices based on characteristics such as capacity, organization, speed, volatility, and cost.

- There are two addressing schemes, fully decoded and partial linear. Fully decoded addressing decodes all address bus lines, whereas partial linear does not; rather, partial linear addressing assigns some address bus lines to enable memory devices via the $\overline{CS}$ pin.

- Either addressing scheme may have two separate addressing spaces, one for memory addresses and the other for I/O addresses (an isolated I/O system) or just one address space containing both memory and I/O addresses (a memory mapped I/O system).

- Memory ICs of the same memory system with differing numbers of address pins can produce nonexistent addresses unless paging is used.

- Address foldback exists unless all address bus lines are used in determining memory addresses. Foldback results in multiple addresses for the same location.

- Dynamic memories have address pins that are shared among functions (multiplexed). They first serve as row address input pins and then as column address input pins. Dynamic memories also require refresh cycles for the device to remember the data stored. A DRAM controller can be used to demultiplex the address pins as well as to provide the required refresh cycle.

- To static test a memory, static level generators are used to input addresses, control signals, and data when a write operation is performed.

- Static testing is limited to testing for component failure and wiring problems.

- Static logic level detectors are used to detect the logic levels of the nodes under static testing.

- As with any digital troubleshooting procedure, a truth table can be used to assist in determining the correct logic level for an output for any given input. The truth table enables the diagnostician to identify an incorrect response. Once the error has been located, the diagnostician can retrace through the logic circuits until the source of the error is located.

- Dynamic testing is conceptually the same as static testing. The difference in implementation is that dynamic testing requires logic level generators that have changing

logic levels. This implies test equipment that can detect changing logic levels and display this data so that it can be observed and interpreted by the user. The signature analyzer and logic analyzer are two pieces of dynamic test equipment.

- Signature analysis is well suited for troubleshooting memory failures when nodal signatures are known.

- Logic analyzers can present data in several formats, of which a timing diagram is one. The diagnostician can use this timing diagram much like a truth table and from it identify correct and incorrect logic levels on the test system's nodes.

Problems

Section 9.2   Applications of semiconductor memories

1. What is the primary function of a scratch pad memory?

2. Why is a scratch pad memory usually an integral part of an MPU rather than a designated portion of main memory, as is a stack memory?

3. Using Figure 9.1a as a model, design a scratch pad memory. Assume each register (A, B, C, and D) to be an 8-bit register.

4. Why are buffer memories accessed using FIFO techniques?

5. Design a $4 \times N$ buffer memory.

6. Because a stack memory is often a portion of main memory, how do the two differ?

7. Could a stack memory be accessed using FIFO techniques?

8. With reference to Figure 9.1 and the explanation given of a stack memory, what determines how a stack memory is accessed?

9. Explain how using a queue memory can enable more efficient use of the system data bus insofar as the MPU is concerned.

10. Can a queue memory be likened to a small-capacity buffer memory?

11. When is a look-up table memory a viable solution to a programming problem?

12. Explain how look-up table memories are implemented using ROMs and encoders.

Table 9.12   Translation table for hex to seven segment

Hexadecimal value	Binary code	Seven-segment code
0	0000	00111111
1	0001	00000110
2	0010	01011011
3	0011	01001111
4	0100	01100110
5	0101	01101101
6	0110	01111101
7	0111	00000111
8	1000	01111111
9	1001	01101111
A	1010	01110111
B	1011	01111100
C	1100	00101001
D	1101	01011110
E	1110	01111001
F	1111	01110001

13. Design a translation device to convert hexadecimal values to seven-segment LED code. The code for hexadecimal and its corresponding seven-segment code is given in Table 9.12. Assume the look-up table is to be stored at beginning address 0300_{16}.

14. Design a look-up table that will square an integer with values of 0 to 9 and give the answer as the binary equivalent.

Section 9.3 Memory device architecture and timing characteristics

15. Which of the technologies described in this section is the fastest?

16. Which technology has the highest RAM bit density?

17. Why is MOS technology so popular?

18. What characterizes the main memory of a PC (personal computer)?

19. How would the main memory of a PC differ from the main memory of a microcontroller?

20. If a semiconductor memory device has an organization of 8192×8 bits, how many address and data pins does it have?

21. From the row and column address pin assignments of the 8128 architecture of Figure 9.7, determine how many rows and columns there are. Explain the discrepancy between the number of columns stated in the architecture (128) and the number calculated from the number of address pins.

22. According to Figure 9.5, when will a 2148 output the contents of an accessed memory location?

23. Why are memory device AC parameters often referenced with respect to an address being input to the memory device via the address pins?

24. From the write timing diagram of Figure 9.6, explain the AC parameter t_{DW}. When the MPU is writing to a 2148, how does t_{DW} relate to MPU timing?

25. What is the functional difference between $\overline{CE}$ and $\overline{OE}$ of an 8128?

26. What is the functional difference between t_{ACE} and t_{OE} of Figure 9.8?

27. Notice in the memory matrix function-block of Figure 9.9 that the memory organization is listed as $128 \times 16 \times 8$. Relate this method of notation to Figure 8.11 and the answer to Problem 21.

28. Explain the significance of the time interval specified by AC parameters $t_{AS} + t_{ACS}$ in Figure 9.10a.

29. Explain the meaning of data output, as indicated by D_{OUT} in Figure 9.10b.

30. Explain the purpose of the output enable pin and chip enable pin on the 2764.

31. What is the significance of AC parameter t_{ACC} in Figure 9.12? As part of your answer relate the meaning of this AC specification to a memory designer.

32. Name the major advantages and disadvantages of using DRAMs in a memory design.

Section 9.4 AC compatibility

33. Whenever a microprocessor executes a read or write operation there are three events that must take place. Referring to Figure 9.15, determine these three events and briefly explain the purpose of each event. You may also wish to refer to Figure 8.8.

34. Explain the significance of AC parameters t_{AD} and $t_{AC} + t_{CC}$ in Figure 9.15a. Relate your explanation to an MPU-reading main memory.

35. Which memory devices with the following AC specification are, or are not, AC compatible with an MPU with the AC specifications of Table 9.6? Explain your answer.
    a. t_{ACS} of Figure 9.5 is 750 ns.
    b. t_{DW} of Figure 9.6 is 475 ns.
    c. t_{OE} of Figure 9.8 is 470 ns.
    d. t_{DS} of Figure 9.10b is 470 ns.

36. Explain the implications of the time when $t_{AC} + t_{CC}$ occurs in Figure 9.15b, relative to the MPU writing to memory.

Section 9.5 Designing static memories

37. How can the enable pins of a 74LS138 be utilized when expanding the page addresses beyond eight pages?

38. Explain the purpose of creating an address bus table such as Tables 9.7, 9.8, 9.10, and 9.11.

39. Explain the purpose of a page selector in a fully decoded addressing scheme. Why does a partial linear addressing scheme not require a page selector?

40. Address foldback creates multiple addresses for a memory location while using memory devices with an unequal number of address pins can result in addresses for nonexistent memory locations. How can each of these situations be prevented?

41. Can foldback and nonexistent memory locations, as discussed in Problem 40, always be prevented? Explain your answer.

42. What defines a page of memory?

43. How are multiple memory page assignments to a memory device implemented in hardware?

44. How does partial linear addressing differ from fully decoded addressing?

45. When might a memory designer choose to use partial linear addressing versus fully decoded addressing? Explain your answer.

46. Design a fully decoded memory-mapped I/O main memory that is without address foldback or nonexistent addresses in the memory map for a 16-bit address bus. This memory is to have 1 K of nonvolatile memory for the bootstrap loader and 4 K of user RAM. The beginning address of the bootstrap loader is to be 0000H. Use 2716 and 8417 memory devices. The pin configuration of a 2716 is basically the same as that of a 2764 or 2732 except that it has just ten address pins (it is AC compatible with this MPU). As part of your design solution, develop an address bus table and memory map. How would you modify your design to make it an isolated I/O memory?

47. Redesign the memory of Problem 46 using partial linear isolated I/O addressing. Do not concern yourself with the starting address of the bootstrap loader.

48. Some MPUs fetch the first instruction to be executed after a restart or cold start from address zero. This would require the bootstrap loader program to begin at location zero, as specified in Problem 46. Redesign the memory of Problem 47 so that the beginning address of the 2716 is 0000H.

49. There are MPUs that fetch the first instruction to be executed after a cold start or reset from the top portion of memory (the highest possible address). Redesign the memory of Problem 46 for this type of MPU.

Section 9.6   Dynamic memories

50. Redesign the memory of Problem 46, but use 8128's instead of 8417's.

51. From a designer's point of view, what is the difference between using 8417's and 8128's?

Troubleshooting

Section 9.7   Troubleshooting

52. If we were to verify correct operation for each memory location of Figure 9.27 using static testing, how many addresses must be generated?

53. Suppose that the decoder of the memory-mapped I/O memory of Figure 9.17 was suspected of failure. How would you test the decoder using static testing?

54. Devise a static test to verify proper operation of the design in Figure 9.17. Illustrate the test configuration in a manner similar to Figure 9.27.

55. Suppose that the chip enable of the 2764-0 in Figure 9.17 was "stuck" high. Devise a static test to locate the problem.

56. Repeat Problem 54 using dynamic testing. Identify what nodal signal might be used for the trigger source.

57. Repeat Problem 55 but use a logic analyzer. Assume that a MPU is present.

58. Devise a dynamic test to verify proper operation of page 5 of the design in Figure 9.20. Use a nodal signal as the trigger source.

10

INTRODUCTION TO COMPUTER ARCHITECTURE AND MICROPROCESSORS

OBJECTIVES

- The objectives of this chapter are to:
- Describe the architecture of a generic microprocessor
- Explain the function of each function-block within the MPU architecture and how it is controlled by instructions of the instructions set
- Explain the fundamental concepts of MPU AC characteristics
- Introduce the concepts of interfacing an MPU to system hardware, that is, memory and I/O
- Present the fundamentals of instruction coding
- Introduce programming concepts
- Introduce I/O addressing
- Survey some real microprocessors

The personal computer is finding its way into almost all aspects of our everyday lives. Here a personal computer is being used with graphics software to display a map of Europe (Courtesy IBM).

10.1
Introduction

Microprocessor technology is a major application of digital circuits. A microprocessor, or *microprocessor unit* (MPU), is a semiconductor IC that as a minimum is capable of implementing a CPU, such as the one depicted in Figure 8.8a. It provides the pseudointelligence of a system because it manipulates data, for example, by performing arithmetic and logic operations and by reading data from and writing data to memory and I/O. Most significantly, based on data computational results, the MPU is capable of making decisions. The MPU has no real intelligence, but rather a pseudointelligence, because at present it is the human intellect that provides the intelligence of an MPU. Its pseudointelligence is provided by programs that instruct the MPU about the task to be performed by providing a step-by-step procedure.

The IC wafer on which an MPU is manufactured is very small, often slightly less than the size of a thumbnail, and contains thousands of MOSFETs (there are also bipolar MPUs). Because of the small size of an MPU and its pseudointellect — that is, its ability to process data and make decisions, it can be used in most applications requiring data processing and/or mechanism control. MPUs are found in medical equipment, automobiles, home appliances, PCs, electronic games, etc. In fact, one could say that an MPU can be used for any application requiring decision capability and/or process control.

Microprocessor technology can be divided into three areas of study: (1) MPU architecture, (2) hardware, and (3) software. MPU architecture provides a function-block diagram representation of the MPU that illustrates its internal structure. By knowing and understanding the internal structure of an MPU, a user (system designer and/or programmer, who can be the same person) can quickly refer to an architectural diagram to be reminded of the capabilities and limitations of an MPU as well as use it as an aid in understanding program instructions.

Hardware considerations are the physical aspects of the system, such as the pin configuration and system connections to those pins. From the memory designs of

Chapter 9 you should already have a feel for hardware considerations such as assignments and connections of address pins, data pins, and the control bus signal pins for read and write operations. As stated above, the architecture diagram aids in system hardware design and implementation by serving as a quick reference.

Software describes the programs and MPU instructions that make up a program. The software controls the hardware via the MPU. To write microprocessor programs for an MPU, the programmer must use instructions from a specific set of instructions the MPU is designed to recognize and execute, which is the *instruction set* for that particular MPU. The programmer must understand and have a general feel for the repertoire of instructions in the instruction set for that MPU. As already stated, the MPU architecture diagram can be a visual aid to programmers in helping them remember the software capabilities of an MPU as well as the general characteristics of the instruction set.

MPU architecture is a major focal point for understanding a specific type of MPU and the workings of the instructions within its instruction set. For this reason much of this chapter will be devoted to developing the concepts of MPU architecture. There are various types of MPUs available; however, their principal concepts of operation are very similar and to understand the concepts of one MPU is to understand them all. We shall take a somewhat generic approach, but our generic MPU is oriented to two specific microprocessors, the 8085 and 8086. At the end of the chapter three specific MPUs will be surveyed — the 8085, 8086, and 6800 — to which our generic architecture will be applied.

Our investigation and use of software will be limited because comprehensive coverage of programming is beyond the scope of this book. We encourage the reader to continue digital studies by studying microprocessors and their applications.

10.2
Computer architecture

To logically develop MPU architecture computer architecture must first be understood. Fortunately, most of the essential concepts of computers have already been covered in Chapters 2 through 9. Much of this section will review, collect, and organize our knowledge of computer systems.

Figure 10.1 is an illustration of a computer architecture, much like Figure 8.8a but expanded. This architecture has a three-bus structure and also function-blocks MPU, main memory, and I/O. The three busses are the address bus, data bus, and control bus; the control bus is formed by those lines over which MPU-generated signals $\overline{RD}$, $\overline{WR}$, $DMA\ ACK$, $IO/\overline{M}$, and $INT\ ACK$ and I/O-generated signals $DMA\ REQ$ and $INT\ REQ$ are transmitted. The computer architecture of Figure 10.1 shows MPU address pins A_0 through A_n connected to the address bus and the MPU data pins D_0 to D_m connected to the data bus. This MPU can generate 2^n addresses, as indicated by the number of MPU address pins, and can read or write m data bits in parallel for each read or write operation because there are m data pins. The process of accessing a main memory location was explained in Chapter 9.

To review some of the concepts developed in Chapter 9, recall that the MPU outputs the address of the memory location to be accessed on the address bus. The page selector enables the memory IC within main memory that contains that

Figure 10.1

A computer architecture.

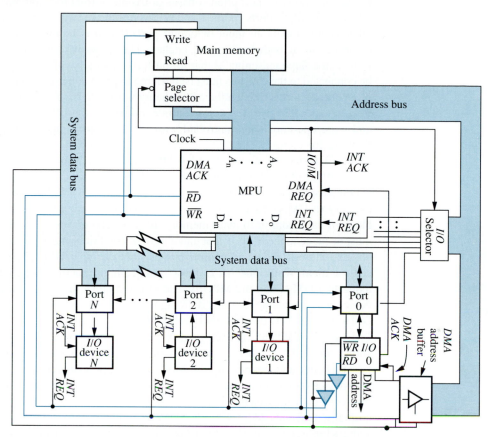

location. The accessed memory location has the datum either read from it or written into it, depending on the program instruction being executed. If the datum is being read from memory, the accessed memory location loads its contents on the data bus when MPU-generated read control signal $\overline{RD}$ is driven active. The MPU latches that datum off the data bus on the positive edge of that read control signal. If the MPU is writing to the addressed location, it outputs the datum on the data bus and drives the write control signal $\overline{WR}$ low at the same time. The positive edge of $\overline{WR}$ serves as a strobe for memory; as a result, memory latches the datum that is on the data bus into the addressed memory location.

Reading or writing to an I/O device is done in a similar fashion to reading or writing to a memory location. As shown in Chapters 8 and 9, the difference between memory and I/O operations concerns the logic state of MPU pin $IO/\overline{M}$.

Figure 10.1 employs a fully decoded addressing scheme as suggested by the presence of the page selector. Also, as in Figure 8.8a, Figure 10.1 indicates that addressed I/O ports (I/O ports are interfaces for I/O devices) are enabled via the I/O selector, which functions much like the page selector except for the method whereby it is enabled. That is, it is enabled when the $IO/\overline{M}$ control signal is a logic high.

Figure 10.1 has $N+1$ I/O devices that are buffered from the data bus by a port with the corresponding identification number. A port is the interfacing device that connects an I/O device to the computer system via the data bus. It is important to understand the difference between an I/O port and an I/O device. The hardware implementation of a port often consists of tristate buffers for input ports (port 2) and latches for output ports (ports 1 and *N*), where the direction of the arrow specifies whether the port is an input or output port. Port 0 is bidirectional and requires a combination of both buffers and latches. The necessity for buffering input ports follows the same reasoning as that for buffering memory cells: all outputs that are to be connected to the data bus must be buffered to prevent bus contention. The purpose of an output port is to latch the datum being written to the I/O device from the data bus. In many instances the port is an integral part of the I/O device. I/O port read and write operations are controlled with MPU control signals $\overline{RD}$ and $\overline{WR}$, just as they are for memory.

I/O device 0 of Figure 10.1 has DMA capability, which could mean that it is a mass storage device (DMA means that the device has direct access to main memory). I/O device 0 can request from the MPU, via the *DMA REQ* pin, a direct access to memory (a DMA); that is, it is requesting to bypass the MPU and read or write a block of data from or to main memory directly. The reason the request for a DMA must be made to the MPU is that under normal conditions the MPU controls the three busses (address, data, and control busses); but during a DMA, I/O device 0 will control them. Thus, the MPU must relinquish its control. When I/O device 0 requests a DMA, and the MPU is ready to relinquish those busses, it notifies I/O device 0 by activating its *DMA ACK* signal, which enables the device to execute the DMA process. I/O device 0 will keep *DMA REQ* in the active logic state during the DMA process. When I/O device 0 drives the *DMA REQ* pin low (inactive) the MPU knows the DMA has been completed and then takes back control of the busses.

To better understand why the MPU must relinquish control of the busses, it is I/O device 0 that must generate the memory addresses where the block of data is to be stored or from which it is to be read, whichever is the case. Because the MPU normally generates addresses, for a DMA it must relinquish control and use of the address bus and I/O device 0 must take control. To implement this process with the hardware, the address pins of the MPU are internally tristated and are put in the Hi-*Z* state during a DMA. The *DMA ACK* signal is also used to turn on the DMA address buffers of Figure 10.1, which allows I/O device 0 to gain control of the address bus for as long as the *DMA ACK* signal is active.

Not only must I/O device 0 control the address bus during a DMA, it must also control the control bus, for I/O device 0 must generate and transmit, via the control bus, the appropriate read or write control signal. As with its address and data pins, during a DMA the MPU puts its control signal pins in the Hi-*Z* state and activates *DMA ACK*, which turns on the two read and write control tristate buffers of I/O device 0. This action enables the I/O device to control the $\overline{RD}$ and $\overline{WR}$ lines of the control bus by becoming the source of those control signals. By controlling the control signals $\overline{RD}$ and $\overline{WR}$, it also controls the data bus.

When I/O device 0 is finished with the DMA, it drives its *DMA REQ* pin to the inactive logic state, which is detected by the MPU. In response, the MPU drives its *DMA ACK* pin to the inactive logic state, putting all of I/O device 0's DMA-related buffers in the Hi-*Z* state, and the MPU again takes control of the busses.

An exchange of request and acknowledge between two devices is known as a *handshake*. In the case just described for a DMA, the handshake is between I/O device 0 and the MPU. The request is made by I/O device 0 for a DMA by activation of its *DMA REQ* pin and the acknowledgment of that request by the MPU by activation of the MPUs *DMA ACK* pin. Specifically, these events are called a *DMA handshake*. Handshaking between two devices is necessary when the two devices share control of one or more busses or when one device needs the attention of the other.

Recall from the discussion on stack memories in Chapter 9 that some I/O devices are able to request of the MPU that it interrupt the program it is currently executing and instead vector (go to) to its service routine and service it. When the MPU acknowledges the interrupt request, which often is as soon as it finishes the instruction execution that was underway at the time of the request (it will not stop in the middle of executing an instruction), it activates its *INT ACK* pin. The *INT ACK* signal is input to the requesting I/O device, acknowledging that its request for service by the MPU has been granted. For illustrative purposes, all I/O devices except 0 were given interrupt capability. The interrupt request and acknowledge exchange between MPU and I/O is an *interrupt handshake*.

The computer architecture of Figure 10.1 and related explanations have provided an understanding of how a computer system operates. Use this knowledge to relate these concepts to the memory designs seen in Chapter 9.

Review Questions

1. What function does the MPU signal $IO/\overline{M}$ serve?
2. What do the page and I/O selectors have in common and how do they differ?
3. What is the function of an I/O port?
4. Explain how the MPU controls read and write operations.
5. What is meant by the term "handshaking"?
6. Why is handshaking essential?
7. Why does the MPU electronically isolate itself during a DMA?
8. Why does the DMA-requesting I/O device take control of the system busses during a DMA?

10.3
Microprocessor architecture

Our approach to developing and understanding MPU architecture is to divide it into sections according to function(s). We will partition the architecture into five sections:

1. Addressing section
2. Instruction section
3. Data storage section
4. Arithmetic, logic, and decision section
5. DMA and interrupt section

These five sections should be sufficient for describing the essence of all MPU architectures. Fortunately, from an educational point of view, understanding the architecture and operation of any MPU provides insight into the architecture and operation of all of them.

Addressing section

There are three different registers that can be used to address memory. These are the program counter (PC), data counter (DC), and stack pointer (SP). The PC and SP were discussed in Chapter 9 (*see* Fig. 9.1) but are briefly reviewed here. Each of these three registers addresses memory for a specific type of memory content. The contents of the PC are output on the address bus by the MPU when the MPU is fetching a program instruction from memory. The contents of the SP are output on the address bus by the MPU when the MPU is addressing the stack for the purpose of either reading or writing the datum from or to it as the result of a *POP, RET,* or *PUSH* instruction. Recall that when it acknowledges an interrupt request, the MPU automatically pushes the return address on the stack, which is popped off the stack using the *RET* instruction. Review Figures 9.1 and 9.2 as well as Example 9.3 if this material has been forgotten, that is, if your memory has become dynamic and you must therefore enter a refresh cycle. The DC is used as the addressing register when the MPU is accessing memory for the purpose of storing or retrieving data.

The only method for addressing I/O is via the I/O pointer (I/O P) register. Any time I/O is addressed, this register supplies the address of the I/O port being addressed.

Figure 10.2 is an illustration of the address section. The address latch, which has buffered outputs, serves as a buffer between the internal address bus of the MPU and the external system address bus. The output pins of the address latch are A_0 through A_n, which are represented in Figure 10.1. The address latch also serves as a latch in which the desired address can be stored by the MPU via its control unit. That is, the outputs of the PC, DC, SP, and I/O P are tristated, and the MPU, via its internal control unit, can enable the output buffers of that register whose contents are to be latched by the address latch. When the appropriate enable line is driven active by the control unit, the output tristated buffers of that address register (PC, DC, SP, or I/O P) output its contents on the MPU's internal address bus. While that address is

Figure 10.2

Address section of an MPU architecture.

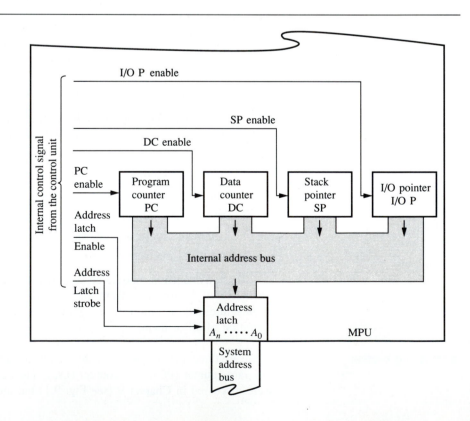

on the internal address bus, the *control unit* (CU) within the MPU (to be explained) generates and transmits a strobe signal to the address latch so that it latches the address off the internal data bus. The CU then activates the address latch enable line, which turns on the output buffers of the address latch, causing the content of the address latch to be loaded on the address bus.

As the CU outputs the memory or I/O address on the address bus, it also drives the $IO/\overline{M}$ pin either high (for I/O) or low (for memory) to indicate whether the address being output is an I/O or memory address. As seen in earlier chapters and in Figure 10.1, this signal is used to enable the appropriate selector so as to isolate the memory and I/O address spaces from each other.

Instruction section

When an instruction is fetched from memory it is stored in an MPU register so that the binary code representing the instruction can be decoded. Decoding the instruction's binary code tells the MPU which specific instruction of its instruction set it has fetched from memory. When it identifies the instruction to be executed, the MPU knows what steps to take to execute that instruction. As discussed in Chapter 9 and illustrated in Figure 9.3, to make a more efficient use of the data bus, some MPUs have an instruction queue within which instructions can be prefetched and stored while awaiting execution. Our model MPU will have an instruction queue.

Figure 10.3 illustrates the instruction section. The first observation to be made from Figure 10.3 is that the last function-block in the instruction fetch and execute stream is the CU. The CU controls the operation of all other function-blocks within the MPU as well as controlling external devices (memory ICs and I/O ports) via the control signals that function as its input and output signals. This means that the CU also controls the function-blocks of Figure 10.2. The CU may be likened to a special-purpose computer that is dedicated to the fetching and execution of instructions. The purpose of the instruction fetch and execute stream is to instruct the CU as to which instruction to execute. If the CU is truly a computer, it must have memory, I/O, and busses over which it can control and communicate with its memory and I/O (refer to Figs. 8.8a and 10.1). Figures 10.2 and 10.3 show that the CU does indeed have three busses and I/O, which are the various internal registers. The CU's memory is internal to the CU and is used to hold the program that contains the procedures and steps necessary to execute the instructions in the MPU's instruction set. This program is known as the *microprogram*, not to be confused with *macroprograms*, which are stored in main memory and executed by the MPU. A microprogram is written by the MPU designer and stored in the CU's internal memory (MROM) at the time of manufacture. A microprogram is executed by the CU. For most MPUs the microprogram cannot be altered by the user.

Some discussions will make reference to instructions from both the microprogram and the macroprogram. To distinguish between the two instructions, microinstruction will be used to mean an instruction from the microprogram, and instruction or macroinstruction for clarity, to identify one from the macroprogram.

Now that we understand the purpose of the instruction fetch and execute stream as well as the CU, let us start at the beginning and discuss each of its function-blocks. The data bus buffer/latch of Figure 10.3 is composed of tristate buffers and latches. It buffers the internal data bus from the system data bus, as well as latches the datum on the system data bus for MPU read operations and loads the datum on the data bus for MPU write operations. The CU controls these read/write operations and therefore controls the data bus buffer/latch. The data bus buffer/latch has data pins D_0 through D_m, that are connected to the system's data bus as indicated in Figure 10.1. The CU drives the tristate buffers in the Hi-Z state when the MPU is not using the data bus, such as during a DMA.

Figure 10.3

The instruction section of an
MPU architecture.

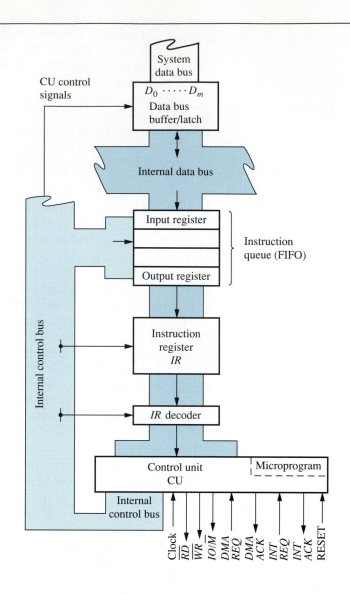

When an instruction is being fetched from memory, the CU begins by loading
the address in the PC of Figure 10.2 to the address latch, also shown in that same
figure. It then outputs that memory address on the address bus via the address latch
(refer to Fig. 9.15 for MPU timing). The CU then drives the read control pin $\overline{RD}$
active, which (recall from Chapters 8 and 9) causes the addressed memory location
to load its content (the instruction) onto the system data bus. While the instruction is
on the data bus, the CU sends a strobe pulse to the data bus buffer / latch of Figure
10.3, causing it to latch the instruction off the system data bus. The CU loads the
content of the buffer latch (the instruction) on the internal data bus and then strobes
the instruction queue (a FIFO memory), which causes it to latch the instruction off
the internal data bus. This instruction is shifted in the queue to the first empty
location. This fetching of instructions continues until the queue is filled. While filling
the queue, the CU may also be executing instructions. From the queue the CU loads
the instruction into the *instruction register* (IR), where it is decoded and identified
for CU execution.

The instruction now resides in the IR and can be decoded by the IR decoder. Decoding results in activating outputs at the IR decoder, which are inputs to the CU. These active CU inputs impact on the CU in two ways. They cause the CU to vector to that portion of the microprogram that pertains to the execution of the instruction in the IR, and they enable the appropriate CU circuitry required for the execution of that instruction. When the CU completes execution of the last microinstruction of the microprogram, pertaining to the execution of the instruction in the IR, it fetches the next instruction to be executed from the queue, loads it into the IR, and repeats the instruction execution cycle. This fetch and execute cycle of instruction continues until the macroprogram has been executed.

As pointed out in Chapter 9, the presence of a queue allows more efficient use of the data bus and faster instruction fetching and execution.

Timing for CU operations is provided by the clock. The clock signal is a repetitive series of pulses of fixed rate and duty cycle. It provides not only CU timing but also system read and write timing, except during a DMA, via CU control signals $\overline{RD}$ and $\overline{WR}$.

Because many MPUs do not have an instruction queue, consider the instruction section without one. Without the instruction queue the CU must fetch the instructions from main memory. The CU proceeds as it did with the queue; that is, it increments the contents of the PC (the address of the next instruction) and loads it on the address bus. It then activates the read control signal $\overline{RD}$, which enables the output pin of the memory device being addressed. The addressed memory location outputs its content (the instruction) on the data bus, from which the CU causes the data bus buffer / latch to latch it. The CU then loads the instruction from the data bus buffer / latch to the IR via the internal data bus (recall that the queue is absent). Once the macroinstruction is in the IR, it is decoded and executed via microinstructions, just as if an instruction queue were present. After execution of the instruction is completed by the CU, it increments the PC and fetches and executes the next instruction.

It is interesting to speculate how an MPU designer might implement an IR decoder. It might be designed from logic gates, as with any combinational logic design. Another approach might be to use a ROM that functions as a look-up table or encoder. The input to the IR decoder (ROM), which is the binary code of the macroinstruction, serves as an address. The content of that addressed location would have been programmed with the binary bit size and pattern required to enable the CU logic circuitry to execute the instruction and vector the CU to the appropriate portion of the microprogram, as discussed previously. If an MPUs instructions are 8 bits, there could be as many as 256 (2^8) instructions in its instruction set if there is an instruction for each possible binary code. Some MPU instruction sets use almost every binary code possible to create an instruction set, whereas others do not. This means that MPUs that use all or most binary codes have a relatively large instruction set compared to those that have the same bit size IR but do not have an instruction for many of the binary codes. MPUs intended to function in a wide range of applications need an instruction set that is general and large, and hence utilize all or most of the binary patterns. An MPU intended for a limited variety of applications would not need as many instructions and would use a smaller number of the binary codes for its instruction set. An example of an MPU with a limited instruction set is a microcontroller.

Data storage section

The data section of an MPU was first discussed in Chapter 9 and is illustrated in a portion of Figure 9.1. Figure 10.4 is an illustration of the data storage section. Registers B, C, D, and E constitute the internal data storage for the MPU, and hence these registers function as a scratch pad memory. A datum is written into or read from one of these registers via the internal data bus and the read and write signals

Figure 10.4

The data section of an MPU
architecture.

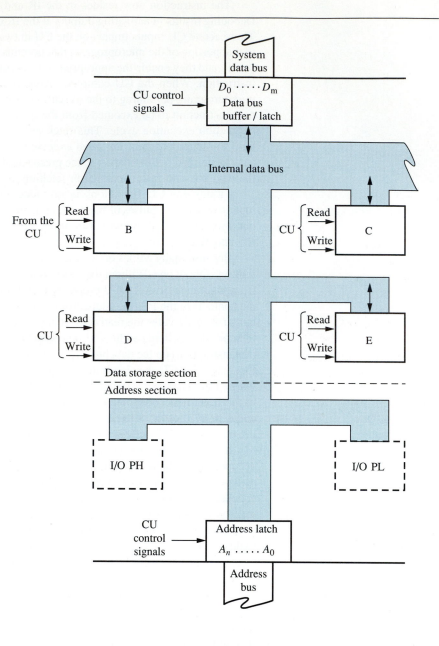

originating from the CU. The CU read and write signals of Figure 10.4 are not to be
confused with the external read and write control signals $\overline{RD}$ and $\overline{WR}$. These are for
internal use only.

As an example of how the data section functions, suppose that in the process of
executing an instruction, the MPU is to read a data byte from memory and store it in
register B. The CU will have addressed the memory location and loaded its content
on the data bus by driving the read control signal $\overline{RD}$ active. On the positive edge of
$\overline{RD}$ the CU latches that data off the data bus and into the data bus buffer / latch, as
previously described. It then loads that data byte onto the internal data bus, from
which register B can latch it by means of a strobe signal from the CU to the read input
of register B. If the MPU is a 16-bit MPU, then the internal data bus and register are
also 16 bits, however, the process is basically the same.

For the MPU to write the content of one of its scratch pad registers to memory, the CU drives the write input of that register active, which loads the content of that register onto the internal data bus by turning on its output buffers. Recall that all outputs connected to a bus must be buffered to prevent bus contention. While the CU is holding that datum on the internal data bus (by keeping the write input of the register active), it also activates the buffers of the data bus buffer/latch, which electronically connects the internal data bus to the system data bus and thereby loads the datum on the internal data bus onto the system data bus. While these actions are occurring internally, the CU has also output the address of the memory location on the address bus and t_{AC} seconds later (see Fig. 9.15) activates its external write control signal $\overline{WR}$. With the address of the memory location on the address bus and the datum to be stored in memory on the system data bus, the CU deactivates $\overline{WR}$, creating the positive edge needed to strobe the datum on the system data bus into the addressed memory location.

Refer to the discussion of the addressing section and Figure 10.2, recall that the DC is used by the CU to furnish memory addresses for data storage or retrieval. The DC in this architecture is not a specific register but rather the pairing of scratch pad memory registers (pairing is required, because memory addresses are 16 bits but scratch pad memory registers are 8 bits). Under macroprogram control the CU is instructed to create a DC by pairing either registers B and C or registers D and E. The addresses created by those register pairs are such that registers C and E form the low byte of the addresses and registers B and D form the high bytes. When the address of the DC (a register pair) is to be output on the address bus for an 8-bit MPU, the CU loads the content of the low byte of the register pair (either) C or E on the internal data bus and latches that byte into the address latch. The CU then loads the content of the high-byte register (either B or D) into the address latch via the internal data bus. The CU then loads the content of the address latch on the address bus. For a 16-bit MPU, the content of the DC (register pair) is loaded on the internal data bus from where it is latched by the address latch.

From the above discussion we can state that in order to make maximum use of registers B, C, D, and E, the MPU has been designed so that they can serve in either of two functions — to serve as scratch pad memory registers, or as the low and high byte of a DC. Which function it serves depends on instructions within the macroprogram. A register pair could be acting as a DC for one macroinstruction and as a scratch pad memory register for another instruction.

How are I/O addresses formed? They are also the result of registers much like the scratch pad register pairs. The difference is that these registers are used exclusively to store I/O addresses. Because these I/O address function-blocks (*I/O PH* and *I/O PL*, for high and low bytes of the address) are part of the architecture and are not programmable, that is, they are not programmable by the user (there are no instructions that can be used to program their content), their function-blocks are phantomed in using dotted lines. These function-blocks have been included in the address section of Figure 10.4 and are configured much the same as scratch pad registers to remind us that although they function much like a DC, they differ from a DC in that they cannot store data.

Arithmetic, logic, and decision section

Figure 10.5 is an illustration of the arithemetic, logic, and decision sections of an MPU architecture. The ALU is the function-block that performs all arithmetic (add, substract, etc.) and logic (*AND, OR*, etc.) operations on data. The data to be operated on is input to the ALU via the two registers that are identified as *ACC*1 and *ACC*2, ACC being an abbreviation for *accumulator*. These registers are called accumulators because they collect, or accumulate, ALU results.

Figure 10.5

Arithmetic, logic, and
decision section of an MPU
architecture.

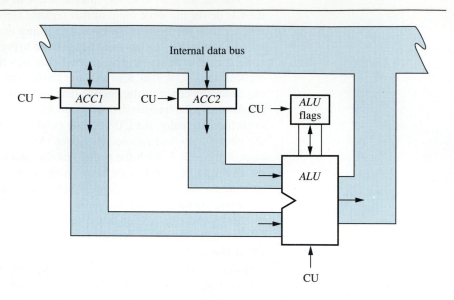

The ALU flags are flip-flops whose logic state depends on the result of an ALU operation. There can be any number of ALU flags, but imagine that there are four for our generic MPU. These four flags are the sign-flag (S-flag), zero-flag (Z-flag), carry-flag (C-flag), and the parity-flag (P-flag). If an ALU-related instruction is executed whose result could be either positive or negative, such as when a subtraction is performed between the two accumulators, the logic state of the S-flag is determined accordingly. If the result of the ALU operation is negative, the S-flag is set (1); if the result is positive, the S-flag is reset (0). The logic state of the S-flag will be the same as the logic state of the MSB of the ALU result, which corresponds to the mechanics of complement arithmetic. The logic state of the Z-flag is set if the result of the ALU operation is a zero; otherwise it is reset. The C-flag is set if a carry or borrow resulted from the MSB of the ALU operation; otherwise it is reset. The P-flag is set if the ALU result is even parity; otherwise it is reset.

Example 10.1

For the indicated ALU operations, determine the logic states of the S-, Z-, and C-flags. All data is 4 bits in length.

(a) 1000 + 0001   (b) 1000 + 1001

(c) 0100 − 1000   (d) 1101 *AND* 0010

(e) 1101 *AND* 0001   (f) 0100 *OR* 0111

Solution

(a) 1000
　　0001
　　‾‾‾‾
　　1001

　　S-flag is reset
　　Z-flag is reset
　　C-flag is reset (no carry)

(b) 　1000
　　　1001
　　　‾‾‾‾
　1　0001

　　S-flag is reset
　　Z-flag is reset
　　C-flag is set

(c)   0100
     − 1000
   1 ‾1100

      S-flag is set
      Z-flag is reset
      C-flag is set

(d)   1101
AND 0010
    ‾0000

      S-flag is reset
      Z-flag is set
      C-flag is not affected for
      an *AND* operation

(e)   1100
AND 0001
    ‾0001

      S-flag is reset
      Z-flag is reset
      C-flag is not affected

(f)   0100
OR 0111
   ‾0111

      S-flag is reset
      Z-flag is reset
      C-flag is not affected

The *C*-flag has no meaning in logic operations (*AND, OR*, etc.) and is not affected. Sometimes flags are affected by an ALU operation that seems to have no logical reason for being affected, such as the *S*-flag for logic operations.

The ALU flags are the decision-making mechanisms of an MPU. There are MPU macroinstructions that enable the logic state of a flag to be checked, based on its logic state, the programmer can choose whether the MPU should execute an instruction or not — hence a decision is made. These are known as *conditional instructions*. Usually conditional instructions are conditional branch instructions, that is, if the conditional branch instruction is executed, the MPU branches from one part of the program to another part, allowing selective parts of a program to be executed.

Example 10.2

Suppose that an instruction from an MPU's instruction set is a conditional jump (*J*) instruction, which is dependent on the logic state of the *Z*-flag. This instruction has two forms, one based on the *Z*-flag being set (*Z*) and the other based on its being reset (*NZ*). The instruction for execution if the *Z*-flag is set is written as *JZ ADDR* (jump to *ADDR* if ALU result is zero); it is written as *JNZ ADDR* if ALU result is nonzero. To determine whether it is to execute the instruction *JZ ADDR*, the CU first checks the logic state of the *Z*-flag. If the *Z*-flag is set (a logic 1) the CU will cause the MPU to jump to the 16-bit memory address following *JZ*, which is symbolized by *ADDR*, by loading that address in the PC; otherwise it increments the content of the PC and continues to the next instruction. The execution of *JNZ ADDR* is similar to that of the *JZ* instruction except that it jumps to memory location *ADDR* if the *Z*-flag is a logic 0.

If the contents of *ACC*1 and *ACC*2 are as given, what is the result when the MPU executes the three-instruction program shown?

        *AND ACC*1, *ACC*2   (*AND ACC*1 and *ACC*2)
        JNZ   0200H
        JZ   0500H
        END   (end of program)

(a) *ACC*1 = 00001000
     *ACC*2 = 00001110

(b) *ACC*1 = 00011111
     *ACC*2 = 10000000

Solution

The datum of registers $ACC1$ and $ACC2$ is ANDed.

(a)   00001000
AND 00001110
        ‾‾‾‾‾‾‾‾
        00001000
        Z-flag is reset

Therefore, when instruction JNZ is loaded into the IR as the next instruction, it is executed since $Z = 0$, and the MPU jumps to memory location 0200H.

(b)   00011111
AND 10000000
        ‾‾‾‾‾‾‾‾
        00000000
        Z-flag $= 1$

The JNZ instruction is not executed. The MPU fetches the next instruction, which is JZ. It is executed and the MPU branches to memory location 0500H.

An application of computer decision making via the ALU flags is the writing of letters that are personalized to the recipient (Example 10.3). Suppose a company wishes to mail out marketing promotional letters that are tailored to the recipient. The company would have a file, a collection of data stored on disk, on each individual to whom it intends to send a letter. A file can be very short and simple, or it can be very long and complex; in either case the file contains a marketing profile of the recipient. Many of these characteristics can be specified with the logic state of a single bit, such as gender. If the binary code that contains these marketing characteristics comes in, for example, a byte, there are eight marketing characteristics. To interrogate the data byte for a market characteristic, the programmer must write a program that isolates the sought-after information from other information in that byte of data. The technique often used is called *masking*. Masking uses a binary bit pattern, called the *mask*, that either ANDs or ORs the data and the mask. When the mask is ANDed with the data byte being interrogated, all bits are masked out except the bit of interest, that is, all bits being masked are forced to a known logic state, 0. Hence, those bits to be masked out are 0's in the mask. Those unwanted bits could be masked to a logic 1 by ORing the data byte with the mask-byte, so that the mask-byte would have logic 1's in all unwanted bit positions. Either logic operation affects the ALU flags, which can be detected and affect the execution (or not) of a conditional instruction.

Example 10.3

Using an approach similar to the program of Example 10.2, write a program that vectors to memory location 0200H if the recipient is female and to address 0500H if the recipient is male. The market profile of this recipient was stored at main memory location 0800H. This market profile data byte is structured so that the recipient's gender is indicated by bit position D_3, where the LSB position is identified as D_0. If the recipient is female, D_3 is a logic 1; otherwise D_3 is a logic 0. The instruction $MOV\ ACC1,[ADDR]$ can be used to move the data byte (the marketing profile) stored at location $ADDR$ into register $ACC1$, and instruction $MVI\ ACC2$ can be used to move the binary value of the mask-byte into $ACC2$.

Solution

$MOV\ ACC1,[0800H]$
$MVI\ ACC2, 08H$

$AND\ ACC1, ACC2$
$JNZ\ 0200H$
$JZ\ 0500H$

END

The first instruction moves (*MOV*) the data at memory location 0800H into *ACC*1. If the recipient is female bit $D_3 = 1$, and bit $D_3 = 0$ if the recipient is male. The other 7 bits are of no interest at this time. The instruction *MVI ACC*2,08H moves the immediate data (*MVI*), which is the mask-byte, into *ACC*2. Mask-byte 08H (00001000_2) will mask the *AND*ed result bits to a logic 0 except for D_3, which is the bit that indicates gender. After *ACC*1 and *ACC*2 are *AND*ed, and if the recipient is female, the Z-flag is reset to 0 (nonzero result), and if the recipient is male the Z-flag is set to a logic 1. Hence, if the recipient is female ($Z = 0$) the MPU vectors to location 0200H, and if male ($Z = 1$) the MPU vectors to 0500H. Memory location 0200H is the beginning of the program that personalizes the letters to women, and beginning at memory location 0500H personalizes those to men.

Examples 10.2 and 10.3 show why the ALU flags are the function-blocks on which programming decisions are made.

DMA and interrupt section

As illustrated in Figure 10.3, the DMA and interrupt section has been included as a part of the CU. They could be represented with separate architectural function-blocks, as was done for the other sections, but each of them would be found to be no more than a single function-block with external handshaking pins (request and acknowledge) and internal output and input going to the CU to provide DMA and interrupt control.

The purpose of the DMA and interrupt handshaking pins was thoroughly discussed in Section 10.2, so they will be reviewed only briefly here. The *DMA REQ* input pin is the request pin on which an I/O device with DMA capability can request the MPU to release control and use of the system data bus so that it may take control and use the data bus for a DMA. When the MPU is ready to release the data bus, it notifies the I/O device by activating its *DMA ACK* output pin, which signals the I/O device to take control. The I/O device holds the *DMA REQ* pin active until the DMA is completed. The CU monitors the logic state of that pin, and during the time that the I/O device holds it active, the CU holds the *DMA ACK* pin active and puts all of the MPU bus-related pins in the Hi-Z state, which electronically isolates the MPU from the busses, thereby allowing the I/O device to control the system busses. When the I/O device has completed the DMA, it drives the *DMA REQ* pin inactive, which is sensed by the CU. The CU then drives its *DMA ACK* pin inactive and takes the MPU bus-related pins out of the Hi-Z state. At that point the MPU takes back control and use of the data bus for the normal operation of fetching and executing instructions.

Interrupt handshaking involves an I/O device requesting an interrupt of the MPU via the *INT REQ* input pin. When an interrupt request has been made of the MPU and it completes the execution of the current instruction, via the CU, it notifies the I/O device that it is ready to service its request by activating its *INT ACK* pin and vectoring to a prespecified memory location, which contains the first instruction of the program subroutine (service program) that services the device. Prior to vectoring to this subroutine, the MPU stores the return address on the stack. The first instructions of the subroutine are *PUSH* instructions, which save the contents of scratch pad memory by pushing (storing) their contents on the stack. At the end of the subroutine *POP* instructions restore the original scratch pad data pushed on the stack by writing it to the scratch pad registers. The last instruction of the subroutine is return (*RET*), which pops the return address off the stack and writes it into the PC, enabling the MPU to return to that address.

The partial MPU architecture of Figure 10.3 shows just one interrupt request pin; however, some MPUs have multiple interrupt request pins. If there is just one interrupt request pin serving more than a single I/O device, there must be external logic to establish interrupt request identities and priorities. This I/O identification and interrupt priority logic is not shown in Figure 10.1 and is beyond the scope of this book.

Figure 10.6 is a composite of Figures 10.2 through 10.5. For convenience it shows two CU control busses when in fact there is just one. Also, the CU control bus signals going to registers C, E, I/O PL, and SP are not shown connected to the CU control bus although they actually are. The CU control bus signals symbolize the control that the CU has over all other function-blocks in carrying out the process of instruction fetch and execution. Although just one line is shown coming from the CU control bus and going to a function-block, there may be more than one.

Figure 10.6

A microprocessor architecture.

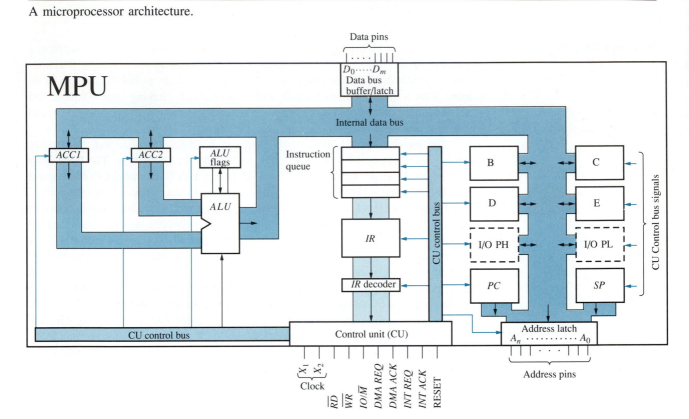

ALU results are output on the internal data bus, as the ALU arrow direction of Figure 10.6 indicates. Once this resultant datum is on the internal data bus, the CU can cause any programmable register to latch that datum off the data bus. Registers *ACC*1, *ACC*2, B, C, D, and E are programmable. Under this condition, the resultant ALU output can be stored under program control in any scratch pad register or in register *ACC*2 or *ACC*1. However, there are MPUs that do not allow a choice of location of data storage. These MPUs automatically store the results in *ACC*1. For this type of MPU the original datum in *ACC*1 is destroyed (it is written over by the ALU results). If the datum in *ACC*1 is not to be lost as the result of an ALU operation, it must be stored in scratch pad memory before execution of the ALU-related instruction.

Rather than having a single clock input to the CU, as illustrated in Figure 10.3, an MPU would probably have two input pins for connecting a crystal (X_1 and X_2) as shown in Figure 10.6. The crystal is the source of MPU clock pulses, but the output signal of the crystal must be wave shaped into pulses by the internal circuitry known as the clock (not shown) before it can be used for MPU timing. Some MPUs do not have an on-board clock. If this is the case, there must be external circuitry to support this function. As one can imagine, the amount of MPU support hardware can be reduced by selecting the proper MPU for an application.

Review Questions

1. How many different registers can be used by the MPU architecture of Figure 10.6 to address memory, and when is each of these used?

2. How many registers can be used to address I/O?

3. What is the function of the address latch?

4. What is an instruction queue and what purpose does it serve?

5. Explain the purpose of the IR and IR decoder.

6. In general terms, define the CU and explain its function.

7. What is a microprogram, and how does it differ from a macroprogram?

8. Explain handshaking and why it is necessary.

9. What is the dual function of scratch pad memory registers B, C, D, and E?

10. How are the logic states of the ALU flags altered?

11. What purpose does the ALU serve?

12. Under what condition would the contents of $ACC1$ be destroyed as a result of an ALU operation?

10.4
MPU timing

The basic external timing of an MPU was introduced in Chapter 9 (Fig. 9.15) and is again illustrated in Figure 10.7. As discussed in Chapter 9, any time the MPU is to access either memory or I/O, the read or write cycle begins with the MPU outputting an address on the address bus of Figure 10.1, which is indicated by the timing diagram of Figure 10.7. During the time that the address is being output, the MPU also drives its $IO/\overline{M}$ pin to logic 1 if the address is an I/O address and to logic 0 if it is a memory address, enabling the appropriate selector (page or I/O) in Figure 10.1. The MPU then activates the appropriate control bus pin—$\overline{RD}$ for a read and $\overline{WR}$ for a write. If an MPU read is being performed, the read control signal enables the addressed memory location or I/O port to load its datum on the system data bus, making that datum available to the MPU data pins via the system data bus of Figure 10.1. As indicated in Figure 10.7, the MPU latches that data off the data bus on the positive edge of $\overline{RD}$, which completes the read cycle. If the MPU had been performing a write operation, it would activate its write control pin $\overline{WR}$ at the same time it outputs the datum on the data bus. The write control pin is held active as long as the MPU holds the datum on the data bus. Just before the MPU removes that datum from the data bus (t_{CL} seconds), it drives $\overline{WR}$ inactive, from which a positive edge is formed. As indicated in Figure 10.7, this positive edge serves as a strobe to the addressed memory location or I/O port.

The timing diagram for an interrupt or reset is basically that of a read. Recall that for either of these cases the MPU vectors to a specified memory location and fetches

Figure 10.7

A representative timing
diagram for an MPU.

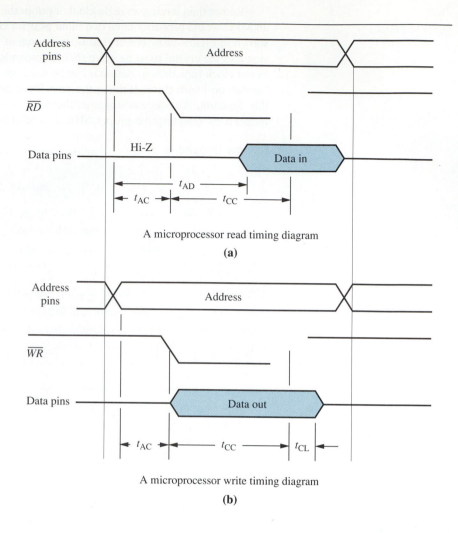

A microprocessor read timing diagram

(a)

A microprocessor write timing diagram

(b)

(reads) the first instruction of the subroutine program that is to be executed. The only
thing that sets an interrupt and reset apart from a normal memory read is how the
vector address is obtained. For an interrupt or reset, the CU loads the vector address
into the PC as a result of the interrupt request or reset. For a normal instruction fetch,
the memory address is established by the CU incrementing the PC.

Review Questions

1. What two events must occur for every MPU read and write operation?
2. What purpose does the MPU read control signal serve to the system?
3. What purpose does the MPU write control signal serve to the system?
4. Rationalize the timing of the $IO/\overline{M}$ pin.
5. When does the MPU actually read the datum from the data bus?
6. When does memory or I/O normally latch the datum being written by
   the MPU?

10.5
MPU instruction set

A hypothetical instruction set will be developed for the MPU architecture of Figure 10.6, which closely resembles that of the 8085 and 8086. The first thing to note is an instruction set is valid for a specific architecture. The programmer must know the instruction set of an MPU as well as its architecture to properly apply those instructions.

Before developing an instruction set for an MPU, decide on the capabilities of the architecture, that is, which registers are to be programmable and between which registers data transfers can occur. The capabilities of the ALU must also be known. Because this hypothetical instruction set's only purpose is to serve as a vehicle for explaining concepts, keep it simple. Suppose that the ALU can perform just four arithmetic and logic functions — addition, subtraction, *AND*ing, and *OR*ing. Register data manipulation will be allowed but will be restricted to simply moving data between the programmable registers B, C, D, E, *ACC*1, *ACC*2, *SP*, and memory. The SP must be programmable because it points to the stack, which is a defined portion of main memory and may have different locations in different systems. To summarize the capabilities of the architecture thus far, it must be capable of ALU operations and the transfer of data between registers, memory, and the SP.

From previous discussions on decision making and vectoring, this architecture must also be capable of branching from one place in memory to another. The branch instructions already looked at are *JZ* and *JNZ* (jump on zero and jump on nonzero, respectively), which are conditional branch instructions. Also required are some unconditional branch instructions whose execution is independent of the ALU flags. Adding the ability of branching to this architecture expands its capabilities to include ALU operations, data transfer, and branching.

Last, this architecture must allow transfer of data between register *ACC*1 and I/O. The exchange of I/O data has been limited to *ACC*1 and I/O to simplify coding of the instruction set. The I/O data transfers have not been grouped under the general heading of data transfers because of the $IO/\overline{M}$ pin. Because this pin partitions the address space into I/O and memory (recall the I/O and page selectors of Fig. 10.1), there must also be a similar partitioning of instructions in the instruction set. When those memory-related data transfer instructions are executed the CU drives the $IO/\overline{M}$ pin low, and when an I/O data transfer instruction is executed the CU drives the $IO/\overline{M}$ pin high.

This completes the capabilities of this simple instruction set. Any instruction within the instruction set can be grouped into one of the following four categories:

1. ALU
2. Branching
3. Memory data transfer
4. I/O data transfer

The next task is to create a detailed listing of the instructions that comprise the instruction set, along with an explanation of the MPU operation to be performed. A sampling of that listing is given in Table 10.1. These samplings of instructions have been grouped according to the four MPU operations — ALU, branching, memory data transfer, and I/O data transfer.

In Table 10.1, the term *mnemonic* means memory aid. The mnemonic used to represent an instruction is an abbreviation that symbolizes the MPU operation to be performed. Mnemonics are a shorthand method for writing instructions, but they are for the programmer's convenience, because the MPU understands only binary code, which it receives via the IR and IR decoder. For each mnemonic of an instruction there must also be a corresponding binary code.

Table 10.1   The instruction set for the MPU architecture of Figure 10.6

	Mnemonic	Explanation of instruction
ALU	ADD R1,R2	Add contents of registers R1 and R2 and deposit the sum in register R1, where R1 is ACC1 but R2 can be register ACC2, B, C, D, or E. Set ALU flags accordingly.
	AND R1,R2	AND registers R1 and R2. Deposit the results in R1. R1 and R2 as above. Set ALU flags.
Branch	JMP ADDR	Jump to memory location whose 16-bit address is specified by ADDR.
	CALL ADDR	Jump to the memory location specified by the 16-bit address ADDR and push the return address on the stack.
	JNZ ADDR	Jump to ADDR if Z=0; otherwise do not execute.
	JZ ADDR	Jump to ADDR if Z=1; otherwise do not execute.
	CNZ ADDR	Jump to ADDR if Z=0 and push return address on the stack; otherwise do not execute.
	CZ ADDR	Jump to ADDR if Z=1 and push return address on the stack.
Data transfer	MOV R1,R2	Move the data in register R2 into register R1. R1 and R2 can be registers ACC1, ACC2, B, C, D, or E.
	MOV R2,M	Move the data from memory location M, whose address is in register pair D and E, and load that data into register R2.
	MOV M,R2	Move the data from register R2 into memory location M. R2 can be either register ACC1, ACC2, B, C, D, or E. The address of location M is the contents of register pair D and E.
	MVI R,I	Move the immediate data I into register R. R can be ACC1, ACC2, B, C, D, or E.
I/O	IN ADDR	Read the port specified by ADDR and load the data read into ACC1
	OUT ADDR	Write the content of ACC1 to port ADDR.

To design binary codes for each instruction of the MPU instruction set, binary bit pattern templates would be devised for each instruction. These templates are partitioned into bits, or groups of bits, so that when decoded by the IR decoder the output from the decoder directs the MPU's CU as to the precise operations to be executed. The output signals of the IR decoder:

1. Enables the appropriate MPU internal logic circuits necessary for execution of the instruction.
2. Vectors the MPU's CU to the starting address within the microprogram where the series of microinstructions for executing the (macro) instruction reside. The microinstructions instruct the CU as to the precise operations that are to be done and in what order.

The template binary patterns must specify:

1. The general type of MPU operation (ALU, branch, data transfer, or I/O) to be performed.

2. The specific operation to be executed within the general type of MPU operation. That is, if the general type MPU operation to be performed requires the ALU, the specific ALU operation (add, subtract, *AND*, or *OR*) must also be specified.

3. If registers are involved in the execution of the instruction, which ones and whether they are to be the source or destination for the data.

Table 10.2 Binary code for encoding MPU operation types

Binary code	Microprocessor operation
00	ALU
01	Branch
10	Data transfer
11	I/O

To design the instruction set templates the bit size of the IR must be known. For simplicity, assume that it is an 8-bit MPU for which the instructions of Table 10.1 apply. Hence, the bit size of the IR is 8 bits, thus also requiring the instruction templates to be 8 bits. To identify each bit position of the 8-bit template the designations T_0 to T_7 will be used, where T_0 is the LSB.

Next, how will the template bits be grouped and what will be their relative position within the template? First, specify the binary codes and their bit position for the general type operation to be executed by the MPU. Table 10.1 shows that all instructions fall into one of four groups (ALU, branch, data transfer, or I/O). To code four identities requires 2 bits. The four binary codes and their corresponding MPU operations are shown in Table 10.2. Realize that combinations different from those of Table 10.2 could have been chosen. The only absolutes are the four 2-bit codes and the four MPU operations. Template bits T_7 and T_6 have been designated for coding of the MPU operation.

Template bits T_5 and T_4 will specify the exact MPU operation to be executed within the general operation specified by bits T_7 and T_6. Within the two general type MPU operations of ALU and branch there are four specific operations. Again, to provide binary codes for four identities requires 2 bits. The 2-bit codes and their corresponding operations are shown in Table 10.3.

Some of the instructions in Table 10.1 require the use of MPU registers and memory as well as loading immediate (I) data. There are eight sources or destinations for MPU data transfers. To code eight identities requires 3 bits. The 3-bit codes and their corresponding identities are shown in Table 10.4. Template bits T_2 to T_0 will be used to code source or destination for data transfers, completing the assignment of template bit positions.

The binary codes for the instructions of Table 10.1 will now be determined. Be reminded that this exercise is to develop concepts and not to imply that these instruction set templates are actual or even well designed. Rather, they are relatively simple and are intended to emphasize how binary codes are determined and the steps by which they execute an operation.

Table 10.3 Codes for encoding operations

Code	ALU operation
00	*ADD*
01	*SUB* (subtract)
10	*AND*
11	*OR*

Code	Branch
00	Conditional jump
01	Conditional call
10	Jump
11	Call

Table 10.4 Codes for encoding registers

Code	Register
000	Immediate data
001	[ACC]1
010	[ACC]2
011	B
100	C
101	D
110	E
111	M (D and E)

Consider the instruction *ADD R1, R2* of Table 10.1. From the description of the instruction, *R1* is always register *ACC1*. Because there is no other choice for *R1* it need not be coded. Codes are needed for the general type MPU operation (bits T_7 and T_6), the specific operation (bits T_5 and T_4), and the generic register *R2*. For a specific example of the generic instruction *ADD R1, R2*, determine the binary code for the instruction *ADD ACC1, B*, the template for which is shown in Figure 10.8a. The binary code for bits T_7 and T_6 are shown in Table 10.2. Because this is an ALU operation bits T_7 and T_6 are both 0. This ALU operation is an *ADD*, so from Table 10.3 the code is 00. Hence, 00 is put in bit positions T_5 and T_4. All that remains to be coded is register B, which from Table 10.4 requires the 3-bit code 011, therefore bits T_2 to T_0 are used to identify register B. T_3 is an extra bit (a don't care). A logic 0 will

Figure 10.8

Example encoding templates
for the instruction set of
Table 10.1.

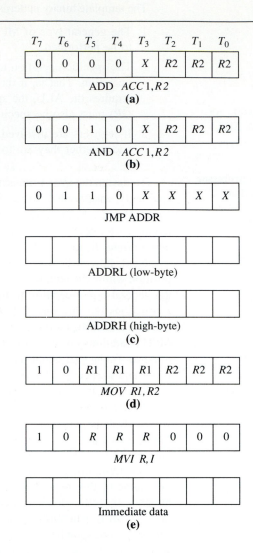

be arbitrarily assigned to all don't cares. Combining the template bits the binary code for instruction *ADD ACC*1,B is 00000011.

To ensure full understanding of the purpose of instruction templates and the binary codes derived from them, trace the instruction *ADD ACC*1,B through the IR decoding process. The MPU fetches binary code 00000011 from memory, which is the code for instruction *ADD ACC*1,B. The MPU loads this binary code in the IR, which decodes it by:

1. Decoding bits T_7 and T_6 for the MPU operation to be performed. Because these 2 bits are 00, which is an ALU operation according to Table 10.2, the ALU is enabled.

2. Decoding bits T_5 and T_4 for the specific ALU operation. These bits are mode select inputs (recall the mode select for the 74181, or refer to Appendix C for review) to the ALU.

According to Table 10.3, the mode selected is an *ADD* for bits 00. So far the ALU has been enabled and instructed to add the data at its two inputs. Next, data to be added will be specified.

3. The architecture of Figure 10.6 shows that there are only two sources of data for all ALU operations, registers $ACC1$ and $ACC2$. Therefore, all data to be operated on by the ALU must be in these two registers. In agreement with the architecture, the explanation for the instruction ADD $ACC1$,B, provided in Table 10.1, states that the data byte of register $ACC1$ is to serve as one input, but it also states that the second data byte can be the contents of either register $ACC2$, B, C, D, or E. Bits T_2 to T_0 will identify which of these registers is to provide the second byte of data. Decoding bits T_2 to T_0 (011), and referring to Table 10.4, identifies register B as the source for the second data byte. To comply with the architecture and the instruction code, the CU loads the content of register B on the internal data bus and strobes it into register $ACC2$. The ALU then adds the contents of $ACC1$ and $ACC2$ and stores the result in register $ACC1$. The condition flags are set according to the arithmetic result. This completes the execution of the ADD $ACC1$,B instruction. The CU then increments the content of the PC by one, which is the address of the next instruction, and the next instruction is fetched and executed.

Continuing with the instruction templates, consider the AND template of Figure 10.8b. According to Table 10.2 bits T_7 and T_6 are both to be logic 0's because an AND is an ALU operation. The template of Figure 10.8b shows this to be the case. Bits T_5 and T_4 are to be 10, respectively, according to Table 10.3, as are also shown in the template of Figure 10.8b. Explanation of the AND instruction provided in Table 10.1 states that register R1 is always register $ACC1$ but register R2 can be either register B, C, D, or E, just as was the case for the ADD instruction. Then, the binary code for the AND instruction will have to specify which register is to serve as the source register for the second byte of data to the ALU. Bits T_2 through T_0 specify the second source register. Refer to Table 10.4 for that binary code. If the AND instruction is AND $ACC1$,D then the binary code would be 00100101 (let $X = 0$).

The template of Figure 10.8c states that jump instructions require 3 bytes. The first byte is the op-code (operation code), which specifies that a jump is to be executed. The next 2 bytes are known as *operands*; these are the low byte ($ADDRL$) and high byte ($ADDRH$) of the address that the MPU is to jump to. The op-code template states that bits T_7 and T_6 are 01, which agrees with Table 10.2 for a branch instruction. Bits T_5 and T_4 specify the branch operation, which is coded as a jump according to the code of Table 10.3. If the MPU were to jump to address 0506H, the mnemonic would be written as JMP 0506H and the 3-byte binary code would be 10100011 ($X=0$) for the op-code, 00000110 for $ADDRL$, and 00000101 for $ADDRH$.

The remaining two templates of Figure 10.8 are data transfers; hence T_7 and T_6 of the op-code are 10 in both cases as specified in Table 10.2. For the template of Figure 10.8d the source register R2 and destination register R1 are coded according to the codes of Table 10.4. For instance, the binary code of instruction MOV C,B is 10100011, since according to Table 10.4 the binary code is 100 for register C and 011 for register B.

The MVI R,I instruction of Figure 10.8e is a 2-byte instruction. The first byte is the op-code and specifies the instruction to be executed. The second byte is an operand and is the (immediate) data to be operated on, which in this case will be copied into the register specified in the op-code. Bits T_5 through T_3 specify the destination register, using the codes of Table 10.4, and bits T_2 through T_0 specify the data to be immediate, again using the codes of Table 10.4. For the mnemonic MVI C,08H the binary code is 10100000 for the op-code and 00001000 for the immediate data 08H. Verify that the binary code 000 from Table 10.4 is used to identify this instruction as having immediate data, meaning that the data is encoded as part of the instruction.

Example 10.4

From the architecture of Figure 10.6 there must be an instruction with which to program the SP. Can an *MVI SP,I* instruction be properly coded using the template of Figure 10.8e? Justify your answer. If the answer is no, what might be some possible solutions?

Solution

No, because there is no code for SP in Table 10.4.

Some solutions are the following: (1) Assign the 000 code of Table 10.4 to identify SP and devise another method to code immediate data. This would also mean increasing the *MVI SP,I* instruction to 3 bytes, because the SP is a 16-bit register. (2) The best solution is to create another subset of data transfer instructions in addition to those shown in Table 10.1. The mnemonic might appear as *LD SP,ID*, which when executed would load (*LD*) the immediate 16 bits of data (*ID*) into register SP. Table 10.2 bits T_7 and T_6 must be 10, but the remainder of the template is yet to be defined.

The purpose of this example is to demonstrate the options that the designer of an MPU encounters.

Example 10.5

For the following mnemonics encode their corresponding binary codes using the templates of Figure 10.8 and the instruction definitions of Table 10.1.
(a) *ADD ACC1,D*   (b) *AND ACC1,C*

(c) *JMP* 12ABH   (d) *MOV C,D*

(e) *MVI ACC2,50H*

Solution

Let $X=0$ for all templates.
(a) Using the template of Figure 10.8a

$$ADD \quad ACC1,D = 00000101$$

because D = 101 from Table 10.4.

(b) *AND ACC1, C* = 00100100

(c) *JMP* 12ABH
   JMP = 01100000
   ABH = 10101011
   12H = 00010010

(d) *MOV C,D* = 10100101
   because C = 100 and D = 101 from Table 10.4.

(e) *MVI ACC2,50H*
   MVI ACC2 = 10010000
   50H = 01010000

With an 8-bit IR there are 256 binary combinations, and therefore there could be as many mnemonics. If the architecture of an MPU has the capability and versatility to require 256 instructions, there would of course be an instruction for each 8-bit binary combination. Most MPUs do not use all binary combinations. Of those binary combinations used, almost no manufacturer would put don't cares in the templates, but rather would put in either a 1 or a 0. For 16- and 32-bit MPUs the instruction set size is potentially 2^{16} and 2^{32} instructions, respectively. Even though the instruction sets are never this large, they become more varied and complex as the bit size of the MPU is increased.

Information concerning the instruction set of an MPU is supplied by its manufacturer. Fortunately, the user does not have to write programs using the binary code format of the instruction set but instead uses the more convenient mnemonic form. However, the mnemonic program must be translated into the binary code. The program can then be stored in main memory and executed by the MPU.

The process of translating a program from mnemonics to binary code is known as *assembly*. If a person does the assembly, the translation process is known as *hand assembly*. Computer programs that assemble mnemonics into binary code are called *assemblers*.

A program in mnemonic form is identified as a *source program* or a *source file*, because it is the source file of the assembler. The file created by the assembly process is known as the *object file* or *machine code* file, because the object of translation (assembly) is to create a binary file that can be executed by an MPU (an MPU is sometimes referred to as a machine).

Example 10.6

Using the mnemonics of Table 10.1, write a program that will read a data byte from ports 3H and 4H and determine their sum. This sum is to be written to port 0006H and stored at memory location 2134H.

Solution

$$MVI\ E,34H$$
$$MVI\ D,21H$$

$$IN\ 0003H$$
$$MOV\ B,ACC1$$

$$IN\ 0004H$$
$$ADD\ ACC1,B$$

$$OUT\ 0006H$$
$$MOV\ M,ACC1$$

$$END$$

Registers D and E are to serve as the register pair making up the DC for this program. The first two instructions initialize registers E and D with the address of memory location 2134H (see *M* of Table 10.4). The *IN* instruction reads the data byte of port 3 into *ACC1*. This data byte must be stored in scratch pad memory because port 4 is also to be read and that data will overwrite any previous data in register *ACC1*. The *MOV B,ACC1* saves the data byte read from port 3 into register B. Next, the data byte of port 4 is read and loaded in *ACC1*. The 2 data bytes can be added using *ADD ACC1,B*. The sum of this addition is automatically written into register *ACC1* and will then be written out to port 6 using the instruction *OUT* 0006H. Because the *OUT* instruction is a nondestructive instruction (it did not alter the content of *ACC1*), *MOV M,ACC1* can be used to write the content of *ACC1* to memory, where register pair D and E serves as the DC.

As already stated, before an MPU can execute the program of Example 10.6 it must first be assembled and stored in main memory. To execute a program the MPU would repeat the fetch and execute cycle discussed earlier. An important part of the execute cycle is the decoding of the binary code in the IR, which enables the appropriate circuitry and vectors the CU to the appropriate location within the microprogram. From the microprogram the CU reads and executes the microinstructions sequence for executing the instruction.

Review Questions

1. What determines the capability of an instruction set?
2. What determines the maximum number of instructions within an instruction set?
3. What are some considerations in determining the instruction set templates?
4. When might the machine code of an instruction exceed a single byte?
5. What are mnemonics, and what purpose do they serve?
6. Can an MPU execute a mnemonic?
7. What is assembly?
8. What is an assembler?

10.6

Memory and I/O interfacing: the creation of a microcomputer

To interface two or more devices means to join them together with what is required to make the joining compatible. The required hardware is referred to as glue and includes chips, resistors, and capacitors. Often the glue is buffers and latches, such as the I/O ports of Figure 10.1 that interface I/O devices to the MPU via the data bus.

Once memory and I/O have been interfaced to an MPU, a microcomputer has been created, as suggested by the computer architecture of Figure 10.1. In Chapter 9 memories were designed that are compatible with the MPU of Figure 10.6.

Figure 10.9 shows the glue for interfacing I/O devices N and M to the system data bus. I/O device N requires an input port, and therefore a buffer will serve as its interface. When the MPU is instructed to read data from I/O device N, the MPU accesses that device by the usual accessing method by first outputting the address on the address bus and then activating the read control signal $\overline{RD}$. Because the buffer (74LS244) does not have a CS pin for address selection and an output enable for read timing, the selection and timing must be done via the output enable pins ($\overline{OE}$) of the 74LS244. To accomplish this the I/O selector is enabled by the MPU read and write control signals by *AND*ing them and having the output of that *AND* gate drive either of the active-low enable pins ($\overline{E}_1$ or $\overline{E}_2$). The other active-low enable pin can be used to prevent address foldback if desired. For isolated I/O systems, $IO/\overline{M}$ is used to enable the I/O address decoder by being connected to E_3, because $IO/\overline{M} = 1$ if the address is an I/O address. Address pins A_0 to A_2 of the 74LS138 go to corresponding address bus lines so that they can decode the I/O port address on the address bus. If more than eight I/O ports are to be interfaced, the I/O selector must be expanded, as was done for the page selector of Figure 9.22.

The output port of Figure 10.9 is a latch. When the MPU writes a data byte to this port, the 74LS138 decodes the address and activates the addressed port (0 in this case) when the write enable is driven active. The MPU loads and holds the data byte on the data bus and during that period of time it also holds $\overline{WR}$ low. Just before removing the data byte from the data bus it drives $\overline{WR}$ inactive, which forms a positive-edge strobe as $\overline{WR}$ goes from a logic 0 to a logic 1. This positive edge is duplicated at the output of the I/O selector by disabling the 74LS138 when $\overline{WR} = 1$.

Figure 10.9

Input and output ports.

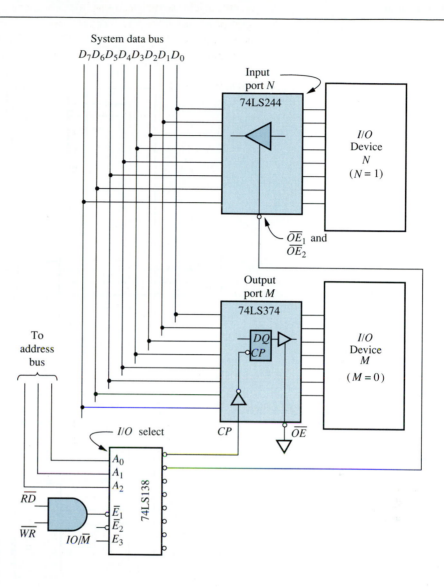

This positive-edge strobe is input to the 74LS374 latch, which latches the data byte off the data bus and completes the write operation.

10.7
Isolated I/O memory systems

Figure 10.10 is an illustration of a microprocessor-based computer sometimes referred to as a microcomputer. The memory design of Figure 9.20 was used to implement main memory, and the I/O ports of Figure 10.9 are used to implement the I/O ports of Figure 10.10. As stated in Chapter 8, to partition memory addresses from I/O addresses the control signal $IO/\overline{M}$ is used (Fig. 9.20 was modified for isolated I/O), as is depicted in Figure 10.1. $IO/\overline{M}$ is input to an enable pin of each 74LS138 address decoder (selector) such that the logic level of $IO/\overline{M}$ will enable the appropriate selector. The page selector for the memory address space is the 74LS138-3 and the I/O selector for the I/O address space is the 74LS138-4. To enable the 74LS138-3 page selector, when memory addresses are on the address bus,

Figure 10.10

A microprocessor-based microcomputer.

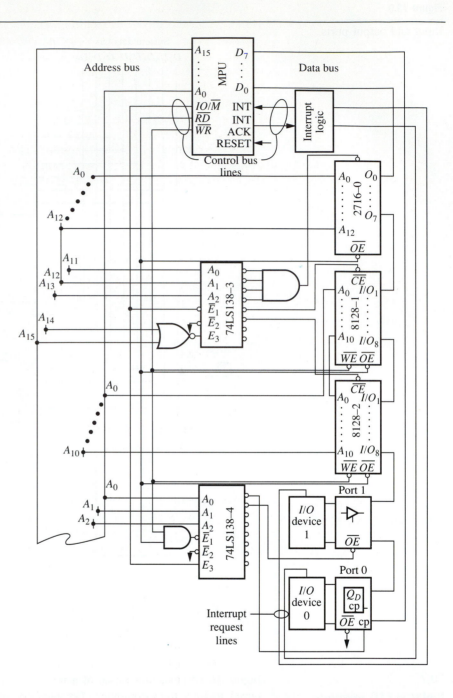

$IO/\overline{M}$ was input to its $\overline{E}_1$ enable pin. To enable the 74LS138-4 I/O selector, $IO/\overline{M}$ is input to its E_3 enable pin. Address pins A_0 through A_2 of both selectors are connected to address bus lines so that when the appropriate selector is enabled these bus lines are decoded for selection of either the addressed memory page or I/O port.

Recall from Chapters 8 and 9 that computer systems that partition address spaces, such as shown in Figure 10.10, are known as isolated I/O systems. By creating two separate address spaces (one for memory and the other for I/O) the number of addresses is greater than if there is just one address space containing both memory and I/O addresses, such as is the case for memory-mapped I/O systems. However, for those systems requiring relative few memory and I/O addresses the two spaces can be combined into just one address space, creating a memory-mapped system. A memory-mapped system does not utilize the $IO/\overline{M}$ control signal. As a result, the system hardware makes no distinction between memory and I/O addresses. This maps (merges) the I/O address space into the memory address space, memory and I/O then share the same address space.

Both I/O devices are given interrupt request capability. These interrupt requests are input to interrupt request logic, which is implemented with logic gates and establishes interrupt priority and interrupt identification. There are commercial chips available that can be used to implement the interrupt request logic. Neither I/O device has DMA capability, so MPU pins *DMA REQ* and *DMA ACK* do not appear in Figure 10.10.

Review Questions

1. What is required to interface memory and I/O to an MPU?
2. What logic is usually used as interface glue for an input I/O device?
3. What logic is usually used as interface glue for an output I/O device?
4. What is the role of the MPU's $IO/\overline{M}$ pin relative to memory and I/O addressing?
5. Why must the I/O selector not only select the addressed port but also provide timing for read and write operations?

There are MPUs that do not have a control signal such as $IO/\overline{M}$. The 6800 MPU is one of those microprocessors. Computer systems that employ this type of MPU can have only memory-mapped I/O systems.

10.8

A Memory-mapped I/O system

The system of Figure 10.11 is representative of a memory-mapped I/O system that uses linear addressing for accessing I/O devices. For the memory portion of this system, the memory of Figure 8.15 was modified for memory-mapped I/O. To accomplish this modification the control signal $IO/\overline{M}$ and accompanying *AND* gates of Figure 8.15 were eliminated, as shown in the inset of Figure 10.11, and four more address lines were added to accomodate the linear addressed I/O. To provide timing for the read/write operations, the *AND* gates of Figure 10.11 (identified as control circuitry) were incorporated. As a result, to access an I/O device the appropriate address line must be active (high) and the appropriate read ($\overline{RD}$) or write ($\overline{WR}$) control must also be low.

As an example of an I/O write operation, if the MPU is to write to I/O port 1 of Figure 10.11, it must drive address line A_9 high and keep it high (note the inverters on the address bus lines); when it has loaded the datum on the data bus, it then drives $\overline{WR}$ low. When these two inputs to the *OR* gate are low, the output of the *OR* gate is driven low, which enables I/O port 1, which latches the datum on the data bus on the negative edge of $\overline{WR}$. As an example of a read operation, when the MPU is to read a datum from an I/O device, say I/O port 3, it first drives A_{11} high and then drive $\overline{RD}$ low at the appropriate time (when it is ready to latch the datum from the data bus). With both A_{11} and $\overline{RD}$ active, the output of the *OR* gate is low, which enables I/O port 3 via its $\overline{CS}$ pin. Enabling I/O port 3 causes it to load its datum onto the data bus, from which the MPU can latch it.

Figure 10.11

A memory-mapped I/O
system with linear
addressing for I/O.

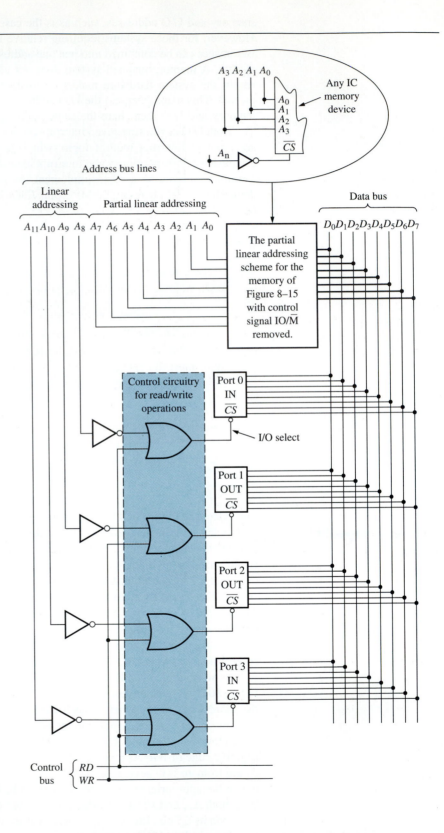

Review Questions

1. What is the conceptual difference between a memory-mapped I/O system and an isolated I/O system?

2. How is an I/O read or write executed relative to MPU control signals $\overline{RD}$ and $\overline{WR}$?

3. What signal determines whether a system is memory-mapped or isolated I/O? Explain how it works.

4. Why are the control signals $\overline{RD}$ and $\overline{WR}$ gated with an address line or I/O select line?

5. Why is the positive edge of $\overline{WR}$ of particular importance?

6. What are the conceptual differences (if any) between addressing memory and I/O?

10.9
A Survey of some real microprocessors

This section will make a cursory study of the architectures of three popular microprocessors — the 8085, 6800, and 8086. The 8085 and 6800 MPUs are 8-bit microprocessors and the 8086 is a 16-bit microprocessor. The architecture of each will be compared to that of Figure 10.6, looking at similarities and differences.

The architecture of the 8085 is illustrated in Figure 10.12a. As compared with the architecture of Figure 10.6, the two are very similar. The number in parentheses

Figure 10.12a

The architectures of the 8085 MPU.

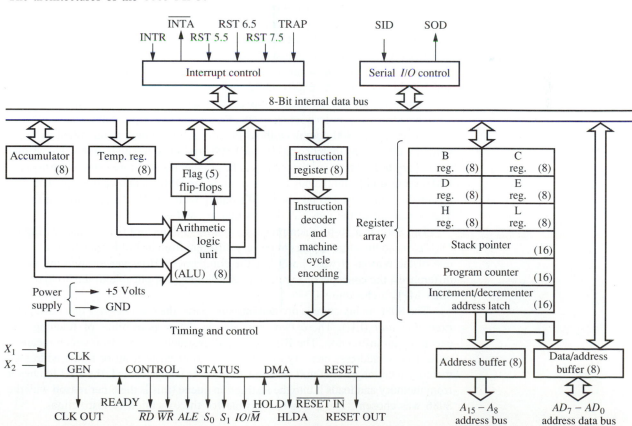

in the function-block of the 8085 architecture indicates the bit size of that function-block. The internal data bus is 8 bits. Register $ACC1$ of Figure 10.6 is labeled as accumulator in the 8085 architecture, and $ACC2$ of Figure 10.6 is identified as being a temporary register in Figure 10.12a. The term ''temporary register'' means that the register is not programmable by the user but is used by the CU to temporarily hold the datum that is to be input to the ALU. Thus, there cannot be an instruction of the mnemonic form $ADD ACC1,ACC2$, because $ACC2$ is not directly programmable. The 8085 add instructions appear as ADD B, ADD C, ADD D, etc.; it is not necessary to specify accumulator in the mnemonic because all ALU operations must have the contents of the accumulator as one of its inputs. The second input to the ALU (the temporary register) will be a copy of the register contents specified in the mnemonic (B, C, D, etc.). Section 10.5 examined such an instruction with the study of instruction $ADD ACC1$,B. There are six scratch pad memory/data counter registers (B, C, D, E, H, and L). To conserve on the number of pins required, the low-byte address pins are multiplexed (time shared) with the data pins. These multiplexed pins are identified as AD_0 through AD_7. The 8085 architecture has multiple interrupt request inputs (*INTR, RST* 5.5, $\overline{RST}$ 6.5, *RST* 7.5, and *TRAP*). The interrupt acknowledge output is identified as $\overline{INTA}$. The DMA handshaking pins are *HOLD* for the request and *HLDA* for the acknowledge. The 8085 has two serial I/O ports (a single-bit port), one of which is an input port (SID) and the other an output port (SOD). The architecture of Figure 10.6 does not have serial ports. The 8085 does not have an instruction queue. Rather, instructions fetched from memory are directly loaded from the system data bus to the IR, via the internal data bus, discussed in section 10.3 on those MPUs without a queue. The 8085 has two I/O addressing registers, which are temporary registers, much like I/O PH and I/O PL in Figure 10.6; however, they are not shown in its architecture.

Figure 10.12b is an illustration of the 6800 microprocessor. It has a PC and SP. The DC is the index register. There is no $IO/\overline{M}$ pin to distinguish between memory and I/O addresses because the 6800 uses memory-mapped I/O addressing. There are two programmable accumulators identified as A and B. The architecture of Figure 10.6 identified these registers as $ACC1$ and $ACC2$. There are 8 data bus pins and 16 address pins. The 6800 architecture lumps the IR decoder and CU into one function-block labeled as instruction decode and control. There are no internal scratch pad registers or instruction queue. Rather than having separate read and write control pins, the 6800 combines them in a single pin which is identified as read/write. When reading memory or I/O the 6800 drives that pin high, and when writing to memory or I/O it drives the pin low. The pin identified as three-state control is the DMA request pin, and the bus-available pin is the DMA acknowledge pin. The data bus enable pin is used to put the 6800's data pins in the Hi-Z state during a DMA. The remainder of the pins have labels that are self-explanatory.

The 8086 is a 16-bit microprocessor whose architecture is illustrated in Figure 10.12c. Its architecture appears to be very different from that of Figure 10.6, which is one of the reasons it was chosen for the comparison. However, despite the visual differences, the concepts of Figure 10.6 are valid and applicable to the 8086 or any other MPU architecture.

The 8086 architecture is divided into two parts, the *bus interface unit* (BIU) and *execution unit* (EU). These two units divide the responsibility of fetching and executing an instruction. The BIU fetches all instructions and data, which means it must access (address) memory or I/O and read or write from or to the addressed memory location or I/O port. If an instruction is being fetched, the BIU fetches it from memory and loads it into the instruction queue (this is the other reason why the 8086 was chosen). The EU is responsible for the execution of instructions.

Figure 10.12b

The architecture of the 6800 MPU.

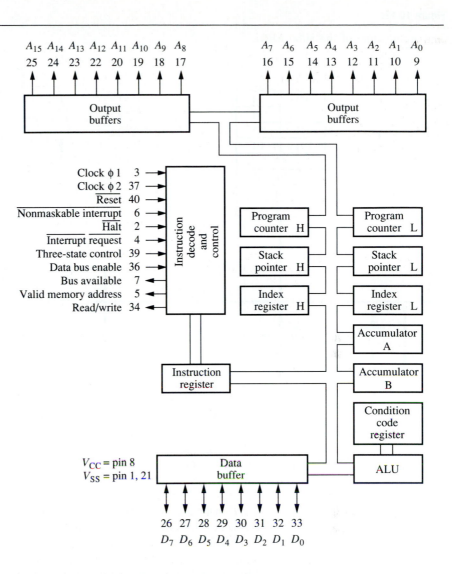

The EU of Figure 10.12c is basically the architecture of Figure 10.6 (or more specifically, a 16-bit version of an 8085) with the instruction queue located in the BIU. With a queue the 8086 employs pipelining and is therefore a pipeline microprocessor. For the sake of simplicity, the EU architecture is simplified to the essentials and similar function-blocks are grouped and merged into one. Because the function of the EU is to execute instructions (not to fetch them . . . the BIU does that) it basically requires the same function-blocks of the generic MPU of Figure 10.6, or those of the 8085 shown in Figure 10.12a. The exceptions are those function-blocks required for interfacing to the system data bus (the data bus buffer/latch) and addressing of memory for fetching instructions (the program counter serves that purpose). These are BIU-type function-blocks and as a result are located in that section.

The instruction stream of Figure 10.6, that is, the IR, IR decoder, and CU are grouped into one function-block in the 8086 architecture illustration of Figure 10.12c and is labeled as EU control system. The registers AH, AL, BH, BL, CH, CL, DH, and DL of the 8086 architecture serve the same purpose as registers *ACC*1, B, C, D,

Figure 10.12c

The architecture of the 8086 MPU.

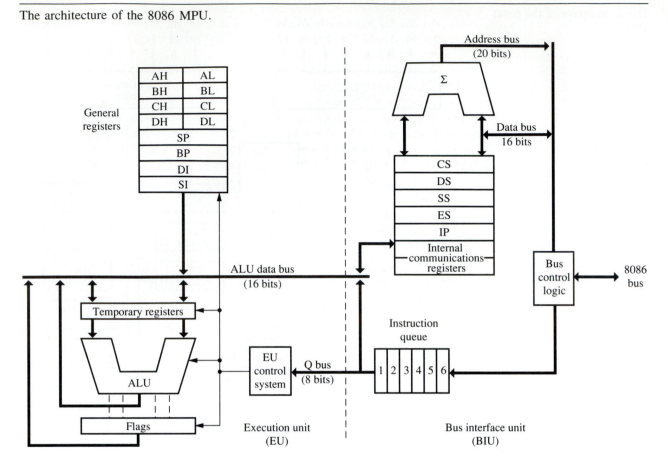

and E in the architecture of Figure 10.6 and the 8085. The main difference is that the registers of the 8086 can be programmed as 8-bit scratch pad registers or they can be paired (BH and BL, CH and CL, and DH and DL) to form 16-bit scratch pad registers. The H in the register designations represent "high-byte" and L signifies "low byte." When an instruction is to pair registers the letter X is used in the instruction mnemonic. For instance, the instruction *MOV* AX,BX would copy the 16-bit data of register BX (BH and BL) into register AX (AH and AL). Instruction *MOV* CL,DL would copy the 8-bit data in register DL into register CL. It should be mentioned that register AX is the accumulator, just as register A is for the 8085. In Figure 10.6 the accumulator is designated as *ACC*1.

In the BIU section of the 8086 architecture of Figure 10.12c there is a function-block labeled "bus control logic." It interfaces the 8086 with all busses (the label "8086 bus" in Fig. 10.12c means all busses, the address bus, control bus, and data bus). As a result, the bus interface function-blocks of Figure 10.6 (the data bus buffer/latch, address latch, and the control signals interface are not specifically shown) are all grouped together and are represented by the 8086 function-block identified as bus control logic. This grouping of function-blocks leaves out some function detail, but the up side is that it simplifies the illustration of the architecture. As MPUs get more complex it is often necessary to omit some of the detail in the architecture illustration to prevent cluttering.

To understand the addressing techniques of the 8086, and the architecture of the BIU section of the 8086, it is necessary to have a basic understanding of modern programming techniques. The addressing technique of the 8086 is based on modular programming. A programming project, or programming task, is modularized into smaller program modules, known as subroutines. This way a large and complex program is broken into smaller and less complex program modules that are easier to program and manage. Also, these modules could be written relatively independent of each other. This could mean that more that one programmer could be assigned to the project with each working in parallel with the other. This could greatly reduce the overall program development time and possibly get the product to market quicker.

As each program module is written it is assigned a *relative address* by its programmer. This is the starting address of the module and is referenced to the starting address of the segment within which it resides. As an example, suppose that there are two large complex programs to be written. Suppose that program no. 1 was modularized into three modules and program no. 2 was divided into four modules. Seven programmers (one per module) can be assigned to the project, all working in parallel. To keep the two programs separated each will be assigned its own segment of program memory or *code segment* (short for program code) as it is known in 8086 terminology. The code segment for program no. 1 begins at, let us say, location 10000H (the 8086 has 20-bit memory addresses) and let us arbitrarily assign address 20000H as the starting address of the code segment for program no. 2. Then code segment 1 begins at location 10000H and will have three modules making up that segment and code segment 2 begins at location 20000H and is composed of four modules. This segmentation is illustrated in Figure 10.13.

As each module is programmed, the programmer assigns an address that is relative to the starting address of the code segment within which it resides. For the memory map of Figure 10.13, if module 2 of code segment 1 was assigned relative

Figure 10.13

Segmentation of program code.

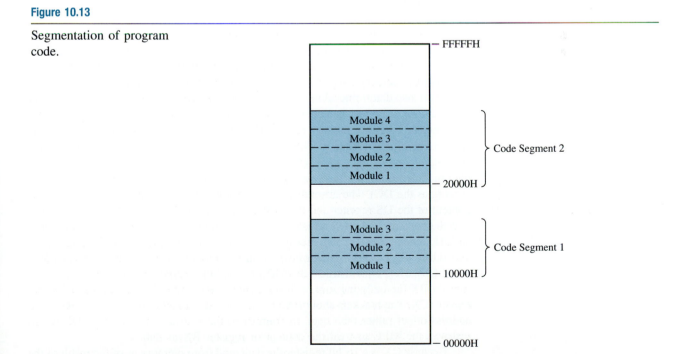

address 500H then the absolute address of module 2 is 10500H (10000H + 500H = 10500H). If module 3 of code segment 2 was assigned beginning address 320H the starting address of module 3 is 20320H (20000H + 320H = 20320H).

From our discussion and examination of Figure 10.13, we can see that the addressing scheme of the 8086 is based on segmentation of programs, which translates into the segmentation of memory. To accommodate various types of segments memory is imagined to be segmented four ways: *code segment*, *data segment*, *stack segment*, and *extra segment*. Except for extra segment, the names of the segments are self-explanatory. The extra segment is also for storing data, as is the data segment, but is viewed as an extra data segment. The starting addresses of the various segments are addressed, or pointed to, by the content of a 20-bit segment register (CS, DS, SS, or ES) in the BIU. If modules from code segment 1 of Figure 10.13 are being addressed the content of the CS register would be 10000H. When modules from code segment 2 are being addressed the content of the CS register is 20000H. Then to switch from one code segment to another requires that the content of the CS register be changed. The same can be said for all the other segments (data, stack, and extra) and their corresponding segment registers (DS, SS, and ES).

The *instruction pointer* (IP) in the BIU of Figure 10.12c serves the same basic function as the PC in both the 8085 and the MPU of Figure 10.6, it is the source register for instruction addresses. The major difference between the PC and IP is the manner in which instruction addresses are determined. In an architecture such as Figure 10.6 the PC contains the absolute memory address of the instruction to be fetched and executed. However, in the 8086 architecture the content of the IP is the relative memory address of the instruction to be fetched and executed. The absolute address from the 8086 is the sum of the contents of the code segment (CS) register and the contents of the IP. If the starting address of a program module within code segment 1 of Figure 10.13 is 426H (recall that this address is relative to the starting address of the code segment within which the module resides) then the absolute starting address of that module is 10426H.

Because the BIU is responsible for fetching instructions it must determine the memory address of the instruction to be fetched and load that address on to the address bus, which is why the CS, IP, and bus control logic are located in the BIU. Note the Greek symbol sigma (signifying summation) located in the ALU within the BIU. It is this ALU that adds the contents of the CS and IP. In addition to determining memory addresses of program instructions, for instruction fetches, the BIU must also determine addresses for reading and writing data from and to memory locations. As previously stated, when data is being read or written from or to memory the content of the data segment register (DS) points to the starting address of the data segment in which the data module resides, just as the content of the CS points to the starting address of the code segment. The relative address, or address offset, of the data module within that data segment is the content of the BX (here register BX is serving as the DC). The absolute address of the memory location is the sum of the content of the DS register and the content of register BX. As an example, suppose that the content of the DS register is 50000H and the content of the BX register is 6346H. When the BIU calculates the memory address from the sum of registers DS and BX it will be 56346H. An instruction mnemonic for this type of addressing is *MOV* CX,[BX]. This instructs the MPU to read the memory location pointed to by register BX (actually the sum of the contents of BX and DS) and load that data into register CX. The brackets about BX instruct the MPU to treat the content of BX as an address offset rather than data, in contrast to the instruction *MOV* CX,BX, which instructs the MPU to treat the content of register BX as data.

Because CX is a 16-bit register the data read from memory as of the result of the instruction *MOV* CX,[BX] is also 16 bits. To read 16 bits from an 8-bit memory two memory locations are read at the same time.

One might wonder why the BX register is not in the BIU because it is used to determine an address in the instance of executing instruction *MOV* CX,[BX]. One of the reasons is that register BX not only serves as a DC but can also provide scratch memory functions that are not BIU functions. More importantly, register BX is programmable, meaning that to execute those instructions requiring its involvement (such as *MOV* BX,CX and *MOV* CX,[BX]) requires that it be part of the instruction execution architecture, the EU.

To address stack segment modules the contents of the stack segment and the SP are added. How stack addresses are determined by the BIU is similar to the explanations given for the other segments.

As a program is written for an 8086 program modules can be assigned to various segments (code, data, and stack). It is easy to see the advantages of this method of writing programs and (as important) managing them. It allows complex programs to be simplified by modularizing them into subroutines (modules). These less complex subroutines can be written by more than one programmer, in parallel, and treated as if they were independent of other subroutines. With modulation also comes ease of program management, that is, it is easier for program troubleshooting, program documentation, and taking care of program maintenance (program up-dates, etc.).

Because the 8086 registers can be programmed as 8-bit registers (which is the size of 8085 registers), and the instruction set mnemonics of both microprocessors are similar, 8085 programs can easily be modified to be executed by an 8086, that is, 8085 programs are *up-ward compatible* with the 8086.

Not only must the instruction set of the 8085 be up-ward compatible with the 8086 in order for the 8086 to execute 8085 instructions but the addressing techniques of both microprocessors must be similar. To determine addressing compatibility let us investigate the address techniques of both microprocessors. The IP of the 8086 and the PC of the 8085 are both 16 bits, meaning they both are limited to addressing 65,536 (2^{16}) contiguous memory locations, or one complete program module within an 8086 code segment. Likewise, the SP in both the 8085 and 8086 is a 16-bit register. Therefore, stack modules for both microprocessors are limited to 65,536 contiguous memory locations and every location within the stack module is address-able by either an 8085 or 8086 SP. The same can be said for 8085 and 8086 DCs. An 8085 program will reside in the same memory segment address spaces (code segment, data segment, and stack segment) that would have been created for a similar 8086 program. This means that each location within each of the segments is addressable by the 8086, via the CS and IP, or the SS and SP, or the DS and DC (content of BX). Hence an 8085 type program can be modified to be executed by an 8086.

Review Questions

1. What registers of the 8085 form its scratch pad memory?
2. The CU is labeled as what 8085 function-block?
3. What are the 8085 serial I/O ports?
4. How does the 8085 differ in interrupt capability from the MPU architecture of Figure 10.6?
5. What is meant by "the low-byte address pins are multiplexed with the data pins on the 8085"?
6. Does the 6800 have an internal scratch pad memory?

7. Does the 6800 have two programmable accumulators?

8. What purpose does the 6800 index register serve?

9. What are the EU and BIU of the 8086?

10. How can the 8086 make more efficient use of the data bus than the 8085 or 6800?

11. Why does the 8086 show an ALU symbol in both the EU and BIU?

12. What similarities do the architectures of the 8085 and 8086 have?

13. How are memory locations addressed by an 8086?

10.10
Troubleshooting

Because microprocessors are on a single IC wafer, if any single MPU function fails, the entire MPU fails catastrophically and must be replaced.

To devise a test for an MPU recall that it is the primary source of all dynamic logic levels. The MPU generates the system's addresses, control signals, and data on the data bus when the MPU is performing a write operation. Measuring dynamic logic levels requires that a logic analyzer be used to monitor those logic levels. The test configuration is nearly the same as those dynamic tests discussed in Chapter 9. There are two differences: (1) The instruction being executed at the time the data samples were taken affects the proper timing diagram for an MPU. (2) The MPU has input pins that, when activated, can modify the operation of the MPU, such as an interrupt or reset. To investigate thoroughly the many complex faults that can arise requires a more in-depth knowledge of an MPU's instruction set and the function of some input pins than is within the scope of this book. Therefore, only some of the more straightforward aspects of troubleshooting an MPU are discussed here.

Test configurations dynamically monitor (1) address bus nodes, (2) control signal bus nodes, (3) data bus nodes, and (4) MPU input nodes that alter the sequence of program execution. Using the MPU of Figure 10.6, we will statically (or dynamically) activate MPU inputs, DMA request, interrupt request, and reset and observe whether the proper MPU response occurs. In addition to the above, check for the obvious, such as proper power supply voltages, grounds, and whether an unused input has been left disconnected (this should never be done).

Example 10.7

Suppose the MPU of Figure 10.10 is to be tested for correctness for each memory location and I/O device. Devise the test.

Solution

To monitor each MPU node requires 29 logic analyzer channels — 16 channels for the address bus, 3 for the control bus, 8 for the data bus, and 2 for the interrupt request and acknowledge. The MPU must execute a program that accesses each memory location and I/O device; that is, each ROM location and input I/O device is read and each RAM location is written into and then read, while output devices are written to. This test is programmed in a 2716 UV PROM, which is then placed in the IC socket for 2716-0.

Once the programmed 2716 is in the system, the MPU's reset pin is activated. Resetting this MPU causes it to vector to address 0000H, from which it fetches and executes the first instruction. Using 0000H as the trigger word, the logic analyzer samples and fills its FIFO memory with nodal data. If the FIFO stores 128 words (128 × 29 bits for 29 channels), then every 128 addresses a new trigger word must be updated and input to the logic analyzer. This updated trigger word is the first address of the next 128 samples to be displayed. Updating the trigger word and resetting the MPU continues until the MPU has accessed every memory and I/O address. The display of the logic analyzer is then checked for correctness.

Seldom would one want, or need, to go through the task of testing every memory location and I/O device via a logic analyzer because there are self-diagnostic programs that can accomplish the task faster and better. Most often the diagnostician is able to narrow the diagnosis to an event or events by means of the symptoms of the failure, as was the case in troubleshooting the system of Chapter 9.

Example 10.8

Suppose that locations within the 8128-1 of Figure 10.10 failed the test, that a program executed by the MPU wrote a known byte of data into the 8128-1's locations, but the data read back was not what had been written. Devise a test to locate the fault.

Solution

Verify that the $\overline{CE}$, $\overline{WE}$, and $\overline{OE}$ have the proper logic levels for the MPU read or write operation. From the $\overline{CE}$ of the 8128-1 of Figure 10.10 the $\overline{O}_4$ output of the page selector 74LS138-3 must be low when 8128-1 is addressed. The truth table of a 74LS138 shows that the inputs A_2, A_1, and A_0 must have logic levels 1, 0, and 0, respectively, to activate output $\overline{O}_4$. In addition to the correct logic levels on the address pins of the 74LS138-3, the enable pins must have the proper logic levels.

As a result of the above analysis connect the channels of the logic analyzer to $\overline{CE}$, $\overline{WE}$, and $\overline{OE}$ of the 8128-1 and to $A_0, A_1, A_2, \overline{E}_1$, and E_3 of the 74LS138-3 ($\overline{E}_2$ is strapped to ground; therefore, it could be tested with a logic probe or VOM).

If the $\overline{CE}$ node of the 8128-1 is high when the 8128-1 is addressed, either the page select *AND* gate has failed or the active output of the 74LS138-3 is at fault. Checking the 74LS138-3 output for proper operation requires that its address and enable pins be checked for proper logic levels. Enable pin $\overline{E}_1$ is directly connected to $IO/\overline{M}$, and E_3 is driven by address lines A_{13} and A_{14}, so these pins can be easily checked for correct logic levels. If E_3 is a logic low and the address bus lines A_{14} and A_{15} have the correct logic level, the *NOR* gate has failed or enable pin E_3 of the 74LS138-3 has shorted internally to ground if the *NOR* gate is operating properly. Disconnecting the *NOR* gate from the 74LS138-3 and determine which is the case and replace either the *NOR* gate or the 74LS138-3.

If the 74LS138-3 checks out, either $\overline{WE}$ or $\overline{OE}$ of the 8128-1 has failed. Their logic levels can also be verified via the display.

If the interrupt logic of the MPU in Figure 10.10 is to be tested, either I/O port 0 or 1 must be forced to request an interrupt. With the logic analyzer monitoring the address bus and MPU inputs *INT* and *INT ACK*, the logic level on *INT* is used as the logic analyzer trigger source. When the interrupt request is made, the logic analyzer begins to sample and store the nodal data. This data can be displayed by the logic analyzer, where it is checked for correctness. If the interrupt request is working properly, the address on the address bus should be the vector address that vectors the MPU to the I/O service routine, and the logic level on interrupt acknowledge (*INT ACK*) should be high.

Although troubleshooting MPUs is involved and is highly dependent on the program being executed, the foregoing discussion should provide basic techniques and insight about methodology.

Review Questions

1. Why can the MPU be termed the system's logic level generator?
2. What MPU output nodes can be monitored for verification of MPU operation?
3. How can the input nodes of an MPU be tested (excluding data pins)?
4. What is the essential difference between testing a memory/IO system and testing an MPU that is part of a total system (i.e., one that contains MPU, memory, and I/O?

Summary

- The computer architecture of Figure 10.1 shows the MPU communicating with memory and I/O via the system data bus. The selection of the memory location or I/O device is made by the MPU over the address bus.

- The control bus is used by the MPU to transmit read and write control signals for controlling use of the data bus and to provide read and write operation timing. Control signal $IO/\overline{M}$ differentiates between I/O and memory addresses and is used to enable the appropriate selector (I/O or page).

- An I/O device, such as mass storage, may have the capability of requesting DMA. When an I/O device has DMA capability, it must control all busses because it must supply the memory location address, control signals, and data.

- An I/O port interfaces an I/O device to the system data bus and is accessed by the MPU.

- The MPU architecture of Figure 10.6 has three registers with which to address memory—the PC for fetching instructions, the SP for accessing the stack, and the DC (a register pair) for accessing data.

- There is just one register pair for addressing I/O—the nonprogrammable register pair I/O PH and I/O PL of Figure 10.6.

- The instruction stream of this architecture is the instruction queue, IR, IR decoder, and CU.

- The CU is a special-purpose computer that is dedicated to the execution of instructions from the instruction set.

- The ALU performs the actual arithmetic and logic operations, and the logic states of the ALU flags are determined by the results of ALU operations. These flags

are used as program decision makers by testing the logic levels of these flags.

- The instruction set of an MPU is determined by the capabilities of the MPU's architecture.

- Instructions have two forms—mnemonics, which are designed to be readable by the programmer, and machine code, which is readable by the MPU and is the binary code equivalent of a mnemonic instruction.

- The MPU can execute only machine code. To translate a program written in mnemonics into machine code, the programmer can hand assemble the mnemonics or use an assembler.

- The 8085 has an architecture very similar to that of Figure 10.6, but without an instruction queue and with serial I/O ports.

- The 6800 mainly differs in architecture from Figure 10.6 by not having an instruction queue, scratch pad memory, and an $IO/\overline{M}$ pin (or one that is similar in function).

- The 8086 is different in many ways from the MPU architecture in Figure 10.6, but the most obvious difference is the existence of the BIU and EU.

- The 8086 employs an addressing technique known as segmented memory addressing.

- Having an architecture that contains a BIU and an EU allows pipelining to be implemented.

- Troubleshooting an MPU is similar to troubleshooting memory or I/O. The major difference is its dependence on program execution and the MPU's ability to have its program sequence altered via instructions and external requests, such as interrupts and DMAs.

Problems

Section 10.2   Computer architecture

1. Explain the purpose of each bus in Figure 10.1.

2. What do bus control signals control?

3. What is the purpose of the two selectors in Figure 10.1, and how are they enabled? Under what condition would there be:
    a. Just one selector for addressing both memory and I/O ($IO/\overline{M}$ not being used)?
    b. No selectors but rather address bus lines are used to select memory and I/O? Again, control signal $IO/\overline{M}$ is not being used.

4. What is the necessity of handshaking?

5. What are the events that take place for
    a. DMA handshake
    b. Interrupt handshake

6. Explain why I/O port 0 of Figure 10.1 is the only I/O device directly connected to the address and control bus via buffers.

Section 10.3   Microprocessor architecture

7. What are the various methods that the MPU architecture of Figure 10.6 has to address (also refer to Fig. 10.2).
    a. Memory
    b. I/O ports

8. What function-block controls the content that is latched by the address latch of Figure 10.6 and what is done with the data (address) latched by the address latch? Also explain the purpose of this process. Refer to Figure 10.4.

9. How are the contents of locations addressed by the PC and SP different?

10. How does the method for determining the next sequential address differ between the PC and SP? Explain the purpose of this difference.

11. Using Figure 10.3 as a model, explain the fetch and execute cycle of an MPU. Begin your explanation with CU loading the address of the instruction on the address bus.

12. What purpose do the IR and IR decoder serve?

13. Define the CU of the architecture of Figure 10.6.

14. What are the three essential function-blocks that a computer must have, and how does the CU meet these criteria?

15. What is a microprogram, and how does it differ from a macroprogram?

16. Why does the architecture of Figure 10.6 show control lines going from the CU to every function-block?

17. All ALU operations must be done with two data at a time. How does the architecture of Figure 10.6 indicate this?

18. What are the ALU flags, and what purpose do they serve?

19. Why can an MPU reset be categorized in the same way as an interrupt and what effect on the MPU do both events have in common?

20. What function does the $IO/\overline{M}$ pin serve?

21. The control signals of some microprocessor-based systems require that the control bus signals be made up of $\overline{I/O\ R}$, $\overline{I/O\ W}$, $\overline{MEM\ R}$, and $\overline{MEM\ W}$, which are separate I/O and memory read and write control signals. Using the control signals of Figure 10.6, design the necessary logic.

22. What is the essential function of the data bus buffer/latch of Figure 10.6?

Section 10.4   MPU timing

23. Suppose that the machine code for instruction *MVI B*,50H begins at memory location 102CH, followed by the instruction *ADD ACC*1,B. Using the timing diagram of Figure 10.7 and the architecture of Figures 10.1 and 10.6, explain in detail the events that are occurring on the various busses and within the MPU architecture to fetch and execute these two instructions. Imagine the instruction queue to be empty, just as if the MPU had vectored to location 102CH as the result of an interrupt.

24. When does the MPU latch the datum being read from the data bus?

25. How does the MPU signal an addressed memory location, or I/O device, that it has the datum to be written on the data bus?

Section 10.5   MPU instruction set

26. Explain how the designer and/or manufacturer of an MPU determines what type of instructions will make up an instruction set.

27. What is a mnemonic and what is the significance of mnemonics?

28. What is a source file and an object file? How are object files different from machine code files?

29. What is an assembler and what does it do?

30. What is an instruction template?

31. How are the configurations of instruction templates arrived at by MPU designers?

32. An IR decoder can be a decoder or a look-up table. Explain each of these decoding techniques relative to this application.

33. Hand assemble the following mnemonic instructions using Table 10.1, the templates of Figure 10.8, and the binary codes of Tables 10.2, 10.3, and 10.4.
    a. *ADD ACC1,E*   b. *ADD ACC1,B*
    c. *AND ACC1,E*   d. *JMP* 0500H
    e. *MOV* B,D   f. *MOV* D,B
    g. *MOV* M,C   h. *MOV* C,M
    i. *MVI* E,7AH   j. *MVI* C,3BH

34. Using your knowledge of the instruction templates of Figure 10.8a and b, hand assemble *SUB ACC1,D* and *OR ACC1,E*.

35. Using the templates of Figure 10.8, why is it that instructions such as *ADD* C,D, or *AND* E,D cannot exist?

36. Is the architecture a limiting factor in creating instructions *ADD* C,D and *AND* E,D? Explain your answer.

Section 10.6   Memory and I/O interfacing: the creation of a microcomputer

37. What is an I/O port, and what purpose does it serve?

38. What type of digital circuit is often used to implement I/O ports? Explain your answer.

39. Redesign the system of Figure 10.10 so that its memory will be a memory-mapped I/O.

40. Redesign the memory of Figure 10.10 so that partial linear addressing is used.

41. Repeat Problem 40 using linear addressing for I/O.

42. To the system of Figure 10.10 add two more I/O ports, an input and an output port.

43. If the system of Figure 10.10 is to have a total of ten I/O ports what changes must be made? Show those changes.

44. To provide I/O read and write timing via pin E_3 of the 74LS138-4 in Figure 10.10, what changes are necessary? Show the changes.

Section 10.8   Memory-mapped I/O systems

45. For the system of Figure 10.11 show the logic circuit(s) that are used to implement the I/O ports. Refer to the ports of Figure 10.10.

46. For a 16-bit address bus, how many more I/O ports can be added to the system of Figure 10.11? Explain your answer.

47. Why are the inverters of Figure 10.11 required. Your answer is to be related to the content of the PC after a MPU reset (PC = 0000H).

48. The 6800 MPU loads its PC with address FFF8H after a reset. Would the system of Figure 10.11 work for a 6800 MPU-based system? Explain your answer and if changes are necessary make them and show your modifications.

49. Relative to MPU control signals and instructions, explain the difference between memory-mapped I/O and isolated I/O systems.

Section 10.9   A survey of some real microprocessors

50. Compare the architectures of the 8085 and 6800.

51. Compare the architecture of the 8086 to that of the 8085 and the 6800.

Troubleshooting

Section 10.10   Troubleshooting

52. Suppose you want to test the DMA operation of a system. How would you configure the test?

53. If the MPU of Figure 10.10 failed when reading I/O port 1, what test procedures would you use?

54. Repeat Problem 52 but with I/O port 0.

55. How would you configure a test for the 2716-0 of Figure 10.10?

11

PROGRAMMABLE LOGIC DEVICES

OBJECTIVES

- The objectives of this chapter are to:
- Introduce the concept of the programmable fuse-link
- Explain PLD notation
- Introduce programmable logic elements (PLE)
- Introduce programmable array logic (PAL)
- Examine procedures to program PAL devices using PALASM4
- Introduce programmable logic arrays (PLA)
- Investigate some combinational and sequential logic applications for PLEs, PALs, and PLAs

The erasable programmable read-only memory is an IC that can be programmed with a sequence of binary numbers to instruct a microcomputer or to replace complex logic circuits (Reprinted by permission of Intel Corporation. © Intel Corporation 1989).

11.1 Introduction

In the introductory chapters of this book design techniques were developed for combinational and sequential logic circuit solutions to design problems. The microprocessor was introduced as a possible alternate solution. The microprocessor has the advantage of being programmable, so that changes in design circumstances can be accommodated by programming changes. A design change for combinational and sequential logic designs often requires scrapping much of the old design and hardware and beginning anew. This suggests that microprocessor technology has made combinational and sequential logic circuits obsolete; however, this is not the case. In fact, most microprocessor-based systems require some combinational logic circuits to interconnect the various parts of the system, that is, it is the combinational logic that "glues" a microprocessor system together. Logic circuit designs that are tailored for a particular application are often faster than a microprocessor solution, but they are less flexible and often occupy more printed circuit board area because of the many ICs required to implement them. To reduce the chip count, these logic circuits are sometimes integrated on a single IC wafer. These types of ICs are known as Application-Specific Integrated Circuits (ASICs). Because of the manufacturing process, their use can be expensive (for low volumes) and time consuming.

This chapter introduces a third possible solution for a design problem— *programmable logic devices* (PLDs). These are ICs that contain many logic gates, and in some cases registers also, that can be configured (programmed) by the user into various logic combinations. This approach offers flexibility in the specific combinational or sequential logic design to be implemented by the PLD, with the additional benefit that the logic design is contained in one 20- or 24-pin IC rather than several ICs. Three types of PLDs will be discussed: (1) *programmable logic elements* (PLE), (2) *programmable array logic* (PAL), and (3) *programmable logic arrays* (PLA).

11.2
The programmable fuse-link

Regardless of the PLD, the essence of its programmability is the *fuse-link*, which connects two points electrically. If a connection is desired between the two points, the fuse is left intact; if no connection is needed between the two points, the fuse is removed by "blowing" it. The ability of the user to select which fuses remain and which are removed implements the programmability of a PLD. A programmer (a PROM programmer or similar apparatus) is used to "burn" those fuses to be blown.

Figure 11.1 illustrates the concept of the fuse-link. The logic gate of Figure 11.1a is a four-input *AND* gate that has fuse-links at each input. As stated, the user can selectively blow these fuse-links, implementing the programmability of the *AND* gate. The *AND* gate of Figure 11.1b has the no. 2 fuse-link blown, which has the same effect as if the input logic was a 1 ($B = 1$). As a result, the Boolean expression for output E is

$$E = A \cdot 1 \cdot C \cdot D = ACD.$$

Observe that an open input is equivalent to a logic 1 for an *AND* gate. The reason for this will be explained in Chapter 12.

Figure 11.1

Fuse-link examples.

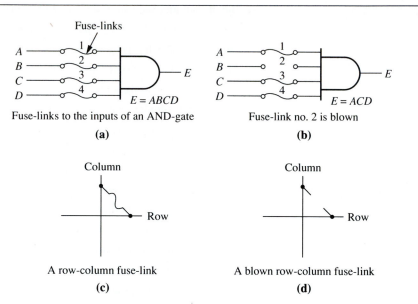

Fuse-links to the inputs of an AND-gate

(a)

Fuse-link no. 2 is blown

(b)

A row-column fuse-link

(c)

A blown row-column fuse-link

(d)

The row and column lines of Figure 11.1c are connected via an intact fuse-link. Being connected the two have the same logic level, whereas blowing the connecting fuse-link makes the row and column lines independent of each other. Figure 11.1d illustrates a blown fuse-link for a row and column connection.

Figure 11.2a is an illustration of how fuse-links might be utilized in a PLD to implement a sum-of-product (SOP) Boolean expression. The Boolean independent variables are the inputs at pins I_0 to I_2. The user may select (program) which of those variables will act as inputs to *AND* gates *A*, *B*, and *C* via the row-column fuse-links. The *OR* gate fuse-links are used to program which *AND* gate outputs will be *OR*ed together.

The logic circuit of Figure 11.2b is the PLD of Figure 11.2a that has been programmed. The only fuse-link of *AND* gate *A* that has been left intact is the *Z* input; hence, the output of *AND* gate *A* is *Z* (recall that an open input is the same as a

Figure 11.2

The fuse-links of a PLD.

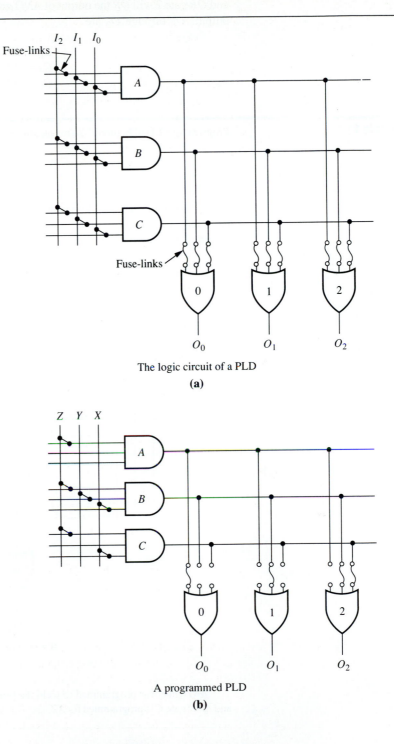

The logic circuit of a PLD

(a)

A programmed PLD

(b)

logic 1 applied at that input). *AND* gate *B* has all fuse-links intact, so its output is equal to the Boolean expression *XYZ*. The output of *AND* gate *C* can be expressed by the Boolean expression *XZ*. *OR* gate 0 will *OR* just the output of *AND* gate *A*, because that output is the only one with its fuse-link intact. Note that an open input of an *OR* gate is equivalent to a logic 0. Similarly, *OR* gate 1 will *OR* the output of *AND* gate *B*, and *OR* gate 2 will *OR* the output of *AND* gates *B* and *C*. The Boolean expressions for outputs O_0 through O_2 are

$$O_0 = Z$$
$$O_1 = XYZ$$
$$O_2 = XYZ + ZX$$

Example 11.1

Program the PLD of Figure 11.2a to yield the SOP Boolean expression

$$W = XY + YZ + XZ$$

Solution

As in Figure 11.2b, in Figure 11.3 arbitrarily input variable X at input I_0, Y at I_1, and Z at I_2. Because only one output is required, the output of *OR* gate 0 is used; however, note that if other *OR* gate fuses are left intact, any one of the outputs will yield the desired Boolean expression.

Figure 11.3

A PLD programmed for the Boolean expression of Example 11.1.

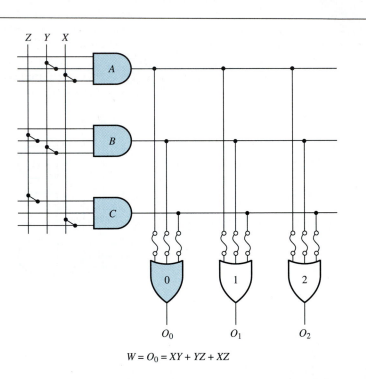

$$W = O_0 = XY + YZ + XZ$$

AND gate *A* is programmed to yield the product *XY*, *AND* gate *B* is programmed for *YZ*, and *AND* gate *C* is programmed for *XZ*.

PLDs offer the user the capability to implement logic circuit designs and to minimize the number of ICs, because many of the logic circuits can be programmed within a single 20- or 24-pin PLD.

Review Questions

1. What relationship exists between programmability and fuse-links?
2. What is the resultant logic state for a blown fuse at the input of an *AND* gate?
3. What is the resultant logic state for a blown fuse at the input of an *OR* gate?
4. What is a PLD?
5. How can PLDs be programmed to implement combinational logic?

11.3

Programmable logic elements (PLE) and PLD symbolic notation

To develop PLD notation, it is best to have a model that can be referenced for examples. Because PLEs are the most fundamental PLDs, they will be used to develop some basic concepts and notation.

As stated in the Introduction, there are basically three types of PLDs — PLE, PAL, and PLA. Except for the placement of the fuse-link, all three types of PLDs have essentially the same conceptual logic gate architecture. It is their fuse-link placement and hence their programming flexibility that differentiate one type from another. A PLE has a fixed *AND* gate array (it is not programmable) whose outputs serve as inputs to programmable (fused) *OR* gates. Figure 11.4 is the logic architecture of a simplified two-input, four-output PLE.

The two inputs of the PLE of Figure 11.4 are identified as I_0 and I_1. If this device were to function as a PROM, these inputs would be used as address pins. However, in a PLE think of these input variables in a more generic sense; that is, these inputs are just pins to which Boolean independent variables are input to the PLE. The two input

Figure 11.4

The logic architecture of a PLE.

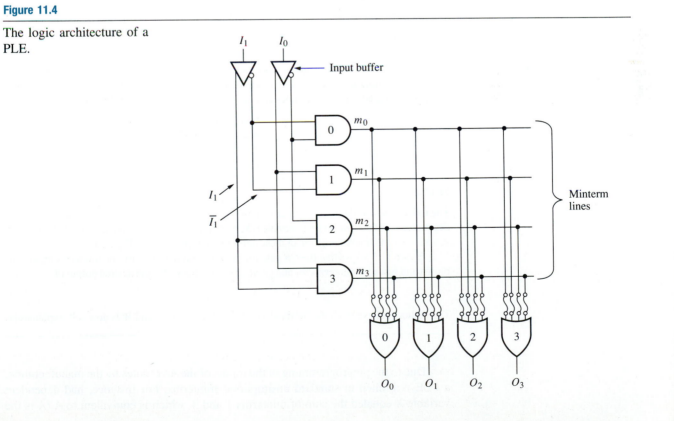

buffers have as outputs the input variables and their complements. Hence, the output of *AND* gate number 0 (m_0) is minterm $\bar{I}_1\bar{I}_0$, and the output of *AND* gate 1 is minterm m_1 ($\bar{I}_1 I_0$), etc. The output of each *AND* gate is a minterm of a (SOP) Boolean equation. Table 11.1 shows the Boolean expression for each minterm as a function of the generic input variables I_1 and I_0 and the truth table for those variables.

Table 11.1 Boolean expressions for the minterms of the PLD of Figure 11.4

Minterm	Boolean expression	Input logic states
m_0	$\bar{I}_1\bar{I}_0$	00
m_1	$\bar{I}_1 I_0$	01
m_2	$I_1\bar{I}_0$	10
m_3	$I_1 I_0$	11

In using a PLE, its flexibility is the ability to program which minterm(s) will serve as inputs to the *OR* gate of Figure 11.4. If a fuse-link is left intact, the respective minterm is input to that *OR* gate. In contrast, when a fuse-link is blown, the minterm served by that fuse-link is not input to that *OR* gate; instead an equivalent logic 0 is applied. The Boolean expression for the output of each *OR* gate of Figure 11.4 has the potential of being equal to

$$O_n = m_0 + m_1 + m_2 + m_3 = \sum_{n=0}^{3} m_n$$

The reader should recognize that the above equation is in the form of an SOP Boolean expression.

Example 11.2

Use a PLE, similar to Figure 11.4, to implement the given Boolean expression, where A and B are the independent (input) variables and the dependent (outputs) variables are X, Y, Z, and W.

$$X = \Sigma_m 1, 2$$
$$Y = m_3$$
$$Z = \Sigma_m 0, 3$$
$$W = \Sigma_m 0, 1, 2$$

Solution

The independent variables (the input variables) A and B are defined as $A = I_0$ and $B = I_1$. The dependent variables (the output variables) X, Y, Z, and W are defined as $O_0 = X$, $O_1 = Y$, $O_2 = Z$, and $O_3 = W$. Implementation of these definitions is illustrated in Figure 11.5.

Figure 11.5 shows that the *OR* gate for output X has the fuses for the minterms m_0 and m_3 blown and fuses for minterms m_1 and m_2 still intact. This yields the desired output of

$$X = \Sigma_m 1, 2$$

The authors believe the logic for the dependent variables Y, Z, and W is now self-explanatory.

Due to the preprogramming of the inputs of the *AND* gates by the manufacturers, a PLE is limited to standard unsimplified minterms. For instance, had dependent variable X equaled the sum of minterms 1 and 3, which is equivalent to A (A is the

Figure 11.5

A PLE programmed for the Boolean expressions of Example 11.2.

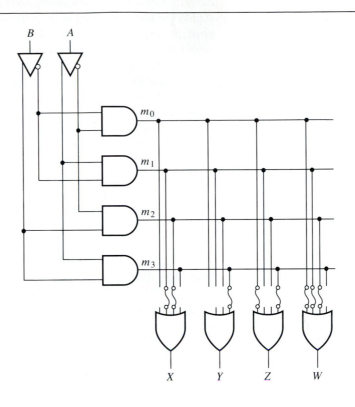

LSB) when simplified, it could not be implemented in simplified form. As will be shown, PLAs and PALs do not have this restriction.

The number of minterms of a PLE and, of course, the number of *AND* gates are directly related to the number of input variables. Previous chapters showed that for *n* independent variables there are 2^n binary combinations of those independent variables. Because there is an *AND* gate for each combination (a minterm), as shown in Table 11.1, there are also 2^n *AND* gates. The outputs of these *AND* gates are the minterms of a SOP-type Boolean equation. To reiterate, these same *AND* gates would be labeled as address decoders if the PLE were being used to implement a PROM.

The number of *OR* gates of a PLE can be varied at the time of manufacture, just as can the bit width (the number of bits in each memory location) of a PROM. For a PLE created PROM, the number of *OR* gates determines the bit width of the PROM. For instance, if the PLE of Figure 11.4 were being used as a PROM, each memory location would contain 4 bits, 1 bit per *OR* gate output. The logic level of each bit position of a memory location is determined by the logic level of the corresponding *OR* gate output. For instance, if the PLE of Figure 11.5 were considered to be a PROM, the output of *OR* gate W would be the LSB of the memory locations, the output of *OR* gate Z is the next highest bit position, *OR* gate Y outputs the logic level for the next highest bit position, and the logic level output from *OR* gate X is the MSB for those memory locations. The output logic levels of those *OR* gates are determined by the logic state produced by the minterm *AND* gates fuse-linked to their inputs and the logic 0s produced by those blown fuse-links.

The number of minterm *AND* gates, which is the same as the number of inputs to the *OR* gates, determine the number of memory locations. There are 4 minterm *AND* gates in Figure 11.5, as well as 4 inputs to each of the 4 *OR* gates, hence there are 4 memory locations. These minterm *AND* gates serve two purposes. First, their outputs supply the logic 1s stored in the various memory locations. Second, they serve the same purpose as address decoders, to locate the addressed memory location, in standard IC memories. The addresses of the memory locations are the logic level inputs to the minterm *AND* gates. From Figure 11.5 it can be seen that input variables A and B, and their complements, are applied to the minterm AND gates. As shown in Table 11.2, for every address there is a corresponding minterm AND gate whose output (minterm line m_0, m_1, m_2, or m_3) is active (a logic 1). The minterm *AND* gate whose output is a logic 1 is used to supply the logic 1s stored at that address. As a result, the logic 1 output from minterm m_0 is used to supply the logic 1s stored at address 00, and the logic 1s stored at address 01 are supplied by minterm AND gate m_1. This logic continues for the rest of the addresses and the contents of their memory locations. The *OR* gate fuse-links are left intact when a logic 1 is stored in a bit position of a memory location and the fuse-link is blown when a logic 0 is to be stored.

Table 11.2   The PLE of Figure 11.5 used as a PROM

Address		Active address line (product-line)	Programmed bit stored			
B	*A*		*X*	*Y*	*Z*	*W*
0	0	m_0 (P_0)	0	0	1	1
0	1	m_1 (P_1)	1	0	0	1
1	0	m_2 (P_2)	1	0	0	1
1	1	m_3 (P_3)	0	1	1	0

As an example of how this works, consider address 00 input (via inputs *A* and *B*) to the PROM of Figure 11.5. Minterm line m_0 goes high and the other minterm lines are low. The *OR* gate of output X has its fuse-link to m_0 blown; therefore, the X output is a logic 0, as indicated by the first row of Table 11.2. The connection from minterm m_0 to the input of the *OR* gate with output Y is also blown, so Y is also a logic 0, as is also indicated in Table 11.2. The connections from minterm line m_0 to *OR* gate outputs Z and W are intact; as a result, these outputs go high when the minterm m_0 goes high, which is also indicated by the first row of Table 11.2.

The number of *OR* gate outputs required for a PLE varies according to the number of dependent variables (outputs) to be implemented. If a PLE is used to implement a design with two dependent variables, two outputs are all that is required. If there are to be five dependent variables, there must be five *OR* gates, etc. There is no strong relationship between the number of *OR* gates and the number of independent variables (inputs) for a PLE, just as there is no direct relationship between the number of input variables (the number of address bits) of a PROM and its bit size. What can be said with certainty is that for *n* input variables there are 2^n *AND* gates, where the output of each *AND* gate is a minterm of a SOP Boolean expression. The number of *OR* gate outputs can be any number that a designer and/or manufacturer wishes to include as part of the IC. The size of the silicon wafer and the number of pins on the IC have a limiting effect.

A practical upper limit on the number of storage bits within a PROM is 64 K (65,536) bits (recall the 2764 PROMs of Chapter 9). Because microprocessors have data bus structures that are multiples of 8 bits, these 64 K PROMs are often configured as an 8 K × 8 memory. With 8 K locations, there are 13 inputs (address pins). However, when a PLD is being fabricated for use as a PLE, rather than as a PROM, the number of inputs is decreased, but the number of outputs is increased. A PLE may have 6 inputs and as many as 16 outputs. This redistribution of input variables versus output variables seems reasonable, because seldom would a logic design have more than six independent variables, but there could be many dependent variables that are to be generated from those input variables.

Example 11.3

For a PLE with 6 inputs and 16 outputs, what is the maximum number of minterms that an SOP Boolean expression can have for any one of those outputs?

Solution

For six inputs there are 2^6 (64) minterms. If the connections between all minterm lines (m_0, $m_1, \ldots, m_{63}$) and the fuse-links to an *OR* gate are left intact, that *OR* gate would *OR* each minterm of the *AND* gate array. The resulting Boolean expression for this output would be

$$O_n = \Sigma_m 0, 1, 2, 3, 4, 5, 6, 7, 8, 9, \ldots, 63.$$

Imagine having to make an architectural drawing, similar to Figure 11.4, for the PLE of Example 11.3. There would be six inputs (I_0 through I_5) and input buffers, 64 *AND* gates, and 16 *OR* gates. Each of the 64 *AND* gates would have six inputs, these inputs being combinations of the input variables and their complements, which form the minterms of the resulting SOP equations. Each of the 16 *OR* gates would have 64 programmable (fuse-linked) inputs. (If you were not discouraged from making this drawing with the 64 *AND* gates, surely these 16 *OR* gates with 64 inputs each would do the job.)

PLD symbolic notation

The large number of logic gate inputs and fuse-links that, in part, make up a PLD architectural drawing requires from a practical point of view that a simpler symbolic representation be adopted. The following PLD notation has been adopted by industry:

1. Show a single input line for all logic gates. For the minterm *AND* gates these lines are known as *product lines*.

2. Put an × at any programmable (fuse-linked) connection where the fuse is intact.

3. Use a dot, as is standard practice in electronic schematics, to indicate a fixed (nonprogrammable) electrical connection.

Figure 11.6 is a logic architectural drawing of the PLE of Figure 11.5 using this PLD symbolic notation. The minterm *AND* gates in Figure 11.6 have a single input that is identified as a product line, as stated above. Those inputs (*A* and *B*) to the *AND* gate are identified as fixed (nonprogrammable) connections by the use of a dot. The progammable connections of the *OR* gate inputs that are fuse-linked (the fuse-link is intact) to a minterm line (m_0, m_1, m_2, and m_3) are indicated by an × at the intersection of the *OR* gate input and the minterm line. Fuse-links that are blown make no connection to the minterm line and OR gate input, which are indicated by the absence of a × at the intersection of the minterm line and the *OR* gate input.

Figure 11.6

The PLE of Figure 11.5 using PLD symbolic notation.

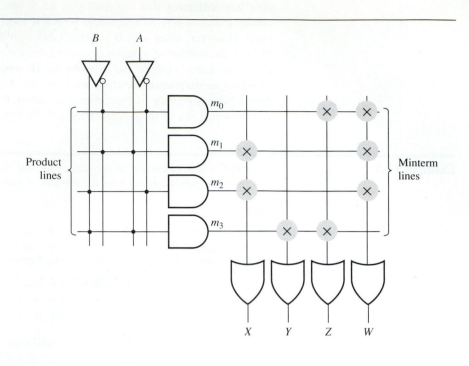

Example 11.4

Using a PLE with four inputs and four outputs, similar in architecture to Figure 11.4, design the 2-bit parallel binary adder of Figure 4.15. Show the programmed logic architecture using PLD symbolic notation.

Solution

As in Chapter 4, first construct a truth table, which relates the logic states of the independent variables (inputs) and the dependent variables (outputs). From this truth table determine the SOP Boolean expression for each of the outputs. These equations are simplified and then implemented. Follow the same steps except do not simplify the Boolean equations because PLE devices implement only minterms.

Minterm	B_1	A_1	B_0	A_0	C	S_1	S_0
0	0	0	0	0	0	0	0
1	0	0	0	1	0	0	1
2	0	0	1	0	0	0	1
3	0	0	1	1	0	1	0
4	0	1	0	0	0	1	0
5	0	1	0	1	0	1	1
6	0	1	1	0	0	1	1
7	0	1	1	1	1	0	0
8	1	0	0	0	0	1	0
9	1	0	0	1	0	1	1
10	1	0	1	0	0	1	1
11	1	0	1	1	1	0	0
12	1	1	0	0	1	0	0
13	1	1	0	1	1	0	1
14	1	1	1	0	1	0	1
15	1	1	1	1	1	1	0

The design equations are

$$S_0 = \Sigma_m\, 1, 2, 5, 6, 9, 10, 13, 14$$
$$S_1 = \Sigma_m\, 3, 4, 5, 6, 8, 9, 10, 15$$
$$C = \Sigma_m\, 7, 11, 12, 13, 14, 15$$

The PLE must have at least 2^4 (16) *AND* gates, product lines, and minterm lines, because there are four independent variables (A_0, B_0, A_1, and B_1). Because there are three dependent variables (S_0, S_1, and C), the chosen PLE must have a minimum of three outputs. As indicated in Figure 11.7, the selected PLE has four outputs, one of which is not used.

Figure 11.7

A 2-bit parallel binary adder implemented with a PLE.

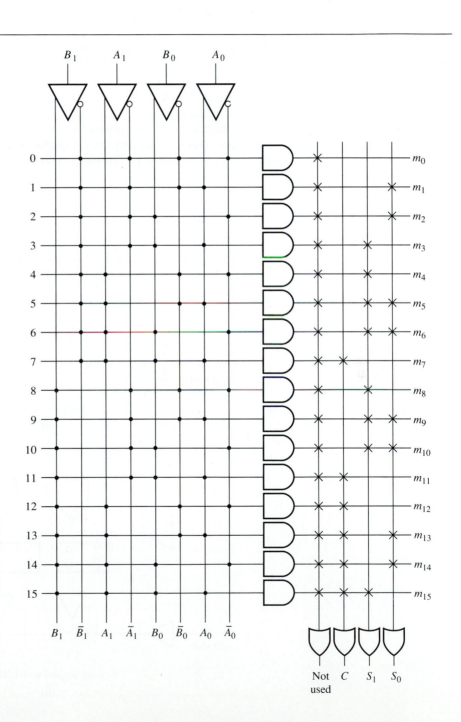

The Boolean equations are programmed into the PLE of Figure 11.7. To verify proper programmed connections, refer to the Boolean design equation, the input line of *OR* gate S_0 must be connected to minterm lines m_1, m_2, m_5, etc, as indicated by the connection ($\times$) at the intersection of the *OR* gate input and minterm lines. The *OR* gate for output S_0 in Figure 11.7 has been properly programmed. Repeat program verification for *OR* gates S_1 and C and find them also to be programmed according to their design equations. The *OR* gate that is not used indicates that all fuse-links were left intact. Because it is not used, and by leaving the fuses intact, that *OR* gate can be considered a spare that may be programmed at a later time if needed.

Refer to the input variable vertical lines ($A_0, \overline{A}_0, B_0$, etc.) and notice the pattern of connections with the product lines (*AND* gate inputs). This pattern is the same as counting in binary; that is, the connections appear in a pattern equivalent to the binary input of each minterm *AND* gate.

The actual programming of a PLE is done with a PROM programmer. Fortunately, there are computer programs available that take Boolean design equations and format them so that they are compatible with PROM programmers. One such program was developed by Monolithic Memories, Inc. (MMI) and can be identified

Figure 11.8

Output variations of a PLE.

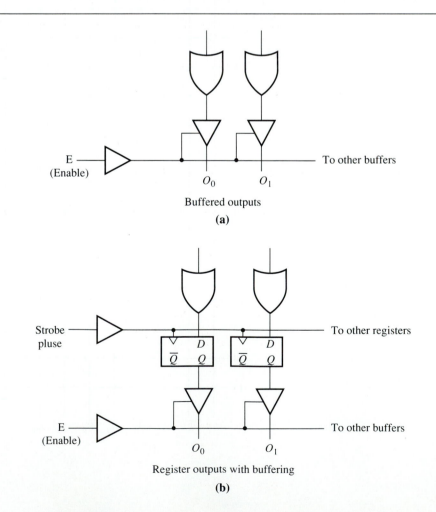

Buffered outputs

(a)

Register outputs with buffering

(b)

by the trademark PLEASM (PLE Assembler). Programming with an assembler will be shown later in the chapter.

Programmable logic elements are manufactured with various numbers of inputs and outputs — 5 inputs and 8 outputs, 6 inputs and 16 outputs, 9 inputs and 4 outputs, and 12 inputs and 8 outputs. In addition, they also have different output capability. Buffered outputs are available as well as outputs with buffered registers. Figure 11.8

Figure 11.9

Representative *AND/OR* logic array of a PLE.

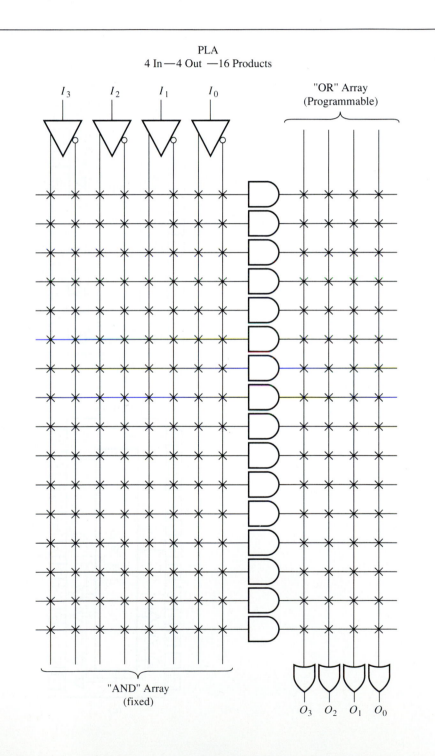

shows a sampling of available outputs. As already known, buffering is required for PLE applications whose outputs are connected to a bus, and registers are necessary any time the output logic state must be retained (remembered) even though there has been a change in the logic state of one or more input variables.

Consider the logic architecture of Figure 11.9 to be representative of PLE logic arrays. If the number of inputs is increased, the number of *AND* gates increases to 2^n,

Figure 11.10

Samples of PLE devices. (Copyright © Advanced Micro Devices, Inc., 1992. Reprinted with permission of copyright owner. All rights reserved.)

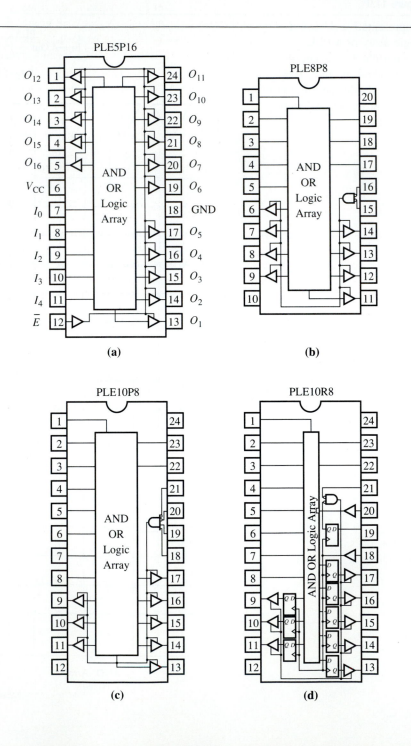

where *n* is the number of inputs, and there will be one *OR* gate per output. If the outputs are buffered they appear as shown in Figure 11.8a, and if they have buffered registers they appear as shown in Figure 11.8b.

Figure 11.10 is an illustration of some commercially available PLEs. Figure 11.10a is the logic symbol and pin configuration of a PLE5P16. The *AND / OR* logic array symbol is greatly simplified by being represented by a rectangle. There are five inputs that are identified as I_0, I_1, I_2, I_3, and I_4 (pins 7 through 11), which means there are 32 (2^5) *AND* gates. There are 16 buffered outputs identified as $O_1, O_2, O_3, \ldots$, O_{16}. The buffer enable input is identified as $\overline{E}$ and is pin 12. The power supply V_{CC} is 5 volts DC and is pin 6, and the ground is pin 18.

Figure 11.10b is another PLE with buffered outputs; however, there are two enable inputs, pins 15 and 16, as can be reasoned from the *AND* gate that controls the state of the tri-state buffers. Because the inputs to this *AND* gate are active lows, the enable signals are also active lows. The inputs are not labeled but are pins 1, 2, 3, 4, 5, 17, 18, and 19. The outputs are pins 6, 7, 8, 9, 11, 12, 13, and 14. They are not labeled but are obvious because of the buffers.

Figure 11.10c is very similar to the PLE of Figure 11.10b except that there are 10 inputs rather than eight, and four enable inputs are required. Two of those inputs are active low (pins 20 and 21) and two are active high (pins 18 and 19).

Figure 11.10d shows a PLE10R8; it has eight easily recognized buffered register outputs and 10 inputs. Output buffers are enabled via an active low at pin 21 and the *D*-type flip-flop, which is the synchronous enable flip-flop, set at pin 19. The strobe pulse for the flip-flops is input at pin 18. Pin 20 is an input pin to which an initialization input signal can be applied. When low, this allows one of 16-bit patterns to be set into the output registers, where the desired pattern is determined by the logic levels of input pins 5, 6, 7, and 8. This feature allows the flip-flops to be set to a desired binary combination, which can be useful in some sequential or feedback designs. If this feature is not to be used, pin 20 is connected to a logic high.

There are a variety of PLDs available. Regardless of the source, the identifying nomenclature for a PLD is similar. Figure 11.11 illustrates a common approach to PLD identification. The leftmost symbols identify the PLD type (PLE, PLA, or PAL). PLE is a registered trademark of MMI; therefore, they use PLE for their product line of programmable logic elements. For a five-input PLE the number 5 is in the *II* position of the product identification symbol. The *TOP* symbol indicates whether the outputs are registers (R), nonregistered (P), etc. The number of outputs is designated by *OO*. The PLE5P16 in Figure 11.10a has 5 inputs and 16 outputs, which are nonregistered (P). The other PLEs of Figure 11.10 are identified by the same notation as indicated in Figure 11.11.

Figure 11.11

PLD product identification nomenclature.

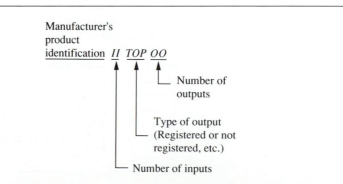

Review Questions

1. Relative to the *AND/OR* logic array of a PLE or PROM, which gates have fixed inputs and which have programmable inputs?

2. What relationship exists between the number of inputs and the number of *AND* gates?

3. What relationship exists between the number of inputs and the number of outputs?

4. What part of an SOP Boolean equation does the *AND* gate yield?

5. What part of an SOP Boolean equation does the *OR* gate implement?

6. Given the logic architecture of a PLE and the SOP equation to be programmed, how can a person methodically program (on paper) a PLE?

7. What is the nomenclature for identifying a PLE?

11.4
Programmable array logic (PAL)

A PAL device is a PLD that is similar to a PLE, but differs from a PLE in that it has a programmable *AND* gate array with a fixed *OR* gate array. PAL is an acronym for programmable array logic and is a registered trademark of MMI. Figure 11.1a illustrates the programmable *AND* gate. Figure 11.12 illustrates the general format of a PAL using conventional PLD symbols as indicated by the ×'s. The product lines are programmable but the inputs to the *OR* gate are fixed as signified by the dots. Each *OR* gate *OR*s four product terms. Until programmed, the output of each *AND* gate (the product term) is a logic 0, because it is the product of each input variable and its complement. Hence, if a product line is not programmed (all of its fuse-links are left intact), the output of that *AND* gate is a permanent logic 0. The symbolic notation used to indicate a product line with all independent variable (input variables) fuse-links intact is a "×" inside the *AND* gate logic symbol of that product line (*see* Fig. 11.13).

Example 11.5

Use the PAL of Figure 11.12 to implement the dependent variables Y and Z if the independent variables are A, B, and C and the Boolean design equations are

$$Y = BC + AC + \overline{A}B\overline{C}$$
$$Z = \Sigma_m 0, 3, 6 \quad (C \text{ is the LSB position})$$

Solution

Only two outputs are being used. We choose output O_0 for the dependent variable Y and output O_1 for dependent variable Z. Outputs O_2 and O_3 are not used; as a result, the *AND* gates that these two *OR* gates *OR* together do not need to be programmed. Product lines 0 through 7 of Figure 11.13 show an X within the respective *AND* gate symbol to indicate this. These outputs can serve as spares for later circuit revisions if needed.

From Figure 11.13 it can be reasoned that to program a PAL pick a product term from the Boolean equation and implement it with one of the *AND* gates being connected to the *OR* gate that was chosen to represent the dependent variable. For dependent variable Y product line 12 was used to implement the product term BC, product line 13 to implement AC, and product line 14 to implement the product term $\overline{A}B\overline{C}$. Product line 15 is not used, so all its fuse-links are left intact. The procedure is repeated for dependent variable Z.

To reiterate, the unused inputs have their fuse-links intact for product lines 0 to 7, 11, and 15.

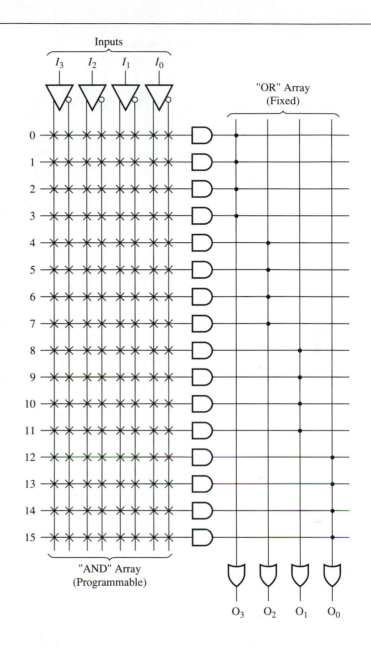

Just as there are a variety of PLEs, there are also various types of PALs. The basic nomenclature of Figure 11.11 also applies to PALs. There are PALs that have inverted buffered outputs (the *TOP* designation is an L) as well as noninverted buffered outputs (the *TOP* is H), buffered outputs that can be programmed to be either inverted or noninverted (the *TOP* is P), and registered outputs (the *TOP* is R). There is also a V (versatile) *TOP* designation that identifies PALs that can be programmed for active high (H) outputs, active low (L) outputs, and registered (R) outputs. Figure 11.14 shows a sampling of these variations of the basic PAL.

Figure 11.14a is a logic array cell of a PAL, which is conceptually the same as the logic array of one OR gate (without the inverter) from Figure 11.12. The major difference between the two is that they are drawn slightly different and the OR gate

Figure 11.13

The programmed PAL for
Example 11.5.

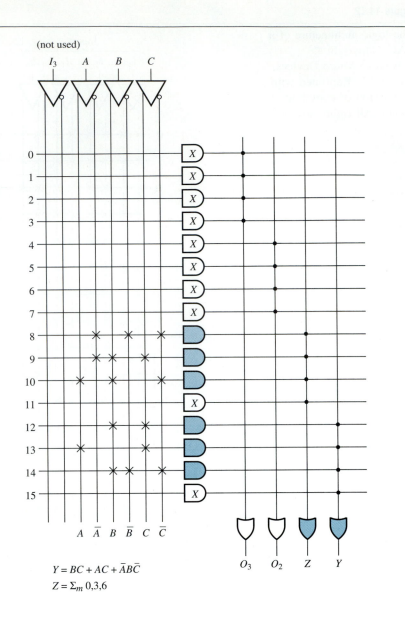

$$Y = BC + AC + \bar{A}B\bar{C}$$
$$Z = \Sigma_m \, 0,3,6$$

output is buffered. To control the state of the tri-state buffer, *AND* gate 1 has been added. *AND* gates 2, 3, 4, and 5 of Figure 11.14a serve the same function as any four *AND* gates connected to the same *OR* gate input line in Figure 11.12.

AND gate 1 also allows the user to program a pin so that it may be used as an input or output pin. That is, if an output is not required but additional inputs are needed, the fuse-links to *AND* gate 1 can be left intact to drive its output to a permanent logic 0, which puts the output buffer in the Hi-Z state. As a result of this action, the output buffer is disabled and that pin can be used as input pin I_n. On the other hand, if all the input fuse-links of *AND* gate 1 are blown, it permanently outputs a logic 1, which enables the buffer, and the pin is then an output pin O_n. *AND* gate 1 can also be programmed so that one or more of the input variables can serve as an enable input to it, thereby allowing the inputs to control the state of the output buffer. For instance, if *AND* gate 1 was programmed so that only input I_0 was an input to it, the logic level of

Figure 11.14

Variations of the basic PAL
logic array.

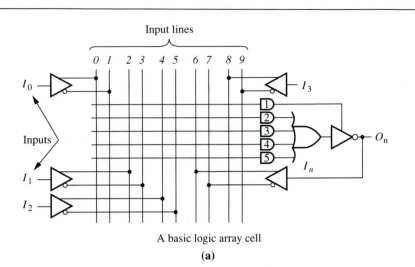

A basic logic array cell

(a)

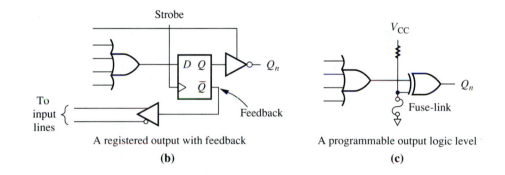

A registered output with feedback

(b)

A programmable output logic level

(c)

I_0 would control state of the output buffer of output O_n. For *AND* gate 1 to have been
programmed in this manner, input I_0 is used as an output enable for output O_n. Also
note from Figure 11.14a that when a pin is functioning as an output pin (O_n), the attached input (I_n) becomes a feedback line ($O_n = I_n$), which works out nicely for sequential logic designs.

Observe from Figure 11.14a that the number of *AND* gates for a PAL is not necessarily equal to 2^n as it is for PLEs. A PAL has a number of product terms available
to the user that may or may not be minterms, because the *AND* gate inputs are programmable. Also, be aware of the inverted output for this logic array cell, which
means that the DeMorgan theorem must be used to properly format SOP Boolean design equations.

Figure 11.14b is basically the same logic array as Figure 11.14a except that the
logic state of the output is latched by the register (flip-flop). The logic array cell of
Figure 11.14b has a register output similar to the PLE register outputs already discussed. Feedback is provided via the feedback buffer, which is connected to $\overline{Q}$.

Figure 11.14c illustrates, in concept, how the active state of an output of a *V*-type
PAL is programmable. The Boolean equation for an *X-OR* gate can be expressed as

$$Q_n = A\overline{B} + \overline{A}B.$$

If input variable B is the variable that is the grounded input, and the fuse is left intact,

Figure 11.15

Logic architecture of the
PAL16L8. (Copyright ©
Advanced Micro Devices,
Inc., 1992. Reprinted with
permission of copyright
owner. All rights reserved.)

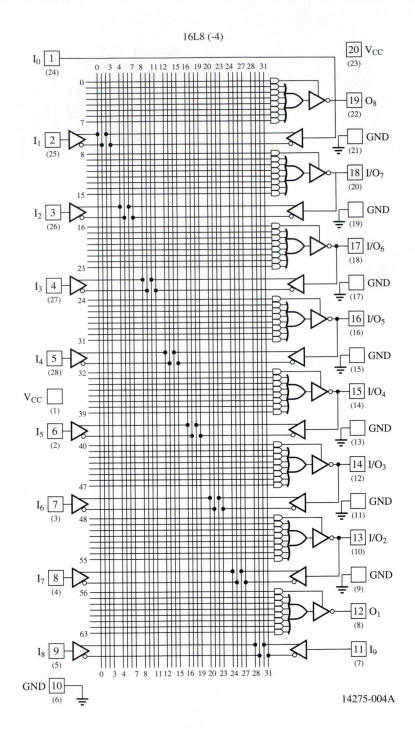

16L8 (-4)

14275-004A

then

$$Q_n = A\overline{0} + \overline{A}0 = A$$

and the output is not inverted. However, if the fuse-link is blown, $B = 1$ and

$$Q_n = A\overline{1} + \overline{A}1 = \overline{A}$$

which states that the output of the *OR* gate is A inverted.

Figures 11.15 and 11.17 are samples of PAL devices. Figure 11.15 is the logic diagram for a PAL16L8, and from the nomenclature of Figure 11.11 we know that it is a PAL with 16 inputs and 8 active low (L) buffered outputs. The block diagram and pinout for the PAL16L8 is shown in Figure 11.16. There are 10 fixed (or dedicated) input pins labeled I_0 through I_9 and two fixed output pins O_1 and O_8. In addition, there are six others that are user programmable as either input or output by means of the technique discussed for Figure 11.14a. The pins labeled I/O_2 through I/O_7 are the programmable pins. With reference to Figure 11.15 observe that the vertical lines

Figure 11.16

Block diagram and pinout for PAL16L8.

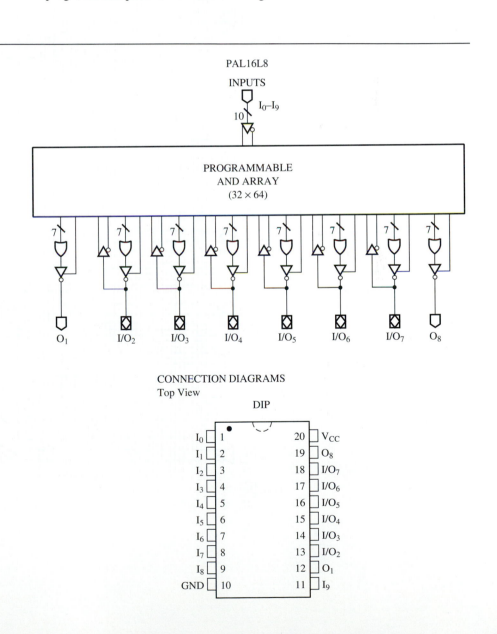

PAL16L8

INPUTS

I_0–I_9

10

PROGRAMMABLE
AND ARRAY
(32×64)

O_1 I/O_2 I/O_3 I/O_4 I/O_5 I/O_6 I/O_7 O_8

CONNECTION DIAGRAMS
Top View

DIP

I_0	1		20	V_{CC}
I_1	2		19	O_8
I_2	3		18	I/O_7
I_3	4		17	I/O_6
I_4	5		16	I/O_5
I_5	6		15	I/O_4
I_6	7		14	I/O_3
I_7	8		13	I/O_2
I_8	9		12	O_1
GND	10		11	I_9

Figure 11.17

Logic architecture of the PAL16R4. (Copyright © Advanced Micro Devices, Inc., 1992. Reprinted with persmission of copyright owner. All rights reserved.)

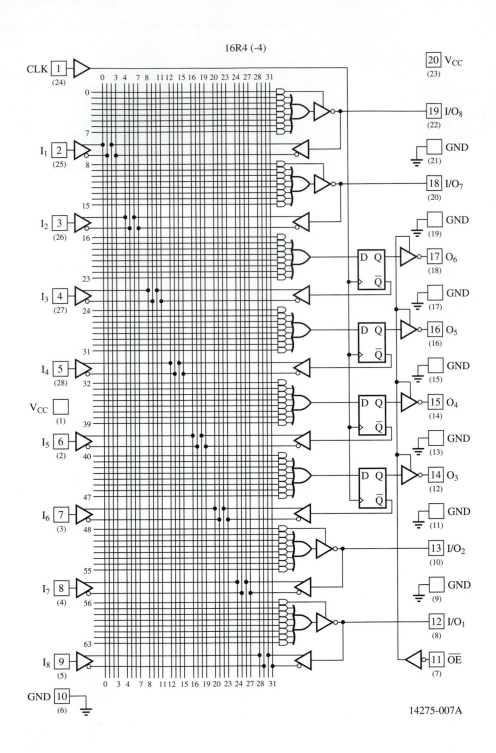

14275-007A

0 to 31 are the independent variables and their complements, and the horizontal lines 0 through 63 are the product lines.

The PAL16R4 shown in Figures 11.17 and 11.18 is very similar to the PAL16L8 except that it has four registered *L*-type outputs (O_3 to O_6). It also has eight fixed input pins (I_1 to I_8), four programmable *I/O* pins ($I/O_1, I/O_2, I/O_7, I/O_8$), a clock input ($CLK$), and an output enable input ($\overline{OE}$).

Figure 11.18

Block diagram and pinout for PAL16R4.

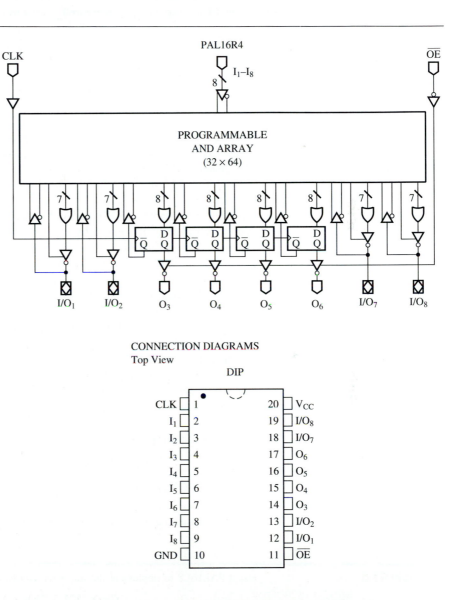

CONNECTION DIAGRAMS
Top View

PALs are manufactured in various configurations to suit different applications. The actual selection of a PAL depends on the input and output requirements of the circuit to be designed as well as on the type of circuit (combinational or sequential). Table 11.3 lists some industry standard PALs and their charateristics. Generally TTL (or bipolar) PALs once programmed cannot be reprogrammed because the internal fuses are actually destroyed during the programming process and therefore cannot be restored. Incorrectly programmed PALs must be discarded. Manufacturers also

Table 11.3 Industry-standard PAL devices. (Copyright © Advanced Micro Devices, Inc., 1992. Reprinted with permission of copyright owner. All rights reserved.)

		Standard PAL Devices										
		Functional Description							Commerical specifications			
		Array Inputs			Array outputs							
Part Number	Pin Count	bidir.	dedcid.	reg. fdbk.	reg.	comb.	macrocell	Prod. terms per output	Spd/Pwr Options	tPD (ns)	fMAX* (MHz)	ICC (mA)
PAL16L8	20	6	10	—	—	8	—	7	B	15	—	180
PAL16R8	20	—	8	8	8	—	—	8	B-2	25	25	90
PAL16R6	20	2	8	6	6	2	—	7–8	A	25	25	180
PAL16R4	20	4	8	4	4	4	—	7–8	B-4	35	16	55
PAL20L8	24	6	14	—	—	8	—	7	B	15	—	210
PAL20R8	24	—	12	8	8	—	—	8	B-2	25	25	105
PAL20R6	24	2	12	6	6	2	—	7–8	A	25	25	210
PAL20R4	24	4	12	4	4	4	—	7–8				

*fMAX is defined as $1/(t_s + t_{CO})$ for the external feedback.

produce CE (erasable CMOS) or GAL* (generic array logic) versions of their PALs that can be electrically erased by exposure to UV light and then reprogrammed. These versions are generally pin compatible with their TTL counterparts and provide much more flexibility and less wastage during the design process.

Look at the following two examples of logic design using PALs. The first design is a combinational logic circuit using the PAL16L8 and the second is a sequential logic circuit using a PAL16R4. These designs will be implemented by modifying the arrangement of fuses inside the PAL chip. Section 11.5 will show how this is actually done through the process of PAL programming.

If a designer uses inverted output (L) type PALs it is better to work with product-of-sum (POS)-type Boolean design equations. This can be verified from the basic logic array cell of Figure 11.14a. The Boolean equation for output O_n is

$$O_n = \overline{m_2 + m_3 + m_4 + m_5}$$

Applying DeMorgan's law

$$O_n = \overline{m_2} \cdot \overline{m_3} \cdot \overline{m_4} \cdot \overline{m_5} = M_2 \cdot M_3 \cdot M_4 \cdot M_5$$

which is a POS-type Boolean expression.

*GAL is a registered trademark of Lattice Semiconductor Corporation.

Example 11.6

Use a PAL16L8 to implement the following Boolean design equations:

$$X = \Sigma_m 1, 2, 3, 7 \text{ where } X(A, B, C) \tag{11.1}$$

$$Y = \Sigma_m 0, 3, 5, 6 \text{ where } Y(D, E, F) \tag{11.2}$$

$$Z = \Sigma_m 2, 3, 4, 5, 6 \text{ where } Z(G, H, J) \tag{11.3}$$

Solution

Because PAL16L8 has inverted outputs, first convert the Boolean equations to equivalent product-of-sums Boolean equations. We then apply DeMorgan's theorem to properly format the Boolean equations for implementation. To convert from SOP to POS, form the POS equation from those terms not present in the SOP equation. For variable X minterms 0, 4, 5 and 6 are not used in its SOP equation. Therefore, the maxterms 0, 4, 5, and 6 make up its POS equation. For variable Y minterms 1, 2, 4, and 7 are not used, thus maxterms 1, 2, 4, and 7 make up its POS equation. Therefore

$$X = \Sigma_m 1, 2, 3, 7 = \Pi_M 0, 4, 5, 6 \text{ where } X(A, B, C)$$

$$Y = \Sigma_m 0, 3, 5, 6 = \Pi_M 1, 2, 4, 7 \text{ where } Y(D, E, F)$$

$$Z = \Sigma_m 2, 3, 4, 5, 6 = \Pi_M 0, 1, 7 \text{ where } Z(G, H, J)$$

After minimization, where possible

$$X = (\bar{A} + B)(\bar{A} + C)(B + C)$$

$$Y = \Pi_M 1, 2, 4, 7$$

$$= (D + E + \bar{F})(D + \bar{E} + F)$$
$$(\bar{D} + E + F)(\bar{D} + \bar{E} + \bar{F})$$

$$Z = (G + H)(\bar{G} + \bar{H} + \bar{J})$$

Applying DeMorgan's law

$$\bar{X} = \overline{(\bar{A} + B)(\bar{A} + C)(B + C)}$$

$$= \overline{(\bar{A} + B)} + \overline{(\bar{A} + C)} + \overline{(B + C)} = A\bar{B} + A\bar{C} + \bar{B}\,\bar{C}$$

Negating the equation one more time, so as to return the $\bar{X}$ variable to its original logic state of X,

$$\bar{\bar{X}} = X = \overline{A\bar{B} + A\bar{C} + \bar{B}\,\bar{C}} \tag{11.4}$$

Using the same conversion principles for Y and Z,

$$Y = \overline{\overline{D}EF + \overline{D}E\overline{F} + D\overline{E}F + DEF} \tag{11.5}$$

$$Z = \overline{\overline{G}\overline{H} + GHJ} \tag{11.6}$$

Figure 11.19 shows the PAL16L8 of Figure 11.15 programmed to implement X, Y, and Z. For example, the independent variables, A, B, C, and D were input at pins 1, 2, 3, and 4, respectively, as indicated in the table below. Due to these pin assignments, the variables and their complements are available on the vertical lines indicated in the table.

Independent variable	Vertical input line
A(pin 1)	2
$\bar{A}$	3
B (pin 2)	0
$\bar{B}$	1
C (pin 3)	4
$\bar{C}$	5
D (pin 4)	8
$\bar{D}$	9
etc.	

As an academic exercise we decided to program pin 15 to function as an input pin for the independent variable J rather than use a standard input pin. This was accomplished by leaving all fuse-links of the tristate buffer control line intact (as indicated by the X within the control *AND* gate), which causes that *AND* gate to output a fixed logic 0, and as a result, drive that output buffer in the Hi-Z state.

Because only three outputs are required (plus the one that is programmed as an input pin), the four outputs not used were not connected (NC) and all fuse-links were left intact. The outputs that were used did not require the use of all of their *AND* gate arrays so these *AND* gates were left fully fused.

Figure 11.19

A PAL16L8 programmed for the Boolean equations of Example 11.6.

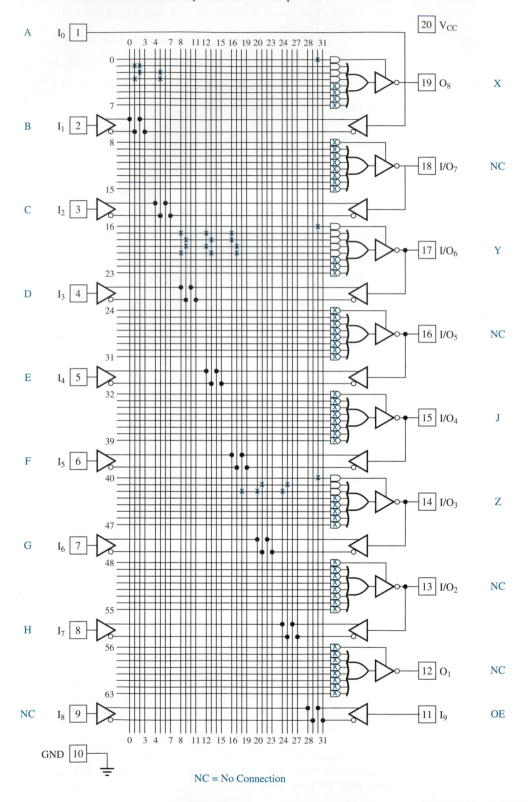

NC = No Connection

Figure 11.20

A PAL16R4 programmed as a 2-bit down-counter.

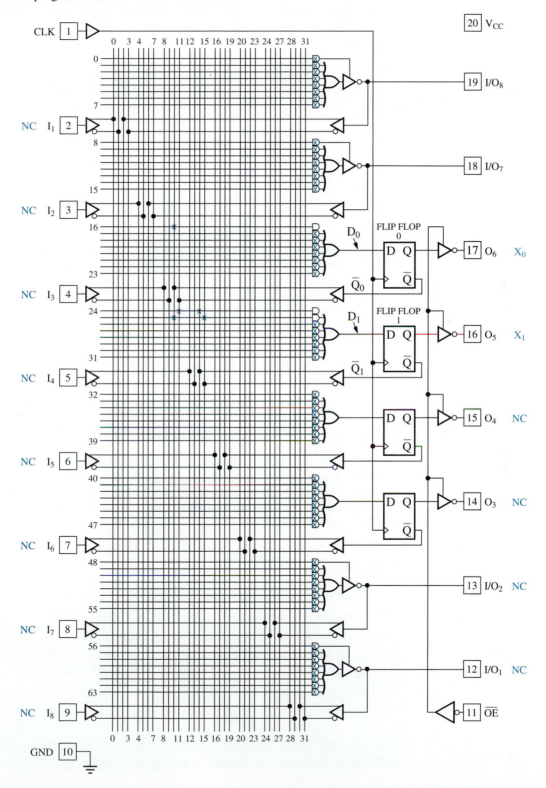

To program the PAL follow the product line that is to be used for a product term and put an × at each intersection of the product line and the vertical lines that are to be connected to that product line. For instance, because output pin 19 was chosen for the dependent variable X, horizontal product lines 0 to 7 must be programmed. The second *AND* gate from the top (product line 1) was chosen to implement the product $A\bar{B}$ of the Boolean equation

$$\overline{A\bar{B} + A\bar{C} + \bar{B}\bar{C}}.$$

Then product line 1 is connected to vertical lines 1, which is $\bar{B}$, and 2, which is A. Product line 2 is used to implement the product $A\bar{C}$, so product line 2 is connected to vertical lines 2 and 5. Lastly, product line 3 is used to implement the product $\bar{B}\,\bar{C}$, which requires that vertical lines 1 and 5 be connected to that product line.

Input pin 11 is used as an output enable pin; therefore, the product line of each control *AND* gate (product lines 0, 16, and 40) is connected to the vertical line 30. If an active-low output enable were required, the product lines of those control *AND* gates would be connected to the vertical line 31, which is $\overline{OE}$.

Example 11.7

Implement a 2-bit binary down-counter using a PAL16R4 (*see* Figure 11.17).

Solution

All register outputs of a PAL16R4 are inverted; that is, the Q output of each flip-flop is inverted and made available at the output pins of the PAL16R4. Recall from Chapter 6 that a binary *UP* counter can be made to count *DOWN* (and vice versa) by simply inverting the outputs of the counter. Therefore, if the PAL16R4 is used to implement the design equations for the 2-bit binary *UP* counter shown in Figure 6.25 (Chapter 6), a 2-bit binary *DOWN* counter should be obtained. The design equations from Figure 6.25 are:

$$D_0 = \bar{Q}_0 \tag{11.7}$$

$$D_1 = Q_1 \oplus Q_0$$
$$= \bar{Q}_1 Q_0 + Q_1 \bar{Q}_0 \tag{11.8}$$

Assign the D-type flip-flop output pins 17 and 16 to implement the outputs of the counter X_0 and X_1, respectively, as indicated in Figure 11.20. Note that Q_0 is available at vertical line 11 ($\bar{Q}_0$ is available at vertical line 10) and Q_1 is available at vertical line 15 ($\bar{Q}_1$ is available at vertical line 14).

To implement the design equations for the D input of flip-flop 0 (D_0), horizontal product line 16 is connected to the vertical line 10, which is $\bar{Q}_0$. The D input of flip-flop 1 (D_1) uses the horizontal product line 24 for the product $\bar{Q}_1 Q_0$ and product line 25 for the product $Q_1 \bar{Q}_0$. Hence, product line 24 is connected to vertical lines 11 and 14, and product line 25 is connected to vertical lines 10 and 15.

Review Questions

1. What is the essential difference between a PLE and a PAL?
2. What function does the *AND* gate array of a PLE serve?
3. What is the symbolic meaning of a product line?
4. Is the quantity of *AND* gates in the *AND* gate array of a PAL necessarily equal to 2^n, where n is the number of input variables?
5. What is the significance of the *OR* gate inputs being fixed?
6. How can a PAL's programmable I/O pin be programmed to be either an input or an output?
7. From PLD product identification nomenclature, how are the number of inputs and outputs designated as well as the output type?

8. When is it that SOP-type Boolean equations rather than POS equations are best used to implement a logic design, and how does that relate to a PAL?

9. Why is a PALIIROO-type PAL required for sequential logic design?

10. What is the essence of how a programmable *X-OR* gate implements the active logic state (*L* or *H*) of an output?

11. What techniques are used to program a PAL by hand, as was done for the PAL of Figures 11.3, 11.19, and 11.20?

11.5

PAL Programming and simulation — combinational logic

In the previous section we saw that a PAL could be programmed to implement various combinational or sequential logic designs by selecting an appropriate PAL device and then blowing fuses in the *AND* gate array to implement the design equations. The actual process of blowing these fuses is accomplished by using an instrument known as a PAL programmer. Many PROM programmers are also capable of programming PALs. The PAL programmer is given a fuse-map of the PAL to be programmed that identifies the fuses to be blown and the ones to be left intact. It then uses this information to program the PAL.

PAL programming can be very tedious if the designer has to explicitly specify the individual fuses to be blown for a given design. This technique was used in the previous section for two rather simple designs and the process was very time consuming and error prone. For more complex designs, this technique is not practical and a better solution is needed.

There are several industry-standard PAL design tools available that automate the process of designing digital circuits with PALs. These tools are available in the form of software packages that run on various platforms such as MSDOS, UNIX, etc. In all these tools, the designer simply specifies the design equations or truth tables for the circuit to be designed and the software generates the appropriate fuse map information. Many of these packages also provide simulation to verify the proper operation of a digital design before a PAL is programmed. Some of the more popular software packages for PAL design are PLDshell from Intel Corp., The Universal Compiler for Programmable Logic (CUPL) from Logical Devices, Inc., ABEL-4 from Data I/O Corp., PLDesigner from Minc Inc., and PALASM from Advanced Micro Devices (AMD) Inc. By far the most popular package for PAL design is PALASM because it was developed by the inventors of the PAL device, MMI now a subsidiary of AMD. Most other software packages maintain compatibility with the PALASM syntax and its files. This section deals with PAL programming and simulation using PALASM4, the latest version of PALASM distributed by AMD. The software runs on an IBM-compatible PC and is available at no charge by contacting AMD.

Figure 11.21 illustrates the flow for PAL design using PALASM4. The chart also includes the path for verifying digital logic designs using simulation (shown in blue). To get acquainted with the process of PAL design briefly examine each step shown in Figure 11.21; each step will be investigated in detail later in this section.

The first step in the PAL design/simulation process is to create a PAL design file (PDS file). This file can be created using any ASCII text editor and contains a description of the design, the design equations, and optionally, simulation commands. The name given to the file should be given the extension ''.PDS''. The PDS

Figure 11.21

Programming and simulation flow using PALASM4.

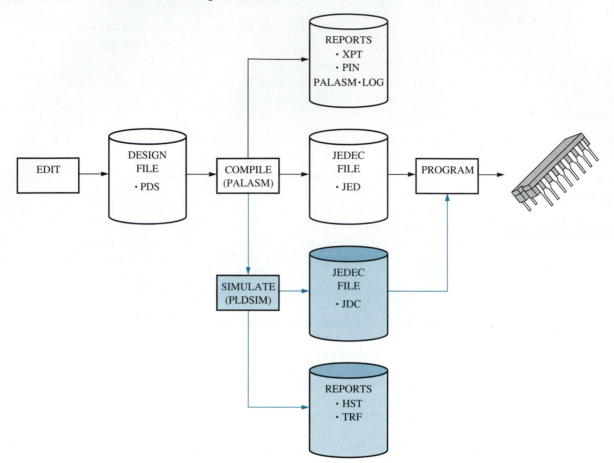

file is fed into the PALASM compiler, which reads the design equations, converts and simplifies (minimizes) the equations if necessary, and produces a file containing the fuse-map. The fuse-map is in a standard format that can be read by most PAL programming units—the JEDEC (Joint Electronic Devices Engineering Council) format. This file is given the extension ".JED" and contains all the information necessary for the PAL programmer to program the PAL.

The PALASM4 compiler also produces several report files for the benefit of the designer. Although these files are not required by the PAL programming unit, they are very helpful to the designer for troubleshooting and documenting a design. The compiler produces a listing of the fuse-map that corresponds to the logic diagram of the PAL being programmed; this file is known as the fuse-plot file and is given the extension ".XPT". The XPT file can be quite useful for verifying that the proper fuses have been blown. The compiler can optionally also produce a file illustrating the pinouts of the PAL device being used with the pins assigned to the dependent and independent variables; this file is given the extension ".PIN". Finally, the PALASM compiler produces a file containing a log of the compilation process (PALASM.LOG) that can be used for diagnostic purposes in case any errors were encountered during the compilation of the PDS file.

The PALASM4 software also provides the designer with optional simulation and testing facilities. Various simulation commands can be used to simulate and thus verify the proper operation of the logic circuit being designed before the PAL device is actually programmed. This can greatly reduce the development time and eliminate wastage if bipolar (disposable) PAL devices are being used. Furthermore, the simulator can generate special information known as *test vectors* in the JEDEC fuse-map file that can be used by the PAL programmer unit to test the PAL device *after* it has been programmed. In Figure 11.21, the blocks drawn in blue represent the flow of the design process if simulation is to be performed. If the PDS file contains any simulation commands, the PLDSIM simulator can be executed. PLDSIM also produces a JEDEC fuse-map file that contains not only the fuse information for the PAL programming unit but also test vectors for testing the device; to distinguish this file from the file produced by the compiler it is given in extension ''.JDC''. PLDSIM produces two output report files — the ''.HST'' file that contains a history of all the input and output pins of the PAL device during simulation, and the ''.TRF'' file that contains a trace of the logic levels at selected input and output pins. These two files are extremely useful in verifying the proper operation of the logic circuit being implemented.

To illustrate the procedure for programming a PAL, the three combinational logic equations that were implemented with the PAL16L8 in Example 11.6 will be used. The fuse-plot (XPT file) produced by PALASM4 will be compared with the circuit shown in Figure 11.19 to verify that the correct fuses have been blown.

Figure 11.22 shows a listing of the PDS file to implement the three equations from Example 11.6 and recreate the design using PALASM4. The file name for the PDS file shown in Figure 11.22 is COMBLOGA.PDS. The PDS file is divided into three segments — the declaration segment, the PIN declarations, and the Boolean equation segment. Note that any line that begins with a semicolon (;) is treated by PALASM4 as a comment and is ignored. Comments can be interspersed throughout a PALASM4 design file at the discretion of the designer and are used to assist the reader in understanding the contents of the file.

The declaration segment of a PDS file begins with some preliminary information that is used to identify the design file. The keywords TITLE, PATTERN, REVISION, AUTHOR, COMPANY, and DATE can be used to document the design project. The information that is placed following these keywords is left entirely to the user and like comments is simply ignored by PALASM4. In fact these keywords are optional and are not required in the PDS file. However, it is good practice to start every PALASM4 design file with this information. The only required keyword in the declarations segment is CHIP. This keyword identifies the actual PAL chip being programmed. The syntax for the CHIP declaration is:

```
CHIP    name_of_chip    part_number
```

Because we are customizing a PAL device by programming it, the resulting PAL device can be named. This name is for convenience of identification and can be any name the designer chooses to give the chip. The CHIP declaration must also contain a standard manufacturer's part number. For example in Figure 11.22 the following CHIP declaration was used:

```
CHIP    COMBLOG    PAL16L8
```

The chip was called COMBLOG for combinational logic and identified as a PAL16L8 for programming.

Figure 11.22

Listing of PAL design file (PDS) for Example 11.6.

```
;PALASM Design Description

;-------------------------------- Declaration Segment -----------

TITLE     PAL Implementation of Example 11-6
PATTERN   COMBLOGA.PDS (design file)
REVISION  A
AUTHOR    Ismail & Rooney
COMPANY   University of Dayton
DATE      10/12/92

CHIP   COMBLOG   PAL16L8 ◄
```
                         ← *Part number*
                         ← *Name & chip*

```
;--------------------------------- PIN Declarations ---------------

PIN   1          A              ;INDEPENDENT VARIABLES (INPUTS)
PIN   2          B              ;          "            "        "
PIN   3          C              ;          "            "        "
PIN   4          D              ;          "            "        "
PIN   5          E              ;          "            "        "
PIN   6          F              ;          "            "        "
PIN   7          G              ;          "            "        "
PIN   8          H              ;          "            "        "
PIN   9          NC
PIN  10          GND
;
PIN  11          OE             ;OUTPUT ENABLE
PIN  12          NC
PIN  13          NC
PIN  14          Z              ;DEPENDENT VARIABLE (OUTPUT)
PIN  15          J              ;INDEPENDENT VARIABLE (INPUT)
PIN  16          NC
PIN  17          Y              ;DEPENDENT VARIABLE (OUTPUT)
PIN  18          NC
PIN  19          X              ;DEPENDENT VARIABLE (OUTPUT)
PIN  20          VCC

;--------------------------------- Boolean Equation Segment ------

EQUATIONS ◄
```
               ← *Identifies start of equation segment*

```
;SIMPLIFIED SOP EQUATIONS (INVERTED)

/X = A * /B  +  A * /C  +  /B * /C ◄
```
              ← *Equation 11-4*

```
X.TRST = OE ◄
```
              ← *Controls Tri-State output*

```
/Y = /D * /E * F  +  /D * E * /F  +  D * /E * /F  +  D * E * F ◄
```
              ← *Equation 11-5*

```
Y.TRST = OE

/Z = /G * /H  +  G * H * J ◄
```
              ← *Equation 11-6*

```
Z.TRST = OE
```

The PIN declaration segment always follows the declaration segment. Here is listed each pin of the specified PAL device and the name of the signal assigned to it

$$\begin{array}{ll} \text{PIN\#} & \text{signal\_name} \\ \text{PIN\#} & \text{signal\_name} \\ \quad . & \quad . \\ \quad . & \quad . \end{array}$$

The order in which the pin assignments are specified is not important but it is convenient to organize them in some logical manner. In Figure 11.22 the 20 pins of the PAL16L8 were divided into two groups of 10 pins corresponding to the left and right sides of the chip. Pins 1 through 8 have been assigned to the independent variables A through H, respectively, pin 11 is the (active-high) output enable line OE, pin 15 is the independent variable J, and the dependent variables X, Y, and Z have been assigned to pins 19, 17, and 14, respectively. Note that the pin assignments are the same as those as in Example 11.6. The power connections to the PAL device are identified by VCC and GND on pins 20 and 10, respectively. Any unconnected (or unused) pin is labeled *NC* for no connection. Comments placed on the right of most of the pin assignments shown in Figure 11.22 make it convenient for the reader to easily identify the characteristics of certain signals.

The third segment of the PDS file is the Boolean equation segment that always begins with the keyword EQUATIONS. This segment is used to specify the logic equations to be implemented on the PAL device. Because the PAL16L8 is an active-low device the equations were implemented for $\overline{X}$, $\overline{Y}$, and $\overline{Z}$. From Example 11.6, the minimized equation for X (Eq. 11.4) is:

$$X = \overline{A\overline{B} + A\overline{C} + \overline{B}\,\overline{C}}$$

Therefore, inverting the left-hand side of the equation

$$\overline{X} = A\overline{B} + A\overline{C} + \overline{B}\,\overline{C}$$

The above equation is written in PALASM4 syntax and entered into the *PDS* file as follows:

$$/X = A * /B + A * /C + /B * /C$$

Because most computer keyboards are incapable of generating an overbar ($\overline{}$) on a symbol, the slash (/) is used to represent complemented variables. Also, the asterisk (*) is used for the *AND* operator.

Similarly from Example 11.6, Equation 11.5 is solved for $\overline{Y}$ as follows:

$$\overline{Y} = \overline{D}\,\overline{E}F + \overline{D}E\overline{F} + D\overline{E}\,\overline{F} + DEF$$

and written in PALASM4 syntax as:

$$/Y = /D * /E * F + /D * E * /F + D * /E * /F + D * E * F$$

Equation 11.6 is solved for $\overline{Z}$ as follows:

$$\overline{Z} = \overline{G}\,\overline{H} + GHJ$$

and written in PALASM4 syntax as:

$$/Z = /G * /H + G * H * J$$

The equations for $/X$, $/Y$, and $/Z$ are then entered in the Boolean equations segment of the PDS file as shown in Figure 11.22. There are three other equations listed in this segment:

```
X.TRST = OE
Y.TRST = OE
Z.TRST = OE
```

These three equations are used to program the tristate outputs of X, Y, and Z. Recall that the active-low outputs of the PAL16L8 have tristate buffers. The enable lines of these tristate buffers are connected to product lines. In Example 11.6 (Fig. 11.19) these product lines (0, 16, and 40) were connected to vertical input line 30 (*OE*). Thus, when $OE = 1$, the outputs X, Y, and Z were enabled. In the three PALASM equations shown above, the ".TRST" extension to the output variable symbol identifies the *TRISTATE* control for the output. Thus, the equation:

$$X.TRST = OE$$

indicates that the tristate control for output X is to be enabled by *OE*. If the output Y were to be enabled by $\overline{OE}$, the equation would be:

$$Y.TRST = /OE$$

And, if the output Z were enabled when *OE* and *C* were both high, the equation would be:

$$Z.TRST = OE * C$$

Because the tristate gate at each output of the PAL device is controlled by a product term, the right-hand side of the TRST equation can have any Boolean term or expression.

Once the PDS file shown in Figure 11.22 has been created, it is then compiled by the PALASM4 compiler. The compiler produces a report of the compilation as shown in Figure 11.23. The PALASM4 compiler always gives this file the name PALASM.LOG and is used only for diagnostics purposes. It lists the various steps that the compiler goes through during the compilation process, identifies any errors, warnings, or conflicts, and indicates the amount of minimization performed on each equation. The annotations in Figure 11.23 explain some of the reports in the PALASM.LOG file.

The PALASM4 compiler also produces a fuse-plot file shown in Figure 11.24. This file is given the name COMBLOGA.XPT. The file shows the vertical input lines and the horizontal product lines corresponding to the manufacturer's logic architecture (*see* Fig. 11.19). Fuses left intact are represented by "X" while blown fuses are represented by a "-". The total number of fuses blown (or programmed) is listed at the end of the file (358 in this example). Notice how the fuse-plot shown in Figure 11.24 corresponds exactly to the PAL16L8 programmed in Example 11.6 and shown in Figure 11.19. For instance, in Figure 11.19 row 0 of the PAL device shows only the fuse of vertical line 30 intact. The fuse-plot in Figure 11.24 has the same connections. Also, row 1 in Figure 11.19 is connected to vertical lines 1 and 2. The fuse-plot in Figure 11.24 confirms this by placing Xs in columns 1 and 2.

The JEDEC file produced by PALASM4 for this example (COMBLOGA.JED) is very similar to the fuse plot (XPT) and contains the necessary information for a PAL programming unit to blow the appropriate fuses. The format in which this information is arranged is an industry standard so that it can be read by programming units manufactured by a variety of manufacturers. The designer usually does not have to be concerned about the content of this file because the XPT file provides the same information in an easy-to-read format. However, to become familiar with the contents of the JEDEC file it is shown in Figure 11.25. Some of the information contained in the JEDEC file can be easily identified as shown in the annotations in Figure 11.25. As was stated earlier, the contents of this file are used by the programming unit and is not important to the designer.

For convenience, PALASM4 can also (optionally) produce a file illustrating the pinouts of the PAL device along with the pin assignments specified by the designer. This file (COMBLOGA.PIN) is shown in Figure 11.26 and simply contains a graphical representation of the pins of the PAL16L8 — the pins assigned to the

Figure 11.23

Log file produced by
PALASM4.

```
combloga.pds

PALASM4  PARSER    - MARKET RELEASE 1.5 (7-10-92)
(C) - COPYRIGHT ADVANCED MICRO DEVICES INC., 1992

                   ***********************************
                   *      PALASM PARSER LISTING      *
                   ***********************************

LINE #  |----+----1----+----2----+----3----+----4----+----5----+----6----+
   1    |;PALASM Design Description
   2    |
   3    |;------------------------------- Declaration Segment ------------
   4    |
   5    |TITLE    PAL Implementation of Example 11-6
   6    |PATTERN  COMBLOGA.PDS (design file)
   7    |REVISION A
   8    |AUTHOR   Ismail & Rooney
   9    |COMPANY  University of Dayton
  10    |DATE     10/12/92
  11    |
  12    |CHIP  COMBLOG  PAL16L8
  13    |
  14    |;------------------------------- PIN Declarations --------------
  15    |
  16    |PIN  1        A              ;INDEPENDENT VARIABLES (INPUTS)
  17    |PIN  2        B              ;     "          "        "
  18    |PIN  3        C              ;     "          "        "
  19    |PIN  4        D              ;     "          "        "
  20    |PIN  5        E              ;     "          "        "
  21    |PIN  6        F              ;     "          "        "
  22    |PIN  7        G              ;     "          "        "
  23    |PIN  8        H              ;     "          "        "
  24    |PIN  9        NC
  25    |PIN 10        GND
  26    |;
  27    |PIN 11        OE             ;OUTPUT ENABLE
  28    |PIN 12        NC
  29    |PIN 13        NC
  30    |PIN 14        Z              ;DEPENDENT VARIABLE (OUTPUT)
  31    |PIN 15        J              ;INDEPENDENT VARIABLE (INPUT)
  32    |PIN 16        NC
  33    |PIN 17        Y              ;DEPENDENT VARIABLE (OUTPUT)
  34    |PIN 18        NC
  35    |PIN 19        X              ;DEPENDENT VARIABLE (OUTPUT)
  36    |PIN 20        VCC
  37    |
  38    |;------------------------------- Boolean Equation Segment ------
  39    |
  40    |EQUATIONS
  41    |
  42    |;SIMPLIFIED SOP EQUATIONS (INVERTED)
  43    |
  44    |/X = A * /B  +  A * /C  +  /B * /C
  45    |
  46    |X.TRST = OE
  47    |
  48    |/Y = /D * /E * F  +  /D * E * /F  +  D * /E * /F  +  D * E * F
  49    |
  50    |Y.TRST = OE
  51    |
  52    |/Z = /G * /H  +  G * H * J
  53    |
  54    |Z.TRST = OE
  55    |
```

Listing of PDS file for Reference purposes

```
%% PARSE %%    No errors.  No warnings.
```

Error and warning count for syntax errors in PDS file

```
%% PARSE %%    File processed successfully.   File: combloga.pds
PALASM4  MINIMIZE   - MARKET RELEASE 1.5 (7-10-92)
(C) - COPYRIGHT ADVANCED MICRO DEVICES INC., 1992

Processing equation ====> /X        MINIMIZE_ON  (3 pt ---> 3 pt)
Processing equation ====> X.TRST     MINIMIZE_ON  (1 pt ---> 1 pt)
Processing equation ====> /Y        MINIMIZE_ON  (4 pt ---> 4 pt)
Processing equation ====> Y.TRST     MINIMIZE_ON  (1 pt ---> 1 pt)
Processing equation ====> /Z        MINIMIZE_ON  (2 pt ---> 2 pt)
Processing equation ====> Z.TRST     MINIMIZE_ON  (1 pt ---> 1 pt)
```

Minimization of each equation (pt = Product Term) none performed here

```
%%  MINIMIZE  %% Maximum memory allocated was: 4904 bytes.

%%  MINIMIZE  %% File Processed Successfully. File: combloga.pds
%%  MINIMIZE  %% ERROR count: 0  WARNING count: 0
```

Error and warning count for minimization

independent variables, *A* through *J*, the dependent variables *X*, *Y*, and *Z*, the output enable line, OE, the power supplies, VCC and GND, and the unused pins, NC.

PALASM4 has the capability to minimize and automatically convert logic equations for active-low PAL devices. We did not have to minimize Equations 11.1, 11.2, and 11.3 in Example 11.6 and convert them to active-low form before entering

Figure 11.24

Fuse-plot file (XPT)
produced by PALASM4.

```
PALASM4  PAL ASSEMBLER   - MARKET RELEASE 1.5a (8-20-92)
 (C) - COPYRIGHT ADVANCED MICRO DEVICES INC., 1992

TITLE    :PAL Implementation of Example 11-6    AUTHOR :Ismail & Rooney
PATTERN :COMBLOGA.PDS (design file)             COMPANY:University of Dayton
REVISION:A                                      DATE   :10/12/92

PAL16L8
COMBLOG


                  11   1111 1111 2222 2222 2233
        0123 4567 8901 2345 6789 0123 4567 8901

0       ---- ---- ---- ---- ---- ---- ---- --X-
1       -XX- ---- ---- ---- ---- ---- ---- ----
2       --X- -X-- ---- ---- ---- ---- ---- ----
3       -X-- -X-- ---- ---- ---- ---- ---- ----
4       XXXX XXXX XXXX XXXX XXXX XXXX XXXX XXXX
5       XXXX XXXX XXXX XXXX XXXX XXXX XXXX XXXX
6       XXXX XXXX XXXX XXXX XXXX XXXX XXXX XXXX
7       XXXX XXXX XXXX XXXX XXXX XXXX XXXX XXXX

8       XXXX XXXX XXXX XXXX XXXX XXXX XXXX XXXX
9       XXXX XXXX XXXX XXXX XXXX XXXX XXXX XXXX
10      XXXX XXXX XXXX XXXX XXXX XXXX XXXX XXXX
11      XXXX XXXX XXXX XXXX XXXX XXXX XXXX XXXX
12      XXXX XXXX XXXX XXXX XXXX XXXX XXXX XXXX
13      XXXX XXXX XXXX XXXX XXXX XXXX XXXX XXXX
14      XXXX XXXX XXXX XXXX XXXX XXXX XXXX XXXX
15      XXXX XXXX XXXX XXXX XXXX XXXX XXXX XXXX

16      ---- ---- ---- ---- ---- ---- ---- --X-
17      ---- ---- X--- X--- X--- ---- ---- ----
18      ---- ---- -X-- -X-- X--- ---- ---- ----
19      ---- ---- -X-- X--- -X-- ---- ---- ----
20      ---- ---- X--- -X-- -X-- ---- ---- ----
21      XXXX XXXX XXXX XXXX XXXX XXXX XXXX XXXX
22      XXXX XXXX XXXX XXXX XXXX XXXX XXXX XXXX
23      XXXX XXXX XXXX XXXX XXXX XXXX XXXX XXXX

24      XXXX XXXX XXXX XXXX XXXX XXXX XXXX XXXX
25      XXXX XXXX XXXX XXXX XXXX XXXX XXXX XXXX
26      XXXX XXXX XXXX XXXX XXXX XXXX XXXX XXXX
27      XXXX XXXX XXXX XXXX XXXX XXXX XXXX XXXX
28      XXXX XXXX XXXX XXXX XXXX XXXX XXXX XXXX
29      XXXX XXXX XXXX XXXX XXXX XXXX XXXX XXXX
30      XXXX XXXX XXXX XXXX XXXX XXXX XXXX XXXX
31      XXXX XXXX XXXX XXXX XXXX XXXX XXXX XXXX

32      XXXX XXXX XXXX XXXX XXXX XXXX XXXX XXXX
33      XXXX XXXX XXXX XXXX XXXX XXXX XXXX XXXX
34      XXXX XXXX XXXX XXXX XXXX XXXX XXXX XXXX
35      XXXX XXXX XXXX XXXX XXXX XXXX XXXX XXXX
36      XXXX XXXX XXXX XXXX XXXX XXXX XXXX XXXX
37      XXXX XXXX XXXX XXXX XXXX XXXX XXXX XXXX
38      XXXX XXXX XXXX XXXX XXXX XXXX XXXX XXXX
39      XXXX XXXX XXXX XXXX XXXX XXXX XXXX XXXX

40      ---- ---- ---- ---- ---- ---- ---- --X-
41      ---- ---- ---- ---- ---- -X-- -X-- ----
42      ---- ---- ---- ---- --X- X--- X--- ----
43      XXXX XXXX XXXX XXXX XXXX XXXX XXXX XXXX
44      XXXX XXXX XXXX XXXX XXXX XXXX XXXX XXXX
45      XXXX XXXX XXXX XXXX XXXX XXXX XXXX XXXX
46      XXXX XXXX XXXX XXXX XXXX XXXX XXXX XXXX
47      XXXX XXXX XXXX XXXX XXXX XXXX XXXX XXXX

48      XXXX XXXX XXXX XXXX XXXX XXXX XXXX XXXX
49      XXXX XXXX XXXX XXXX XXXX XXXX XXXX XXXX
50      XXXX XXXX XXXX XXXX XXXX XXXX XXXX XXXX
51      XXXX XXXX XXXX XXXX XXXX XXXX XXXX XXXX
52      XXXX XXXX XXXX XXXX XXXX XXXX XXXX XXXX
53      XXXX XXXX XXXX XXXX XXXX XXXX XXXX XXXX
54      XXXX XXXX XXXX XXXX XXXX XXXX XXXX XXXX
55      XXXX XXXX XXXX XXXX XXXX XXXX XXXX XXXX

56      XXXX XXXX XXXX XXXX XXXX XXXX XXXX XXXX
57      XXXX XXXX XXXX XXXX XXXX XXXX XXXX XXXX
58      XXXX XXXX XXXX XXXX XXXX XXXX XXXX XXXX
59      XXXX XXXX XXXX XXXX XXXX XXXX XXXX XXXX
60      XXXX XXXX XXXX XXXX XXXX XXXX XXXX XXXX
61      XXXX XXXX XXXX XXXX XXXX XXXX XXXX XXXX
62      XXXX XXXX XXXX XXXX XXXX XXXX XXXX XXXX
63      XXXX XXXX XXXX XXXX XXXX XXXX XXXX XXXX

SUMMARY
-------

     TOTAL FUSES BLOWN    = 358
```

Vertical input lines

Fuses left intact "x"

Blown fuses "–"

Product term

Figure 11.25

JEDEC file produced by
PALASM4.

```
PALASM4  PAL ASSEMBLER    - MARKET RELEASE 1.5a (8-20-92)
(C) - COPYRIGHT ADVANCED MICRO DEVICES INC., 1992


TITLE    :PAL Implementation of Example 11-6     AUTHOR :Ismail & Rooney
PATTERN :COMBLOGA.PDS (design file)              COMPANY:University of Dayton
REVISION:A                                       DATE   :10/12/92


PAL16L8
COMBLOG*
QP20*
QF2048*                                                   Number of pins
GO*FO*                                                    Total no. of fuses
L0000 111111111111111111111111111111101*
L0032 100111111111111111111111111111111*
L0064 110110111111111111111111111111111*
L0096 101110111111111111111111111111111*
L0128 000000000000000000000000000000000*
L0160 000000000000000000000000000000000*
L0192 000000000000000000000000000000000*
L0224 000000000000000000000000000000000*
L0256 000000000000000000000000000000000*
L0288 000000000000000000000000000000000*
L0320 000000000000000000000000000000000*
L0352 000000000000000000000000000000000*
L0384 000000000000000000000000000000000*
L0416 000000000000000000000000000000000*
L0448 000000000000000000000000000000000*
L0480 000000000000000000000000000000000*    "0" fuse left intact
L0512 111111111111111111111111111111101*
L0544 111111110111011011111111111111111*
L0576 111111110101101101111111111111111*
L0608 111111110110111011111111111111111*
L0640 111111101111011011111111111111111*
L0672 000000000000000000000000000000000*
L0704 000000000000000000000000000000000*    "1" blown fuse
L0736 000000000000000000000000000000000*
L0768 000000000000000000000000000000000*
L0800 000000000000000000000000000000000*
L0832 000000000000000000000000000000000*
L0864 000000000000000000000000000000000*
L0896 000000000000000000000000000000000*
L0928 000000000000000000000000000000000*
L0960 000000000000000000000000000000000*
L0992 000000000000000000000000000000000*
L1024 000000000000000000000000000000000*
L1056 000000000000000000000000000000000*
L1088 000000000000000000000000000000000*
L1120 000000000000000000000000000000000*
L1152 000000000000000000000000000000000*
L1184 000000000000000000000000000000000*
L1216 000000000000000000000000000000000*
L1248 000000000000000000000000000000000*
L1280 111111111111111111111111111111101*
L1312 111111111111111111110110111111*
L1344 111111111111111110101101111111*
L1376 000000000000000000000000000000000*
L1408 000000000000000000000000000000000*
L1440 000000000000000000000000000000000*
L1472 000000000000000000000000000000000*
L1504 000000000000000000000000000000000*
L1536 000000000000000000000000000000000*
L1568 000000000000000000000000000000000*
L1600 000000000000000000000000000000000*
L1632 000000000000000000000000000000000*
L1664 000000000000000000000000000000000*
L1696 000000000000000000000000000000000*
L1728 000000000000000000000000000000000*
L1760 000000000000000000000000000000000*
L1792 000000000000000000000000000000000*
L1824 000000000000000000000000000000000*
L1856 000000000000000000000000000000000*
L1888 000000000000000000000000000000000*
L1920 000000000000000000000000000000000*
L1952 000000000000000000000000000000000*
L1984 000000000000000000000000000000000*
L2016 000000000000000000000000000000000*
C2E21*
E9DC
```

Figure 11.26

PIN file produced by
PALASM4.

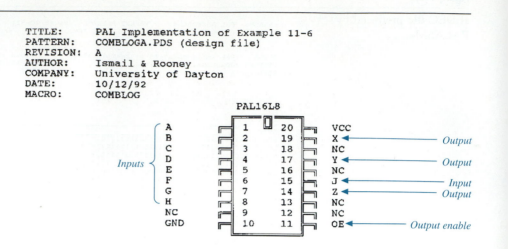

```
TITLE:      PAL Implementation of Example 11-6
PATTERN:    COMBLOGA.PDS (design file)
REVISION:   A
AUTHOR:     Ismail & Rooney
COMPANY:    University of Dayton
DATE:       10/12/92
MACRO:      COMBLOG
```

them into the PDS file. The Boolean equations segment of the PDS file shown in Figure 11.22 was modified to include the original design equations (Eq. 11.1, 11.2, and 11.3) of Example 11.6 as follows:

```
EQUATIONS
;UNSIMPLIFIED SOP EQUATIONS (NON-INVERTED)
X = /A * /B * C + /A * B * /C + /A * B * C + A * B * C
X.TRST = OE
Y = /D * /E * /F + /D * E * F + D * /E * F + D * E * /F
Y.TRST = OE
Z = /G * H * /J + /G * H * J + G * /H * /J + G * /H * J + G * H * /J
Z.TRST = OE
```

If the resulting *PDS* file is recompiled, the PALASM4 compiler will automatically minimize the equations, convert them to match the active-low outputs of the PAL16L8 device, and produce a fuse-plot that is exactly the same as shown in Figure 11.24. This eliminates the need for a lot of additional work for the designer.

Example 11.8

Write the PALASM PDS file to implement the following design equation on a PAL16L8:

$$P = \overline{W}X\overline{Y}Z + WX\overline{Y}Z + \overline{W}XYZ$$

Examine the fuse-plot file (XPT) produced by the compiler and verify that the PAL has been programmed properly.

Solution

Figure 11.27 shows the PDS file (EX11-8.PDS) for the given design equation. The independent variables, *W*, *X*, *Y*, and *Z* have been arbitrarily assigned to pins 2 through 5, respectively. The dependent variable *P* has been assigned to pin 19. The given design equation is converted to PALASM4 syntax and entered in the EQUATIONS segment of the PDS file:

$$P = /W * X * /Y * Z + W * X * /Y * Z + /W * X * Y * Z$$

Because this design does not require an output enable signal, we have specified that the tristate control of output *P* be permanently enabled by connecting it to a logic high (VCC):

$$P.TRST = VCC$$

Figure 11.27

PALASM4 PDS file for
Example 11.8.

```
;PALASM Design Description

;-------------------------------- Declaration Segment ------------

TITLE      Example 11-8
PATTERN    EX11-8.PDS (design file)
REVISION A
AUTHOR     Ismail & Rooney
COMPANY    University of Dayton
DATE       10/31/92

CHIP  ex11_8  PAL16L8

;-------------------------------- PIN Declarations ---------------

PIN   1          NC
PIN   2          W              ; INDEPENDENT VARIABLES (INPUTS)
PIN   3          X              ;    "         "        "
PIN   4          Y              ;    "         "        "
PIN   5          Z              ;    "         "        "
PIN   6          NC
PIN   7          NC
PIN   8          NC
PIN   9          NC
PIN   10         GND
;
PIN   11         NC
PIN   12         NC
PIN   13         NC
PIN   14         NC
PIN   15         NC
PIN   16         NC
PIN   17         NC
PIN   18         NC
PIN   19         P              ;DEPENDENT VARIABLE (OUTPUT)
PIN   20         VCC

;-------------------------------- Boolean Equation Segment ------

EQUATIONS

P   =   /W * X * /Y * Z  +  W * X * /Y * Z  +  /W * X * Y * Z

P.TRST = VCC
```

The tristate control equation shown above is not absolutely necessary. If this equation is omitted, the PALASM compiler will assume that the output P is to be permanently enabled. It is good practice, however, to explicitly specify this through an equation in case the assumption made by the compiler is changed in a later version of the PALASM software.

Figure 11.28a shows a partial listing of the fuse-plot produced by PALASM4. Based on this information and the logic architecture of the PAL16L8, a circuit diagram that represents the fuse-plot information has been constructed; this diagram is shown in Figure 11.28b. All the fuses on horizontal line 0 have been blown thus applying logic 1's (V_{CC}) to the inputs of the tristate buffer control *AND* gate. As a result, the output of this *AND* gate is a logic 1 and will permanently enable the tristate buffer at output P. Horizontal lines 1, 2, and 3 are used to apply three product terms, $\overline{X}$, $\overline{Z}$, and WY to the *OR* gate. The output equation therefore is

$$\overline{P} = \overline{X} + \overline{Z} + WY.$$

To verify the validity of this equation, solve the equation in terms of the variable P. If both sides of the equation are inverted

$$\overline{\overline{P}} = \overline{\overline{X} + \overline{Z} + WY}$$

Applying DeMorgan's law:

$$P = \overline{\overline{P}} = \overline{\overline{X}} \cdot \overline{\overline{Z}} \cdot \overline{WY}$$
$$P = X \cdot Z \cdot \overline{WY}$$
$$P = X \cdot Z \cdot (\overline{W} + \overline{Y})$$
$$P = XZ(\overline{W} + \overline{Y})$$
$$P = \overline{W}XZ + X\overline{Y}Z$$

Figure 11.28

Fuse-plot and circuit
diagram for Example 11.8

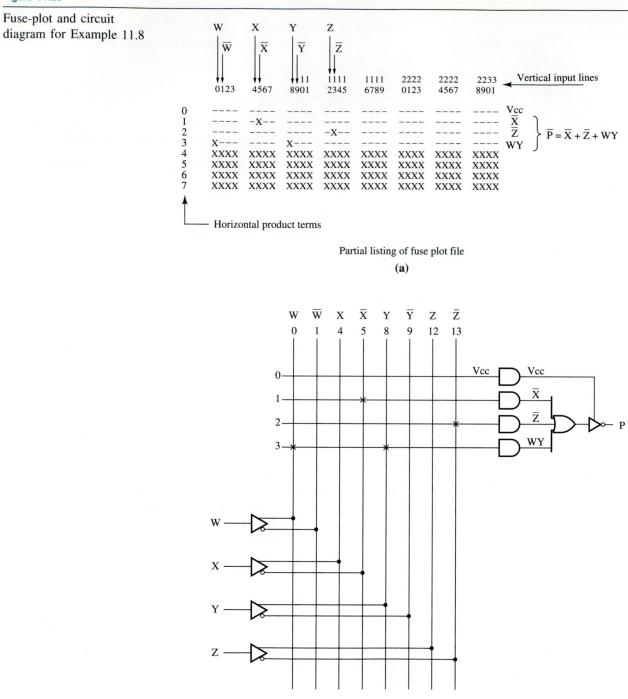

Partial listing of fuse plot file

(a)

Circuit diagram representing fuse plot information

(b)

To verify the above equation, simplify the original (given) design equation and see if it is the same:

$$P = \overline{W}X\overline{Y}Z + WX\overline{Y}Z + \overline{W}XYZ$$
$$P = X\overline{Y}Z(\overline{W} + W) + \overline{W}XYZ$$
$$P = X\overline{Y}Z + \overline{W}XYZ$$
$$P = XZ(\overline{Y} + Y\overline{W})$$
$$P = XZ(\overline{Y} + \overline{W})$$
$$P = \overline{W}XZ + X\overline{Y}Z$$

The simplified equation shown above is equivalent to the equation implemented by PALASM4.

PALASM4 has not only generated a fuse-map to implement the design equation specified but has also simplified the logic and converted the equation to suit the active-low output of the PAL16L8 device.

Simulation

Example 11.8 showed that one way to verify a design is to examine the fuse-plot file produced by PALASM4 and then recreate the logic equations or circuit for the design. This is acceptable for fairly simple designs but for more complex ones, a better solution is needed. PALASM4 provides this solution through *simulation*.

The PALASM4 software package includes a simulator (PLDSIM) that can be used to verify that a PAL design works before programming the PAL. Through the use of various simulation commands, the designer can instruct the simulator to apply logic levels at specified pins and simulate the operation of the logic circuit to be implemented. The outputs can then be examined to verify that the circuit does indeed work as intended.

It was stated at the beginning of this section (with reference to Fig. 11.21) that if the PDS file contains any simulation commands, the PLDSIM simulator can be executed. These simulation commands are placed after the Boolean equation segment of the PDS file, in the *simulation segment*.

Take the PDS file for Example 11.6 that is shown in Figure 11.22 and append to it a simulation segment. The resulting PDS file is shown in Figure 11.29. Note that (for variety) the Boolean equation segment has been modified to include Equations 11.1, 11.2, and 11.3 from Example 11.6 rather than the simplified inverted equations (Eq. 11.4, 11.5, and 11.6); recall from earlier discussion that doing so will not affect the design in any way and will cause the PALASM4 compiler to automatically perform simplification and inversion of the output.

Each logic equation implemented on a PAL device must be separately tested. The simulation commands shown in the simulation segment will only test the proper PAL implementation of Equation 11.1. We shall later see how the other equations can also be tested. The segment begins with the SIMULATION keyword. The first simulation command initiates a TRACE for the inputs, *OE*, *A*, *B*, and *C*, and the output *X*. The SETF commands that follow are used to SET an input pin to either a logic 1 or a logic 0. For example, the command,

SETF   OE

sets the *OE* line to a logic 1 state.
The command,

SETF / A / B / C

sets *A*, *B*, and *C* to the logic 0 state
and the command,

SETF / A / B   C

sets *A* and *B* low and *C* high.

Figure 11.29

PDS file for Example 11.6
with simulation commands.

```
;PALASM Design Description

;-------------------------------- Declaration Segment ------------

TITLE     PAL Implementation of Example 11-6
PATTERN   COMBLOGC.PDS (design file)
REVISION  C
AUTHOR    Ismail & Rooney
COMPANY   University of Dayton
DATE      10/12/92

CHIP   COMBLOG  PAL16L8

;---------------------------- PIN Declarations ---------------

PIN  1        A                    ;INDEPENDENT VARIABLES (INPUTS)
PIN  2        B                    ;      "          "        "
PIN  3        C                    ;      "          "        "
PIN  4        D                    ;      "          "        "
PIN  5        E                    ;      "          "        "
PIN  6        F                    ;      "          "        "
PIN  7        G                    ;      "          "        "
PIN  8        H                    ;      "          "        "
PIN  9        NC
PIN  10       GND
;
PIN  11       OE                   ;OUTPUT ENABLE
PIN  12       NC
PIN  13       NC
PIN  14       Z                    ;DEPENDENT VARIABLE (OUTPUT)
PIN  15       J                    ;INDEPENDENT VARIABLE (INPUT)
PIN  16       NC
PIN  17       Y                    ;DEPENDENT VARIABLE (OUTPUT)
PIN  18       NC
PIN  19       X                    ;DEPENDENT VARIABLE (OUTPUT)
PIN  20       VCC

;-------------------------------- Boolean Equation Segment ------

EQUATIONS

;UNSIMPLIFIED SOP EQUATIONS (NON-INVERTED)

X = /A * /B * C  +  /A * B * /C  +  /A * B * C  +  A * B * C          ◄─── Equation 11-1

X.TRST = OE

Y = /D * /E * /F  +  /D * E * F  +  D * /E * F  +  D * E * /F          ◄─── Equation 11-2

Y.TRST = OE

Z = /G * H * /J  +  /G * H * J  +  G * /H * /J  +  G * /H * J  +  G * H * /J   ◄─── Equation 11-3

Z.TRST = OE

;-------------------------------- Simulation Segment ------------

SIMULATION  ◄──────────────────────────────── Identifies the start of the simulation segment

TRACE_ON OE A B C X ◄─────────────────────── Initiates a trace for selected variables

SETF OE ◄────── Sets OE=1

SETF /A /B /C ┐
SETF /A /B  C │
SETF /A  B /C │
SETF /A  B  C │  Try all possible
SETF  A /B /C │  combinations
SETF  A /B  C │
SETF  A  B /C │
SETF  A  B  C ┘

SETF /OE ◄────── Sets OE=0

SETF /A /B /C ┐
SETF /A /B  C │
SETF /A  B /C │
SETF /A  B  C │  Try all possible
SETF  A /B /C │  combinations
SETF  A /B  C │
SETF  A  B /C │
SETF  A  B  C ┘

TRACE_OFF ◄────── Turn off trace listing
```

Following the TRACE_ON command, the *OE* line is first set high (SETF OE)
and all possible combinations are applied to the *A*, *B*, and *C* inputs. The *OE* line is
then set low (SETF /OE) and once again all possible combinations are applied to the
three inputs. The last command is the TRACE_OFF command that turns off the
trace.

After the PDS file shown in Figure 11.29 is compiled by PALASM4, the file is
run through the simulator (PLDSIM). PLDSIM executes each command in the

simulation segment and produces two output files — a trace file (*TRF*) that contains a trace of the logic levels at inputs *A*, *B*, *C*, and *OE*, and output *X*, and a history file that contains a trace of the logic levels of all defined pins in the PAL device.

The trace file for the *PDS* file of Figure 11.29 is shown in Figure 11.30. This file contains a selective listing of only the inputs and outputs specified in the *TRACE_ON* command. When the *OE* line is set high, the circuit output (*X*) is enabled and for each combination applied to the input the simulator determines the logic level at the output *X*. The following truth table for *X* can be obtained from Figure 11.30.

A	B	C	X
0	0	0	0
0	0	1	1
0	1	0	1
0	1	1	1
1	0	0	0
1	0	1	0
1	1	0	0
1	1	1	1

The logic equation for *X* from the above truth table is:

$$X = \Sigma_m\ 1,2,3,7$$

This equation is the same as Equation 11.1, thus verifying the design.

Figure 11.30

Trace file (*TRF*) for Example 11.6.

```
PALASM4  PLDSIM   - MARKET RELEASE 1.5 (7-10-92)
(C) - COPYRIGHT ADVANCED MICRO DEVICES INC., 1992

PALASM SIMULATION SELECTIVE TRACE LISTING

Title   : PAL Implementation of Example 11-6    Author : Ismail & Rooney
Pattern : COMBLOGC.PDS (design file)            Company: University of Dayton
Revision: C                                     Date   : 10/12/92

PAL16L8
Page : 1

       gggggggggggggggggggg               ── Each "g" represents a SETF command
    OE HHHHHHHHHLLLLLLLLLL
    A  LLLLLHHHHHLLLLLHHHH
    B  LLLHHLLHHHLLHHLLHH
    C  LLHLHLHLHHLHLHLHLH
    X  LLHHHLLLLHZZZZZZZZZ
       ↳   OE=1   ↳  OE=0
          output     output
          enabled    disabled
```

Observe in Figure 11.30 that when the *OE* line is set low, the output *X* is disabled and regardless of the logic levels applied to *A*, *B*, and *C*, the output *X* is in a Hi-Z state (identified by the letter *Z*).

The simulation history listing (*HST* file) produced by the simulator is very similar to the trace listing and is shown in Figure 11.31. The only difference is that the history listing shows the logic levels at all defined pins on the PLD device. It is preferable to use the more selective trace listing (*TRF* file) rather than the history listing because the former is easier to interpret and less cluttered.

History file (HST) for
Example 11.6.

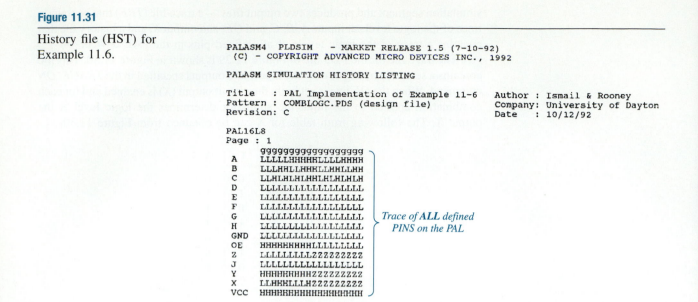

```
PALASM4  PLDSIM   - MARKET RELEASE 1.5 (7-10-92)
(C) - COPYRIGHT ADVANCED MICRO DEVICES INC., 1992

PALASM SIMULATION HISTORY LISTING

Title   : PAL Implementation of Example 11-6    Author : Ismail & Rooney
Pattern : COMBLOGC.PDS (design file)            Company: University of Dayton
Revision: C                                     Date   : 10/12/92

PAL16L8
Page : 1
        ggggggggggggggggggg
    A   LLLLLHHHHHLLLLHHHH
    B   LLLHHLLHHHLLHHLLHH
    C   LLHLHLHLHHLHLHLHLH
    D   LLLLLLLLLLLLLLLLLL
    E   LLLLLLLLLLLLLLLLLL
    F   LLLLLLLLLLLLLLLLLL
    G   LLLLLLLLLLLLLLLLLL
    H   LLLLLLLLLLLLLLLLLL
   GND  LLLLLLLLLLLLLLLLLL
    OE  HHHHHHHHHLLLLLLLLL
    Z   LLLLLLLLLZZZZZZZZZ
    J   LLLLLLLLLLLLLLLLLL
    Y   HHHHHHHHHZZZZZZZZZ
    X   LLHHHLLLHZZZZZZZZZ
   VCC  HHHHHHHHHHHHHHHHHH
```

*Trace of **ALL** defined
PINS on the PAL*

Example 11.9   Verfiy the proper operation of the logic circuit designed in Example 11.8 through simulation.

Solution

The logic equation that was implemented on the PAL16L8 in Example 11.8 is

$$P = \overline{W}X\overline{Y}Z + WX\overline{Y}Z + \overline{W}XYZ.$$

The equation can be written in short-form notation as:

$$P = \Sigma_m 5,7,13$$

A truth table for the logic equation can be constructed as shown below:

W	X	Y	Z	P
0	0	0	0	0
0	0	0	1	0
0	0	1	0	0
0	0	1	1	0
0	1	0	0	0
0	1	0	1	1
0	1	1	0	0
0	1	1	1	1
1	0	0	0	0
1	0	0	1	0
1	0	1	0	0
1	0	1	1	0
1	1	0	0	0
1	1	0	1	1
1	1	1	0	0
1	1	1	1	0

To verify the truth table shown above through simulation, the following simulation segment can be appended to the PDS file that was shown in Figure 11.27:

```
SIMULATION
TRACE_ON W X Y Z P
SETF / W / X / Y / Z
SETF / W / X / Y   Z
SETF / W / X   Y / Z
SETF / W / X   Y   Z
SETF / W   X / Y / Z
SETF / W   X / Y   Z
SETF / W   X   Y / Z
SETF / W   X   Y   Z
SETF   W / X / Y / Z
SETF   W / X / Y   Z
SETF   W / X   Y / Z
SETF   W / X   Y   Z
SETF   W   X / Y / Z
SETF   W   X / Y   Z
SETF   W   X   Y / Z
SETF   W   X   Y   Z
TRACE_OFF
```

The simulation commands shown above initiate a trace of the inputs W, X, Y, and Z, and the output P. All possible combinations of the inputs are then applied to the simulated PAL device using the *SETF* command. The trace listing produced by the simulator (shown below) corresponds to the truth table for the original design equation:

```
  gggggggggggggggg
W LLLLLLLLHHHHHHHH
X LLLLHHHHLLLLHHHH
Y LLHHLLHHLLHHLLHH
Z LHLHLHLHLHLHLHLH
P LLLLLHLHLLLLLHLL
```

Notice in the trace listing shown above that the dependent variable P is a high (logic 1) for combinations 0101, 0111, and 1101 (minterms 5, 7, and 13). Thus

$$P = \Sigma_m 5,7,13$$

It was stated earlier in this section that the PLDSIM simulator also produces a JEDEC output file containing fuse information for the PAL programming unit (*see* Fig. 11.21). This file is exactly the same as the JEDEC file produced by the PALASM4 compiler (JED file) except that in addition to a fuse-map, it also contains test vectors that are generated during simulation. The test vectors can be used by the PAL programming unit to test the proper operation of the PAL device after program-ming. A test vector is simply a binary number that the PAL programming unit uses to apply logic levels at the inputs of the PAL device and verify that the PAL produces the proper output. Test vectors generated by PLDSIM correspond to the simulation commands used in the simulation segment. Unlike a trace, however, test vectors verify the proper operation of the actual PAL device rather than simulate its operation. The JEDEC file produced by PLDSIM is given the extension .JDC to distinguish it from the JEDEC file produced by PALASM (*see* Fig. 11.21). The JDC file produced for Figure 11.29 is shown in Figure 11.32. The fuse-map information is the same as in Figure 11.25 and for brevity a partial listing is shown. However, at the end of the fuse-map, PLDSIM has appended 18 test vectors (labeled V0001 through

Figure 11.32

JEDEC output file (JDC)
with test vectors.

```
PALASM4   PAL ASSEMBLER    - MARKET RELEASE 1.5a (8-20-92)
(C) - COPYRIGHT ADVANCED MICRO DEVICES INC., 1992


Title   : PAL Implementation of Example 11-6   Author : Ismail & Rooney
Pattern : COMBLOGC.PDS (design file)           Company: University of Dayton
Revision: C                                    Date   : 10/12/92


PAL16L8
COMBLOG*
QV0018*
QP20*
QF2048*
G0*F0*
L0000 11111111111111111111111111111101*
L0032 10011111111111111111111111111111*
L0064 11011011111111111111111111111111*
L0096 10111011111111111111111111111111*
L0128 00000000000000000000000000000000*
L0160 00000000000000000000000000000000*
      .    .    .    .    .    .
      .    .    .    .    .    .
      .    .    .    .    .    .
      .    .    .    .    .    .
L1856 00000000000000000000000000000000*
L1888 00000000000000000000000000000000*
L1920 00000000000000000000000000000000*
L1952 00000000000000000000000000000000*
L1984 00000000000000000000000000000000*
L2016 00000000000000000000000000000000*
X0*
V0001 XXXXXXXXXN1XXLXXHXLN*
V0002 000XXXXXXN1XXLXXHXLN*
V0003 001XXXXXXN1XXLXXHXHN*
V0004 010XXXXXXN1XXLXXHXHN*
V0005 011XXXXXXN1XXLXXHXHN*
V0006 100XXXXXXN1XXLXXHXLN*
V0007 101XXXXXXN1XXLXXHXLN*
V0008 110XXXXXXN1XXLXXHXLN*
V0009 111XXXXXXN1XXLXXHXHN*
V0010 111XXXXXXN0XXZXXZXZN*
V0011 000XXXXXXN0XXZXXZXZN*
V0012 001XXXXXXN0XXZXXZXZN*
V0013 010XXXXXXN0XXZXXZXZN*
V0014 011XXXXXXN0XXZXXZXZN*
V0015 100XXXXXXN0XXZXXZXZN*
V0016 101XXXXXXN0XXZXXZXZN*
V0017 110XXXXXXN0XXZXXZXZN*
V0018 111XXXXXXN0XXZXXZXZN*
C2E21*
7580
```

*Fuse map
(same as Figure 11-25)*

Test vectors

V0018) corresponding to the 18 SETF commands in the simulation segment of the PDS file (Fig. 11.29). As was stated earlier, the contents of the JEDEC files are not of importance to the designer but just to investigate how these test vectors work, examine, for example, test vector V0003:

$$V0003 \quad 001XXXXXXN1XXLXXHXHN*$$

The test vector was generated by the SETF /A /B C command, the third SETF command in the simulation segment. Each bit in the test vector corresponds to a pin on the PLD device as follows:

```
PIN: 1 2 3 4 5 6 7 8 9 10 11 12 13 14 15 16 17 18 19 20
     0 0 1 X X X X X X N  1  X  X  L  X  X  H  X  H  N
```

After the PAL device is programmed, the PLD programming unit applies the logic combination 0 0 1 to pins 1, 2, and 3 (*A*, *B*, and *C*), respectively, and a logic 1 on pin 11 (*OE*). It then examines pin 14 (*Z*) for a logic 0, and pins 17 (*Y*) and 19 (*X*) for a logic 1. An *N* in the test vector identifies the power supply pins whereas an *X* indicates that no logic level is applied or examined. If the PLD programming unit detects a discrepancy between the PAL device outputs and the expected outputs in the test vector, it signals a verification error to the designer.

SETF and TRACE commands can be used to trace through more than one logic equation implemented on a PAL device. For example, to trace the outputs *X* and *Z* for all possible combinations of inputs use the following simulation segment for the PDS file shown in Figure 11.29:

```
SIMULATION
SETF OE
TRACE_ON A B C X   G H J Z
SETF / A / B / C    / G / H / J
SETF / A / B   C    / G / H   J
SETF / A   B / C    / G   H / J
SETF / A   B   C    / G   H   J
SETF   A / B / C      G / H / J
SETF   A / B   C      G / H   J
SETF   A   B / C      G   H / J
SETF   A   B   C      G   H   J
TRACE_OFF
```

The TRACE command shown above initiates a trace of the inputs *A*, *B*, and *C*, and the output *X*, the inputs *G*, *H*, and *J*, and the output *Z*. Notice that the input *OE* is not specified in the trace and that *OE* is set high before the TRACE_ON command is issued. The resulting trace will look like the following:

```
    gggggggg
A LLLLHHHH
B LLHHLLHH
C LHLHLHLH
X LHHHLLLH
G LLLLHHHH
H LLHHLLHH
J LHLHLHLH
Z LLHHHHHL
```

Example 11.10

The PAL20V8 is a popular PAL device that can be used for many combinational or sequential logic designs. It has 20 inputs and 8 outputs. Because it is a *V* or versatile-type PAL device its outputs are fully programmable — active high or active low. The block diagram and pinout for the PAL20V8 is shown in Figure 11.33. The outputs of the PAL shown in the block diagram have macrocells that can be configured as active high/low registered outputs, combinational inputs/outputs, or dedicated inputs, giving the designer tremendous flexibility.

Implement the hexadecimal to seven-segment decoder that was designed in Chapter 4 (Fig. 4.50) on the PAL20V8. Verify that the design produces the correct outputs by simulating all possible input combinations and comparing the simulation trace listing to the truth table shown in Table 4.17.

Solution

The PDS file shown in Figure 11.34 contains the pin definitions for the inputs and outputs of the decoder. Pins 1 through 4 have been arbitrarily assigned to the decoder inputs *A3*, *A2*, *A1*,

Block diagram and pinout
for the PAL20V8.

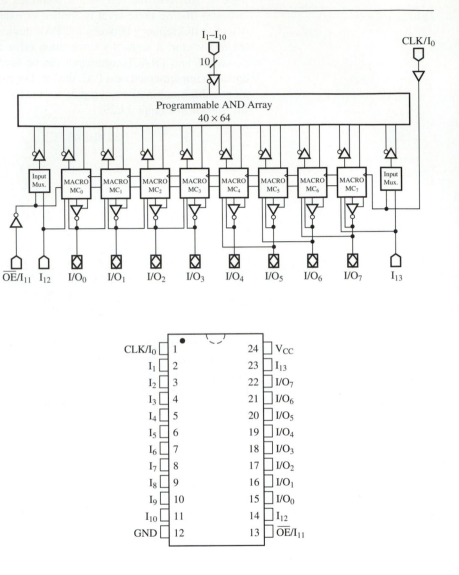

and *A0*, respectively. The decoder outputs (*a* through *g*) have been assigned to pins 15 through 21. The EQUATIONS segment contain Equations 4.37 through 4.43 in PALASM4 syntax. Recall that these equations are the simplified equations. The SIMULATION segment initiates a trace of the inputs (*A0* to *A3*) and the outputs (*a* to *g*) and then using the SETF command all possible combinations are applied to the inputs.

The simulation selective trace listing file (TRF) is shown in Figure 11.35. For each of the eight binary combinations applied to the inputs (*A0* to *A3*) the logic levels produced at the outputs (*a* to *g*) correspond to the truth table shown in Table 4.17.

It is apparent in this example that a PAL can greatly reduce the number of components required to implement a logic circuit. The hexadecimal to seven-segment decoder designed in Chapter 4 would have required more than 15 ICs when implemented. With PAL technology only one IC is required.

Figure 11.34

PDS file for Example 11.10.

```
TITLE      Hex to 7-segment Decoder
PATTERN    7segB.pds (design file)
REVISION B
AUTHOR     Ismail & Rooney
COMPANY    University of Dayton
DATE       10/31/92

CHIP   7seg   PAL20V8

PIN  1            A3             ;binary inputs
PIN  2            A2
PIN  3            A1
PIN  4            A0
PIN  5            NC
PIN  6            NC
PIN  7            NC
PIN  8            NC
PIN  9            NC
PIN  10           NC
PIN  11           NC
PIN  12           GND
;
PIN  13           NC
PIN  14           NC
PIN  15           g              ;7-segment outputs
PIN  16           f
PIN  17           e
PIN  18           d
PIN  19           c
PIN  20           b
PIN  21           a
PIN  22           NC
PIN  23           NC
PIN  24           VCC

EQUATIONS

a =      A2 * A1  +  /A3 * A1  +  A3 * /A0  +  /A2 * /A0
    +  /A3 * A2 * A0  +  A3 * /A2 * /A1

b =      /A3 * /A2  +  A3 * /A1 * A0  +  /A2 * /A0  +  /A3 * /A1 * /A0
    +  /A3 * A1 * A0

c =      A3 * /A2  +  /A3 * A2  +  /A1 * A0  +  /A3 * /A1  +  /A3 * A0

d =      A3 * /A1 * /A0  +  A2 * /A1 * A0  +  A2 * A1 * /A0
    +  /A2 * A1 * A0  +  /A3 * /A2 * /A0

e =      /A2 * /A0  +  A3 * A2  +  A1 * /A0  +  A3 * A1

f =      /A1 * /A0  +  A3 * /A2  +  A2 * /A0  +  A3 * A1  +  /A3 * A2 * /A1

g =      A3 * /A2  +  A3 * A0  +  /A2 * A1  +  A1 * /A0  +  /A3 * A2 * /A1

SIMULATION

TRACE_ON A3 A2 A1 A0 a b c d e f g

SETF /A3 /A2 /A1 /A0
SETF /A3 /A2 /A1  A0
SETF /A3 /A2  A1 /A0
SETF /A3 /A2  A1  A0
SETF /A3  A2 /A1 /A0
SETF /A3  A2 /A1  A0
SETF /A3  A2  A1 /A0
SETF /A3  A2  A1  A0
SETF  A3 /A2 /A1 /A0
SETF  A3 /A2 /A1  A0
SETF  A3 /A2  A1 /A0
SETF  A3 /A2  A1  A0
SETF  A3  A2 /A1 /A0
SETF  A3  A2 /A1  A0
SETF  A3  A2  A1 /A0
SETF  A3  A2  A1  A0

TRACE_OFF
```

Figure 11.35

TRF file for Example 11.10.

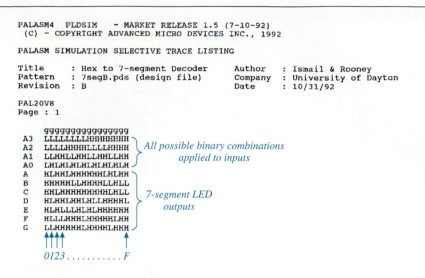

```
PALASM4  PLDSIM   - MARKET RELEASE 1.5 (7-10-92)
(C) - COPYRIGHT ADVANCED MICRO DEVICES INC., 1992

PALASM SIMULATION SELECTIVE TRACE LISTING

Title    : Hex to 7-segment Decoder     Author  : Ismail & Rooney
Pattern  : 7segB.pds (design file)      Company : University of Dayton
Revision : B                            Date    : 10/31/92

PAL20V8
Page : 1
```

```
        gggggggggggggggg
A3      LLLLLLLLHHHHHHHH
A2      LLLLHHHHLLLLHHHH   ⎫  All possible binary combinations
A1      LLHHLLHHLLHHLLHH   ⎬       applied to inputs
A0      LHLHLHLHLHLHLHLH   ⎭
A       HLHHLHHHHHHLHLHH   ⎫
B       HHHHHLLHHHHLLHLL   ⎪
C       HHLHHHHHHHHHLHLL   ⎬  7-segment LED
D       HLHHLHHLHLLHHHHL   ⎪      outputs
E       HLHLLLHLHLHHHHHH   ⎪
F       HLLLLHHHLHHHHHLHH  ⎪
G       LLHHHHHLHHHHHLHHH  ⎭
        ↑↑↑↑             ↑
        0123 .......... F
```

Review Questions

1. Describe the content and purpose of each file processed and produced by PALASM4.

2. How many segments can a PDS file have? Describe the purpose of each one.

3. What is the purpose of simulation? Describe contents of the files produced by PLDSIM.

4. How can simulation be used to verify that a PAL device has been programmed properly?

11.6
PAL programming and simulation — sequential logic

Sequential logic design using PALASM4 is very similar to the design of combinational logic circuits. PALASM4, however, allows the designer to define the sequential logic circuit using equations (as was done for combinational logic circuits) or by using state transitions. Both of these techniques will be examined in this section.

To examine the manner in which PALASM4 accomplishes the design of a sequential logic circuit, implement the 2-bit synchronous binary down counter from Example 11.7. Recall that this design had been implemented on a PAL16R4 and we had manually determined the fuse arrangement (*see* Fig. 11.20) from a given set of design equations. Also remember that because the PAL16R4 had active-low registered outputs the design equations for an *UP* counter were implemented; the *UP* counter with inverted outputs would count *DOWN*.

The PDS file for the implementation of the down counter is shown in Figure 11.36. The PDS file is similar to the PDS file for a combinational logic design in that it contains the same four segments — the declarations segment, the pin declarations, the Boolean equations segment, and the simulation segment. The same pin assignments as those in Example 11.7 were used and therefore, in the PIN declarations, pin 1 has been assigned to the clock *CP*, pin 11 is the nonprogrammable (active-low) output enable line, and pins 16 and 17 are the active-low outputs of the counter /*X*1

Figure 11.36

PAL implementation of Example 11.7.

```
;PALASM Design Description

;----------------------------- Declaration Segment ------------
TITLE    PAL Implementation of Example 11-7
PATTERN  ex11-7C.pds (design file)
REVISION C
AUTHOR   Ismail & Rooney
COMPANY  University of Dayton
DATE     10/12/92

CHIP  ex11_7C  PAL16R4

;------------------------------- PIN Declarations --------------
PIN  1        CP         ;Clock input
PIN  2        NC
PIN  3        NC
PIN  4        NC
PIN  5        NC
PIN  6        NC
PIN  7        NC
PIN  8        NC
PIN  9        NC
PIN  10       GND
;
PIN  11       OE         ;Output enable
PIN  12       NC
PIN  13       NC
PIN  14       NC
PIN  15       NC
PIN  16       /X1        ;Counter output ⎫ active low
PIN  17       /X0        ;Counter output ⎭
PIN  18       NC
PIN  19       NC
PIN  20       VCC

;------------------------------- Boolean Equation Segment ------
EQUATIONS

X0 := /X0  ◄─────────────────────────── Equation 11-7

X1 := X1 :+: X0  ◄────────────────────── Equation 11-8

;------------------------------- Simulation Segment ------------
SIMULATION

TRACE_ON CP OE /X1 /X0

SETF /OE  ◄────── OE=0  output enabled
CLOCKF CP ⎫
CLOCKF CP
CLOCKF CP
CLOCKF CP
CLOCKF CP  ⎬ Apply 8 clock pulses
CLOCKF CP
CLOCKF CP
CLOCKF CP ⎭
SETF OE  ◄────── OE=1  output disabled
CLOCKF CP ⎫
CLOCKF CP
CLOCKF CP
CLOCKF CP
CLOCKF CP  ⎬ Apply 8 clock pulses
CLOCKF CP
CLOCKF CP
CLOCKF CP ⎭

TRACE_OFF
```

and $/X0$, respectively. The reason why these two outputs have been identified as active low (using the "/" symbol before the pin name) will be seen shortly.

Equations 11.7 and 11.8 must first be converted to the proper PALASM4 syntax for sequential logic circuits before they can be placed in the PDS file. Equation 11.7

$$D_0 = \overline{Q_0}$$

becomes

$$X0 := /X0$$

in PALASM4 syntax. All equations must be written in terms of the output of each flip-flop.

Figure 11.37

Fuse-plot for PAL
implementation of Example
11.7.

```
PALASM4  PAL ASSEMBLER   - MARKET RELEASE 1.5a (8-20-92)
  (C) - COPYRIGHT ADVANCED MICRO DEVICES INC., 1992

TITLE   :PAL Implementation of Example 11-7    AUTHOR :Ismail & Rooney
PATTERN :ex11-7C.pds (design file)             COMPANY:University of Dayton
REVISION:C                                     DATE   :10/12/92

PAL16R4
EX11_7C

               11   1111  1111  2222  2222  2233
     0123  4567 8901  2345  6789  0123  4567  8901

0    XXXX  XXXX XXXX  XXXX  XXXX  XXXX  XXXX  XXXX
1    XXXX  XXXX XXXX  XXXX  XXXX  XXXX  XXXX  XXXX
2    XXXX  XXXX XXXX  XXXX  XXXX  XXXX  XXXX  XXXX
3    XXXX  XXXX XXXX  XXXX  XXXX  XXXX  XXXX  XXXX
4    XXXX  XXXX XXXX  XXXX  XXXX  XXXX  XXXX  XXXX
5    XXXX  XXXX XXXX  XXXX  XXXX  XXXX  XXXX  XXXX
6    XXXX  XXXX XXXX  XXXX  XXXX  XXXX  XXXX  XXXX
7    XXXX  XXXX XXXX  XXXX  XXXX  XXXX  XXXX  XXXX

8    XXXX  XXXX XXXX  XXXX  XXXX  XXXX  XXXX  XXXX
9    XXXX  XXXX XXXX  XXXX  XXXX  XXXX  XXXX  XXXX
10   XXXX  XXXX XXXX  XXXX  XXXX  XXXX  XXXX  XXXX
11   XXXX  XXXX XXXX  XXXX  XXXX  XXXX  XXXX  XXXX
12   XXXX  XXXX XXXX  XXXX  XXXX  XXXX  XXXX  XXXX
13   XXXX  XXXX XXXX  XXXX  XXXX  XXXX  XXXX  XXXX
14   XXXX  XXXX XXXX  XXXX  XXXX  XXXX  XXXX  XXXX
15   XXXX  XXXX XXXX  XXXX  XXXX  XXXX  XXXX  XXXX

16   ----  ---- --X-  ----  ----  ----  ----  ----     ◄─── Fuse at vertical input line 10
17   XXXX  XXXX XXXX  XXXX  XXXX  XXXX  XXXX  XXXX            left intact
18   XXXX  XXXX XXXX  XXXX  XXXX  XXXX  XXXX  XXXX
19   XXXX  XXXX XXXX  XXXX  XXXX  XXXX  XXXX  XXXX
20   XXXX  XXXX XXXX  XXXX  XXXX  XXXX  XXXX  XXXX
21   XXXX  XXXX XXXX  XXXX  XXXX  XXXX  XXXX  XXXX
22   XXXX  XXXX XXXX  XXXX  XXXX  XXXX  XXXX  XXXX
23   XXXX  XXXX XXXX  XXXX  XXXX  XXXX  XXXX  XXXX

24   ----  ---- ---X  --X-  ----  ----  ----  ----     ◄─── Fuses at vertical input lines
25   ----  ---- --X-  ---X  ----  ----  ----  ----            11 and 14 left intact
26   XXXX  XXXX XXXX  XXXX  XXXX  XXXX  XXXX  XXXX
27   XXXX  XXXX XXXX  XXXX  XXXX  XXXX  XXXX  XXXX      Fuses at vertical input lines 10 and 15
28   XXXX  XXXX XXXX  XXXX  XXXX  XXXX  XXXX  XXXX                  left intact
29   XXXX  XXXX XXXX  XXXX  XXXX  XXXX  XXXX  XXXX
30   XXXX  XXXX XXXX  XXXX  XXXX  XXXX  XXXX  XXXX
31   XXXX  XXXX XXXX  XXXX  XXXX  XXXX  XXXX  XXXX

32   XXXX  XXXX XXXX  XXXX  XXXX  XXXX  XXXX  XXXX
33   XXXX  XXXX XXXX  XXXX  XXXX  XXXX  XXXX  XXXX
34   XXXX  XXXX XXXX  XXXX  XXXX  XXXX  XXXX  XXXX
35   XXXX  XXXX XXXX  XXXX  XXXX  XXXX  XXXX  XXXX
36   XXXX  XXXX XXXX  XXXX  XXXX  XXXX  XXXX  XXXX
37   XXXX  XXXX XXXX  XXXX  XXXX  XXXX  XXXX  XXXX
38   XXXX  XXXX XXXX  XXXX  XXXX  XXXX  XXXX  XXXX
39   XXXX  XXXX XXXX  XXXX  XXXX  XXXX  XXXX  XXXX

40   XXXX  XXXX XXXX  XXXX  XXXX  XXXX  XXXX  XXXX
41   XXXX  XXXX XXXX  XXXX  XXXX  XXXX  XXXX  XXXX
42   XXXX  XXXX XXXX  XXXX  XXXX  XXXX  XXXX  XXXX
43   XXXX  XXXX XXXX  XXXX  XXXX  XXXX  XXXX  XXXX
44   XXXX  XXXX XXXX  XXXX  XXXX  XXXX  XXXX  XXXX
45   XXXX  XXXX XXXX  XXXX  XXXX  XXXX  XXXX  XXXX
46   XXXX  XXXX XXXX  XXXX  XXXX  XXXX  XXXX  XXXX
47   XXXX  XXXX XXXX  XXXX  XXXX  XXXX  XXXX  XXXX

48   XXXX  XXXX XXXX  XXXX  XXXX  XXXX  XXXX  XXXX
49   XXXX  XXXX XXXX  XXXX  XXXX  XXXX  XXXX  XXXX
50   XXXX  XXXX XXXX  XXXX  XXXX  XXXX  XXXX  XXXX
51   XXXX  XXXX XXXX  XXXX  XXXX  XXXX  XXXX  XXXX
52   XXXX  XXXX XXXX  XXXX  XXXX  XXXX  XXXX  XXXX
53   XXXX  XXXX XXXX  XXXX  XXXX  XXXX  XXXX  XXXX
54   XXXX  XXXX XXXX  XXXX  XXXX  XXXX  XXXX  XXXX
55   XXXX  XXXX XXXX  XXXX  XXXX  XXXX  XXXX  XXXX

56   XXXX  XXXX XXXX  XXXX  XXXX  XXXX  XXXX  XXXX
57   XXXX  XXXX XXXX  XXXX  XXXX  XXXX  XXXX  XXXX
58   XXXX  XXXX XXXX  XXXX  XXXX  XXXX  XXXX  XXXX
59   XXXX  XXXX XXXX  XXXX  XXXX  XXXX  XXXX  XXXX
60   XXXX  XXXX XXXX  XXXX  XXXX  XXXX  XXXX  XXXX
61   XXXX  XXXX XXXX  XXXX  XXXX  XXXX  XXXX  XXXX
62   XXXX  XXXX XXXX  XXXX  XXXX  XXXX  XXXX  XXXX
63   XXXX  XXXX XXXX  XXXX  XXXX  XXXX  XXXX  XXXX

SUMMARY
-------

    TOTAL FUSES BLOWN   = 91
```

The ": =" identifies $X0$ to be a registered output (the output of the flip-flop whose input is D_0). Recall that equations that describe combinational outputs use the " = " operator.

Equation 11.8

$$D_1 = Q_1 \oplus Q_0$$

becomes

$$\text{X1 : = X1 : + : X0}$$

Here, ": + :" is used to identify the exclusive-*OR* operator $\oplus$.

Because the PAL16R4 has active-low registered outputs, PALASM4 would normally convert the two equations to match the active-low outputs of the PAL device, as was seen in the previous section. However, in this example we do not want to do that because Equations 11.7 and 11.8 are actually the equations for an *UP* counter and we are depending on the fact that the outputs of the flip-flops are inverted so that the counter counts down. By identifying the two outputs (pins 16 and 17) in the pin declaration segment as active low outputs $/X0$ and $/X1$, PALASM4 will not change the equations in any way because it assumes that we intend to keep the outputs configured as active low. If we did not identify these two outputs as active-low outputs, PALASM4 would invert the right-hand side of each equation to match the active-low outputs; this would make the counter an *UP* counter as described by the two equations.

The SIMULATION segment initiates a trace of the clock input *CP*, the output enable line *OE*, and the outputs of the counter X_1, X_0. *OE* is set low to enable the counter's outputs using the SETF command. Next the CLOCKF command is used eight times to apply eight positive-going clock pulses to the clock input *CP* of the counter. Then the *OE* line is set high to disable the outputs of the counter and again eight clock pulses are applied to the counter's clock input.

Figure 11.37 shows a listing of the fuse-plot file (XPT file) produced by PALASM4 after the PDS file in Figure 11.36 is compiled. Compare this fuse-plot information with the programmed PAL16R4 shown in Figure 11.20. Figure 11.20 shows that horizontal product line 16 is connected to vertical line 10. This is verified in Figure 11.37, which shows that the fuse connecting these two lines is kept intact while the other fuses on product line 16 are blown. Similarly, in Figure 11.20, horizontal product line 24 is connected to vertical lines 11 and 14. This is verified in the fuse-plot file that shows that the fuses between product line 24 and vertical lines 11 and 14 are left intact. Also, in Figure 11.20, horizontal product line 25 is connected to vertical lines 10 and 15. This is verified in the fuse-plot file that shows that the fuses between product lines 10 and 15 are left intact. The fuse-plot listing thus verifies that the design equations have been implemented properly.

The trace file (TRF) produced by PLDSIM is shown in Figure 11.38. The previous section showed that this file could be examined to determine proper operation of the designed circuit. Each transition of the clock pulse applied to "CP" from a low (*L*) to a high (*H*) state is identified by the letter "*c*" in the trace listing. For the first eight clock pulses we specified that *OE* be set low. If the outputs of the counter *X*1 and *X*0 are examined during this time the counter counts down in the following sequence — 3, 2, 1, 0, 3, ... For the next eight clock pulses applied to *CP*, *OE* is set high. This disables the outputs of the counter and places them in a Hi-*Z* state.

The PLDSIM simulator also produces a graphical representation of the trace known as a *waveform display*. This is basically a timing diagram of the pins being

Figure 11.38

Trace listing for PAL
implementation of Example
11.7.

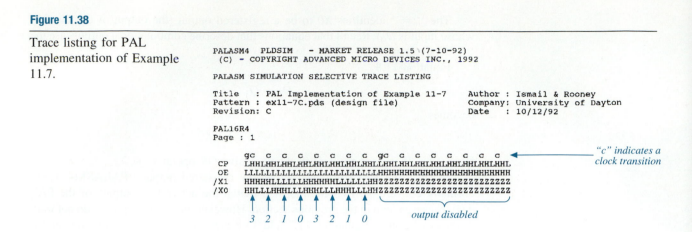

"c" indicates a clock transition

output disabled

Figure 11.39

Waveform display for PAL
implementation of Example
11.7.

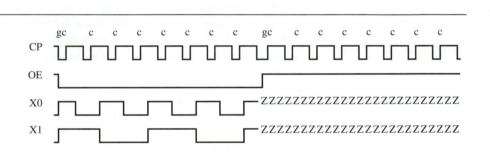

traced during the simulation and is very useful for sequential logic designs. The
waveform display for the 2-bit down counter is shown in Figure 11.39.

Example 11.11

Implement the 3-bit synchronous counter shown in Figure 6.33 on a PAL16R4. Simulate the
design and verify that the circuit counts in the following sequence: 6, 3, 5, 0, 6, 3, . . .

Solution

The equations for the circuit shown in Figure 6.33 are

$$D_0 = \overline{Q_1}Q_0 + Q_2Q_1$$
$$D_1 = \overline{Q_0}$$
$$D_2 = \overline{Q_2}Q_1$$

This time implement the equations in terms of $X0$, $X1$, and $X2$ and let PALASM4 convert the
equations to match the active-low outputs of the PAL16R4. Then the equations in PALASM4
syntax become

```
X0 := X1 * X0 + X2 * X1
X1 := / X0
X2 := / X2 * X1
```

These equations are incorporated into the Boolean equation segment of the following PDS file

```
;PALASM Design Description

;-------------------Declaration Segment-------------------
TITLE     PAL Implementation of Figure 6.33
PATTERN   3BIT.PDS (design file)
REVISION A
AUTHOR    Ismail & Rooney
COMPANY   University of Dayton
DATE      11 / 05 / 92

CHIP  3BIT  PAL16R4

;-------------------PIN Declarations-------------------
PIN 1      CP       ;Clock Input
PIN 2      NC
PIN 3      NC
PIN 4      NC
PIN 5      NC
PIN 6      NC
PIN 7      NC
PIN 8      NC
PIN 9      NC
PIN 10     GND
;
PIN 11     OE       ;Output Enable
PIN 12     NC
PIN 13     NC
PIN 14     NC
PIN 15     X2       ;Counter Outputs
PIN 16     X1
PIN 17     X0
PIN 18     NC
PIN 19     NC
PIN 20     VC

;-----------------Boolean Equation Segment----------------
EQUATIONS
X0 := X1 * X0 + X2 * X1
X1 := / X0
X2 := / X2 * X1

;------------------Simulation Segment------------------
SIMULATION
SETF / OE
TRACE_ON CP X0 X1 X2
CLOCKF CP
CLOCKF CP
CLOCKF CP
CLOCKF CP
CLOCKF CP
CLOCKF CP
CLOCKF CP
CLOCKF CP
CLOCKF CP
TRACE_OFF
;-------------------------------------------------------
```

The trace file listing after simulation of the PDS file is shown below:

```
PALASM4  PLDSIM   -MARKET RELEASE 1.5 (7-10-92)
(C)-COPYRIGHT ADVANCED MICRO DEVICES INC., 1992
PALASM SIMULATION SELECTIVE TRACE LISTING

Title : PAL Implementation of Fig. 6.33    Author : Ismail & Rooney
Pattern  : 3BIT.PDS (design file)  Company : University of Dayton
Revision : A                       Date     : 11/05/92


PAL16R4
Page : 1


     C  C  C  C  C  C  C  C  C
CP  HHLHHLHHLHHLHHLHHLHHLHHLHHL
X0  HHHHLLLLLLLLLLLHHHHHHHLLLLLLLLLL
X1  HLLLLLLLHHHHHHHHHHLLLLLLLHHHHH
X2  HLLLLLLLLLLHHHLLLHHHLLLLLLLHH
```

From the trace of the outputs of the counter $X2$, $X1$, and $X0$, the following sequence can be obtained:

$$7, 1, 0, 2, 6, 3, 5, 0, 2,$$

verifying the proper operation of the circuit because when the counter enters state 0 it follows the sequence, 2, 6, 3, 5, ...

In both the sequential logic circuits that have been implemented so far, the SIMULATION segment contained multiple *CLOCKF* statements that were used to apply clock pulses to the clock inputs of the counters. In fact typical simulations require many more such statments to completely simulate a design. As a convenience, PALASM4 allows the designer to construct a loop with a single *CLOCKF* statement. For example, instead of including eight *CLOCKF* statements in the SIMULATION segment, the following loop would produce the same results:

```
FOR  K := 1 TO 8 DO
        BEGIN
            CLOCKF CP
        END
```

In fact any simulation command placed between the *BEGIN* and *END* keywords would be executed eight times.

The PDS file shown in Figure 11.40 illustrates how this feature can be used to simulate the operation of a 2-bit synchronous up/down counter. This PDS file implements the up/down counter shown in Figure 6.40. The counter has two outputs, $X0$ and $X1$ that have been assigned to pins 21 and 20 of a PAL20V8, respectively, and a clock input that has been assigned to pin 1. The counter's direction select input S is assigned to pin 2 of the PAL device; when $S = 1$ the counter counts *UP* and when $S = 0$, the counter counts *DOWN*. The EQUATIONS segment of the PDS file contains Equations 6.7 and 6.8 in PALASM4 syntax.

We simulate the operation of the counter in the SIMULATION segment by first initiating a trace of the clock, CP, the select line, S, and the outputs of the counter, $X1$ and $X0$. The SETF command is then used to initialize S to a logic 1 for an *UP* count.

Figure 11.40

PDS file for a 2-bit up/down counter.

```
;PALASM Design Description

;------------------------------- Declaration Segment ------------
TITLE     PAL Implementation of an UP/DOWN Counter
PATTERN   updwnB.pds (design file)
REVISION  B
AUTHOR    Ismail & Rooney
COMPANY   University of Dayton
DATE      10/12/92

CHIP   updwn  PAL20V8

;------------------------------------ PIN Declarations --------------
PIN   1          CP            ;CLOCK INPUT
PIN   2          S             ;SELECT DIRECTION  1=UP  0=DOWN
PIN   3          NC
PIN   4          NC
PIN   5          NC
PIN   6          NC
PIN   7          NC
PIN   8          NC
PIN   9          NC
PIN   10         NC
PIN   11         NC
PIN   12         GND
;
PIN   13         NC
PIN   14         NC
PIN   15         NC
PIN   16         NC
PIN   17         NC
PIN   18         NC
PIN   19         NC
PIN   20         X1            ;OUTPUT
PIN   21         X0            ;OUTPUT
PIN   22         NC
PIN   23         NC
PIN   24         VCC

;------------------------------------ Boolean Equation Segment ------
EQUATIONS

X0 := /X0                                                           ← Equation 6-7

X1 := /X1 * /X0 * /S  +  /X1 * X0 * S  +  X1 * X0 * /S  +  X1 * /X0 * S  ← Equation 6-8

;------------------------------------ Simulation Segment ------------
SIMULATION

TRACE_ON CP S X0 X1

SETF S                    ← set S=1 (up count)

FOR J := 1 TO 8 DO
  BEGIN
     CLOCKF CP            } Apply 8 clock pulses
  END

SETF /S                   ← set S=0 (down count)

FOR J := 1 TO 8 DO
  BEGIN
     CLOCKF CP            } Apply 8 clock pulses
  END

TRACE_OFF
```

Eight clock pulses are then applied to the *CP* input to determine if the counter goes through all its states in the proper sequence and then recycles. Next the *SETF* command is used to initialize *S* to a logic 0. Again eight clock pulses are applied to *CP* to verify that the counter works properly during the *DOWN* counting sequence.

The trace listing and the waveform display for the simulation are shown in Figures 11.41 and 11.42, respectively. From the annotated trace listing when *S* is high the counter follows the sequence — 0, 1, 2, 3, . . . (*UP* count) and when *S* is low the counter follows the sequence — 3, 2, 1, 0, . . . (*DOWN* count) thus verifying the design.

State Machine Design

In all the counter implementations seen so far, the designer must specify the design equations (excitation equations) of the counter to be designed. This involves a considerable amount of work because the procedures of synthesis covered in Chapter 6 would have to be used to obtain these design equations. PALASM4 provides the designer with a much simpler alternative for the design of sequential logic circuits —

Figure 11.41

Trace listing for a 2-bit up/down counter.

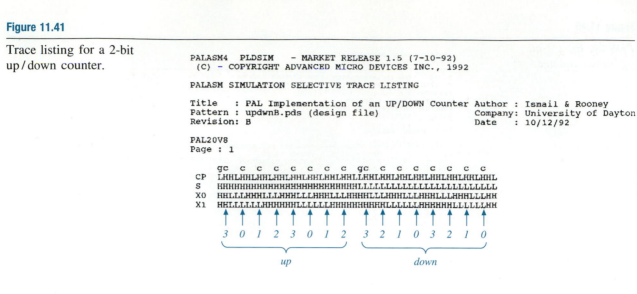

```
PALASM4  PLDSIM    - MARKET RELEASE 1.5 (7-10-92)
 (C) - COPYRIGHT ADVANCED MICRO DEVICES INC., 1992

PALASM SIMULATION SELECTIVE TRACE LISTING

Title   : PAL Implementation of an UP/DOWN Counter  Author : Ismail & Rooney
Pattern : updwnB.pds (design file)                  Company: University of Dayton
Revision: B                                         Date   : 10/12/92

PAL20V8
Page : 1

      gc c  c  c  c  c  c  c  gc c  c  c  c  c  c  c
 CP   LHHLHHLHHLHHLHHLHHLHHLHHLLHHLHHLHHLHHLHHLHHLHHL
 S    HHHHHHHHHHHHHHHHHHHHHHHHHHLLLLLLLLLLLLLLLLLLLLLL
 X0   HHLLLHHHLLLLHHHLLLABHHLLLHHHHLLLHHHLLLHHHLLLHHHLLLLHH
 X1   HHLLLLLLHHHHHHLLLLLLHHHHHHHHHHLLLLLLHHHHHHLLLLLLHH
```

```
      3 0 1 2 3 0 1 2   3 2 1 0 3 2 1 0
           up                down
```

Figure 11.42

Waveform display for a 2-bit up/down counter.

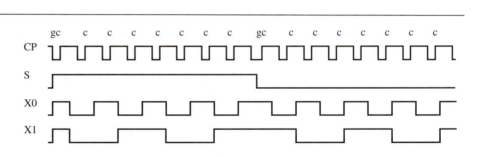

state machine design, where the operation of the circuit is described to PALASM4 in terms of a state diagram (or state table) rather than the excitation equations.

To illustrate the procedures for state machine design, implement the 2-bit *UP/DOWN* counter (*see* Figs. 11.40 through 11.42) by only specifying the state transitions rather than the design equations. The result of this design will be the same. The annotated PDS file for the state machine design is shown in Figure 11.43.

By comparing Figure 11.43 with Figure 11.40, the only difference between the two is that the PDS file in Figure 11.43 does not contain a Boolean equation segment but contains a *state segment* instead. The state segment begins with the keyword ''STATE''. Following this keyword is a statement that identifies the type of state machine to be designed—MEALY or MOORE. A Mealy machine, or a machine with ''Mealy outputs'' has outputs that depend on the present state of the circuit as well as on some external inputs. A Moore machine, or a machine with ''Moore outputs'' has outputs that are only dependent on the present state of the circuit and are independent of any external inputs. Typically, circuits whose outputs are the direct outputs of the flip-flops are Moore circuits, whereas circuits whose outputs are ''gated'' with flip-flop outputs and other external inputs are Mealy circuits. The counters examined in this book are all categorized as Moore circuits because the outputs of the counters are the outputs of the flip-flops. The statement MOORE_MACHINE in the PDS file identifies the circuit as a Moore circuit.

The state assignments follow the MOORE_MACHINE statement. Here each state, S0, S1, S2, and S3 are assigned to the flip-flop outputs, 00, 01, 10, and 11, respectively. Next the designer must list the state transitions required to implement

Figure 11.43

PDS file for a 2-bit up/down counter.

```
;PALASM Design Description

;-------------------------------- Declaration Segment ------------
TITLE     2-bit UP/DOWN Counter (State Machine)
PATTERN   updwnC.pds (design file)
REVISION C
AUTHOR    Ismail & Rooney
COMPANY   University of Dayton
DATE      11/05/92

CHIP  updwn  PAL20V8

;--------------------------------- PIN Declarations ---------------
PIN  1          CP          ;CLOCK INPUT
PIN  2          S           ;SELECT DIRECTION 1=UP  0=DOWN
PIN  3          NC
PIN  4          NC
PIN  5          NC
PIN  6          NC
PIN  7          NC
PIN  8          NC
PIN  9          NC
PIN  10         NC
PIN  11         NC
PIN  12         GND
;
PIN  13         NC
PIN  14         NC
PIN  15         NC
PIN  16         NC
PIN  17         NC
PIN  18         NC
PIN  19         NC
PIN  20         X1              ;OUTPUT
PIN  21         X0              ;OUTPUT
PIN  22         NC
PIN  23         NC
PIN  24         VCC
```

Same as figure 11-40

```
;-------------------------------State Segment------------------------
STATE
```
identifies beginning of state segment
```
MOORE_MACHINE
```
Moore type circuit
```
;STATE ASSIGNMENTS

S0 =  /X1 * /X0
S1 =  /X1 *  X0
S2 =   X1 * /X0
S3 =   X1 *  X0
```
Assign each state to outputs of counter S0=00, S1=01, S2=10, S3=11

```
; STATE TRANSITIONS

S0 := UP    -> S1
   + DOWN -> S3
S1 := UP    -> S2
   + DOWN -> S0
S2 := UP    -> S3
   + DOWN -> S1
S3 := UP    -> S0
   + DOWN -> S2
```

```
; INPUT CONDITIONS

CONDITIONS

UP = S
DOWN = /S
```
If S = 1 Up count
If S = 0 Down count

```
;-------------------------------- Simulation Segment ------------
SIMULATION

TRACE_ON CP S X0 X1

SETF S

FOR J := 1 TO 8 DO
   BEGIN
      CLOCKF CP
   END

SETF /S

FOR J := 1 TO 8 DO
   BEGIN
      CLOCKF CP
   END

TRACE_OFF
```
Same as figure 11-40

Figure 11.44

PDS file for a BCD
up/down counter.

```
;PALASM Design Description

;--------------------------------------- Declaration Segment ------------
TITLE      BCD up/down counter (State Machine Design)
PATTERN    bcd.pds (design file)
REVISION A
AUTHOR     Ismail & Rooney
COMPANY    University of Dayton
DATE       10/12/92

CHIP  bcd  PAL16V8

;--------------------------------- PIN Declarations ---------------
PIN  1        CP            ;CLOCK
PIN  2        UPDOWN        ;DIRECTION CONTROL INPUT
PIN  3        NC
PIN  4        NC
PIN  5        NC
PIN  6        NC
PIN  7        NC
PIN  8        NC
PIN  9        NC
PIN  10       GND
;
PIN  11       NC
PIN  12       NC
PIN  13       NC
PIN  14       X0            ;BCD outputs
PIN  15       X1
PIN  16       X2
PIN  17       X3
PIN  18       NC
PIN  19       NC
PIN  20       VCC

;-------------------------------State Segment-------------------------
STATE

MOORE_MACHINE

;STATE ASSIGNMENTS

S0  =   /X3 * /X2 * /X1 * /X0
S1  =   /X3 * /X2 * /X1 *  X0
S2  =   /X3 * /X2 *  X1 * /X0
S3  =   /X3 * /X2 *  X1 *  X0
S4  =   /X3 *  X2 * /X1 * /X0
S5  =   /X3 *  X2 * /X1 *  X0
S6  =   /X3 *  X2 *  X1 * /X0
S7  =   /X3 *  X2 *  X1 *  X0
S8  =    X3 * /X2 * /X1 * /X0
S9  =    X3 * /X2 * /X1 *  X0
```

Assign states S0 through S9 to outputs 0000 through 1001 respectively

```
; STATE TRANSITIONS

S0  :=  UP    -> S1
      + DOWN -> S9
S1  :=  UP    -> S2
      + DOWN -> S0
S2  :=  UP    -> S3
      + DOWN -> S1
S3  :=  UP    -> S4
      + DOWN -> S2
S4  :=  UP    -> S5
      + DOWN -> S3
S5  :=  UP    -> S6
      + DOWN -> S4
S6  :=  UP    -> S7
      + DOWN -> S5
S7  :=  UP    -> S8
      + DOWN -> S6
S8  :=  UP    -> S9
      + DOWN -> S7
S9  :=  UP    -> S0
      + DOWN -> S8
```

```
; INPUT CONDITIONS

CONDITIONS

UP = UPDOWN
DOWN = /UPDOWN

;--------------------------------------- Simulation Segment ------------
SIMULATION

TRACE_ON CP UPDOWN X0 X1 X2 X3

SETF UPDOWN

FOR J := 1 TO 15 DO
    BEGIN
      CLOCKF CP
    END

TRACE_OFF

TRACE_ON CP UPDOWN X0 X1 X2 X3

SETF /UPDOWN

FOR J := 1 TO 15 DO
    BEGIN
      CLOCKF CP
    END

TRACE_OFF
```

the design. This information can be obtained from a state diagram (or state table) of the counter. For example the first transition listed,

$$S0 := UP \quad -> S1$$
$$+ DOWN -> S3$$

can be interpreted as follows. If the present state is S0, and the condition UP is true, the next state is S1. If the present state is S0 and the condition DOWN is true, the next state is S3. After the transition for each state is defined in a similar manner, the designer needs to define the ''conditions'' specified in the state transitions. Following the CONDITIONS keyword, the following two statements are listed:

$$UP = S$$
$$DOWN = / S$$

These two statements are interpreted by PALASM4 as follows. If $S = 1$, the condition UP is true, and if $S = 0$, the DOWN condition is true.

After compiling the PDS file, PALASM4 will generate the appropriate circuit to implement the specified state transitions. In fact the operation of the circuits produced by the PDS files in Figures 11.40 and 11.43 will be identical. After simulation the trace listing and waveform display for the PDS file in Figure 11.43 will be the same as that shown in Figures 11.41 and 11.42, respectively.

Example 11.12

Design a BCD up/down counter on a PAL16V8. The BCD counter will have a direction control input *UPDOWN* that will cause the counter to count *UP* if *UPDOWN* = 1 and count down if *UPDOWN* = 0.

Solution

The PAL16V8 is a *V*-type device that has 16 programmable inputs at pins 1 through 9, and 11 through 19. Pins 12 through 19 can also serve as either eight registered or combinational outputs whereas pin 1 functions as the clock input for registered applications. The PDS file for the design of the BCD counter is shown in Figure 11.44. Pin 1 has been assigned to *CP* the clock input of the BCD counter and pin 2 to *UPDOWN,* the direction control input of the counter. Pins 14 through 17 have been assigned to the counter outputs, *X0, X1, X2,* and *X3*, respectively.

The state segment of the PDS file begins with the STATE and MOORE_MACHINE keywords followed by the state assignments. The state assignments assign the states *S0* through *S9* to the counter outputs 0000 through 1001, respectively. The state transitions then define the transition each state must make if *UPDOWN* = 1 or if *UPDOWN* = 0 as specified in the input conditions section of the state segment.

For this example, two separate traces are initiated—one with *UPDOWN* = 1 and the other with *UPDOWN* = 0. This will make the simulator produce a more compact version of the trace listing and waveform display as will be seen shortly. First, a trace of the clock pulse *CP, UPDOWN,* and the counter outputs (*X0* to *X3*) is initiated. *UPDOWN* is then set to a logic 1 and 15 clock pulses are applied to *CP*. The trace is then turned off. A new trace is then initiated with the same pins specified before. This time *UPDOWN* is set to a logic 0 and 15 clock pulses are applied to *CP*. The trace is then turned off again to complete the simulation session.

The annotated trace listing is shown in Figure 11.45. The first trace shows the counting sequence when *UPDOWN* is high. We can see by examining the outputs of the counter that the counting sequence is a BCD up count. Notice that the counter starts off in an undefined state (outputs are all high) and makes a transition to *S8* which is in the BCD counting sequence. Recall that a BCD counter counts through 10 of 16 possible states. The remaining six states are undefined and if the counter should enter into one of these six states PALASM4 will ensure that the counter makes a transition into the defined counting sequence. The second trace

Figure 11.45

Trace listing for BCD
up/down counter.

```
PALASM4  PLDSIM   - MARKET RELEASE 1.5 (7-10-92)
(C) - COPYRIGHT ADVANCED MICRO DEVICES INC., 1992

PALASM SIMULATION SELECTIVE TRACE LISTING

Title   : BCD up/down counter (State Machine)   Author : Ismail & Rooney
Pattern : bcd.pds (design file)                 Company: University of Dayton
Revision: A                                     Date   : 10/12/92

PAL16V8
Page : 1

        gc c  c  c  c  c  c  c  c  c  c  c  c  c  c
CP      LHHLHHLHHLHHLHHLHHLHHLHHLHHLHHLHHLHHLHHLHHL    ◄── UPDOWN = 1
UPDOWN  HHHHHHHHHHHHHHHHHHHHHHHHHHHHHHHHHHHHHHHHHHH        Up count
X0      HHLLLHHHLLLHHHLLLHHHLLLHHHLLLHHHLLLHHHLLLHHHLL
X1      HHLLLLLLLLLLLLLHHHHHHLLLLLLLHHHHHLLLLLLLLLLLLHH
X2      HHLLLLLLLLLLLLLLHHHHHHHHHHHLLLLLLLLLLLLLLLLL
X3      HHHHHHHHLLLLLLLLLLLLLLLLLLLLLLLLHHHHHHLLLLLLLLL

            ↑  ↑  ↑  ↑  ↑  ↑  ↑  ↑  ↑  ↑  ↑  ↑  ↑  ↑

            ? 8  9  0  1  2  3  4  5  6  7  8  9  0  1


PAL16V8
Page : 2

        gc c  c  c  c  c  c  c  c  c  c  c  c  c  c
CP      LHHLHHLHHLHHLHHLHHLHHLHHLHHLHHLHHLHHLHHLHHL    ◄── UPDOWN = 0
UPDOWN  LLLLLLLLLLLLLLLLLLLLLLLLLLLLLLLLLLLLLLLLLLL        Down count
X0      LLHHHLLLHHHLLLHHHLLLHHHLLLHHHLLLHHHLLLHHHLLLHH
X1      HHLLLLLLLLLLLLLHHHHHHLLLLLLLHHHHHHLLLLLLLLLLLLLLHH
X2      LLLLLLLLLLLLLLLHHHHHHHHHHHLLLLLLLLLLLLLLLLLLHH
X3      LLLLLLLLHHHHHHLLLLLLLLLLLLLLLLLLLLLLHHHHHHLL

            ↑  ↑  ↑  ↑  ↑  ↑  ↑  ↑  ↑  ↑  ↑  ↑  ↑  ↑

            2  1  0  9  8  7  6  5  4  3  2  1  0  9  8
```

continues where the first left off. *UPDOWN* is now low and the counting sequence is a BCD
down count. The waveform display representing this data in the form of a timing diagram is
shown in Figure 11.46.

State machine designs using PALASM4 can also be done for simple counters
that unconditionally count from one state to another. For example to implement the
counter designed in Example 11.11 using state machine design instead of design
from excitation equations, the Boolean equation segment of the PDS file shown in
Example 11.11 could be replaced with the following state segment:

```
STATE
MOORE_MACHINE
DEFAULT_BRANCH S0

;State assignments
S0 = / X2 * / X1 * / X0
S2 = / X2 *   X1 * / X0
S3 = / X2 *   X1 *   X0
S5 =   X2 * / X1 *   X0
S6 =   X2 *   X1 * / X0
```

```
;State Transitions
S0 := VCC - > S2
S2 := VCC - > S6
S6 := VCC - > S3
S3 := VCC - > S5
S5 := VCC - > S0
```

The state segment shown above defines the five states of the counter, *S*0, *S*2, *S*3, *S*5, and *S*6, and assigns them to the outputs, 000, 010, 011, 101, and 110, respectively. Because the counter is to count in the sequence — 6, 3, 5, 0, 6, 3, . . ., without any input conditions, the state transitions have no conditions associated with them. For example, the state transition,

$$S0 := VCC -> S2$$

indicates that if the present state is *S*0 the next state should be *S*2. Because the syntax of this statement requires a condition, V_{CC} (logic 1) was used, which is always true. Therefore, the transition will occur unconditionally.

If the PDS file in Example 11.11 is compiled and simulated with this state segment replacing the Boolean equations segment, the trace listing will be identical

Figure 11.46

Waveform display for BCD up/down counter.

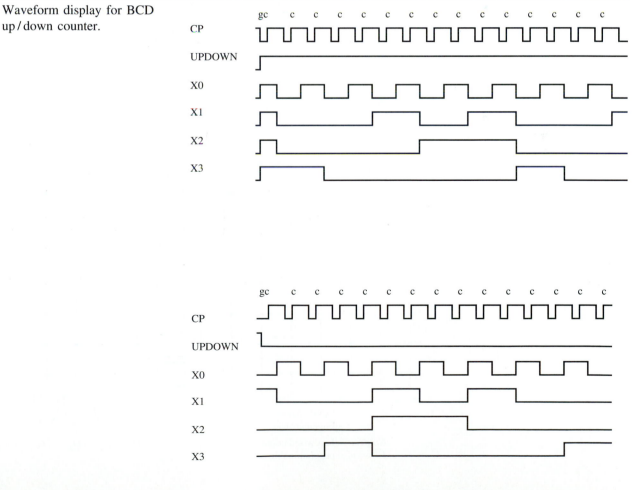

to the one shown in Example 11.11. This indicates that a PAL device programmed from either PDS file will function the same. State machine design provides the designer with a much simpler means of designing sequential logic circuits because it does not involve the derivation of excitation equations.

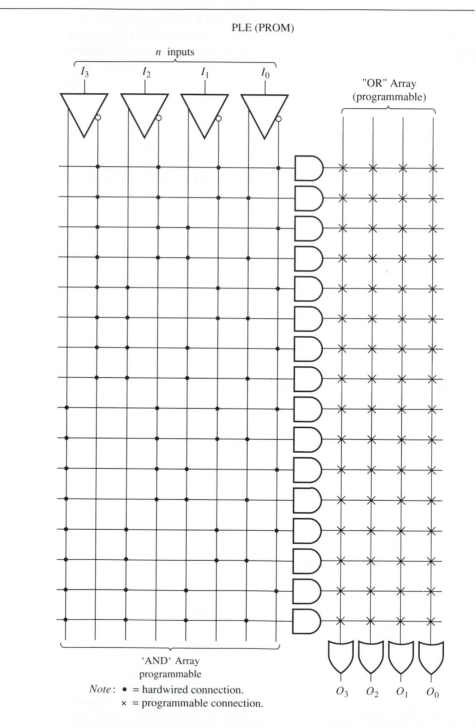

PLE (PROM)

n inputs

I_3 I_2 I_1 I_0

"OR" Array (programmable)

'AND' Array programmable

Note: • = hardwired connection.
× = programmable connection.

O_3 O_2 O_1 O_0

Review Questions

1. How are equations in a PDS file for a sequential logic design different from the equations in a PDS file for a combinational logic design?

2. Which simulation command is used to apply clock pulses to the clock input pin of a PAL device?

3. What does a waveform display produced by the simulator contain?

4. What is the difference between sequential logic design using equations and sequential logic design using a state machine design?

11.7
Programmable logic array (PLA)

A programmable logic array is similar to a PAL, except that for a PLA both the *AND* gate and *OR* gate arrays are programmable. The PLD of Figure 11.2a is a PLA. Having the ability to program both arrays yields PLDs with the greatest flexibility. The price paid for this increased flexibility is a slight loss in performance speed (an increase in propagation delay time — see Chapter 12) and a decrease in the number of logic gates on the silicon wafer due to the increased number of fuse-links. Figure 11.47 shows a logic architectural diagram of a PLA.

Comparing the PLA of Figure 11.47 with that of Figure 11.12, the PLA is not limited to four product terms per output, but rather can be programmed to have as many product terms as there are *AND* gates in the *AND* gate array. Comparison of the PLE of Figure 11.9 and the PAL of Figure 11.12 with the PLA of Figure 11.47 shows that a PLA has the programming capability of both PLE and PAL logic architectures. As with PAL, the number of *AND* gates in the *AND* gate array does not have to be equal to 2^n (where n equals the number of inputs) unless the designer of the PLA intends that the user have these *AND* gates available for *OR*ing all minterms.

Example 11.13

Using the PLA of Figure 11.47, implement the Boolean design equations

$$Y = \overline{A}\,\overline{C} + \overline{A}BD + \overline{B}C + A\overline{B}C + AC\overline{D}$$
$$W = \overline{A}\,\overline{C} + AB\overline{C} + \overline{A}BD + \overline{B}C$$

Solution

Program the product lines by taking each nonrepeated product term in the Boolean design equation and connecting the indicated variables to that product line, just as was done for the PAL. The first product line of Figure 11.48 is connected to variables $\overline{A}$ and $\overline{C}$; its product $\overline{A}\,\overline{C}$ is the first product term of the equation for Y. For convenience, when programming the *OR* gate array, the authors identified the product $\overline{A}\,\overline{C}$ at the product-term line (the output of the *AND* gate). The next product line was programmed for the second product $(\overline{A}BD)$ of the equation for Y. The product-term line for that *AND* gate was identified as $\overline{A}BD$. This procedure is repeated for all products in the Boolean equations for Y and W. Those *AND* gates that are not needed are not programmed. Notice that equations Y and W have two products in common.

Once the *AND* gate array has been programmed, the *OR* gate can be programmed by putting an $\times$ at the intersection of each horizontal product-term line in the Boolean equation and the vertical input line for the *OR* gate being programmed. The vertical input line for *OR* gate Y is connected to the five products indicated by the Boolean design equation. The *OR* gate with outputs O_1 and O_0 are not being used.

Figure 11.48

A programmed PLA for
Example 11.13.

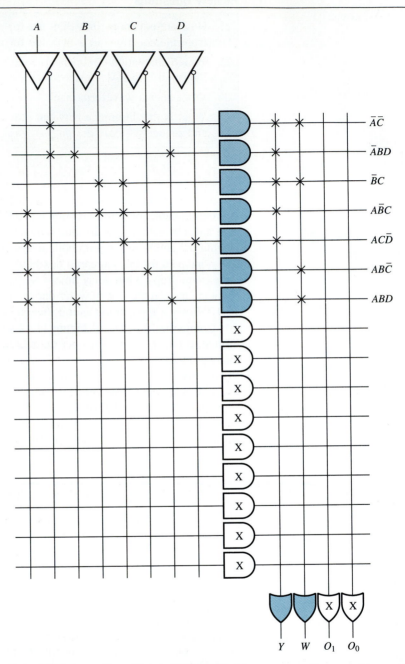

The nomenclature for a PLA is indicated in Figure 11.11. Hence, a PLA with 16
inputs and 8 registered outputs would be identified as a PLA16R8.

Review Questions

1. How are PLAs and PLEs alike, and how do they differ?
2. How are PLAs and PALs alike, and how do they differ?
3. What are the major advantages and disadvantages of PLAs compared to PLEs
   and PALs?

11.8
Troubleshooting

PLDs, whether PLEs, PLAs, or PALs, are programmable combinational or sequential logic circuits contained within a single semiconductor IC and are therefore viewed as a system function-block rather than individual logic circuits for troubleshooting purposes. To troubleshoot a PLD for a fault is a very straightforward process. By means of a schematic of the programmed PLD and/or a truth table of the input and output variables, the diagnostician knows the correct output responses for any combination of logic levels for the input variables. With this information the diagnostician can devise a test to evaluate the PLD either statically or dynamically. That is, either static or dynamic logic level generators are used to generate logic levels for the input variables and either static or dynamic logic level detectors are used to detect the resultant output logic levels. If an output logic level is not correct, the PLD has failed and must be replaced. Of course, to verify failure this output must be isolated from the input(s) it is driving to ensure that it is not the input of the circuit being driven that has failed, for example, an input that has become a permanent short to ground due to an electronic failure.

Review Questions

1. Why must PLDs be viewed as function-block diagrams relative to troubleshooting?
2. If a fault is found with a PLD, whether the fault is with an input or output, what is the remedy?
3. What is the essence of any PLD static or dynamic test?

Summary

- Programmable logic devices are structured primarily of *AND* and *OR* gate arrays. The inputs to either the *AND* gates, the *OR* gates, or both are programmable via a fuse-link. These programmable fuse-links afford the PLD a flexibility that ordinary combinational logic circuits do not have. A 20- or 24-pin PLD can often replace up to six ICs containing typical logic gates.

- The type of PLD is determined by which logic gate array is programmable. If the logic of the *AND* gate array is fixed (not programmable) and the *OR* gate array is programmable, the PLD is a PLE. If the logic of the *AND* gate array is programmable and that of the *OR* gate array is fixed, the PLD is a PAL device. When the logic of both arrays is programmable, the PLD is a PLA. PLEs are the least flexible of the PLDs, PLAs are the most flexible, and PALs are in-between.

- Programmability extracts a price of increased propagation delay time and reduced logic density on the silicon wafer. PAL is a convenient compromise between flexibility and performance. Using a PAL is seldom a sacrifice, because most logic designs do not require the program flexibility of a PLA.

- PLDs have a variety of output combinations. There are buffered outputs that are noninverting (active-high [*H*]) or inverting (active-low [*L*]). Some PLDs have registers as their outputs (*D*-type flip-flops), with or without feedback. Lastly, some PLDs have programmable I/O

pins. The user must decide which PLD best accommodates the design.

- PLD nomenclature readily identifies a PLD type. The first symbols usually identify the type of device (PLE, PAL, or PLA), and in some instances the manufacturer as well. The number following the type identification indicates the number of inputs, the next symbol specifies the output configuration, and the next number indicates the number of outputs available.

- To symbolically program a PLD, an *x-y* coordinate system is used. A × is placed at any horizontal and vertical intersection that is to be connected to indicate that the fuse-link is to remain intact.

- One of the software packages that is used to program PLDs is known as PALASM4. PALASM4 allows the designer to specify the logic equations of the circuit(s) to be designed and automatically generates a fuse-map identifying the fuses to be blown. PALASM4 also provides the designer the facility to test a design through the use of PLDSIM, a simulator.

- One of the outputs produced by PALASM4 is a file containing the fuse-map information in a standard format known as the JEDEC format. This file is used by the PAL programming unit to blow the appropriate fuses in the PAL device.

- The type of PAL device to be programmed, the logic equations to be implemented, and the simulation commands are specified in a file known as the PDS (PAL design) file. This file is compiled by the PALASM4 compiler into fuse-map information for the PAL programming unit.

- Besides using simulation commands to simulate the operation of the circuit being designed, PALASM4 also generates test vectors in the JEDEC file that can be used by the PAL programming unit to test the operation of the PAL device after programming.

- The PDS files for sequential logic designs are similar to those for combinational logic designs except that logic equations are described in terms of the outputs of the flip-flops and the ":=" assignment operator is used to indicate that the outputs are registered. The logic

equations implemented for a sequential logic design are the excitation equations.

- State machine design allows the designer to describe the operation of a sequential logic circuit in terms of state transitions rather than excitation equations. State machine design requires less work on the part of the designer since the excitation equations need not be derived.

- In troubleshooting, the PLD is viewed as an IC that is either properly functioning or has failed and must be replaced, rather than as individual logic circuits.

- To troubleshoot a PLD, the logic levels of the input variables are applied and the output variables are tested for correct responses. An incorrect response constitutes a failure of the PLD.

Problems

Section 11.2   The programmable fuse-link

1. Explain the concept of PLD programmability and explain what a fuse-link is.

2. Use the PLD of Figure 11.2a to implement the Boolean design equations

$$X = AB + BC$$
$$Y = AC$$

Show the fuse-links that are intact and those that are blown in the logic diagram of Figure 11.2a.

3. Use the PLD of Figure 11.2a to program the Boolean design equation $X = \overline{A}B + AB$. Make a drawing of the logic circuit.

4. Modify the PLD of Figure 11.2a so that the Boolean design equation

$$Y = A\overline{B} + \overline{A}B$$

can be implemented. Show the logic diagram of the modified circuit as it appears when programmed.

5. Show the logic circuit, similar to the logic circuit of Figure 11.2a, that is capable of implementing a design with three independent variables (inputs) and four dependent variables (outputs). The logic circuit is to accommodate the independent variables and their complements.

6. Use the logic circuit of Problem 5 to program the Boolean equations.

$$X = \Sigma_m 0, 3, 5, 6$$
$$Y = \Sigma_m 1, 3, 5, 6, 7$$
$$Z = \Sigma_m 0, 6, 7$$

The independent variables are A, B, and C. For the truth table assign A to be the MSB and C to be the LSB. Show the resultant programmed logic circuit for the simplified Boolean equations.

7. To accommodate all possible minterms, how many *AND* gates must there be in a PLD?

Section 11.3   Programmable logic elements (PLE) and PLD symbolic notation

8. What are the conceptual differences between the PLE of Figure 11.4 and the PLD of Figure 11.2a?

9. If a PLE has six input pins for inputting independent variables how many *AND* gates are there?

10. What relationship is there between the number of *AND* gates and *OR* gates for a PLE? Explain your answer.

11. How many inputs does each *OR* gate of a PLE have? Explain your answer.

12. Use the PLE of Figure 11.4 to implement the Boolean design equations

$$A = \Sigma_m 0, 3$$
$$B = \Sigma_m 1, 2$$
$$D = m_0$$
$$E = \Sigma_m 0, 2, 3$$

Show the resultant programmed logic diagram, which should be similar to Figure 11.5.

13. Repeat Problem 2 using the PLD symbolic notation for indicating a fuse-link that is to remain intact and single lines to indicate the inputs to the *AND* and *OR* gate arrays, as in Figures 11.6 and 11.7.

14. Use a PLE as a look-up table to convert binary hexadecimal values 0 through F into their equivalent ASCII binary codes. Design the PLE required. Show the programmed version of the PLE, which will be similar to the programmed PLE of Figure 11.7.

15. Modify the outputs of Problem 14 so that they may be connected to the data bus of a microprocessor-based system.

16. Modify the outputs of Problem 14 so that their output logic states can be latched (remembered) and then read at some later time.

17. What is the major restriction to the flexibility of a PLE?

18. Can the PLE of Figure 11.9 be used to implement the simplified version of the Boolean equation

$$A = \Sigma_m 1, 3, 5, 7, 10, 11, 12, 13$$
$$\text{where } A(X, Y, W, Z)?$$

Explain your answer.

19. What would be the PLD product identification for:
    a. The PLE of Problem 15?
    b. The PLE of Problem 16?

Section 11.4 Programmable array logic (PAL)

20. How does a PAL differ from a PLE?

21. Why is a PAL more flexible than a PLE?

22. What relationship exists between the number of input variables and logic gates in the *AND* and *OR* gate array of a PAL?

23. Use the PAL of Figure 11.12 to implement the simplified Boolean equation of Problem 18.

24. How many inputs and outputs are required for a PAL to be used to implement the Boolean design equations given below?
    a. $Y(A, B, C)$
       $Z(A, B, C)$
    b. $A(X, Y, W, Z)$ $D(X, Y, W, Z)$
       $B(X, Y, W, Z)$ $E(X, Y, W, Z)$
       $C(X, Y, W, Z)$ $H(X, Y, W, Z)$

25. Construct a PAL that will accommodate the implementation of the Boolean design equations given below.

$$W = \Sigma_m 0, 1, 2, 7$$
$$X = \Sigma_m 3, 5, 6, 7$$
$$Y = \Sigma_m 0, 3, 4, 6$$
$$Z = \Sigma_m 0, 3$$

where the independent variables are *A*, *B*, and *C* with *A* = MSB and *C* = LSB of the truth table.

26. How would you modify the PAL of Problem 25 if it were to be interfaced with an MPU-based system?

27. Explain how an output pin can be programmed to function as an input pin. Use the basic logic array cell of Figure 11.14a as a model for your explanation.

28. Explain how an *X-OR* gate can be programmed to determine if a logic state is inverted or not. Use Figure 11.14c as a model for your explanation.

29. Repeat Problem 25, but use the PAL of Figure 11.15 to implement the design.

30. Repeat Problem 25, but use the PAL of Figure 11.17 to implement the design.

31. Design a three-stage down-counter and implement the design using a PAL16R4 (see Fig. 11.17).

Section 11.5 PAL programming and simulation — combinational logic

32. Write the following logic equations in PALASM4 syntax:
    a. $X = A\bar{B} + \bar{A}BC + D$
    b. $P = X(Y + \bar{Z})$
    c. $T = \Sigma_m 3,5,6$ (independent variables: *P*, *Q*, *R*)
    d. $Z = \Pi_m 2,6,7$ (independent variables: *A*, *B*, *C*)

33. A partial listing of the fuse-plot file produced by PALASM4 is shown below. Assuming that all fuses not shown are left intact, draw the PLD circuit corresponding to this fuse plot.

```
PAL16L8
PROB33
                  11 1111 1111 2222 2222 2233
      0123 4567 8901 2345 6789 0123 4567 8901
 0 - - - - - - - - X - - - - - - - - - - - - - - - - - - - -
 1 - X X - - - - - - - - - - - - - - - - - - - - - - - -
 2 X - - X - - - - - - - - - - - - - - - - - - - - - -
 3 X - - - - X - - - - - - - - - - - - - - - - - - - -
 4 XXXX XXXX XXXX XXXX XXXX XXXX XXXX XXXX
 5 XXXX XXXX XXXX XXXX XXXX XXXX XXXX XXXX
 6 XXXX XXXX XXXX XXXX XXXX XXXX XXXX XXXX
 7 XXXX XXXX XXXX XXXX XXXX XXXX XXXX XXXX
     TOTAL FUSES BLOWN = 121
```

34. A partial listing of the fuse-plot file produced by PALASM4 is shown below. Assuming that all fuses not shown are left intact, draw the PLD circuit corresponding to this fuse plot.

```
PAL16L8
PROB34
                   11 1111 1111 2222 2222 2233
      0123 4567 8901 2345 6789 0123 4567 8901
 0  XXXX XXXX XXXX XXXX XXXX XXXX XXXX XXXX
 1  XXXX XXXX XXXX XXXX XXXX XXXX XXXX XXXX
 2  XXXX XXXX XXXX XXXX XXXX XXXX XXXX XXXX
 3  XXXX XXXX XXXX XXXX XXXX XXXX XXXX XXXX
 4  XXXX XXXX XXXX XXXX XXXX XXXX XXXX XXXX
 5  XXXX XXXX XXXX XXXX XXXX XXXX XXXX XXXX
 6  XXXX XXXX XXXX XXXX XXXX XXXX XXXX XXXX
 7  XXXX XXXX XXXX XXXX XXXX XXXX XXXX XXXX
 8  - - - - - - - - - - - - - - - - - - - - - - - - - - - -
 9  X - X - X - - - - - - - - - - - - - - - - - - - - - -
10 X - X - - - - - X - - - - - - - - - - - - - - - - -
11 XXXX XXXX XXXX XXXX XXXX XXXX XXXX XXXX
12 XXXX XXXX XXXX XXXX XXXX XXXX XXXX XXXX
13 XXXX XXXX XXXX XXXX XXXX XXXX XXXX XXXX
14 XXXX XXXX XXXX XXXX XXXX XXXX XXXX XXXX
15 XXXX XXXX XXXX XXXX XXXX XXXX XXXX XXXX
     TOTAL FUSES BLOWN = 90
```

35. Implement the equations given in Problem 25 on a PAL16L8 using PALASM4. Write the PDS file for the design. Examine the XPT file and compare it to the results obtained in Problem 29.

36. Simulate the design in Problem 35 using PALASM4 and verify that the equations are implemented correctly.

37. Implement the 2-bit magnitude comparator shown in Figure 4.28 on a PAL20V8. Create a PDS file for the design and then use simulation to verify that the circuit produces the correct outputs for all possible combinations of inputs.

38. Implement a 1-of-4 decoder with active-low outputs on a PAL20V8. The decoder should also have an active-low enable line that will disable the outputs of the decoder (all high) when at the logic 1 state. Create a PDS file for the design and then use simulation to verify that the circuit produces the correct outputs for all possible combinations of inputs.

Section 11.6   PAL programming and simulation — sequential logic

39. Write the following excitation equations in PALASM4 syntax. Use X to identify the output pins of the counter and assume that the counter outputs are the flip-flop outputs.
   a. $D_3 = \overline{Q_1} Q_0 + Q_1 Q_3$
   b. $D_0 = Q_0 \oplus (Q_1 \oplus Q_0)$
   c. $D_1 = Q_2 (Q_1 + \overline{Q_3})$
   d. $D_2 = \overline{Q_1}$

40. A partial listing of the fuse-plot file produced by PALASM4 is shown below. Assuming that all fuses not shown are left intact, draw the PLD circuit corresponding to this fuse plot.

```
PAL16R4
PROB40
                11 1111 1111 2222 2222 2233
    0123 4567 8901 2345 6789 0123 4567 8901
31 XXXX XXXX XXXX XXXX XXXX XXXX XXXX XXXX
32 - - - - - - - - - - - - - - - -X - - - - - - - - - - - -
33 - - - - - - - - - - - - - - - - - -X - - - - - - - -
34 XXXX XXXX XXXX XXXX XXXX XXXX XXXX XXXX
35 XXXX XXXX XXXX XXXX XXXX XXXX XXXX XXXX
36 XXXX XXXX XXXX XXXX XXXX XXXX XXXX XXXX
37 XXXX XXXX XXXX XXXX XXXX XXXX XXXX XXXX
38 XXXX XXXX XXXX XXXX XXXX XXXX XXXX XXXX
39 XXXX XXXX XXXX XXXX XXXX XXXX XXXX XXXX
40 - - - - - - - - - - - - - - -X- - - - - - - - - - - -
41 XXXX XXXX XXXX XXXX XXXX XXXX XXXX XXXX
42 XXXX XXXX XXXX XXXX XXXX XXXX XXXX XXXX
    TOTAL FUSES BLOWN = 93
```

41. Implement a 2-bit up counter on a PAL16R4. A diagram of such a counter is shown in Figure 6.25. Create a PDS file for the design and then use simulation to verify that the circuit produces the correct counting sequence.

42. Implement a counter to count in the following sequence: 7, 5, 3, 1, 7, 5, . . . on a PAL20V8. Implement the design using the excitation equations for the circuit. Create a PDS file for the design and then use simulation to verify that the circuit produces the correct counting sequence.

43. Repeat Problem 42 using state machine design.

44. Using state machine design, create a PDS file for a counter that will count up through all possible odd 3-bit combinations if the *ODD* input line is a logic 1 and count up through all possible even 3-bit combinations if the *ODD* input line is a logic 0. For example,
   If *ODD* = 1, counting sequence is 1,3,5,7,1,3,5,7,1
   . . .
   If *ODD* = 0, counting sequence is   0,2,4,6,0,2,4,6,0
   . . .
   Use a PAL20V8 and use simulation to verify your design.

Section 11.7   Programmable logic array (PLA)

45. What is the relationship between the number of independent variables and the number of *AND* and *OR* gates in the logic gate arrays of a PLA?

46. Use the PLA of Figure 11.47 to implement the following Boolean design equations:

   $G = \Sigma_m 0, 5, 10, 11, 14, 15$
   $H = \Sigma_m 4, 7, 10, 11, 14, 15$
   $J = \Sigma_m 1, 3, 9, 10, 11, 14, 15$
   $K = \Sigma_m 1, 3, 5, 10, 11, 14, 15$

   where the independent variables are A, B, C, and D (A = MSB).

Troubleshooting

Section 11.8   Troubleshooting

47. Devise a static test to troubleshoot the PLE of Figure 11.7.

48. How would you dynamically test the PLE of Figure 11.7?

49. Devise a dynamic test for the PAL of Figure 11.13 and state those input logic levels for which there should be a logic 1 at output Y.

50. Devise a static test for the PAL of Figure 11.19.

51. What logic levels would be input to test output Z of Figure 11.19?

52. Devise a test configuration for the PAL of Figure 11.20 and show the expected output for $\overline{Q}_1$ and $\overline{Q}_2$.

12

THE TECHNOLOGY OF LOGIC FAMILIES

OBJECTIVES

The objectives of this chapter are to:

• Explain the operating characteristics of bipolar and MOSFET electronic switches

• Describe the structuring of electronic switches that form logic gates

• Analyze the voltage and current requirements of logic families

• Explain loading rules for various logic families

• Investigate the propagation delay time characteristic of various logic families

• Examine the memory cell for different technologies

• Investigate some parameters that are specified in manufacturers' data sheets

• Apply the knowledge learned in this chapter to some representative designs

Various designs or technologies are used to fabricate highly complex digital circuits onto a piece of silicon inside an IC. The photograph shows a typical memory chip capable of storing approximately 16,000 bits of information (Reprinted by permission of Intel Corporation. © Intel Corporation 1989).

12.1 Introduction

Many different logic families may be used to implement a digital design. Logic families include *transistor-transistor logic* **(TTL),** *resistor-transistor logic* **(RTL),** *diode-transistor logic* **(DTL),** *emitter-coupled logic* **(ECL),** *N-channel MOS* **(NMOS), and** *complementary MOS* **(CMOS). This chapter will cover the more popular of these logic families, TTL, NMOS, and CMOS. Each of these logic families has characteristics that are better suited for a particular application than are other families. By studying the electronics of these logic families we shall understand why logic families have specific characteristics and have more insight into the loading characteristics introduced in Chapter 2. Among the characteristics of particular interest are power consumption, voltage levels, current requirements, loading rules, and propagation delay time.**

This chapter will investigate which logic families are directly compatible and which need interfacing circuitry. This will enable a mix and match of logic families when necessary or desirable.

The electronic principles of memory cells will also be examined and will be related to the design concepts developed in Chapters 8 and 9. In addition to understanding the whys of memory access time, we will be able to improve the memory designs and I/O port interfacing dealt with in Chapter 9 by considering loading criteria.

12.2
The electronic switch

Regardless of which logic family is being studied, the electronic switch is the key to the technology in question. An electronic switch is a transistor (bipolar or MOS) that is always in one of two states — on or off. All logic families are constructed of either bipolar switches (TTL, RTL, DTL, or ECL) or *metal oxide semi-conductor field effect transistor* (MOSFET) switches (PMOS, NMOS, or CMOS).

Figure 12.1a is an illustration of an NPN *bipolar junction transistor* (BJT) configured as an electronic switch. Figure 12.1b and c are equivalent circuits of that electronic switch for the on and off states. Figure 12.1a shows the transistor voltage parameters V_{BE} and V_{CE} as well as the input voltage V_I, which is the input voltage, and the output voltage V_O. Note that the transistor voltage parameter V_{CE} is the same as the output voltage V_O. The purpose in labeling the same voltage with two different symbols has to do with our prior knowledge and our present interest. Those who have studied BJTs in other courses, such as an electronic device course, are accustomed to using the collector-to-emitter voltage V_{CE} and are familiar with the characteristic curve of Figure 12.1a. However, in digital electronics the interest focuses on the electronic switch parameters V_I and V_O. Both notations will be used as we study the electronic switch.

To understand the BJT electronic switch we must know the cause-and-effect relationship of its parameters V_{BE}, I_B, I_C, and V_{CE}. To turn the NPN transistor of Figure 12.1a on, the base-to-emitter voltage V_{BE} must have the polarity shown and

Figure 12.1

An *NPN* bipolar electronic switch.

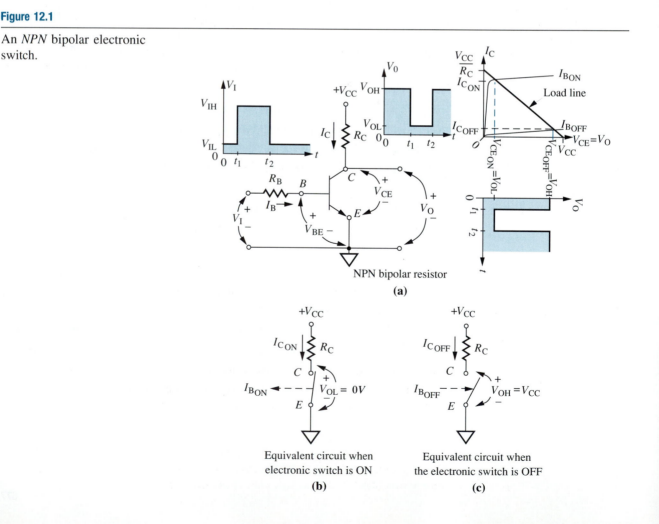

be greater than approximately 0.6 volts. When this condition occurs, the base-to-emitter junction of the transistor is forward biased (forward biasing turns on the base-to-emitter junction like a switch), and base current I_B flows through the base-to-emitter junction to ground. The electronic switch is designed so that R_B has a magnitude that allows a relatively large amount of base current to flow when the base-to-emitter junction is turned on. In the characteristic curves of the BJT in Figure 12.1a (I_C vs. V_{CE}) this base current has been labeled $I_{B\ ON}$. Whenever a base current that is equal to or greater than $I_{B\ ON}$ is present it turns the collector-to-emitter junction on ''hard'' so that its equivalent circuit is a closed switch and collector current $I_{C\ ON}$ flows, as depicted in Figure 12.1b. When the electronic switch is on, the output voltage is zero — the voltage level for a logic 0. As a result, when the electronic switch of Figure 12.1a is turned on, its output is driven to the logic low (0) level, symbolically designated V_{OL} (V for voltage, O for output, and L for logic low, as recalled from Section 2.9).

To relate the events and magnitudes of the electronic switch for the on condition to the characteristic curve (I_C vs. V_{CE}) of the BJT, refer to Figure 12.1a. When base current $I_{B\ ON}$ is flowing through the base-to-emitter junction, the point of intersection with the load line when projected vertically to the V_{CE} axis indicates that an output voltage of $V_{CE\ ON}$ is present at V_{CE}. The magnitude of $V_{CE\ ON}$ is the same as V_{OL}. The collector current is at a maximum and is identified as $I_{C\ ON}$, which also is shown in Figure 12.1b.

If the base-to-emitter junction of Figure 12.1a is not forward biased, the BJT is turned off and its collector-to-emitter equivalent circuit is that of an open circuit, as illustrated in Figure 12.1c. For the BJT to be turned off, V_{BE} must be less than approximately 0.6 volts. Under this condition the base-to-emitter junction is an equivalent open circuit, meaning that there is no base current. Without base current there is no collector current I_C, so there is an equivalent open circuit between the collector and the emitter, which is shown in the case Fig. 12.1c. With the collector-to-emitter circuit open, the collector-to-emitter voltage of Figure 12.1c is at voltage level V_{CC}. V_{CE} is the output voltage V_O, and because its voltage level is V_{CC}, which is a logic high (1), the output voltage is a logic high. The output voltage of V_O is designated V_{OH} for the logic high state (V for voltage, O for output and H for logic high).

To relate the off condition to the BJT's characteristic curve (I_C vs. V_{CE}) of Figure 12.1a, the condition of no base current (or almost none — there is always leakage current) is labeled $I_{B\ OFF}$. The intersection of $I_{B\ OFF}$ with the load line results in a voltage magnitude of $V_{CE\ OFF}$ for V_{CE}, which is a logic high level designated V_{OH}.

Note in each equivalent circuit of Figure 12.1b and c that base current I_B is symbolically shown to control the position of the toggle. If the base current is large enough, it closes the switch; otherwise, the switch is open.

In Figure 12.1a there is an input (V_I) and output (V_O) voltage waveform. Review the events of the electronic switch using these voltage waveforms. During the interval $t_1 < t < t_2$, $V_I = V_{IH}$ (the input voltage is a logic high) and its magnitude and polarity are correct for forward biasing the base-to-emitter junction of the BJT. This turns on the collector-to-emitter junction, producing output voltage V_{OL}. However, when the input voltage is at the logic 0 level (V_{IL}) the magnitude is not sufficient to turn the base-to-emitter junction on and there is little to no base current, which is indicated by $I_{B\ OFF}$ (only a little leakage current flows). Without base current there is no collector current, and the collector-to-emitter voltage is at a logic high level (V_{OH}), as indicated in the output waveform of Figure 12.1a. Note from these two waveforms that the electronic switch is an inverter.

It is important to understand that bipolar technologies are current driven; that is, in addition to a base-to-emitter voltage of sufficient magnitude and proper polarity, a base-to-emitter current is required to turn a BJT switch on.

There are a variety of MOSFETs — the junction FET, the channel depletion mode, and the channel enhancement mode — and each of the modes can be either a negative (N) or a positive (P) channel. They are all voltage-controlled devices. The more popular mode is the N-channel enhancement mode, which will be used to discuss the workings of a MOSFET electronic switch.

The N-channel enhancement mode MOSFET of Figure 12.2a has areas at each end that are heavily doped with negative-(N) yielding.type material. This doping

Figure 12.2

An *N*-channel enhancement mode MOSFET electronic switch.

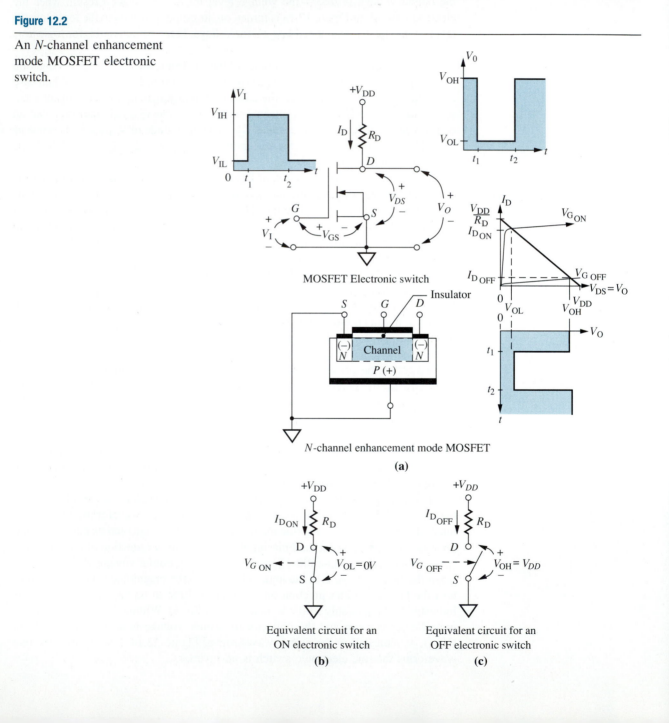

MOSFET Electronic switch

N-channel enhancement mode MOSFET

(a)

Equivalent circuit for an
ON electronic switch

(b)

Equivalent circuit for an
OFF electronic switch

(c)

results in either end, the drain (D) and the source (S), being N-type semiconductor material. Between the source and the drain is the channel, which has been doped with a material that yields a P-type semiconductor. Before current can flow from source to drain, the channel must be changed to a pseudo–N-type material. That is, while the channel doping is not physically changed, there must be a method that has the same net result. To change the channel so that it appears to be an N-type channel, a positive voltage is placed on the gate (G). As illustrated in Figure 12.2a, the gate is physically placed over the channel but is insulated from it by a thin layer of metal oxide. The schematic symbol for a MOSFET shows the gate insulated from the drain and source by a space as well as the space existing between source, channel, and drain. This positive gate voltage attracts electrons (opposites attract) throughout the channel material and concentrates them beneath the gate, which runs the length of the channel. At the same time this positive gate voltage repels positive charges (holes) from the channel. As a result, the concentration of N-type and P-type charges within the channel changes in favor of the N-type charges (the channel has been enhanced with current carrier electrons), which has the effect of changing the channel to an N-type material, causing electron current to flow between the source and drain. As the positive gate voltage is increased, it attracts more electrons and repels more holes in the channel, increasing the current flow between the source and drain (I_D).

The characteristic curve of Figure 12.2a supports what has been said. At a gate voltage of $V_{G\ OFF}$, which is insufficient to attract enough electrons within the channel for current flow, there is no drain current. However, from the characteristic curve even with a gate voltage of $V_{G\ OFF}$ there is a small value of gate leakage current, $I_{D\ OFF}$. Ideally $I_{D\ OFF}$ is zero, meaning that the electronic switch is open circuited, as illustrated in Figure 12.2c. With the electronic switch open, the output voltage V_O of the switch is V_{DD}, which is a logic high, and the output voltage for this condition is identified as V_{OH}.

As indicated by the characteristic curve, as the gate voltage is increased to $V_{G\ ON}$, the drain current increases to $I_{D\ ON}$. When the drain current has a value of $I_{D\ ON}$, the electronic switch is considered closed, as illustrated in Figure 12.2b. The characteristic curve also shows that the output voltage of the MOSFET is nearly at zero volts and, consistent with our notation, is designated V_{OL}.

Like the bipolar electronic switch, the MOSFET electronic switch is an inverter. The difference between the two is that the logic state of a MOSFET is voltage controlled (gate voltage), whereas that of the bipolar switch is current controlled (base current).

Review Questions

1. For a logic high input voltage, what is the logic level of the output of an electronic switch?

2. What input parameter of a bipolar switch controls the logic level of its output?

3. What input parameter of a MOSFET switch controls the logic level of its output?

4. Both bipolar and MOSFET characteristic curves indicate that for the off condition (open circuit) there is a small collector ($I_{C\ OFF}$), or drain ($I_{D\ OFF}$) current. Why are these currents present?

5. What do the symbols V_{OL} and V_{OH} represent? Explain the meaning of each letter of the symbols.

12.3

Fundamental electronic principles of logic gates

As seen in Chapter 2, electronic switches can be used to construct logic gates. Figure 12.3 shows how two electronic switches form an *AND* gate when connected in series and an *OR* gate when connected in parallel. The position of the toggle of each switch is controlled by base currents I_{B1} and I_{B2} for bipolar electronic switches and gate voltages V_{G1} and V_{G2} for MOSFET switches. The accompanying truth tables verify the logic of each circuit, where the inputs I_B and V_G are simply labeled no. 1 and no. 2. An input logic 0 means no base current ($I_{B\ OFF}$) for the bipolar switch and no gate voltage ($V_{G\ OFF}$) for the MOSFET switches, whereas a logic 1 input means $I_{B\ ON}$ or $V_{G\ ON}$. Although actual logic gates are more complicated than those of Figure 12.3, they do function in a similar fashion.

Figure 12.3

Logic gates constructed from electronic switches.

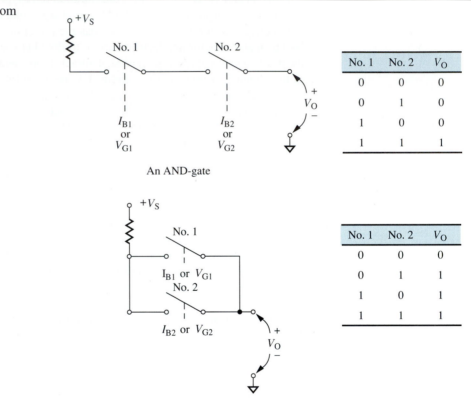

An AND-gate

No. 1	No. 2	V_O
0	0	0
0	1	0
1	0	0
1	1	1

An OR-gate

No. 1	No. 2	V_O
0	0	0
0	1	1
1	0	1
1	1	1

Transistor-transistor logic

The circuits of Figure 12.4 are two TTL logic circuits, an inverter and a tristate buffer with inverted output. We will discuss Figure 12.4a now but defer the discussion of Figure 12.4d until multiple emitters (Q_1) have been discussed.

Begin by examining the inverter of Figure 12.4a. The diodes D_1 and D_2 are not part of the electronic switches. D_1 serves to protect Q_1 from a reverse polarity input voltage. D_2 ensures that Q_3 is off when Q_2 is on. As we trace through the electronic switches take note of their relationship to one another. Transistor Q_1 controls the base current of Q_2 and hence determines whether Q_2 is on or off. Transistor Q_2 in turn determines whether transistor Q_3 or Q_4 is on or off. If Q_2 is off, Q_3 is on and Q_4 is off; however, if Q_2 is on, Q_3 is off and Q_4 is on. The series arrangement of Q_3 and

Figure 12.4

Basic TTL logic circuits.

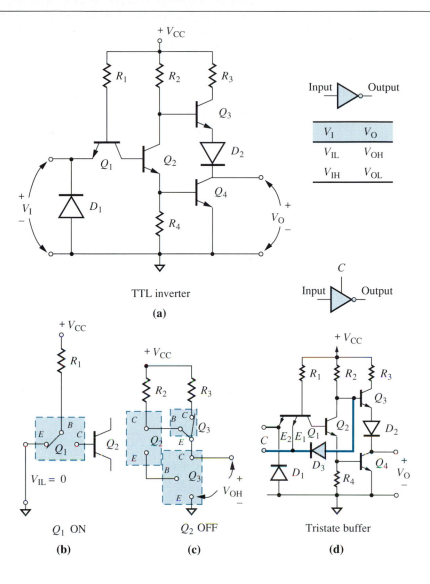

TTL inverter

(a)

Q_1 ON

(b)

Q_2 OFF

(c)

Tristate buffer

(d)

Q_4 is known as a *totem pole*. When Q_4 is off the output voltage is a logic high, and when Q_4 is on the output voltage is a logic low.

Before analyzing the inverter of Figure 12.4a, examine the equivalent circuit of the electronic switch circuit for Q_1. This equivalent circuit is an single-pole-double-throw (SPDT) switch, as indicated in Figure 12.4b. The reason for this type of equivalent circuit is that current can flow in one of two paths, it will either be flowing through the base-to-emitter junction (B-E path is closed) or the base-to-collector junction (the B-C path is closed).

Apply a logic 0 and then a logic 1 to the input V_I of the TTL inverter of Figure 12.4a and determine the resultant output logic level for each case. Applying a logic 0 to the input, which is designated V_{IL} in the truth table of Figure 12.4a, pulls the emitter of Q_1 to ground (or close to it). With the emitter at approximately zero volts (ground) and the base pulled to V_{CC} ($+5$ volts), via R_1, the base-to-emitter junction is forward biased and Q_1 turns on and base-to-emitter current flows, as shown in Figures 12.4b and 12.5a. Even though Q_1 is on, there is effectively no collect-to-emitter current (I_C), because there is no source for that current. That is, the current would have to flow through the base-collector junction of Q_2 in order to provide a closed path from ground to the power supply V_{CC}. As shown in Figure 12.5a, the base-to-collector of Q_2 is an equivalent open circuit. Without sufficient base-to-emitter current to Q_2 it is off, and it has an equivalent open collector-to-emitter circuit, as shown in Figures 12.4c and 12.5a. With Q_2 acting as an equivalent open circuit there can be no base-to-emitter current to Q_4, but R_2 provides a direct path through which base-to-emitter current can be supplied by the power supply to Q_3. R_2 also pulls the base potential of Q_3 up toward V_{CC} volts. With approximately V_{CC}

Figure 12.5

Equivalent circuits for the
TTL inverter of Figure 12.4.

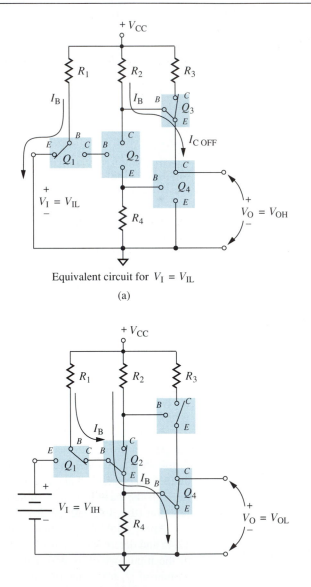

Equivalent circuit for $V_I = V_{IL}$

(a)

Equivalent circuit for $V_I = V_{IH}$

(b)

volts on the base of Q_3 and a source for a base-to-emitter current, Q_3 turns on. It should be noted that I_B of Q_3 is small, because Q_4 is off, but even with Q_4 off $I_{C\,OFF}$ amperes can flow through Q_4, and Q_3 is designed so that this much base-to-emitter current will turn Q_3 on, as illustrated in Figure 12.5a. When Q_3 is on, its collector-emitter is an effective short circuit; it pulls the collector of Q_4 toward V_{CC}, which is the logic 1 state that is designated V_{OH} in the truth table of Figure 12.4a. Therefore $V_O = V_{OH}$, as shown in Figure 12.5a.

Inputting V_{IH} (a logic 1) volts at the emitter of Q_1 does not forward bias Q_1's base-to-emitter junction. As a result, that junction is an equivalent open circuit, as shown in Figure 12.5b. The emitter of Q_2 is referenced to ground via R_4. This has the effect of referencing the collector of Q_1 to ground. With the collector of Q_1 pulled toward ground potential and its base near V_{CC} volts, the base-to-collector junction of Q_1 is forward biased, acting as a forward biased diode that is designed to turn on and act as an equivalent short circuit, as shown in Figure 12.5b. This provides a source of base-to-emitter current for Q_2, which turns Q_2 on, as indicated in Figure 12.5b. There is now a base-to-emitter current path for Q_4, which turns Q_4 on and pulls the collector of Q_4 to approximately zero volts. Pulling V_O to near ground potential results in a logic level of V_{OL}.

Compare the equivalent circuits of Figure 12.5a and b and note that Q_2 controls whether Q_3 or Q_4 is on (both cannot be on at the same time), and Q_1 controls the state of Q_2.

Why not use the simple inverter of Figure 12.1a rather than the more complex one of Figure 12.4? The answer is the reduced current drain at the output. The inverter of Figure 12.4 is designed so that there is always a high impedance between V_{CC} and ground, because either Q_3 or Q_4 is off.

The inverter of Figure 12.4a can be modified so that it becomes the *NAND* gate of Figure 12.7. To make this modification, Q_1 is designed and manufactured with multiple emitters, one emitter for each input. Figure 12.6a shows the schematic symbol of a multiemitter transistor, and Figure 12.6b is its diode equivalent. The base-to-emitter diodes are connected so that if one or more of these diodes are forward biased, base-to-emitter current can flow and turn Q_1 on; otherwise, the base-to-collector diode is forward biased and a base-to-collector current will flow. Because Q_1 of Figure 12.6a has either a base-to-emitter current flow or a base-to-collector current flow, the equivalent circuit of a multiemitter transistor is also a SPDT switch. As a result, the equivalent circuits of Figure 12.5 are also valid for analyzing the *NAND* gate of Figure 12.7.

When all inputs to Q_1 of Figure 12.7 are a logic high, the base-to-emitter junction of each equivalent diode is not forward biased, resulting in no base-emitter current for Q_1. With this condition, the equivalent circuit of Figure 12.5b is valid and the output voltage V_O is a logic 0, which agrees with the truth table of Figure 12.7. Anytime a logic low (V_{IL}) is input to one or more emitters of Q_1, a path exists for base-to-emitter current to flow, and Q_1 is turned on. With Q_1 on, the equivalent circuit of Figure 12.5a is valid, and the output voltage is a logic high ($V_O = V_{OH}$). From the truth table of Figure 12.7 any time an input is at a logic low the equivalent circuit of Figure 12.5a is valid and V_O is a logic 1 (V_{OH}).

Return now to the tristate buffer of Figure 12.4d. If the control line C is a logic high, D_3 and emitter E_1 are reverse biased and the circuit functions as the inverter of Figure 12.4a. However, if control C is pulled to a logic low, diode D_3 is forward biased, which pulls the base of Q_3 low, turning Q_3 off. Emitter E_1 of transistor Q_1 is also forward biased, which turns Q_1 on and turns Q_2 off. As a result of diode D_3 and transistor Q_1 being on, both transistors Q_3 and Q_4 are off. Hence, the output

Figure 12.6

Multiemitter transistor Q_1.

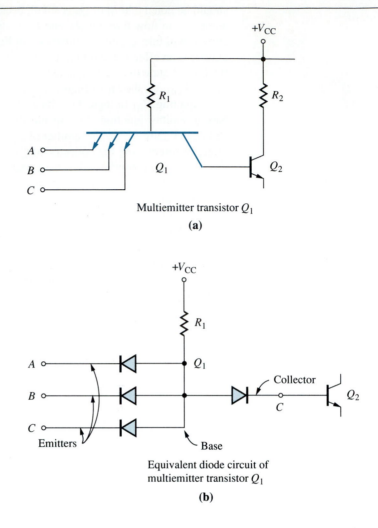

Multiemitter transistor Q_1

(a)

Equivalent diode circuit of
multiemitter transistor Q_1

(b)

impedance is very high (Hi-Z) and very little current can be sinked or sourced (the output has no current drive capability). As a result, the Hi-Z output can be easily driven to either a logic 0 or 1 by any other output connected to it.

Normally the outputs of logic gates cannot be connected because this may result in damage to the output transistor Q_4 as well as bus contention. One type of TTL technology is designed so that outputs can be connected: *open-collector logic*. The top half of the output totem pole is removed, as illustrated for the *NAND* gate of Figure 12.8a. In place of the removed portion of the totem pole is an external *pull-up resistor R* (Fig. 12.8b), which configures Q_4 as an electronic switch similar to Figure 12.1a. Figure 12.8b shows that by connecting the outputs of open-collector logic gates, *NAND* gates in this case, the output transistors (Q_4) of each *NAND* gate have been paralleled. Paralleling electronic switches implements *OR* logic, but owing to negation the resultant logic is an *AND*. Thus, the logic at the connection is the *AND*ing of the three *NAND* gates whose outputs are connected, which is why the resultant logic from wiring the outputs of open-collected logic is known as *wired-AND logic*. Figure 12.8c shows the symbolic representation for wired-*AND* logic.

The pull-up resistor derives its name from the fact that it pulls the collector(s) of Q_4 to V_{CC} when Q_4 is off. It also serves to limit the current from the power supply

Figure 12.7

TTL three-input *NAND* gate.

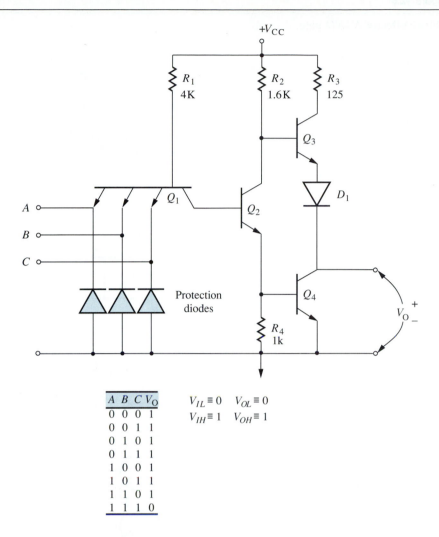

A	B	C	V_O
0	0	0	1
0	0	1	1
0	1	0	1
0	1	1	1
1	0	0	1
1	0	1	1
1	1	0	1
1	1	1	0

$V_{IL} \equiv 0 \quad V_{OL} \equiv 0$
$V_{IH} \equiv 1 \quad V_{OH} \equiv 1$

when Q_4 is on. The magnitude of this current must be less than the magnitude specified by the manufacturer as damaging to Q_4.

There are various open-collector TTL logic functions available, such as *NAND*, *NOR*, *AND*, and *OR* gates, multiplexers, registers, and memory chips. Regardless of the logic function, its output operates as described for the *NAND* gate of Figure 12.8.

The on-to-off switching time of the bipolar transistors can be shortened by not allowing those electronic switch transistors to be turned on so hard, that is, by not letting them be driven into saturation. To accomplish this the base current must be limited as the transistor approaches $I_{B\,ON}$, which is shown on the characteristic curve of Figure 12.1a. To limit the base current a specially designed bypass diode, known as a Schottky diode, is placed from the base to the collector of a bipolar transistor, as shown in Figure 12.9a. The cathode of a Schottky diode is drawn as an ∫ in the schematic. This diode has a very low forward biased voltage drop of approximately 0.2 volt. As a result, with the base-to-emitter junction of the transistor at approximately 0.6 volts (forward biased voltage drop) and the collector voltage V_{CE} for an on transistor at about 0.4 volts ($V_{CE\,ON}$), the Schottky diode is forward biased and turns on, which bypasses any excessive base current to the collector. This action

Figure 12.8

Open-collector *NAND* gate.

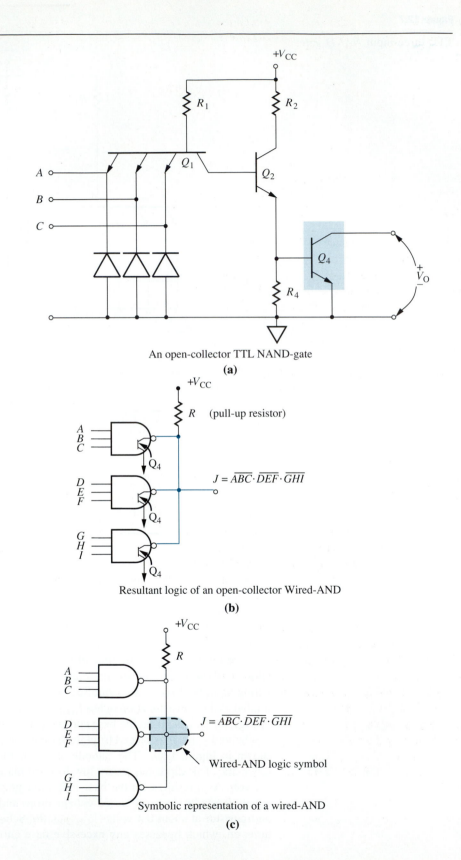

An open-collector TTL NAND-gate

(a)

$$J = \overline{ABC} \cdot \overline{DEF} \cdot \overline{GHI}$$

Resultant logic of an open-collector Wired-AND

(b)

$$J = \overline{ABC} \cdot \overline{DEF} \cdot \overline{GHI}$$

Wired-AND logic symbol

Symbolic representation of a wired-AND

(c)

Figure 12.9

Schottky electronics and logic gate.

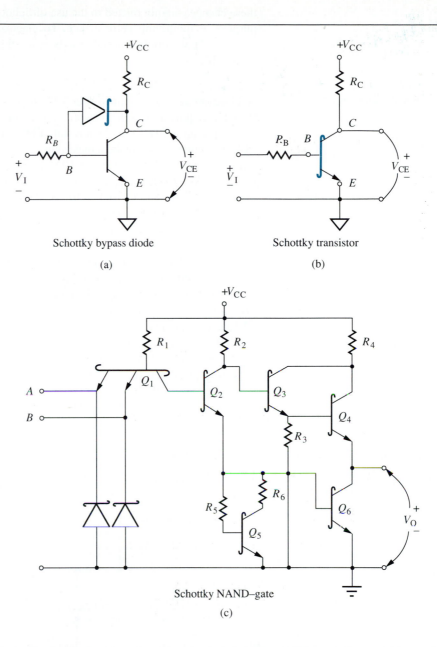

Schottky bypass diode

(a)

Schottky transistor

(b)

Schottky NAND–gate

(c)

limits the base current to the amount needed to turn on the transistor and keeps it from being driven into saturation. When the transistor is to be turned off using this technique, there are fewer excessive charges in the base-to-emitter region of the transistor (NP junctions act like capacitors in that they collect charges) that must first be drained off. A Schottky transistor is created when the Schottky diode of Figure 12.9a is manufactured as an integral part of the transistor, and its logic symbol is that of Figure 12.9b. Schottky transistors are of the bipolar TTL family, and therefore logic gates that utilize Schottky transistors are still TTL devices. Schottky-based logic gates are designated with an ∫ to differentiate them from others of the same family.

Figure 12.9c is a Schottky *NAND* gate. Although there are some circuit changes in the *NAND* gate of Figure 12.9c, when compared to the *NAND* gate of Figure 12.7, these changes are not related to the use of Schottky transistors but rather are for the purpose of improving performance. To electronically analyze the *NAND* gate of Figure 12.9c we would proceed exactly as we did for the standard TTL *NAND* gate of Figure 12.7.

N-channel MOS logic

A disadvantage to manufacturing TTL logic is the implementation of resistors. Resistors require a lot of the IC waffer area, which meaning relatively few logic gates can be put on a TTL chip. As a result, TTL is restricted to *small-scale integration* (SSI), meaning that there are a maximum of 10 logic gates on a single chip, *medium-scale integration* (MSI), meaning 10 to 100 gates, or *large-scale integration* (LSI), meaning 100 to 1000 gates. Metal oxide semiconductor technology ICs use MOSFETs as biasing resistors; that is, these MOSFETs are substituted for R_D of Figure 12.2a. This technique allows the number of logic gates per chip to be greatly increased. At present, the upper limit is approximately 200,000 logic gates per chip, which is designated as *very large-scale integration* (VLSI). MOS technology is used primarily for VLSI applications, such as microprocessors and memory chips.

The MOSFET identified as Q_R in Figure 12.10a and b serves as bias resistor R_D. The gate of Q_R is tied to V_{DD}, which biases it to conduct (partially turns it on) so as to offer an impedance of R_D ohms. The two MOSFET electronic switches of Figure 12.10a, Q_A and Q_B, are connected in parallel. From Figure 12.3 we know that paralleling of switches is an *OR* configuration. However, because electronic switches invert the logic (negative logic), the paralleling of electronic switches Q_A and Q_B must result in a negated *OR*, which is a *NOR* gate.

To electronically verify the type of logic gate of Figure 12.10a input the logic levels as indicated in the truth table. In keeping with our notation, V_{IL} and V_{OL} are

Figure 12.10

NMOS logic gates.

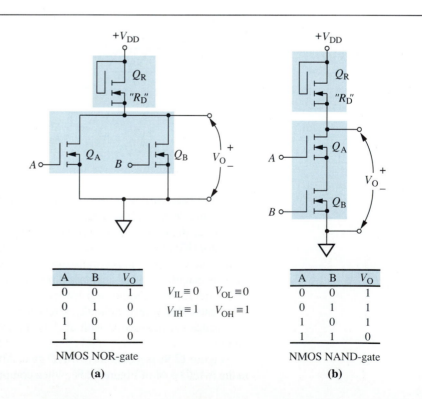

A	B	V_O
0	0	1
0	1	0
1	0	0
1	1	0

$V_{IL} \equiv 0$ $V_{OL} \equiv 0$
$V_{IH} \equiv 1$ $V_{OH} \equiv 1$

NMOS NOR-gate

(a)

A	B	V_O
0	0	1
0	1	1
1	0	1
1	1	0

NMOS NAND-gate

(b)

logic 0's, and V_{IH} and V_{OH} are logic 1's. Input a logic 0 to inputs A and B. Both Q_A and Q_B will be off, thereby pulling the output up to a logic 1, which agrees with the first row of the truth table. Applying a logic 0 to input A and a logic 1 to input B results in Q_A being turned off and Q_B being turned on. With Q_B on, the output is pulled to ground potential, a logic 0. This agrees with the second row of the truth table. Because Q_A and Q_B are in parallel, any time either or both electronic switches of Figure 12.10a are on, the output is a logic 0, as verified by the truth table. This completes the verification.

When electronic switches are connected in series, they configure a logic *AND* according to Figure 12.3. Because electronic switches Q_A and Q_B of Figure 12.10b are in series and electronic switches invert, the resultant logic will be a negated *AND* gate, or *NAND* gate. This can be verified electronically by inputting a logic 0 to both switches at inputs A and B. Under this condition both switches are off and the output is pulled toward V_{DD}, a logic 1. In fact, owing to the series connection of the switches, any time either or both switches are off, the output is pulled to V_{DD}. The output is other than a logic 1 only when both switches are on — that is, both inputs A and B are a logic 1 — because then the output is pulled to ground (a logic 0). This is the truth table of a *NAND* gate.

Complementary metal oxide semiconductor logic

To reduce power consumption the resistance of Q_R of Figure 12.10 could be increased. In fact, if the resistance value of Q_R were in the megaohm range, the current drawn from the power supply V_{DD} would be greatly reduced. A MOS technology that accomplishes this goal is referred to as complementary MOS (CMOS). CMOS uses the complement channel type of electronic switch to structure Q_R. As shown in Figure 12.11, the bias resistors Q_{RA} and Q_{RB} are P-channel MOSFETs, and the electronic switches Q_A and Q_B are N-channel MOSFETs. With complementary channels connected so that their gates are in parallel, as are MOSFETs Q_A and Q_{RA} and MOSFETs Q_B and Q_{RB}, turning on an electronic switch (Q_A or Q_B) turns off the corresponding bias resistor (Q_{RA} or Q_{RB}), which means there is always a large impedance from the power supply to ground.

Figure 12.11a is configured as a *NOR* gate and Figure 12.11b as a *NAND* gate, as is evident from the electronic switch arrangements. Electronic switches Q_A and Q_B are in parallel in Figure 12.11a, which is an *OR* arrangement; however, because electronic switches negate the logic, this is really a *NOR* gate. The two electronic switches of Figure 12.11b are in series, which is an *AND* configuration, but the resultant logic is a *NAND* configuration due to negation. To make the electronic principles clearer, apply a logic 0 (which is a few tenths of a volt) to both inputs A and B of Figure 12.11a. The logic 0 on the gates of Q_A and Q_B is not large enough to enhance their channels; therefore, both electronic switches are off. However, those logic 0's are also applied to the gates of the P-channel MOSFETs Q_{RA} and Q_{RB}. With the source of Q_{RA} at a positive V_{DD} potential, by writing a loop equation it can be determined that the magnitude and polarity of the gate-to-source voltage are negative V_{DD} volts (for simplicity assume a logic 0 to be 0 volts). A negative potential of magnitude V_{DD} is sufficient to enhance the P-channel of Q_{RA}, which turns it on. Turning Q_{RA} on pulls the source of Q_{RB} to $+V_{\text{DD}}$, which then biases the gate of Q_{RB} with $-V_{\text{DD}}$ volts and also turns on Q_{RB}. With Q_{RA} and Q_{RB} on and electronic switches Q_A and Q_B off, the output is pulled up to potential V_{DD}, which is a logic 1. This agrees with the logic states shown in the truth table.

Because the electronic switches of Figure 12.11a are connected in parallel, if either or both switches are turned on the output is shorted to ground and the output is a logic 0, which agrees with the truth table. To walk through the electronics for an example of a logic 1 input, apply a logic 1 to input B and a logic 0 to input A. For the

Figure 12.11

CMOS logic gates.

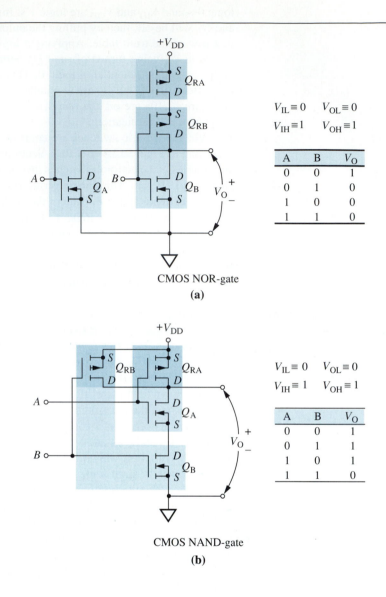

A	B	V_O
0	0	1
0	1	0
1	0	0
1	1	0

$V_{IL} \equiv 0$ $V_{OL} \equiv 0$
$V_{IH} \equiv 1$ $V_{OH} \equiv 1$

CMOS NOR-gate

(a)

$V_{IL} \equiv 0$ $V_{OL} \equiv 0$
$V_{IH} \equiv 1$ $V_{OH} \equiv 1$

A	B	V_O
0	0	1
0	1	1
1	0	1
1	1	0

CMOS NAND-gate

(b)

logic 0 input to A, Q_A is off and Q_{RA} is on, which pulls the source of Q_{RB} to V_{DD}. The logic 1 (near $+V_{DD}$ volts) on the gate of Q_B sufficiently enhances its N-channel so that it turns on; hence Q_B will be on. Because that same logic 1 ($+V_{DD}$) is on the gate of P-channel Q_{RB}, its gate-to-source potential is zero volts (recall that Q_{RA} is on), which does not enhance the channel of Q_{RB} and it therefore is turned off. With Q_B on, the output is a logic 0, which agrees with the truth table, but little current is drawn from the power supply V_{DD} because Q_{RB} is off. Verify the remaining logic conditions of the truth table.

To trace through an electronic state of Figure 12.11b, input logic 0's to both A and B. The channels of both Q_A and Q_B are not sufficiently biased to enhance them for source-to-drain current conduction. Thus, both Q_A and Q_B are off and there is an equivalent open circuit from the drain of Q_A to the source of Q_B. The logic 0's input at A and B result in a bias potential of $-V_{DD}$ volts between the source-to-gate of Q_{RA} and Q_{RB}, and this situation is sufficient biasing to turn them on. With either Q_{RA} or

Q_{RB} turned on, and with Q_A and Q_B off, the drain of Q_A is pulled to $+V_{DD}$ volts, which is a logic 1. This agrees with the first row of the truth table of Figure 12.11b.

The series configuration of Figure 12.11b shows that the only way to get a logic 0 at V_O is to turn both Q_A and Q_B on. This requires that logic 1's be input to A and B, resulting in both Q_{RA} and Q_{RB} being turned off and again greatly reduces any current flow from the power supply V_{DD} to ground. With inputs A and B at a logic 1 level, Q_A and Q_B are on and the output voltage is V_{OL}, which agrees with the truth table, and there is very little current drain from V_{DD} because Q_{RA} and Q_{RB} are off.

We have not investigated each and every logic gate configuration for the technologies discussed (standard TTL, open-collector TTL, Schottky TTL, NMOS, and CMOS), nor have we discussed all the technologies available (RTL, DTL, ECL, I^2L, PMOS, or HMOS). We have examined only the essential concepts of the most prominent technologies to establish a solid background for reasoning our way through an analysis of those technologies and logic gates not studied. More importantly, from the fundamental knowledge we have developed, we will have a better understanding of manufacturers' data sheet specifications so as to interpret their significance relative to design requirements.

Review Questions

1. What is required to turn on a bipolar electronic switch?
2. What controls the toggle position of a MOSFET electronic switch?
3. When either a bipolar or MOSFET electronic switch is off, what is the logic level of the output voltage?
4. When the output voltage of an electronic switch is V_{OL}, what is the input logic level of that switch, whether it is a bipolar or MOSFET electronic switch?
5. What must be the gate-to-source voltage polarity of an NMOS and PMOS electronic switch in order to turn them on?
6. How are enhancement-type MOSFETs turned on?
7. What is significant about the totem pole output of TTL technology?
8. What purpose does Q_2 serve in Figures 12.4 and 12.7?
9. How does a multiemitter bipolar transistor function?
10. Is input current I_{IL} required of a TTL gate for the logic condition V_{IL}?
11. What is a wired-*AND* logic gate?
12. What is a Schottky transistor, and how does it improve the performance of the standard TTL technology?
13. How does MOS technology improve the logic gate density of an IC?
14. For the logic condition of V_{IL}, does an NMOS device require current I_{IL} in order to function?
15. In what circuit configuration are the electronic switches of a *NOR* gate connected?
16. In what circuit configuration are the electronic switches of an *AND* gate connected?
17. What is meant by complementary channel?
18. How does CMOS reduce power consumption?

12.4

Electronic memory cells

Figure 12.12 is the schematic of a static bipolar R/W memory cell, which is a flip-flop. There are two multiemitter transistors — Q_1 and Q_2. The generic Boolean variable Q was arbitrarily chosen to be represented by the output of Q_1, and the output of Q_2 represents the complement of the variable, which is the negation of Q ($\overline{Q}$). A static memory cell can have a logic 1 written into it ($Q=1$ and $\overline{Q}=0$), or a logic 0 ($Q=0$ and $\overline{Q}=1$), whichever logic state is to be remembered by the memory cell. As explained in Chapter 9, a static memory cell retains the logic of the bit stored in it so long as power is applied to the cell (it is volatile).

Figure 12.12

Static bipolar R/W memory cell.

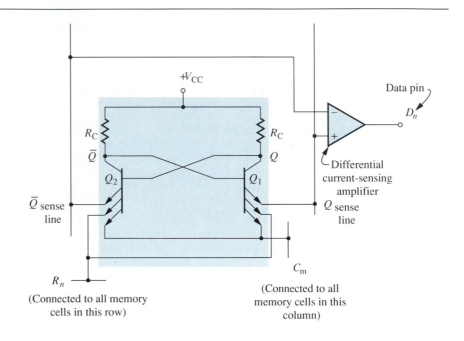

Transistors Q_1 and Q_2 of Figure 12.12 are electronic switches that are dependent on each other for the needed forward biased base-to-emitter potential in order to be turned on. That is, the base of Q_1 is connected to the collector of Q_2, and the base of Q_2 is connected to the collector of Q_1; therefore, if Q_1 is off, its collector is pulled up to V_{CC} by R_C, and this voltage is applied to the base of Q_2, which will forward bias Q_2 and turn it on. With Q_2 on, its collector voltage is pulled low and as a result the base voltage is not of sufficient magnitude to forward bias Q_1 and it remains off. The memory cell remains in that state ($Q=1$ and $\overline{Q}=0$) as long as power is applied to the cell, or until Q_1 is turned on or Q_2 is turned off.

When the memory cell of Figure 12.12 is not being addressed, the row (R_n) and/or column (C_m) lines are inactive, which is a logic low. With either or both R_n and C_m low, the base-to-emitter current of the on transistor has a low-resistance path to ground, which keeps that transistor on. The off transistor has no base current and therefore does not use the row or column line as a low-resistance path to ground.

When a memory cell is addressed to be read, the row and column lines are first driven to a logic high by decoding the address being input to the memory chip (*see* Fig. 8.7a). When both row and column lines go high, this reduces the base-to-emitter forward biased voltage so that the four emitters connected to R_n and C_m are no longer forward biased. However, the sense lines are at a logic low, which means that there is a base-to-emitter forward biased voltage present to keep the on transistor forward biased and conducting. This base-to-emitter current is conducted to the sense line. The off transistor is not forward biased and therefore does not conduct base-to-emitter current to its sense line. The sense line with base current is detected by a current-sensing differential amplifier, which outputs the proper logic voltage level V_{OL} at the data pin. The sense line connected to the off transistor would not have a base-to-emitter current flowing in it, which of course would be sensed by the current-sensing amplifier and as a result its output would be V_{OH}.

To write to the memory cell of Figure 12.12, again the R_n and C_m lines are driven high via row and column address decoders within the memory chip. As we know from the timing diagram of an MPU, the datum being written to memory is loaded on the data bus and then latched by the memory chip on the positive edge of $\overline{WR}$. The memory chip then applies the latched logic level to the Q sense line (the emitter of Q_1), and the complement of that logic level is applied to the sense line of $\overline{Q}$ (the emitter of Q_2). If the logic level to be stored is a logic 1, Q_1 is turned off but Q_2 is turned on, because a logic 0 would be applied to the emitter of Q_2. The address is then removed from the address pins by the MPU, causing the row and column lines to go to a logic low and thereby provide a path for the base current of the on transistor to continue to flow. This keeps Q_2 on and Q_1 off after the logic level 1 being written into the memory cell has been removed from the sense lines.

In summary, to read the data bit stored in the memory cell of Figure 12.12, or to write a data bit into it, the row and column lines must be driven to a logic high state. If the memory cell is being read, this action results in a base-to-emitter current flowing in the sense line that is connected to the emitter of the on transistor. The sense line connected to the off transistor does not have a current flowing in it. A current-sensing differential amplifier detects the presence or absence of current in these sense lines and outputs to the data pin the appropriate logic level. When writing to the memory cell, the logic level being stored is applied to Q, and the complement of that logic level is applied to $\overline{Q}$ by way of their respective sense lines. This turns on the transistor whose emitter is connected to the sense line with the logic low, and the other transistor is turned off. This completes the write operation. The row and column lines then go inactive (low), and the logic state written into the cell remains.

The configuration of the static R/W NMOS memory cell of Figure 12.13 is similar to that of the static bipolar memory cell of Figure 12.12. The MOSFETs Q_1 and Q_2 are cross-coupled so that each enhancement gate voltage is dependent on the other's drain voltage. Whenever electronic switch Q_1 or Q_2 is on, its drain-to-source voltage is pulled to a logic low and is therefore inadequate to enhance the channel of the other to support conduction, and as a result it is turned off. The MOSFETs identified as Q_B are the biasing resistors, just as R_D is in Figure 12.10. When the cell is to be read or written to, its row and column lines are activated, as they were in Figure 12.12. This action turns on the MOSFETs identified as Q_X (for row) and Q_Y (for column). If a read is to occur, the gate of Q_R is driven high (it can be designed for an active low or high), which turns Q_R on. With Q_X, Q_Y, and Q_R on, the logic state of Q_1 is transferred to the data pin $\overline{D}_n$. If the cell is to be written into, the data bit to be stored is input at data pin D_n, and the write input is driven high (again, it could be designed to be low), which turns Q_W on. Since Q_Y and Q_X are already on, this loads the data bit on the write line and applies it to the gate of Q_1. If the data bit is a logic 1 it turns on Q_1 and pulls the drain of Q_1 low, which turns off Q_2. If the data bit is a logic 0, Q_1 is turned off and Q_2 is turned on. Notice that the logic level at the drain of Q_1 is the inverse of the logic level input at pin D_n. This is the reason for the negation symbol of the read output pin $\overline{D}_n$. Of course, the read output can be inverted, and then the write data pin and read data pin could be made to be one pin.

Figure 12.13

An NMOS static R/W
memory cell.

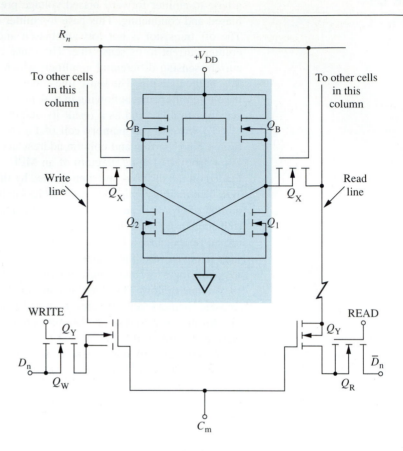

The sampling of R/W memory cells should give a feeling for static R/W memory cells, regardless of the logic family technology. The only memory cell that is fundamentally different is that of a dynamic memory, which will be examined next.

There are a variety of dynamic memory cell arrangements, but the one illustrated in Figure 12.14 is representative of the principle of operation of all of them. The MOSFETs are standard MOSFETs but are drawn in a more conventional and convenient fashion. The storage mechanism is the capacitor C, a parasitic capacitance of the MOSFET Q_2. The logic level being read from the cell appears at C_O and is the inverse logic level stored on C. Because C is a parasitic capacitance, it has a relatively large amount of leakage current. This leakage is large enough to discharge C within 2 to 3 ms, so it must be refreshed, as discussed in Chapter 9. Because of the reduced number of components per cell, dynamic memories have the largest bit capacity.

When a data bit is to be written into the dynamic memory cell of Figure 12.14, it is input at D_{IN} and the gate of Q_1 is activated so that Q_1 turns on. With Q_1 on, the capacitor C is charged (or discharged) to the logic level of the data bit. Once the data bit has been stored on C, the write control signal is driven inactive, Q_1 is turned off, and C is buffered (no current is drained from it) between Q_1 and the gate of Q_2.

When a data bit is to be read from the dynamic memory cell of Figure 12.14, the output capacitor C_O is precharged to a logic 1. The read control signal is then driven active, turning Q_3 on. If a logic 1 is stored on C, Q_2 is turned on and the logic 1 stored

Figure 12.14

A dynamic memory cell.

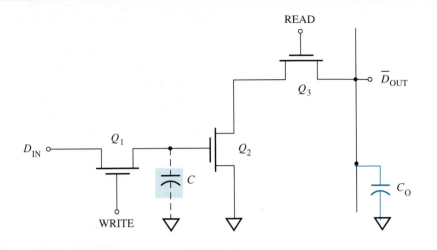

on C_O is discharged to a logic 0. The logic level of the data bit output at $\overline{D}_{OUT}$, via C_O, is the inverse of that stored, which means that the output logic bit must be inverted. If a logic 0 is stored on C, Q_2 is off, and even with Q_3 on there is no discharge path for the logic 1 stored on C_O and the output at $\overline{D}_{OUT}$ is a logic 1. Again, the logic level output is the inverse of the logic bit stored.

Figure 12.15 conceptually represents a typical ROM of any type (MROM, PROM, or EPROM). The ROM of Figure 12.15 is an $N \times 4$-bit ROM; that is, it has N rows of memory locations and 4 bits per location. The boxes represent the memory

Figure 12.15

A typical ROM.

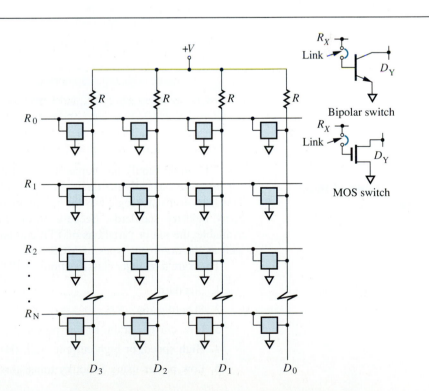

cells implemented with electronic switches. Representatives of those switches are shown for the bipolar and MOS family. Bipolar switches have a fuse-link in the base, whereas MOS switches have a fuse-link in the gate. If a logic 1 is to be programmed into a memory cell, this link is removed; otherwise it remains. If the ROM is an MROM, these links are put in or removed as part of the manufacturing process. If the ROM is a PROM, all links are manufactured in the electronic switches but are removed at the discretion of the user. A PROM programmer is used to cause an excessive base current to flow in those bipolar transistors for which a logic 1 is to be programmed, thereby blowing the fuse-link in the base, or applying the gate fuse-link with excessive gate voltage for a MOS switch. For either family, the link is removed for those memory cells that are to be programmed with a logic 1 and left intact for those to be programmed with a 0.

When a memory location is to be read, the appropriate row is activated (driven high) via the address decoder of that ROM. When that address row goes high it turns on all electronic switches connected to it that have their links still intact. When these switches are turned on, they pull the data line (D_n) connected to them to ground, resulting in a logic 0. Those electronic switches whose links are removed and therefore are always off do not short their data line to ground when the row line goes high, and as a result these links are pulled to $+V$ volts, which is a logic 1. The pull-up resistors identified as R are resistors if the family is TTL but are MOSFETs if the ROM is from the MOS family.

Review Questions

1. What do all static memory cells have in common?

2. How is a bipolar static R/W memory cell addressed, and how does that affect the electronic switches of that cell?

3. How is a bit written into a bipolar memory cell?

4. What electronic switches are turned on as the result of an address in an NMOS static R/W memory cell?

5. What function do the electronic switches Q_W and Q_R of Figure 12.13 serve?

6. What makes dynamic memory cells dynamic?

7. What is a refresh cycle, and why is it necessary for DRAMs?

12.5
Designations of logic families

The TTL logic family has numerous variations, whereas the MOS logic family is composed of basically the three already introduced—NMOS, PMOS, and CMOS. Transistor-transistor logic was the first major logic family, and as a result established many industry standards. Because of the importance and variety of TTL logic available, the major variations of TTL and their designations warrant discussion as well as the dominant types of MOS logic—NMOS and CMOS.

There are five major classifications of TTL logic:

1. Standard
2. Schottky (S)
3. Low current drain, or low power (L)
4. High speed, or high current drain (H)
5. Low power using Schottky transistors (LS)

The first TTL logic available is now known as *standard TTL*. Its designation is either 54XX or 74XX, the numbers indicating the operating temperature range of the IC. Those ICs with designations of 74 operate from 0 to 70°C, which meets most commercial requirements and those designated 54 operate in a temperature range of − 55 to 125°C, which meets military specifications. The two or three digits following the number specify the logic. For instance, 7400 and 5400 indicate that the IC contents are quad two-input *NAND* gates (four 2-input *NAND* gates per chip), and 54181 and 74181 are 4-bit ALUs. Standard TTL logic ICs are designated the 54/74 family. Figure 12.7 shows a standard TTL three-input *NAND* gate, and an IC that contains three of these *NAND* gates is designated either 5410 or 7410, depending on the operating temperature range.

Variations of the 54/74 series were implemented to improve performance, for example, by increased operating speed, reduced power consumption, or both. Many of these performance improvements were accomplished by simply changing the values of resistors, such as R_2, R_3, and R_4 of Figure 12.7, whereas others required circuit modifications such as replacing R_4 of Figure 12.7 with circuitry composed of R_5, R_6, and Q_5 of Figure 12.9. Some speed improvements were made if Schottky transistors were used instead of the standard bipolar transistor.

The S (Schottky) series is faster than the standard 54/74 and uses Schottky transistors to achieve the improved performance. Its designation is 54S/74S. That same triple three-input *NAND* gate that is designated 5410 or 7410 for standard TTL is designated 54S10 or 74S10 for the Schottky version (they are pin-for-pin compatible).

In the low-power (L) series the values of the resistors are increased. For example, in Figure 12.7, R_2 would be increased to 20 K ohms, R_3 to 500 ohms, and R_4 to 12 K ohms. With these increased resistances less current is drained from the power supply. The disadvantage is that the operating time also increases. This low-power series is designated 54L/74L. Hence, the previously mentioned triple three-input *NAND* gate IC would be designated 54L10 or 74L10.

In the high-power (H) series, the biasing resistance values are reduced, which increases the current drain, in order to achieve greater output current as well as a reduction in operating speed. For instance, R_3 of Figure 12.7 would be reduced to 50 ohms. This high-power and higher-speed series is designated 54H/74H. Therefore, the triple three-input *NAND* gate IC would be designated either 54H10 or 74H10.

The last circuit modification designation to be considered is the low-power Schottky (LS) variety. This series uses a combination of Schottky transistors and circuit modifications, as in Figure 12.9. In fact, Figure 12.9 is a two-input *NAND* gate of the LS variety. If there are four of these *NAND* gates per IC, the designation is 54LS00 or 74LS00.

If a CMOS IC is pin compatible with that of a 54/74 series, it is designated with a prefix 74C; otherwise it has a 4000 (4001, 4002, etc.) designation. Our interest is restricted to the TTL pin-compatible 54/74 series.

There are five different types of 74 CMOS series logic devices: 74C, 74HC, 74HCT, 74AC, and 74ACT. The 74C was the first TTL pin-compatible CMOS. Its major disadvantage is its relatively low value for I_{OL} (0.36 mA), which means it could not sink the input current I_{IL} of a single 74 series TTL load (− 1.6 mA). If 74C CMOS were to drive a TTL logic device, the latter would have to be from the 74LS series, and even then it would be incompatible by 0.04 mA.

The 74HC series is a high-voltage CMOS (it can be operated up to + 15 volts). It is a counterpart to TTL open-collector logic devices. 74HC logic devices are open-drain devices and are used in wired-*AND* configurations. A disadvantage to the 74HC series is its minimum input voltage level for a logic 1 (V_{IH}), which is 3.5 volts;

this is also true for the 74C series. Because TTL has a V_{OH} of 2.4 volts, interface circuitry must be used if 74 series TTL logic devices are to drive 74HC CMOS logic devices.

The 74HCT CMOS was designed to correct the incompatibility of the 74HC series with TTL logic devices. For the 74HCT series V_{IH} is 2.0 volts.

The 74AC series is similar to the 74HC series — that is, it has open-drain technology — but differs in the amount of current it can sink in the logic low state ($I_{OL} = 24$ mA). The AC series is also faster than the HC series.

The 74ACT series is the AC series but is directly compatible with TTL logic because it has an input voltage requirement for a logic high (V_{IH}) of 2.0 volts.

After load currents have been examined, Table 12.5 will be presented, which allows easy reference for the voltage level specifications of the various logic families discussed. At that time examples will also be given.

Review Questions

1. What designation is used to signify standard TTL logic ICs?
2. What is the difference between 54 and 74 logic ICs?
3. What performance characteristic is improved with use of the Schottky transistor?
4. Over the 54/74 series, what performance improvements are noted with (a) 54L/74L series? (b) 54H/74H series? (c) 54LS/74LS series?
5. How is a CMOS IC designated if there is a 54/74 IC with which it is pin compatible?
6. What do the designations given below indicate?
   (a) 74C   (b) 74HC   (c) 74HCT   (d) 74AC   (e) 74ACT

12.6

Definition of logic voltage levels and loading specifications

Recall from Section 2.9 that logic voltage levels are simply the specified magnitudes of input and output voltages that define the maximum voltage of a logic 0 and the minimum voltage of a logic 1. Loading is the act of connecting outputs to inputs. Of primary interest are the resultant currents and output capacitance due to loading. Load currents are the result of connecting an output of a digital device to one or more inputs. Load capacitance is the capacitance of the output due to the input capacitance of the driven circuits. Various consequences can result from loading an output. For example, the output voltage levels may not meet specifications for a logic 1 or 0; there could be damage to the output circuitry due to excessive current demands; or propagation time delays may become unacceptable owing to the loading capacitance (the adding of parasitic capacitance). It is these effects of loading that are of interest.

Two terms that will be used to develop concepts that have evolved from loading considerations: *driver* and *driven*. A driver is an output that must drive the inputs connected to it to the appropriate logic level, either a logic 0 or 1. Figure 12.16 illustrates this concept. Device A has two outputs, O_1 and O_2, and device B has one output, O_1. Device A has one input, as do devices B and C, labeled I_1 for each device. These inputs and outputs are accessible via pins on the IC. The outputs of Figure 12.16 are the driver pins and the inputs are the driven pins. A device may have both driver and driven pins. The memory device of Figure 9.23 provides confirmation of this, because $\overline{CS}$, $\overline{RD}$, $\overline{WE}$, and the address inputs A_0 through A_9 are examples of driven pins, whereas I/O pins function as both driver and driven pins. The I/O pins are driven during write operations and are driver pins during read operations.

Figure 12.16

Illustration of driver and driven.

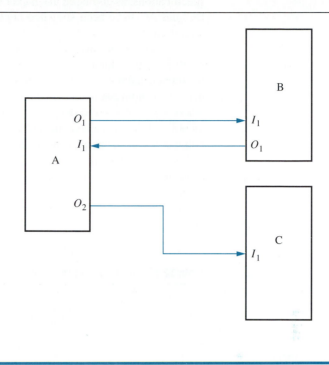

Example 12.1

Identify whether the pins of the page select decoder and the ROMs of Figure 9.20 are drivers, driven, or both.

Solution

Input pins are driven pins, output pins are driver pins, and those that serve as both input and output are both driven and driver, depending on the operation being executed.

The page selector (74LS138) has driver pins $\overline{O}_0$ through $\overline{O}_5$ and driven pins A_0 to A_2 and E_3. The address pins of the 2764-0 are driven pins, as are pins $\overline{CE}$ and $\overline{OE}$. The output pins O_0 through O_7 of the 2764-0 are driver pins. The address pins of both 8128's are driven pins, as are pins $\overline{CE}$, $\overline{OE}$, $\overline{WE}$, and the I/O data pins for write operations. The driver pins for the 8128's are the I/O data pins in a read mode of operation. The input pins of the *NOR* gate are driven, and its output pin is a driver.

Inputs of semiconductor digital-type devices have specifications that must be met by their drivers in order for the driver and driven to be compatible. If driver and driven devices are incompatible, some circuitry must be placed between them to make them compatible. Recall that this circuitry is known as an interface, or glue, because it lies between the two devices. Interface circuitry can be as simple as a resistor or fairly complex, with multiple transistors.

To determine if two digital devices are electronically compatible, use the manufacturer's data specifications for the input and output pins of these devices. The driver of every driven pin must be checked for compatability of both voltage levels and loading current. One need know very little of the internal electronics to make these comparisons. Usually just knowing which families of technology (TTL, MOS, CMOS, etc.) are involved is sufficient.

Manufacturers categorize the data specification of a digital device into two basic groups, which are listed as DC or AC characteristics. DC characteristics are concerned with voltage and current specifications, and AC characteristics provide timing specifications, as discussed in Chapter 9. The reader should note that the meanings of DC and AC have been stretched beyond their original meanings when applied to circuit analysis.

To review the symbolic notation used: The data sheet of a digital device, whether it is a logic gate or a memory device, provides output and input DC parameters using an O subscript to denote an output parameter and an I subscript to denote an input parameter. If the parameter is a voltage, a V is used to identify it, whereas an I is used to denote a current. For example, if a manufacturer wishes to identify an output voltage, the notation V_O is used, and V_I is used to denote an input voltage. I_I is used to represent an input current and I_O to indicate an output current. In addition, manufacturers use the subscripts L and H to indicate the specified logic level, where L (low) designates a logic 0 and H (high) indicates a logic 1. Thus V_{IH} signifies an input (I) voltage (V) which is a logic 1 (H), and I_{OL} signifies an output (O) current (I) in the 0 logic state (L). Table 12.1 shows all possible symbolic representations of input and output voltages and currents in either logic state.

Table 12.1   Symbolic representations of input and output DC parameters

DC parameter	Explanation of parameter
V_{OH}	Output voltage in the logic 1 state
V_{OL}	Output voltage in the logic 0 state
V_{IH}	Input voltage in the logic 1 state
V_{IL}	Input voltage in the logic 0 state
I_{OH}	Output current in the logic 1 state
I_{OL}	Output current in the logic 0 state
I_{IH}	Input current in the logic 1 state
I_{IL}	Input current in the logic 0 state

Figure 12.17

Parameters of comparison.

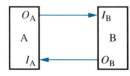

Driven	Driver
A (I_A)	B (O_B)
V_{IH}	V_{OH}
V_{IL}	V_{OL}
I_{IH}	I_{OH}
I_{IL}	I_{OL}
B (I_B)	A (O_A)
V_{IH}	V_{OH}
V_{IL}	V_{OL}
I_{IH}	I_{OH}
I_{IL}	I_{OL}

Figure 12.17 is an illustration of two devices that have both input and output pins, the input pins represented by an I and the output pins by an O. The outputs are drivers and the inputs are driven by these drivers. Figure 12.17 also contains a table that indicates which parameters of devices *A* and *B* are to be compared for compatibility. Recalling that the driver output parameters must be compatible with the input parameters of the driven, a comparison must be made of both voltage and current parameters for the logic 0 and logic 1 states. As shown, O_A is the driver and I_B is the driven. In the first row of that table, V_{IH} and V_{OH} are compared for voltage magnitude and polarity compatibility, as are V_{IL} and V_{OL} in the second row. The third and fourth rows contain current parameters I_{IH}, I_{OH}, I_{IL}, and I_{OL}, which must be compatible in magnitude and direction. A comparison must also be made for O_B as the driver pin and I_A as the driven pin, which is done in the next four rows.

Voltage compatibility

To understand how to compare the input and output voltage parameters of Figure 12.17, the driven input parameters dictate criteria for the driver (output). That is, if an input pin has a minimum voltage level for a logic 1 (V_{IH}), the driver (V_{OH}) must meet or exceed this level if the input and output are to be voltage compatible for the logic 1 state. Also, the driven pin has a maximum acceptable voltage for a logic 0 (V_{IL}), and V_{OL} must be equal to or less than this maximum value in order for V_{IL} and V_{OL} to be compatible. Polarities must be the same, which for most modern semiconductors is not a problem, since TTL logic has set a pseudostandard of +5 volts DC for the power supply voltage (V_{CC}). The required relationships between the input and output voltages are given in Table 12.2.

To visualize the relationships of input voltages V_I and output voltages V_O of Table 12.2, refer to Figure 12.18. As previously stated, voltage level criteria are determined by the input voltage levels of the driven device; hence Figure 12.18 shows V_{IH} and V_{IL} as establishing baseline requirements that the driver must meet or even exceed. These input and output voltage levels are guaranteed specifications that are supplied by the manufacturer. If the logic 1 level of the output voltage V_{OH} is theoretically equal to or greater than V_{IH}, the two devices are voltage compatible for the logic 1 level; otherwise they are not compatible and interfacing must be done. The equal-to condition is too close for a good design and is ignored as acceptable. Therefore, Figure 12.18 represents V_{OH} as exceeding V_{IH} by

$$V_{OH} - V_{IH} = V_{NH} \tag{12.1}$$

As illustrated in Figure 12.18, V_{NH} is a voltage safety factor, which is the amount of voltage variation that can exist between output and input under worst-case conditions (minimum voltage levels). This voltage safety factor is known as the noise margin for the logic 1 state. Similarly, V_{IL} is a guaranteed maximum voltage that can be input to the device and be interpreted by the device as a logic 0, which establishes a baseline maximum voltage requirement for V_{OL}, as indicated in Figure 12.18. V_{NL} is the noise margin for the logic 0 state and is equal to

$$V_{IL} - V_{OL} = V_{NL} \tag{12.2}$$

Table 12.2 V_O and V_I criteria

V_{IH} (min)	<	V_{OH} (min)
V_{IL} (max)	>	V_{OL} (max)

Figure 12.18

V_I and V_O voltage level specifications.

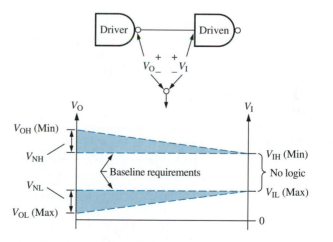

Table 12.3   TTL voltage specifications

V_{IH}	=	2.0 volts min
V_{IL}	=	0.8 volt max
V_{OH}	=	2.4 volts min
V_{OL}	=	0.4 volt max

Understand the reasoning of Table 12.2 well enough to duplicate it without reliance on memory.

Voltage specifications for standard TTL (54/74 series) are given in Table 12.3.

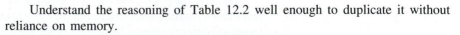

Example 12.2

Suppose that the voltage specifications for devices *A* and *B* of Figure 12.17 are as follows.

1. Devices *A* and *B* are TTL devices; therefore, refer to Table 12.3.
2. $V_{IH} = 3.2$ volts, $V_{IL} = 0.2$ volts
   $V_{OH} = 2.7$ volts, $V_{OL} = 0.4$ volts

Are these devices voltage compatible for conditions 1 and 2?

Solution

Applying the inequalities of Table 12.2 for the specifications of condition 1,

1.

$$V_{IH} = 2.0 \text{ volts} < 2.4 \text{ volts} = V_{OH}$$
$$V_{IL} = 0.8 \text{ volt} > 0.4 \text{ volt} = V_{OL}$$

Because all of the inequalities of Table 12.2 agree with the specifications of condition 1, devices *A* and *B* are voltage compatible. The current specifications for devices *A* and *B* must also be compared for compatibility.

2. Filling in the values of Table 12.2 with the specified values for condition 2,

$$V_{IH} = 3.2 \text{ volts} < 2.7 \text{ volts} = V_{OH}$$

Because this inequality is not true, devices *A* and *B* are not compatible in the logic 1 state. Testing for the logic 0 state

$$V_{IL} = 0.2 \text{ volt} > 0.4 \text{ volt} = V_{OL}$$

which is also not true. As a result of these comparisons, devices *A* and *B* are not voltage compatible in either logic state. Therefore, if the two devices are to be used together, an interface must be used to make the two devices voltage compatible.

Current compatibility

To understand the meaning of current specifications, it is necessary to first understand the meanings of the terms *current sink* and *current source*. The *NAND* gates of Figure 12.19 (*NAND* gates are the dominant logic gate used in industry and for that reason are often used in a generic sense) are used to illustrate the meaning of these two terms. Current source describes an input or output that acts as a source of current, such as the output of *NAND* gate *A* in Figure 12.19a, or the input of *NAND* gate *B* in Figure 12.19b. A current sink is an input or output that acts as receiver (a sink) of a source current, such as input *B* in Figure 12.19a and the output of device *A* in Figure 12.19b. The direction and magnitude of a current depend on the logic level and the technology of the device (TTL, MOS, etc.).

Figure 12.19

Examples of sourcing and sinking of current.

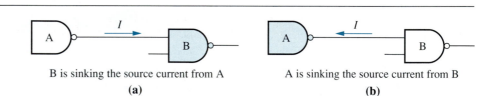

B is sinking the source current from A   A is sinking the source current from B

(a)   (b)

Because the pin of a device (either input or output) may be a sink for one logic level and a source for the other, industry has established a common direction for all input and output currents, regardless of the logic level. As indicated in Figure 12.20, all currents, I_O or I_I, are assumed to flow into the device, which also says that

Figure 12.20

The assumed current directions for output current I_O and input current I_I.

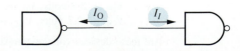

industry assumes all pins to be current sinks. Intuitively, if not in fact, this cannot be because there must be one or more current sources and current sinks whenever two or more pins are connected (if there is at least one current source, there must be at least one current sink because what goes in must come out). Figure 12.21 illustrates the physical logic of this statement; in the figure there are many current sinks but not one source of current for those sink currents, which is illogical. When a current is actually a source current (i.e., one that flows in the opposite direction from that assumed in Figure 12.20), industry assigns a negative sign to its magnitude in the DC specifications. For instance, a statement in the DC specification table of a device that $I_{IL} = -1.6$ mA means that for this input, the logic 0 level current (I_{IL}) is a source current (the negative sign states it is flowing out of the pin) with a magnitude of 1.6 mA. Of course, it is implied that there must be one or more current-sinking outputs (I_{OL}) connected to this input pin that are capable of handling (sinking) this source current of 1.6 mA.

Figure 12.21

The assumed current direction for multiple device connections.

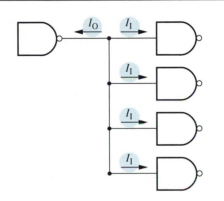

Current considerations are most important if the technology (bipolar or MOSFET) used to implement a design requires source and/or sink current to operate. Bipolar (TTL, RTL, ECL, etc.) technology is a current-controlled technology and therefore requires certain currents to operate, whereas MOSFET technology (NMOS, PMOS, or CMOS) is a voltage-controlled technology and does not require either a sink or a source current. For instance, recall that TTL technology requires an input source current (to turn on Q_1) for the logic 0 state (I_{IL}) of an input. This implies that there must also be an output, or outputs, connected to this input that can sink that source current to ground; otherwise, where would the source current go? Hence, the specified magnitude of the sink current (I_{OL}) of the output must be larger in magnitude than that of I_{IL}, or the sum of I_{IL}s if the output is driving more than one input, and the sink and source currents must be of opposite sign. In contrast, I_{IH} for TTL is an input sink current that is not required for operation but flows if it is provided with a path to ground. This path is provided via the driver output. There is a problem for this condition if the output has to drive too many inputs, such that the sum of the individual input sink currents exceeds that of the source current (I_{OH}) specification. If this occurs, the voltage drops internal to the driver device are large enough that the output voltage cannot reach the specified minimum voltage level to satisfy the guaranteed minimum voltage (V_{OH}) for a logic 1. In summary, for TTL technology we must be concerned with both the sinking and the sourcing of current for both logic states. For the logic 0 condition ensure that the I_{OL} sink current exceeds the source current I_{IL}. For the logic 1 state ensure that the source current I_{OH} is greater than the sink current I_{IH} so that the output is at least minimum voltage V_{OH}. Table 12.4 gives the current load specifications for standard TTL technology.

Example 12.3

If the logic circuit of Figure 12.22 is implemented with standard TTL *NAND* gates (7400), is there current compatibility?

Figure 12.22

NAND gate *A* driving *NAND* gates *B*, *C*, and *D*.

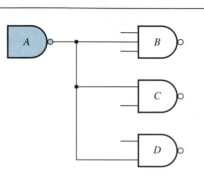

Solution

To test for current compatibility consider both the logic 0 and logic 1 states; if compatibility fails for either logic state, the logic gates are not compatible. Testing the logic 1 state for current compatibility, compare I_{IH} of *NAND* gates *B*, *C*, and *D* of Figure 12.22 with I_{OH} of *NAND* gate *A*. The current specifications of Table 12.4 state that I_{IH} of each TTL input is a sink current (positive sign) and its magnitude is guaranteed not to exceed 40 μA. Because there are three inputs (gates *B*, *C*, and *D*), 120 μA (3 $\times$ 40 μA) of sink current is drawn from the source current I_{OH} of gate *A*. $I_{OH} = -400$ μA, which means it can reliably source up to 400 μA (note that I_{IH} and I_{OH} are opposite in sign) and therefore is capable of sourcing the 120 μA. These gates are compatible for the logic 1 state.

Testing the logic 0 state, the current specifications of Table 12.4 are $I_{IL} = -1.6$ mA, which means that in the logic 0 state a TTL input sources 1.6 mA of current. This source current must be present in order for the *NAND* gate to operate. Because there are three of these inputs connected to the output of gate *A* in Figure 12.22, I_{OL} of gate *A* must exceed 3 I_{IL} (3 $\times$ 1.6 mA = 4.8 mA). For current direction compatibility, the source and sink currents must be of opposite sign. From the current specifications, $I_{OL} = 16$ mA, which exceeds the source current magnitude requirement of 4.8 mA and is also sign compatible with I_{IL}. Because there is current compatibility for both logic states, the TTL-implemented logic design of Figure 12.22 is current compatible.

Table 12.4   Current specifications for TTL technology

I_{IH}	=	40 μA max
I_{IL}	=	-1.6 mA max
I_{OH}	=	-400 μA max
I_{OL}	=	16 mA max

Example 12.4

Can a TTL output drive 12 TTL inputs relative to current requirements?

Solution

With 12 inputs the output must sink 12 $\times$ I_{IL} (12 $\times$ [-1.6 mA] $= -19.2$ mA) of source current for the logic 0 state. Because $I_{OL} = 16$ mA, one TTL output cannot sink 19.2 mA and the logic design would not be compatible in the logic 0 state, which means the choice of technology must be altered or the design changed.

Even though current compatibility has failed based on the logic 0 state analysis, for practice also determine current compatibility (or noncompatibility) for the logic 1 state. $I_{IH} = 40$ μA maximum; therefore, under worst case conditions (all 12 inputs sinking the maximum current) I_{OH} must be capable of sourcing 480 μA (12 $\times$ 40 μA) of current without degradation of the voltage level required for V_{OH}. Table 12.4 shows that the manufacturer guarantees just 400 μA ($I_{OH} = -400$ μA) of source current, which is not sufficient. Hence, the logic circuit also fails current compatibility for the logic 1 state.

To provide a more convenient method of evaluating load currents, TTL circuit manufacturers have devised the *unit load* (UL). A unit load is a standardized load that is representative of maximum TTL input loading on a TTL output. The *NAND* gate was chosen to be representative of TTL logic. The load currents of Table 12.4 are for a TTL *NAND* gate, and therefore the unit load is based on these values. Because the input parameters set the criteria to be met by the driver, it is logical that the input load currents be used to define the UL. Because there are two logic states and the input load current is different for each, the UL must be defined for both logic states. The UL is defined to be

$$\text{UL (logic 1 state)} = 40\ \mu\text{A} \qquad (12.3)$$

$$\text{UL (logic 0 state)} = 1.6\ \text{mA} \qquad (12.4)$$

Note that the UL definitions are based on magnitudes, and therefore the designer must be conscious of sign (current direction) when mixing logic technologies, such as TTL and CMOS.

Example 12.5

The current DC characteristics of a TTL logic device are listed in the manufacturer's data specifications as

$$I_{IH} = 1.25\ \text{UL and } I_{IL} = 2\ \text{UL}$$

1. How much sink and source current does this represent, and what minimum requirements do these specifications impose on the driver?
2. How many of these types of input can a standard TTL output drive?

Solution

1. To determine the amount of load current, multiply the number of ULs by the magnitude of the current standard for the input logic level in question. Then

$$I_{IH} = (\text{UL}) \times 40\ \mu\text{A} = 1.25(40)\ \mu\text{A} = 50\ \mu\text{A}$$

and

$$I_{IL} = -(\text{UL}) \times 1.6\ \text{mA} = -2 \times 1.6\ \text{mA} = -3.2\ \text{mA}$$

2. $I_{IH} = 50$ μA states that this input could sink as much as 50 μA from the driver. Therefore, the driver's I_{OH} specification must not exceed 50 μA of source current in order that degradation of V_{OH} does not occur. For a standard TTL output $I_{OH} = -400$ μA; because $I_{IH} = 50$ μA, eight (400 μA / 50 μA) such inputs could be driven for the logic 1 state. For the logic 0 state $I_{IL} = -3.2$ mA, meaning that this input will source 3.2 mA, which the driver output must be capable of sinking. Determining the number of inputs that can be driven by a standard TTL output for the logic 0 state, $I_{OL} / I_{IL} = 16$ mA / 3.2 mA = five inputs. A standard TTL output can drive eight inputs based on logic 1 load current specifications and five inputs based on logic 0 load current specifications. Obviously it cannot do both, and therefore the worst case situation must be used, which is five inputs.

The terminology used to describe the number of inputs an output of the same technology can drive is termed fan-out. The fan-out of an output is determined by dividing the output current for both logic states by the input current of the same logic state. The integer of that quotient is the fan-out for that output. For instance, the fan-out for a standard TTL output driving a standard TTL input is

$$\frac{\text{Fan-out}}{\text{for } H} = \frac{I_{OH}}{I_{IH}} = \frac{400\ \mu A}{40\ \mu A} = 10$$

and

$$\frac{\text{Fan-out}}{\text{for } L} = \frac{I_{OL}}{I_{IL}} = \frac{16\ mA}{1.6\ mA} = 10$$

For this case the fan-out is the same, regardless of the logic state.

Example 12.6

Determine the approximate value of I_{OL} for a TTL output that has a fan-out of 15.

Solution

This output can drive 15 UL inputs and because 1 UL = 1.6 mA, I_{OL} = 15 × I_{IL} = 15 × 1.6 mA = 24 mA.

Having studied logic voltage levels and load currents and their relevance to compatibility, let us summarize them for various TTL and CMOS families in table form. Table 12.5 is a listing of the logic voltage levels and load currents for some of the logic families previously mentioned.

The *NAND* gate has become the pseudostandard, which is the reason for using the *NAND* gate as the gate of comparison. Many of the logic devices (*OR, AND, NOR*, etc.) have these same specifications, so they are representative of the entire series. The specifications of Table 12.5 represent the series rather than just the *NAND* gate of that series.

Table 12.5   Voltage levels and load currents for various logic families

	Series	V_{IL} max (volts)	V_{IH} min (volts)	V_{OL} max (volts)	V_{OH} min (volts)	I_{IL} max	I_{IH} max (μA)	I_{OL} max (mA)	I_{OH} max (mA)
TTL family	74	0.8	2.0	0.4	2.4	−1.6 mA	40	16	−0.4
	74L	0.7	2.0	0.4	2.4	−0.18 mA	10	3.6	−0.2
	74H	0.8	2.0	0.4	2.4	−2.0 mA	50	20	−0.5
	74S	0.8	2.0	0.5	2.7	−2.0 mA	50	20	−1.0
	74LS	0.8	2.0	0.5	2.7	−0.4 mA	20	8.0	−0.4
CMOS family	74C	1.5	3.5	0.33	4.5	−1.0 μA	1.0	0.36	−0.36
	74HC	1.2	3.5	0.33	3.84	−1.0 μA	1.0	4.0	−4.0
	74HCT	0.8	2.0	0.33	3.84	−1.0 μA	1.0	4.0	−4.0
	74AC	0.8	3.5	0.33	3.84	−1.0 μA	1.0	24	−4.0
	74ACT	0.8	2.0	0.33	3.84	−1.0 μA	1.0	24	−4.0

Table 12.5 clearly shows those technologies that are current controlled and those that are voltage controlled. That is, the input current I_{IL} for the current-dependent TTL family is in the milliampere range, whereas that for the CMOS family is in the microampere range. Input current for a CMOS device is leakage current. Also notice those series that have the greatest noise margins V_{NH} and V_{NL}, according to Equations 12.1 and 12.2, and the largest output current sinking capability for a logic low (I_{OL}).

Example 12.7

From the specifications of Table 12.5 and Equations 12.1 and 12.2, determine the noise margins for the following: (a) standard TTL   (b) 74LS   (c) 74C   (d) 74ACT

Solution

(a)
$$V_{NH} = V_{OH} - V_{IH} = 2.4 - 2.0 = 0.4 \text{ volts}$$
$$V_{NL} = V_{IL} - V_{OL} = 0.8 - 0.4 = 0.4 \text{ volts}$$

(b)
$$V_{NH} = 2.7 - 2.0 = 0.7 \text{ volts}$$
$$V_{NL} = 0.8 - 0.5 = 0.3 \text{ volts}$$

(c)
$$V_{NH} = 4.5 - 3.5 = 1.0 \text{ volts}$$
$$V_{NL} = 1.5 - 0.33 = 1.17 \text{ volts}$$

(d)
$$V_{NH} = 3.84 - 2.0 = 1.84 \text{ volts}$$
$$V_{NL} = 0.8 - 0.33 = 0.47 \text{ volts}$$

Recall that noise margin is the measure of acceptable voltage variation within a logic level. It can then be said that noise margin is a voltage tolerance. If this variation in voltage were due to noise, such as an unwanted voltage spike, the noise margin for a logic level (*H* or *L*) would specify how large that spike could be before the input voltage level of the driven device would be out of the specified voltage limits for a logic 0 (V_{IL}) or logic 1 (V_{IH}). This tolerance for noise is referred to as *noise immunity*. Example 12.7 shows that standard TTL has the least immunity to noise for a logic high (0.4 volts) and the 74ACT series has the largest immunity to noise (1.84 volts). The 74LS series has the least immunity to logic 0 level noise, and the 74C series has the largest. High values of noise immunity are a most desirable characteristic, especially if the logic circuits are to be placed in a noisy environment (e.g., in close proximity to electric motors that create power surges).

Example 12.8

Suppose you had to design a logic circuit that was to be placed in an environment where 0.8-volt spikes often appeared on the connecting lines (printed circuit lines) of the logic circuit. From the logic series of Example 12.7, which would you use?

Solution

Because this 0.8-volt spike could appear in either logic level, V_{NL} and V_{NH} must be equal to or greater than 0.8 volts. Only the 74C series meets this design criterion.

It is sometimes desirable to mix logic families within a design. To do this the voltage levels and load currents must be checked for compatibility using the same methods already mentioned.

Example 12.9

Suppose that an older 74 series logic circuit design is to be interfaced with low-power Schottky (LS) and CMOS 74AC logic circuits. Check for both input and output compatibility and determine whether the 74LS and 74AC logic can be interfaced directly or interfacing circuitry must be used.

Solution

See the table of Figure 12.17 for the voltage levels and current levels that are to be compared for compatibility, and Table 12.5 for the criteria for voltage level compatibility. The criteria for load current compatibility are the following.

1. Are input currents required?
2. If input currents are required, are they necessary in the logic low state, the logic high, or both?
3. Are the required sink and source currents compatible in both magnitude and sign (direction)?

Create a table similar to the one in Figure 12.17. The column to the left contains driver specifications and column to the right contain specifications for those inputs that are to be driven.

Comparing voltage logic levels of the 74 and 74LS series,

$$V_{OL} = 0.4 \text{ volts} < V_{IL} = 0.8 \text{ volts}$$

which agrees with the criteria of Table 12.2, and

$$V_{OH} = 2.4 \text{ volts} > V_{IH} = 2.0 \text{ volts}$$

which also agrees with the criteria of Table 12.2. Therefore, the 74 and 74LS logic circuits are voltage compatible for 74 logic driving 74LS logic.

Now compare the 74 and 74LS series for current compatibility. From the stated criteria the answer to the first question, which asks if input current is required, is yes, because both series are from the TTL family. Now that we know that an input current is required, determine for which logic state. From previous discussion and the analysis of Figure 12.5, we know that input current is necessary for just the logic low state and in this logic state the driven input acts as a current source and the driver output must be capable of sinking this source current. Then the condition $I_{OL} > I_{IL}$ must be met. From Table 12.6 we find for the condition when the 74 series is driving 74LS

$$I_{OL} = 16 \text{ mA} > I_{IL} = 0.4 \text{ mA}$$

which means that the 74 series can sink up to 16 mA and the 74LS series requires a source (the negative sign) current of just 0.4 mA. Hence, the two series are compatible for the logic low state.

Figure 12.5b shows that input current is not required for the logic high state. However, the specifications of Table 12.6 state that the 74 can source up to a magnitude of 0.4 mA in the logic high state if required ($I_{OH} = -0.4$ mA) and specifications for the 74LS series state that it will drain (sink) a maximum of 20 μA (leakage) if the source current is available. Therefore, the 74LS series would draw 20 μA from the 74 series, but it is not necessary for operation. As a result, the logic low state specifications must be met for load current compatibility. Because the 74 and 74LS series are compatible for both logic voltage levels and load currents, the two series can be directly connected when a 74 series device drives a 74LS device, if they have the specifications of Table 12.5 (check actual data sheets of logic devices in question).

Checking compatibility for the 74 series driving the 74AC series, the first and second rows of Table 12.6 show

$$V_{OL} = 0.4 \text{ volts} < V_{IL} = 0.8 \text{ volts}$$

and

$$V_{OH} = 2.4 \text{ volts} < V_{IH} = 3.5 \text{ volts}$$

Table 12.6   Voltage logic levels and load current comparisons

Driver 74		Driven 74LS		Driven 74AC	
V_{OL} =	0.4 volts	V_{IL} =	0.8 volts	V_{IL} =	0.8 volts
V_{OH} =	2.4 volts	V_{IH} =	2.0 volts	V_{IH} =	3.5 volts
I_{OL} =	16.0 mA	I_{IL} =	−0.4 mA	I_{IL} =	−1.0 μA
I_{OH} =	−0.4 mA	I_{IH} =	20.0 μA	I_{IH} =	1.0 μA
74LS		**74**			
V_{OL} =	0.5 volts	V_{IL} =	0.8 volts		
V_{OH} =	2.7 volts	V_{IH} =	2.0 volts		
I_{OL} =	8.0 mA	I_{IL} =	−1.6 mA		
I_{OH} =	−0.4 mA	I_{IH} =	40.0 μA		
74AC		**74**			
V_{OL} =	0.33 volts	V_{IL} =	0.8 volts		
V_{OH} =	3.84 volts	V_{IH} =	2.0 volts		
I_{OL} =	24.0 mA	I_{IL} =	−1.6 mA		
I_{OH} =	−4.0 mA	I_{IH} =	40.0 μA		

Table 12.2 shows that logic voltage level compatibility fails for the high state, and hence the two series are not directly compatible. To interface the two series, where a 74 series is driving a 74AC series, there must be an interface circuit that corrects the logic high voltage level.

Checking for compatibility when 74LS series is driving 74 series logic,

$$V_{OL} = 0.5 \text{ volts} < V_{IL} = 0.8 \text{ volts}$$

and

$$V_{OH} = 2.7 \text{ volts} > V_{IH} = 2.0 \text{ volts}$$

which agrees with the inequalities of Table 12.2. The two series are logic voltage level compatible. Current specifications state that a 74 series logic device must source 1.6 mA of current in the logic low state and the 74LS can sink up to 8.0 mA. Hence, they are compatible for the logic low state, and, current is not required for a logic high, and the 7400 sinks (40 μA) less than the 74LS can source (0.4 mA) without degradation of the output voltage specified for V_{OH}, the two series are also current compatible.

When 74AC drives 74 series logic the two are voltage and current compatible.

In Example 12.9 we checked for load current compatibility, but did not determine the fan-out for each configuration and we know that an output may have to drive many inputs. Load current specifications limit the number of inputs that can be connected to an output. For instance, when the 74 series logic is driving 74LS logic, the fan-out is 40 (16 mA / 0.4 mA). Any design that has multiple inputs connected to an output must be checked for current loading to ensure that load current specifications have not been exceeded. If they are, a driver must be used as an interfacing circuit between the output and inputs. Drivers are buffers without the control for Hi-Z state and with large output load current specifications.

Example 12.10

Suppose that the NMOS MPU of Figure 10.10 has DC load current specifications

$$I_{OL} = 2 \, \text{mA}, I_{OH} = -400 \, \mu\text{A}$$

and the input current for both logic levels is just leakage current (recall its NMOS technology) with values of $\pm 10 \, \mu\text{A}$. Is this MPU load current compatible with its memory?

The logic level input currents for the 8128's (NMOS) are leakage currents I_{LI} and I_{LO}, with values of $\pm 10 \, \mu\text{A}$. Output load current specifications are $I_{OL} = 2.1 \, \text{mA}$ and $I_{OH} = -1$ mA. The 2716 has input leakage currents that are $\pm 10 \, \mu\text{A}$, and its output load currents are $I_{OL} = 2 \, \text{mA}$ and $I_{OH} = -400 \, \mu\text{A}$.

Solution

When a 8128 is being read by the MPU the 8128 data pins are driving the MPU data pins. Compare the output load current of the 8128 to the input load current of the MPU for both logic levels. The specifications indicate that the 8128 can sink 2.1 mA in the logic low state ($I_{OL} = 2.1 \, \text{mA}$), but the MPU requires no input current, just leakage values of $\pm 10 \, \mu\text{A}$. This is a fan-out of 210 (2.1 mA / 10 μA). For the logic high state, the 8128 can source up to 1 mA ($I_{OH} = -1 \, \text{mA}$) and the MPU again requires no input current but draws a leakage current of 10 μA. Hence, the fan-out of the 8128 for a logic high is 100. Then the worst case is a fan-out of 100. Hence, one hundred 8128 inputs can be connected to this MPU.

For reading the 2716, its data pins drive the data pins of the MPU. The 2716 can sink 2 mA of current in the logic low state and source 400 μA in the logic high state. The MPU requires no input current, but 10 μA of leakage current will flow. Hence, the worst case fan-out is 40.

The MPU is writing to the 8128 when the MPU outputs an address to its address pins, data to its data pins, or control signals to its enable pins. Checking the fan-out of the MPU relative to its writing to an 8128, the MPU can sink 2 mA of current in the logic low state and source 400 μA in the high state. The input current to the 8128 is a leakage current of $\pm 10 \, \mu\text{A}$ values; hence the worst case fan-out is 40. The 2716 has the same write conditions as the 8128 (except the MPU writes only to the address and enable pins) and therefore also has a worst case fan-out of 40.

Considering all worst case fan-outs, this MPU can have a total of forty 2716's and / or 8128's connected to it, which is well within the three memory chips of this design. However, the I / O 74LS138 *NOR* gate loading must be also considered.

Reader's should make certain that they understand the criteria for DC compatibility and are able to evaluate the compatibility of various logic families. In addition to understanding compatibility, readers must be able to determine the fan-out of an output relative to the inputs it is driving.

Review Questions

1. What are the symbolic designations for the voltage logic levels and load currents? Explain the reasoning of those symbols used.

2. What are the requirements for compatibility relative to logic voltage levels and load currents?

3. What is fan-out, and what role does it play in the design of logic circuits?

4. What is a UL?

5. Why is it that MOS family devices do not require current to operate?

6. Why do bipolar family devices require current to operate? Are both input and output currents required, and for which logic state?

12.7

Propagation delay and load capacitance

No voltage or current can change its magnitude, polarity, or direction instantaneously, as is the case of all things with mass. There is a delay from the time the input of a logic device theoretically causes a change in the output logic voltage level to the time when it actually occurs. Figure 12.23 illustrates this concept.

Figure 12.23

Logic level transition delay.

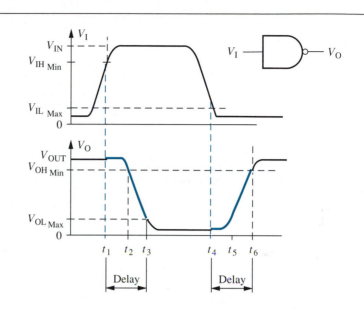

Figure 12.23 shows the input logic level going from a logic 0 to a logic 1 and the input of the logic circuit (a *NAND* gate is shown for the purpose of demonstration) detects the logic 1 level at the time t_1. The *NAND* gate begins to drive its output to the logic 0 level, but there is a delay from time t_1 to t_2. This delay is due to internal parasitic capacitance of the electronic switches requiring time to charge or discharge. Once they are charged or discharged to the proper voltage level, the electronic switches begin to make the transition to the opposite state, which occurs in the interval t_2 to t_3. At time t_3 the output voltage is at a logic 0 voltage level and the transition from a logic 1 to a logic 0 has been completed. Hence, the total delay time, as measured from when the input detected a logic 1 to when the output reached a logic 0, is $t_3 - t_1$.

As the input voltage level of Figure 12.23 traverses from a logic 1 to a logic 0, there is also a delay in the output voltage going from a logic 0 to a logic 1 due to the same parasitic capacitors. When the input voltage reaches the logic 0 level, at time t_4, the electronic switch begins to change states. At time t_5 the output voltage actually begins to change from a logic 0 to a logic 1. At time t_6 the output voltage has reached the logic 1 state and the transition is complete. The total delay time is $t_6 - t_4$.

These delay times are known as propagation delay time and are symbolically represented as t_p. Propagation delay time was first introduced in Chapter 2. If the propagation delay time is the delay time in going from a low (0) to a high (1), it is designated t_{pLH}, and t_{pHL} is used to signify the propagation delay time in going from a high to a low.

Propagation delays, both t_{pLH} and t_{pHL}, are measured with respect to the 50% level between a logic 1 (V_{OH}) and a logic 0 (V_{OL}). Using the same waveforms as Figure 12.23, propagation delay parameters are defined as illustrated in Figure 12.24.

Figure 12.24

Propagation delay times.

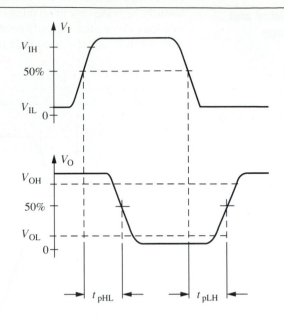

In addition to parasitic capacitance, propagation delay times are affected by capacitance that appears at the output of a logic device. As a result, every output pin must not have its specified load capacitance exceeded in order that the propagation delay time remain within the manufacturer's specifications. That is, each input pin has an input capacitance (C_{IN}) associated with it, and when connected in parallel these input capacitances are additive. Hence, as inputs are connected to an output, these input capacitances add and increase the output load capacitance. However, their sum must remain below the specified load capacitance (C_L) of the output pin. Most often manufacturers provide a graph of propagation delay versus load capacitance. From this graph the user can determine the allowable load capacitance for an acceptable propagation delay time.

Example 12.11

If the load capacitance of an 8128 output pin is 100 pF and the input capacitance of a 74C00 is 6 pF, how many 74C00 inputs can be connected to an 8128?

Solution

Considering just capacitive loading, sixteen (100/6) 74C00 inputs can be connected to an 8128 output.

As a pulse is propagated through logic circuits, the total propagation delay of the pulse is the sum of propagation delay time for each logic circuit through which it passes.

Example 12.12

If a pulse passes through five 74C00 *NAND* gates, what is the resultant propagation delay time? Assume a load capacitance of 60 pF, which results in a propagation time of 58 ns.

Solution

The total delay time is 290 ns (5 × 58 ns).

Propagation delay times are most critical when two events must occur simultaneously, such as when two logic levels are *AND*ed. If these two events begin from their source with proper synchronization but each must pass through different logic circuits, they can arrive at their destination out of synchronization. This misalignment is due to their different propagation delay times.

Example 12.13

From the MPU read timing diagram of Figure 10.7a the address and read control signals originate from the MPU according to the timing indicated. When this timing diagram is applied to the system of Figure 10.10, how is it altered?

Solution

Some address bits are applied instantaneously to the 2716-0 (A_0 through A_{12}) and 8128's (A_0 through A_{10}), whereas others are delayed owing to the propagation delay of the 74LS138-3 and the *AND* gate. But even the enabling of the 74LS138-3 is delayed by the propagation delay of the *NOR* gate (approximately 5 ns). Thus, the total delay of the effects of address lines A_{11} through A_{15} is the sum of the propagation delay time of the *NOR* gate, the 74LS138-3, and the *AND* gate. If the propagation delay time for both gates is 5 ns and the 74LS138 has a propagation delay of 27 ns, the total is 37 ns. This is acceptable if $t_{AC} + t_{CC} - t_{AD}$ of Figure 9.15 is greater than the delay time. Table 9.6 shows that 250 ns + 450 ns − 600 ns = 100 ns is greater than 37 ns. Therefore, the delay is acceptable.

Review Questions

1. What is propagation delay, and why does it exist?
2. Why can propagation delay time be a problem?
3. Why does increasing the load capacitance increase the propagation delay?
4. Why does increasing the number of inputs an output must drive also increase the propagation delay time?
5. When logic circuits are cascaded, what effect does that have on the overall propagation delay time, and why is it affected in this manner?

12.8
Troubleshooting

This chapter has investigated the whys of logic level voltage specification, current, and capacitance loading. In Chapter 2 and this chapter these specifications were to ensure that the designs used were within specification limits for fan-out and interfacing various technologies. To troubleshoot on the basis of fan-out and the mixing of technologies, drive the output under test to the two logic states and measure them to verify that V_{OH} and V_{OL} are within specifications. If they are, there is no fault with that output. If an output fails to meet the output voltage specifications, it may be that the output is overloaded and is either sinking or sourcing too much current, as discussed in this chapter. The input voltage must also be verified as being within specifications.

To test for capacitance load compare the input pulse that gates the logic circuit on and off and measure the resultant propagation delay times. If these are greater than allowed, the amount of capacitance load must be reduced. To reduce the capacitance loading attempt to replace the driven logic circuit with logic circuits that have less input capacitance. Alternatively, buffer the output of the driver, therefore seeing just a single input.

The troubleshooting techniques relative to the material of this chapter are similar in concept to those techniques previously developed; that is, input a known logic state and measure its resultant output state. From this, determine if input and output parameters are within specifications. If they are not in specification, check for incompatibility and loading requirements via the manufacturer's data sheets. If compatibility and loading are within specifications, then this specific device is out of specification due to an internal fault and must be replaced.

Review Questions

1. What are the results of loading problems?
2. How are loading problems detected?
3. How can loading problems be corrected?

Summary

- Digital circuitry contains electronic switches that operate in one of two states, on and off. There are bipolar and MOSFET electronic switches.

- The state of a bipolar electronic switch is controlled by the presence or absence of base current.

- The state of a MOSFET electronic switch is controlled by the presence or absence of a gate voltage.

- There are various bipolar-based logic circuits, but the dominant family is TTL. Also, there are various types of TTL logic, such as 54/74, 74S, and 74LS. These variations are the result of changing biasing resistance values and/or using the faster Schottky transistors.

- MOS family logic circuits also come in various alterations of the basic MOS circuit. There are MOS logic circuits available in NMOS, PMOS, and CMOS. NMOS and CMOS are the dominant MOS technologies.

- MOS technology enabled VLSI ICs to become a reality.

- Logic voltage levels and load current symbols are standardized, and manufacturers specify these voltage levels and load currents in data sheets. In order for two logic families to be compatible the logic voltage levels at the output of the driver must be of proper magnitude and polarity, so that these levels are properly interpreted as 1's and 0's at the input of the driven logic circuit.

- For load current compatibility, if load currents are required for either logic level there must be a current source and sink. The current sink must be capable of sinking the required source current.

- Noise margin is a measure of voltage tolerance that exists between output and input logic voltage levels for either logic level.

- When noise margin is applied to logic families' tolerance of noise, it is known as noise immunity.

- A high noise immunity is desired.

- Owing to the parasitic capacitance of electronic switches, they do not turn on or off instantaneously. This results in a propagation delay time.

- As a pulse propagates through a number of cascaded logic circuits, its propagation delay time is the sum of propagation delays of each individual logic circuit it passes through.

- Propagation delay time is a potential problem when synchronization must occur between two or more pulses, such as in MPU read and write operations.

- To test a node for loading and/or mismatched technologies, the output voltage parameters are measured.

- When a fault due to loading is detected (including too much input capacitance), a buffer (driver) could be the solution.

Problems

Section 12.2   The electronic switch

1. Explain why the voltage logic level V_{IL} cannot turn on the electronic switch of Figure 12.1.

2. Why does the gate voltage of the electronic switch of Figure 12.2 have to be of magnitude $V_{G\,ON}$, which is the same as V_{IH}, in order to turn it on?

3. What logic gate is formed by connecting two or more switches in series?

4. When switches are paralleled, what logic gate is formed?

Section 12.3 Fundamental electronic principles of logic gates

5. What is the primary function of Q_1 of the TTL inverter that is shown in Figure 12.4? Explain how it achieves that purpose.

6. What is the function of Q_2 in the schematic of Figure 12.4?

7. TTL logic has a totem pole output. Explain what is meant by this term and what its advantage is over the output of the electronic switch of Figure 12.1.

8. What is the essential difference in operation between the circuits of Figures 12.4 and 12.7?

9. When any emitter input (*A, B,* or *C*) of Figure 12.7 is at ground potential, what is the approximate magnitude of that emitter current if V_{CC} is 5 volts?

10. If the inputs of the *NAND* gate of Figure 12.7 were not connected — that is, they were left floating — what would be the resultant logic level at the output? This illustrates why inputs should always be connected — never leave an input unconnected.

11. What is the advantage of open-collector logic?

12. Explain why connecting open-collector outputs together results in *AND*ing the output logic of those gates.

13. Explain how the diodes at the inputs of TTL logic protect the circuit. Also explain why protection is required.

14. What is the essential difference between a standard bipolar transistor and a Schottky bipolar transistor?

15. Why does a Schottky transistor have a faster transition time than a standard bipolar transistor?

16. Why do the NMOS logic circuits drain less current from their power supply than does a TTL gate?

17. Why is MOS logic used to produce VLSI ICs and not TTL logic?

18. Explain why by just a quick examination of the circuits of Figure 12.10 the logic gates are easily identified as *NOR* and *NAND* gates.

19. Why do the circuits of Figure 12.10 justify the statement that the input currents of those circuits are only leakage currents?

20. Using switches, create equivalent circuits that can be used to verify the truth tables for the circuits of Figure 12.10.

21. Repeat Problem 20 for the circuits of Figure 12.11.

22. Why does CMOS drain less current than either TTL or NMOS?

23. What is significant in the wiring connections of Q_A and Q_B of Figure 12.11 relative to determining the resultant type of logic gate?

24. Using circuit analysis, prove that when Q_A is on, Q_{RA} is off and vice versa.

Section 12.4 Electronic memory cells

25. How does the static memory cell ensure that when one of its electronic switches is on the other is off?

26. Why must both row and column of both Figures 12.12 and 12.13 be high in order to access the cell?

27. Explain how read/write operations are performed on the memory cell of Figure 12.12.

28. Repeat Problem 27 for the memory cell of Figure 12.13.

29. What is the purpose of Q_W and Q_R in the memory cell of Figure 12.13?

30. Explain how a dynamic memory cell functions.

31. How is a memory cell programmed in the ROM memory cells of Figure 12.15?

32. What constitutes a memory location in the ROM of Figure 12.15, and how is it read?

Section 12.5 Designations of logic families

33. List designations used to identify the TTL logic families, and explain the difference in electronics from the TTL standard.

34. List designations used to identify the various types within the MOS family.

Section 12.6 Definition of logic voltage levels and loading specifications

35. What is meant by driver and driven?

36. When referring to the parameters of a driver logic circuit, are its input or output parameters of interest? Explain your answer.

37. Relative to compatibility, what is the significance of the table in Figure 12.17? Explain the significance in terms of voltage and current magnitudes and signs (polarity or direction).

38. Explain the significance of Table 12.2. Use the graph of Figure 12.18 as the basis of your explanation.

39. How does a manufacturer indicate a source or sink current in the DC specifications?

40. What is fan-out and how is it determined?

41. Check these logic families for compatibility.
    a. 74L driven by 74C
    b. 74S driven by 74HC
    c. 74H driven by 74AC
    d. 74HC driven by 74S
    e. 74C driven by 74

42. Check the I/O section of Figure 10.10 for compatibility if the I/O ports of Figure 10.9 are used.

43. Check the memory of Figure 9.17 for load current compatibility if the 2764 has the same DC specifications as a 2716 (see Example 12.10).

44. What would be the resultant propagation delay time of any one of the 74LS138 decoders of Figure 9.22 measured with respect from when the address was loaded on the address bus? Explain your answer(s). Assume the inverter is a 74LS04.

Section 12.7   Propagation delay and load capacitance

45. What causes a logic gate to have a propagation delay?

46. Explain what would constitute the resultant propagation delay time of the chip enable input signal ($\overline{CE}$) of the 2732 of Figure 9.23?

47. Explain how the resultant propagation delay time of the chip enable and output enable of any of the memory ICs of Figure 9.25 is determined.

48. When can propagation delay be a problem?

Troubleshooting

Section 12.8   Troubleshooting

49. If the fan-out of a TTL *NAND* gate is exceeded, what is the expected output response?

50. How would you know that the capacitance being driven by an *OR* gate has been exceeded?

51. If exceeding fan-out produces logic levels out of specification, what are the possible problems and how would you test for them?

52. What is a remedy for overloading an output?

53. If the pull-up resistor of an open-collector *NAND* gate were too large, what would be the symptom detected?

54. How could you detect a failed tristate buffer? State your reasoning.

13

INTERFACING PARALLEL, SERIAL, AND ANALOG I/O DEVICES

OBJECTIVES

- The objectives of this chapter are to:

- Explain how to interface parallel and serial I/O devices

- Study some devices that are specifically for interfacing parallel and serial I/O devices

- Introduce operational amplifiers as they apply to digital-to-analog and analog-to-digital converters

- Examine the essential concepts of digital-to-analog conversion

- Explain the electronic principles involved in the operation of a digital-to-analog converter

- Examine the essential concepts of analog-to-digital conversion

- Explain the electronic principles involved in the operation of an analog-to-digital converter

- Implement the concepts of digital-to-analog converters and analog-to-digital converters as interfacing between digital and analog systems

A digital computer is designed to interface with various components. The photograph shows an instrument known as an emulator/analyzer that assists in the design and development of the components that make-up and interact with the digital computer (Photo courtesy Hewlett-Packard Company).

13.1
Introduction

As already discussed, various types of I/O devices must be connected to the data bus of the system. Sometimes an I/O device can be directly connected to the data bus, whereas other I/O devices require additional circuitry, such as tristate buffers or latches. Whatever the case, the process of connecting I/O devices to the data bus is known as *interfacing the I/O device to the system.* The interface circuitry may range from a very simple circuit (made up of wires) to one that is quite complex.

Each I/O interfacing is a path over which the MPU can communicate with an I/O device, and the interface is known as an I/O port. An I/O port is circuitry that provides access to the data bus.

Input/output devices can be classified into one of three groups: (1) parallel, (2) serial, or (3) analog. Parallel I/O devices transfer or receive data bits in parallel. An 8-bit parallel I/O device reads to or writes from the 8-bit data bus lines in parallel. Parallel I/O devices are the type shown in the memory and I/O designs of Chapter 10. Serial I/O devices can transfer, or receive, just 1 bit at a time. Analog I/O devices do not digitize data (they do not put the data in a format of 1's and 0's) but rather process data in analog form. The data is a continuous waveform and there-fore has no sudden jumps (like going from a 1 to a 0). An audio amplifier is an example of an analog device.

There are many applications in which a digital computer is used to control a process, such as adjusting the fuel/air mixture ratio of an automobile engine or the speed of an electric motor. Controlling a process digitally may require that the digi-tal output, which is composed of bits, be converted to a single analog like output voltage whose magnitude is proportional to the digital value being converted. This type of conversion is known as a digital-to-analog conversion. The circuitry that per-forms the conversion is known as a *digital-to-analog converter* (DAC). There are also applications in which the reverse conversion must occur, known as analog-to-digital conversion; in this case, the conversion is performed by an *analog-to-digital converter* (ADC).

Parallel, serial, and analog interfacing will be discussed in this chapter. Digital-to-analog conversion and analog-to-digital conversion will be studied from both a conceptual and an electronic point of view. The electronics of DACs and ADCs will be approached analytically; however, those not interested in a detailed analysis of the electronics may omit that material without any loss of essential concepts.

13.2
Interfacing parallel I/O devices

Chapter 10 discussed the need for buffering input (to prevent bus contention) and output devices (to reduce loading). To interface a parallel I/O device, tristate buffers and latches are used, as illustrated in Figure 13.1, which shows input device *N* being interfaced with the data bus via tristate buffers. When I/O device *N* is addressed, its I/O select goes to the active state (high) and the buffers turn on, via the *CS* pin, which in turn loads the datum of input device *N* on the data bus. On the other hand, when output device *M* is written to by the MPU the datum written on the data bus is latched by the latch when its I/O select line goes from a 0 to a 1 (positive edge). Both the tristate buffers and the latches provide access to the data bus and are therefore

Figure 13.1

I/O port interfacing.

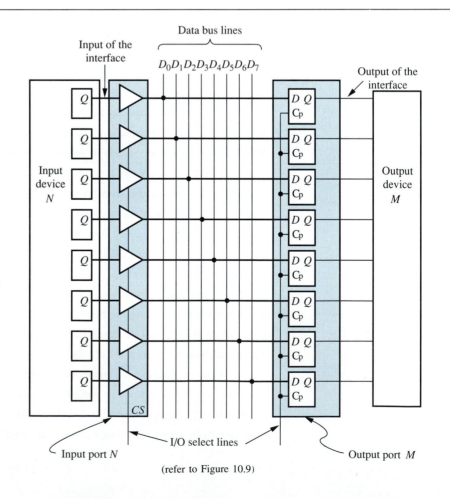

(refer to Figure 10.9)

I/O ports. For specific I/O ports, refer to devices such as the 74125, 74126, and 74LS374 in Appendix C. Buffers and latches were also examined in Chapter 2.

There is a large demand for devices that can serve as parallel I/O ports. Industry has provided a somewhat universal parallel I/O interface that is applicable to almost any parallel I/O port. The configuration of this interfacing device is programmable, which enables it to be tailored to a particular application. To program the interfacing device, the MPU writes a control word to it on start up of the system. The control word (often an 8-bit code) is read by the device and stored in its control register. The control word is then decoded by the interfacing device. It then configures itself according to the dictates of the control word and remains in that configuration so long as power is applied to the interfacing device.

Figure 13.2 is representative of a programmable interface. It has three I/O ports (A, B, and C) to which I/O devices 0, 1, and 2 are connected. The arrows show that, in this instance, port A has been programmed to be bidirectional, port B has been programmed to function as an output port, and port C is an input port. The address pins A_0 and A_1 permit each port to be addressed, as well as the control register. That is, logic level 00 addressed port A, logic level 01 addresses port B, logic level 10 addresses port C, and logic level 11 addresses the control register. The control register is the internal register that contains the control word that configures the interfacing device. As already stated, the MPU writes the control word to the control register. The write ($\overline{WR}$) and read ($\overline{RD}$) pins control read/write operations to or from the ports, as well as to or from the control register of the interfacing device. The chip select pin $\overline{CS}$ enables the interfacing device via an I/O select line from the I/O selector decoder.

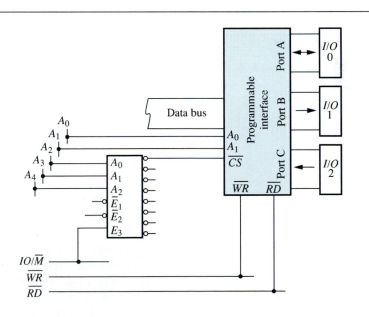

One specific programmable I/O interfacing device is an 8255. It is identified by the acronym PPI, which stands for *programmable peripheral interface*. Figure 13.3 illustrates its architecture and pin configuration. There are three 8-bit ports. Port A pins are identified as PA_7 through PA_0 and port B pins as PB_7 through PB_0. Port C

Figure 13.3

The architecture and pin configuration of an 8255 PPI. (Courtesy of INTEL Corp.)

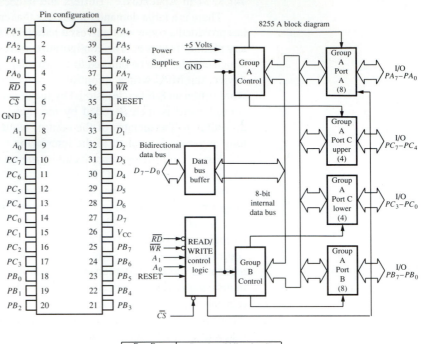

D_7-D_0	Data bus (bidirectional)
RESET	Reset input
$\overline{CS}$	Chip select
$\overline{RD}$	Read input
$\overline{WR}$	Write input
$A_0\,A_1$	Port address
PA_7-PA_0	Port A (bit)
PB_7-PB_0	Port B (bit)
PC_7-PC_0	Port C (bit)
V_{CC}	+ 5 Volts
GND	0 Volts

pins have been divided into two groups of four and are identified as PC_7 through PC_4 and PC_3 through PC_0. Port C pins are divided so that they can provide either of two functions: an 8-bit port, as illustrated in Figure 13.2, or handshaking capability for ports A and B. Because handshaking is a request and acknowledge action, two pins per port are required, or a total of four pins.

To program the 8255 the MPU outputs the address of the 8255 on the address bus and the control word on the data bus a short time later in accordance with the MPU timing diagram of Figure 9.15. The address selects the 8255 via its $\overline{CS}$ pin and also addresses the control register within the 8255 (which is similar to addressing a memory location within a semiconductor memory). To access the control register both A_1 and A_0 must be a logic high. The MPU then activates the write control signal of the control bus, to which $\overline{WR}$ of the 8255 is connected, while it is holding the datum (the control word) on the data bus. The control signal causes the control word to be latched off the data bus by the control register. The control word is decoded and the 8255 then configures itself accordingly.

To access a port, its address must enable the $\overline{CS}$ pin and input the proper logic levels to the address pins of the 8255. To address port A requires that A_1 and A_0 be a logic 0. To address port B, A_1 must be low and A_0 high, and port C requires $A_1 = 1$

and $A_0 = 0$. Whether a read or write operation is performed on the port is a function of the logic level on the read and write pins.

There are many more features to the 8255, but these functions and basic operations are sufficient for now. There are other parallel interfacing devices besides the 8255, but conceptually its purpose and operation are representative of all of them.

Review Questions

1. What is parallel interfacing?
2. What two types of logic circuits are used to interface parallel I/O devices?
3. What is an I/O port, and how does it relate to the system and I/O device?
4. What is a PPI and how does it work?

13.3
Interfacing serial I/O devices

Serial I/O devices output or input 1 bit of data at a time. When a serial I/O device is interfaced to a data bus (data bus lines are read or written to in parallel), there must be an interfacing device between the serial I/O device and the parallel data bus. When the serial I/O device is an input device the interfacing device must assemble the serial data being written by the serial I/O device into a parallel format before loading it on the data bus. When an output serial I/O device is interfaced to the data bus, the interface must latch the parallel data being written to the serial I/O device and then disassemble that data and transfer it to the serial I/O device 1 bit at a time. Shift registers can be used to implement both types of interfaces.

Figure 13.4 illustrates shift registers from Chapter 7 used as serial-to-parallel and parallel-to-serial interfaces. The serial-to-parallel interface shifts data into the interfacing unit 1 bit at a time. The shift rate is controlled by the clock, which may be synchronous or asynchronous with the I/O device. After the 8 bits of data have been shifted to the interface, the MPU activates control signal $\overline{RD}$, which turns on the buffers and loads the datum on the 8-bit data bus in parallel.

The parallel-to-serial interface of Figure 13.4b functions opposite to the serial-to-parallel circuit of Figure 13.4a. The parallel datum on the data bus is written into the parallel-to-serial interface device when its write pin is driven low. The datum is then shifted to the output serial I/O device 1 bit at a time, where the rate of shift is determined by the clock rate.

As with parallel-to-parallel interfacing devices, there is also demand for serial interfacing devices. One such device is a *universal synchronous/asynchronous receiver/transmitter* (USART). Within a single IC both serial-to-parallel and parallel-to-serial circuits exist. Figure 13.5 is conceptually representative of a USART. The system data bus is connected to pins D_7 through D_0. An input serial I/O device sends its data over the receive line, which is input at the *RXD* (receive data) pin. The USART transmits serial data to an output I/O device over the transmit line, which is connected to output *TXD* (transmit data). The rate of serial data transmission is determined by the clock rate input at *TXC*, and the clock rate input at *RXC* determines the rate of serial data reception by the USART. The read and write operations discussed in Figure 13.4a and b are controlled by the logic levels on pins $\overline{RD}$ and $\overline{WR}$. The $\overline{CS}$ pin is activated as a result of the USART being addressed. The $C/\overline{D}$ pin determines whether the USART is reading or writing data ($C/\overline{D} = 0$) or a control word ($C/\overline{D} = 1$). The USART can be programmed for certain data formats.

Figure 13.4

Serial I/O interfacing.

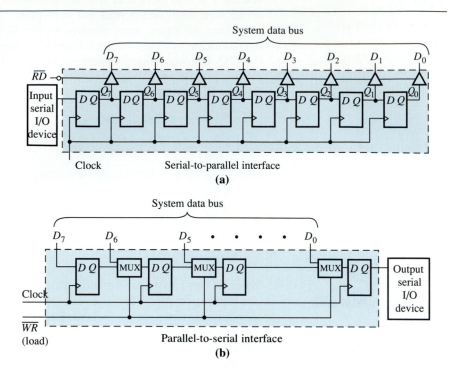

(a)

(b)

Figure 13.5

A representation of a USART.

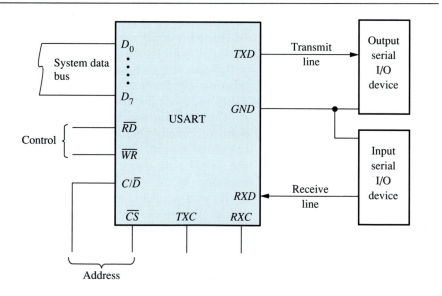

The 8251 is a popular USART. Its architecture and pin configuration are shown in Figure 13.6. The additional pins of the 8251 show that it is much more complex than this discussion has indicated; however, we have set forth the essence of an 8251 and of USARTs in general.

The discussion of serial data interfacing devices is not complete without mention of a standard known as *RS-232C*, to which many serial I/O devices adhere.

Figure 13.6

The architecture and pin configuration of the 8251 USART. (Courtesy of INTEL Corp.)

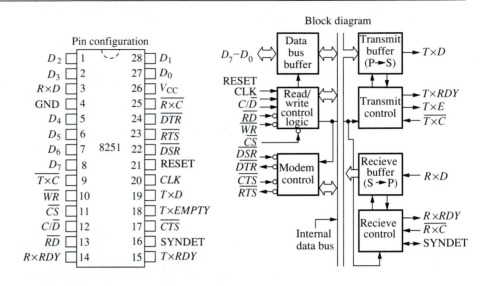

The formal title of this standard is "Interface Between Data Terminal Equipment and Data Communication Equipment Employing Serial Binary Data Interchange." Among other things, it defines the voltage levels necessary to represent a logic 1 and 0, which are from $+3$ to $+25$ volts for a logic 0 and from -3 to -25 volts for a logic 1. The reason for such large voltage levels is that an RS-232C line is often used for the exchange of data over long distances; with large voltage levels noise is less of a problem. Most modern ICs operate with a 5-volt DC power supply, so there must be an interface between a data bus operating over a range of 0 to 5 volts and an RS-232C serial I/O. These interfaces are available.

Review Questions

1. How can a shift register function as a parallel-to-serial interface?
2. How can a shift register function as a serial-to-parallel interface?
3. What is a USART, and how do such devices function?
4. What is an RS-232C standard, and why does it require interfacing?

13.4
The operational amplifier

The *operational amplifier* (*op-amp*) is an essential component of many DACs and ADCs. Therefore, the principles of op-amps necessary for understanding the operation of DACs and ADCs will be examined.

As with any active electronic component, such as a transistor, op-amps can be operated in either a nonlinear or a linear mode. In the nonlinear mode they operate in either extreme of their operating range; that is, they are either on or off. An amplifier operated as a nonlinear device is essentially an electronic switch, which is another way of saying it is operated as a digital device. The electronic switches of Figures 12.1a and 12.2a are examples of amplifiers that were designed to operate in a nonlinear mode.

Linear amplifiers are designed to operate within the range bounded by the extremes defined by "fully turned on" and "fully turned off." This range of operation is known as its *active region* or linear range of operation. As examples, the linear range of operation for the characteristic curve of Figure 12.1a can be defined by the inequality

$$V_{CE\,ON} < V_{CE} < V_{CE\,OFF}$$

The inequality

$$V_{OL} < V_{DS} < V_{OH}$$

defines the linear range of operation for the characteristic curve of Figure 12.2a. Linear amplifiers are what we normally think of as an amplifier; that is, the output has the same waveform appearance as the input, but amplified.

An op-amp is a differential amplifier in that it amplifies the difference of the voltage signals applied at its two inputs. Depending on its external circuit configuration, the op-amp amplifies the difference signal e_d in either a linear or a nonlinear fashion.

The symbol for an op-amp is shown in Figure 13.7a, and Figure 13.7b shows the characteristic curve of an op-amp. The symbol shows two power supplies (V), one with positive and the other with negative polarity. The magnitude and polarity of the output voltage v_o are bounded by the magnitudes and polarities of the power supplies. The maximum magnitude of the output voltage, positive or negative, is approximately 0.5 volts less than the magnitude of the power supply with the corresponding polarity. These maximum magnitudes are identified as saturation

Figure 13.7

The symbol and characteristic curves of an op-amp.

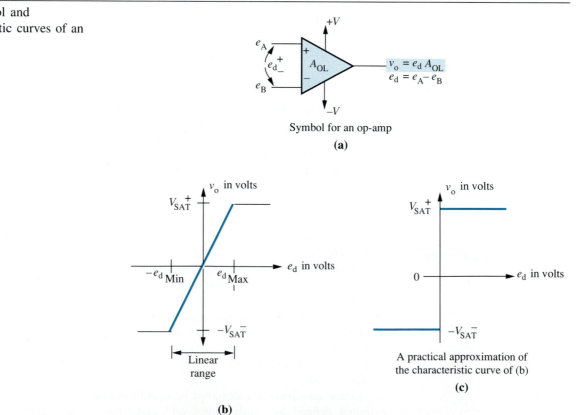

Symbol for an op-amp

(a)

A practical approximation of the characteristic curve of (b)

(c)

(b)

voltages, V^+_{SAT} for the magnitude of the positive saturation voltage and V^-_{SAT} for the magnitude of the negative output saturation voltage, which is illustrated in Figure 13.7b.

The schematic of Figure 13.7a has two inputs, one positive and the other negative. The positive sign indicates that any voltage input at this input will be in phase with the amplified voltage at the output. In contrast, the amplified output voltage will be 180 degrees out of phase with the voltage applied at the negative input. The gain of the op-amp is designated A_{OL} (open-loop gain), and any voltage at either input is amplified by this gain. To determine the resultant output voltage from voltages being applied at both inputs, use the discussion above and the superposition theorem. The output voltage resulting from input voltage e_A of Figure 13.7a is

$$v'_{\text{o}} = e_A A_{\text{OL}}$$

The output voltage resulting from e_B is

$$v''_{\text{o}} = e_B(-A_{\text{OL}}) = -e_B A_{\text{OL}}$$

Using the superposition theorem to determine the resultant output,

$$v_{\text{o}} = v_{\text{o}}' + v''_{\text{o}}$$

or

$$v_{\text{o}} = (e_A - e_B)A_{\text{OL}} \tag{13.1}$$

Now, the difference voltage at the inputs can be expressed as

$$e_d = e_A - e_B \tag{13.2}$$

Equation 13.1 can be written as

$$v_{\text{o}} = e_d A_{\text{OL}}. \tag{13.3}$$

Equation 13.3 confirms that this is a difference amplifier, because it amplifies the difference of two input voltages. The polarity of e_d shown in Figure 13.7a is an assumed polarity and is valid if $e_A > e_B$.

The characteristic curve of Figure 13.7b shows that an op-amp is operating in a linear mode so long as its input voltage is within the range of

$$-e_{d\,\text{min}} < e_d < e_{d\,\text{max}} \tag{13.4}$$

If e_d is outside this range, the output voltage is driven into saturation and the op-amp is functioning in a nonlinear mode. If the negative power supply is removed, and its place of connection is grounded, then $-V^-_{\text{SAT}}$ is 0 volts, which is a logic 0. In addition, if a $+5$ volt power supply is used for the positive power supply, then V^+_{SAT} is a logic 1 ($V_{\text{OH}} = 5 - 0.5 = 4.5$ volts). By grounding the negative power supply connection and using a $+5$ volt power supply to implement $+V$, the output is voltage compatible for all TTL and CMOS families, as can be verified from Table 12.5.

Nonlinear mode of operation

As stated by Equation 13.4 and illustrated in Figure 13.7b, if the difference voltage is greater than $e_{d\,\text{max}}$ or less than $-e_{d\,\text{min}}$, the op-amp is operating in a nonlinear mode. Using a realistic value of 10^5 for A_{OL}, let us get a feel for the value of $e_{d\,\text{max}}$. Using Equation 13.3 solve for e_d

$$e_d = \frac{v_{\text{o}}}{A_{\text{OL}}} \tag{13.5}$$

To solve for maximum conditions, let the output voltage be at $V^+{}_{SAT}$. Then Equation 13.5 becomes

$$e_{d\,max} = \frac{V^+{}_{SAT}}{A_{OL}}. \tag{13.6}$$

If the positive polarity power supply is $+5$ volts DC (V_{CC}), $V^+{}_{SAT}$ is equal to $+4.5$ volts, as discussed previously. Substituting values into Equation 13.6,

$$e_{d\,max} = \frac{4.5}{10^5} = 45 \ \mu V$$

If the same reasoning is used for the negative voltage range ($V^- = -5$ volts DC),

$$e_{d\,min} = \frac{-4.5}{10^5} = -45 \ \mu V$$

The linear range for e_d is thus $\pm 45 \ \mu V$ (a 90 μV swing). This voltage range is so small that it may be considered to be zero, which yields the characteristic curve of Figure 13.7c. From this figure we reason that the magnitude and polarity of the output voltage are determined by the polarity of voltage e_d. From Figure 13.7c we can mathematically state that if

$$e_d > 0 \text{ volts}, \ v_o = V^+{}_{SAT} \tag{13.7}$$

and if

$$e_d < 0 \text{ volts}, \ v_o = -V^-{}_{SAT} \tag{13.8}$$

The characteristic curve of Figure 13.7c is that of a binary digital device; that is, it operates in one of two states.

Nonlinear applications

The practical characteristic curve of Figure 13.7c is that of a nonlinear device and represents the output voltage as a function of the input voltage. The most common application of an op-amp operated in a nonlinear mode is a comparator.

Example 13.1

For the circuit of Figure 13.8a derive equations that can be used to predict the op-amp output voltage as a function of the input voltage $e(t)$. Then apply those equations to the input voltage waveform of Figure 13.8b and plot the resultant output voltage.

Solution

Using Equations 13.7 and 13.8 to predict the output voltage, find out the polarity of e_d as a function of $e(t)$. To find this relationship, write a loop equation at the input circuit of Figure 13.8a. Hence,

$$e(t) - e_d = 0.$$

Solving for dependent variable e_d (the polarity and magnitude of e_d depend on e),

$$e_d = e(t) \tag{13.9}$$

By substituting Equation 13.9 into Equation 13.7,

$$e(t) > 0 \text{ volts}, \ v_o = V^+{}_{SAT}$$

From the plot of $e(t)$ in Figure 13.8b this inequality is true for the time range 0 to t_1 and as a result $v_o = V^+{}_{SAT}$ over this same interval of time.

Figure 13.8

An op-amp comparator.

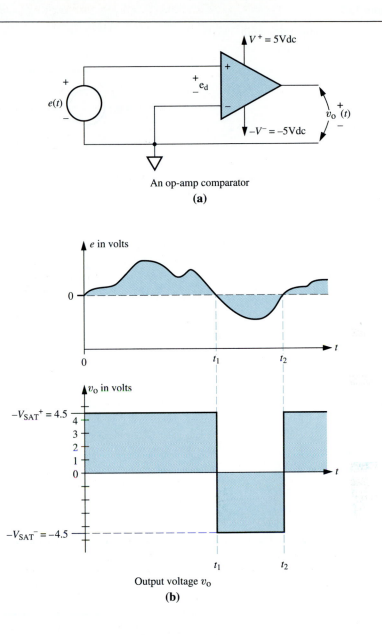

An op-amp comparator

(a)

Output voltage v_o

(b)

When $e(t)$ is negative, Equation 13.8 applies and the output is -4.5 volts, as shown in the interval of time from t_1 to t_2 in Figure 13.8b.

From the mathematical analysis and the plot of v_o in Figure 13.8b, it can be stated that an op-amp operating in an open-loop (OL) mode, that is, without negative feedback, is a comparator. In the case of the circuit of Figure 13.8a the input voltage $e(t)$ is compared with the reference voltage (the voltage at the negative input), which is 0 volts. As a result, when $e(t)$ becomes greater than the reference voltage (0 volts) the output is $+4.5$ volts, and when $e(t)$ drops below the reference voltage the output voltage becomes -4.5 volts.

Figure 13.9

An op-amp comparator with
a reference of 0 volts.

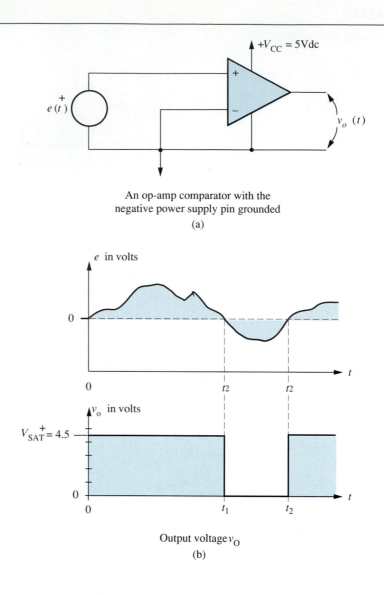

An op-amp comparator with the
negative power supply pin grounded

(a)

Output voltage v_O

(b)

Example 13.2

For the circuit of Figure 13.9 apply the input voltage of Figure 13.8b and plot the output voltage.

Solution

The circuit of Figure 13.9a has the negative power supply pin connected to ground. As a result, when $e_d < 0$, the output voltage goes to 0 volts rather than $-V^-_{SAT}$. The solution of Example 13.1 is valid for this circuit, except that $-V^-_{SAT}$ is equal to 0 volts, as indicated by the output voltage plot of Figure 13.9b.

Linear applications

Figure 13.7b and Equation 13.5 show that it is the large magnitude of open-loop gain (A_{OL}) that causes an open-loop op-amp to be a nonlinear device, as graphically indicated in Figure 13.7c. To use an op-amp in a linear mode of operation requires

Figure 13.10

An adder.

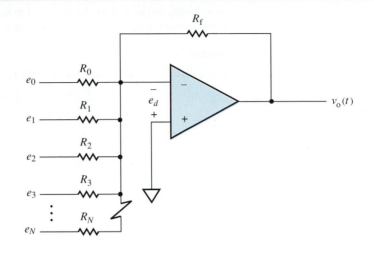

that the effects of open-loop gain be reduced in the circuit. To accomplish this, negative feedback is used. Figure 13.10 shows negative feedback being implemented with the resistor R_f.

The circuit of Figure 13.10 is an analog adder; that is, the output voltage is proportional to the sum of the input voltages e_0, e_1, e_2, etc. To analytically verify that this is true, begin the analysis by writing a nodal equation at the negative input of the op-amp. Those not interested in the electronic and mathematical details of the analysis may progress directly to the answer (Eq. 13.12), which is printed in color.

$$\frac{e_o - (-e_d)}{R_0} + \frac{e_1 - (-e_d)}{R_1}$$

$$+ \frac{e_2 - (-e_d)}{R_2} + \dots + \frac{e_N - (-e_d)}{R_N}$$

$$+ \frac{v_o - (-e_d)}{R_f} = 0 \tag{13.10}$$

For all practical purposes $e_d = 0$. Equation 13.10 can then be written as

$$\frac{e_0}{R_0} + \frac{e_1}{R_1} + \frac{e_2}{R_2} + \dots + \frac{e_N}{R_N} + \frac{v_o}{R_f} = 0 \tag{13.11}$$

Solving Equation 13.11 for v_o gives the equation

$$v_o = - \left[\left(\frac{R_f}{R_0}\right) e_0 + \left(\frac{R_f}{R_1}\right) e_1 + \dots + \left(\frac{R_f}{R_N}\right) e_N \right] \tag{13.12}$$

Equation 13.12 states that the output voltage is 180 degrees out of phase (the negative sign) with the sum of the products of coefficients and input voltages, where the coefficients (R_f/R_0, R_f/R_1, etc.) are gains of the individual input voltages. These gains can be used as weighting factors; that is, their values will assign a "weight of influence" to each input voltage when determining the magnitude of the output voltage.

Example 13.3

Suppose that the adder of Figure 13.10 had three input voltages with values of $e_0 = 2$ volts, $e_1 = -3$ volts, and $e_2 = 1.5$ volts. If $R_f = 100$ K, $R_0 = 50$ K, $R_1 = 30$ K, and $R_2 = 25$ K, what is the output voltage for power supplies equal to $+$ and $-$ 25 volts DC?

Solution

Substituting these values into Equation 13.12

$$v_o = -\left[\left(\frac{100}{50}\right)2 + \left(\frac{100}{30}\right)(-3) + \left(\frac{100}{25}\right)1.5\right]$$

$$= -(4 - 10 + 6) = 0 \text{ volts}$$

Note the weighting factor of the coefficients.

Understanding op-amp–based comparators and adders is sufficient for understanding op-amps as they relate to DACs and ADCs.

Review Questions

1. What is the meaning of the positive and negative sign at the input of an op-amp?
2. Why can an op-amp be classified as a difference amplifier?
3. In what two modes of operation can an op-amp operate?
4. Why is the open-loop gain of an op-amp responsible for its nonlinear characteristic?
5. Why does negative feedback enable an op-amp to function in a linear mode?
6. When an op-amp is used without negative feedback, why is it a voltage comparator?
7. What is meant by weighting factor as it applies to Equation 13.12?

13.5
Digital-to-analog conversion

An analog signal is continuous; that is, it has no points of discontinuity. The input voltage $e(t)$ of Figure 13.8b is an analog signal. All digital signals are discontinued signals. The output voltage v_o of that same figure is a digital signal. This waveform is discontinuous because it changes abruptly at t_1 and t_2.

The "real world" functions in an analog mode. However, people have created a digital world through digital devices and computers. Often it is convenient, or even necessary to cross back and forth between these two worlds, which is the reason for DACs and ADCs.

To develop the concepts of DACs two basic techniques of converting a digital signal to an analog signal will be studied. These two techniques are (1) a resistor ladder network known as an *R-2R* ladder, and (2) an *R-2R* ladder that is buffered with an op-amp circuit that is either a voltage-follower, an adder, or an inverting amplifier.

R-2R ladder network

Figure 13.11 is a schematic of a three-input *R-2R* ladder network. If additional inputs are required, the pattern of Figure 13.11 is repeated for each additional input. The inputs e_0, e_1, and e_2 are digital inputs, so their voltage level is either V_{OL} (0) or V_{OH}

Figure 13.11

R-2R resistor ladder network.

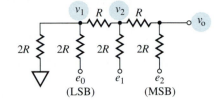

(1). To analyze the circuit of Figure 13.11 write nodal equations at each node. The nodal voltages are identified by v_1, v_2, and v_o. Solve for v_o, the converted analog signal. To skip the analysis of Figure 13.11 advance to the answer of Equation 13.13. As with the analysis of the adder, the answer is printed in color for those wishing to omit the analysis. This format will be used for all circuit analyses.

Writing a nodal equation at node 1 of the *R-2R* ladder network,

$$0 = -\frac{v_1}{2R} + \frac{e_0 - v_1}{2R} + \frac{v_2 - v_1}{R}$$

At node 2

$$0 = \frac{v_1 - v_2}{R} + \frac{e_1 - v_2}{2R} + \frac{v_o - v_2}{R}$$

At node 3, the output node

$$0 = \frac{v_2 - v_o}{R} + \frac{e_2 - v_o}{2R}$$

Dividing the *R*s out of the three nodal equations and setting the knowns (e_1, e_2, and e_3) equal to the unknowns (v_1, v_2, and v_3)

$$\frac{e_0}{2} = 2v_1 \qquad\qquad - v_2$$

$$\frac{e_1}{2} = -v_1 + \left(\frac{5}{2}\right) v_2 \qquad\qquad - v_o$$

$$\frac{e_2}{2} = \qquad\qquad - v_2 + \left(\frac{3}{2}\right) v_o$$

From these three equations structure the determinate *D*

$$D \quad \begin{vmatrix} 2 & -1 & 0 \\ -1 & \dfrac{5}{2} & -1 \\ 0 & -1 & \dfrac{3}{2} \end{vmatrix}$$

$$D = 2\left[\left(\frac{5}{2}\right)\left(\frac{3}{2}\right) - 1 \right] + 1\left[-1\left(\frac{3}{2}\right) \right] = 4$$

Solving for Dv_o

$$Dv_o = \begin{vmatrix} 2 & -1 & \dfrac{e_0}{2} \\[2ex] -1 & \dfrac{5}{2} & \dfrac{e_1}{2} \\[2ex] 0 & -1 & \dfrac{e_2}{2} \end{vmatrix}$$

$$= 2\left[\frac{(5/2)e_2}{2} + \frac{e_1}{2}\right] + \left[\frac{-e_2}{2} + \frac{e_0}{2}\right]$$

$$= 2e_2 + e_1 + \frac{e_0}{2}$$

Divide Dv_o by the determinate $(D = 4)$

$$v_o = \frac{e_2}{2} + \frac{e_1}{4} + \frac{e_0}{8}$$

Put these quantities under a common denominator

$$v_o = \frac{1}{8}(4e_2 + 2e_1 + e_0) \tag{13.13}$$

Generalize Equation 13.13 so that it is valid for any number of inputs by recognizing a pattern of the coefficients. Using n to represent the nth input and N to represent the total number of inputs, Equation 13.13 can be written as

$$v_o = \frac{1}{2^N}\left(\sum_{n=0}^{N-1} 2^n e_n\right). \tag{13.14}$$

Note from Equations 13.13 and 13.14 that each input voltage has a weighting function, which is the coefficient of each input voltage and is equal to 2^{n-N}; as a result input voltage e_2 is given more weight than e_1 or e_0. This means that the MSB of, say, a 3-bit binary input would be connected to input e_2 and the LSB would be connected to e_0, because the MSB should have more weight when determining the output voltage than any of the other bits and e_0 should have the least weight.

Example 13.4

For the number of inputs specified, draw the appropriate schematic for an *R-2R* ladder network and from Equation 13.14 determine the equation for the output voltage v_o.
(a) $N = 4$ (b) $N = 8$

Solution

Figure 13.12 shows the schematics for parts (a) and (b).
(a) For $N = 4$, Equation 13.14 becomes

$$v_o = \frac{1}{2^4}\left(\sum_{n=0}^{4-1} 2^n e_n\right)$$

$$= \frac{1}{16}(2^0 e_0 + 2^1 e_1 + 2^2 e_2 + 2^3 e_3)$$

$$v_o = \frac{1}{16}(e_0 + 2e_1 + 4e_2 + 8e_3)$$

Figure 13.12

Four- and 8-bit *R-2R* ladder networks.

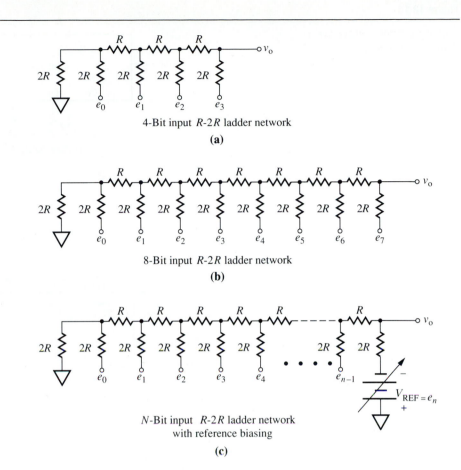

4-Bit input *R-2R* ladder network

(a)

8-Bit input *R-2R* ladder network

(b)

N-Bit input *R-2R* ladder network
with reference biasing

(c)

(b) For $N = 8$

$$v_o = \frac{1}{256}(e_0 + 2e_1 + 4e_2 + 8e_3$$

$$+ 16e_4 + 32e_5 + 64e_6 + 128e_7)$$

Example 13.5

Suppose that the output of a 74LS series 4-bit up-counter is to be converted to an analog signal. Use an *R-2R* ladder network as shown in Figure 13.13a to make the conversion. Graph the analog output signal as the digital counter progresses from count 0000 to 1111.

Solution

Using the equation from Example 13.4a,

$$v_o = \frac{1}{16}(e_0 + 2e_1 + 4e_2 + 8e_3)$$

calculate the analog output voltage from the 16 binary numbers that make up the count. Table 12.5 shows that a logic 0 output voltage (V_{OL}) is 0.5 volts and a logic 1 (V_{OH}) is 2.7 volts. For the first count of 0000 all the input voltages are $V_{OL} = 0.5$ volts. Then,

$$v_o = \frac{1}{16}[0.5 + 2(0.5) + 4(0.5) + 8(0.5)] = 0.469 \text{ volts}$$

Figure 13.13

Analog output voltage for a
4-bit *R*-2*R* ladder DAC.

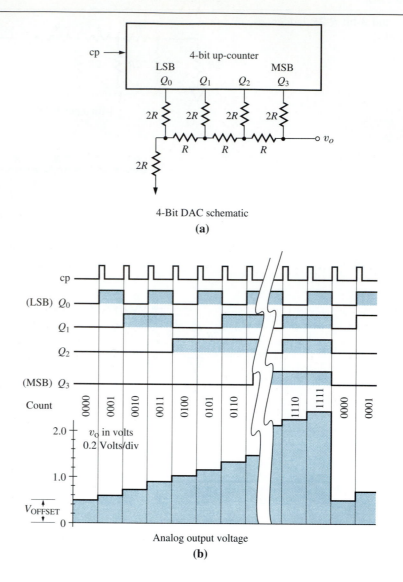

4-Bit DAC schematic

(a)

Analog output voltage

(b)

When the count is 0001

$$v_o = \frac{1}{16}[2.7 + 2(0.5) + 4(0.5) + 8(0.5)] = 0.606 \text{ volts}$$

Count = 0010

$$v_o = \frac{1}{16}[0.5 + 2(2.7) + 4(0.5) + 8(0.5)] = 0.744 \text{ volts}$$

Count = 0011

$$v_o = \frac{1}{16}[2.7 + 2(2.7) + 4(0.5) + 8(0.5)] = 0.881 \text{ volts}$$

Count = 0100

$$v_o = \frac{1}{16}[0.5 + 2(0.5) + 4(2.7) + 8(0.5)] = 1.018 \text{ volts}$$

Count = 0101

$$v_o = \frac{1}{16}[2.7 + 2(0.5) + 4(2.7) + 8(0.5)] = 1.156 \text{ volts}$$

Count = 0110

$$v_o = \frac{1}{16}[0.5 + 2(2.7) + 4(2.7) + 8(0.5)] = 1.294 \text{ volts}$$

Count = 1110

$$v_o = \frac{1}{16}[0.5 + 2(2.7) + 4(2.7) + 8(2.7)] = 2.394 \text{ volts}$$

Count = 1111

$$v_o = \frac{2.7}{16}[1 + 2 + 4 + 8] = 2.531 \text{ volts}$$

The output voltages for each corresponding count are plotted in Figure 13.13b.

Figure 13.13b shows that the analog output voltage has an offset voltage of V_{OFFSET} when the binary input is 0000 and that the output voltage is not a true analog signal in that it is not a smooth, continuous curve. However, it can be considered close enough to be an analog signal; if a smoother (more continuous) output voltage is desired, a capacitive load can be put at the output. To get rid of the offset voltage, a reference biasing voltage source V_{REF} can be added to the *R*-2*R* ladder network, as illustrated in Figure 13.12c. Using this technique the negative reference voltage can be viewed as the *n* input voltage; that is, $e_n = -V_{\text{REF}}$. Making this addition to the ladder of Figure 13.11 would modify Equation 13.13 to appear as

$$v_o = \frac{1}{8}(-4V_{\text{REF}} + 2e_1 + e_0)$$

because $e_2 = -V_{\text{REF}}$. To incorporate this change into Equation 13.14,

$$v_o = \frac{1}{2^N}\left(\sum_{n=0}^{N-2} 2^n e_n - 2^{N-1}V_{\text{REF}}\right) \tag{13.15}$$

To cancel the offset voltage V_{OFFSET} of Figure 13.13b would require that

$$\sum_{n=0}^{N-2} 2^n e_n - 2^{N-1}V_{\text{REF}} = 0$$

when $e_n = V_{\text{OL}}$, which is the first count. Solving for V_{REF}

$$V_{\text{REF}} = \frac{1}{2^{N-1}}\left(\sum_{n=0}^{N-2} 2^n e_n\right)$$

where $e_n = V_{\text{OL}}$. This equation states that V_{REF} must be adjusted to a magnitude of V_{OFFSET} volts, which is equal to

$$\frac{1}{2^{N-1}}\left(\sum_{n=0}^{N-2} 2^n e_n\right) \text{ volts}$$

where $e_n = V_{\text{OL}}$.

More importantly, the DACs of Figure 13.13 will convert a digital signal into a single output analog voltage whose voltage level is directly proportional to the binary value of the input.

Buffered *R-2R* ladder network

The *R-2R* ladder network has one major disadvantage — the input resistance of the load must be at least ten times larger than the value *R* of the network. This requirement is necessary so as not to load down the ladder network so that Equation 13.14 or 13.15 is no longer valid. There are three op-amp solutions to this problem. One is to use the adder circuit of Figure 13.10 when it is properly designed to yield an output voltage equal to the magnitude of Equation 13.14 or 13.15. To take care of the sign difference use an inverting amplifier at the output of the adder.

Using the same technique to express the adder op-amp output voltage of Equation 13.12 in a general fashion, as was used to express the *R-2R* ladder output voltage of Equation 13.14, Equation 13.12 can be written as

$$v_o = -\sum_{n=0}^{N-1} \left(\frac{R_f}{R_n}\right) e_n \tag{13.16}$$

Because Equation 13.14 can be written as

$$v_o = \sum_{n=0}^{N-1} 2^{n-N} e_n \tag{13.17}$$

equating weighting functions (coefficients)

$$\frac{R_f}{R_n} = 2^{n-N}$$

or

$$R_n = 2^{-(N-n)} R_f \tag{13.18}$$

Example 13.6

If the feedback resistor R_f of Figure 13.10 has a value of 100 K, what values should the remaining resistors have, according to Equation 13.18, for a 4-bit digital input?

Solution

Evaluating Equation 13.18 for $N = 4$

$$R_0 = 2^{-(4-0)} \times 100\,\text{K} = \frac{1}{16}(100\,\text{K}) = 6.25\,\text{K}$$

$$R_1 = 2^{-(4-1)} \times 100\,\text{K} = \frac{1}{8}(100\,\text{K}) = 12.50\,\text{K}$$

$$R_2 = 2^{-2} \times 100\,\text{K} = 25\,\text{K}$$

$$R_3 = 50\,\text{K}$$

In addition to having an inverted output (the negative sign of Equation 13.16), the solution of Example 13.6 requires the use of five different values of resistors in the manufacture of this design. Because of the price break given for quantity purchases, it would be more desirable if the number of different values of resistors could be reduced. The R-$2R$ ladder is desirable because it requires only two values of resistors. Incorporating the R-$2R$ ladder network with an op-amp that has an input resistance in the megaohm range, so that the op-amp buffers the R-$2R$ ladder from the load, would be a suitable solution.

The voltage follower of Figure 13.14 can serve as buffer between the R-$2R$ ladder network and the load, which is represented by resistor R_L. Voltage followers have a voltage gain of one, so that Equation 13.14 or 13.15 would be valid to describe the output voltage of the voltage follower.

Figure 13.14

An R-$2R$ ladder buffered with an op-amp voltage follower.

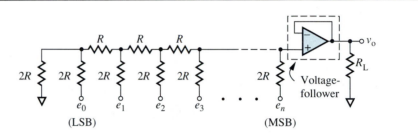

Another way to buffer the R-$2R$ ladder network from the load is to modify the adder of Figure 13.10 so that it is an inverting amplifier and then use it as a buffer. The inverting amplifier can be designed to simply duplicate the output voltage described by Equations 13.14 and 13.15, or it can be designed to amplify it by some scaling factor, depending on the gain designed into the amplifier.

To determine how to modify Figure 13.10 so that it is an inverting amplifier and to derive its output voltage equation, notice from Equation 13.12 that if all inputs but one, say e_0, are forced to 0 volts (shorted to ground) then

$$v_o = -\left(\frac{R_f}{R_0}\right) e_0$$

Shorting these inputs to ground allows removal of them and their associated resistor R_n. The resulting circuit is the inverting amplifier of Figure 13.15a with a gain of $-(R_f/R_0)$. Figure 13.15b is the inverting amplifier of Figure 13.15a with the output voltage of the R-$2R$ ladder serving as its input voltage, where the Ds are used to indicate digital inputs.

To derive the equation for the output voltage of Figure 13.15b, use the output voltage equation for the inverting amplifier of Figure 13.15a and Thevenin's equivalent circuit for the R-$2R$ ladder network of Figure 13.15b. To understand the reasoning for deriving the output voltage equation of Figure 13.15b in this manner, recall from circuit analysis that a Thevenin's equivalent circuit are a Thevenin's equivalent voltage source and a Thevenin's equivalent resistance in series with each other. Hence, the series circuit of voltage source e_0 and resistance R_0 of Figure 13.15a can be thought of as a Thevenin's equivalent circuit. To find Thevenin's equivalent circuit for the ladder network of Figure 13.15b the ladder network of Figure 13.12c shall be analyzed, which is the same ladder network of Figure 13.15b if the reference voltage is considered to be input voltage D_{N-1}. This will result in the reference voltage being part of the output voltage equation. Applying Thevenin's theorem to the ladder network of Figure 13.12c, the equivalent voltage source is described by Equation 13.15 and Thevenin's equivalent resistance is R. Substituting

Figure 13.15

An inverting amplifier
buffering an *R-2R* ladder.

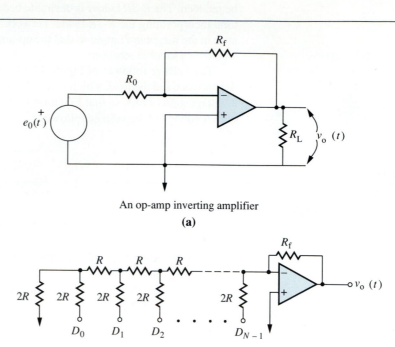

An op-amp inverting amplifier

(a)

An *R-2R* Ladder buffered with an
inverting op-amp amplifier

(b)

these two quantities into the output voltage equation for the inverting amplifier of
Figure 13.15a the output voltage equation for Figure 13.15b is

$$v_o = - \left(\frac{R_f}{R}\right) \frac{1}{2^N} \left[\sum_{n=0}^{N-1} 2^n e_n + 2^{N-1} V_{REF}\right]. \qquad (13.19)$$

Example 13.7

If resistor *R* of Figure 13.15b has a value of 20 K

(a) Design the circuit of Figure 13.15b so that Equation 13.19 is equal to Equation 13.15.

(b) Design the circuit so that Equation 13.15 is scaled (amplified) by a factor of 2.

· (c) Adjust V_{REF} to null V_{OFFSET} for the design of part (a) if the digital input is an 8-bit CMOS
counter.

Solution

(a) Equation 13.19 shows that R_f/R must equal 1; because $R = 20$ K, R_f must also
equal 20 K.

(b) For a gain of 2, Equation 13.19 states that $R_f/R = 2$, or $R_f = 2R$. Then $R_f = 40$ K.

(c) From the previously derived equation

$$V_{REF} = \frac{1}{2^{N-1}} \left(\sum_{n=0}^{N-1} 2^n e_n\right)$$

CMOS inputs (V_{OL} = 0.33 volts) when N = 9 (V_{REF} is an input),

$$V_{REF} = \frac{1}{2^8}(e_0 + 2e_1 + 4e_2 + \ldots + 128e_7)$$

$$= \frac{1}{256}V_{OL}(1 + 2 + 4 + 8 + 16 + 32 + 64 + 128)$$

$$= \left(\frac{1}{256}\right)(0.33)(255) = 0.329 \text{ volts}$$

Other techniques are available to implement a DAC, but most are variations of the concepts presented.

Figure 13.16a shows the schematic symbol for a DAC and Figure 13.16b illustrates a representative application of a DAC in which the converted digital signal is driving an analog load, such as an electrical motor. The symbol of Figure 13.16a shows that the inputs are on the left and the outputs are on the right, with the power supply and reference voltage at the top. The output labeled ''R_{FB}'' is an external pin for access to the feedback resistor labeled ''R_f'' in Figure 13.15b. There is a feedback resistor manufactured within the DAC, but because its value may not be large enough the manufacturer has made it possible to increase the feedback resistance by adding external resistor R_2 in series with internal resistor R_{FB}. If the DAC user does not want an inverting amplifier, such as that in Figure 13.16, then the pin for R_{FB} is not connected and the voltage follower of Figure 13.14 is used instead.

Figure 13.16a also shows that for this DAC there are 8 bits of digital data that can be input via data pins DB_0 through DB_7. Figure 13.16b shows that these pins are connected to the source of the digital data to be converted, which in this case is an MPU via the data bus (the MPU writes the data byte to be converted to the DAC). The data byte on the data bus is latched by the DAC under control of the write control signal, which is input at the $\overline{WR}$ pin. The $\overline{CS}$ pin serves the same function as the chip selects of memory ICs, which is to enable the DAC chip. Figure 13.16b shows the $\overline{CS}$ pin being activated via the I/O selector, which employs the same design concepts that select the I/O ports of Figure 10.9. The truth table of Figure 13.16a shows that the $\overline{CS}$ pin must be at a logic low for the DAC to convert data. When the write enable pin is activated, that is, is driven low by the MPU when it has written a data byte on the data bus, and if $\overline{CS}$ is also low at this time, the DAC converts the data byte on the data bus and outputs the converted analog signal at output OUT_1. As shown in Figure 13.16b, the write enable pin of the DAC is connected to the write control signal of the control bus. We realize from the connections of Figure 13.16b that a DAC is just another I/O port to the MPU.

Some DACs are designed so that the reference voltage V_{REF} serves as a multiplier of the output OUT_1. For this type of DAC the output voltage of the op-amp of Figure 13.16a would be a modification of Equation 13.14 such that V_{REF} is a multiplier. Hence,

$$v_o = -V_{REF}\left(\frac{R_{FB} + R_2}{R}\right)\frac{1}{2^N}\sum_{n=0}^{N-1}2^n e_n \tag{13.20}$$

Note from Equation 13.20 that if V_{REF} is made a negative voltage, the output voltage of the op-amp is no longer negative.

Figure 13.16

A DAC system application.

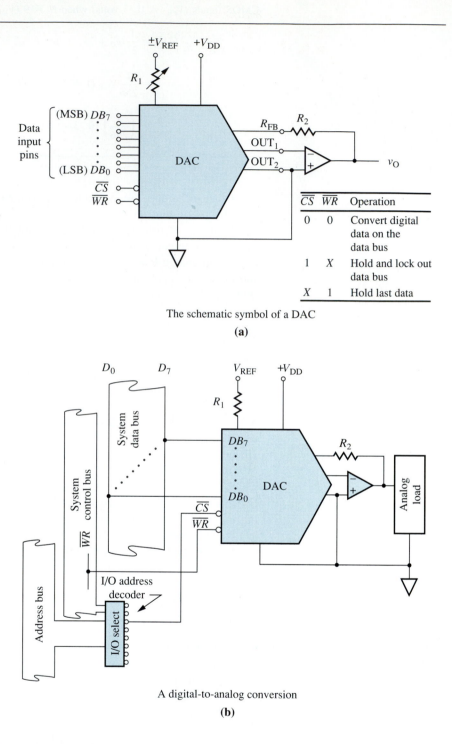

The schematic symbol of a DAC

(a)

$\overline{CS}$	$\overline{WR}$	Operation
0	0	Convert digital data on the data bus
1	X	Hold and lock out data bus
X	1	Hold last data

A digital-to-analog conversion

(b)

Example 13.8

Suppose that the DAC of Figure 13.16b has an $R_{FB} = R$ (of the internal R-$2R$ ladder) $= 50$ K and $R_2 = 25$ K. What polarity and magnitude must V_{REF} have for the output of the op-amp to be equal to

$$v_o = 15\left(\frac{1}{2^N}\right)\sum_{n=0}^{N-1} 2^n e_n$$

Solution

Equating the above equation to Equation 13.20 and using the stated resistance values, this would require that

$$-V_{REF}\left(\frac{75 \times 10^3}{50 \times 10^3}\right) = 15$$

or

$$V_{REF} = -\frac{2}{3}(15) = -10 \text{ volts}$$

Following are some definitions associated with DACs. These are:

1. Resolution. The resolution of a magnitude is the smallest change that can be detected or resolved. For a DAC it is a change in output voltage that must be resolved. Because any change in output voltage is the result of a corresponding change in the digital input, the smallest change in output voltage is due to the weight assigned to the LSB. As an example, Figure 13.14 shows that $n = 0$ for the LSB, and from Equation 13.14 the change in the analog output voltage for this equation is

$$Resolution = \frac{1}{2^n}e_0 \text{ volts} \tag{13.21}$$

where $e_0 = V_{OH}$ and the gain of the voltage follower is unity.

2. Quantization error. As with any quantity that is expressed with a finite number (a limited number of digit positions), there is an inherent error. That is, if it is stated that an object is 0.42 cm long, the error exists in the unstated third position to the right of the decimal point. The length may be closer to 0.41723 cm, if the magnitude was truncated. However, now the 3 is in question, for it may also be the result of truncation. Of course a digital number can be truncated just like an analog number. Then the weight of the LSB of a digital number is the bit in question and it may be positive or negative by one half of the weight of that LSB. For example, the quantization error (QE) of Equation 13.14 can be expressed as

$$QE = \pm\left[\frac{1}{2}\left(\frac{1}{2^N}\right)e_0\right] = \pm\left[\left(\frac{1}{2^{N+1}}\right)e_0\right] \tag{13.22}$$

3. Accuracy. The accuracy is the difference between the calculated analog output voltage and the measured value. These differences can be attributed to noise, drift, difference in calculated and actual gain, offset potential, etc.

Example 13.9

What are the resolution, QE, and accuracy of the DAC of Example 13.8 for a 74LS 8-bit digital input if the measured output voltage was 15.03 volts for an input of 00111010?

Solution

For this DAC $N = 8$

1. Resolution is determined from the analog output voltage equation for $n = 0$. From Example 13.8 the weight of e_0 is
   Resolution $= (15/2^8)(2^0e_0) = (15/256)(2.7)$
   $= 0.158$ volt (truncated) where $V_{OH} = 2.7$ volts.
   Then 0.158 volts is the smallest detectable change in the analog signal.

2. QE $= \pm \frac{1}{2}(0.158)$ volts $= \pm 0.079$ volts, because the weight of e_0 is as stated above.

3. The calculated value of the output voltage is

$$v_o = \frac{15}{256}(e_0 + 2e_1 + 4e_2 + 8e_3 + 16e_4 + 32e_5 + 64e_6 + 128e_7)$$

$$= 0.0586\,[0.5 + 2(2.7) + 4(0.5) + 8(2.7) + 16(2.7) + 32(2.7)$$
$$+ 64(0.5) + 128(0.5)]$$

$$= 0.0586(255.10) = 14.95 \text{ volts (truncated)}$$

Because the measured voltage is 15.03 volts, the difference between the calculated and measured value is -0.08 volts. As a result,

$$\text{Accuracy} = -0.08\,/\,14.95 \times 100 = -0.54\% \text{ (truncated)}$$

Now apply these concepts and equations to a commercially available DAC, the DAC-0800 of Figure 13.18. The equations previously developed must be modified to fit the format required by the DAC-0800. On examination of Equation 13.20 if the series resistors R_{FB} and R_2 are chosen so that their sum is equal to $2R$ then Equation 13.20 becomes

$$v_o = -V_{REF}\ \frac{1}{2^{N-1}}\sum_{n=0}^{N-1} 2^n e_n \qquad (13.23)$$

To understand how the reference voltage V_{REF} of Equation 13.23 can act as a multiplier refer to Figure 13.17 and Equation 13.23. Figure 13.17 shows that the binary bit input voltage levels (D_0, D_1, etc.) to the DAC have been standardized to be either V_{REF} volts for a logic 1, or 0 volts for a logic 0. This standardization is accomplished via the electronic switches (transitors) represented by SPDT switch. When an input to the DAC is a logic 0 (V_{OL} volts) the corresponding SPDT switch is driven to the 0 position (to the left) and 0 volts are input to the DACs R-$2R$ ladder network. When a logic 1 (V_{OH} volts) is input to the DAC the corresponding SPDT switch is driven to position 1 and V_{REF} volts are input to the R-$2R$ ladder network. With this standardization of logic 0s and 1s voltage levels to the R-$2R$ ladder network, the output voltage v_o will be independent of those variances in voltage levels of V_{OL} and V_{OH}. Also, there will no longer be an offset voltage due to binary 0 inputs being something other than zero volts.

Figure 13.17

An R-$2R$ buffered ladder with standardized input voltages.

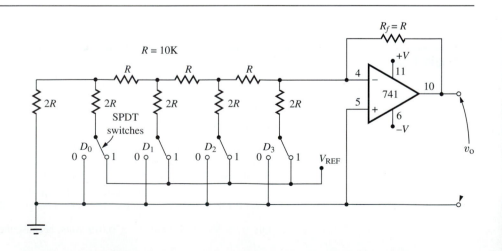

As discussed in Chapter 1, each binary bit position has a decimal weighted value of

Decimal value of the nth bit = binary value of the n bit (a 0 or a 1) $\times 2^n$.   (13.24)

The decimal weight of the nth bit position with a 1 can be expressed as $1 \times 2^n = 2^n$, according to Equation 13.24. If the binary value of that bit is 0 then its decimal weight is $0 \times 2^n = 0$.

Then the decimal equivalent of any binary number with $N-1$ bits according to Equation 13.24 is

$$\text{Decimal equivalent} = D_{N-1} \times 2^{N-1} + D_{N-2} \times 2^{N-2} + \ldots$$
$$+ D_1 \times 2^1 + D_0 \times 2^0 = \sum_{n=0}^{N-1} 2^n D_n, \tag{13.25}$$

where D_n is the binary input to the DAC, such as the DAC of Figure 13.17, and is the same as the e_n of Figure 13.14 and Equation 13.20.

For the sake of simplicity, express the decimal equivalent of Equation 13.25 as N_{10}. Then Equation 13.25 becomes

$$N_{10} = \sum_{n=0}^{N-1} 2^n D_n, \tag{13.26}$$

Substitute Equation 13.26 into Equation 13.23, which then becomes

$$v_{\text{o}} = - V_{\text{REF}} \frac{N_{10}}{2^{N-1}} \text{ volts} \tag{13.27}$$

Equation 13.27 is the form required by the DAC-0800. As seen from Equation 13.27, to determine the output voltage of a DAC divide the decimal equivalent of the binary number being input to the DAC by 2^{N-1}, where N is the number of inputs to the DAC ($N = 8$ for the DAC-0800). Then the product of this value and V_{REF} yields the negative of the equivalent decimal value in volts, which is the analog equivalent voltage of the binary input. If the reference voltage V_{REF} is negative then v_{o} is positive.

Using the proper format for the analog conversion equation for output voltage v_{o}, the DAC-0800 of Figure 13.18 shall be investigated to understand the proper use of Equation 13.27. The DAC-0800 is packaged in a 16-pin DIP and is an 8-bit current-output DAC. The eight digital inputs are identified as B_1 through B_8 with B_8

Figure 13.18

Wiring configuration for the DAC-0800.

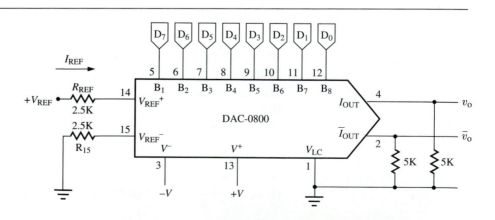

being the LSB, which is why Figure 13.18 shows binary bit D_0 input to the B_8 pin. The DAC-0800 is a TTL compatible device. It requires an input reference current (I_{REF}) of 2.0 mA. The magnitudes of the output currents (pins 2 and 4) are proportioned by the magnitude of the binary input at inputs B_0 through B_8 and reference current I_{REF}. For instance, if the binary input is one half its maximum input value (01111111) then the output current will be one half (1.0 mA) of the reference current I_{REF}, or -5 volts at the output of pin 4 with a 5 K ohm resistor at the output v_o ($1 \times 10^{-3} \times 5 \times 10^3 = 5$). As seen, the accuracy of I_{REF} is critical to the accuracy of the conversion, for it establishes the proportional current value of I_{REF} to be output at I_{OUT}. $\overline{I_{OUT}}$ is the complement of I_{OUT}. By complement the manufacturer means the difference between the maximum current possible (2 mA for an I_{REF} of 2 mA) at output pin 4 minus the actual value of current at pin 4. As an example, if the current at pin 4 is 0.027343 mA then the current at pin 2 is 1.9726 mA ([2.0000 $-$ 0.027343] mA $=$ 1.9726 mA). The more accurate the resistor R_{REF}, the more accurate the current I_{REF} will be. Equation 13.28 can be used to determine the value of R_{REF}.

$$R_{REF} = \frac{V_{REF}}{I_{REF}} = \frac{V_{REF}}{2 \times 10^{-3}} \text{ ohms} \tag{13.28}$$

By connecting external resistors to output pins 2 and 4 the output parameters are converted from current to volts, v_o volts, as is illustrated in Figure 13.18. Connecting 5 K ohm resistors to the outputs, as is the case shown in Figure 13.18, will provide a 10 volt swing for the negative and positive polarities (a 20 volt total). This 10 volt swing can be verified by finding the product of the maximum output current (2 mA ... if I_{REF} is 2 mA) and the output resistor (5 K in this case). Equation 13.27 states mathematically that the output voltage may be either positive or negative, depending on the polarity of V_{REF}. To achieve both polarities positive and negative power supplies are required. These power supplies may be set at voltages within range of from 4.5 to 18 volts and are connected to the DAC via pins 3 and 13.

Example 13.10

Suppose that the DAC of Figure 13.18 was constructed with V_{REF} set to -5 volts and the supply voltages $-V$ and $+V$ are each 15 volts.

1. What is the output voltage at pin 4 if the binary number input to the DAC is

    (a) 00000111?

    (b) 00111111?

2. Repeat part (a), but for $V_{REF} = 5$ volts.

3. What is the resolution of this DAC?

Solution

The binary numbers of parts a and b are first converted to decimal so that Equation 13.27 can be used. Hence,

(a) $N_{10} = 7$. Substituting 7 into Equation 13.27, where N $=$ 8

$$v_o = -(-5)\frac{7}{2^7} = 0.2734375 \text{ volts} = 273.4375 \text{ mV}$$

(b) $v_0 = -5\,\frac{7}{2^7} = -0.2734375 \text{ volts} = -273.4375 \text{ mV}$

(c) From Equation 13.21 determine the smallest change in output voltage for a change in the LSB of the input. The smallest change the input is capable of is when the LSB goes from a 0 to a 1, or in this case, going from a binary input of 00000000 to an input of 00000001. Applying this to Equation 13.27 the smallest change is determined to be

$$v_0 = + \text{ or } - 5\frac{1}{2^7} = + \text{ or } - 39.0625 \text{ mV, or simply } 39.0625 \text{ mV}$$

because only the absolute magnitude is of interest. Observe that for this application R_{REF} of pins 14 and 15 are 2.5 K according to Equation 13.28. Another way to determine the resolution is to divide the maximum output voltage (10 volts for an I_{REF} of 2 mA and a 5 K ohm resistor connected to the output) by 2^8, because for an 8-bit input the number of binary numbers that the output voltage is to be proportioned is 2^8. Then the resolution for this example is

$$\text{Resolution} = \frac{10 \text{ volts}}{2^8} = 39.0625 \text{ mV per binary number.}$$

Review Questions

1. What function does a weighting factor serve?
2. Which DAC input should have the largest weighting factor and why?
3. Why is the output of an R-$2R$ ladder buffered?
4. When would a voltage follower be used to buffer an R-$2R$ ladder network?
5. When would an amplifier be used to buffer an R-$2R$ ladder?
6. What function does the write enable pin of a DAC serve?
7. Why is it often desirable that a DAC have latching capability?
8. What function does the chip enable of a DAC pin serve, and how is it utilized in a system design?
9. What is the function of the R_{FB} pin of a DAC, and under what circumstance is it used?
10. What signal is at the output OUT_1 of the DAC in Figure 13.16?
11. What benefits are there to standardizing the input voltages to the R-$2R$ ladder network of a DAC?
12. Why is the accuracy of the DAC-0800 current I_{REF} important?
13. How is the magnitude of the output current, available at pin 4 of a DAC-0800, determined as a function of the input binary number?
14. How is the resolution of the DAC-0800 determined?

13.6
Analog-to-digital conversion

There are many applications that require that an analog signal be converted to a digital signal. Such a conversion is necessary when the amplitude of an analog signal is to be detected by a digital computer or some other digital device and then the amplitude of that analog signal is to be processed by the computer. For example, computers control the environment of many residences and workplaces. To control such parameters as temperature, humidity, and air flow, the computer must know the temperature and humidity of its source for fresh air (the air outside) as well as the required volume of fresh air (how many people are in the room). Because parameters such as temperature, humidity, and the air flow of a room are analog signals, they must be converted to digital signals before a digital computer can process them. The magnitude of the converted digital signal must be directly proportional to the amplitude of the analog signal. In other words, it is the reverse of a digital-to-analog conversion.

As with a digital-to-analog conversion, there are various methods to achieve analog-to-digital conversion. Regardless of the technique, the device that performs the conversion is the ADC. There are three techniques of conversion that will demonstrate the basic concepts of all ADCs, that is, simultaneous, tracking, and successive approximation ADC.

Simultaneous analog-to-digital converter

Figure 13.19 is a schematic of a 3-bit simultaneous ADC. As an overview, the analog signal e_a is applied to the positive inputs of the seven op-amps, that are serving as comparators, because they are being used in a nonlinear mode of operation (open loop). The comparison voltage (the reference voltage) for each comparator is the voltage input at the negative input, which is identified as V_1, V_2, V_3, etc. These reference voltages are the result of the voltage divider formed by the DC voltage V_{REF} and the series resistors identified as R (recall that an op-amp has a large input impedance — consider it to be an open circuit). As the analog signal is input to the comparators, their output reflects whether the analog signal is greater than or less than their reference voltage by outputting either a logic 1 or 0. These logic levels are input to the priority encoder that encodes the appropriate binary number for the magnitude of analog signal being input.

To analyze the ADC of Figure 13.19, determine how the comparators function. Figure 13.19 has an inset that is a representative model of the comparators, which is identified as the nth op-amp ($n = 1, 2, 3$, etc.). If $e_d > 0$ volts the output voltage of O_n is a logic 1 (V_{SAT}); otherwise it is a logic 0 (the negative power supply input of the op-amp is connected to ground). Because it is the polarity of the input voltage e_d that controls the output logic level of the comparator, write a loop equation at the input and solve for e_d. Hence

$$V_n + e_d - e_a = 0$$

from which

$$e_d = e_a - V_n \tag{13.29}$$

Solving for the condition $e_d > 0$

$$e_d = e_a - V_n > 0 \text{ and if}$$

$$e_a > V_n \tag{13.30}$$

then $O_n = 1$; otherwise $O_n = 0$.

From Equation 13.30 it can be reasoned that for various voltage ranges of e_a:

$e_a < V_1$	then the outputs of all comparators are 0.
$V_1 < e_a < V_2$	then $O_1 = 1$ and the rest are 0.
$V_2 < e_a < V_3$	then $O_1 = O_2 = 1$ and the rest are 0.
$V_3 < e_a < V_4$	then $O_1 = O_2 = O_3 = 1$ and the rest are 0, etc.

These ranges of e_a and the resultant logic level of the comparators are given in Table 13.1. From Figure 13.19 note that the outputs of the comparators are connected to the corresponding inputs of the priority encoder, or $O_n = I_n$.

As Table 13.1 shows, as e_a increases above a reference voltage V_n (V_1, V_2, V_3, etc.), the output of the corresponding comparator is a logic 1 as well as those comparator outputs with preceding reference voltages. That is, for the first row of

Figure 13.19

A simultaneous ADC.

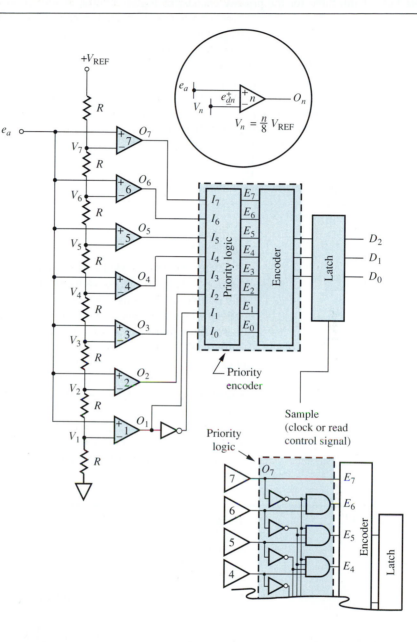

Table 13.1 $e_a < V_1$, the magnitude of e_a is less than all reference voltages and as a result the outputs of all comparators are a logic 0. For the second row e_a increases above V_1 but remains below V_2, and as a result output $O_1 = 1$ ($I_1 = 1$). For the third row e_a increases above V_2 but remains below V_3, and both outputs O_1 and O_2 are a logic 1. This continues until e_a increases above v_7 and the outputs of all the comparators are a logic 1.

To derive a general equation for the reference voltages, apply the voltage divider equation,

$$V_n = \left(\frac{n}{N}\right) V_{REF} = \left(\frac{n}{8}\right) V_{REF} \tag{13.31}$$

where N is the number of resistors R ($N = 8$ for the ADC of Figure 13.19).

Table 13.1　Truth table for the priority encoder of Figure 13.19

Voltage range of e_a	Priority logic								Encoder								ADC output		
	I_7	I_6	I_5	I_4	I_3	I_2	I_1	I_0	E_7	E_6	E_5	E_4	E_3	E_2	E_1	E_0	D_2	D_1	D_0
$e_a < V_1$	0	0	0	0	0	0	0	1	0	0	0	0	0	0	0	1	0	0	0
$V_1 < e_a < V_2$	0	0	0	0	0	0	1	0	0	0	0	0	0	0	1	0	0	0	1
$V_2 < e_a < V_3$	0	0	0	0	0	1	1	0	0	0	0	0	0	1	0	0	0	1	0
$V_3 < e_a < V_4$	0	0	0	0	1	1	1	0	0	0	0	0	1	0	0	0	0	1	1
$V_4 < e_a < V_5$	0	0	0	1	1	1	1	0	0	0	0	1	0	0	0	0	1	0	0
$V_5 < e_a < V_6$	0	0	1	1	1	1	1	0	0	0	1	0	0	0	0	0	1	0	1
$V_6 < e_a < V_7$	0	1	1	1	1	1	1	0	0	1	0	0	0	0	0	0	1	1	0
$e_a > V_7$	1	1	1	1	1	1	1	0	1	0	0	0	0	0	0	0	1	1	1

$V_n = (n/N)\, V_{\text{REF}} = (n/8)\, V_{\text{REF}}$.

After the analog signal has been digitized via the comparator outputs, the digitized output must be converted to a binary number. Use an encoder to accomplish this portion of the conversion. Encoders require that one, and only one, input be active (a logic 1 in this case) at any given time; hence, it must be ensured that this occurs for all ranges of e_a. As seen from Table 13.1, many outputs are active at the same time once e_a is greater than V_2 volts. A solution is to input these outputs to priority logic, which is illustrated as an inset in Figure 13.19. The priority logic outputs a logic 1 for the comparator with the highest priority whose output is a logic 1, and all other comparator outputs are disabled via *AND* gates in the priority logic (see inset). V_7 has the highest priority and V_1 has the lowest. Table 13.1 confirms the function of the priority logic, because the outputs of the priority logic, which are the inputs of the encoder, have only one active output at any given time. The encoder encodes the binary inputs, which are listed as E_0, E_1, E_2, etc., in Table 13.1, to the ADC output binary number shown in columns D_2, D_1, and D_0 of Table 13.1.

Example 13.11

In Figure 13.19 the analog signal shown in Figure 13.20 is applied to its input. If V_{REF} is 16 volts DC, determine the corresponding converted digital output of e_a and superimpose this digital plot on the analog plot of e_a.

Solution

From Equation 13.31 the reference voltages of each comparator can be determined.

$$V_n = \left(\frac{n}{8}\right) 16 \text{ volts}$$

so

$$V_1 = \left(\frac{1}{8}\right) 16 = 2 \text{ volts}$$

$$V_2 = \left(\frac{2}{8}\right) 16 = 4 \text{ volts}$$

and so on. As a guide for determining the various ranges of e_a, a vertical axis for V_n was created beside the axis for e_a in Figure 13.20.

Applying the values for V_1, V_2, V_3, etc., to Table 13.1, determine the various voltage ranges for e_a and the corresponding ADC output. For instance, $V_1 = 2$ volts and $V_2 = 4$ volts; from Table 13.1 the range 2 volts $< e_a < 4$ volts is in the second row from the top, which

Figure 13.20

The ADC output of Figure 13.19.

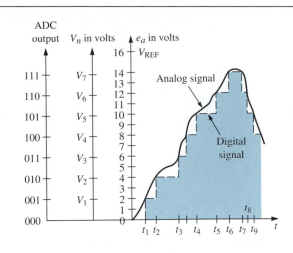

results in ADC output 001. From the plot of e_a in Figure 13.20 this range of e_a is valid over the time interval t_1 to t_2. When e_a increases to 4 volts, then 4 volts $= V_2 < e_a < V_3 = 6$ volts. The ADC output is 010 according to Table 13.1 and is in the time interval of t_2 to t_3 according to Figure 13.20. The digital magnitude of the dotted line of Figure 13.20 corresponds to the vertical scale of the ADC output and is the superimposed digital plot of e_a.

As can be seen from the digital output shown in Figure 13.20, the digital ADC output is an approximation of the analog signal. To improve the approximation would require more comparators, which would increase the number of comparison voltage ranges of the analog signal. It is the number of comparators required that is the major disadvantage to this type of ADC. Because most MPUs have a minimum data bus size of 8 bits, to encode the analog signal into bytes would require 255 (256 − 1) op-amps and associated priority logic. Two hundred fifty-five comparators would yield 256 analog signal ranges, just as the seven comparators of Figure 13.19 yield eight analog ranges. This increase in the number of analog ranges and ADC output bit size would reduce the QE and improve the resolution.

Example 13.12

Apply the concepts of resolution and QE to Example 13.11.

Solution

By definition, the resolution is equal to the weight assigned to the LSB. For this application the LSB changes for every 2 volt change of the analog signal; hence,

$$\text{Resolution} = 2 \text{ volts}$$

The QE is plus or minus one half the weight of the LSB, which for this application is one half of 2 volts. Then

$$\text{QE} = \pm 1 \text{ volt}$$

Example 13.13

What would be the resolution and QE for the ADC output of Example 13.11 if the ADC contained 255 comparators?

Solution

As already discussed, with 255 comparators there would be 256 analog ranges. To determine the size of each range use Equation 13.31. Thus,

$$V_n = \left(\frac{n}{256}\right) 16 = n62.5 \text{ mV}$$

So each analog range is 62.5 mV; therefore, the LSB will represent 62.5 mV. Then

$$\text{Resolution} = 62.5 \text{ mV}$$

and the $\text{QE} = \pm \left(\frac{1}{2}\right) 62.5 \text{ mV} = \pm 31.25 \text{ mV}.$

An ADC with less than 8 bits can be used with an 8-bit system. It is the required resolution of the output that determines the bit size of the ADC.

Tracking analog-to-digital converter

The digital output of a tracking ADC follows the magnitude of the analog input signal. Figure 13.21 is a block diagram of a tracking ADC. The up/down counter is the heart of this type of ADC, for its output is used as the voltage of comparison to the analog input voltage, where equality is being sought. As shown, a clock pulse (a continuous stream of pulses with a fixed repetition rate) is input to the up/down counter, causing the counter to either increase or decrease its count based on the logic level of its up/$\overline{\text{down}}$ input. If the logic level is a high at input up/$\overline{\text{down}}$, it will count up and increase its count; if it is a logic low, it counts down and decreases its count. It is the output of the comparator that determines the logic level of this input. When $e_d > 0$ volts, the output of the comparator is a high; when $e_d < 0$ volts, the output of the comparator is low.

The output of the up/down counter is converted to an analog voltage, e'_a, which is proportional to the digital count. Identifying the signal at the negative input of the comparator is e'_a and the input signal at the positive input as e_{aH}, then by writing a loop equation

$$e_d = e_{aH} - e'_a$$

Figure 13.21

A tracking ADC block diagram.

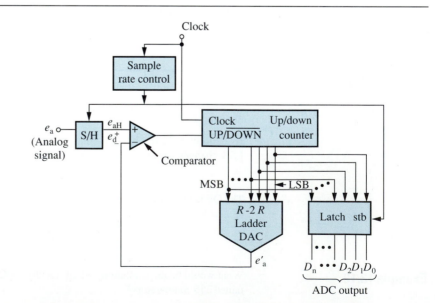

From this equation, if

$$e_{aH} > e'_a \qquad (13.32)$$

the counter counts up, and if

$$e_{aH} < e'_a \qquad (13.33)$$

the counter counts down.

The analog signal e_a is sampled at a fixed rate and its amplitude is held at the value sampled by the sample/hold (S/H) circuitry of Figure 13.21. The output of the S/H circuit is the sampled-and-held value of the analog signal and is identified in Figure 13.21 as e_{aH}, which is input to the positive input of the comparator. The counter digital output is converted to an analog signal e'_a. The comparator compares these two voltages. When Equation 13.32 is valid, the digital number of the counter is less than the sampled analog signal and therefore the digital count must be increased, which is why the counter acts as an up counter. In contrast, when Equation 13.33 is valid, the digital value of the counter is greater than the sampled value of the analog signal and therefore must be decreased, which is why the counter must act as a down counter for this condition. The digital number of the counter never equals the amplitude of the sampled analog signal but rather follows it. The accuracy of the conversion depends on how fast the counter increases or decreases its count versus how fast a new sample is taken of the analog signal e_a. The faster the count, relative to the sample time, the better the accuracy. It is hoped that the digital value can track the analog signal with no more than a difference (error) of one count. This being true, the QE would be plus or minus the weight of the LSB.

As each new sample is taken, the counter contains the binary equivalent of the analog amplitude of the previous sample. This binary number is to be latched by the latch of Figure 13.21, usually at the same sample rate as that of the analog signal.

Figure 13.22 illustrates these concepts. At t_1 a sample of e_a was made and held by the S/H circuitry. The amplitude of the analog signal at that sample time was e_{aH1}. The output of the DAC (e'_a) is less than e_{aH1}; hence, the counter up counts until time t_2. At t_2 the voltage e'_a is greater by one count than e_{aH1}, so the counter down counts. From t_2 to t_3 the counter increases and decreases by one count about the e_{aH1} value, and it is this digital value that would be considered equivalent to e_{aH1}.

Figure 13.22

A tracking ADC digital signal for the ADC of Figure 13.21.

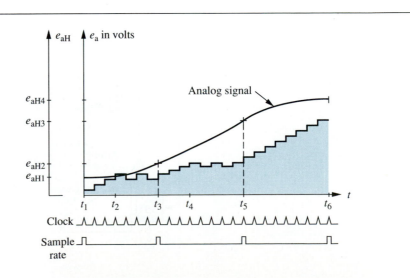

The equivalent digital value was essentially reached for the analog signal sampled at t_1 at time t_2 and oscillated about that value until the new sample was taken at t_3. The latch would latch this value and output it at pins D_0 through D_n. A new sample e_{aH2} is taken at t_3 and the counter increases until t_4, at which time the count is too large by 1. From t_4 to t_5 the counter again oscillates about the sampled value e_{aH2}. A new sample is taken at t_5. The digital equivalent has not enough time (clock pulses) in the time interval t_5 to t_6 for the counter to reach a value equivalent to e_{aH3}; hence, this design would fail. There are several solutions.

1. Increase the weight of the LSB so that for each count, e'_a will increase faster. This solution sacrifices resolution and causes a higher QE.

2. Lengthen the sample time. This means fluctuations in e_a that appear between samples could be missed, which is not a good idea and implies that to determine the sample rate we must have some idea of how rapidly the analog signal may change.

3. Increase the repetition rate of the clock pulses per sample. This is probably the best solution.

Successive approximation analog-to-digital converter

Figure 13.23 is an illustration of the essential function-blocks of a successive approximation ADC. Compare Figures 13.23 and 13.21 and note that a successive approximation ADC utilizes the content of a *successive approximation register* (SAR) for comparison with the sampled-and-held analog signal e_{aH} via the DAC output, rather than the content of an up/down counter. Other than that, the two are essentially the same.

The SAR is a register (the series of rectangles within the SAR of Figure 13.23 represents flip-flops) whose contents are successively adjusted and compared to the sampled-and-held value of the analog signal e_{aH}. To successively adjust and compare the content of the SAR and e_{aH}, each bit position of the SAR is successively set to a logic 1, beginning with the MSB, and then this new adjusted digital value is converted to e'_a, via the DAC, and compared to e_{aH}. If the most recent bit that was set to a logic 1 causes the content of the SAR to be larger than e_{aH} ($e'_a > e_{aH}$) that bit is reset to a logic 0, because the content of the SAR is too large; otherwise it remains a logic 1. Specifically, when the start input of the SAR is activated, the SAR sets its

Figure 13.23

Successive approximation ADC.

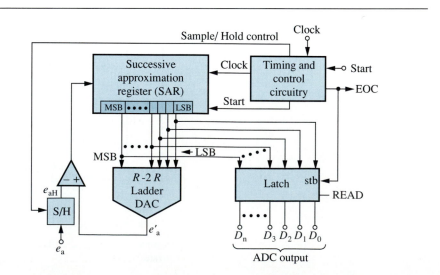

MSB to a logic 1 and all other bits to a logic 0, which is the mid-value of the SAR, as the first approximation to e_{aH}. The content of the SAR is converted to an analog signal e'_a and compared to e_{aH}. As can be determined by writing a loop equation at the input of the comparator, if $e'_a > e_{aH}$ then the output voltage of the comparator will be a logic 1, and if $e'_a < e_{aH}$ the comparator output voltage will be a logic 0. When $e'_a > e_{aH}$ the content of the SAR is too large and must be reduced, whereas if $e'_a < e_{aH}$ the content of the SAR is too small and therefore must be increased. It is the logic level of the comparator output that makes the adjustments to the content of the SAR. If the output of the comparator is a logic 1 ($e'_a > e_{aH}$) the MSB is reset to a logic 0, whereas if the output of the comparator is a logic 0 ($e'_a < e_{aH}$) the MSB logic level is left as a logic 1. After the MSB has been set to a logic 1 and then tested and adjusted, the SAR sets its second MSB to a logic 1 and repeats the test-and-adjust procedure for this bit position. The process is then repeated for the third, fourth, etc. positions until the LSB has been tested and adjusted, at which time the successive approximation is completed and the content of the SAR is the digital approximation of the sampled analog voltage e_{aH}. This digital value is latched and output via pins D_0 through D_n and at the same time an *end-of-conversion* (EOC) is output. The EOC signal can be used to request an interrupt of an MPU, because the ADC would have digital data for the MPU at that time.

Example 13.14

Suppose that an 8-bit ADC, such as that shown in Figure 13.23, is to be used to convert the analog signal of Figure 13.24. Following Figure 13.22, superimpose the resulting ADC output on the same graph as the analog signal. Assume that the SAR is 74ACT technology.

Solution

Because there are eight flip-flops within the SAR, there are eight comparisons and adjustments before EOC occurs. Hence, a new sample of the analog is taken every eight clock pulses, as indicated in Figure 13.24.

Figure 13.24

Successive approximation of an analog signal.

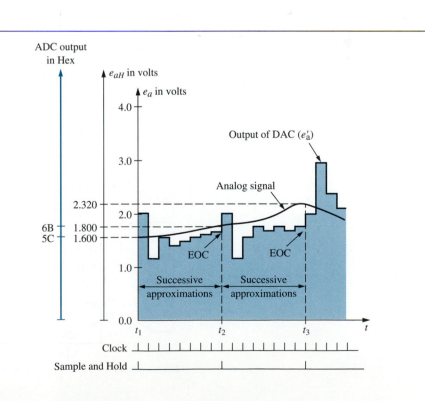

Assuming that the DAC does not have amplification, as could be the case if the DAC of Figure 13.15 were used, Equation 13.14 can be used to determine the magnitude of e'_a. Hence, for an 8-bit DAC

$$e'_a = \frac{1}{256}(128e_7 + 64e_6 + 32e_5 + 16e_4 + 8e_3 + 4e_2 + 2e_1 + e_0)$$

At t_1 the first sample of e_a is made and held. Suppose that the sampled-and-held value is 1.600 volts. While the sample of e_a is being taken, the start command is issued to the SAR, which causes the MSB of the SAR (e_7) to be set to a logic 1 and the other 7 bits to be reset to a logic 0. Because for 74ACT logic $V_{OH} = 3.84$ volts and $V_{OL} = 0.33$ volts, when SAR = 10000000

$$e'_a = \frac{1}{256}[128(3.84) + 127(0.33)] = 2.083 \text{ volts}$$

Because $e'_a = 2.083$ volts > 1.600 volts $= e_{aH}$, the MSB is reset and then the second MSB is set. For SAR = 01000000 the DAC output voltage is

$$e'_a = \frac{1}{256}[128(0.33) + 64(3.84) + 63(0.33)] = 1.206 \text{ volts}$$

For this adjustment $e'_a < 1.600$ volts; hence, the logic 1 remains and the next bit is set. When SAR = 01100000

$$e'_a = (1/256)[128(0.33) + 64(3.84) + 32(3.84) + 31(0.33)] = 1.630$$

volts, which is greater than 1.600 volts so the bit is reset and the next MSB is set. When SAR = 01010000, $e'_a = 1.426$ volts which is less than 1.600 volts; therefore, the bit remains a 1 and the next MSB is set.

SAR = 01011000
$e'_a = 1.535$ volts $< e_{aH}$; set the next bit.
SAR = 01011100
$e'_a = 1.590$ volts $< e_{aH}$; set the next bit.
SAR = 01011110
$e'_a = 1.618$ volts $> e_{aH}$; reset the bit and set the next bit.
SAR = 01011101

Figure 13.25

An 8-bit successive approximation ADC.

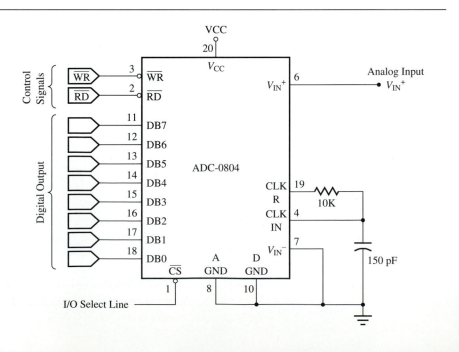

$e'_a = 1.604$ volts > 1.600 volts. Therefore, the LSB is reset and SAR $= 01011100 = 5$CH and $e'_a = 1.590$ volts, which differs by 0.010 volts from the actual sampled-and-held value of 1.600 volts.

Because all registers of the SAR have been adjusted (set or reset), the successive approximation has been completed. The content of the SAR is latched and output via the EOC signal. Figure 13.24 shows the digital output (ADC output) in hexadecimal values. Note that the resolution is 15 mV and the QE is ± 7.5 mV. For this sample the accuracy is -0.625%.

At t_2 another sample of e_a is made and held. For this sample suppose that the sample-and-held value is $e_{aH} = 1.800$ volts. Again the MSB of the SAR is set and the process is again repeated.

$$SAR = 10000000$$
$$e'_a = 2.083 \text{ volts} > e_{aH}$$
$$SAR = 01000000$$
$$e'_a = 1.206 \text{ volts} < e_{aH}$$
$$SAR = 01100000$$
$$e'_a = 1.630 \text{ volts} < e_{aH}$$
$$SAR = 01110000$$
$$e'_a = 1.864 \text{ volts} > e_{aH}$$
$$SAR = 01101000$$
$$e'_a = 1.755 \text{ volts} < e_{aH}$$
$$SAR = 01101100$$
$$e'_a = 1.809 \text{ volts} > e_{aH}$$
$$SAR = 01101010$$
$$e'_a = 1.782 \text{ volts} < e_{aH}$$
$$SAR = 01101011$$
$$e'_a = 1.796 \text{ volts} < e_{aH}$$

The conversion is completed; the EOC output is driven high (requesting and interrupt of the MPU) and the content of the SAR (SAR $= 01101011 = 6$BH) is latched and output to the data pins. At t_3 another sample is taken and the process is repeated. As before, for each EOC the content of the SAR is latched and output at pins D_0 through D_7. The output of the ADC can be read by activating its read pin.

Note from Figure 13.24 that there is a delay time that is eight clock pulses from the time a sample of the analog signal is made until the EOC occurs and the digital output is displayed. As in the case of the tracking ADC, how much delay time exists is directly related to the repetition rate of the clock and, of course, the sampling rate.

There are many types of ADCs. The one to be studied is the ADC-0804 of Figure 13.25, an 8-bit successive approximation ADC. For most applications an ADC is an I/O device for an MPU that converts analog data to digital data. The pins DB_o through DB_7 (DB_7 is the MSB) is where the converted data is output and are connected to the system data bus. These output pins are buffered and are enabled via the chip enable pin $\overline{CS}$. In an MPU-based system the $\overline{CS}$ pin is connected to either an I/O select line, if address decoding is the addressing technique, or an address bus line if linear addressing is used.

The ADC-0804 executes a conversion of the analog signal input at pin $V_{IN}{}^+$ when it receives a positive edge (0 to 1) trigger on the write enable ($\overline{WR}$) pin. The MPU control signal $\overline{WR}$ is connected to the $\overline{WR}$ pin of the ADC-0804. Recalling the MPU timing diagram of Figure 10.7b, the MPU pulls the $\overline{WR}$ pin low when it has written a data byte to the ADC-0804, which is then converted on the positive edge of $\overline{WR}$. When the conversion has been completed the ADC-0804 pulls its $\overline{INTR}$ pin (interrupt request) low for a few microseconds. This can serve as an interrupt request signal to the MPU by being input to the $INTR$ pin of the MPU (after being inverted). As a result of the interrupt request the MPU will read the binary output of the ADC-0804 via the system data bus.

To read the datum created by the conversion, the MPU will drive its read control signal $\overline{RD}$ low, which is connected to the $\overline{RD}$ pin of the ADC-0804. When the $\overline{RD}$ pin of the ADC-0804 is pulled low the equivalent binary value is output to data pins DB_0 through DB_7.

The number of times the input analog signal is sampled for conversion is a function of how many times the $\overline{WR}$ pin and $\overline{RD}$ are activated. For each conversion (sample) the $\overline{WR}$ pin must be pulled low and then high (to create the positive edge) followed by pulling $\overline{RD}$ low so that the converted data can be read. This is accomplished under program control using the *OUT* and *IN* instruction, respectively.

Because the ADC-0804 is a successive approximation ADC it requires a clock. Figure 13.25 shows the clock source to be an R-C circuit composed of a 10 K ohm resistor and a 150 pF capacitor.

The input voltage range is from 0 to 5 volts. There are actually two inputs, $V_{IN}{}^+$ and $V_{IN}{}^-$. It is the difference between the voltages applied to these two pins that is converted. This allows a variety of input voltages to be converted. That is, negative or positive voltages can be converted (ground the input that has the opposite polarity of that being converted), by establishing a reference voltage on one of the inputs and converting the difference between the analog signal and the reference voltage. As always, V_{CC} is $+5$ volts.

Example 13.15

For the ADC-0804 of Figure 13.25, what is the binary output for an analog voltage of:
(a) 0.500 volts

(b) 1.500 volts

(c) 3.500 volts

Solution

The solution requires the resolution of the ADC-0804. Because this is an 8-bit converter with a maximum input voltage of 5 volts, the resolution is

$$\text{Resolution} = \frac{5 \text{ volts}}{2^8 \text{ binary numbers}} = 19.53125 \text{ mV}/\text{binary number}$$

and the decimal equivalent of the analog input voltage is

$$\text{Decimal equivalent} = \frac{\text{Analog input voltage}}{\text{Resolution}} = \frac{V_{IN}{}^+}{19.53125 \text{ mV}/\text{binary number}}$$
$$= 51.20 \, V_{IN}{}^+ \text{ binary number,} \qquad (13.34)$$

which is a base 10 value and must be converted to base 2 (binary).

(a) For an input analog voltage of 0.500 volts the output is

$$\frac{0.500}{19.53125 \text{ mV}/\text{binary number}} = 25.6_{10} = 00011001_2$$

according to Equation 13.34. There was some rounding off because $11001_2 = 25_{10}$.

(b) Again, using Equation 13.34,

$$\text{Decimal equivalent} = 51.20 \times 1.500 = 76.8_{10} = 01001100_2$$

(c) Decimal equivalent $= 51.20 \times 3.5 = 179.2 = 10110011_2$

Review Questions

1. What function do the comparators of a simultaneous ADC serve?

2. Why must a priority encoder be employed in a simultaneous ADC?

3. What is the difficulty in fabricating a simultaneous ADC?

4. How do a tracking ADC and a successive approximation ADC function?

5. What function does the up/down counter serve in a tracking ADC?

6. What function does the successive approximation register serve in a successive approximation ADC?

7. What are resolution and QE?

8. How is the ADC-0804 triggered so that the analog signal at its input is converted to an equivalent digital value?

9. How is the converted data of an ADC-0804 read?

10. How can one calculate the converted value of the input analog signal of an ADC?

11. What determines the sample rate of an ADC?

13.7
Troubleshooting

To troubleshoot parallel I/O device interfaces, use either static or dynamic testing. For either case address the I/O port under test and then either read data from it or write to it. If the I/O port is an input port, such as input port N of Figure 13.1, a static logic level generator or the input device itself may be used to input the datum to be read. When the read control signal is activated (by the MPU or a logic level generator) the CS pin of the interface buffer (I/O port) also goes active and in turn activates those buffers, which loads the datum of input I/O device N on the data bus. Monitoring the address, control, and data busses dynamically, the logic analyzer samples the logic levels on these busses. To capture the data samples of interest, the logic analyzer can be triggered on the address of I/O device N. The captured data can then be displayed and checked to see that the datum being input to port N is also on the data bus while the read control signal is active.

To test an output port dynamically, such as port M in Figure 13.1, again use a logic analyzer to monitor the address, control, data busses, and the output of the interface (I/O port). Using the address of output device M as the trigger word, when it appears, the logic analyzer captures the data on the busses as well as the output datum of port M. Examining the display of this sampled data, check to see that while the write control signal is active the datum on the data bus also appears at the output of port M. A static test of an output port is conducted in a similar fashion.

To troubleshoot a PPI dynamically, use a logic analyzer to monitor the busses as well as I/O ports' nodal connections to the I/O devices. Using the I/O port address as the trigger word, the datum captured from the data bus would be that being read or written to the port. Once the captured data is displayed, verification of proper operation can be made using the same methods for the I/O ports previously discussed. That is, if an input port is being tested, such as port C of Figure 13.2, then, with its address as the trigger word, the displayed datum from the data bus can be checked to see that the datum being input to port C is the same as that on the data bus during the interval that the read pin $\overline{RD}$ is active. For an output port, such as port B of Figure 13.2, the datum being output by the port must be the same as that on the data bus when port B of the PPI is addressed and the write pin $\overline{WR}$ is activated. The bidirectional port A would be tested for a read and then a write.

Testing serial I/O interfaces requires that for a serial-to-parallel interface, the assembled serial input datum be the same as the parallel datum output and the reverse for a parallel-to-serial interface. To test the serial-to-parallel interface of Figure 13.4a, use a logic analyzer that is triggered on the address of that serial I/O port. Monitoring the input node to the interface and the data bus would enable us to verify assembly of the serial datum. For a parallel-to-serial interface monitor the data bus and the output of the interface.

To dynamically test the USART of Figure 13.5, we use its address as the trigger word and monitor nodes $\overline{CS}$, $\overline{RD}$, $\overline{WE}$, TXD, and RXD. Similar to the test of those serial interfaces in Figure 13.4, verify that for a read the serial bits input to the USART via RXD (and assembled) contain the same datum as on the data bus while the read pin is active. For a write, the datum on the data bus and the serial data output at TXD are monitored to verify that their bit compositions are the same.

To test a DAC or ADC statically input the static data and measure the resultant output data. For example, to statically test the R-$2R$ buffered ladder of Figure 13.14, connect static logic level generators to inputs e_0 through e_n and measure output voltage v_o with a VOM. To statistically test the simultaneous ADC of Figure 13.17 use a DC supply to simulate the input analog voltage e_a. This input voltage is varied from zero volts to the maximum allowed for the design. Static logic level detectors or a logic probe are used to detect the digital output at D_2 to D_0.

To test DACs and ADCs dynamically, use test instruments that can display both digital and analog signals. For instance, for a 4-bit R-$2R$ ladder DAC use the display of Figure 13.13. If a particular logic analyzer has the ability to display both digital and analog values, our testing problems are solved; otherwise improvise by using a standard dual-trace oscilloscope.

To test a DAC dynamically with a logic analyzer, monitor the digital inputs and analog output and use the address of the DAC as the trigger word. Using the DAC of Figure 13.16b as the test model, monitor the $\overline{CS}$ node as well as the write node ($\overline{WR}$). We would have a display of the sampled data similar to Figure 13.13b, but for eight digital inputs instead of three, during the time that the write node was active. To produce the output voltage of Figure 13.13 requires that the digital input data begin at zero and increase to the maximum count. To input the desired digital data the inputs of the DAC could be interfaced to the system data bus and the MPU programmed to perform this count. Otherwise an up counter could be used.

If a dual-trace oscilloscope is used to monitor the output voltage of a DAC, one input channel of the oscilloscope should be connected to the LSB of the digital input of the DAC and the other channel to the DAC output. Monitoring the LSB of the input provides us with a reference for each change in the output voltage, as can be imagined by viewing Figure 13.13 without the presence of Q_1, Q_2, and Q_3.

To test the ADC of Figure 13.20, the amplitude of the analog input signal must be monitored as well as the digital output D_2 through D_0. When the read control signal is active, the amplitude of the analog input voltage e_a can be compared to the digital output for correctness. The address of the ADC or the read control signal can be used to trigger the logic analyzer. The ADCs of Figures 13.21 and 13.23 are tested in much the same manner.

Review Questions

1. What is the procedure for testing a parallel I/O port?
2. What is the procedure for testing a serial I/O port?
3. What is the procedure for testing a DAC?
4. What is the procedure for testing an ADC?

Summary

- The data written to or read from a system bus is written in parallel fashion.

- There are two parallel I/O interfaces: tristate buffers and latches. Tristate buffers are used to implement input ports, and latches are used to implement output ports.

- A PPI is a parallel interface that can be programmed so that its ports are either input or output. Port A can be programmed to be a bidirectional port.

- Serial I/O devices require interfacing to a parallel data bus. For a serial input port the serial data is assembled by the interface before being loaded in parallel on the data bus. A serial output port requires that the interface disassemble parallel data bus data into a stream of serial data. Shift registers are used to implement both types of interfaces.

- It is often necessary to interface digital systems with analog systems. For a digital system to output to an analog system, the digital data must be converted to an equivalent analog signal. For an analog system to input its data to a digital system, the analog signal must be converted to digital data. The interface for a digital system to output data to an analog system is a DAC. The interface for the reverse data flow is an ADC.

- There are various DACs, which are based on either the *R*-2*R* ladder network or an op-amp adder. All operate by assigning a weight to each bit position of the digital number to be converted. The MSB is given the greatest weight and the LSB is assigned the least.

- There are also various ADCs, which either compare the analog signal directly, as does the simultaneous ADC, or make a comparison with the content of a digital circuit (either a counter or a register). As these comparisons are made, the resulting digital signal is adjusted to be within one count (the resolution) of equivalent value to the analog signal. In the case of the simultaneous ADC, the adjustment is done via a priority encoder. The adjustments for the tracking ADC are implemented with an up/down counter, and the adjustment is accomplished with a successive approximation register for the successive approximation ADC.

- Regardless of the conversion being performed, there is a quantization error, and by the very nature of quantization the resolution of the answer is limited by the number of bits used to represent the signal. It is the LSB that determines the resolution and the QE.

- To test a DAC the digital input is compared to the analog output voltage while the write control signal is active.

- To test an ADC the analog input amplitude is compared to the digital output while the read control signal is active.

Problems

Section 13.2 Interfacing parallel I/O devices

1. How would an input port differ when implemented with a 74125 rather than a 74126?

2. How could an output port be implemented with *J-K* flip-flops?

3. Design an I/O system that has six I/O ports. Use programmable interfaces such as illustrated in Figure 13.2 to implement the six ports.

Section 13.3 Interfacing serial I/O devices

4. Using the serial-to-parallel interface of Figure 13.4a, design a serial-to-parallel interface and show it integrated into a system similar to the programmable interface of Figure 13.2.

5. Repeat Problem 4 for the parallel-to-serial interface of Figure 13.4b.

6. What is a USART and essentially how does it work?

7. Why are clocks necessary for a USART?

Section 13.4 The operational amplifier

8. Why is an op-amp used as an open-loop "amplifier" considered a nonlinear device? Also explain why open-loop amplifiers are comparators.

9. Configure the op-amp comparator of Figure 13.8a so that $e(t)$ is input at its inverting input ($-$) and the positive input is grounded. Then analytically determine the output voltage of the op-amp for the signal $e(t)$.

10. Configure the comparator of Figure 13.9a so that the positive power supply pin is grounded and apply -5 volts DC to the negative power supply pin. Analytically determine the resultant output voltage for the input signal e of Figure 13.9b.

11. What is the maximum output voltage magnitude of an op-amp and what will be its polarity?

12. For linear applications of an op-amp, why must negative feedback be added?

13. Analytically show that the difference voltage e_d is approximately equal to zero voltage for linear applications of an op-amp.

14. Design the adder of Figure 13.10 so that its output voltage is equal to

$$v_o = -(10e_2 + 5e_1 + 2.5e_0)$$

Let $R_f = 150$ K.

15. Design the adder of Figure 13.10 for the output voltage stated in Problem 14 with the restriction that the maximum feedback current I_f will be approximately 0.1 mA. For ease of analysis, assume that the output voltage is V_{SAT}, with the input voltages approximately equal to zero volts. The power supply voltages are $+$ and -19.5 volts DC.

16. For the adder of Problem 15, what is the output voltage if
    a. $e_0 = 2.0$ volts
       $e_1 = -1.6$ volts
       $e_2 = 0.3$ volts
    b. $e_0 = 0.1$ volts
       $e_1 = 0.5$ volts
       $e_2 = 3.0$ volts

Section 13.5   Digital-to-analog conversion

17. Determine the output voltage of a voltage-follower–buffered R-$2R$ ladder network if the digital input has standard TTL technology and is equal to
    a. 1001
    b. 1010
    c. 10000001
    d. 10000010

18. Determine the value of V_{REF} of Figure 13.12 if the digital input has 74C technology and has
    a. 4 bits.
    b. 8 bits.
    c. 16 bits.

19. Plot the analog output of the DAC of Figure 13.13a if the input digital sequence is 0000, 1000, 1100, 1110, and 1111 and is from the 74C logic family. Take note of the decrease in voltage change as the bit position decreases.

20. If R_f of Figure 13.15 is 150 K and it is desired that the voltage output of the R-$2R$ ladder network be amplified by 10, what must be the value of R?

21. Design the R-$2R$ amplifier of Figure 13.15b such that a gain of 10 is achieved and the maximum feedback current I_f (the current flowing through R_f) is approximately 0.1 mA. For simplicity of analysis, assume that the output voltage is V_{SAT} and the input voltages are zero. The negative power supply is 15 volts DC, and the positive power supply input is grounded.

22. Repeat Problem 21 but use resistor values in stock, which are 10 K, 12 K, 13 K, 15 K, 100 K, 120 K, and 150 K. What is the value of I_f using these standard values?

23. For $R = 15$ K and $R_f = 120$ K of Figure 13.15b, what is the magnitude of the actual worst case feedback current I_f?

24. What is the largest digital number that can be converted by the DAC of Figure 13.15b?

25. If the DAC of Figure 13.15b is designed for a gain of -5, what is the maximum output voltage of the op-amp if the digital technology is 74LS and the number of bits is
    a. $N = 4$
    b. $N = 8$

26. For the 8-bit DAC of Figure 13.15b, with 74LS logic driving it, what is the maximum gain that the amplifier can have? Assume that the negative power supply is 20 volts DC.

27. What are the resolution and QE for Problem 25?

28. What is the accuracy of the DAC of Problem 25a and b if the measured output voltage for part (a) is -12.45 volts and that for (b) is -13.21 volts, with all inputs a logic 1?

29. For the DAC-0800 of Figure 13.18, determine the output voltage at pin 4 if:
    a. The binary input is 00001111
    b. The binary input is 11111111

30. What would be the value of R_{REF} if the reference voltage V_{REF} is 10 volts?

31. How would you modify the circuit of Figure 13.18 to accommodate a negative V_{REF}?

32. For an I_{REF} of 2 mA, what are the voltages at pins 2 and 4 of Figure 13.18 if the current at pin 4 is 1.2 mA?

Section 13.6   Analog-to-digital conversion

33. For a 4-bit simultaneous ADC, how many op-amps and resistors R are there?

34. For a 4-bit simultaneous ADC, how many inputs are there to the priority encoder?

35. For a 4-bit simultaneous ADC with $V_{REF} = 20$ volts DC, what magnitude is the maximum comparator reference voltage? Mathematically describe the relationship between V_{REF} and V_n.

36. For the ADC of Problem 35, what is the maximum amplitude of an analog signal that can be converted?

37. Derive an equation for the ADC of Figure 13.19 that can serve as a guide in selecting a value for V_{REF} for a given analog signal.

38. For an analog voltage that is a ramp, which peaks to 5 volts in 5 ms, plot the digital output similar to Figure 13.20. The ADC is to be a 4-bit simultaneous ADC.

39. If the DAC of Figure 13.21 is a 4-bit DAC, how many clock pulses must there be before a sample and hold of the analog can be made?

40. For the 4-bit tracking ADC of Figure 13.21, what is the approximate upper limit of an analog signal amplitude that can be converted? Assume that the DAC does not have an amplified output and the DAC has 74C technology.

41. The 4-bit tracking ADC of Figure 13.21 is to convert a 5 volt ramp that peaks in 5 ms. Plot the results using Figure 13.22 as a guide.

42. Explain the difference in process if an 8-bit successive approximation ADC would have been used for Problem 41.

43. For a 4-bit successive approximation ADC, using Figure 13.24 as a guide describe the conversion process.

44. When converting a digital value to an equivalent analog value, how much error can there be in the conversion, and what is the source of that error?

45. For the ADC-0804 of Figure 13.25, what would be the analog output if the binary input is:
    a. 0.020 volts
    b. 1.000 volts
    c. 4.000 volts
    d. 5.000 volts

Troubleshooting

Section 13.7 Troubleshooting

46. Devise a static test for the parallel ports of Figure 13.1.

47. Devise a static test for the programmable I/O interface of Figure 13.2.

48. Devise a dynamic test for the programmable interface of Figure 13.2.

49. Devise a dynamic test for the USART of Figure 13.5.

50. Devise a static test for the *R*-2*R* buffered ladder of Figure 13.15b.

51. Devise a dynamic test that uses an up counter to generate the digital input to a three-input DAC (similar to that of Fig. 13.14) and a dual-trace oscilloscope to detect the resultant output voltage.

52. If the up counter of Problem 50 has 74C technology, produce a drawing of the expected output voltage and LSB of the input voltage.

53. Changing the digital input to 3 bits, make a timing diagram for the DAC of Figure 13.16b similar to the one displayed by a logic analyzer.

54. Devise a dynamic test for the ADC of Figure 13.21.

A

POWERS OF 2

Powers of Two

2^n	n	2^{-n}
1	0	1.0
2	1	0.5
4	2	0.25
8	3	0.125
16	4	0.062 5
32	5	0.031 25
64	6	0.015 625
128	7	0.007 812 5
256	8	0.003 906 25
512	9	0.001 953 125
1 024	10	0.000 976 562 5
2 048	11	0.000 488 281 25
4 096	12	0.000 244 140 625
8 192	13	0.000 122 070 312 5
16 384	14	0.000 061 035 156 25
32 768	15	0.000 030 517 578 125
65 536	16	0.000 015 258 789 062 5
131 072	17	0.000 007 629 394 531 25
262 144	18	0.000 003 814 697 265 625
524 288	19	0.000 001 907 348 632 812 5
1 048 576	20	0.000 000 953 674 316 406 25
2 097 152	21	0.000 000 476 837 158 203 125
4 194 304	22	0.000 000 238 418 579 101 562 5
8 388 608	23	0.000 000 119 209 289 550 781 25
16 777 216	24	0.000 000 059 604 644 775 390 625
33 554 432	25	0.000 000 029 802 322 387 695 312 5
67 108 864	26	0.000 000 014 901 161 193 847 656 25
134 217 728	27	0.000 000 007 450 580 596 923 828 125
268 435 456	28	0.000 000 003 725 290 298 461 914 062 5
536 870 912	29	0.000 000 001 862 645 149 230 957 031 25
1 073 741 824	30	0.000 000 000 931 322 574 615 478 515 625
2 147 483 648	31	0.000 000 000 465 661 287 307 739 257 812 5
4 294 967 296	32	0.000 000 000 232 830 643 653 869 628 906 25
8 589 934 592	33	0.000 000 000 116 415 321 826 934 814 453 125
17 179 869 184	34	0.000 000 000 058 207 660 913 467 407 226 562 5
34 359 738 368	35	0.000 000 000 029 103 830 456 733 703 613 281 25
68 719 476 736	36	0.000 000 000 014 551 915 228 366 851 806 640 625
137 438 953 472	37	0.000 000 000 007 275 957 614 183 425 903 320 312 5
274 877 906 944	38	0.000 000 000 003 637 978 807 091 712 951 660 156 25
549 755 813 888	39	0.000 000 000 001 818 989 403 545 856 475 830 078 125
1 099 511 627 776	40	0.000 000 000 000 909 494 701 772 928 237 915 039 062 5
2 199 023 255 552	41	0.000 000 000 000 454 747 350 886 464 118 957 519 531 25
4 398 046 511 104	42	0.000 000 000 000 227 373 675 443 232 059 478 759 765 625
8 796 093 022 208	43	0.000 000 000 000 113 686 837 721 616 029 739 379 882 812 5
17 592 186 044 416	44	0.000 000 000 000 056 843 418 860 808 014 869 689 941 406 25
35 184 372 088 832	45	0.000 000 000 000 028 421 709 430 404 007 434 844 970 703 125
70 368 744 177 664	46	0.000 000 000 000 014 210 854 715 202 003 717 422 485 351 562 5
140 737 488 355 328	47	0.000 000 000 000 007 105 427 357 601 001 858 711 242 675 781 25
281 474 976 710 656	48	0.000 000 000 000 003 552 713 678 800 500 929 355 621 337 890 625
562 949 953 421 312	49	0.000 000 000 000 001 776 356 839 400 250 464 677 810 668 945 312 5
1 125 899 906 842 624	50	0.000 000 000 000 000 888 178 419 700 125 232 338 905 334 472 656 25
2 251 799 813 685 248	51	0.000 000 000 000 000 444 089 209 850 062 616 169 452 667 236 328 125
4 503 599 627 370 496	52	0.000 000 000 000 000 222 044 604 925 031 308 084 726 333 618 164 062 5
9 007 199 254 740 992	53	0.000 000 000 000 000 111 022 302 462 515 654 042 363 166 809 082 031 25
18 014 398 509 481 984	54	0.000 000 000 000 000 055 511 151 231 257 827 021 181 583 404 541 015 625
36 028 797 018 963 968	55	0.000 000 000 000 000 027 755 575 615 628 913 510 590 791 702 270 507 812 5
72 057 594 037 927 936	56	0.000 000 000 000 000 013 877 787 807 814 456 755 295 395 851 135 253 906 25
144 115 188 075 855 872	57	0.000 000 000 000 000 006 938 893 903 907 228 377 647 697 925 567 676 950 125
288 230 376 151 711 744	58	0.000 000 000 000 000 003 469 446 951 953 614 188 823 848 962 783 813 476 562 5
576 460 752 303 423 488	59	0.000 000 000 000 000 001 734 723 475 976 807 094 411 924 481 391 906 738 281 25

EXPLANATION OF THE ANSI/IEEE 91-1984 LOGIC SYMBOLS

1.0

Introduction

The *International Electrotechnical Commission* (IEC) has been developing a very powerful symbolic language that can show the relationship of each input of a digital logic circuit to each output without showing explicitly the internal logic. At the heart of the system is dependency notation, which will be explained in Section 4.

The system was introduced in the United States in a rudimentary form in IEEE/ANSI Standard Y32.14-1973. Lacking at that time a complete development of dependency notation, it offered little more than a substitution of rectangular shapes for the familiar distinctive shapes for representing the basic functions of *AND*, *OR*, negation, etc. This is no longer the case.

Internationally, Working Group 2 of IEC Technical Committee TC-3 has prepared a new document (Publication 617-12) that consolidates the original work started in the mid 1960s and published in 1972 (Publication 117-15) and the amendments and supplements that have followed. Similarly for the United States, IEEE Committee SCC 11.9 has revised the publication IEEE Std 91/ANSI Y32.14. Now numbered simply IEEE Std 91-1984, the IEEE standard contains all of the IEC work that has been approved, and also a small amount of material still under international consideration. Texas Instruments (TI) is participating in the work of both organizations and this document introduces new logic symbols in accordance with the new standards. When changes are made as the standards develop, future editions will take those changes into account.

The following explanation of the new symbolic language is necessarily brief and greatly condensed from what the standards publications will contain. This is not intended to be sufficient for those people who will be developing symbols for new devices. It is primarily intended to make possible the understanding of the symbols used in this data book and is somewhat briefer than the explanation that appears in several of TI's data books on digital logic. However, it includes a new section (6.0) that explains in detail several symbols for actual devices. This has proven to be a powerful learning aid.

2.0 Symbol Composition

A symbol comprises an outline or a combination of outlines together with one or more qualifying symbols. The shape of the symbols is not significant. As shown in Figure 1, general qualifying symbols are used to tell exactly what logical operation is performed by the elements. Table 1 shows general qualifying symbols defined in the new standards. Input lines are placed on the left and output lines are placed on the right. When an exception is made to that convention, the direction of signal flow is indicated by an arrow (*see* Fig. 9).

Figure 1

Symbol composition

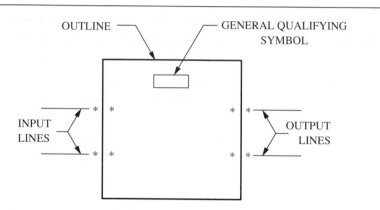

*Possible positions for qualifying symbols relating to inputs and outputs

3.0 Qualifying symbols

Table 1 shows general qualifying symbols defined by IEEE Standard 91. These characters are placed near the top center or the geometric center of a symbol or symbol element to define the basic function of the device represented by the symbol or of the element.

3.1 General qualifying symbols

X/Y is the general qualifying symbol for identifying coders, code converters, and level converters. *X* and *Y* may be used in their own right to stand for some code or either or both may be replaced by some other indication of the code or level such as BCD or TTL. As might be expected, interface circuits make frequent use of this set of qualifying symbols.

3.2 Qualifying symbols for inputs and outputs

Qualifying symbols for inputs and outputs are shown in Table 2 and will be familiar to most users with the possible exception of the logic polarity and analog signal indicators. The older logic negation indicator means that the external 0 state produces the internal 1 state. The internal 1 state means the active state. Logic negation may be used in pure logic diagrams; to tie the external 1 and 0 logic states to the levels H (high) and L (low), a statement of whether positive logic (1 = H, 0 = L) or negative logic (1 = L, 0 = H) is being used is required or must be assumed. Logic

Table 1   General qualifying symbols

Symbol	Description
&	*AND* gate or function.
≥ 1	*OR* gate or function. The symbol was chosen to indicate that at least one active input is needed to activate the output.
$= 1$	Exclusive-*OR*. One and only one input must be active to activate the output.
1	The one input must be active.
$\triangleright$ or $\triangleleft$	A buffer or element with more than usual output capability (symbol is oriented in the direction of signal flow).
⎍	Schmitt trigger; element with hysteresis.
X/Y	Coder, code converter, level converter.

The following are examples of subsets of this general class of qualifying symbols used in this book:

BCD / 7-SEG	BCD to seven-segment display driver.
TTL / MOS	TTL to MOS level converter.
CMOS / PLASMA DISP	Plasma-display driver with CMOS-compatible inputs.
MOS / LED	Light-emitting-diode driver with MOS-compatible inputs.
CMOS / VAC FLUOR DISP	Vacuum-fluorescent display driver with CMOS-compatible inputs.
CMOS / EL DISP	Electroluminescent display driver with CMOS-compatible inputs.
TTL / GAS DISCH DISPLAY	Gas-discharge display driver with TTL-compatible inputs.

SRGm	Shift register. m is the number of bits.

polarity indicators eliminate the need for calling out the logic convention and are used in this data book in the symbology for actual devices. The presence of the triangle polarity indicator indicates that the L logic level will produce the internal 1 state (the active state) or that, in the case of an output, the internal 1 state will produce the external L level. Note how the active direction of transition for a dynamic input is indicated in positive logic, negative logic, and with polarity indication.

When nonstandardized information is shown inside an outline, it is usually enclosed in square brackets [like these]. The square brackets are omitted when associated with a nonlogic input, which is indicated by an X superimposed on the connection line outside the symbol.

3.3 Symbols inside the outline

Table 3 shows some symbols used inside the outline. Note particularly that open-collector (open-drain), open-emitter (open-source), and three-state outputs have distinctive symbols. Also note that an *EN* input affects all of the outputs of the element and has no effect on inputs. An *EN* input affects all the external outputs of the element in which it is placed, plus the external outputs of any elements shown to be influenced by that element. It has no effect on inputs. When an enable input affects only certain outputs, affects outputs located outside the indicated influence of the element in which the enable input is placed, or affects one or more inputs, a form of dependency notation will indicate this (*see* Section 4.9). The effects of the *EN* input on the various types of outputs are shown.

Table 2 Qualifying symbols for inputs and outputs

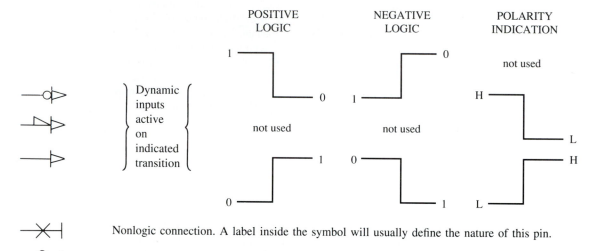

Logic negation at input. External 0 produces internal 1.

Logic negation at output. Internal 1 produces external 0.

Active-low input. Equivalent to ⊸◁ in positive logic.

Active-low output. Equivalent to ▷⊸ in positive logic.

Active-low input in the case of right-to-left signal flow.

Active-low output in the case of right-to-left signal flow.

Signal flow from right to left. If not otherwise indicated, signal flow is from left to right.

Bidirectional signal flow.

POSITIVE LOGIC   NEGATIVE LOGIC   POLARITY INDICATION

Dynamic inputs active on indicated transition

Nonlogic connection. A label inside the symbol will usually define the nature of this pin.

Input for analog signals (on a digital symbol) (*see* Fig. 11).

Input for digital signals (on an analog symbol) (*see* Fig. 11).

It is particularly important to note that a *D* input is always the data input of a storage element. At its internal 1 state, the *D* input sets the storage element to its 1 state, and at its internal 0 state it resets the storage element to its 0 state.

3.4 Combinations of outlines and internal connections

When a circuit has one or more inputs that are common to more than one element of the circuit, the common-control block may be used. This is the only distinctively shaped outline used in the IEC system. Figure 2 shows that unless otherwise qualified by dependency notation, an input to the common-control block is an input to each of the elements below the common-control block.

The outlines of elements may be embedded within one another or abutted to form complex elements, in which case the following rules apply. There is no logic connection between elements when the line common to their outlines is in the direction of signal flow. There is at least one logic connection when the line common

Table 3 Symbols inside the outline

	Bi-threshold input (input with hysteresis).
	N-P-N open-collector or similar output that can supply a relatively low-impedance L level when not turned off. Requires external pull-up. Capable of positive-logic wired-AND connection.
	Passive-pull-up output is similar to N-P-N open-collector output but is supplemented with a built-in passive pull-up.
	N-P-N open-emitter or similar output that can supply a relatively low-impedance H level when not turned off. Requires external pull-down. Capable of positive-logic wired-OR connection.
	Passive pull-down output is similar to N-P-N open-emitter output but is supplemented with a built-in passive pull-down.
	3-state output.
	Output with more than usual output capability (symbol is oriented in the direction of signal flow).
EN	Enable input When at its internal 1-state, all outputs are enabled. When at its internal 0-state, open-collector, open-emitter, and three-state outputs are at external high-impedance state, and all other outputs (i.e., totem-poles) are at the internal 0-state.
J, K, R, S, T	Usual meanings associated with flip-flops (e.g., R = reset, T = toggle)
D	Data input to a storage element equivalent to:
→m ←m	Shift right (left) inputs, m = 1, 2, 3, etc. If m = 1, it is usually not shown.
	Binary grouping. m is highest power of 2. Produces a number equal to the sum of the weights of the active inputs.
	Input line grouping . . . indicates two or more terminals used to implement a single logic input, e.g., differential inputs.

Figure 2

Common-control block.

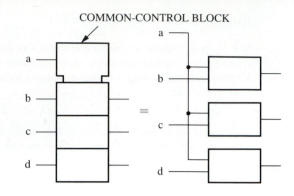

to two outlines is perpendicular to the direction of signal flow. If no indications are shown on either side of the common line, it is assumed that there is only one logic connection. If more than one internal connection exists between adjacent elements, the number of connections will be clarified by the use of one or more of the internal connection symbols from Table 4 and/or appropriate qualifying symbols or dependency notation.

Table 4 shows symbols that are used to represent internal connection with specific characteristics. The first is a simple noninverting connection, the second is inverting, the third is dynamic. As with this symbol and an external input line, the transition from 0 to 1 on the left produces a momentary 1 state on the right. The fourth symbol is similar except that the active transition on the left is from 1 to 0.

Table 4 Symbols for internal connections

	Internal connection. 1 state on left produces 1 state on right.
	Negated internal connection. 1 state on left produces 0 state on right.
	Dynamic internal connection. Transition from 0 to 1 on left produces transitory 1 state on right.
	Dynamic internal connection. Transition from 1 to 0 on left produces transitory 1 state on right.

Only logic states, not levels, exist inside symbols. The negation symbol (○) is used internally even when direct polarity indication ($\triangleright$) is used externally.

The binary grouping symbol will be explained in more detail in Section 6.11. Binary-weighted inputs are arranged in order and the binary weights of the least-significant and the most-significant lines are indicated by numbers. In this document weights of input and output lines will be represented by powers of two usually only when the binary grouping symbol is used, otherwise decimal numbers will be used. The grouped inputs generate an internal number on which a mathematical function can be performed or that can be an identifying number for dependency notation. This number is the sum of the weights $(1, 2, 4 \ldots 2^n)$ of those inputs standing at their 1 states. A frequent use is in addresses for memories.

Reversed in direction, the binary grouping symbol can be used with outputs. The concept is analogous to that for the inputs and the weighted outputs will indicate the internal number assumed to be developed within the circuit.

In an array of elements, if the same general qualifying symbol and the same qualifying symbols associated with inputs and outputs would appear inside each of the elements of the array, these qualifying symbols are usually shown only in the first element. This is done to reduce clutter and to save time in recognition. Similarly, large identical elements that are subdivided into smaller elements may each be represented by an unsubdivided outline. The MC3446 symbol illustrates this principle.

4.0 Dependency Notation

Some readers will find it more to their liking to skip this section and proceed to Section 6.0 for the explanation of the symbols for a few actual devices. Reference will be made there to various parts of this section as it is needed. If this procedure is followed, it is recommended that Section 5.0 be read after Section 6.0 and then all of Section 4.0 be reread.

4.1 General explanation

Dependency notation is the powerful tool that sets the IEC symbols apart from previous systems and makes compact, meaningful symbols possible. It provides the means of denoting the relationship between inputs, outputs, or inputs and outputs without actually showing all the elements and interconnections involved. The information provided by dependency notation supplements that provided by the qualifying symbols for an element's function.

In the convention for the dependency notation, use will be made of the terms "affecting" and "affected." In cases where it is not evident which inputs must be considered as being the affecting or the affected ones (e.g., if they stand in an *AND* relationship), the choice may be made in any convenient way.

So far, 11 types of dependency have been defined but only the 8 used in this book are explained. They are listed below in the order in which they are presented and are summarized in Table 5 following Section 4.10.2.

Section	Dependency type or other subject
4.2	*G, AND*
4.3	General rules for dependency notation
4.4	*V, OR*
4.5	*N*, Negate (Exclusive-*OR*)
4.6	*Z*, Interconnection
4.7	*X*, Transmission
4.8	*C*, Control
4.9	*EN*, Enable
4.10	*M*, Mode

4.2 *G (AND)* dependency

A common relationship between two signals is to have them *AND*ed together. This has traditionally been shown by explicitly drawing an *AND* gate with the signals connected to the inputs of the gate. The 1972 IEC publication and the 1973 IEEE/ANSI standard showed several ways to show this *AND* relationship using dependency notation. Although 10 other forms of dependency have since been defined, the ways to invoke *AND* dependency are now reduced to one.

In Figure 3 input b is *AND*ed with input a and the complement of b is *AND*ed with c. The letter *G* has been chosen to indicate *AND* relationships and is placed at input **b**, inside the symbol. A number considered appropriate by the symbol designer (1 has been used here) is placed after the letter *G* and also at each affected input. Note the bar over the 1 at input c.

Figure 3

G dependency between inputs.

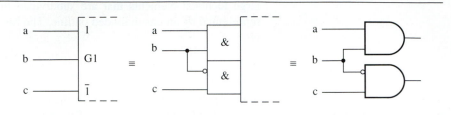

In Figure 4, output b affects input a with an *AND*ed relationship. The lower example shows that it is the internal logic state of b, unaffected by the negation sign, that is *AND*ed. Figure 5 shows input a to be *AND*ed with a dynamic input b.

The rules for *G* dependency can be summarized as follows:

When a *Gm* input or output (*m* is a number) stands at its internal 1 state, all inputs and outputs affected by *Gm* stand at their normally defined internal logic states. When the *Gm* input or output stands at its 0 state, all inputs and outputs affected by *Gm* stand at their internal 0 states.

Figure 4

G dependency between outputs and inputs.

Figure 5

G dependency with a dynamic input.

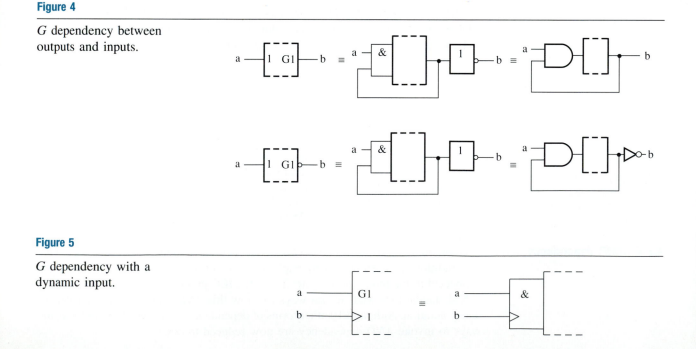

4.3 Conventions for the application of dependency notation in general

The rules for applying dependency relationships in general follow the same pattern as was illustrated for *G* dependency.

Application of dependency notation is accomplished by:

1. Labeling the input (or output) affecting other inputs or outputs with the letter symbol indicating the relationship involved (e.g., *G* for *AND*) followed by an identifying number, appropriately chosen
2. Labeling each input or output affected by that affecting input (or output) with that same number.

If it is the complement of the internal logic state of the affecting input or output that does the affecting, then a bar is placed over the identifying numbers at the affected inputs or outputs (Fig. 3).

If two affecting inputs or outputs have the same letter and same identifying number, they stand in an *OR* relationship to each other (Fig. 6).

Figure 6

*OR*ed affecting inputs.

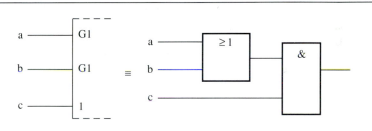

If the affected input or output requires a label to denote its function (e.g., ''D''), this label will be prefixed by the identifying number of the affecting input (*see* Fig. 12).

If an input or output is affected by more than one affecting input, the identifying numbers of each of the affecting inputs will appear in the label of the affected one, separated by commas. The normal reading order of these numbers is the same as the sequence of the affecting relationships (*see* Fig. 12).

4.4 V (OR) dependency

The symbol denoting *OR* dependency is the letter *V* (Fig. 7).

Figure 7

V (OR) dependency.

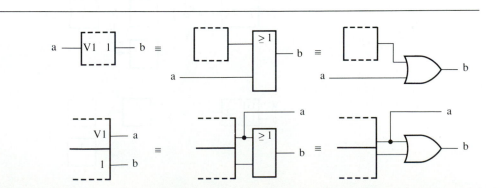

When a *Vm* input or output stands at its internal 1 state, all inputs and outputs affected by *Vm* stand at their internal 1 states. When the *Vm* input or output stands at its internal 0 state, all inputs and outputs affected by *Vm* stand at their normally defined internal logic states.

4.5 *N* (negate) (Exclusive-*OR*) dependency

The symbol denoting negate dependency is the letter *N* (Fig. 8). Each input or output affected by an *Nm* input or output stands in an Exclusive-*OR* relationship with the *Nm* input or output.

Figure 8

N (negate) (Exclusive-*OR*) dependency.

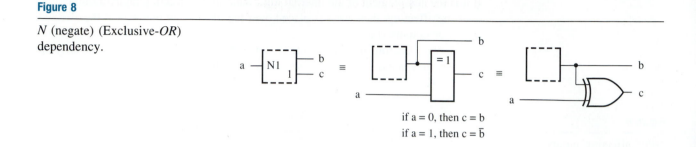

if a = 0, then c = b
if a = 1, then c = $\bar{b}$

When an *Nm* input or output stands at its internal 1 state, the internal logic state of each input and each output affected by *Nm* is the complement of what it would otherwise be. When an *Nm* input or output stands at its internal 0 state, all inputs and outputs affected by *Nm* stand at their normally defined internal logic states.

Figure 9

Z (interconnection) dependency.

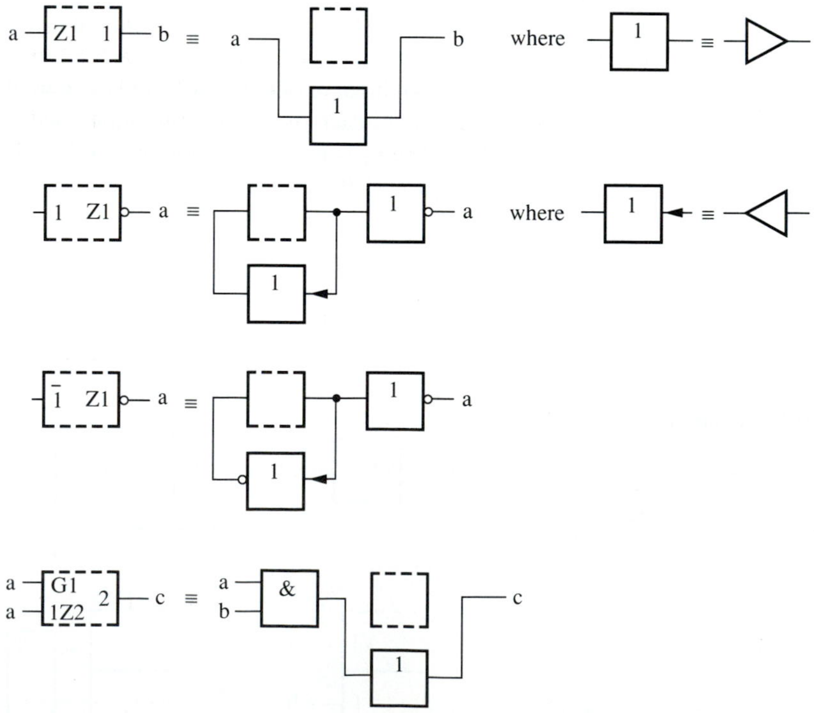

4.6 *Z* (interconnection) dependency

The symbol denoting interconnection dependency is the letter *Z*.

Interconnection dependency is used to indicate the existence of internal logic connections between inputs, outputs, internal inputs, and/or internal outputs.

The internal logic state of an input or output affected by a *Zm* input or output will be the same as the internal logic state of the *Zm* input or output, unless modified by additional dependency notation (Fig. 9).

4.7 *X* (transmission) dependency

The symbol denoting transmission dependency is the letter *X*.

Transmission dependency is used to indicate controlled bidirectional connections between affected input/output ports (Fig. 10).

Figure 10

X (transmission) dependency.

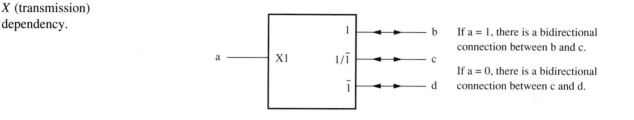

If a = 1, there is a bidirectional connection between b and c.

If a = 0, there is a bidirectional connection between c and d.

When an *Xm* input or output stands at its internal 1 state, all input-output ports affected by this *Xm* input or output are bidirectionally connected together and stand at the same internal logic state or analog signal level. When an *Xm* input or output stands at its internal 0 state, the connection associated with this dependency notation does not exist.

Although the transmission paths represented by *X* dependency are inherently bidirectional, use is not always made of this property. This is analogous to a piece of wire, that may be constrained to carry current in only one direction. If this is the case in a particular application, then the directional arrows shown in Figures 10 and 11 would be omitted.

Figure 11

Analog data selector (multiplexer/demultiplexer).

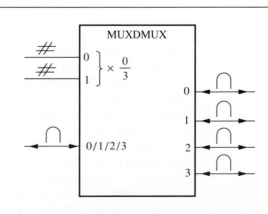

4.8 *C* (control) dependency

The symbol denoting control dependency is the letter *C*.

Control inputs are usually used to enable or disable the data (*D, J, K, R,* or *S*) inputs of storage elements. They may take on their internal 1 states (be active) either statically or dynamically. In the latter case the dynamic input symbol is used as shown in the second example of Figure 12.

Figure 12

C (control) dependency.

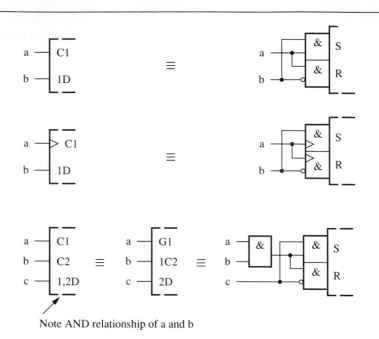

Note AND relationship of a and b

When a *Cm* input or output stands at its internal 1 state, the inputs affected by *Cm* have their normally defined effect on the function of the element, that is, these inputs are enabled. When a *Cm* input or output stands at its internal 0 state, the inputs affected by *Cm* are disabled and have no effect on the function of the element.

4.9 *EN* (enable) dependency

The symbol denoting enable dependency is the combination of letters *EN*.

An *ENm* input has the same effect on outputs as an *EN* input, (*see* Section 3.3), but it affects only those outputs labeled with the identifying number *m*. It also affects those inputs labeled with the identifying number *m*. By contrast, an *EN* input affects all outputs and no inputs. The effect of an *ENm* input on an affected input is identical to that of a *Cm* input (Fig. 13).

When an *ENm* input stands at its internal 1 state, the inputs affected by *ENm* have their normally defined effects on the function of the element and the outputs affected by this input stand at their normally defined internal logic states, that is, these inputs and outputs are enabled.

When an *ENm* input stands at its internal 0 state, the inputs affected by *ENm* are disabled and have no effect on the function of the element, and the outputs affected by *ENm* are also disabled. Open-collector outputs are turned off, three-state outputs stand at their Hi-2 state, and all other outputs (e.g., totem-pole outputs) stand at their internal 0 states.

Figure 13

EN (enable) dependency.

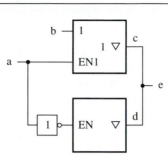

If a = 0, input b and output c are disabled and e = d
If a = 1, output d is disabled and e = c

4.10 *M* (mode) dependency

The symbol denoting mode dependency is the letter *M*.

Mode dependency is used to indicate that the effects of particular inputs and outputs of an element depend on the mode in which the element is operating.

If an input or output has the same effect in different modes of operation, the identifying numbers of the relevant affecting *Mm* inputs will appear in the label of that affected input or output between parentheses and separated by solidi, for example, $(1/2)CT = 0 \equiv 1CT = 0/2CT = 0$ where 1 and 2 refer to *M*1 and *M*2.

4.10.1 *M* dependency affecting inputs

M dependency affects inputs the same as *C* dependency. When an *Mm* input or output stands at its internal 1 state, the inputs affected by this *Mm* input or output have their normally defined effect on the function of the element, that is, the inputs are enabled.

When an *Mm* input or output stands at its internal 0 state, the inputs affected by this *Mm* input or output have no effect on the function of the element. When an affected input has several sets of labels separated by solidi (e.g., $C4/2 \rightarrow /3+$), any set in which the identifying number of the *Mm* input or output appears has no effect and is to be ignored. This represents disabling of some of the functions of a multifunction input.

The circuit in Figure 14 has two inputs, b and c, that control which one of four modes (0, 1, 2, or 3) will exist at any time. Inputs d, e, and f are *D* inputs subject to dynamic control (clocking) by the a input. The numbers 1 and 2 are in the series chosen to indicate the modes so inputs e and f are only enabled in mode 1 (for parallel loading) and input d is only enabled in mode 2 (for serial loading). Note that input a has three functions. It is the clock for entering data. In mode 2, it causes right

Figure 14

M (mode) dependency affecting inputs.

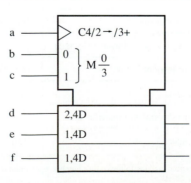

Note that all operations are synchronous.

In MODE 0 (b = 0, c = 0), the outputs remain at their existing states as none of the inputs has an effect.

In MODE 1 (b = 1, c = 0), parallel loading takes place thru inputs e and f.

In MODE 2 (b = 0, c = 1), shifting down and serial loading thru input d take place.

In MODE 3 (b = c = 1), counting up by increment of 1 per clock pulse takes place.

shifting of data, which means a shift away from the control block. In mode 3, it causes the contents of the register to be incremented by one count.

4.10.2 *M* dependency affecting outputs

When an *Mm* input or output stands at its internal 1 state, the affected outputs stand at their normally defined internal logic states, that is, the outputs are enabled.

When an *Mm* input or output stands at its internal 0 state, at each affected output any set of labels containing the identifying number of that *Mm* input or output has no effect and is to be ignored. When an output has several different sets of labels separated by solidi (e.g., 2,4/3,5), only those sets in which the identifying number of this *Mm* input or output appears are to be ignored.

Figure 15 shows a symbol for a device whose output can behave like either a three-state output or an open-collector output depending on the signal applied to input a. Mode 1 exists when input a stands at its internal 1 state and, in that case, the three-state symbol applies and the open-element symbol has no effect. When a = 0,

Figure 15

Type of output determined by mode

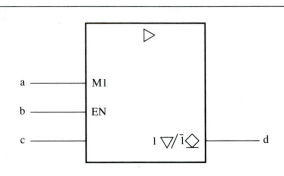

Table 5   Summary of dependency notation

Type of dependency	Letter Symbol*	Affecting input at its 1 state	Affecting input at its 0 state
Control	*C*	Permits action	Prevents action
Enable	*EN*	Permits action	Prevents action of inputs ◇ outputs turned off ▽ outputs at external high impedance Other outputs at internal 0 state
AND	*G*	Permits action	Imposes 0 state
Mode	*M*	Permits action (mode selected)	Prevents action (mode not selected)
Negate (Ex-*NOR*)	*N*	Complements state	No effect
OR	*V*	Imposes 1 state	Permits action
Transmission	*X*	Bidirectional connection exists	Bidirectional connection does not exist
Interconnection	*Z*	Imposes 1 state	Imposes 0 state

*These letter symbols appear at the affecting input (or output) and are followed by a number. Each input (or output) affected by that input is labeled with that same number.

mode 1 does not exist so the three-state symbol has no effect and the open-element symbol applies.

5.0
Bistable Elements

The dynamic input symbol and dependency notation provide the tools to identify different types of bistable elements and make synchronous and asynchronous inputs easily recognizable (Fig. 16).

Figure 16

Latches and flip-flops

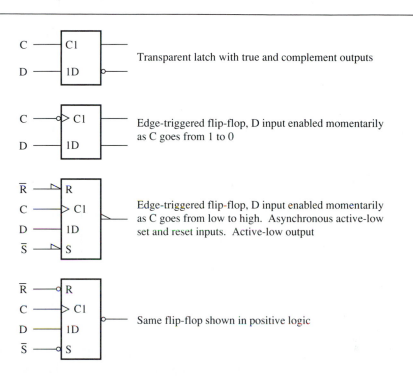

Transparent latch with true and complement outputs

Edge-triggered flip-flop, D input enabled momentarily as C goes from 1 to 0

Edge-triggered flip-flop, D input enabled momentarily as C goes from low to high. Asynchronous active-low set and reset inputs. Active-low output

Same flip-flop shown in positive logic

Transparent latches have a level-operated control input. The *D* input is active as long as the *C* input is at its internal 1 state. The outputs respond immediately. Edge-triggered elements accept data from *D, J, K, R,* or *S* inputs on the active transition of *C*.

Synchronous inputs can be readily recognized by their dependency labels (a number preceding the functional label, 1D in these examples) compared to the asynchronous inputs (*S* and *R*), which are not dependent on the *C* inputs. Of course if the set and reset inputs were dependent on the *C* inputs, their labels would be similarly modified (e.g., 1S, 1R).

6.0
Examples of Actual Device Symbols

The symbols explained in this section include some of the most complex in this book. These were chosen, not to discourage, but to illustrate the amount of information that can be conveyed. It is likely that if these explanations are read and followed reasonably well, most of the other symbols will seem simple indeed. The explanations are intended to be independent of each other so they may seem somewhat repetitious. However, each illustrates new principles. They are arranged more or less in the order of complexity.

6.1 SN75430 dual peripheral positive-*AND* driver

There are two identical sections. The symbology is complete for the first element; the absence of any symbology for the second element indicates it is identical. The two elements share pin 1 as an input. Had there been more than two elements this would have been indicated more conveniently with a common input block.

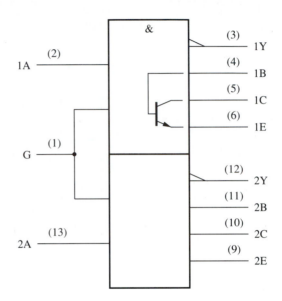

Each of the two elements is indicated by the "&" to be an *AND* gate. The output, pin 3, is active low so to this point the device is what would commonly be called a *NAND* gate. An extension of symbology used for analog devices has been used to show a floating transistor. Its emitter, base, and collector are shown lined up with the terminals to which they are connected; they are not connected internally to anything else. The device is usually used with pin 3 connected to pin 4 providing an inverting driver output and converting the *NAND* to *AND*.

6.2 SN75437 quadruple peripheral driver

There are four identical sections. The symbology is complete for the first element; the absence of any symbology for the other elements indicates they are identical. The top two elements share a common output clamp, pin 2. This is shown to be a nonlogic connection by the superimposed *X* on the line. The function for this type of connection is indicated briefly and not necessarily exactly by a small amount of text within the symbol. The bottom two elements likewise share a common clamp.

Each element is shown to be an inverter with amplification (indicated by ▷). Taking TTL as a reference, this means that either the input is sensitive to lower level signals, or the output has greater drive capability than usual. The latter applies in this case. The output is shown by ◇ to be open collector.

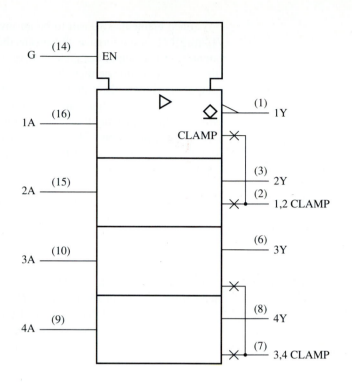

All the outputs share a common *EN* input, pin 14. See Figure 2 for an explanation of the common-control block. When *EN* = 0 (pin 14 is low), the outputs, being open-collector types, are turned off and go high.

6.3 SN75128 eight-channel line receiver

There are eight identical sections. The symbology is complete for the first element; the absence of any symbology for the next three elements indicates they are identical. Likewise the symbology is complete for the fifth element; the absence of any symbology for the next three elements indicates they are identical to the fifth.

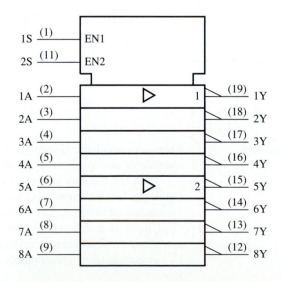

Each element is shown to be an inverter with amplification (indicated by $\triangleright$). Taking TTL as a reference, this means that either the input is sensitive to lower level signals, or the output has greater drive capability than usual. The former applies in this case. Because neither the symbol for open-collector ($\Diamond$) or three-state (∇) outputs is shown, the outputs are of the totem-pole type.

The top four outputs are shown to be affected by affecting input number 1, which is *EN*1, meaning they will be enabled if *EN*1 = 1 (pin 1 is high). *See* Section 4.9 for an explanation of *EN* dependency. If pin 1 is low, *EN*1 = 0 and the affected outputs will go to their inactive (high) levels. Similarly, the lower four outputs are controlled by pin 11.

6.4 SN75122 triple line drivers

There are two identical sections. The symbology is complete for the first section; the absence of any symbology for the next section indicates it is identical. Likewise the symbology is complete for the third section, which is similar, but not identical, to the first and second.

The top section may be considered to be an OR element (≥ 1) with two embedded *AND*s (&), one of which has an active-low amplified input ($\triangleright$) with hysteresis ($\boxed{}$), pin 14. This is *AND*ed with pin 15 and the result is *OR*ed with the *AND* of pins 1 and 2. The output of the *OR*, pin 13, is active low.

The third section is identical to the first except that pin 12 has no input *AND*ed with it. Because neither the symbol for open-collector ($\Diamond$) or three-state (∇) outputs is shown, the outputs are of the totem-pole type.

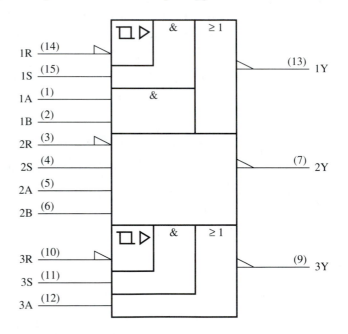

6.5 SN75142 dual line receivers

There are two identical sections. The symbology is complete for the first section; the absence of any symbology for the second section indicates it is identical.

Each section may be considered to be a three-input *OR* element (≥ 1). The first input, pin 2, is common to both *OR* elements. See Figure 2 for an explanation of the common-control block. The second input in the case of the first section is pin 3. The third input is a differential pair shown coming into an embedded element with amplification ($\triangleright$). Because neither the symbol for open-collector ($\Diamond$) or three-state (∇) outputs is shown, the outputs are of the totem-pole type. The outputs are active low.

The common-control block is sometimes used as a point of placement for an output that originates with either more than one or, as in this case, none of the elements in the array. Pin 9 is shown to be a nonlogic connection by the superimposed X on the line. The function for this type of connection is indicated briefly and not necessarily exactly by a small amount of text within the symbol.

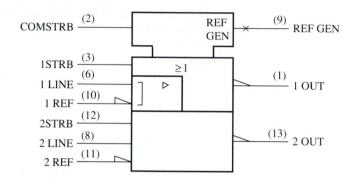

6.6 DS8831 quad single-ended or dual differential line drivers

There are four similar elements in the array. Each element is shown to be non-inverting with amplification (indicated by $\triangleright$). Taking TTL as a reference, this means that either the input is sensitive to lower level signals or the output has greater drive capability than usual. The latter applies in this case. The outputs are shown by $\triangledown$ to be of the three-state type.

The top two outputs are shown to be affected by affecting input number 2, which is $EN2$, meaning they will be enabled if $EN2 = 1$. *See* Section 4.9 for an explanation of EN dependency. If $EN2 = 0$, the affected outputs will go to their Hi-Z state (off). $EN2$ is the output of an AND gate (indicated by $\&$) whose active-low inputs are pins 1 and 2. Both pins 1 and 2 must be low to enable pins 3 and 5. Likewise, both pins 14 and 15 must be low to enable pins 11 and 13 through $EN3$.

Input pins 6 and 10 are shown to be affected by affecting input number 1, which is $N1$, meaning they will be negated if $N1 = 1$. *See* Section 4.5 for an explanation of N (negate or exclusive-OR) dependency. If $N1 = 0$, the input signals are not negated. $N1$ is the output of an OR gate (indicated by ≥ 1) whose active-high inputs are pins 7 and 9. Thus, if either of these pins are high, then the second and third elements become inverters.

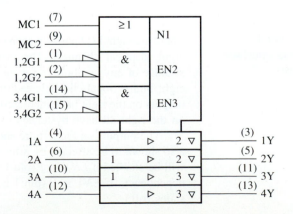

6.7 SN75113 differential line drivers with split three-state outputs

There are two similar elements in the array. The first is a two-input *AND* element (indicated by &); the second has only a single input. Both elements are shown to have special amplification (indicated by ▷). Taking TTL as a reference, this means that either the input is sensitive to lower level signals, or the output has greater drive capability than usual. The latter applies in this case.

Each element has four outputs. Pins 4 and 3 are a pair consisting of one open-emitter output (◇) and one open-collector output (◇). Relative to the *AND* function, both are active high. Pins 1 and 2 are a similar pair but relative to the *AND* function, both are active low. All outputs of a single, unsubdivided element always have identical internal logic states determined by the function of the element except when otherwise indicated by an associated symbol or label inside the element. Here there is no such contrary indication. All four outputs are shown to be affected by affecting input number 1, which is *EN*1, meaning they will all be enabled if *EN*1 = 1. *See* Section 4.9 for an explanation of *EN* dependency. If *EN*1 = 0, all the affected outputs will be turned off. *EN*1 is the output of an *AND* gate (indicated by &) whose active-high inputs are pins 7 and 9. Both pins 7 and 9 must be high to enable the outputs of the top element. Assuming they are enabled and that pins 5 and 6 are both high, the internal state of all four outputs will be a 1. Pins 4 and 3 will both be high, pins 1 and 2 will both be low. The part is designed so that pins 3 and 4 may be connected together creating an active-high three-state output. Likewise, pins 1 and 2 may be connected together to create an active-low three-state output.

All that has been said about the first element regarding its outputs and their enable inputs also applies to the second element. Pins 9 and 10 are the enable inputs in this case.

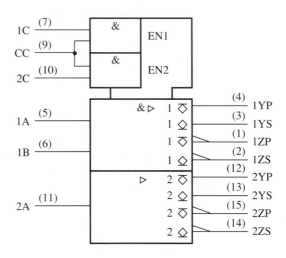

6.8 SN75163A octal general-purpose interface bus transceiver

There are eight I/O ports on each side, pins 2 through 9 and 12 through 19. There are eight identical channels. The symbology is complete for the first channel; the absence of any symbology for the other channels indicates they are identical. The eight bidirectional channels each have amplification from left to right, that is, the outputs on the right have increased drive capability (indicated by ▷), and the inputs on the right all have hysteresis (indicated by ⎍).

The outputs on the left are shown to be three-state outputs by the ▽. They are also shown to be affected by affecting input number 4, which is *EN*4, meaning they will be enabled if *EN*4 = 1 (pin 1 is low). *See* Section 4.9 for an explanation of *EN* dependency. If *EN*4 = 0 (pin 1 is high), the affected outputs will go to their Hi-Z states (off).

The labeling at pin 2, which applies to all the outputs on the right, is unusual because the outputs themselves have an unusual feature. The label includes both the symbol for a three-state output (∇) and for an open-collector output ($\diamondsuit$), separated by a slash indicating that these are alternatives.

The symbol for the three-state output is shown to be affected by affecting input number 1, which is $M1$, meaning the ∇ label is valid when $M1 = 1$ (pin 11 is high), but is to be ignored when $M1 = 0$ (pin 11 is low). *See* Section 4.10 for an explanation of M (mode) dependency. Likewise, the symbol for the open-collector output is shown to be affected by affecting input number 2, which is $M2$, meaning the $\diamondsuit$ label is valid when $M2 = 1$ (pin 11 is low), but is to be ignored when $M2 = 0$ (pin 11 is high). These labels are enclosed in parentheses (used as in algebra); the numeral 3 indicates that in either case the output is affected by $EN3$. Thus the right-hand outputs will be off if pin 1 is low. It can now be seen that pin 1 is the direction control and pin 11 is used to determine whether the outputs are of the three-state or open-collector variety.

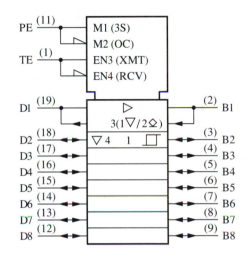

6.9 SN75161 octal general-purpose interface bus transceiver

There are eight I/O ports on each side, pins 2 through 9 and 12 through 19. Pin 13 is not only an I/O port; the lines running into the common-control block (*see* Fig. 2) indicate that it also has control functions. Pins 1 and 11 are also controls. The eight bidirectional channels each have amplification from left to right, that is, the outputs on the right have increased drive capability (indicated by $\triangleright$), and the inputs on the right all have hysteresis (indicated by $\square$). All of the outputs are shown to be of the three-state type by the ∇ symbol except for the outputs at pins 9, 4, and 5, which are shown to have passive pullups by the $\diamondsuit$ symbol.

Starting with a typical I/O port, pin 18, the output portion is identified by an arrow indicating right-to-left signal flow and the three-state output symbol (∇). This output is shown to be affected by affecting input number 1, which is $EN1$, meaning it will be enabled as an output if $EN1 = 1$ (pin 11 is high). *See* Section 4.9 for an explanation of EN dependency. If pin 11 is low, $EN1 = 0$ and the output at pin 18 will be in its Hi-Z state (off). This also applies to the three-state outputs at pins 13 and 19 and to the passive-pullup output at pin 9. On the other hand, the outputs at pins 8, 2, 3, and 12 all are affected by the complement of $EN1$. This is indicated by the bar over the 1 at each of those outputs. They are enabled only when pin 11 is low. Thus one function of pin 11 is to serve as direction control for the first, third, fourth, and fifth channels.

Similarly, it can be seen that pin 1 serves as direction control for the sixth, seventh, and eighth channels. If pin 1 is high, transmission will be from left to right in the sixth channel, right to left in the seventh and eighth. These transmissions are reversed if pin 1 is low.

The direction control for the second channel, *EN3*, is more complex. *EN3* is the output of an *OR* (≥1) function. One of the inputs to this *OR* is the active-high signal on pin 13. This signal is shown to be affected at the input to the *OR* gate by affecting input number 5, which is *G5*, meaning that pin 13 is *AND*ed with pin 1 before entering the *OR* gate. *See* Section 4.2 for an explanation of *G* (*AND*) dependency. The other input to the *OR* is the active-low signal on pin 13. This signal is *AND*ed with the complement of pin 11 before entering the *OR* gate. This is indicated by the *G4* at pin 1 and the 4 with a bar over it at pin 13. Thus, for *EN3* to stand at the 1 state, which would enable transmission from pin 14 to pin 7, both pins 13 and 1 must be high or both pins 13 and 11 must be low.

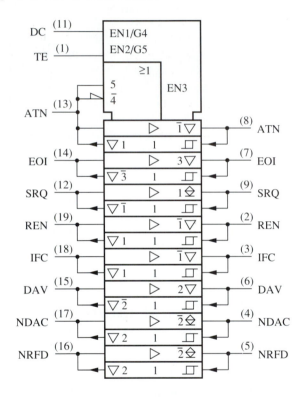

6.10 SN75584 seven-segment cathode driver with latched inputs

The heart of this device and its symbol is a BCD to seven-segment decoder. Supplying the inputs to the decoder are four elements with *D* inputs. A *D* input indicates a storage element. All four *D* inputs are shown to be affected by affecting input number 10, which is *C10*, meaning they will all be enabled if *C10* = 1. *See* Section 4.8 for an explanation of *C* dependency and Section 5.0 for a discussion of bistable elements. Because the *C* input is not dynamic, the storage element is a transparent latch. While *C* = 1, meaning in this case while pin 6 is high, the outputs of the latches (and hence the inputs of the decoder) will follow the input pins 8, 2, 3, and 7. When *C* = 0 (pin 6 goes low), those inputs are latched.

The BCD inputs to the decoder are labeled 1, 2, 4, and 8 corresponding to the weights of the inputs. The outputs are labeled a through g corresponding to the accepted segment designations for seven-segment displays. When the decoder is in operation, an internal number is produced that is equal to the sum of the weights of the BCD inputs that stand at their 1 states. This causes those outputs corresponding to the segments needed to display that number to take on their 1 states. For example, if pins 8 (weight 1) and 7 (weight 8) were high and pins 2 and 3 were low while pin 6 was high, the internal number would be the sum of 8 and 1. All the segment outputs except e would be active (low).

The remaining input to the decoder, pin 5, is an *EN* input. An *EN* input affects all the outputs of the element in which it is placed. When $EN = 0$, all the segment outputs take on their 0 states. Being active low, that means they are forced high.

Located below the decoder is another transparent latch. In this case its active-low output is brought out to a terminal. This latch is also under the control of $C10$ (pin 6). The output, pin 17, is shown to be affected by affecting input number 11, which is $EN11$, meaning it will be enabled if $EN11 = 1$ (pin 5 is high). *See* Section 4.9 for an explanation of *EN* dependency. Although the effect of pin 5 is the same for the latch output as for the decoder outputs, it is necessary to use *EN* dependency for the latter because an *EN* input affects all outputs of the element in which it is placed and any other elements shown to be affected by that element. The latch is shown to have no logic connection to the decoder making it necessary to use dependency notation to show that its output is also controlled by pin 5.

Located below that latch is another element whose function is defined by its single input, pin 4. This is shown to be a nonlogic connection by the superimposed *X* on the line. The function for this type of connection is indicated briefly and not necessarily exactly by a small amount of text within the symbol. In this case the function of the element is to adjust the output current of the decoder, but not that of the latch.

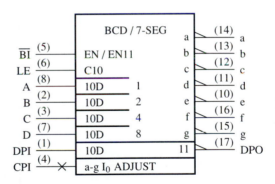

6.11 SN75500A AC plasma display driver with CMOS-compatible inputs

The heart of this device and its symbol is an 8-bit shift register. It has a single *D* input, pin 2, which is shown to be affected by affecting input number 9, which is $C9$, meaning it will be enabled if $C9 = 1$. *See* Section 4.8 for an explanation of *C* dependency and Section 5.0 for a discussion of bistable elements. Because the *C* input is dynamic, the storage elements are edge-triggered flip-flops. While $C = 1$, which in this case will occur on the transition of pin 3 from low to high, the state of the *D* input will be stored. Pin 2 is shown to be active low so to store a 1, pin 2 must be low.

In addition to controlling the *D* input, pin 3 is shown by $/\rightarrow$ to have an additional function. As pin 3 goes from low to high, data stored in the shift register is shifted one position. The right-pointing arrow means that the data is shifted away from the control block (down).

On the right side of the symbol an abbreviation technique has been used that is practical only when the internal labels and the pin numbers are both consecutive. Thus, it should be clear that the input of the element whose output is pin 5 is affected by affecting input number 2, just as the input of the element whose output is pin 4 is affected by affecting input number 1. Affecting inputs 1 through 8 are Z inputs ($Z1$ through $Z8$), which means their signals are transferred directly to the output elements. *See* Section 4.6 for an explanation of Z dependency.

The inputs of the 32 implicitly shown output elements are also shown to be affected by affecting inputs numbers 11, 12, 13, and 14 in four blocks of eight each. These inputs will be found in the common-control block preceded by a letter G and a brace. The brace is called the binary grouping symbol. It is equivalent to a decoder with outputs in this case driving four G inputs ($G11$, $G12$, $G13$, and $G14$). The weights of the inputs to the coder are shown to be 2^0 and 2^1 for pins 1 and 39, respectively. The decoder has four outputs corresponding to the four possible sums of the weights of the activated decoder inputs. If pins 1 and 39 are both low, the sum of the weights = 0 and $G11 = 1$. If pin 1 is low while pin 39 is high, the sum = 2 and $G13 = 1$, G indicates *AND* dependency (*see* Section 4.2). Only one of the four affecting G inputs at a time can take on the 1 state. The block of eight output elements affected by that G input are enabled; the 0 state is imposed on the other 24 output elements and externally those output pins are low.

Because of their high-current, high-voltage characteristics, the outputs are labeled with the amplification symbol $\triangleright$. All the outputs share a common *EN* input, pin 38. *See* Figure 2 for an explanation of the common-control block. When $EN = 0$ (pin 38 is high), the outputs take on their internal 0 states. Being active high, that means they are forced low.

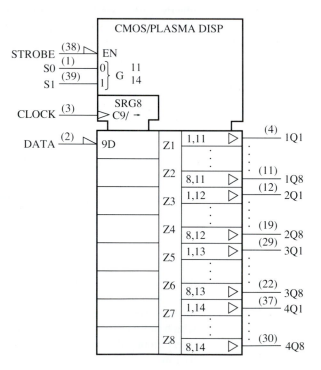

6.12 SN75551 electroluminescent row driver with CMOS-compatible inputs

The heart of this device and its symbol is a 32-bit shift register. It has a single D input, pin 24, which is shown to be affected by affecting input number 40, which is $C40$, meaning it will be enabled if $C40 = 1$. *See* Section 4.8 for an explanation of C dependency and Section 5.0 for a discussion of bistable elements. Because the C input is dynamic, the storage elements are edge-triggered flip-flops. While $C = 1$, which in this case will occur on the transition of pin 20 from high to low, the state of the *D* input will be stored. Pin 24 is shown to be active high so to store a 1, pin 24 must be high.

In addition to controlling the *D* input, pin 20 is shown by $/\rightarrow$ to have an additional function. As pin 20 goes from high to low, data stored in the shift register is shifted one position. The right-pointing arrow means that the data is shifted away from the control block (down). The internal outputs of the shift register are all shown to be affected by affecting inputs 41 and 42. Affecting input 41 is *G*41, meaning that pin 19 is *AND*ed with each of the internal register outputs. If pin 19 is high, the affected outputs are enabled. If pin 19 is low, the 0 state is imposed on the affected outputs. *See* Section 4.2 for an explanation of *G* (*AND*) dependency. Affecting input 42 if *V*42, meaning that pin 23 (active low) is *OR*ed with each of the internal register outputs. If pin 23 is high, *V*42 = 0 and the affected outputs are enabled. If pin 23 is low, *V*42 = 1 and the 1 state is imposed on the affected outputs. *See* Section 4.4 for an explanation of *V* (*OR*) dependency. The affect of *V*42 is taken into account after that of *V*41 because of the order in which the labels appear. This means that the imposition of the 1 state by pin 23 would take precedence over the imposition of the 0 state by pin 19 in case both inputs were active. Pin 18 is shown to be an output directly from the 32nd stage of the shift register. The dependency label 41,42 does not apply to this output, so pins 19 and 23 do not affect it.

An abbreviation technique has been used for the shift register elements, the output lines, and the associated dependency notation. This technique is practical only when the internal labels and the pin numbers are both consecutive. Thus, it should be clear that the output at pin 28 is affected by affecting input number 3, just as the output at pin 27 is affected by affecting input number 2. Affecting inputs 1 through 32 are *X* inputs (*X*1 through *X*32). If one of these *X* inputs stands at the 1 state, there is a connection established between the ports labeled with the number of the *X* input. In the case of *X*2, there would be a connection between pin 27 and pin 21. Pin 21 (labeled $1/2\ldots/31/32$) is the common point for all the connections indicated by *X* dependency in this symbol. *See* Section 4.7 for an explanation of *X* dependency.

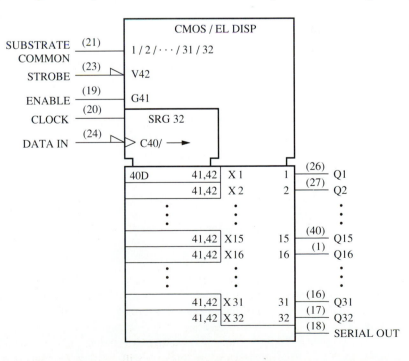

MANUFACTURER'S IC DATA SHEETS

*The material on pages C.3–C.82 has been reprinted courtesy of Signetics Company, a division of North American Philips Corporation.

Signetics

Logic Products

7400, LS00, S00
Gates

Quad Two-Input NAND Gate
Product Specification

TYPE	TYPICAL PROPAGATION DELAY	TYPICAL SUPPLY CURRENT (TOTAL)
7400	9ns	8mA
74LS00	9.5ns	1.6mA
74S00	3ns	15mA

ORDERING CODE

PACKAGES	COMMERCIAL RANGE $V_{CC} = 5V \pm 5\%$; $T_A = 0°C$ to $+70°C$
Plastic DIP	N7400N, N74LS00N, N74S00N
Plastic SO	N74LS00D, N74S00D

NOTE:
For information regarding devices processed to Military Specifications, see the Signetics Military Products Data Manual.

INPUT AND OUTPUT LOADING AND FAN-OUT TABLE

PINS	DESCRIPTION	74	74S	74LS
A, B	Inputs	1ul	1Sul	1LSul
Y	Output	10ul	10Sul	10LSul

NOTE:
Where a 74 unit load (ul) is understood to be 40μA I_{IH} and -1.6mA I_{IL}, a 74S unit load (Sul) is 50μA I_{IH} and -2.0mA I_{IL}, and 74LS unit load (LSul) is 20μA I_{IH} and -0.4mA I_{IL}.

FUNCTION TABLE

INPUTS		OUTPUT
A	B	Y
L	L	H
L	H	H
H	L	H
H	H	L

H = HIGH voltage level
L = LOW voltage level

PIN CONFIGURATION

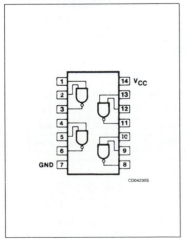

LOGIC SYMBOL

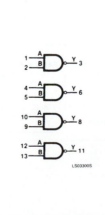

LOGIC SYMBOL (IEEE/IEC)

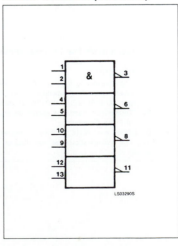

Gates

ABSOLUTE MAXIMUM RATINGS (Over operating free-air temperature range unless otherwise noted.)

PARAMETER		74	74LS	74S	UNIT
V_{CC}	Supply voltage	7.0	7.0	7.0	V
V_{IN}	Input voltage	−0.5 to +5.5	−0.5 to +7.0	−0.5 to +5.5	V
I_{IN}	Input current	−30 to +5	−30 to +1	−30 to +5	mA
V_{OUT}	Voltage applied to output in HIGH output state	−0.5 to +V_{CC}	−0.5 to +V_{CC}	−0.5 to +V_{CC}	V
T_A	Operating free-air temperature range		0 to 70		°C

RECOMMENDED OPERATING CONDITIONS

PARAMETER		74			74LS			74S			UNIT
		Min	Nom	Max	Min	Nom	Max	Min	Nom	Max	
V_{CC}	Supply voltage	4.75	5.0	5.25	4.75	5.0	5.25	4.75	5.0	5.25	V
V_{IH}	HIGH-level input voltage	2.0			2.0			2.0			V
V_{IL}	LOW-level input voltage			+0.8			+0.8			+0.8	V
I_{IK}	Input clamp current			−12			−18			−18	mA
I_{OH}	HIGH-level output current			−400			−400			−1000	μA
I_{OL}	LOW-level output current			16			8			20	mA
T_A	Operating free-air temperature	0		70	0		70	0		70	°C

TEST CIRCUITS AND WAVEFORMS

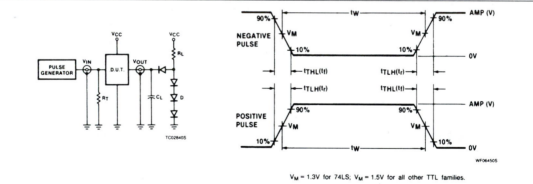

V_M = 1.3V for 74LS; V_M = 1.5V for all other TTL families.

Test Circuit For 74 Totem-Pole Outputs

Input Pulse Definition

DEFINITIONS
R_L = Load resistor to V_{CC}; see AC CHARACTERISTICS for value.
C_L = Load capacitance includes jig and probe capacitance;
  see AC CHARACTERISTICS for value.
R_T = Termination resistance should be equal to Z_{OUT}
  of Pulse Generators.
D = Diodes are 1N916, 1N3064, or equivalent.
t_{TLH}, t_{THL} Values should be less than or equal to the table
entries.

FAMILY	INPUT PULSE REQUIREMENTS				
	Amplitude	Rep. Rate	Pulse Width	t_{TLH}	t_{THL}
74	3.0V	1MHz	500ns	7ns	7ns
74LS	3.0V	1MHz	500ns	15ns	6ns
74S	3.0V	1MHz	500ns	2.5ns	2.5ns

Gates   7400, LS00, S00

DC ELECTRICAL CHARACTERISTICS (Over recommended operating free-air temperature range unless otherwise noted.)

PARAMETER		TEST CONDITIONS[1]		7400			74LS00			74S00			UNIT
				Min	Typ[2]	Max	Min	Typ[2]	Max	Min	Typ[2]	Max	
V_{OH}	HIGH-level output voltage	V_{CC} = MIN, V_{IH} = MIN, V_{IL} = MAX, I_{OH} = MAX		2.4	3.4		2.7	3.4		2.7	3.4		V
V_{OL}	LOW-level output voltage	V_{CC} = MIN, V_{IH} = MIN	I_{OL} = MAX		0.2	0.4		0.35	0.5			0.5	V
			I_{OL} = 4mA (74LS)					0.25	0.4				V
V_{IK}	Input clamp voltage	V_{CC} = MIN, I_I = I_{IK}				−1.5			−1.5			−1.2	V
I_I	Input current at maximum input voltage	V_{CC} = MAX	V_I = 5.5V			1.0						1.0	mA
			V_I = 7.0V						0.1				mA
I_{IH}	HIGH-level input current	V_{CC} = MAX	V_I = 2.4V			40							μA
			V_I = 2.7V						20			50	μA
I_{IL}	LOW-level input current	V_{CC} = MAX	V_I = 0.4V			−1.6			−0.4				mA
			V_I = 0.5V									−2.0	mA
I_{OS}	Short-circuit output current[3]	V_{CC} = MAX		−18		−55	−20		−100	−40		−100	mA
I_{CC}	Supply current (total)	V_{CC} = MAX	I_{CCH} Outputs HIGH		4	8		0.8	1.6		10	16	mA
			I_{CCL} Outputs LOW		12	22		2.4	4.4		20	36	mA

NOTES:
1. For conditions shown as MIN or MAX, use the appropriate value specified under recommended operating conditions for the applicable type.
2. All typical values are at V_{CC} = 5V, T_A = 25°C.
3. I_{OS} is tested with V_{OUT} = +0.5V and V_{CC} = V_{CC} MAX + 0.5V. Not more than one output should be shorted at a time and duration of the short circuit should not exceed one second.

AC WAVEFORM

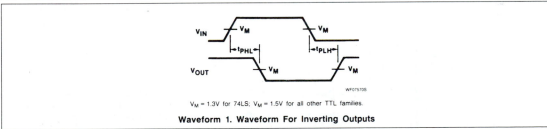

V_M = 1.3V for 74LS; V_M = 1.5V for all other TTL families.

Waveform 1. Waveform For Inverting Outputs

AC ELECTRICAL CHARACTERISTICS T_A = 25°C, V_{CC} = 5.0V

PARAMETER		TEST CONDITIONS	74 C_L = 15pF, R_L = 400Ω		74LS C_L = 15pF, R_L = 2kΩ		74S C_L = 15pF, R_L = 280Ω		UNIT
			Min	Max	Min	Max	Min	Max	
t_{PLH} t_{PHL}	Propagation delay	Waveform 1		22 15		15 15		4.5 5.0	ns

Signetics

Logic Products

7406, 07
Inverter/Buffer/Drivers

'06 Hex Inverter Buffer/Driver (Open Collector)
'07 Hex Buffer/Driver (Open Collector)
Product Specification

TYPE	TYPICAL PROPAGATION DELAY	TYPICAL SUPPLY CURRENT (TOTAL)
7406	10ns (t_{PLH}) 15ns (t_{PHL})	31mA
7407	6ns (t_{PLH}) 20ns (t_{PHL})	25mA

ORDERING CODE

PACKAGES	COMMERCIAL RANGE $V_{CC} = 5V \pm 5\%$; $T_A = 0°C$ to $+70°C$
Plastic DIP	N7406N, N7407N
Plastic SO	N7406D, N7407D

NOTE:
For information regarding devices processed to Military Specifications, see the Signetics Military Products Data Manual.

INPUT AND OUTPUT LOADING AND FAN-OUT TABLE

PINS	DESCRIPTION	74
A	Input	1ul
Y	Output	10ul

NOTE:
Where a 74 unit load (ul) is understood to be 40μA I_{IH} and -1.6mA I_{IL}.

FUNCTION TABLE

'06		'07	
INPUT	OUTPUT	INPUT	OUTPUT
A	Y	A	Y
H	L	H	H
L	H	L	L

H = HIGH voltage level
L = LOW voltage level

PIN CONFIGURATION

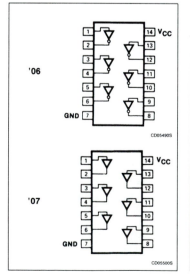

LOGIC SYMBOL

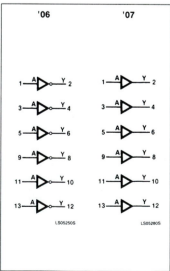

LOGIC SYMBOL (IEEE/IEC)

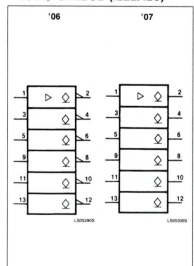

Inverter/Buffer/Drivers

7406, 07

ABSOLUTE MAXIMUM RATINGS (Over operating free-air temperature range unless otherwise noted.)

PARAMETER		74	UNIT
V_{CC}	Supply voltage	7.0	V
V_{IN}	Input voltage	−0.5 to +5.5	V
I_{IN}	Input current	−30 to +5	mA
V_{OUT}	Voltage applied to output in HIGH output state	−0.5 to +30	V
T_A	Operating free-air temperature range	0 to 70	°C

RECOMMENDED OPERATING CONDITIONS

PARAMETER		74			UNIT
		Min	Nom	Max	
V_{CC}	Supply voltage	4.75	5.0	5.25	V
V_{IH}	HIGH-level input voltage	2.0			V
V_{IL}	LOW-level input voltage			+0.8	V
I_{IK}	Input clamp current			−12	mA
V_{OH}	HIGH-level output voltage			30	V
I_{OL}	LOW-level output current			40	mA
T_A	Operating free-air temperature	0		70	°C

TEST CIRCUITS AND WAVEFORMS

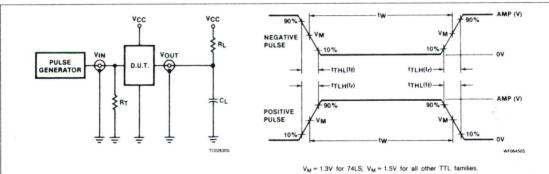

V_M = 1.3V for 74LS; V_M = 1.5V for all other TTL families.

**Test Circuit For 74 Open Collectors
Outputs**

Input Pulse Definition

DEFINITIONS
R_L = Load resistor to V_{CC}; see AC CHARACTERISTICS for value.
C_L = Load capacitance includes jig and probe capacitance;
    see AC CHARACTERISTICS for value.
R_T = Termination resistance should be equal to Z_{OUT}
    of Pulse Generators.
D = Diodes are 1N916, 1N3064, or equivalent.
t_{TLH}, t_{THL} Values should be less than or equal to the table
entries.

FAMILY	INPUT PULSE REQUIREMENTS				
	Amplitude	Rep. Rate	Pulse Width	t_{TLH}	t_{THL}
74	3.0V	1MHz	500ns	7ns	7ns
74LS	3.0V	1MHz	500ns	15ns	6ns
74S	3.0V	1MHz	500ns	2.5ns	2.5ns

Inverter/Buffer/Drivers   7406, 07

DC ELECTRICAL CHARACTERISTICS (Over recommended operating free-air temperature range unless otherwise noted.)

PARAMETER		TEST CONDITIONS[1]		7406, 7407			UNIT
				Min	Typ[2]	Max	
I_{OH}	HIGH-level output current	V_{CC} = MIN, V_{IH} = MIN, V_{IL} = MAX, V_{OH} = 30V				250	μA
V_{OL}	LOW-level output voltage	V_{CC} = MIN, V_{IH} = MIN, V_{IL} = MAX	I_{OL} = 16mA			0.4	V
			I_{OL} = 30mA			0.7	V
			I_{OL} = 40mA			0.7	V
V_{IK}	Input clamp voltage	V_{CC} = MIN, I_I = I_{IK}				−1.5	V
I_I	Input current at maximum input voltage	V_{CC} = MAX, V_I = 5.5V				1.0	mA
I_{IH}	HIGH-level input current	V_{CC} = MAX, V_I = 2.4V				40	μA
I_{IL}	LOW-level input current	V_{CC} = MAX, V_I = 0.4V				−1.6	mA
I_{CC}	Supply current (total)	V_{CC} = MAX	I_{CCH} Outputs HIGH '06		30	48	mA
			I_{CCL} Outputs LOW		32	51	mA
			I_{CCH} Outputs HIGH '07		29	41	mA
			I_{CCL} Outputs LOW		21	30	mA

NOTES:
1. For conditions shown as MIN or MAX, use the appropriate value specified under recommended operating conditions for the applicable type.
2. All typical values are at V_{CC} = 5V, T_A = 25°C.

AC WAVEFORMS

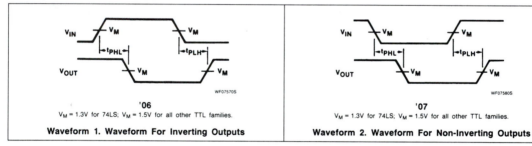

'06
V_M = 1.3V for 74LS; V_M = 1.5V for all other TTL families.

Waveform 1. Waveform For Inverting Outputs

'07
V_M = 1.3V for 74LS; V_M = 1.5V for all other TTL families.

Waveform 2. Waveform For Non-Inverting Outputs

AC ELECTRICAL CHARACTERISTICS T_A = 25°C, V_{CC} = 5.0V

PARAMETER		TEST CONDITIONS	7406		7407		UNIT
			C_L = 15pF, R_L = 110Ω		C_L = 15pF, R_L = 110Ω		
			Min	Max	Min	Max	
t_{PLH}	Propagation delay	Waveform 1, '06		15		10	ns
t_{PHL}		Waveform 2, '07		23		30	

Signetics

7414, LS14
Schmitt Triggers

Hex Inverter Schmitt Trigger
Product Specification

Logic Products

DESCRIPTION

The '14 contains six logic inverters which accept standard TTL input signals and provide standard TTL output levels. They are capable of transforming slowly changing input signals into sharply defined, jitter-free output signals. In addition, they have greater noise margin than conventional inverters.

Each circuit contains a Schmitt trigger followed by a Darlington level shifter and a phase splitter driving a TTL totem-pole output. The Schmitt trigger uses positive feedback to effectively speed-up slow input transition, and provide different input threshold voltages for positive and negative-going transitions. This hysteresis between the positive-going and negative-going input thresholds (typically 800mV) is determined internally by resistor ratios and is essentially insensitive to temperature and supply voltage variations.

TYPE	TYPICAL PROPAGATION DELAY	TYPICAL SUPPLY CURRENT (TOTAL)
7414	15ns	31mA
74LS14	15ns	10mA

ORDERING CODE

PACKAGES	COMMERCIAL RANGE V_{CC} = 5V ±5%; T_A = 0°C to +70°C
Plastic DIP	N7414N, N74LS14N
Plastic SO	N74LS14D

NOTE:
For information regarding devices processed to Military Specifications, see the Signetics Military Products Data Manual.

INPUT AND OUTPUT LOADING AND FAN-OUT TABLE

PINS	DESCRIPTION	74	74LS
A	Inputs	1ul	1LSul
Y	Output	10ul	10LSul

NOTE:
Where a 74 unit load (ul) is understood to be 40μA I_{IH} and −1.6mA I_{IL}, and 74LS unit load (LSul) is 20μA I_{IH} and −0.4mA I_{IL}.

PIN CONFIGURATION

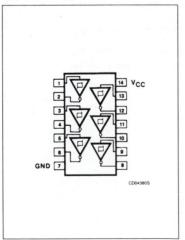

CD04380S

LOGIC SYMBOL

LS03480S

LOGIC SYMBOL (IEEE/IEC)

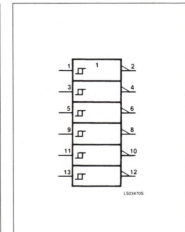

LS03470S

Schmitt Triggers   7414, LS14

ABSOLUTE MAXIMUM RATINGS (Over operating free-air temperature range unless otherwise noted.)

	PARAMETER	74	74LS	UNIT
V_{CC}	Supply voltage	7.0	7.0	V
V_{IN}	Input voltage	−0.5 to +5.5	−0.5 to +7.0	V
I_{IN}	Input current	−30 to +5	−30 to +1	mA
V_{OUT}	Voltage applied to output in HIGH output state	−0.5 to +V_{CC}	−0.5 to +V_{CC}	V
T_A	Operating free-air temperature range	0 to 70		°C

RECOMMENDED OPERATING CONDITIONS

	PARAMETER	74			74LS			UNIT
		Min	Nom	Max	Min	Nom	Max	
V_{CC}	Supply voltage	4.75	5.0	5.25	4.75	5.0	5.25	V
I_{IK}	Input clamp current			−12			−18	mA
I_{OH}	HIGH-level output current			−800			−400	μA
I_{OL}	LOW-level output current			16			8	mA
T_A	Operating free-air temperature	0		70	0		70	°C

TEST CIRCUITS AND WAVEFORMS

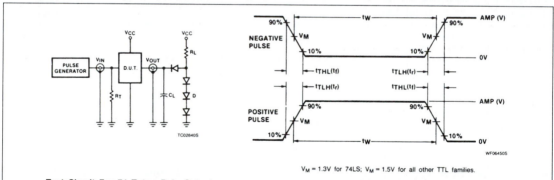

V_M = 1.3V for 74LS; V_M = 1.5V for all other TTL families.

Test Circuit For 74 Totem-Pole Outputs   **Input Pulse Definition**

DEFINITIONS
R_L = Load resistor to V_{CC}; see AC CHARACTERISTICS for value.
C_L = Load capacitance includes jig and probe capacitance;
    see AC CHARACTERISTICS for value.
R_T = Termination resistance should be equal to Z_{OUT}
    of Pulse Generators.
D = Diodes are 1N916, 1N3064, or equivalent.
t_{TLH}, t_{THL} Values should be less than or equal to the table
entries.

FAMILY	INPUT PULSE REQUIREMENTS				
	Amplitude	Rep. Rate	Pulse Width	t_{TLH}	t_{THL}
74	3.0V	1MHz	500ns	7ns	7ns
74LS	3.0V	1MHz	500ns	15ns	6ns
74S	3.0V	1MHz	500ns	2.5ns	2.5ns

Schmitt Triggers   7414, LS14

DC ELECTRICAL CHARACTERISTICS (Over recommended operating free-air temperature range unless otherwise noted.)

PARAMETER		TEST CONDITIONS[1]		7414			74LS14			UNIT
				Min	Typ[2]	Max	Min	Typ[2]	Max	
V_{T+}	Positive-going threshold	$V_{CC} = 5.0V$		1.5	1.7	2.0	1.4	1.6	1.9	V
V_{T-}	Negative-going threshold	$V_{CC} = 5.0V$		0.6	0.9	1.1	0.5	0.8	1.0	V
ΔV_T	Hysteresis ($V_{T+} - V_{T-}$)	$V_{CC} = 5.0V$		0.4	0.8		0.4	0.8		V
V_{OH}	HIGH-level output voltage	$V_{CC} = MIN, V_I = V_{T-MIN}, I_{OH} = MAX$		2.4	3.4		2.7	3.4		V
V_{OL}	LOW-level output voltage	$V_{CC} = MIN, V_I = V_{T+MAX}$	$I_{OL} = MAX$		0.2	0.4		0.35	0.5	V
			$I_{OL} = 4mA$ (74LS)					0.25	0.4	V
V_{IK}	Input clamp voltage	$V_{CC} = MIN, I_I = I_{IK}$				-1.5			-1.5	V
I_{T+}	Input current at positive-going threshold	$V_{CC} = 5.0V, V_I = V_{T+}$			-0.43			-0.14		mA
I_{T-}	Input current at negative-going threshold	$V_{CC} = 5.0V, V_I = V_{T-}$			-0.56			-0.18		mA
I_I	Input current at maximum input voltage	$V_{CC} = MAX$	$V_I = 5.5V$			1.0				mA
			$V_I = 7.0V$						0.1	mA
I_{IH}	HIGH-level input current	$V_{CC} = MAX$	$V_I = 2.4V$			40				µA
			$V_I = 2.7V$						20	µA
I_{IL}	LOW-level input current	$V_{CC} = MAX, V_I = 0.4V$				-1.2			-0.4	mA
I_{OS}	Short-circuit output current[3]	$V_{CC} = MAX$		-18		-55	-20		-100	mA
I_{CC}	Supply current (total)	$V_{CC} = MAX$	I_{CCH} Outputs HIGH		22	36		8.6	16	mA
			I_{CCL} Outputs LOW		39	60		12	21	mA

NOTES:
1. For conditions shown as MIN or MAX, use the appropriate value specified under recommended operating conditions for the applicable type.
2. All typical values are at $V_{CC} = 5V$, $T_A = 25°C$.
3. I_{OS} is tested with $V_{OUT} = +0.5V$ and $V_{CC} = V_{CC}$ MAX + 0.5V. Not more than one output should be shorted at a time and duration of the short circuit should not exceed one second.

FUNCTION TABLE

INPUT	OUTPUT
A	Y
0	1
1	0

AC WAVEFORM

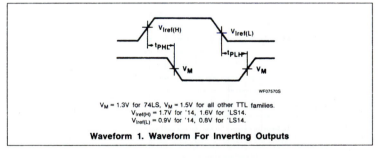

WF07570S

V_M = 1.3V for 74LS, V_M = 1.5V for all other TTL families.
$V_{Iref(H)}$ = 1.7V for '14, 1.6V for 'LS14.
$V_{Iref(L)}$ = 0.9V for '14, 0.8V for 'LS14.

Waveform 1. Waveform For Inverting Outputs

Schmitt Triggers   7414, LS14

AC ELECTRICAL CHARACTERISTICS $T_A = 25°C$, $V_{CC} = 5.0V$

PARAMETER	TEST CONDITIONS	74		74LS		UNIT
		$C_L = 15pF$, $R_L = 400\Omega$		$C_L = 15pF$, $R_L = 2k\Omega$		
		Min	Max	Min	Max	
t_{PLH} Propagation delay t_{PHL}	Waveform 1		22 22		22 22	ns

TYPICAL PERFORMANCE CHARACTERISTICS

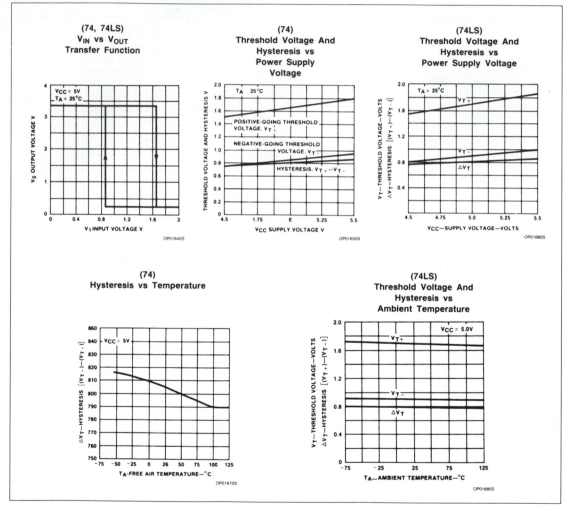

Signetics

Logic Products

7474, LS74A, S74
Flip-Flops

Dual D-Type Flip-Flop
Product Specification

DESCRIPTION

The '74 is a dual positive edge-triggered D-type flip-flop featuring individual Data, Clock, Set and Reset inputs; also complementary Q and $\overline{Q}$ outputs.

Set ($\overline{S}_D$) and Reset ($\overline{R}_D$) are asynchronous active-LOW inputs and operate independently of the Clock input. Information on the Data (D) input is transferred to the Q output on the LOW-to-HIGH transition of the clock pulse. The D inputs must be stable one set-up time prior to the LOW-to-HIGH clock transition for predictable operation. Although the Clock input is level-sensitive, the positive transition of the clock pulse between the 0.8V and 2.0V levels should be equal to or less than the clock-to-output delay time for reliable operation.

TYPE	TYPICAL f_{MAX}	TYPICAL SUPPLY CURRENT (TOTAL)
7474	25MHz	17mA
74LS74A	33MHz	4mA
74S74	100MHz	30mA

NOTE:
For information regarding devices processed to Military Specifications, see the Signetics Military Products Data Manual.

ORDERING CODE

PACKAGES	COMMERCIAL RANGE $V_{CC} = 5V \pm 5\%$; $T_A = 0°C$ to $+70°C$
Plastic DIP	N7474N, N74LS74AN, N74S74N
Plastic SO	N741S74A, N74S74D

NOTE:
For information regarding devices processed to Military Specifications, see the Signetics Military Products Data Manual.

INPUT AND OUTPUT LOADING AND FAN-OUT TABLE

PINS	DESCRIPTION	74	74S	74LS
D	Input	1ul	1Sul	1LSul
$\overline{R}_D$	Input	2ul	3Sul	2LSul
$\overline{S}_D$	Input	1ul	2Sul	2LSul
CP	Input	2ul	2Sul	1LSul
Q,$\overline{Q}$	Outputs	10ul	10Sul	10LSul

NOTE:
Where a 74 unit load (ul) is understood to be 40μA I_{IH} and -1.6mA I_{IL}, a 74S unit load (Sul) is 50μA I_{IH} and -2.0mA I_{IL}, and 74LS unit load (LSul) is 20μA I_{IH} and -0.4mA I_{IL}.

PIN CONFIGURATION

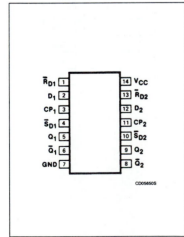

LOGIC SYMBOL

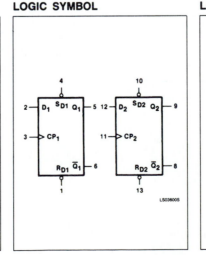

LOGIC SYMBOL (IEEE/IEC)

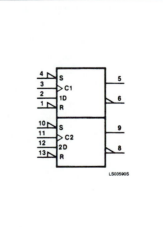

Flip-Flops   7474, LS74A, S74

LOGIC DIAGRAM

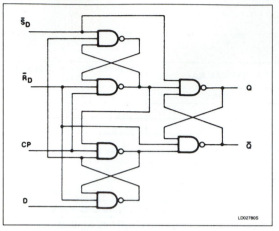

LD02780S

MODE SELECT — FUNCTION TABLE

OPERATING MODE	INPUTS				OUTPUTS	
	$\overline{S}_D$	$\overline{R}_D$	CP	D	Q	$\overline{Q}$
Asynchronous Set	L	H	X	X	H	L
Asynchronous Reset (Clear)	H	L	X	X	L	H
Undetermined[1]	L	L	X	X	H	H
Load "1" (Set)	H	H	↑	h	H	L
Load "0" (Reset)	H	H	↑	l	L	H

H = HIGH voltage level steady state.
h = HIGH voltage level one set-up time prior to the LOW-to-HIGH clock transition.
L = LOW voltage level steady state.
l = LOW voltage level one set-up time prior to the LOW-to-HIGH clock transition.
X = Don't care.
↑ = LOW-to-HIGH clock transition.

NOTE:
(1) Both outputs will be HIGH while both $\overline{S}_D$ and $\overline{R}_D$ are LOW, but the output states are unpredictable if $\overline{S}_D$ and $\overline{R}_D$ go HIGH simultaneously.

ABSOLUTE MAXIMUM RATINGS (Over operating free-air temperature range unless otherwise noted.)

	PARAMETER	74	74LS	74S	UNIT
V_{CC}	Supply voltage	7.0	7.0	7.0	V
V_{IN}	Input voltage	−0.5 to +5.5	−0.5 to +7.0	−0.5 to +5.5	V
I_{IN}	Input current	−30 to +5	−30 to +1	−30 to +5	mA
V_{OUT}	Voltage applied to output in HIGH output state	−0.5 to +V_{CC}	−0.5 to +V_{CC}	−0.5 to +V_{CC}	V
T_A	Operating free-air temperature range	0 to 70			°C

RECOMMENDED OPERATING CONDITIONS

	PARAMETER	74			74LS			74S			UNIT
		Min	Nom	Max	Min	Nom	Max	Min	Nom	Max	
V_{CC}	Supply voltage	4.75	5.0	5.25	4.75	5.0	5.25	4.75	5.0	5.25	V
V_{IH}	HIGH-level input voltage	2.0			2.0			2.0			V
V_{IL}	LOW-level input voltage			+0.8			+0.8			+0.8	V
I_{IK}	Input clamp current			−12			−18			−18	mA
I_{OH}	HIGH-level output current			−400			−400			−1000	μA
I_{OL}	LOW-level output current			16			8			20	mA
T_A	Operating free-air temperature	0		70	0		70	0		70	°C

Flip-Flops

7474, LS74A, S74

DC ELECTRICAL CHARACTERISTICS (Over recommended operating free-air temperature range unless otherwise noted.)

PARAMETER		TEST CONDITIONS[1]		7474			74LS74A			74S74			UNIT
				Min	Typ[2]	Max	Min	Typ[2]	Max	Min	Typ[2]	Max	
V_{OH}	HIGH-level output voltage	V_{CC} = MIN, V_{IH} = MIN, V_{IL} = MAX, I_{OH} = MAX		2.4	3.4		2.7	3.4		2.7	3.4		V
V_{OL}	LOW-level output voltage	V_{CC} = MIN, V_{IH} = MIN, V_{IL} = MAX	I_{OL} = MAX		0.2	0.4		0.35	0.5			0.5	V
			I_{OL} = 4mA (74LS)					0.25	0.4				V
V_{IK}	Input clamp voltage	V_{CC} = MIN, I_I = I_{IK}				−1.5			−1.5			−1.2	V
I_I	Input current at maximum input voltage	V_{CC} = MAX	V_I = 5.5V			1.0						1.0	mA
			V_I = 7.0V — D input						0.1				mA
			$\overline{R}_D$ input						0.2				mA
			$\overline{S}_D$ input						0.2				mA
			CP input						0.1				mA
I_{IH}	HIGH-level input current	V_{CC} = MAX	V_I = 2.4V — D input			40							µA
			$\overline{R}_D$ input			120							µA
			$\overline{S}_D$ input			80							µA
			CP input			80							µA
			V_I = 2.7V — D input						20			50	µA
			$\overline{R}_D$ input						40			150	µA
			$\overline{S}_D$ input						40			100	µA
			CP input						20			100	µA
I_{IL}	LOW-level input current[5]	V_{CC} = MAX	V_I = 0.4V — D input			−1.6			−0.4				mA
			$\overline{R}_D$ input			−3.2			−0.8				mA
			$\overline{S}_D$ input			−1.6			−0.8				mA
			CP input			−3.2			−0.4				mA
			V_I = 0.5V — D input									−2	mA
			$\overline{R}_D$ input									−6	mA
			$\overline{S}_D$ input									−4	mA
			CP input									−4	mA
I_{OS}	Short-circuit output current[3]	V_{CC} = MAX		−18		−57	−20		−100	−40		−100	mA
I_{CC}	Supply current[4] (total)	V_{CC} = MAX			17	30		4	8		30	50	mA

NOTES:
1. For conditions shown as MIN or MAX, use the appropriate value specified under recommended operating conditions for the applicable type.
2. All typical values are at V_{CC} = 5V, T_A = 25°C.
3. I_{OS} is tested with V_{OUT} = +0.5V and V_{CC} = V_{CC} MAX + 0.5V. Not more than one output should be shorted at a time and duration of the short circuit should not exceed one second.
4. Measure I_{CC} with the Clock inputs grounded and all outputs open, with the Q and $\overline{Q}$ outputs HIGH in turn.
5. Set is tested with reset HIGH and reset is tested with set HIGH.

Flip-Flops   7474, LS74A, S74

AC ELECTRICAL CHARACTERISTICS $T_A = 25°C$, $V_{CC} = 5.0V$

PARAMETER		TEST CONDITIONS	74 $C_L = 15pF$, $R_L = 400\Omega$		74LS $C_L = 15pF$, $R_L = 2k\Omega$		74S $C_L = 15pF$, $R_L = 280\Omega$		UNIT
			Min	Max	Min	Max	Min	Max	
f_{MAX}	Maximum clock frequency	Waveform 1	15		25		75		MHz
t_{PLH} t_{PHL}	Propagation delay Clock to output	Waveform 1		25 40		25 40		9 9	ns
t_{PLH} t_{PHL}	Propagation delay Set or Reset to output	Waveform 2		25		25		6	ns
		Waveform 2 CP = HIGH		40		40		13.5	
t_{PHL}	Set or Reset to output	Waveform 2 CP = LOW		40		40		8	ns

NOTE:
Per industry convention, f_{MAX} is the worst case value of the maximum device operating frequency with no constraints on t_r, t_f, pulse width or duty cycle.

AC SET-UP REQUIREMENTS $T_A = 25°C$, $V_{CC} = 5.0V$

PARAMETER		TEST CONDITIONS	74		74LS		74S		UNIT
			Min	Max	Min	Max	Min	Max	
$t_W(H)$	Clock pulse width (HIGH)	Waveform 1	30		25		6		ns
$t_W(L)$	Clock pulse width (LOW)	Waveform 1	37				7.3		ns
$t_W(L)$	Set or reset pulse width (LOW)	Waveform 2	30		25		7		ns
$t_s(H)$	Set-up time (HIGH) data to clock	Waveform 1	20		20		3		ns
$t_s(L)$	Set-up time (LOW) data to clock	Waveform 1	20		20		3		ns
t_h	Hold time data to clock	Waveform 1	5		5		2		ns

AC WAVEFORMS

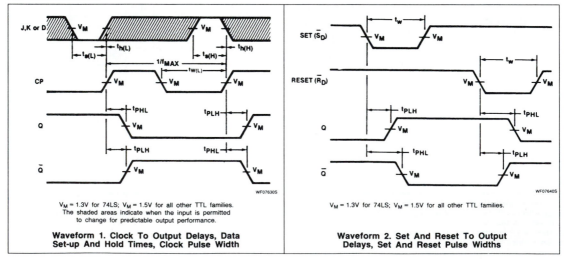

$V_M = 1.3V$ for 74LS; $V_M = 1.5V$ for all other TTL families.
The shaded areas indicate when the input is permitted
to change for predictable output performance.

**Waveform 1. Clock To Output Delays, Data
Set-up And Hold Times, Clock Pulse Width**

$V_M = 1.3V$ for 74LS; $V_M = 1.5V$ for all other TTL families.

**Waveform 2. Set And Reset To Output
Delays, Set And Reset Pulse Widths**

Flip-Flops

7474, LS74A, S74

TEST CIRCUITS AND WAVEFORMS

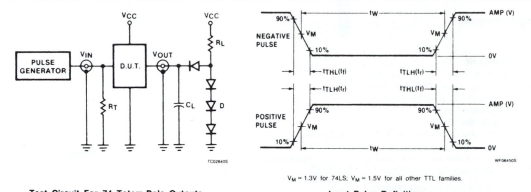

V_M = 1.3V for 74LS; V_M = 1.5V for all other TTL families.

Test Circuit For 74 Totem-Pole Outputs

Input Pulse Definition

DEFINITIONS
R_L = Load resistor to V_{CC}; see AC CHARACTERISTICS for value.
C_L = Load capacitance includes jig and probe capacitance;
    see AC CHARACTERISTICS for value.
R_T = Termination resistance should be equal to Z_{OUT}
    of Pulse Generators.
D = Diodes are 1N916, 1N3064, or equivalent.
t_{TLH}, t_{THL} Values should be less than or equal to the table
entries.

FAMILY	INPUT PULSE REQUIREMENTS				
	Amplitude	Rep. Rate	Pulse Width	t_{TLH}	t_{THL}
74	3.0V	1MHz	500ns	7ns	7ns
74LS	3.0V	1MHz	500ns	15ns	6ns
74S	3.0V	1MHz	500ns	2.5ns	2.5ns

Signetics

7476, LS76
Flip-Flops

Dual J-K Flip-Flop
Product Specification

Logic Products

DESCRIPTION

The '76 is a dual J-K flip-flop with individual J, K, Clock, Set and Reset inputs. The 7476 is positive pulse-triggered. JK information is loaded into the master while the Clock is HIGH and transferred to the slave on the HIGH-to-LOW Clock transition. The J and K inputs must be stable while the Clock is HIGH for conventional operation.

The 74LS76 is a negative edge-triggered flip-flop. The J and K inputs must be stable only one set-up time prior to the HIGH-to-LOW Clock transition.

The Set ($\overline{S}_D$) and Reset ($\overline{R}_D$) are asynchronous active LOW inputs. When LOW, they override the Clock and Data inputs, forcing the outputs to the steady state levels as shown in the Function Table.

TYPE	TYPICAL f_{MAX}	TYPICAL SUPPLY CURRENT (TOTAL)
7476	20MHz	10mA
74LS76	45MHz	4mA

ORDERING CODE

PACKAGES	COMMERCIAL RANGE $V_{CC} = 5V \pm 5\%$; $T_A = 0°C$ to $+70°C$
Plastic DIP	N7476N, N74LS76N

NOTE:
For information regarding devices processed to Military Specifications, see the Signetics Military Products Data Manual.

INPUT AND OUTPUT LOADING AND FAN-OUT TABLE

PINS	DESCRIPTION	74	74LS
$\overline{CP}$	Clock input	2ul	2LSul
$\overline{R}_D$, $\overline{S}_D$	Reset and Set inputs	2ul	2LSul
J, K	Data inputs	1ul	1LSul
Q, $\overline{Q}$	Outputs	10ul	10LSul

NOTE:
Where a 74 unit load (ul) is understood to be 40μA I_{IH} and -1.6mA I_{IL}, and a 74LS unit load (LSul) is 20μA I_{IH} and -0.4mA I_{IL}.

PIN CONFIGURATION

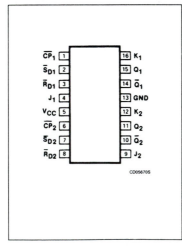

LOGIC SYMBOL

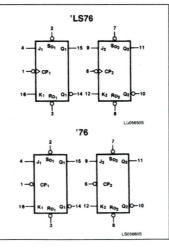

LOGIC SYMBOL (IEE/IEC)

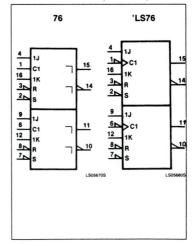

Flip-Flops   7476, LS76

LOGIC DIAGRAM

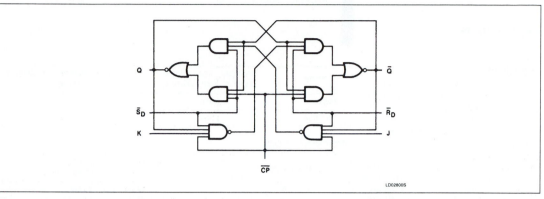

LD02800S

FUNCTION TABLE

OPERATING MODE	INPUTS					OUTPUTS	
	$\overline{S}_D$	$\overline{R}_D$	$\overline{CP}^{(2)}$	J	K	Q	$\overline{Q}$
Asynchronous set	L	H	X	X	X	H	L
Asynchronous reset (Clear)	H	L	X	X	X	L	H
Undetermined[1]	L	L	X	X	X	H	H
Toggle	H	H	⊓	h	h	$\overline{q}$	q
Load "0" (Reset)	H	H	⊓	l	h	L	H
Load "1" (Set)	H	H	⊓	h	l	H	L
Hold "no change"	H	H	⊓	l	l	q	$\overline{q}$

H   = HIGH voltage level steady state.
h   = HIGH voltage level one set-up time prior to the HIGH-to-LOW Clock transition.[3]
L   = LOW voltage level steady state.
l   = LOW voltage level one set-up time prior to the HIGH-to-LOW Clock transition.[3]
q   = Lower case letters indicate the state of the referenced output prior to the HIGH-to-LOW Clock transition.
X   = Don't care.
⊓ = Positive Clock pulse.

NOTES:
1. Both outputs will be HIGH while both $\overline{S}_D$ and $\overline{R}_D$ are LOW, but the output states are unpredictable if $\overline{S}_D$ and $\overline{R}_D$ go HIGH simultaneously.
2. The 74LS76 is edge triggered. Data must be stable one set-up time prior to the negative edge of the Clock for predictable operation.
3. The J and K inputs of the 7476 must be stable while the Clock is HIGH for conventional operation.

Signetics

7483, LS83A
Adders

4-Bit Full Adder
Product Specification

Logic Products

FEATURES
- High speed 4-bit binary addition
- Cascadeable in 4-bit increments
- LS83A has fast internal carry lookahead
- See '283 for corner power pin version

DESCRIPTION
The '83 adds two 4-bit binary words (A_n plus B_n) plus the incoming carry. The binary sum appears on the Sum outputs ($\Sigma_1 - \Sigma_4$) and the outgoing carry (C_{OUT}) according to the equation:

$$C_{IN} + (A_1 + B_1) + 2(A_2 + B_2) + 4(A_3 + B_3) + 8(A_4 + B_4) = \Sigma_1 + 2\Sigma_2 + 4\Sigma_3 + 8\Sigma_4 + 16C_{OUT}$$

Where (+) = plus.

Due to the symmetry of the binary add function, the '83 can be used with either all active-HIGH operands (positive logic) or with all active-LOW operands (negative logic). See Function Table. With active-HIGH inputs, C_{IN} cannot be left open; it must be held LOW when no "carry in" is intended. Interchanging inputs of equal weight does not affect the operation, thus C_{IN}, A_1, B_1, can arbitrarily be assigned to pins 10, 11, 13, etc.

TYPE	TYPICAL ADD TIMES (TWO 8 – BIT WORDS)	TYPICAL SUPPLY CURRENT (TOTAL)
7483	23ns	66mA
74LS83A	25ns	19mA

ORDERING CODE

PACKAGES	COMMERCIAL RANGE $V_{CC} = 5V \pm 5\%$; $T_A = 0°C$ to $+70°C$
Plastic DIP	N7483N, N74LS83AN
Plastic SO	N74LS83AD

NOTE:
For information regarding devices processed to Military Specifications, see the Signetics Military Products Data Manual.

INPUT AND OUTPUT LOADING AND FAN-OUT TABLE

PINS	DESCRIPTION	74	74LS
A_1, B_1, A_3, B_3, C_{IN}	Inputs	2ul	
A_2, B_2, A_4, B_4	Inputs	1ul	
A, B	Inputs		2LSul
C_{IN}	Input		1LSul
Sum	Outputs	10ul	10LSul
Carry	Output	5ul	10LSul

NOTE:
Where a 74 unit load (ul) is understood to be 40μA I_{IH} and – 1.6mA I_{IL}, and a 74LS unit load (LSul) is 20μA I_{IH} and – 0.4mA I_{IL}.

PIN CONFIGURATION

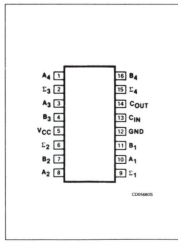

LOGIC SYMBOL

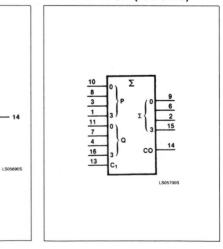

LOGIC SYMBOL (IEEE/IEC)

Adders

7483, LS83A

LOGIC DIAGRAM

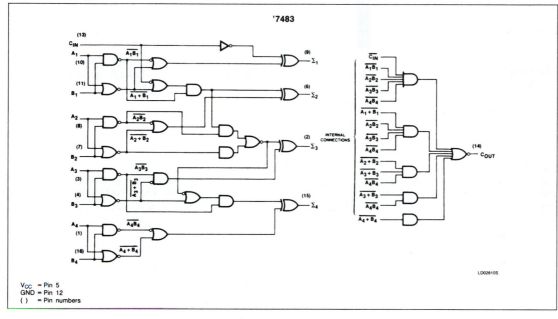

'7483

V_{CC} = Pin 5
GND = Pin 12
( ) = Pin numbers

LOGIC DIAGRAM

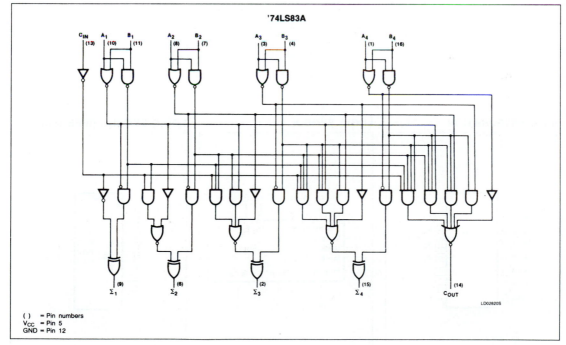

'74LS83A

( ) = Pin numbers
V_{CC} = Pin 5
GND = Pin 12

Signetics

Logic Products

7485, LS85, S85
Comparators

4-Bit Magnitude Comparator
Product Specification

FEATURES
- **Magnitude comparison of any binary words**
- **Serial or parallel expansion without extra gating**
- **Use 74S85 for very high speed comparisons**

DESCRIPTION
The '85 is a 4-bit magnitude comparator that can be expanded to almost any length. It compares two 4-bit binary, BCD, or other monotonic codes and presents the three possible magnitude results at the outputs. The 4-bit inputs are weighted $(A_0 - A_3)$ and $(B_0 - B_3)$, where A_3 and B_3 are the most significant bits.

The operation of the '85 is described in the Function Table, showing all possible logic conditions. The upper part of the table describes the normal operation under all conditions that will occur in a single device or in a series expansion scheme.

TYPE	TYPICAL PROPAGATION DELAY	TYPICAL SUPPLY CURRENT (TOTAL)
7485	23ns	55mA
74LS85	23ns	10mA
74S85	12ns	73mA

ORDERING CODE

PACKAGES	COMMERCIAL RANGE $V_{CC} = 5V \pm 5\%$; $T_A = 0°C$ to $+70°C$
Plastic DIP	N7485N, N74LS85N, N74S85N
Plastic SO	N74LS85D, N74S85D

NOTE:
For information regarding devices processed to Military Specifications, see the Signetics Military Products Data Manual.

INPUT AND OUTPUT LOADING AND FAN-OUT TABLE

PINS	DESCRIPTION	74	74S	74LS
$A_0 - A_3$, $B_0 - B_3$, $I_{A = B}$	Inputs	3ul	3Sul	3LSul
$I_{A < B}$, $I_{A > B}$	Inputs	1ul	1Sul	1LSul
$A = B$, $A < B$, $A > B$	Outputs	10ul	10Sul	10LSul

NOTE:
Where a 74 unit load (ul) is understood to be $40\mu A$ I_{IH} and $-1.6mA$ I_{IL}, a 74S unit load (Sul) is $50\mu A$ I_{IH} and $-2.0mA$ I_{IL}, and 74LS unit load (LSul) is $20\mu A$ I_{IH} and $-0.4mA$ I_{IL}.

PIN CONFIGURATION

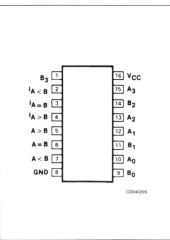

LOGIC SYMBOL

LOGIC SYMBOL (IEEE/IEC)

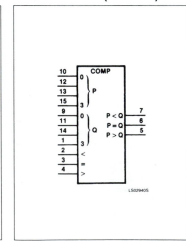

Comparators   7485, LS85, S85

LOGIC DIAGRAM

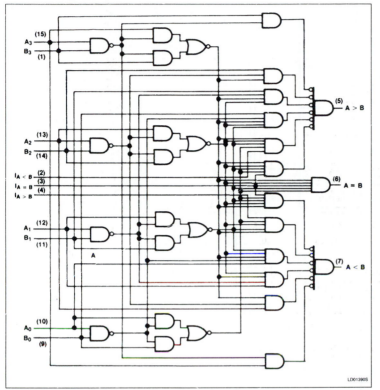

In the upper part of the table the three outputs are mutually exclusive. In the lower part of the table, the outputs reflect the feed-forward conditions that exist in the parallel expansion scheme.

The expansion inputs $I_{A>B}$, $I_{A=B}$, and $I_{A<B}$ are the least significant bit positions. When used for series expansion, the A > B, A = B and A < B outputs of the least significant word are connected to the corresponding $I_{A>B}$, $I_{A=B}$, and $I_{A<B}$ inputs of the next higher stage. Stages can be added in this manner to any length, but a propagation delay penalty of about 15ns is added with each additional stage. For proper operation the expansion inputs of the least significant word should be tied as follows: $I_{A>B}$ = LOW, $I_{A=B}$ = HIGH, and $I_{A<B}$ = LOW.

The parallel expansion scheme shown in Figure 1 demonstrates the most efficient general use of these comparators. In the parallel expansion scheme, the expansion inputs can be used as a fifth input bit position except on the least significant device which must be connected as in the serial scheme. The expansion inputs are used by labeling $I_{A>B}$ as an "A" input, $I_{A<B}$ as a "B" input and setting $I_{A=B}$ LOW. The '85 can be used as a 5-bit comparator only when the outputs are used to drive the ($A_0 - A_3$) and ($B_0 - B_3$) inputs of another '85 device. The parallel technique can be expanded to any number of bits as shown in Table 1.

LD01390S

FUNCTION TABLE

COMPARING INPUTS				CASCADING INPUTS			OUTPUTS		
A_3, B_3	A_2, B_2	A_1, B_1	A_0, B_0	$I_{A>B}$	$I_{A<B}$	$I_{A=B}$	A > B	A < B	A = B
$A_3 > B_3$	X	X	X	X	X	X	H	L	L
$A_3 < B_3$	X	X	X	X	X	X	L	H	L
$A_3 = B_3$	$A_2 > B_2$	X	X	X	X	X	H	L	L
$A_3 = B_3$	$A_2 < B_2$	X	X	X	X	X	L	H	L
$A_3 = B_3$	$A_2 = B_2$	$A_1 > B_1$	X	X	X	X	H	L	L
$A_3 = B_3$	$A_2 = B_2$	$A_1 < B_1$	X	X	X	X	L	H	L
$A_3 = B_3$	$A_2 = B_2$	$A_1 = B_1$	$A_0 > B_0$	X	X	X	H	L	L
$A_3 = B_3$	$A_2 = B_2$	$A_1 = B_1$	$A_0 < B_0$	X	X	X	L	H	L
$A_3 = B_3$	$A_2 = B_2$	$A_1 = B_1$	$A_0 = B_0$	H	L	L	H	L	L
$A_3 = B_3$	$A_2 = B_2$	$A_1 = B_1$	$A_0 = B_0$	L	H	L	L	H	L
$A_3 = B_3$	$A_2 = B_2$	$A_1 = B_1$	$A_0 = B_0$	L	L	H	L	L	H
$A_3 = B_3$	$A_2 = B_2$	$A_1 = B_1$	$A_0 = B_0$	X	X	H	L	L	H
$A_3 = B_3$	$A_2 = B_2$	$A_1 = B_1$	$A_0 = B_0$	H	H	L	L	L	L
$A_3 = B_3$	$A_2 = B_2$	$A_1 = B_1$	$A_0 = B_0$	L	L	L	H	H	L

H = HIGH voltage level
L = LOW voltage level
X = Don't care

Comparators   7485, LS85, S85

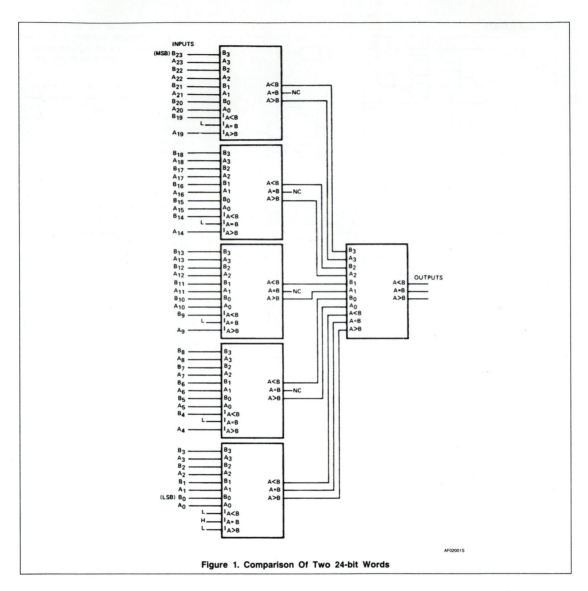

Figure 1. Comparison Of Two 24-bit Words

Table 1

WORD LENGTH	NUMBER OF PACKAGES	TYPICAL SPEEDS		
		74	74S	74LS
1 – 4 Bits	1	23ns	12ns	23ns
5 – 25 Bits	2 – 6	40ns	22ns	46ns
25 – 120 Bits	8 – 31	63ns	34ns	69ns

Signetics

Logic Products

7486, LS86, S86
Gates

Quad Two-Input Exclusive-OR Gate
Product Specification

TYPE	TYPICAL PROPAGATION DELAY	TYPICAL SUPPLY CURRENT (TOTAL)
7486	14ns	30mA
74LS86	10ns	6.1mA
74S86	7ns	50mA

ORDERING CODE

PACKAGES	COMMERCIAL RANGE $V_{CC} = 5V \pm 5\%$; $T_A = 0°C$ to $+70°C$
Plastic DIP	N7486N, N74LS86N, N74S86N
Plastic SO	N74LS86D, N74S86D

NOTE:
For information regarding devices processed to Military Specifications, see the Signetics Military Products Data Manual.

FUNCTION TABLE

INPUTS		OUTPUT
A	B	Y
L	L	L
L	H	H
H	L	H
H	H	L

H = HIGH voltage level
L = LOW voltage level

INPUT AND OUTPUT LOADING AND FAN-OUT TABLE

PINS	DESCRIPTION	74	74S	74LS
A, B	Inputs	1ul	1Sul	1LSul
Y	Output	10ul	10Sul	10LSul

NOTE:
Where a 74 unit load (ul) is understood to be $40\mu A$ I_{IH} and $-1.6mA$ I_{IL}, a 74S unit load (Sul) is $50\mu A$ I_{IH} and $-2.0mA$ I_{IL}, and a 74LS unit load (LSul) is $20\mu A$ I_{IH} and $-0.4mA$ I_{IL}.

PIN CONFIGURATION

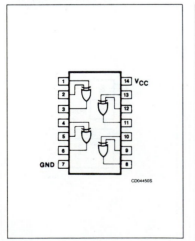

LOGIC SYMBOL

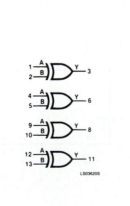

LOGIC SYMBOL (IEEE/IEC)

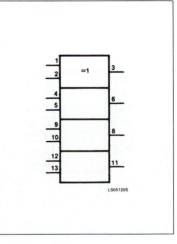

Signetics

7490, LS90
Counters

Decade Counter
Product Specification

Logic Products

DESCRIPTION

The '90 is a 4-bit, ripple-type Decade Counter. The device consists of four master-slave flip-flops internally connected to provide a divide-by-two section and a divide-by-five section. Each section has a separate Clock input to initiate state changes of the counter on the HIGH-to-LOW clock transition. State changes of the Q outputs do not occur simultaneously because of internal ripple delays. Therefore, decoded output signals are subject to decoding spikes and should not be used for clocks or strobes.

A gated AND asynchronous Master Reset ($MR_1 \cdot MR_2$) is provided which overrides both clocks and resets (clears) all the flip-flops. Also provided is a gated AND asynchronous Master Set ($MS_1 \cdot MS_2$) which overrides the clocks and the MR inputs, setting the outputs to nine (HLLH).

Since the output from the divide-by-two section is not internally connected to the succeeding stages, the device may be operated in various counting modes. In a BCD (8421) counter the $\overline{CP}_1$ input must be externally connected to the Q_0 output. The $\overline{CP}_0$ input receives the incoming count producing a BCD count sequence. In a symmetrical Bi-quinary divide-by-ten

counter the Q_3 output must be connected externally to the $\overline{CP}_0$ input. The input count is then applied to the CP_1 input and a divide-by-ten square wave is obtained at output Q_0. To operate as a divide-by-two and a divide-by-five count-

er no external interconnections are required. The first flip-flop is used as a binary element for the divide-by-two function ($\overline{CP}_0$ as the input and Q_0 as the output). The $\overline{CP}_1$ input is used to obtain a divide-by-five operation at the Q_3 output.

TYPE	TYPICAL f_{MAX}	TYPICAL SUPPLY CURRENT
7490	30MHz	30mA
74LS90	42MHz	9mA

ORDERING CODE

PACKAGES	COMMERCIAL RANGE $V_{CC} = 5V \pm 5\%$; $T_A = 0°C$ to $+70°C$
Plastic DIP	N7490N, N74LS90N

NOTE:
For information regarding devices processed to Military Specifications, see the Signetics Military Products Data Manual.

INPUT AND OUTPUT LOADING AND FAN-OUT TABLE

PINS	DESCRIPTION	74	74LS
$\overline{CP}_0$	Input	2ul	6LSul
$\overline{CP}_1$	Input	4ul	8LSul
MR, MS	Inputs		1ul
$Q_0 - Q_3$	Outputs	10ul	10LSul

NOTE:
Where a 74 unit load (ul) is understood to be $40\mu A$ I_{IH} and $-1.6mA$ I_{IL}, and a 74LS unit load (LSul) is $20\mu A$ I_{IH} and $-0.4mA$ I_{IL}.

PIN CONFIGURATION

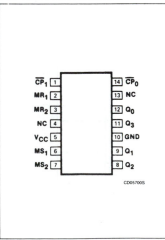

LOGIC SYMBOL

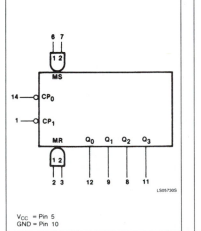

V_{CC} = Pin 5
GND = Pin 10

LOGIC SYMBOL (IEEE/IEC)

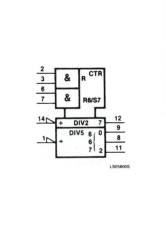

Counters   7490, LS90

LOGIC DIAGRAM

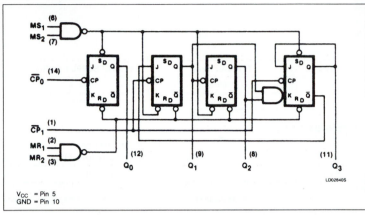

V_{CC} = Pin 5
GND = Pin 10

MODE SELECTION — FUNCTION TABLE

RESET/SET INPUTS				OUTPUTS			
MR_1	MR_2	MS_1	MS_2	Q_0	Q_1	Q_2	Q_3
H	H	L	X	L	L	L	L
H	H	X	L	L	L	L	L
X	X	H	H	H	L	L	H
L	X	L	X	Count			
X	L	X	L	Count			
L	X	X	L	Count			
H	L	L	X	Count			

H = HIGH voltage level
L = LOW voltage level
X = Don't care

BCD COUNT SEQUENCE — FUNCTION TABLE

COUNT	OUTPUTS			
	Q_0	Q_1	Q_2	Q_3
0	L	L	L	L
1	H	L	L	L
2	L	H	L	L
3	H	H	L	L
4	L	L	H	L
5	H	L	H	L
6	L	H	H	L
7	H	H	H	L
8	L	L	L	H
9	H	L	L	H

NOTE:
Output Q_0 connected to input $\overline{CP}_1$.

ABSOLUTE MAXIMUM RATINGS (Over operating free-air temperature range unless otherwise noted.)

	PARAMETER	74	74LS	UNIT
V_{CC}	Supply voltage	7.0	7.0	V
V_{IN}	Input voltage	−0.5 to +5.5	−0.5 to +7.0	V
I_{IN}	Input current	−30 to +5	−30 to +1	mA
V_{OUT}	Voltage applied to output in HIGH output state	−0.5 to +V_{CC}	−0.5 to +V_{CC}	V
T_A	Operating free-air temperature range	0 to 70		°C

NOTE:
V_{IN} is limited to +5.5V on $\overline{CP}_0$ and $\overline{CP}_1$ inputs on the 74LS90 only.

RECOMMENDED OPERATING CONDITIONS

	PARAMETER	74			74LS			UNIT
		Min	Nom	Max	Min	Nom	Max	
V_{CC}	Supply voltage	4.75	5.0	5.25	4.75	5.0	5.25	V
V_{IH}	HIGH-level input voltage	2.0			2.0			V
V_{IL}	LOW-level input voltage			+0.8			+0.8	V
I_{IK}	Input clamp current			−12			−18	mA
I_{OH}	HIGH-level output current			−800			−400	µA
I_{OL}	LOW-level output current			16			8	mA
T_A	Operating free-air temperature	0		70	0		70	°C

Signetics

7492, LS92
Counters

Divide-By-Twelve Counter
Product Specification

Logic Products

DESCRIPTION

The '92 is a 4-bit, ripple-type Divide-by-12 Counter. The device consists of four master-slave flip-flops internally connected to provide a divide-by-two section and a divide-by-six section. Each section has a separate Clock input to initiate state changes of the counter on the HIGH-to-LOW clock transition. State changes of the Q outputs do not occur simultaneously because of internal ripple delays. Therefore, decoded output signals are subject to decoding spikes and should not be used for clocks or strobes.

A gated AND asynchronous Master Reset ($MR_1 \cdot MR_2$) is provided which overrides both clocks and resets (clears) all the flip-flops.

TYPE	TYPICAL f_{MAX}	TYPICAL SUPPLY CURRENT
7492	28MHz	28mA
74LS92	42MHz	9mA

ORDERING CODE

PACKAGES	COMMERCIAL RANGE $V_{CC} = 5V \pm 5\%$; $T_A = 0°C$ to $+70°C$
Plastic DIP	N7492N, N74LS92N

NOTE:
For information regarding devices processed to Military Specifications, see the Signetics Military Products Data Manual.

INPUT AND OUTPUT LOADING AND FAN-OUT TABLE

PINS	DESCRIPTION	74	74LS
MR	Master reset inputs	1ul	1LSul
CP_0	Input	2ul	6LSul
CP_1	Input	4ul	8LSul
$Q_0 - Q_3$	Outputs	10ul	10LSul

NOTE:
Where a 74 unit load (ul) is understood to be 40µA I_{IH} and −1.6mA I_{IL}, and a 74LS unit load (LSul) is 20µA I_{IH} and −0.4mA I_{IL}.

PIN CONFIGURATION

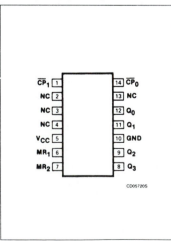

LOGIC SYMBOL

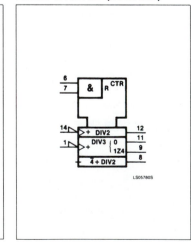

LOGIC SYMBOL (IEEE/IEC)

Counters 7492, LS92

LOGIC DIAGRAM

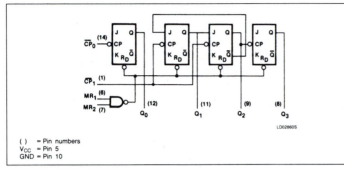

```
        (14)
CP₀ ──○CP
          J Q        J Q        J Q        J Q
          CP         CP         CP         CP
          K R_D Q̄    K R_D Q̄    K R_D Q̄    K R_D Q̄
CP₁ (1)
MR₁ (6)
MR₂ (7)
          (12)       (11)       (9)        (8)
          Q₀         Q₁         Q₂         Q₃
                                        LD02860S
```

( ) = Pin numbers
V_{CC} = Pin 5
GND = Pin 10

Since the output from the divide-by-two section is not internally connected to the succeeding stages, the device may be operated in various counting modes. In a Modulo-12, Divide-by-12 Counter the $\overline{CP}_1$ input must be externally connected to the Q_0 output. The $\overline{CP}_0$ input receives the incoming count and Q_3 produces a symmetrical divide-by-12 square wave output. In a divide-by-six counter no external connections are required. The first flip-flop is used as a binary element for the divide-by-two function. The $\overline{CP}_1$ input is used to obtain divide-by-three operation at the Q_1 and Q_2 outputs and divide-by-six operation at the Q_3 output.

FUNCTION TABLE

COUNT	OUTPUTS			
	Q_0	Q_1	Q_2	Q_3
0	L	L	L	L
1	H	L	L	L
2	L	H	L	L
3	H	H	L	L
4	L	L	H	L
5	H	L	H	L
6	L	L	L	H
7	H	L	L	H
8	L	H	L	H
9	H	H	L	H
10	L	L	H	H
11	H	L	H	H

NOTE:
Output Q_0 connected to input $\overline{CP}_1$.

MODE SELECTION

RESET INPUTS		OUTPUTS			
MR_1	MR_2	Q_0	Q_1	Q_2	Q_3
H	H	L	L	L	L
L	H		Count		
H	L		Count		
L	L		Count		

H = HIGH voltage level
L = LOW voltage level
X = Don't care

ABSOLUTE MAXIMUM RATINGS (Over operating free-air temperature range unless otherwise noted.)

PARAMETER		74	74LS	UNIT
V_{CC}	Supply voltage	7.0	7.0	V
V_{IN}	Input voltage	−0.5 to +5.5	−0.5 to +7.0	V
I_{IN}	Input current	−30 to +5	−30 to +1	mA
V_{OUT}	Voltage applied to output in HIGH output state	−0.5 to +V_{CC}	−0.5 to +V_{CC}	V
T_A	Operating free-air temperature range	0 to 70		°C

NOTE:
V_{IN} is limited to 5.5V on $\overline{CP}_0$ and $\overline{CP}_1$ inputs only on the 74LS92.

RECOMMENDED OPERATING CONDITIONS

PARAMETER		74			74LS			UNIT
		Min	Nom	Max	Min	Nom	Max	
V_{CC}	Supply voltage	4.75	5.0	5.25	4.75	5.0	5.25	V
V_{IH}	HIGH-level input voltage	2.0			2.0			V
V_{IL}	LOW-level input voltage			+0.8			+0.8	V
I_{IK}	Input clamp current			−12			−18	mA
I_{OH}	HIGH-level output current			−800			−400	μA
I_{OL}	LOW-level output current			16			8	mA
T_A	Operating free-air temperature	0		70	0		70	°C

Signetics

7493, LS93
Counters

4-Bit Binary Ripple Counter
Product Specification

Logic Products

DESCRIPTION

The '93 is a 4-bit, ripple-type Binary Counter. The device consists of four master-slave flip-flops internally connected to provide a divide-by-two section and a divide-by-eight section. Each section has a separate Clock input to initiate state changes of the counter on the HIGH-to-LOW clock transition. State changes of the Q outputs do not occur simultaneously because of internal ripple delays. Therefore, decoded output signals are subject to decoding spikes and should not be used for clocks or strobes.

A gated AND asynchronous Master Reset ($MR_1 \cdot MR_2$) is provided which overrides both clocks and resets (clears) all the flip-flops.

Since the output from the divide-by-two section is not internally connected to the succeeding stages, the device may be operated in various counting modes. In a 4-bit ripple counter the output Q_0 must be connected externally to input $\overline{CP}_1$.

TYPE	TYPICAL f_{MAX}	TYPICAL SUPPLY CURRENT (TOTAL)
7493	40MHz	28mA
74LS93	42MHz	9mA

ORDERING CODE

PACKAGES	COMMERCIAL RANGE $V_{CC} = 5V \pm 5\%$; $T_A = 0°C$ to $+70°C$
Plastic DIP	N7493N, N74LS93N
Plastic SO	N74LS93D

NOTE:
For information regarding devices processed to Military Specifications, see the Signetics Military Products Data Manual.

INPUT AND OUTPUT LOADING AND FAN-OUT TABLE

PINS	DESCRIPTION	74	74LS
MR	Master reset inputs	1ul	1LSul
$\overline{CP}_0$	Input	2ul	6LSul
$\overline{CP}_1$	Input	2ul	4LSul
$Q_0 - Q_3$	Outputs	10ul	10LSul

NOTE:
Where a 74 unit load (ul) is understood to be 40μA I_{IH} and -1.6mA I_{IL}, and a 74LS unit load (LSul) is 20μA I_{IH} and -0.4mA I_{IL}.

PIN CONFIGURATION

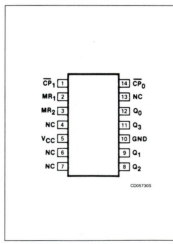

LOGIC SYMBOL

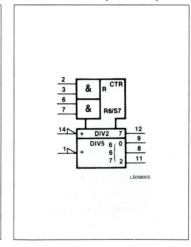

LOGIC SYMBOL (IEEE/IEC)

Counters

7493, LS93

LOGIC DIAGRAM

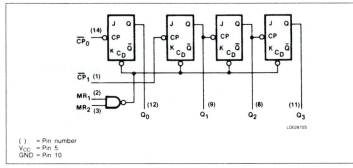

LD02870S

( ) = Pin number
V_{CC} = Pin 5
GND = Pin 10

The input count pulses are applied to input $\overline{CP}_0$. Simultaneous divisions of 2, 4, 8 and 16 are performed at the Q_0, Q_1, Q_2 and Q_3 outputs as shown in the Function Table.

As a 3-bit ripple counter the input count pulses are applied to input $\overline{CP}_1$. Simultaneous frequency divisions of 2, 4 and 8 are available at the Q_1, Q_2 and Q_3 outputs. Independent use of the first flip-flop is available if the reset function coincides with reset of the 3-bit ripple-through counter.

FUNCTION TABLE

COUNT	OUTPUTS			
	Q_0	Q_1	Q_2	Q_3
0	L	L	L	L
1	H	L	L	L
2	L	H	L	L
3	H	H	L	L
4	L	L	H	L
5	H	L	H	L
6	L	H	H	L
7	H	H	H	L
8	L	L	L	H
9	H	L	L	H
10	L	H	L	H
11	H	H	L	H
12	L	L	H	H
13	H	L	H	H
14	L	H	H	H
15	H	H	H	H

NOTE:
Output Q_0 connected to input $\overline{CP}_1$.

MODE SELECTION

RESET INPUTS		OUTPUTS			
MR_1	MR_2	Q_0	Q_1	Q_2	Q_3
H	H	L	L	L	L
L	H	Count			
H	L	Count			
L	L	Count			

H = HIGH voltage level
L = LOW voltage level
X = Don't care

ABSOLUTE MAXIMUM RATINGS (Over operating free-air temperature range unless otherwise noted.)

	PARAMETER	74	74LS	UNIT
V_{CC}	Supply voltage	7.0	7.0	V
V_{IN}	Input voltage	−0.5 to +5.5	−0.5 to +7.0	V
I_{IN}	Input current	−30 to +5	−30 to +1	mA
V_{OUT}	Voltage applied to output in HIGH output state	−0.5 to +V_{CC}	−0.5 to +V_{CC}	V
T_A	Operating free-air temperature range	0 to 70		°C

RECOMMENDED OPERATING CONDITIONS

	PARAMETER	74			74LS			UNIT
		Min	Nom	Max	Min	Nom	Max	
V_{CC}	Supply voltage	4.75	5.0	5.25	4.75	5.0	5.25	V
V_{IH}	HIGH-level input voltage	2.0			2.0			V
V_{IL}	LOW-level input voltage			+0.8			+0.8	V
I_{IK}	Input clamp current			−12			−18	mA
I_{OH}	HIGH-level output current			−800			−400	μA
I_{OL}	LOW-level output current			16			8	mA
T_A	Operating free-air temperature	0		70	0		70	°C

Signetics

74LS112, S112
Flip-Flops

Dual J-K Edge-Triggered Flip-Flop
Product Specification

Logic Products

DESCRIPTION

The '112 is a dual J-K negative edge-triggered flip-flop featuring individual J, K, Clock, Set and Reset inputs. The Set ($\overline{S}_D$) and Reset ($\overline{R}_D$) inputs, when LOW, set or reset the outputs as shown in the Function Table regardless of the levels at the other inputs.

A HIGH level on the Clock ($\overline{CP}$) input enables the J and K inputs and data will be accepted. The logic levels at the J and K inputs may be allowed to change while the $\overline{CP}$ is HIGH and the flip-flop will perform according to the Function Table as long as minimum setup and hold times are observed. Output state changes are initiated by the HIGH-to-LOW transition of $\overline{CP}$.

TYPE	TYPICAL f_{MAX}	TYPICAL SUPPLY CURRENT (TOTAL)
74LS112	45MHz	4mA
74S112	125MHz	15mA

ORDERING CODE

PACKAGES	COMMERCIAL RANGE $V_{CC} = 5V \pm 5\%$; $T_A = 0°C$ to $+70°C$
Plastic DIP	N74S112N, N74LS112N
Plastic SO	N74LS112D, N74S112D

NOTE:
For information regarding devices processed to Military Specifications, see the Signetics Military Products Data Manual.

INPUT AND OUTPUT LOADING AND FAN-OUT TABLE

PINS	DESCRIPTION	74S	74LS
$\overline{CP}$	Clock input	2Sul	4LSul
$\overline{R}_D$, $\overline{S}_D$	Reset and set inputs	3.5Sul	3LSul
J, K	Data inputs	1Sul	1LSul
Q, $\overline{Q}$	Outputs	10Sul	10LSul

NOTE:
A 74 unit load (ul) is 50μA I_{IH} and -2.0mA I_{IL}, and a 74LS unit load (LSul) is 20μA I_{IH} and -0.4mA I_{IL}.

PIN CONFIGURATION

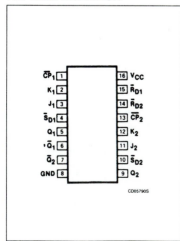

LOGIC SYMBOL

LOGIC SYMBOL (IEEE/IEC)

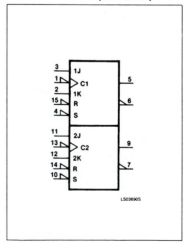

Flip-Flops

74LS112, S112

LOGIC DIAGRAM

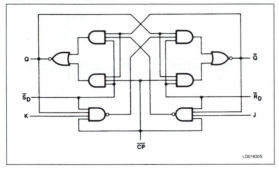

LD01830S

FUNCTION TABLE

OPERATING MODE	INPUTS					OUTPUTS	
	$\overline{S}_D$	$\overline{R}_D$	$\overline{CP}$	J	K	Q	$\overline{Q}$
Asynchronous set	L	H	X	X	X	H	L
Asynchronous reset (clear)	H	L	X	X	X	L	H
Undetermined	L	L	X	X	X	H	H
Toggle	H	H	↓	h	h	$\overline{q}$	q
Load "0" (reset)	H	H	↓	l	h	L	H
Load "1" (set)	H	H	↓	h	l	H	L
Hold "no change"	H	H	↓	l	l	q	$\overline{q}$

H   = HIGH voltage level steady state.
h   = HIGH voltage level one set-up time prior to the HIGH-to-LOW Clock transition.
L   = LOW voltage level steady state.
l   = LOW voltage level one set-up time prior to the HIGH-to-LOW Clock transition.
q   = Lower case letters indicate the state of the referenced output one set-up time prior to the HIGH-to-LOW Clock transition.
X   = Don't care.
↓   = HIGH-to-LOW Clock transition.

NOTE:
Both outputs will be HIGH while both $\overline{S}_D$ and $\overline{R}_D$ are LOW, but the output states are unpredictable if $\overline{S}_D$ and $\overline{R}_D$ go HIGH simultaneously.

ABSOLUTE MAXIMUM RATINGS (Over operating free-air temperature range unless otherwise noted.)

	PARAMETER	74LS	74S	UNIT
V_{CC}	Supply voltage	7.0	7.0	V
V_{IN}	Input voltage	−0.5 to −7.0	−0.5 to +5.5	V
I_{IN}	Input current	−30 to +1	−30 to +5	mA
V_{OUT}	Voltage applied to output in HIGH output state	−0.5 to +V_{CC}	−0.5 to +V_{CC}	V
T_A	Operating free-air temperature range	0 to 70		°C

RECOMMENDED OPERATING CONDITIONS

	PARAMETER	74LS			74S			UNIT
		Min	Nom	Max	Min	Nom	Max	
V_{CC}	Supply voltage	4.75	5.0	5.25	4.75	5.0	5.25	V
V_{IH}	HIGH-level input voltage	2.0			2.0			V
V_{IL}	LOW-level input voltage			+0.8			+0.8	V
I_{IK}	Input clamp current			−18			−18	mA
I_{OH}	HIGH-level output current			−400			−1000	µA
I_{OL}	LOW-level output current			8			20	mA
T_A	Operating free-air temperature	0		70	0		70	°C

Signetics

74121
Multivibrator

Monostable Multivibrator
Product Specification

Logic Products

FEATURES
- Very good pulse width stability
- Virtually immune to temperature and voltage variations
- Schmitt trigger input for slow input transitions
- Internal timing resistor provided

TYPE	TYPICAL PROPAGATION DELAY	TYPICAL SUPPLY CURRENT (TOTAL)
74121	43ns	18mA

ORDERING CODE

PACKAGES	COMMERCIAL RANGE $V_{CC} = 5V \pm 5\%$; $T_A = 0°C$ to $+70°C$
Plastic DIP	N74121 N
Plastic SO	N74121 D

NOTE:
For information regarding devices processed to Military Specifications, see the Signetics Military Products Data Manual.

DESCRIPTION
These multivibrators feature dual active LOW going edge inputs and a single active HIGH going edge input which can be used as an active HIGH enable input. Complementary output pulses are provided.

Pulse triggering occurs at a particular voltage level and is not directly related to the transition time of the input pulse. Schmitt-trigger input circuitry (TTL hysteresis) for the B input allows jitter-free triggering from inputs with transition rates as slow as 1 volt/second, providing the circuit with an excellent noise immunity of typically 1.2 volts. A high immunity to V_{CC} noise of typically 1.5 volts is also provided by internal latching circuitry. Once fired, the outputs are independent of further transitions of the inputs and are a function only of the timing components. Input pulses may be of any duration relative to the output pulse. Output pulse length may be varied from 20 nanoseconds to 28 seconds by choosing appropriate timing components. With no external timing components (i.e., R_{int} connected to V_{CC}, C_{ext} and R_{ext}/C_{ext} open), an output pulse of typically 30 or 35 nanoseconds is achieved which may be used as a dc triggered reset signal. Output rise and fall times are TTL compatible and independent of pulse length.

Pulse width stability is achieved through internal compensation and is virtually independent of V_{CC} and temperature. In most applications, pulse stability will only be limited by the accuracy of external timing components.

Jitter-free operation is maintained over the full temperature and V_{CC} ranges for more than six decades of timing capacitance (10pF to 10μF) and more than one decade of timing resistance (2kΩ to 30kΩ for the 54121 and 2KΩ to 40kΩ for the 74121). Throughout these ranges, pulse width is defined by the relationship: (see Figure 1)

$$t_W(out) = C_{ext}\, R_{ext}\, \ln 2$$
$$t_W(out) \cong 0.7\, C_{ext}\, R_{ext}$$

PIN CONFIGURATION

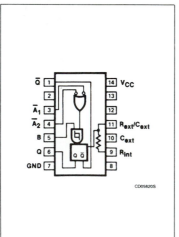

CD05820S

LOGIC SYMBOL

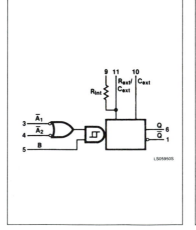

LS05950S

LOGIC SYMBOL (IEEE/IEC)

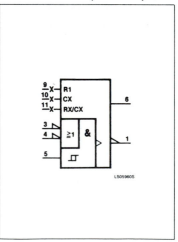

LS05960S

Multivibrator   74121

FUNCTION TABLE

INPUTS			OUTPUTS	
$\overline{A}_1$	$\overline{A}_2$	B	Q	$\overline{Q}$
L	X	H	L	H
X	L	H	L	H
X	X	L	L	H
H	H	X	L	H
H	↓	H	⊓	⊔
↓	H	H	⊓	⊔
↓	↓	H	⊓	⊔
L	X	↑	⊓	⊔
X	L	↑	⊓	⊔

H = HIGH voltage level
L = LOW voltage level
X = Don't care
↑ = LOW-to-HIGH transition
↓ = HIGH-to-LOW transition

INPUT AND OUTPUT LOADING AND FAN-OUT TABLE

PINS	DESCRIPTION	74
$\overline{A}_1$, $\overline{A}_2$	Inputs	1ul
B	Input	2ul
Q, $\overline{Q}$	Outputs	10ul

NOTE:
A 74 unit load (ul) is understood to be 40μA I_{IH} and $-$1.6mA I_{IL}.

In circuits where pulse cutoff is not critical, timing capacitance up to 1000μF and timing resistance as low as 1.4kΩ may be used.

ABSOLUTE MAXIMUM RATINGS (Over operating free-air temperature range unless otherwise noted.)

	PARAMETER	74	UNIT
V_{CC}	Supply voltage	7.0	V
V_{IN}	Input voltage	-0.5 to $+5.5$	V
I_{IN}	Input current	-30 to $+5$	mA
V_{OUT}	Voltage applied to output in HIGH output state	-0.5 to $+V_{CC}$	V
T_A	Operating free-air temperature range	0 to 70	°C

RECOMMENDED OPERATING CONDITIONS

	PARAMETER		74			UNIT
			Min	Nom	Max	
V_{CC}	Supply voltage		4.75	5.0	5.25	V
I_{IK}	Input clamp current				-12	mA
I_{OH}	HIGH-level output current				-400	μA
I_{OL}	LOW-level output current				16	mA
dv/dt	Rate of rise or fall of input pulse	B input	1			V/s
		$\overline{A}_1$, $\overline{A}_2$ inputs	1			V/μs
T_A	Operating free-air temperature		0		70	°C

Multivibrator   74121

DC ELECTRICAL CHARACTERISTICS (Over recommended operating free-air temperature range unless otherwise noted.)

	PARAMETER	TEST CONDITIONS[1]		74121			UNIT
				Min	Typ[2]	Max	
V_{T+}	Positive-going threshold at $\overline{A}$ and B	V_{CC} = MIN				2.0	V
V_{T-}	Negative-going threshold at $\overline{A}$ and B	V_{CC} = MIN		0.8			V
V_{OH}	HIGH-level output voltage	V_{CC} = MIN, V_{IH} = MIN, V_{IL} = MAX, I_{OH} = MAX		2.4	3.4		V
V_{OL}	LOW-level output voltage	V_{CC} = MIN, V_{IH} = MIN, V_{IL} = MAX, I_{OL} = MAX			0.2	0.4	V
V_{IK}	Input clamp voltage	V_{CC} = MIN, I_I = I_{IK}				−1.5	V
I_I	Input current at maximum input voltage	V_{CC} = MAX, V_I = 5.5V				1.0	mA
I_{IH}	HIGH-level input current	V_{CC} = MAX, V_I = 2.4V	$\overline{A}_1$, $\overline{A}_2$ inputs			40	μA
			B input			80	μA
I_{IL}	LOW-level input current	V_{CC} = MAX, V_I = 0.4V	$\overline{A}_1$, $\overline{A}_2$ inputs			−1.6	mA
			B input			−3.2	mA
I_{OS}	Short-circuit output current[3]	V_{CC} = MAX		−18		−55	mA
I_{CC}	Supply current (total)	V_{CC} = MAX	Quiescent		13	25	mA
			Triggered		23	40	mA

NOTES:
1. For conditions shown as MIN or MAX, use the appropriate value specified under recommended operating conditions for the applicable type.
2. All typical values are at V_{CC} = 5V, T_A = 25°C.
3. I_{OS} is tested with V_{OUT} = + 0.5V and V_{CC} = V_{CC} MAX + 0.5V. Not more than one output should be shorted at a time and duration of the short circuit should not exceed one second.

AC ELECTRICAL CHARACTERISTICS T_A = 25°C, V_{CC} = 5.0V

	PARAMETER	TEST CONDITIONS	74		UNIT
			C_L = 15pF, R_L = 400Ω		
			Min	Max	
t_{PLH} t_{PHL}	Propagation delay $\overline{A}$ input to Q & $\overline{Q}$ output	Waveform 1 C_{ext} = 80pF, R_{int} to V_{CC}		70 80	ns
t_{PLH} t_{PHL}	Propagation delay B input to Q & $\overline{Q}$ output	Waveform 2 C_{ext} = 80pF, R_{int} to V_{CC}		55 65	ns
t_W	Minimum output pulse width	C_{ext} = 0pF, R_{int} to V_{CC}	20	50	ns
t_W	Output pulse width	C_{ext} = 80pF, R_{int} to V_{CC}	70	150	ns
		C_{ext} = 100pF, R_{ext} = 10kΩ	600	800	ns
		C_{ext} = 1μF, R_{ext} = 10kΩ	6.0	8.0	ms

Multivibrator   74121

AC SET-UP REQUIREMENTS $T_A = 25°C$, $V_{CC} = 5.0V$

PARAMETER		TEST CONDITIONS	74		UNIT
			Min	Max	
t_W	Minimum input pulse width to trigger	Waveforms 1 & 2	50		ns
R_{ext}	External timing resistor range		1.4	40	kΩ
C_{ext}	External timing capacitance range		0	1000	μF
	Output duty cycle	$R_{ext} = 2kΩ$		67	%
		$R_{ext} = R_{ext}(Max)$		90	%

AC WAVEFORMS

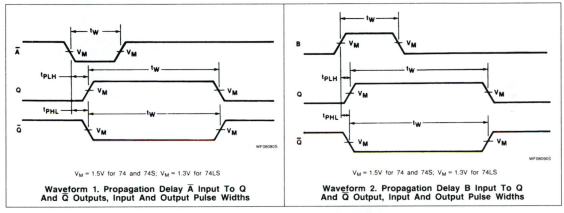

$V_M = 1.5V$ for 74 and 74S; $V_M = 1.3V$ for 74LS

Waveform 1. Propagation Delay Ā Input To Q And Q̄ Outputs, Input And Output Pulse Widths

$V_M = 1.5V$ for 74 and 74S; $V_M = 1.3V$ for 74LS

Waveform 2. Propagation Delay B Input To Q And Q̄ Output, Input And Output Pulse Widths

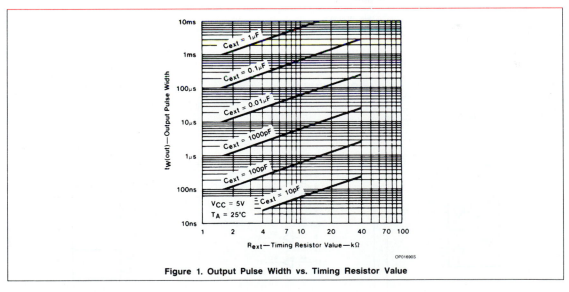

Figure 1. Output Pulse Width vs. Timing Resistor Value

Signetics

Logic Products

74125, 74126, LS125A, LS126A
Buffers

Quad 3-State Buffer
Product Specification

TYPE	TYPICAL PROPAGATION DELAY	TYPICAL SUPPLY CURRENT (TOTAL)
74125	10ns	32mA
74LS125A	8ns	11mA
74126	10ns	36mA
74LS126A	9ns	12mA

FUNCTION TABLE '125, 'LS125A

INPUTS		OUTPUT
C	A	Y
L	L	L
L	H	H
H	X	(Z)

FUNCTION TABLE '126, 'LS126A

INPUTS		OUTPUT
C	A	Y
H	L	L
H	H	H
L	X	(Z)

H = HIGH voltage level
L = LOW voltage level
X = Don't care
(Z) = HIGH impedance (off)

ORDERING CODE

PACKAGES	COMMERCIAL RANGE $V_{CC} = 5V \pm 5\%$; $T_A = 0°C$ to $+70°C$
Plastic DIP	N74125N, N74LS125N N74126N, N74LS126N
Plastic SO	N74LS125AD

NOTE:
For information regarding devices processed to Military Specifications, see the Signetics Military Products Data Manual.

INPUT AND OUTPUT LOADING AND FAN-OUT TABLE

PINS	DESCRIPTION	74	74LS
All	Inputs	1ul	1LSul
All	Outputs	10ul	30LSul

NOTE:
Where a 74 unit load (ul) is understood to be $40\mu A\ I_{IH}$ and $-1.6mA\ I_{IL}$, and a 74LS unit load (LSul) is $20\mu A\ I_{IH}$ and $-0.4mA\ I_{IL}$.

PIN CONFIGURATION

'125
'LS125A

CD04190S

LOGIC SYMBOL

'125
'LS125A

LS03220S

LOGIC SYMBOL (IEEE/IEC)

'125
'LS125A

LS03240S

Buffers   74125, 74126, LS125A, LS126A

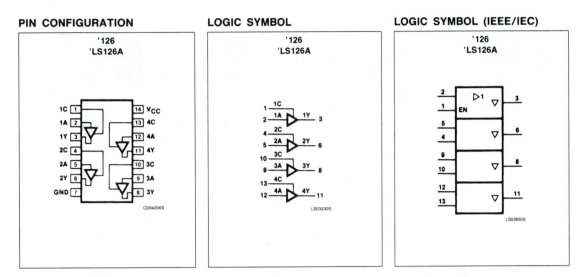

PIN CONFIGURATION

'126 'LS126A

LOGIC SYMBOL

'126 'LS126A

LOGIC SYMBOL (IEEE/IEC)

'126 'LS126A

ABSOLUTE MAXIMUM RATINGS (Over operating free-air temperature range unless otherwise noted.)

PARAMETER		74	74LS	UNIT
V_{CC}	Supply voltage	7.0	7.0	V
V_{IN}	Input voltage	−0.5 to +5.5	−0.5 to +7.0	V
I_{IN}	Input current	−30 to +5	−30 to +1	mA
V_{OUT}	Voltage applied to output in HIGH output state	−0.5 to +V_{CC}	−0.5 to +V_{CC}	V
T_A	Operating free-air temperature range	0 to 70		°C

RECOMMENDED OPERATING CONDITIONS

PARAMETER		74			74LS			UNIT
		Min	Nom	Max	Min	Nom	Max	
V_{CC}	Supply voltage	4.75	5.0	5.25	4.75	5.0	5.25	V
V_{IH}	HIGH-level input voltage	2.0			2.0			V
V_{IL}	LOW-level input voltage			+0.8			+0.8	V
I_{IK}	Input clamp current			−12			−18	mA
I_{OH}	HIGH-level output current			−5.2			−2.6	mA
I_{OL}	LOW-level output current			16			24	mA
T_A	Operating free-air temperature	0		70	0		70	°C

Buffers   74125, 74126, LS125A, LS126A

DC ELECTRICAL CHARACTERISTICS (Over recommended operating free-air temperature range unless otherwise noted.)

PARAMETER		TEST CONDITIONS[1]		74125 74126			74LS125A 74LS126A			UNIT
				Min	Typ[2]	Max	Min	Typ[2]	Max	
V_{OH}	HIGH-level output voltage	V_{CC} = MIN, V_{IH} = MIN, V_{IL} = MAX, I_{OH} = MAX		2.4	3.1		2.4			V
V_{OL}	LOW-level output voltage	V_{CC} = MIN, V_{IH} = MIN, V_{IL} = MAX	I_{OL} = MAX			0.4		0.35	0.5	V
			I_{OL} = 12mA (74LS)					0.25	0.4	V
V_{IK}	Input clamp voltage	V_{CC} = MIN, I_I = I_{IK}				−1.5			−1.5	V
I_{OZH}	Off-state output current, HIGH-level voltage applied	V_{CC} = MAX, V_{IH} = MIN, V_{IL} = MAX, V_O = 2.4V				40			20	μA
I_{OZL}	Off-state output current, LOW-level voltage applied	V_{CC} = MAX, V_{IH} = MIN, V_{IL} = MAX, V_O = 0.4V				−40			−20	μA
I_I	Input current at maximum input voltage	V_{CC} = MAX	V_I = 5.5V			1.0				mA
			V_I = 7.0V						0.1	mA
I_{IH}	HIGH-level input current	V_{CC} = MAX	V_I = 2.4V			40				μA
			V_I = 2.7V						20	μA
I_{IL}	LOW-level input current	V_{CC} = MAX, V_I = 0.4V				−1.6			−0.4	mA
I_{OS}	Short-circuit output current[3]	V_{CC} = MAX		−28		−70	−40		−130	mA
I_{CC}	Supply current (total)	V_{CC} = MAX	'125		32	54		11	20	mA
			'126		36	62		12	22	mA

NOTES:
1. For conditions shown as MIN or MAX, use the appropriate value specified under recommended operating conditions for the applicable type.
2. All typical values are at V_{CC} = 5V, T_A = 25°C.
3. I_{OS} is tested with V_{OUT} = +0.5V and V_{CC} = V_{CC} MAX + 0.5V. Not more than one output should be shorted at a time and duration of the short circuit should not exceed one second.

Buffers

74125, 74126, LS125A, LS126A

AC WAVEFORMS

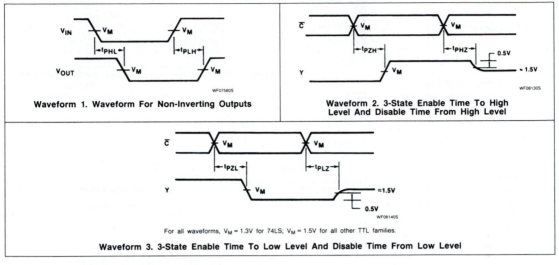

Waveform 1. Waveform For Non-Inverting Outputs

Waveform 2. 3-State Enable Time To High Level And Disable Time From High Level

For all waveforms, V_M = 1.3V for 74LS; V_M = 1.5V for all other TTL families.

Waveform 3. 3-State Enable Time To Low Level And Disable Time From Low Level

AC ELECTRICAL CHARACTERISTICS T_A = 25°C, V_{CC} = 5.0V

PARAMETER		TEST CONDITIONS	74125		74LS125A		74126		74LS126A		UNIT
			C_L = 50pF R_L = 400Ω		C_L = 45pF R_L = 667Ω		C_L = 50pF R_L = 400Ω		C_L = 45pF R_L = 667Ω		
			Min	Max	Min	Max	Min	Max	Min	Max	
t_{PLH}	Propagation delay	Waveform 1		13		15		13		15	ns
t_{PHL}	Data to output			18		18		18		18	
t_{PZH}	Enable to HIGH	Waveform 2		17		20		18		25	ns
t_{PZL}	Enable to LOW	Waveform 3		25		25		25		35	ns
t_{PHZ}	Disable from HIGH	Waveform 2, C_L = 5pF		8.0		20		16		25	ns
t_{PLZ}	Disable from LOW	Waveform 3, C_L = 5pF		12		20		18		25	ns

Signetics

Logic Products

74LS138, S138
Decoders/Demultiplexers

1-Of-8 Decoder/Demultiplexer
Product Specification

FEATURES
- **Demultiplexing capability**
- **Multiple input enable for easy expansion**
- **Ideal for memory chip select decoding**
- **Direct replacement for Intel 3205**

DESCRIPTION
The '138 decoder accepts three binary weighted inputs (A_0, A_1, A_2) and when enabled, provides eight mutually exclusive, active LOW outputs ($\overline{0}-\overline{7}$). The device features three Enable Inputs: two active LOW ($\overline{E}_1$, $\overline{E}_2$) and one active HIGH (E_3). Every output will be HIGH unless $\overline{E}_1$ and $\overline{E}_2$ are LOW and E_3 is HIGH. This multiple enable function allows easy parallel expansion of the device to a 1-of-32 (5 lines to 32 lines) decoder with just four '138s and one inverter.

The device can be used as an eight output demultiplexer by using one of the active LOW Enable inputs as the Data input and the remaining Enable inputs as strobes. Enable inputs not used must be permanently tied to their appropriate active HIGH or active LOW state.

TYPE	TYPICAL PROPAGATION DELAY	TYPICAL SUPPLY CURRENT (TOTAL)
74LS138	20ns	6.3mA
74S138	7ns	49mA

ORDERING CODE

PACKAGES	COMMERCIAL RANGE V_{CC} = 5V ±5%; T_A = 0°C to +70°C
Plastic DIP	N74S138N, N74LS138N
Plastic SO	N74LS138D, N74S138D

NOTE:
For information regarding devices processed to Military Specifications see the Signetics Military Products Data Manual.

INPUT AND OUTPUT LOADING AND FAN-OUT TABLE

PINS	DESCRIPTION	74S	74LS
All	Inputs	1Sul	1LSul
All	Outputs	10Sul	10LSul

NOTE:
Where a 74S unit load (Sul) is 50µA I_{IH} and −2.0mA I_{IL}, and a 74LS unit load (LSul) is 20µA I_{IH} and −0.4mA I_{IL}.

PIN CONFIGURATION

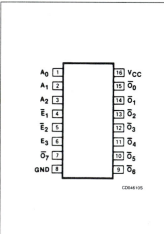

LOGIC SYMBOL

LOGIC SYMBOL (IEEE/IEC)

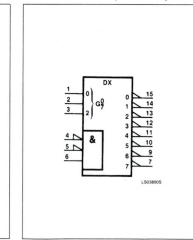

Decoders/Demultiplexers   74LS138, S138

LOGIC DIAGRAM

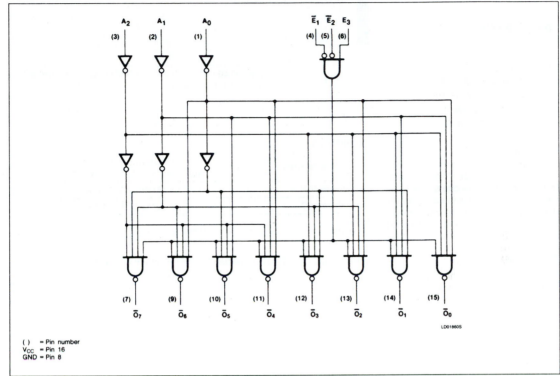

```
( )  = Pin number
Vcc  = Pin 16
GND  = Pin 8
```

FUNCTION TABLE

INPUTS						OUTPUTS							
$\bar{E}_1$	$\bar{E}_2$	E_3	A_0	A_1	A_2	$\bar{0}$	$\bar{1}$	$\bar{2}$	$\bar{3}$	$\bar{4}$	$\bar{5}$	$\bar{6}$	$\bar{7}$
H	X	X	X	X	X	H	H	H	H	H	H	H	H
X	H	X	X	X	X	H	H	H	H	H	H	H	H
X	X	L	X	X	X	H	H	H	H	H	H	H	H
L	L	H	L	L	L	L	H	H	H	H	H	H	H
L	L	H	H	L	L	H	L	H	H	H	H	H	H
L	L	H	L	H	L	H	H	L	H	H	H	H	H
L	L	H	H	H	L	H	H	H	L	H	H	H	H
L	L	H	L	L	H	H	H	H	H	L	H	H	H
L	L	H	H	L	H	H	H	H	H	H	L	H	H
L	L	H	L	H	H	H	H	H	H	H	H	L	H
L	L	H	H	H	H	H	H	H	H	H	H	H	L

```
H = HIGH voltage level
L = LOW voltage level
X = Don't care
```

Signetics

Logic Products

FEATURES
- **Demultiplexing capability**
- **Two independent 1-of-4 decoders**
- **Multifunction capability**
- **Replaces 9321 and 93L21 for higher performance**

DESCRIPTION

The '139 is a high-speed, dual 1-of-4 decoder/demultiplexer. This device has two independent decoders, each accepting two binary weighted inputs (A_0, A_1) and providing four mutually exclusive active LOW outputs ($\overline{0} - \overline{3}$). Each decoder has an active LOW Enable ($\overline{E}$). When $\overline{E}$ is HIGH, every output is forced HIGH. The Enable can be used as the Data input for a 1-of-4 demultiplexer application.

74LS139, S139
Decoders/Demultiplexers

Dual 1-of-4 Decoder/Demultiplexer
Product Specification

TYPE	TYPICAL PROPAGATION DELAY (ENABLE AT 2 LOGIC LEVELS)	TYPICAL SUPPLY CURRENT (TOTAL)
74LS139	19ns	6.8mA
74S139	6ns	60mA

ORDERING CODE

PACKAGES	COMMERCIAL RANGE $V_{CC} = 5V \pm 5\%$; $T_A = 0°C$ to $+70°C$
Plastic DIP	N74S139N, N74LS139N
Plastic SO	N74LS139D, N74S139D

NOTE:
For information regarding devices processed to Military Specifications, see the Signetics Military Products Data Manual.

INPUT AND OUTPUT LOADING AND FAN-OUT TABLE

PINS	DESCRIPTION	74S	74LS
All	Inputs	1Sul	1LSul
All	Outputs	10Sul	10LSul

NOTE:
A 74S unit load (Sul) is 50μA I_{IH} and −2.0mA I_{IL}, and a 74LS unit load (LSul) is 20μA I_{IH} and −0.4mA I_{IL}.

PIN CONFIGURATION

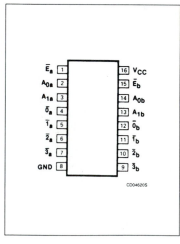

LOGIC SYMBOL

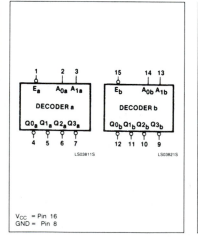

LOGIC SYMBOL (EEE/IEC)

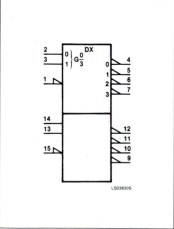

Decoders/Demultiplexers   74LS139, S139

LOGIC DIAGRAM

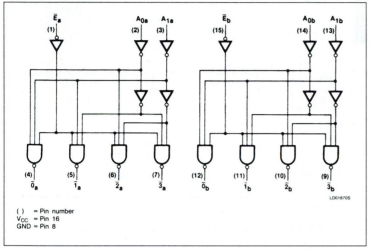

```
( )   = Pin number
Vcc   = Pin 16
GND = Pin 8
```

FUNCTION TABLE

INPUTS			OUTPUTS			
$\overline{E}$	A_0	A_1	$\overline{0}$	$\overline{1}$	$\overline{2}$	$\overline{3}$
H	X	X	H	H	H	H
L	L	L	L	H	H	H
L	H	L	H	L	H	H
L	L	H	H	H	L	H
L	H	H	H	H	H	L

H = HIGH voltage level
L = LOW voltage level

ABSOLUTE MAXIMUM RATINGS (Over operating free-air temperature range unless otherwise noted.)

	PARAMETER	74LS	74S	UNIT
V_{CC}	Supply voltage	7.0	7.0	V
V_{IN}	Input voltage	−0.5 to +7.0	−0.5 to +5.5	V
I_{IN}	Input current	−30 to +1	−30 to +5	mA
V_{OUT}	Voltage applied to output in HIGH output state	−0.5 to +V_{CC}	−0.5 to +V_{CC}	V
T_A	Operating free-air temperature range	0 to 70		°C

RECOMMENDED OPERATING CONDITIONS

	PARAMETER	74LS			74S			UNIT
		Min	Nom	Max	Min	Nom	Max	
V_{CC}	Supply voltage	4.75	5.0	5.25	4.75	5.0	5.25	V
V_{IH}	HIGH-level input voltage	2.0			2.0			V
V_{IL}	LOW-level input voltage			+0.8			+0.8	V
I_{IK}	Input clamp current			−18			−18	mA
I_{OH}	HIGH-level output current			−400			−1000	μA
I_{OL}	LOW-level output current			8			20	mA
T_A	Operating free-air temperature	0		70	0		70	°C

Signetics

74147
Encoder

10-Line-To-4-Line Priority Encoder
Product Specification

Logic Products

FEATURES
- Encodes 10-line decimal to 4-line BCD
- Useful for 10-position switch encoding
- Used in code converters and generators

DESCRIPTION
The '147 9-input priority encoder accepts data from nine active-LOW inputs ($\bar{I}_1 - \bar{I}_9$) and provides a binary representation on the four active-LOW outputs ($A_0 - A_3$). A priority is assigned to each input so that when two or more inputs are simultaneously active, the input with the highest priority is represented on the output, with input line $\bar{I}_9$ having the highest priority.

The device provides the 10-line-to-4-line priority encoding function by use of the implied decimal "zero." The "zero" is encoded when all nine data inputs are HIGH, forcing all four outputs HIGH.

TYPE	TYPICAL PROPAGATION DELAY	TYPICAL SUPPLY CURRENT (TOTAL)
74147	10ns	46mA

ORDERING CODE

PACKAGES	COMMERCIAL RANGE V_{CC} = 5V ±5%; T_A = 0°C to +70°C
Plastic DIP	N74147N

NOTE:
For information regarding devices processed to Military Specifications see the Signetics Military Products Data Manual.

INPUT AND OUTPUT LOADING AND FAN-OUT TABLE

PINS	DESCRIPTION	74
All	Inputs	1ul
All	Outputs	10ul

NOTE:
A 74 unit load (ul) is understood to be 40µA I_{IH} and −1.6mA I_{IL}.

PIN CONFIGURATION

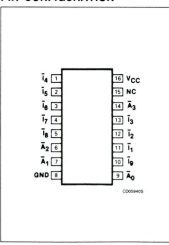

LOGIC SYMBOL

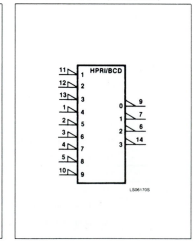

LOGIC SYMBOL (IEEE/IEC)

Encoder   74147

LOGIC DIAGRAM

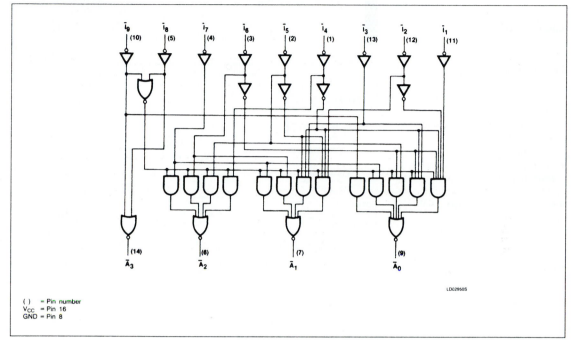

```
( )  = Pin number
VCC  = Pin 16
GND  = Pin 8
```

LD02950S

FUNCTION TABLE

INPUTS									OUTPUTS			
$\bar{I}_1$	$\bar{I}_2$	$\bar{I}_3$	$\bar{I}_4$	$\bar{I}_5$	$\bar{I}_6$	$\bar{I}_7$	$\bar{I}_8$	$\bar{I}_9$	$\bar{A}_3$	$\bar{A}_2$	$\bar{A}_1$	$\bar{A}_0$
H	H	H	H	H	H	H	H	H	H	H	H	H
X	X	X	X	X	X	X	X	L	L	H	H	L
X	X	X	X	X	X	X	L	H	L	H	H	H
X	X	X	X	X	X	L	H	H	H	L	L	L
X	X	X	X	X	L	H	H	H	H	L	L	H
X	X	X	X	L	H	H	H	H	H	L	H	L
X	X	X	L	H	H	H	H	H	H	L	H	H
X	X	L	H	H	H	H	H	H	H	H	L	L
X	L	H	H	H	H	H	H	H	H	H	L	H
L	H	H	H	H	H	H	H	H	H	H	H	L

H = HIGH voltage level
L = LOW voltage level
X = Don't care

Signetics

74148
Encoder

8-Input Priority Encoder
Product Specification

Logic Products

FEATURES
- Code conversions
- Multi-channel D/A converter
- Decimal-to-BCD converter
- Cascading for priority encoding of ''N'' bits
- Input Enable capability
- Priority encoding — automatic selection of highest priority input line
- Output Enable — active LOW when all inputs HIGH
- Group Signal output — active when any input is LOW

DESCRIPTION
The '148 8-input priority encoder accepts data from eight active-LOW inputs and provides a binary representation on the three active-LOW outputs. A priority is assigned to each input so that when two or more inputs are simultaneously active, the input with the highest priority is represented on the output, with input line $\bar{I}_7$ having the highest priority.

TYPE	TYPICAL PROPAGATION DELAY	TYPICAL SUPPLY CURRENT (TOTAL)
74148	10ns	38mA

ORDERING CODE

PACKAGES	COMMERCIAL RANGE $V_{CC} = 5V \pm 5\%$; $T_A = 0°C$ to $+70°C$
Plastic DIP	N74148N
Plastic SO	

NOTES:
For information regarding devices processed to Military Specifications, see the Signetics Military Products Data Manual.

INPUT AND OUTPUT LOADING AND FAN-OUT TABLE

PINS	DESCRIPTION	74
$\bar{I}_0$	Input	1ul
$\bar{I}_1 - \bar{I}_7$	Inputs	2ul
$\overline{EI}$	Input	2ul
All	Outputs	10ul

NOTE:
A 74 unit load (ul) is understood to be 40µA I_{IH} and −1.6mA I_{IL}.

PIN CONFIGURATION

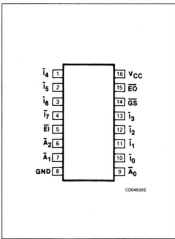

CD04630S

LOGIC SYMBOL

V_{CC} = Pin 16
GND = Pin 8

LS03840S

LOGIC SYMBOL (IEEE/IEC)

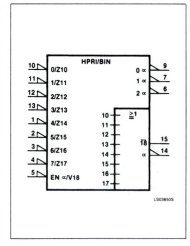

LS03850S

Encoder　　　　　　　　　　　　　　　　　　　　　　　　　　**74148**

A HIGH on the Enable Input ($\overline{EI}$) will force all outputs to the inactive (HIGH) state and allow new data to settle without producing erroneous information at the outputs.

A Group Signal ($\overline{GS}$) output and an Enable Output ($\overline{EO}$) are provided with the three data outputs. The $\overline{GS}$ is active-LOW when any input is LOW; this indicates when any input is active. The $\overline{EO}$ is active-LOW when all inputs are HIGH. Using the Enable Output along with the Enable Input allows priority encoding of N input signals. Both $\overline{EO}$ and $\overline{GS}$ are active-HIGH when the Enable input is HIGH.

LOGIC DIAGRAM

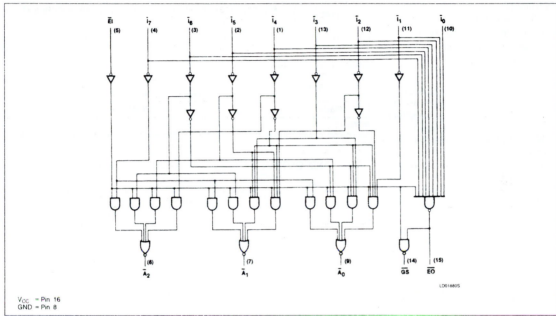

V$_{CC}$ = Pin 16
GND = Pin 8

LD01880S

FUNCTION TABLE

INPUTS									OUTPUTS				
$\overline{EI}$	$\overline{I_0}$	$\overline{I_1}$	$\overline{I_2}$	$\overline{I_3}$	$\overline{I_4}$	$\overline{I_5}$	$\overline{I_6}$	$\overline{I_7}$	$\overline{GS}$	$\overline{A_0}$	$\overline{A_1}$	$\overline{A_2}$	$\overline{EO}$
H	X	X	X	X	X	X	X	X	H	H	H	H	H
L	H	H	H	H	H	H	H	H	H	H	H	H	L
L	X	X	X	X	X	X	X	L	L	L	L	L	H
L	X	X	X	X	X	X	L	H	L	H	L	L	H
L	X	X	X	X	X	L	H	H	L	L	H	L	H
L	X	X	X	X	L	H	H	H	L	H	H	L	H
L	X	X	X	L	H	H	H	H	L	L	L	H	H
L	X	X	L	H	H	H	H	H	L	H	L	H	H
L	X	L	H	H	H	H	H	H	L	L	H	H	H
L	L	H	H	H	H	H	H	H	L	H	H	H	H

H = HIGH voltage level
L = LOW voltage level
X = Don't care

Signetics

74150
Multiplexer

16-Input Multiplexer
Product Specification

Logic Products

FEATURES
- Select data from 16 sources
- Demultiplexing capability
- Active-LOW enable or strobe
- Inverting data output

DESCRIPTION
The '150 is a logical implementation of a single-pole, 16-position switch with the switch position controlled by the state of four Select inputs. S_0, S_1, S_2, S_3. The Multiplexer output ($\overline{Y}$) inverts the selected data. The Enable input ($\overline{E}$) is active-LOW. When $\overline{E}$ is HIGH the $\overline{Y}$ output is HIGH regardless of all other inputs. In one package the '150 provides the ability to select from 16 sources of data or control information.

TYPE	TYPICAL PROPAGATION DELAY	TYPICAL SUPPLY CURRENT (TOTAL)
74150	17ns	40mA

ORDERING CODE

PACKAGES	COMMERCIAL RANGE $V_{CC} = 5V \pm 5\%$; $T_A = 0°C$ to $+70°C$
Plastic DIP	N74150N

NOTE:
For information regarding devices processed to Military Specifications, see the Signetics Military Products Data Manual.

INPUT AND OUTPUT LOADING AND FAN-OUT TABLE

PINS	DESCRIPTION	74
All	Inputs	1ul
$\overline{Y}$	Output	10ul

NOTE:
A 74 unit load (ul) is understood to be 40μA I_{IH} and −1.6mA I_{IL}.

PIN CONFIGURATION

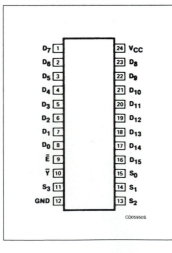

```
D7  [1]        [24]  Vcc
D6  [2]        [23]  D8
D5  [3]        [22]  D9
D4  [4]        [21]  D10
D3  [5]        [20]  D11
D2  [6]        [19]  D12
D1  [7]        [18]  D13
D0  [8]        [17]  D14
Ē   [9]        [16]  D15
Ȳ   [10]       [15]  S0
S3  [11]       [14]  S1
GND [12]       [13]  S2
```
CD05950S

LOGIC SYMBOL

Vcc = Pin 24
GND = Pin 12

LS06180S

LOGIC SYMBOL (IEEE/IEC)

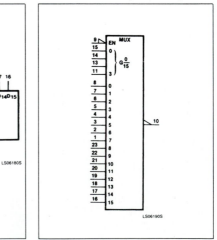

LS06190S

Multiplexer 74150

LOGIC DIAGRAM

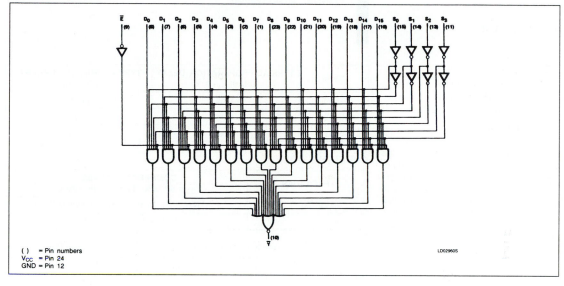

( ) = Pin numbers
V_{CC} = Pin 24
GND = Pin 12

LD02960S

FUNCTION TABLE

S_3	S_2	S_1	S_0	$\bar{E}$	D_0	D_1	D_2	D_3	D_4	D_5	D_6	D_7	D_8	D_9	D_{10}	D_{11}	D_{12}	D_{13}	D_{14}	D_{15}	$\bar{Y}$
										INPUTS											**OUTPUT**
X	X	X	X	H	X	X	X	X	X	X	X	X	X	X	X	X	X	X	X	X	H
L	L	L	L	L	L	X	X	X	X	X	X	X	X	X	X	X	X	X	X	X	H
L	L	L	L	L	H	X	X	X	X	X	X	X	X	X	X	X	X	X	X	X	L
L	L	L	H	L	X	L	X	X	X	X	X	X	X	X	X	X	X	X	X	X	H
L	L	L	H	L	X	H	X	X	X	X	X	X	X	X	X	X	X	X	X	X	L
L	L	H	L	L	X	X	L	X	X	X	X	X	X	X	X	X	X	X	X	X	H
L	L	H	L	L	X	X	H	X	X	X	X	X	X	X	X	X	X	X	X	X	L
L	L	H	H	L	X	X	X	L	X	X	X	X	X	X	X	X	X	X	X	X	H
L	L	H	H	L	X	X	X	H	X	X	X	X	X	X	X	X	X	X	X	X	L
L	H	L	L	L	X	X	X	X	L	X	X	X	X	X	X	X	X	X	X	X	H
L	H	L	L	L	X	X	X	X	H	X	X	X	X	X	X	X	X	X	X	X	L
L	H	L	H	L	X	X	X	X	X	L	X	X	X	X	X	X	X	X	X	X	H
L	H	L	H	L	X	X	X	X	X	H	X	X	X	X	X	X	X	X	X	X	L
L	H	H	L	L	X	X	X	X	X	X	L	X	X	X	X	X	X	X	X	X	H
L	H	H	L	L	X	X	X	X	X	X	H	X	X	X	X	X	X	X	X	X	L
L	H	H	H	L	X	X	X	X	X	X	X	L	X	X	X	X	X	X	X	X	H
L	H	H	H	L	X	X	X	X	X	X	X	H	X	X	X	X	X	X	X	X	L
H	L	L	L	L	X	X	X	X	X	X	X	X	L	X	X	X	X	X	X	X	H
H	L	L	L	L	X	X	X	X	X	X	X	X	H	X	X	X	X	X	X	X	L
H	L	L	H	L	X	X	X	X	X	X	X	X	X	L	X	X	X	X	X	X	H
H	L	L	H	L	X	X	X	X	X	X	X	X	X	H	X	X	X	X	X	X	L
H	L	H	L	L	X	X	X	X	X	X	X	X	X	X	L	X	X	X	X	X	H
H	L	H	L	L	X	X	X	X	X	X	X	X	X	X	H	X	X	X	X	X	L
H	L	H	H	L	X	X	X	X	X	X	X	X	X	X	X	L	X	X	X	X	H
H	L	H	H	L	X	X	X	X	X	X	X	X	X	X	X	H	X	X	X	X	L
H	H	L	L	L	X	X	X	X	X	X	X	X	X	X	X	X	L	X	X	X	H
H	H	L	L	L	X	X	X	X	X	X	X	X	X	X	X	X	H	X	X	X	L
H	H	L	H	L	X	X	X	X	X	X	X	X	X	X	X	X	X	L	X	X	H
H	H	L	H	L	X	X	X	X	X	X	X	X	X	X	X	X	X	H	X	X	L
H	H	H	L	L	X	X	X	X	X	X	X	X	X	X	X	X	X	X	L	X	H
H	H	H	L	L	X	X	X	X	X	X	X	X	X	X	X	X	X	X	H	X	L
H	H	H	H	L	X	X	X	X	X	X	X	X	X	X	X	X	X	X	X	L	H
H	H	H	H	L	X	X	X	X	X	X	X	X	X	X	X	X	X	X	X	H	L

H = HIGH voltage level
L = LOW voltage level
X = Don't care

Signetics

74151, LS151, S151
Multiplexers

8-Input Multiplexer
Product Specification

Logic Products

FEATURES
- Multifunction capability
- Complementary outputs
- See '251 for 3-state version

DESCRIPTION
The '151 is a logical implementation of a single-pole, 8-position switch with the switch position controlled by the state of three Select inputs, S_0, S_1, S_2. True (Y) and Complement ($\overline{Y}$) outputs are both provided. The Enable input ($\overline{E}$) is active LOW. When $\overline{E}$ is HIGH, the $\overline{Y}$ output is HIGH and the Y output is LOW, regardless of all other inputs. The logic function provided at the output is:

$$Y = \overline{E} \bullet (I_0 \bullet \overline{S}_0 \bullet \overline{S}_1 \bullet \overline{S}_2 + I_1 \bullet S_0 \bullet \overline{S}_1 \bullet \overline{S}_2 +$$
$$I_2 \bullet \overline{S}_0 \bullet S_1 \bullet \overline{S}_2 + I_3 \bullet S_0 \bullet S_1 \bullet \overline{S}_2 +$$
$$I_4 \bullet \overline{S}_0 \bullet \overline{S}_1 \bullet S_2 + I_5 \bullet S_0 \bullet \overline{S}_1 \bullet S_2 +$$
$$I_6 \bullet \overline{S}_0 \bullet S_1 \bullet S_2 + I_7 \bullet S_0 \bullet S_1 \bullet S_2$$

In one package the '151 provides the ability to select from eight sources of data or control information. The device can provide any logic function of four variables and its negation with correct manipulation.

TYPE	TYPICAL PROPAGATION DELAY (ENABLE TO $\overline{Y}$)	TYPICAL SUPPLY CURRENT (TOTAL)
74151	18ns	29mA
74LS151	12ns	6mA
74S151	9ns	45mA

ORDERING CODE

PACKAGES	COMMERCIAL RANGE $V_{CC} = 5V \pm 5\%$; $T_A = 0°C$ to $+70°C$
Plastic DIP	N74151N, N74LS151N, N74S151N
Plastic SO	N74LS151D, N74S151D

NOTE:
For information regarding devices processed to Military Specifications, see the Signetics Military Products Data Manual.

INPUT AND OUTPUT LOADING AND FAN-OUT TABLE

PINS	DESCRIPTION	74	74S	74LS
All	Inputs	1ul	1Sul	1LSul
All	Outputs	10ul	10Sul	10LSul

NOTE:
Where a 74 unit load (ul) is understood to be 40μA I_{IH} and -1.6mA I_{IL}, a 74S unit load (Sul) is 50μA I_{IH} and -2.0mA I_{IL}, and 74LS unit load (LSul) is 20μA I_{IH} and -0.4mA I_{IL}.

PIN CONFIGURATION

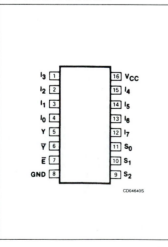

LOGIC SYMBOL

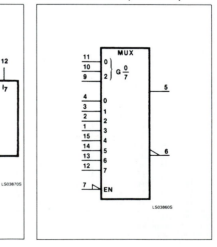

LOGIC SYMBOL (IEEE/IEC)

Multiplexers   **74151, LS151, S151**

LOGIC DIAGRAM

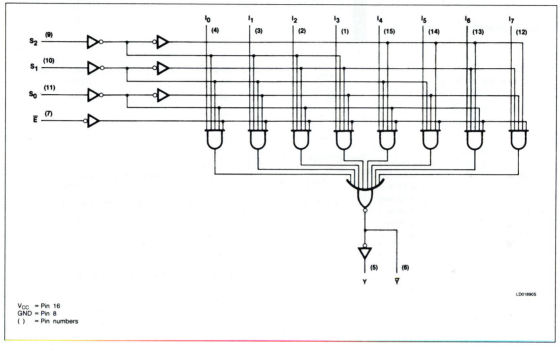

V_{CC} = Pin 16
GND = Pin 8
( ) = Pin numbers

FUNCTION TABLE

		INPUTS										OUTPUTS	
$\bar{E}$	S_2	S_1	S_0	I_0	I_1	I_2	I_3	I_4	I_5	I_6	I_7	$\bar{Y}$	Y
H	X	X	X	X	X	X	X	X	X	X	X	H	L
L	L	L	L	L	X	X	X	X	X	X	X	H	L
L	L	L	L	H	X	X	X	X	X	X	X	L	H
L	L	L	H	X	L	X	X	X	X	X	X	H	L
L	L	L	H	X	H	X	X	X	X	X	X	L	H
L	L	H	L	X	X	L	X	X	X	X	X	H	L
L	L	H	L	X	X	H	X	X	X	X	X	L	H
L	L	H	H	X	X	X	L	X	X	X	X	H	L
L	L	H	H	X	X	X	H	X	X	X	X	L	H
L	H	L	L	X	X	X	X	L	X	X	X	H	L
L	H	L	L	X	X	X	X	H	X	X	X	L	H
L	H	L	H	X	X	X	X	X	L	X	X	H	L
L	H	L	H	X	X	X	X	X	H	X	X	L	H
L	H	H	L	X	X	X	X	X	X	L	X	H	L
L	H	H	L	X	X	X	X	X	X	H	X	L	H
L	H	H	H	X	X	X	X	X	X	X	L	H	L
L	H	H	H	X	X	X	X	X	X	X	H	L	H

H = HIGH voltage level
L = LOW voltage level
X = Don't care

Signetics

74153, LS153, S153
Multiplexers

Dual 4-Line To 1-Line Multiplexer
Product Specification

Logic Products

FEATURES
- **Non-inverting outputs**
- **Separate enable for each section**
- **Common select inputs**
- **See '253 for 3-state version**

DESCRIPTION
The '153 is a dual 4-input multiplexer that can select 2 bits of data from up to eight (8) sources under control of the common Select inputs (S_0, S_1). The two 4-input multiplexer circuits have individual active LOW Enables ($\overline{E}_a$, $\overline{E}_b$) which can be used to strobe the outputs independently. Outputs (Y_a, Y_b) are forced LOW when the corresponding Enables ($\overline{E}_a$, $\overline{E}_b$) are HIGH.

$Y_a = \overline{E}_a \cdot (I_{0a} \cdot \overline{S}_1 \cdot \overline{S}_0 + I_{1a} \cdot \overline{S}_1 \cdot S_0 + I_{2a} \cdot S_1 \cdot \overline{S}_0 + I_{3a} \cdot S_1 \cdot S_2)$

$Y_b = \overline{E}_b \cdot (I_{0b} \cdot \overline{S}_1 \cdot \overline{S}_0 + I_{1b} \cdot \overline{S}_1 \cdot S_0 + I_{2b} \cdot S_1 \cdot \overline{S}_0 + I_{3b} \cdot S_1 \cdot S_2)$

TYPE	TYPICAL PROPAGATION DELAY	TYPICAL SUPPLY CURRENT (TOTAL)
74153	18ns	36mA
74LS153	18ns	6.2mA
74S153	9ns	45mA

ORDERING CODE

PACKAGES	COMMERCIAL RANGE $V_{CC} = 5V \pm 5\%$; $T_A = 0°C$ to $+70°C$
Plastic DIP	N74153N, N74LS153N, N74S153N
Plastic SO	N74LS153D, N74S153D

NOTE:
For information regarding devices processed to Military Specifications, see the Signetics Military Products Data Manual.

INPUT AND OUTPUT LOADING AND FAN-OUT TABLE

PINS	DESCRIPTION	74	74S	74LS
All	Inputs	1ul	1Sul	1LSul
All	Outputs	10ul	10Sul	10LSul

NOTE:
Where a 74 unit load (ul) is understood to be 40μA I_{IH} and −1.6mA I_{IL}, a 74S unit load (Sul) is 50μA I_{IH} and −2.0mA I_{IL}, and 74LS unit load (LSul) is 20μA I_{IH} and −0.4mA I_{IL}.

PIN CONFIGURATION

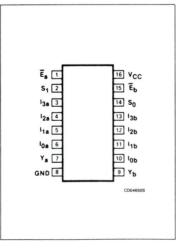

LOGIC SYMBOL

LOGIC SYMBOL (IEEE/IEC)

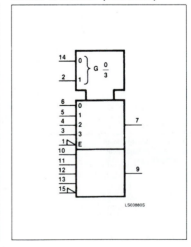

Multiplexers   74153, LS153, S153

The '153 can be used to move data to a common output bus from a group of registers. The state of the Select inputs would determine the particular register from which the data came. An alternative application is as a function generator. The device can generate two functions or three variables. This is useful for implementing highly irregular random logic.

LOGIC DIAGRAM

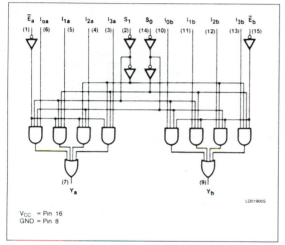

V_{CC} = Pin 16
GND = Pin 8

FUNCTION TABLE

SELECT INPUTS		INPUTS (a or b)					OUTPUT
S_0	S_1	$\bar{E}$	I_0	I_1	I_2	I_3	Y
X	X	H	X	X	X	X	L
L	L	L	L	X	X	X	L
L	L	L	H	X	X	X	H
H	L	L	X	L	X	X	L
H	L	L	X	H	X	X	H
L	H	L	X	X	L	X	L
L	H	L	X	X	H	X	H
H	H	L	X	X	X	L	L
H	H	L	X	X	X	H	H

H = HIGH voltage level
L = LOW voltage level
X = Don't care

ABSOLUTE MAXIMUM RATINGS (Over operating free-air temperature range unless otherwise noted.)

PARAMETER		74	74LS	74S	UNIT
V_{CC}	Supply voltage	7.0	7.0	7.0	V
V_{IN}	Input voltage	−0.5 to +5.5	−0.5 to +7.0	−0.5 to +5.5	V
I_{IN}	Input current	−30 to +5	−30 to +1	−30 to +5	mA
V_{OUT}	Voltage applied to output in HIGH output state	−0.5 to +V_{CC}	−0.5 to +V_{CC}	−0.5 to +V_{CC}	V
T_A	Operating free-air temperature range		0 to 70		°C

RECOMMENDED OPERATING CONDITIONS

PARAMETER		74			74LS			74S			UNIT
		Min	Nom	Max	Min	Nom	Max	Min	Nom	Max	
V_{CC}	Supply voltage	4.75	5.0	5.25	4.75	5.0	5.25	4.75	5.0	5.25	V
V_{IH}	HIGH-level input voltage	2.0			2.0			2.0			V
V_{IL}	LOW-level input voltage			+0.8			+0.8			+0.8	V
I_{IK}	Input clamp current			−12			−18			−18	mA
I_{OH}	HIGH-level output current			−800			−400			−1000	μA
I_{OL}	LOW-level output current			16			8			20	mA
T_A	Operating free-air temperature	0		70	0		70	0		70	°C

Signetics

Logic Products

74154, LS154
Decoder/Demultiplexers

1-of-16 Decoder/Demultiplexer
Product Specification

FEATURES
- 16-line demultiplexing capability
- Mutually exclusive outputs
- 2-input enable gate for strobing or expansion

DESCRIPTION
The '154 decoder accepts four active HIGH binary address inputs and provides 16 mutually exclusive active LOW outputs. The 2-input enable gate can be used to strobe the decoder to eliminate the normal decoding "glitches" on the outputs, or it can be used for expansion of the decoder. The enable gate has two AND'ed inputs which must be LOW to enable the outputs.

The '154 can be used as a 1-of-16 demultiplexer by using one of the enable inputs as the multiplexed data input. When the other enable is LOW, the addressed output will follow the state of the applied data.

TYPE	TYPICAL PROPAGATION DELAY	TYPICAL SUPPLY CURRENT (TOTAL)
74154	21ns	34mA
74LS154	15ns	9mA

ORDERING CODE

PACKAGES	COMMERCIAL RANGE $V_{CC} = 5V \pm 5\%$; $T_A = 0°C$ to $+70°C$
Plastic DIP	N74154N, N74LS154N

NOTE:
For information regarding devices processed to Military Specifications, see the Signetics Military Products Data Manual.

INPUT AND OUTPUT LOADING AND FAN-OUT TABLE

PINS	DESCRIPTION	74	74LS
All	Inputs	1ul	1LSul
All	Outputs	10ul	10LSul

NOTE:
Where a 74 unit load (ul) is understood to be $40\mu A$ I_{IH} and $-1.6mA$ I_{IL}, and a 74LS unit load (LSul) is $20\mu A$ I_{IH} and $-0.4mA$ I_{IL}.

PIN CONFIGURATION

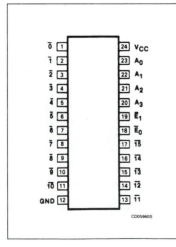

CD05960S

LOGIC SYMBOL

LS06200S

V_{CC} = Pin 24
GND = Pin 12

LOGIC SYMBOL (IEEE/IEC)

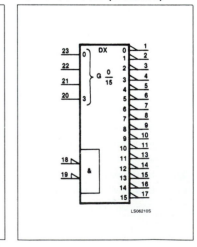

LS06210S

Decoder/Demultiplexers   74154, LS154

LOGIC DIAGRAM

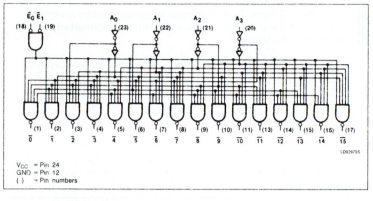

V_{CC} = Pin 24
GND = Pin 12
( ) = Pin numbers

LD029705

FUNCTION TABLE

INPUTS						OUTPUT															
E_0	E_1	A_3	A_2	A_1	A_0	0	1	2	3	4	5	6	7	8	9	10	11	12	13	14	15
L	H	X	X	X	X	H	H	H	H	H	H	H	H	H	H	H	H	H	H	H	H
H	L	X	X	X	X	H	H	H	H	H	H	H	H	H	H	H	H	H	H	H	H
H	H	X	X	X	X	H	H	H	H	H	H	H	H	H	H	H	H	H	H	H	H
L	L	L	L	L	L	L	H	H	H	H	H	H	H	H	H	H	H	H	H	H	H
L	L	L	L	L	H	H	L	H	H	H	H	H	H	H	H	H	H	H	H	H	H
L	L	L	L	H	L	H	H	L	H	H	H	H	H	H	H	H	H	H	H	H	H
L	L	L	L	H	H	H	H	H	L	H	H	H	H	H	H	H	H	H	H	H	H
L	L	L	H	L	L	H	H	H	H	L	H	H	H	H	H	H	H	H	H	H	H
L	L	L	H	L	H	H	H	H	H	H	L	H	H	H	H	H	H	H	H	H	H
L	L	L	H	H	L	H	H	H	H	H	H	L	H	H	H	H	H	H	H	H	H
L	L	L	H	H	H	H	H	H	H	H	H	H	L	H	H	H	H	H	H	H	H
L	L	H	L	L	L	H	H	H	H	H	H	H	H	L	H	H	H	H	H	H	H
L	L	H	L	L	H	H	H	H	H	H	H	H	H	H	L	H	H	H	H	H	H
L	L	H	L	H	L	H	H	H	H	H	H	H	H	H	H	L	H	H	H	H	H
L	L	H	L	H	H	H	H	H	H	H	H	H	H	H	H	H	L	H	H	H	H
L	L	H	H	L	L	H	H	H	H	H	H	H	H	H	H	H	H	L	H	H	H
L	L	H	H	L	H	H	H	H	H	H	H	H	H	H	H	H	H	H	L	H	H
L	L	H	H	H	L	H	H	H	H	H	H	H	H	H	H	H	H	H	H	L	H
L	L	H	H	H	H	H	H	H	H	H	H	H	H	H	H	H	H	H	H	H	L

H = HIGH voltage level
L = LOW voltage level
X = Don't care

ABSOLUTE MAXIMUM RATINGS (Over operating free-air temperature range unless otherwise noted.)

	PARAMETER	74	74LS	UNIT
V_{CC}	Supply voltage	7.0	7.0	V
V_{IN}	Input voltage	−0.5 to +5.5	−0.5 to +7.0	V
I_{IN}	Input current	−30 to +5	−30 to +1	mA
V_{OUT}	Voltage applied to output in HIGH output state	−0.5 to +V_{CC}	−0.5 to +V_{CC}	V
T_A	Operating free-air temperature range	0 to 70		°C

Signetics

74157, 74158, LS157, LS158, S157, S158
Data Selectors/Multiplexers

'157 Quad 2-Input Data Selector/Multiplexer (Non-Inverted)
'158 Quad 2-Input Data Selector/Multiplexer (Inverted)
Product Specification

Logic Products

DESCRIPTION

The '157 is a quad 2-input multiplexer which selects four bits of data from two sources under the control of a common Select input (S). The Enable input ($\overline{E}$) is active LOW. When $\overline{E}$ is HIGH, all of the outputs (Y) are forced LOW regardless of all other input conditions.

Moving data from two groups of registers to four common output busses is a common use of the '157. The state of the Select input determines the particular register from which the data comes. It can also be used as a function generator. The device is useful for implementing highly irregular logic by generating any four of the 16 different functions of two variables with one variable common.

TYPE	TYPICAL PROPAGATION DELAY	TYPICAL SUPPLY CURRENT (TOTAL)
74157	13ns	30mA
74LS157	13ns	9.7mA
74S157	7.4ns	50mA
74158	13ns	30mA
74LS158	13ns	4.8mA
74S158	6ns	40mA

ORDERING CODE

PACKAGES	COMMERCIAL RANGES $V_{CC} = 5V \pm 5\%$; $T_A = 0°C$ to $+70°C$
Plastic DIP	N74157N, N74LS158N, N74S157N N74LS157N, N74S158N, N74LS158N
Plastic SO	N74LS157D, N74S158D

NOTE:
For information regarding devices processed to Military Specifications, see the Signetics Military Products Data Manual.

INPUT AND OUTPUT LOADING AND FAN-OUT TABLE

PINS	DESCRIPTION	74	74S	74LS
S, $\overline{E}$	Inputs	1ul	2Sul	2LSul
Data	Inputs	1ul	1Sul	1LSul
All	Outputs	10ul	10Sul	10LSul

NOTE:
Where a 74 unit load (ul) is understood to be 40μA I_{IH} and -1.6mA I_{IL}, a 74S unit load (Sul) is 50μA I_{IH} and -2.0mA I_{IL}, and a74LS unit load (LSul) is 20μA I_{IH} and -0.4mA I_{IL}.

PIN CONFIGURATION

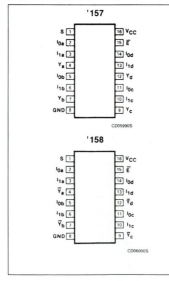

LOGIC SYMBOL

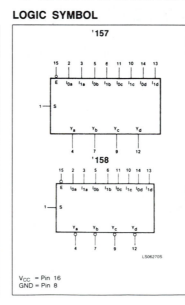

V_{CC} = Pin 16
GND = Pin 8

LOGIC SYMBOL (IEEE/IEC)

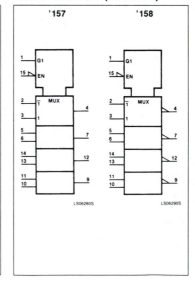

Data Selectors/Multiplexers   74157, 74158, LS157, LS158, S157, S158

LOGIC DIAGRAM, '157

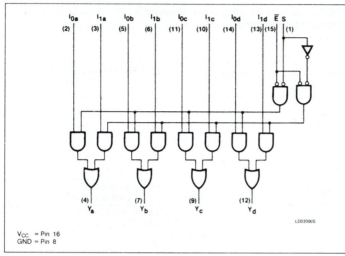

V_{CC} = Pin 16
GND = Pin 8

LD03000S

LOGIC DIAGRAM, '158

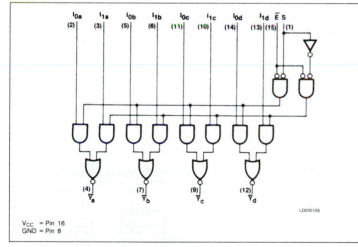

V_{CC} = Pin 16
GND = Pin 8

LD03010S

The device is the logic implementation of a 4-pole, 2-position switch where the position of the switch is determined by the logic levels supplied to the Select input. Logic equations for the outputs are shown below:

$$Y_a = \overline{E} \cdot (I_{1a} \cdot S + I_{0a} \cdot \overline{S})$$
$$Y_b = \overline{E} \cdot (I_{1b} \cdot S + I_{0b} \cdot \overline{S})$$
$$Y_c = \overline{E} \cdot (I_{1c} \cdot S + I_{0c} \cdot \overline{S})$$
$$Y_d = \overline{E} \cdot (I_{1d} \cdot S + I_{0d} \cdot \overline{S})$$

The '158 is similar but has inverting outputs:

$$\overline{Y}_a = \overline{E} \cdot (I_{1a} \cdot S + I_{0a} \cdot \overline{S})$$
$$\overline{Y}_b = \overline{E} \cdot (I_{1b} \cdot S + I_{0b} \cdot \overline{S})$$
$$\overline{Y}_c = \overline{E} \cdot (I_{1c} \cdot S + I_{0c} \cdot \overline{S})$$
$$\overline{Y}_d = \overline{E} \cdot (I_{1d} \cdot S + I_{0d} \cdot \overline{S})$$

FUNCTION TABLE, '157

ENABLE	SELECT INPUT	DATA INPUTS		OUTPUT
$\overline{E}$	S	I_0	I_1	Y
H	X	X	X	L
L	H	X	L	L
L	H	X	H	H
L	L	L	X	L
L	L	H	X	H

H = HIGH voltage level
L = LOW voltage level
X = Don't care

FUNCTION TABLE, '158

ENABLE	SELECT INPUT	DATA INPUTS		OUTPUT
$\overline{E}$	S	I_0	I_1	$\overline{Y}$
H	X	X	X	H
L	L	L	X	H
L	L	H	X	L
L	H	X	L	H
L	H	X	H	L

H = HIGH voltage level
L = LOW voltage level
X = Don't care

ABSOLUTE MAXIMUM RATINGS (Over operating free-air temperature range unless otherwise noted.)

PARAMETER		74	74LS	74S	UNIT
V_{CC}	Supply voltage	7.0	7.0	7.0	V
V_{IN}	Input voltage	−0.5 to +5.5	−0.5 to +7.0	−0.5 to +5.5	V
I_{IN}	Input current	−30 to +5	−30 to +1	−30 to +5	mA
V_{OUT}	Voltage applied to output in HIGH output state	−0.5 to +V_{CC}	−0.5 to +V_{CC}	−0.5 to +V_{CC}	V
T_A	Operating free-air temperature range		0 to 70		°C

Signetics

Logic Products

74LS168A, 74LS169A, S168A, S169A
4-Bit Bidirectional Counters

4-Bit Up/Down Synchronous Counter
Product Specification

FEATURES
- **Synchronous counting and loading**
- **Up/down counting**
- **Modulo 16 binary counter — '169A**
- **BCD decade counter — '168A**
- **Two Count Enable inputs for n-bit cascading**
- **Positive edge-triggered clock**

DESCRIPTION
The '168A is a synchronous, presettable BCD decade up/down counter featuring an internal carry look-ahead for applications in high-speed counting designs. Synchronous operation is provided by having all flip-flops clocked simultaneously so that the outputs change coincident with each other when so instructed by the Count Enable inputs and internal gating. This mode of operation eliminates the output spikes which are normally associated with asynchronous (ripple clock) counters. A buffered Clock input triggers the flip-flops on the LOW-to-HIGH transition of the clock.

TYPE	TYPICAL f_{MAX}	TYPICAL SUPPLY CURRENT (TOTAL)
74LS168A	32MHz	20mA
74S168A	70MHz	100mA
74LS169A	32MHz	20mA
74S169A	70MHz	100mA

ORDERING CODE

PACKAGES	COMMERCIAL RANGE V_{CC} = 5V ±5%; T_A = 0°C to +70°C
Plastic DIP	N74LS168AN, N74S168AN N74LS169AN, N74S169AN
Plastic SO	N74LS169AD, N74LS169AD, N74S169AD

NOTE:
For information regarding devices processed to Military Specifications, see the Signetics Military Products Data Manual.

INPUT AND OUTPUT LOADING AND FAN-OUT TABLE

PINS	DESCRIPTION	74S	74LS
$\overline{PE}$	Input	1Sul	2LSul
$\overline{CET}$	Input	2Sul	1LSul
Other	Inputs	1Sul	1LSul
All	Outputs	10Sul	10LSul

NOTE:
Where a 74S unit load (Sul) is understood to be 50µA I_{IH} and –2.0mA I_{IL} and a 74LS unit load (LSul) is 20µA I_{IH} and –0.4mA I_{IL}.

PIN CONFIGURATION

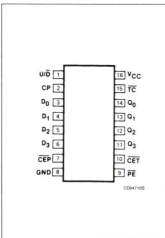

LOGIC SYMBOL

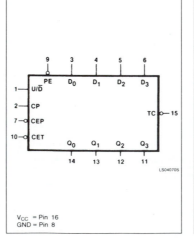

LOGIC SYMBOL (IEEE/IEC)

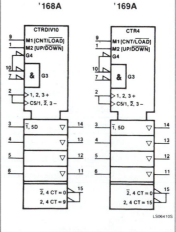

4-Bit Bidirectional Counters   74LS168A, 74LS169A, S168A, S169A

The counter is fully programmable; that is, the outputs may be preset to either level. Presetting is synchronous with the clock and takes place regardless of the levels of the Count Enable inputs. A LOW level on the Parallel Enable ($\overline{PE}$) input disables the counter and causes the data at the D_n input to be loaded into the counter on the next LOW-to-HIGH transition of the clock.

The direction of counting is controlled by the Up/Down (U/$\overline{D}$) input; a HIGH will cause the

count to increase, a LOW will cause the count to decrease.

The carry look-ahead circuitry provides for cascading counters for n-bit synchronous applications without additional gating. Instrumental in accomplishing this function are two Count Enable inputs ($\overline{CET} \cdot \overline{CEP}$) and a Terminal Count ($\overline{TC}$) output. Both Count Enable inputs must be LOW to count. The $\overline{CET}$ input is fed forward to enable the $\overline{TC}$ output. The $\overline{TC}$ output thus enabled will produce a LOW

output pulse with a duration approximately equal to the HIGH level portion of the Q_0 output. This LOW level $\overline{TC}$ pulse is used to enable successive cascaded stages. See Figure A for the fast synchronous multistage counting connections.

The '169A is identical except that it is a Modulo 16 counter.

LOGIC DIAGRAM, '168A

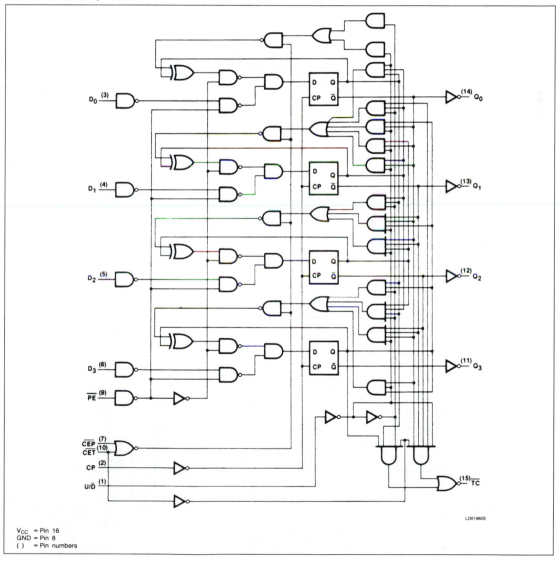

V_{CC} = Pin 16
GND = Pin 8
( ) = Pin numbers

LD01960S

4-Bit Bidirectional Counters   74LS168A, 74LS169A, S168A, S169A

LOGIC DIAGRAM, '169A

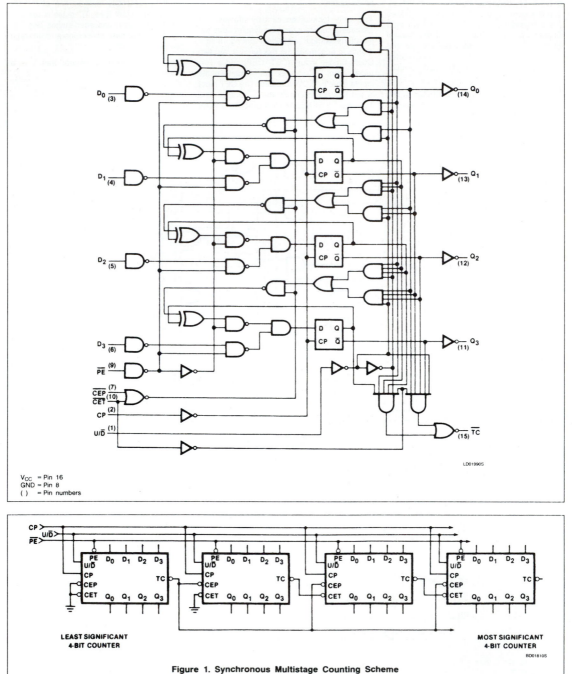

V_{CC} = Pin 16
GND = Pin 8
( ) = Pin numbers

LD01990S

Figure 1. Synchronous Multistage Counting Scheme

BD01810S

4-Bit Bidirectional Counters 74LS168A, 74LS169A, S168A, S169A

MODE SELECT — FUNCTION TABLE

OPERATING MODE	INPUTS						OUTPUTS	
	CP	U/$\overline{\text{D}}$	$\overline{\text{CEP}}$	$\overline{\text{CET}}$	$\overline{\text{PE}}$	D_n	Q_n	$\overline{\text{TC}}$
Parallel Load	↑	X	X	X	l	i	L	(1)
	↑	X	X	X	i	h	H	(1)
Count Up	↑	h	l	l	h	X	Count Up	(1)
Count Down	↑	l	l	l	h	X	Count Down	(1)
Hold (do nothing)	↑	X	h	X	h	X	q_n	(1)
	↑	X	X	h	h	X	q_n	H

H = HIGH voltage level steady state
h = HIGH voltage level one setup time prior to the LOW-to-HIGH clock transition
L = LOW voltage level steady state
l = LOW voltage level one setup time prior to the LOW-to-HIGH clock transition
X = Don't care
q = Lower case letters indicate the state of the referenced output prior to the LOW-to-HIGH clock transition
↑ = LOW-to-HIGH clock transition

NOTE:
1. The $\overline{\text{TC}}$ is LOW when $\overline{\text{CET}}$ is LOW and the counter is at Terminal Count. Terminal Count Up is (HHHH) and Terminal Count Down is (LLLL) for '169A.
   The $\overline{\text{TC}}$ is LOW when $\overline{\text{CET}}$ is LOW and the counter is at Terminal Count. Terminal Count Up is (HLLH) and Terminal Count Down is (LLLL) for '168A.

WAVEFORM (Typical Load, Count, and Inhibit Sequences)

Illustrated below is the following sequence for the '168A. The operation of the '169A is similar.

1. Load (preset) to BCD seven.
2. Count up to eight, nine (maximum), zero, one, and two.
3. Inhibit.
4. Count down to one, zero (minimum), nine, eight, and seven.

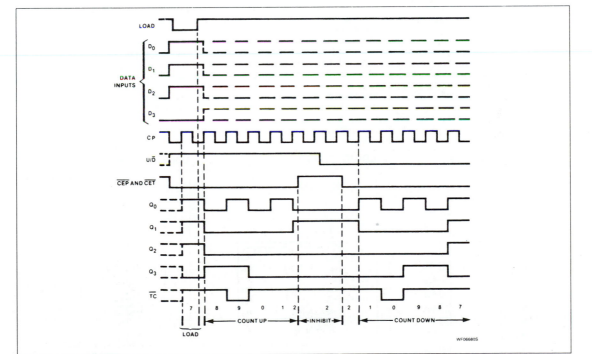

WF06680S

Signetics

74S172
Register File

16-Bit Multiple Port Register File (3-State)
Product Specification

Logic Products

FEATURES
- Simultaneous and independent Read and Write operations
- Expandable to 1024 words on n-bits
- 3-State outputs

DESCRIPTION
The '172 is a high-performance, 16-bit multiport register file with 3-State outputs organized as eight words of two bits each. Multiple address decoding circuitry is used so that the read and write operation can be performed independently on up to three word locations. Data can be written into two word locations through input Port "A" or input Port "C" while data is simultaneously read from both output Port "B" and output Port "C".

Port "A" is an input port which can be used to write two bits of data (D_{A0}, D_{A1}) into one of eight register locations selected by the Address inputs (A_{A0}, A_{A1}, A_{A2}). When the Write Enable ($\overline{WE}_A$) input is LOW one set-up time prior to the LOW-to-HIGH transition of the Clock (CP) input, the data is written into the selected location.

TYPE	TYPICAL f_{MAX}	TYPICAL SUPPLY CURRENT (TOTAL)
74S172	40MHz	160mA

ORDERING CODE

PACKAGES	COMMERCIAL RANGE $V_{CC} = 5V \pm 5\%$; $T_A = 0°C$ to $+70°C$
Plastic DIP	N74S172N

NOTE:
For information regarding devices processed to Military Specifications, see the Signetics Military Products Data Manual.

INPUT AND OUTPUT LOADING AND FAN-OUT TABLE

PINS	DESCRIPTION	74S
All	Inputs	1Sul
All	Outputs	8Sul

NOTE:
A 74S unit load (Sul) is 50μA and I_{IH} –2.0mA I_{IL}.

Port "B" is an output port which can be used to read two bits of data from one of eight register locations selected by the Address inputs (A_{B0}, A_{B1}, A_{B2}). When the Read Enable ($\overline{RE}_B$) is LOW, the selected 2-bit word appears on outputs Q_{B0} and Q_{B1}. When $\overline{RE}_B$ is HIGH, the Q_{B0} and Q_{B1} outputs are in the HIGH impedance "off" state. The read operation is independent of the clock.

Port "C" is a read/write port that has separate Data input and Data output sections, but common Address inputs (A_{C0}, A_{C1}, A_{C2}). Data can be simultaneously written into and read from the same register location. Port "C" can be used to write data into one location while Port "A" is writing into a different location, but data cannot be written reliably into the same location simultaneously.

PIN CONFIGURATION

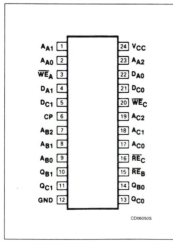

CD06050S

LOGIC SYMBOL

LS06440S

V_{CC} = Pin 24
GND = Pin 12

LOGIC SYMBOL (IEEE/IEC)

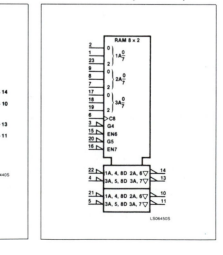

LS06450S

Register File 74S172

If both Ports "A" and "C" are enabled for writing into the same location during the same clock cycle, the LOW data will predominate if there is a conflict.

The register operation is essentially a master-slave flip-flop. Each master acts as a transparent D latch when selected by the "A" or "C" address and the clock and applicable write enable are LOW. The data in the master is transferred to the slave (or output section) following the LOW-to-HIGH transition of the Clock (CP). The Address inputs must be stable while the Clock and Write Enable inputs are LOW to ensure retention of data previously written into the other locations. Any number of masters can be altered while the clock and write enable are LOW, but the new data will not be loaded into the slaves, or be available at the outputs, until the clock goes HIGH.

BLOCK DIAGRAM

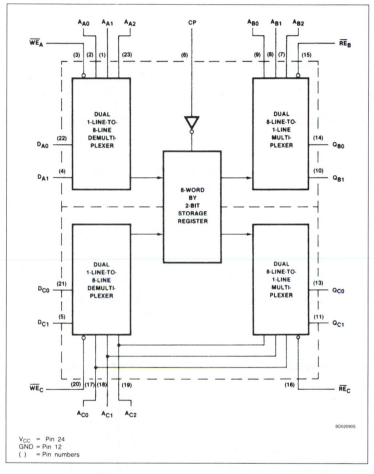

V_{CC} = Pin 24
GND = Pin 12
( ) = Pin numbers

BD02090S

WRITE MODE SELECT TABLE

OPERATING MODE	INPUTS			ADDRESSED REGISTER
	$\overline{CP}$	$\overline{WE}$	D_n	
Write data[a]	↑	l	l	L
	↑	l	h	H
Hold[b]	↓	h	X	no change

READ MODE SELECT TABLE

OPERATING MODE	INPUTS		OUTPUTS
	$\overline{RE}$	Addressed Register	Q_n
Read	L	L	L
	L	H	H
Disabled	H	X	(Z)

H = HIGH voltage level steady state.
= HIGH voltage level one set-up time prior
h to the LOW-to-HIGH or HIGH-to-LOW clock transition.
L = LOW voltage level steady state.
= LOW voltage level one set-up time prior to
l the LOW-to-HIGH clock transition.
X = Don't care.
(Z) = HIGH impedance (off) state.
↑ = LOW-to-HIGH clock transition.
↓ = HIGH-to-LOW clock transition.

NOTES:
a. The Write Address (A_A and A_C) to the "internal register" must be stable while $\overline{WE}$ and CP are LOW for conventional operation.
b. The Write Enable must be HIGH before the HIGH-to-LOW clock transition to ensure that the data in the register is not changed.

ABSOLUTE MAXIMUM RATINGS (Over operating free-air temperature range unless otherwise noted.)

PARAMETER		74S	UNIT
V_{CC}	Supply voltage	7.0	V
V_{IN}	Input voltage	−0.5 to +5.5	V
I_{IN}	Input current	−30 to +5	mA
V_{OUT}	Voltage applied to output in HIGH output state	−0.5 to +V_{CC}	V
T_A	Operating free-air temperature range	0 to 70	°C

Signetics

74173, LS173
Flip-Flops

Quad D-Type Flip-Flop With 3-State Outputs
Product Specification

Logic Products

FEATURES
- Edge-triggered D-type register
- Gated Input enable for hold "do nothing" mode
- 3-State output buffers
- Gated output enable control
- Pin compatible with the 8T10 and DM8551

DESCRIPTION
The '173 is a 4-bit parallel load register with clock enable control, 3-State buffered outputs and master reset. When the two Clock Enable ($\overline{E}_1$ and $\overline{E}_2$) inputs are LOW, the data on the D inputs is loaded into the register synchronously with the LOW-to-HIGH Clock (CP) transition. When one or both $\overline{E}$ inputs are HIGH one set-up time before the LOW-to-HIGH clock transition, the register will retain the previous data. Data inputs and Clock Enable inputs are fully edge triggered and must be stable only one set-up time before the LOW-to-HIGH clock transition.

The Master Reset (MR) is an active HIGH asynchronous input. When the MR is HIGH, all four flip-flops are reset (cleared) independently of any other input condition.

TYPE	TYPICAL f_{MAX}	TYPICAL SUPPLY CURRENT (TOTAL)
74173	35MHz	50mA
74LS173	50MHz	20mA

ORDERING CODE

PACKAGES	COMMERCIAL RANGE $V_{CC} = 5V \pm 5\%$; $T_A = 0°C$ to $+70°C$
Plastic DIP	N74173N, N74LS173N
Plastic SO-16	N74LS173D
Plastic SOL-16	CD7186D

NOTE:
For information regarding devices processed to Military Specifications, see the Signetics Military Products Data Manual.

INPUT AND OUTPUT LOADING AND FAN-OUT TABLE

PINS	DESCRIPTION	74	74LS
All	Inputs	1ul	1LSul
All	Outputs	10ul	30LSul

NOTE:
Where a 74 unit load (ul) is understood to be 40μA I_{IH} and −1.6mA I_{IL} and a 74LS unit load (LSul) is 20μA I_{IH} and −0.4mA I_{IL}.

The 3-State output buffers are controlled by a 2-input NOR gate. When both Output Enable ($\overline{OE}_1$ and $\overline{OE}_2$) inputs are LOW, the data in the register is presented at the Q outputs. When one or both $\overline{OE}$ inputs is HIGH, the outputs are forced to a HIGH impedance "off" state. The 3-State output buffers are completely independent of the register operation; the $\overline{OE}$ transition does not affect the clock and reset operations.

PIN CONFIGURATION

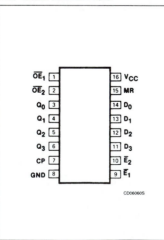

LOGIC SYMBOL

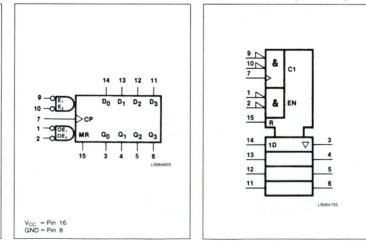

V_{CC} = Pin 16
GND = Pin 8

LOGIC SYMBOL (IEEE/IEC)

Flip-Flops   74173, LS173

LOGIC DIAGRAM

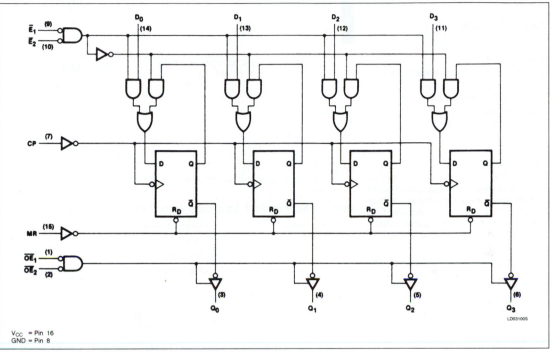

V_{CC} = Pin 16
GND = Pin 8

LD03100S

MODE SELECT — FUNCTION TABLE

REGISTER OPERATING MODES	INPUTS					OUTPUTS
	MR	CP	$\overline{E}_1$	$\overline{E}_2$	D_n	Q_n (Register)
Reset (clear)	H	X	X	X	X	L
Parallel load	L	↑	l	l	l	L
	L	↑	l	l	h	H
Hold (no change)	L	X	h	X	X	q_n
	L	X	X	h	X	q_n

3-STATE BUFFER OPERATING MODES	INPUTS			OUTPUTS
	Q_n (Register)	$\overline{OE}_1$	$\overline{OE}_2$	Q_0, Q_1, Q_2, Q_3
Read	L	L	L	L
	H	L	L	H
Disabled	X	H	X	(Z)
	X	X	H	(Z)

H = HIGH voltage level.
h = HIGH voltage level one set-up time prior to the LOW-to-HIGH clock transition.
L = LOW voltage level.
l = LOW voltage level one set-up time prior to the LOW-to-HIGH clock transition.
q_n = Lower case letters indicate the state of the referenced input (or output) on set-up time prior to the LOW-to-HIGH clock transition.
X = Don't care.
(Z) = HIGH impedance "off" state.
↑ = LOW-to-HIGH clock transition.

Signetics

74180
Parity Generator/Checker

9-Bit Odd/Even Parity Generator/Checker
Product Specification

Logic Products

FEATURES
- Word length easily expanded by cascading
- Generate even or odd parity
- Checks for parity errors
- See '280 for faster parity checker

TYPE	TYPICAL PROPAGATION DELAY, P_O = 0V	TYPICAL SUPPLY CURRENT
74180	36ns	34mA

ORDERING CODE

PACKAGES	COMMERCIAL RANGE V_{CC} = 5V ±5%; T_A = 0°C to +70°C
Plastic DIP	N74180N

NOTE:
For information regarding devices processed to Military Specifications, see the Signetics Military Products Data Manual.

DESCRIPTION
The '180 is a 9-bit parity generator or checker commonly used to detect errors in high speed data transmission or data retrieval systems. Both Even and Odd parity enable inputs and parity outputs are available for generating or checking parity on 8-bits.

True active-HIGH or true active-LOW parity can be generated at both the Even and Odd outputs. True active-HIGH parity is established with Even Parity enable input (P_E) set HIGH and the Odd Parity enable input (P_O) set LOW. True active-LOW parity is established when P_E is LOW and P_O is HIGH. When both enable inputs are at the same logic level, both outputs will be forced to the opposite logic level.

Parity checking of a 9-bit word (8 bits plus parity) is possible by using the two

INPUT AND OUTPUT LOADING AND FAN-OUT TABLE

PINS	DESCRIPTION	74
$I_0 - I_7$	Data inputs	1ul
P_E, P_O	Parity inputs	2ul
Σ_E, Σ_O	Parity outputs	10ul

NOTE:
A 74 unit load (ul) is understood to be 40μA I_{IH} and −1.6mA I_{IL}.

enable inputs plus an inverter as the ninth data input. To check for true active-HIGH parity, the ninth data input is tied to the P_O input and an inverter is connected between the P_O and P_E inputs. To check for true active-LOW parity, the ninth data input is tied to the P_E input and an inverter is connected between the P_E and P_O inputs.

Expansion to larger word sizes is accomplished by serially cascading the '180 in 8-bit increments. The Even and Odd parity outputs of the first stage are connected to the corresponding P_E and P_O inputs, respectively, of the succeeding stage.

PIN CONFIGURATION

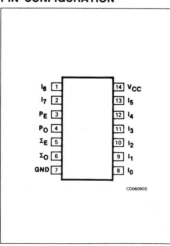

CD06080S

LOGIC SYMBOL

V_{CC} = Pin 14
GND = Pin 7

LS06500S

LOGIC SYMBOL (IEEE/IEC)

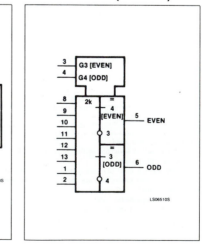

LS06510S

Parity Generator/Checker   74180

LOGIC DIAGRAM

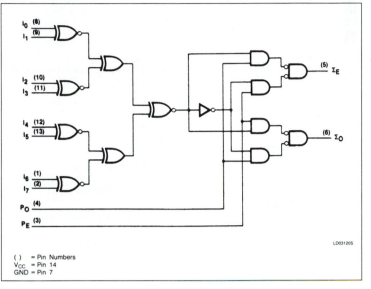

LD03120S

( ) = Pin Numbers
V_{CC} = Pin 14
GND = Pin 7

FUNCTION TABLE

INPUTS			OUTPUTS	
Number of HIGH Data Inputs $(I_0 - I_7)$	P_E	P_O	Σ_E	Σ_O
Even	H	L	H	L
Odd	H	L	L	H
Even	L	H	L	H
Odd	L	H	H	L
X	H	H	L	L
X	L	L	H	H

H = HIGH voltage level
L = LOW voltage level
X = Don't care

ABSOLUTE MAXIMUM RATINGS (Over operating free-air temperature range unless otherwise noted.)

PARAMETER		74	UNIT
V_{CC}	Supply voltage	7.0	V
V_{IN}	Input voltage	−0.5 to +5.5	V
I_{IN}	Input current	−30 to +5	mA
V_{OUT}	Voltage applied to output in HIGH output state	−0.5 to +V_{CC}	V
T_A	Operating free-air temperature range	0 to 70	°C

RECOMMENDED OPERATING CONDITIONS

PARAMETER		74			UNIT
		Min	Nom	Max	
V_{CC}	Supply voltage	4.75	5.0	5.25	V
V_{IH}	HIGH-level input voltage	2.0			V
V_{IL}	LOW-level input voltage			+0.8	V
I_{IK}	Input clamp current			−12	mA
I_{OH}	HIGH-level output current			−800	µA
I_{OL}	LOW-level output current			16	mA
T_A	Operating free-air temperature	0		70	°C

Signetics

74181, LS181, S181
Arithmetic Logic Units

4-Bit Arithmetic Logic Unit
Product Specification

Logic Products

FEATURES
- Provides 16 arithmetic operations: ADD, SUBTRACT, COMPARE, DOUBLE, plus 12 other arithmetic operations
- Provides all 16 logic operations of two variables: Exclusive-OR, Compare, AND, NAND, NOR, OR, plus 10 other logic operations
- Full lookahead carry for high-speed arithmetic operation on long words

DESCRIPTION
The '181 is a 4-bit high-speed parallel Arithmetic Logic Unit (ALU). Controlled by the four Function Select inputs ($S_0 - S_3$) and the Mode Control Input (M), it can perform all the 16 possible logic operations or 16 different arithmetic operations on active HIGH or active LOW operands. The Function Table lists these operations.

TYPE	TYPICAL PROPAGATION DELAY	TYPICAL SUPPLY CURRENT (TOTAL)
74181	22ns	91mA
74LS181	22ns	21mA
74S181	11ns	120mA

ORDERING CODE

PACKAGES	COMMERCIAL RANGE $V_{CC} = 5V \pm 5\%$; $T_A = 0°C$ to $+70°C$
Plastic DIP	N74181N, N74LS181N, N74S181N

NOTE:
For information regarding devices processed to Military Specifications, see the Signetics Military Products Data Manual.

INPUT AND OUTPUT LOADING AND FAN-OUT TABLE

PINS	DESCRIPTION	74	74S	74LS
Mode	Input	1ul	1Sul	1LSul
$\overline{A}$ or $\overline{B}$	Inputs	3ul	3Sul	3LSul
S	Inputs	4ul	4Sul	4LSul
Carry	Input	5ul	5Sul	5LSul
$F_0 - F_3$, = B, C_{n+4}	Outputs	10ul	10Sul	10LSul
$\overline{G}$	Output	10ul	10Sul	40LSul
$\overline{P}$	Output	10ul	10Sul	20LSul

NOTE:
Where a 74 unit load (ul) is understood to be $40\mu A$ I_{IH} and $-1.6mA$ I_{IL}, a 74S unit load (Sul) is $50\mu A$ I_{IH} and $-2.0mA$ I_{IL}, and 74LS unit load (LSul) is $20\mu A$ I_{IH} and $-0.4mA$ I_{IL}.

PIN CONFIGURATION

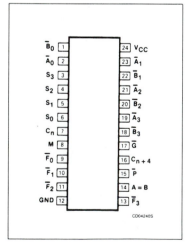

LOGIC SYMBOL

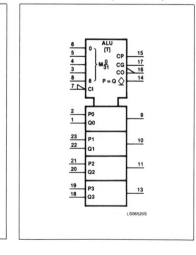

LOGIC SYMBOL (IEEE/IEC)

Arithmetic Logic Units

74181, LS181, S181

When the Mode Control input (M) is HIGH, all internal carries are inhibited and the device performs logic operations on the individual bits as listed. When the Mode Control Input is LOW, the carries are enabled and the device performs arithmetic operations on the two 4-bit words. The device incorporates full internal carry lookahead and provides for either ripple carry between devices using the C_{n+4} output, or for carry lookahead between packages using the signals $\overline{P}$ (Carry Propagate) and $\overline{G}$ (Carry Generate). $\overline{P}$ and $\overline{G}$ are not affected by carry in. When speed requirements are not stringent, it can be used in a simple ripple carry mode by connecting the Carry output (C_{n+4}) signal to the Carry input (C_n) of the next unit. For high-speed operation the device is used in conjunction with the

'182 carry lookahead circuit. One carry lookahead package is required for each group of four '181 devices. Carry lookahead can be provided at various levels and offers high-speed capability over extremely long word lengths.

The A = B output from the device goes HIGH when all four $\overline{F}$ outputs are HIGH and can be used to indicate logic equivalence over 4 bits when the unit is in the subtract mode. The A = B output is open collector and can be wired-AND with other A = B outputs to give a comparison for more than 4 bits. The A = B signal can also be used with the C_{n+4} signal to indicate A > B and A < B.

The Function Table lists the arithmetic operations that are performed without a carry in. An

incoming carry adds a one to each operation. Thus, select code LHHL generates A minus B minus 1 (2s complement notation) without a carry in and generates A minus B when a carry is applied.

Because subtraction is actually performed by complementary addition (1s complement), a carry out means borrow; thus, a carry is generated when there is no underflow and no carry is generated when there is underflow.

As indicated, this device can be used with either active LOW inputs producing active LOW outputs or with active HIGH inputs producing active HIGH outputs. For either case the table lists the operations that are performed to the operands labeled inside the logic symbol.

LOGIC DIAGRAM

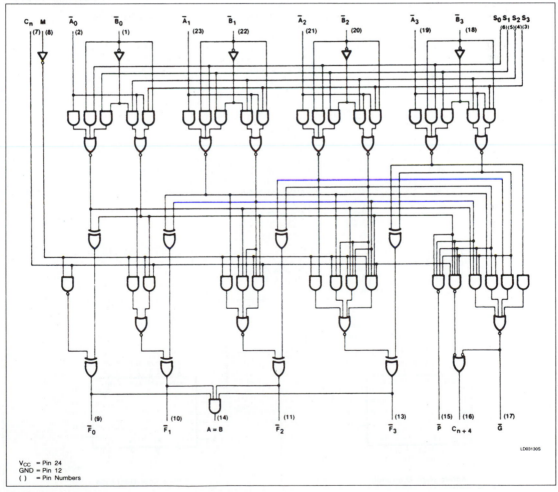

V_{CC} = Pin 24
GND = Pin 12
( ) = Pin Numbers

Arithmetic Logic Units   74181, LS181, S181

MODE SELECT — FUNCTION TABLE

MODE SELECT INPUTS				ACTIVE HIGH INPUTS & OUTPUTS	
S_3	S_2	S_1	S_0	Logic (M = H)	Arithmetic** (M = L) (C_n = H)
L	L	L	L	$\overline{A}$	A
L	L	L	H	$\overline{A + B}$	A + B
L	L	H	L	$\overline{A}B$	A + $\overline{B}$
L	L	H	H	Logical 0	minus 1
L	H	L	L	$\overline{AB}$	A plus A$\overline{B}$
L	H	L	H	$\overline{B}$	(A + B) plus A$\overline{B}$
L	H	H	L	A $\bullet$ B	A minus B minus 1
L	H	H	H	A$\overline{B}$	AB minus 1
H	L	L	L	$\overline{A}$ + B	A plus AB
H	L	L	H	$\overline{A \bullet B}$	A plus B
H	L	H	L	B	(A + $\overline{B}$) plus AB
H	L	H	H	AB	AB minus 1
H	H	L	L	Logical 1	A plus A*
H	H	L	H	A + $\overline{B}$	(A + B) plus A
H	H	H	L	A + B	(A + $\overline{B}$) plus A
H	H	H	H	A	A minus 1

MODE SELECT INPUTS				ACTIVE LOW INPUTS & OUTPUTS	
S_3	S_2	S_1	S_0	Logic (M = H)	Arithmetic** (M = L) (C_n = L)
L	L	L	L	$\overline{A}$	A minus 1
L	L	L	H	$\overline{AB}$	AB minus 1
L	L	H	L	$\overline{A}$ + B	A$\overline{B}$ minus 1
L	L	H	H	Logical 1	minus 1
L	H	L	L	$\overline{A + B}$	A plus (A + $\overline{B}$)
L	H	L	H	$\overline{B}$	AB plus (A + $\overline{B}$)
L	H	H	L	$\overline{A \bullet B}$	A minus B minus 1
L	H	H	H	A + $\overline{B}$	A + $\overline{B}$
H	L	L	L	$\overline{A}B$	A plus (A + B)
H	L	L	H	$\overline{A \bullet B}$	A plus B
H	L	H	L	B	A$\overline{B}$ (A + B)
H	L	H	H	A + B	A + B
H	H	L	L	Logical 0	A plus A*
H	H	L	H	A$\overline{B}$	AB plus A
H	H	H	L	AB	A$\overline{B}$ plus A
H	H	H	H	A	A

L = LOW voltage
H = HIGH voltage level
 *Each bit is shifted to the next more significant position.
**Arithmetic operations expressed in 2s complement notation.

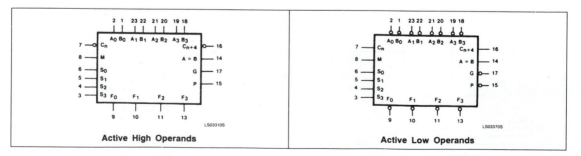

Active High Operands   Active Low Operands

Signetics

74194, LS194A, S194
Shift Registers

4-Bit Bidirectional Universal Shift Register
Product Specification

Logic Products

- **Buffered clock and control inputs**
- **Shift left and shift right capability**
- **Synchronous parallel and serial data transfers**
- **Easily expanded for both serial and parallel operation**
- **Asynchronous Master Reset**
- **Hold (do nothing) mode**

DESCRIPTION

The functional characteristics of the '194 4-Bit Bidirectional Shift Register are indicated in the Logic Diagram and Function Table. The register is fully synchronous, with all operations taking place in less than 20ns (typical) for the 54/74 and 54LS/74LS, and 12ns (typical) for 54S/74S, making the device especially useful for implementing very high speed CPUs, or for memory buffer registers.

TYPE	TYPICAL f_{MAX}	TYPICAL SUPPLY CURRENT (TOTAL)
74194	36MHz	39mA
74LS194A	36MHz	15mA
74S194	105MHz	85mA

ORDERING CODE

PACKAGES	COMMERCIAL RANGE V_{CC} = 5V ±5%; T_A = 0°C to +70°C
Plastic DIP	N74194N, N74LS194AN, N74S194N
Plastic SO-16	N74LS194AD, N745194D

NOTE:
For information regarding devices processed to Military Specifications, see the Signetics Military Products Data Manual.

INPUT AND OUTPUT LOADING AND FAN-OUT TABLE

PINS	DESCRIPTION	74	74S	74LS
All	Inputs	1ul	1Sul	1LSul
$Q_0 - Q_3$	Outputs	10ul	10Sul	10LSul

NOTE:
Where a 74 unit load (ul) is understood to be 40μA I_{IH} and −1.6mA I_{IL}, a 74S unit load (Sul) is 50μA I_{IH} and −2.0mA I_{IL}, and 74LS unit load (LSul) is 20μA I_{IH} and −0.4mA I_{IL}.

PIN CONFIGURATION

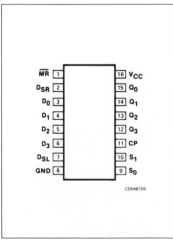

LOGIC SYMBOL

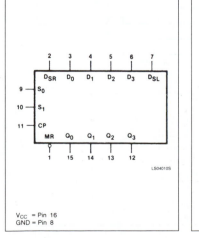

LOGIC SYMBOL (IEEE/IEC)

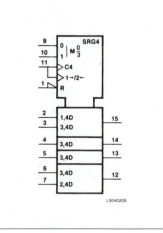

Shift Registers

74194, LS194A, S194

MODE SELECT — FUNCTION TABLE

OPERATING MODE	INPUTS							OUTPUTS			
	CP	$\overline{MR}$	S_1	S	D_{SR}	D_{SL}	D_n	Q_0	Q_1	Q_2	Q_3
Reset (clear)	X	L	X	X	X	X	X	L	L	L	L
Hold (do nothing)	X	H	l(a)	l(a)	X	X	X	q_0	q_1	q_2	q_3
Shift left	↑	H	h	l(a)	X	l	X	q_1	q_2	q_3	L
	↑	H	h	l(a)	X	h	X	q_1	q_2	q_3	H
Shift right	↑	H	l(a)	h	l	X	X	L	q_0	q_1	q_2
	↑	H	l(a)	h	h	X	X	H	q_0	q_1	q_2
Parallel load	↑	H	h	h	X	X	d_n	d_0	d_1	d_2	d_3

H = HIGH voltage level.
h = HIGH voltage level one set-up time prior to the LOW-to-HIGH clock transition.
L = LOW voltage level.
l = LOW voltage level one set-up time prior to the LOW-to-HIGH clock transition.
$d_n(q_n)$ = Lower case letters indicate the state of the referenced input (or output) one set-up time prior to the LOW-to-HIGH clock transition.
X = Don't care.
↑ = LOW-to-HIGH clock transition.
NOTE:
a. The HIGH-to-LOW transition of the S_0 and S_1 inputs on the 74194 should only take place while CP is HIGH for conventional operation.

TYPICAL CLEAR, LOAD, RIGHT-SHIFT, LEFT-SHIFT, INHIBIT AND CLEAR SEQUENCES

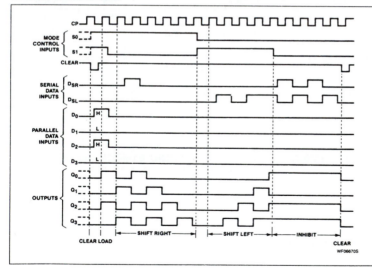

The '194 design has special logic features which increase the range of application. The synchronous operation of the device is determined by two Mode Select inputs, S_0 and S_1. As shown in the Mode Select Table, data can be entered and shifted from left to right (shift right, $Q_0 \rightarrow Q_1$, etc.) or, right to left (shift left, $Q_3 \rightarrow Q_2$, etc.) or, parallel data can be entered, loading all 4 bits of the register simultaneously. When both S_0 and S_1 are LOW, existing data is retained in a hold (do nothing) mode. The first and last stages provide D-type Serial Data inputs (D_{SR}, D_{SL}) to allow multistage shift right or shift left data transfers without interfering with parallel load operation.

Mode Select and Data inputs on the 74S194 and 74LS194A are edge-triggered, responding only to the LOW-to-HIGH transition of the Clock (CP). Therefore, the only timing restriction is that the Mode Control and selected Data inputs must be stable one set-up time prior to the positive transition of the clock pulse. The Mode Select inputs of the 74194 are gated with the clock and should be changed from HIGH-to-LOW only while the Clock input is HIGH.

The four parallel data inputs ($D_0 - D_3$) are D-type inputs. Data appearing on $D_0 - D_3$ inputs when S_0 and S_1 are HIGH is transferred to the $Q_0 - Q_3$ outputs respectively, following the next LOW-to-HIGH transition of the clock. When LOW, the asynchronous Master Reset ($\overline{MR}$) overrides all other input conditions and forces the Q outputs LOW.

Shift Registers

74194, LS194A, S194

LOGIC DIAGRAM

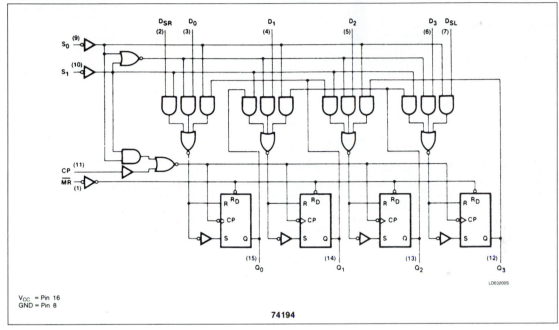

74194

V_{CC} = Pin 16
GND = Pin 8

LOGIC DIAGRAM

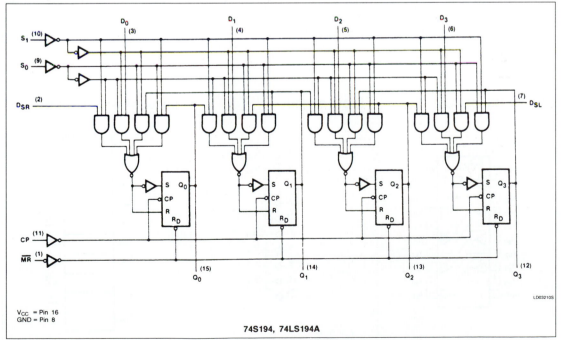

V_{CC} = Pin 16
GND = Pin 8

74S194, 74LS194A

Signetics

Logic Products

74LS244, S244
Buffers

Octal Buffers (3-State)
Product Specification

TYPE	TYPICAL PROPAGATION DELAY	TYPICAL SUPPLY CURRENT (TOTAL)
74LS244	12ns	25mA
74S244	6ns	112mA

FUNCTION TABLE

INPUTS				OUTPUTS	
$\overline{OE}_a$	I_a	$\overline{OE}_b$	I_b	Y_a	Y_b
L	L	L	L	L	L
L	H	L	H	H	H
H	X	H	X	(Z)	(Z)

H = HIGH voltage level
L = LOW voltage level
X = Don't care
(Z) = HIGH impedance (off) state

ORDERING CODE

PACKAGES	COMMERCIAL RANGE $V_{CC} = 5V \pm 5\%$; $T_A = 0°C$ to $+70°C$
Plastic DIP	N74LS244N, 74S244N
Plastic SOL-20	74LS244D

NOTE:
For information regarding devices processed to Military Specifications, see the Signetics Military Products Data Manual.

INPUT AND OUTPUT LOADING AND FAN-OUT TABLE

PINS	DESCRIPTION	74S	74LS
All	Inputs	1Sul	1LSul
All	Outputs	24Sul	30LSul

NOTE:
A 74S unit load (Sul) is 50μA I_{IH} and -2.0mA I_{IL}, and a 74LS unit load (LSul) is 20μA I_{IH} and -0.4mA I_{IL}.

PIN CONFIGURATION

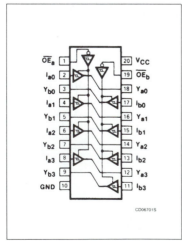

LOGIC SYMBOL

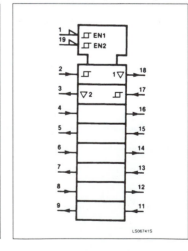

LOGIC SYMBOL (IEEE/IEC)

Buffers   **74LS244, S244**

ABSOLUTE MAXIMUM RATINGS (Over operating free-air temperature range unless otherwise noted.)

	PARAMETER	74LS	74S	UNIT
V_{CC}	Supply voltage	7.0	7.0	V
V_{IN}	Input voltage	−0.5 to +7.0	−0.5 to +5.5	V
I_{IN}	Input current	−30 to +1	−30 to +5	mA
V_{OUT}	Voltage applied to output in HIGH output state	−0.5 to +V_{CC}	−0.5 to +V_{CC}	V
T_A	Operating free-air temperature range	0 to 70		°C

RECOMMENDED OPERATING CONDITIONS

	PARAMETER	74LS			74S			UNIT
		Min	Nom	Max	Min	Nom	Max	
V_{CC}	Supply voltage	4.75	5.0	5.25	4.75	5.0	5.25	V
V_{IH}	HIGH-level input voltage	2.0			2.0			V
V_{IL}	LOW-level input voltage			+0.8			+0.8	V
I_{IK}	Input clamp current			−18			−18	mA
I_{OH}	HIGH-level output current			−15			−15	mA
I_{OL}	LOW-level output current			24			64	mA
T_A	Operating free-air temperature	0		70	0		70	°C

TEST CIRCUITS AND WAVEFORMS

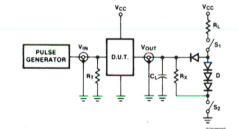

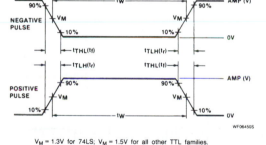

V_M = 1.3V for 74LS; V_M = 1.5V for all other TTL families.

Test Circuit For 3-State Outputs   **Input Pulse Definition**

SWITCH POSITION

TEST	SWITCH 1	SWITCH 2
t_{PZH}	Open	Closed
t_{PZL}	Closed	Open
t_{PHZ}	Closed	Closed
t_{PLZ}	Closed	Closed

FAMILY	INPUT PULSE REQUIREMENTS				
	Amplitude	Rep. Rate	Pulse Width	t_{TLH}	t_{THL}
74	3.0V	1MHz	500ns	7ns	7ns
74LS	3.0V	1MHz	500ns	15ns	6ns
74S	3.0V	1MHz	500ns	2.5ns	2.5ns

DEFINITIONS

R_L = Load resistor to V_{CC}; see AC CHARACTERISTICS for value.

C_L = Load capacitance includes jig and probe capacitance; see AC CHARACTERISTICS for value.

R_T = Termination resistance should be equal to Z_{OUT} of Pulse Generators.

D = Diodes are 1N916, 1N3064, or equivalent.

R_X = 1kΩ for 74, 74S, R_X = 5kΩ for 74LS.

t_{TLH}, t_{THL} Values should be less than or equal to the table entries.

Signetics

Logic Products

74LS373, 74LS374, S373, S374
Latches/Flip-Flops

'373 Octal Transparent Latch With 3-State Outputs
'374 Octal D Flip-Flop With 3-State Outputs
Product Specification

FEATURES
- 8-bit transparent latch — '373
- 8-bit positive, edge-triggered register — '374
- 3-State output buffers
- Common 3-State Output Enable
- Independent register and 3-State buffer operation

DESCRIPTION
The '373 is an octal transparent latch coupled to eight 3-State output buffers. The two sections of the device are controlled independently by Latch Enable (E) and Output Enable ($\overline{OE}$) control gates.

TYPE	TYPICAL PROPAGATION DELAY	TYPICAL SUPPLY CURRENT (TOTAL)
74LS373	19ns	24mA
74S373	10ns	105mA
74LS374	19ns	27mA
74S374	8ns	116mA

ORDERING CODE

PACKAGES	COMMERCIAL RANGE V_{CC} = 5V ±5%; T_A = 0°C to +70°C
Plastic DIP	N74LS373N, N74S373N, N74LS374N, N74S374N
Plastic SOL-20	N74LS373D, N74S373D, N74LS374D, N74S374D

NOTE:
For information regarding devices processed to Military Specifications, see the Signetics Military Products Data Manual.

INPUT AND OUTPUT LOADING AND FAN-OUT TABLE

PINS	DESCRIPTION	74S	74LS
All	Inputs	1Sul	1LSul
All	Outputs	10Sul	30LSul

NOTE:
Where a 74S unit load (Sul) is 50μA I_{IH} and −2.0mA I_{IL}, and a 74LS unit load (LSul) is 20μA I_{IH} and −0.4mA I_{IL}.

PIN CONFIGURATION

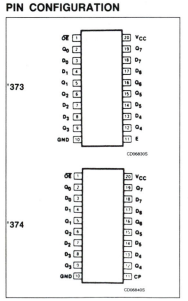

LOGIC SYMBOL

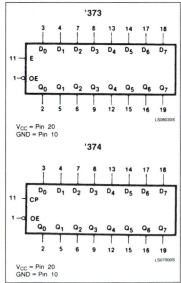

LOGIC SYMBOL (IEEE/EC)

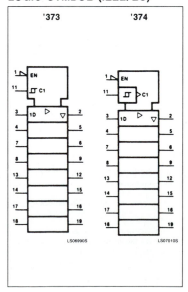

Latches/Flip-Flops   74LS373, 74LS374, S373, S374

The data on the D inputs are transferred to the latch outputs when the Latch Enable (E) input is HIGH. The latch remains transparent to the data inputs while E is HIGH, and stores the data present one set-up time before the HIGH-to-LOW enable transition. The enable gate has hysteresis built in to help minimize problems that signal and ground noise can cause on the latching operation.

The 3-State output buffers are designed to drive heavily loaded 3-State buses, MOS memories, or MOS microprocessors. The active LOW Output Enable ($\overline{OE}$) controls all eight 3-State buffers independent of the latch operation. When $\overline{OE}$ is LOW, the latched or transparent data appears at the outputs. When $\overline{OE}$ is HIGH, the outputs are in the HIGH impedance "off" state, which means they will neither drive nor load the bus.

The '374 is an 8-bit, edge-triggered register coupled to eight 3-State output buffers. The two sections of the device are controlled independently by the Clock (CP) and Output Enable ($\overline{OE}$) control gates.

The register is fully edge triggered. The state of each D input, one set-up time before the LOW-to-HIGH clock transition, is transferred to the corresponding flip-flop's Q output. The clock buffer has hysteresis built in to help minimize problems that signal and ground noise can cause on the clocking operation.

The 3-State output buffers are designed to drive heavily loaded 3-State buses, MOS memories, or MOS microprocessors. The active LOW Output Enable ($\overline{OE}$) controls all eight 3-State buffers independent of the register operation. When $\overline{OE}$ is LOW, the data in the register appears at the outputs. When $\overline{OE}$ is HIGH, the outputs are in the HIGH impedance "off" state, which means they will neither drive nor load the bus.

LOGIC DIAGRAM, '373

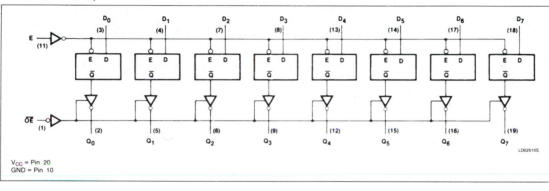

V_{CC} = Pin 20
GND = Pin 10

LOGIC DIAGRAM, '374

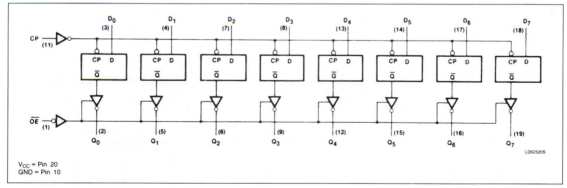

V_{CC} = Pin 20
GND = Pin 10

MODE SELECT — FUNCTION TABLE '373

OPERATING MODES	INPUTS			INTERNAL REGISTER	OUTPUTS
	$\overline{OE}$	E	D_n		$Q_0 - Q_7$
Enable and read register	L L	H H	L H	L H	L H
Latch and read register	L L	L L	l h	L H	L H
Latch register and disable outputs	H H	L L	l h	L H	(Z) (Z)

Signetics

74LS670
Register File

4 x 4 Register File (3-State)
Product Specification

Logic Products

FEATURES
- **Simultaneous and independent Read and Write operations**
- **Expandable to almost any word size and bit length**
- **3-State outputs**
- **See '170 for open collector version**

DESCRIPTION
The '670 is a 16-bit 3-State Register File organized as 4 words of 4 bits each. Separate Read and Write Address and Enable inputs are available, permitting simultaneous writing into one word location and reading from another location. The 4-bit word to be stored is presented to four Data inputs. The Write Address inputs (W_A and W_B) determine the location of the stored word. When the Write Enable ($\overline{WE}$) input is LOW, the data is entered into the addressed location. The addressed location remains transparent to the data while the $\overline{WE}$ is LOW. Data supplied at the inputs will be read out in true (non-inverting) form from the 3-State outputs. Data and Write Address inputs are inhibited when $\overline{WE}$ is HIGH.

TYPE	TYPICAL PROPAGATION DELAY	TYPICAL SUPPLY CURRENT (TOTAL)
74LS670	25ns	30mA

ORDERING CODE

PACKAGES	COMMERCIAL RANGE $V_{CC} = 5V \pm 5\%$; $T_A = 0°C$ to $+70°C$
Plastic DIP	N74LS670N
Plastic SOL-16	N74LS670D

NOTE:
For information regarding devices processed to Military Specifications, see the Signetics Military Products Data Manual.

INPUT AND OUTPUT LOADING AND FAN-OUT TABLE

PINS	DESCRIPTION	74LS
$D_0 - D_3$, W_A, W_B, R_A, R_B	Inputs	1LSul
$\overline{WE}$	Input	2LSul
$\overline{RE}$	Input	3LSul
$Q_0 - Q_3$	Outputs	10LSul

NOTE:
A 74LS unit load (LSul) is 20μA I_{IH} and −0.4mA I_{IL}.

Direct acquisition of data stored in any of the four registers is made possible by individual Read Address inputs (R_A and R_B). The addressed word appears at the four outputs when the Read Enable ($\overline{RE}$) is LOW. Data outputs are in the HIGH impedance "off" state when the Read Enable input is HIGH. This permits outputs to be tied together to increase the word capacity to very large numbers.

PIN CONFIGURATION

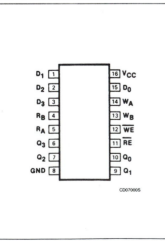

LOGIC SYMBOL

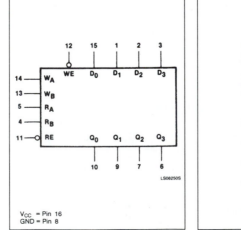

V_{CC} = Pin 16
GND = Pin 8

LOGIC SYMBOL (IEEE/IEC)

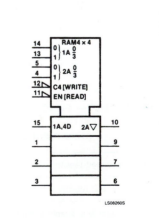

Register File

74LS670

Up to 128 devices can be stacked to increase the word size to 512 locations by tying the 3-State outputs together. Since the limiting factor for expansion is the output HIGH current, further stacking is possible by tying pull-up resistors to the outputs to increase the I_{OH} current available. Design of the Read Enable signals for the stacked devices must ensure that there is no overlap in the LOW levels which would cause more than one output to be active at the same time. Parallel expansion to generate n-bit words is accomplished by driving the Enable and Address inputs of each device in parallel.

LOGIC DIAGRAM

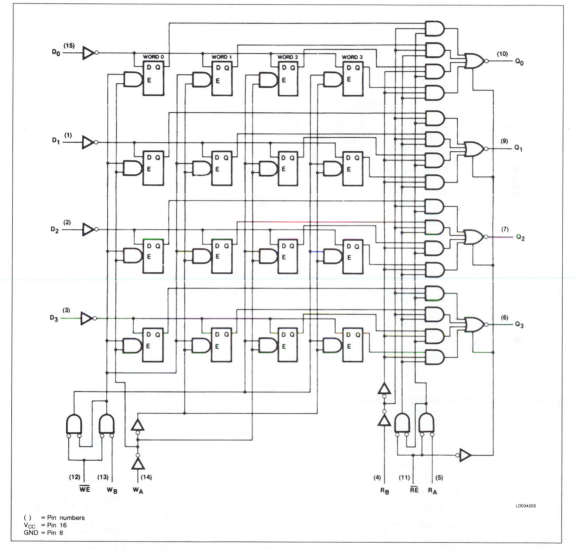

( ) = Pin numbers
V_{CC} = Pin 16
GND = Pin 8

LD03420S

Register File   74LS670

WRITE MODE SELECT TABLE

OPERATING MODE	INPUTS		INTERNAL LATCHES[a]
	$\overline{WE}$	D_n	
Write data	L L	L H	L H
Data latched	H	X	no change

NOTE:

a. The Write Address (W_A and W_B) to the "internal latches" must be stable while $\overline{WE}$ is LOW for conventional operation.

READ MODE SELECT TABLE

OPERATING MODE	INPUTS		OUTPUT Q_n
	$\overline{RE}$	Internal Latches[b]	
Read	L L	L H	L H
Disabled	H	X	(Z)

NOTE:

b. The selection of the "internal latches" by Read Address (R_A and R_B) are not constrained by $\overline{WE}$ or $\overline{RE}$ operation.

H = HIGH voltage level
L = LOW voltage level
X = Don't care
(Z) = HIGH impedance "off" state.

ABSOLUTE MAXIMUM RATINGS (Over operating free-air temperature range unless otherwise noted.)

PARAMETER		74LS	UNIT
V_{CC}	Supply voltage	7.0	V
V_{IN}	Input voltage	−0.5 to +7.0	V
I_{IN}	Input current	−30 to +1	mA
V_{OUT}	Voltage applied to output in HIGH output state	−0.5 to +V_{CC}	V
T_A	Operating free-air temperature range	0 to 70	°C

RECOMMENDED OPERATING CONDITIONS

PARAMETER		74LS			UNIT
		Min	Nom	Max	
V_{CC}	Supply voltage	4.75	5.0	5.25	V
V_{IH}	HIGH-level input voltage	2.0			V
V_{IL}	LOW-level input voltage			+0.8	V
I_{IK}	Input clamp current			−18	mA
I_{OH}	HIGH-level output current			−2.6	mA
I_{OL}	LOW-level output current			8	mA
T_A	Operating free-air temperature	0		70	°C

D

GLOSSARY OF TERMS AND DEFINITIONS

AC characteristics The nomenclature used by manufacturers to describe timing parameters of semiconductor ICs.

Access time The elapsed time from when the address of a memory location is sent to a memory device to when the location is accessible.

Accuracy of a DAC The difference between the calculated analog output voltage and the measured value.

Adder A logic circuit that adds binary numbers.

Address A binary number that is used to reference a particular register in a register array or a particular memory location or I/O port for read or write access.

Address foldback When multiple addresses reference the same memory location or I/O device.

Address map A rectangle that is subdivided into blocks that are identified by starting and ending addresses. These blocks are further identified by programs, data, I/O devices, or whatever occupies those addresses.

ALU Arithmetic logic unit. A logic circuit that implements various arithmetic and logical operations on binary numbers.

Analog The characteristic of being continuously variable.

Analog circuit A circuit that operates on a continuous range of values rather than a discrete (two-stated) set of values.

Analog-to-digital converter (ADC) A semiconductor IC(s) that converts an analog signal to a digital signal.

Analog signal A continuous signal without any discontinuity in the amplitude of the dependent variable.

Analysis A procedure that involves taking apart a circuit or trying to determine the circuit's operation.

AND gate A logic gate that produces a logic 1 at its output if and only if all its inputs are at the logic 1 state.

AND-OR logic Circuits that are made up of *AND* gates, *OR* gates, and inverters, or a combination of these gates.

ANSI American National Standards Institute.

Architecture The symbolic representation of a system in the form of function-blocks and connecting lines that indicate the relationship between function-blocks.

Arithmetic The branch of mathematics that deals with the four basic operations of addition, subtraction, multiplication, and division.

ASCII American Standard Code for Information Interchange.

Assembler A program that translates mnemonics into machine code.

Assembly To translate mnemonics into machine code (object code).

Astable multivibrator An oscillator. A circuit that does not have any stable states but keeps oscillating from one state to another.

Asynchronous An event that is not synchronized with the clock.

Base The number of digits in a numbering system.

BCD Binary coded decimal. A binary code used to represent decimal digits.

Binary A numbering system that is based on a set of two digits.

Binary code A set of binary numbers that are used to represent symbols such as number, letters, etc.

Bistable multivibrator A flip-flop. A circuit that has two stable states.

Bit A binary digit 0 or 1.

Bit bucket A pseudo-device that is used to collect discarded bits.

Boolean algebra The mathematics of logic.

Bootstrap loader A program that is stored in ROM that initially instructs the CPU.

Buffer   A logic gate that provides no logic function but simply electronically isolates the input and output while increasing its output drive.

Buffer memory   A memory (often a FIFO) that acts as a buffer between two devices.

Bus   A line, or lines, that has two or more devices connected in parallel to each line.

Bus contention   A condition whereby two logic circuits put binary data on the bus at the same time resulting in a conflict of logic levels.

Byte   An 8-bit binary number or binary type.

Carry   A digit that is carried over to the next stage in arithmetic operations.

Cascade   A connection of two or more circuits where one circuit drives or triggers another.

Clear   To force the state of a flip-flop or other device to the logic 0 state.

Clock   A stream of pulses used to maintain the basic timing in a logic circuit.

Code   A set of binary numbers that are used to represent symbols such as numbers, letters, etc.

Combinational logic   Logic circuits whose outputs simply depend on the various combinations applied to the inputs.

Comparator   A logic circuit that is used to compare two numbers.

Complement   A term used in Boolean algebra to describe the opposite (or inverse) logic state.

Complementary MOS (CMOS)   A logic family that utilizes both N- and P-channel MOSFETs in a complementary fashion.

Computer   A collection of logic circuits that can be programmed to perform various arithmetic, logic, and data transfer operations.

Control bus   The bus over which control signals are transmitted by the CPU to memory and I/O devices.

Control unit   A special-purpose unit within the CPU that is dedicated to the execution of instructions.

Count   An ordered sequence of numbers.

Counter   A sequential logic circuit that is capable of producing a binary count.

CPU   The central processing unit is the section of a digital computer that processes (executes) program instructions.

Data   Information that is processed by a digital system, generally in binary form.

Data bus   The bus over which the CPU, memory, and I/O transfer data.

DC characteristic   The nomenclature used by manufacturers to describe voltage and current specifications of semiconductor ICs.

Decade counter   A counter that has ten states.

Decimal   A numbering system that is based on a set of ten digits.

Decimal adjust   The procedure used to correct the result of a BCD addition.

Decode   The process of identifying a binary code.

Decoder   A logic circuit that identifies a particular input binary code by activating one of its outputs.

Decrement   The process of decreasing the value of a number by 1.

Demultiplexer   A logic circuit that directs binary data from a single input channel onto one of several output channels.

Difference   The result of a subtraction.

Digit   A symbol in a numbering system that is used to represent a value.

Digital   A circuit, or a field of study, that deals with digits.

Digital-to-analog converter (DAC)   A semiconductor IC(s) that will convert a digital signal to an analog signal.

Digital circuit   A circuit that processes numbers.

Digital signal   A signal that is composed of concatenated binary bits.

DIP   Dual in-line package. A type of package used for ICs.

Disable   To deactivate or make inoperational.

Dividend   The number that is being divided in a division.

Divisor   The number that is divided into the dividend in a division.

Don't care   A logic state that can be a 0 or a 1.

Driven logic circuit   The logic circuit that serves as a load for another logic circuit.

Driver logic circuit   The logic circuit that must drive another logic circuit(s).

Dynamic memory   A memory that has capacitors as storage elements that tend to lose information over a period of time and therefore must be refreshed.

Dynamic one-dimensional memory cell array   A one-dimensional array in which either the content of memory locations or the memory medium is mobile (dynamic).

Dynamic RAM (DRAM)   A R/W memory that must have its contents refreshed. *See also* Dynamic memory.

Dynamic testing   Testing under rapidly changing logic level conditions.

Edge-triggered flip-flop   A type of flip-flop that changes states on detecting a transition of the clock pulse that triggers it.

Electrically erasable PROM (EEPROM)   A PROM that can have its contents erased electronically by the CPU. EEPROMs are also known as read-mostly RAMs and nonvolatile RAMs.

Electronic switch   A transistor (bipolar or MOS) that operates in either the on (saturated) or off state.

Enable   To activate or to make operational.

Encode   The process of producing a binary code.

Encoder   A logic circuit that produces a unique binary code when one of its inputs is activated.

End-around carry   The final carry that is added to the result in unsigned subtraction.

Erasable PROM (EPROM)   A PROM that can have its contents erased using UV light. The EPROM can then be programmed again. Also referred to as a UV-PROM.

Even parity   An indication that a binary number has an even number of 1 bits.

Exclusive-*OR* gate   A logic gate that produces a logic 0 at its output if and only if both inputs are equal.

Fall time   The time it takes for a pulse to change from a high to a low.

Falling edge   The portion of a pulse that makes transition from a 1 to a 0.

Fan-in   Another term used to describe UL.

Fan-out   The number of inputs that can be safely connected to the output of a gate.

First-in first-out (FIFO)   An acronym that describes the method of accessing data. That is, the first data stored is the first data that can be retrieved.

Flag   A flip-flop whose logic state is used to signal the result of an operation or the request for a service.

Flat pack   A type of package used for ICs.

Flip-flop   A bistable sequential logic circuit that has the capability to store a bit.

Floppy disk   A flexible magnetic disk that is used as a medium to store digital data.

Frequency   The rate at which a periodic stream of clock pulses repeats itself.

Full adder   A logic circuit that adds 3 bits.

Fully decoded addressing   An addressing scheme that decodes all address lines of the address bus.

Fuse-link   A fuse in a semiconductor that can be blown (programmed).

Gate   An electronic circuit that implements a basic logic function such as *AND, OR, NOT, NAND,* and *NOR.*

Glitch   An undesired pulse of very short duration.

Glue (VLSI glue)   The ICs used for interfacing various VLSI components and devices in a digital computer circuit.

Half-adder   A logic circuit that adds 2 bits.

Hand assembly   The act of assembling mnemonics (manually) without the use of an assembler.

Handshaking   A technique used to acknowledge the transfer of data between two devices.

Hard disk   A rigid magnetic disk that is used as a medium to store digital data.

Hardware   The electronics of a computer system.

Hexadecimal   A numbering system that is based on a set of 16 digits.

Hi-Z   The high-impedance state of a device with tristate logic. *See also* Tristate logic.

Hold time   The time interval after the clock pulse triggers the flip-flop during which the control inputs must be stable.

IC   Integrated circuit. An electronic package that integrates several circuits and components on a single silicon chip.

I/O device   Devices that are used to either input or output data to or from a computer.

I/O port   The electronic device used to interface an I/O device to a computer system.

I/O space   Those I/O addresses that occupy the address map.

Increment   The process of increasing the value of a number by 1.

Information   Binary data in a digital system.

Input   The signal or collection of signals that go into a circuit.

Instruction register (IR)   The register within the CPU that is used to store the instruction currently being executed.

Instruction set   The set of instructions that a MPU is capable of executing.

Interface   The electronics required to connect (interface) a device to the system.

Interrupt   When the CPU stops the execution of one program at the request of an I/O device in order to service that device.

IR decoder   A decoder that decodes the content of the instruction register.

Isolated I/O   A computer system that differentiates between I/O and memory addresses.

Karnaugh map (K-map)   A variation of the truth table organized to assist in the simplification of logic equations.

Last-in first-out (LIFO)   An acronym that describes the method of accessing data. That is, the last data stored is the first data that can be retrieved.

Latch   A logic circuit that can hold a logic level at its output even after the input stimulus is removed. Also used to describe the capability of a circuit to hold a binary number.

LCD   Liquid crystal display.

LED   Light emitting diode.

Linear addressing   An addressing scheme that uses an address line(s) from the address bus to select and enable either an I/O port or semiconductor memory IC.

Load specifications   Those voltage, current, and capacitance specifications that must be met when one semiconductor device drives another.

Logic   The ability of a circuit to decide what its output(s) will be, based on its inputs and various logical rules.

Logic circuit   A circuit made up of various logic gates.

Logic family   The technology used to implement a group (family) of logic semiconductor ICs.

Logic function   A basic logic operation such as *NOT, AND, OR, NAND,* and *NOR.*

Logic gate   An electronic circuit that implements a basic logic function such as *AND, OR, NOT, NAND,* and *NOR.*

Logic level transition time   The elapsed time required for the logic level output by a device to change from one logic level to the other.

Logic symbol   A simple representation of a logic circuit or gate in terms of its inputs and outputs.

Logic voltage level   The logic level of an input to or an output from a logic circuit.

Logical inversion   The logical operation that converts a 0 (low) to a 1 (high) and a 1 (high) to a 0 (low).

Look-ahead carry   A method of binary addition whereby carries from preceding stages are anticipated, thus avoiding propagation delays.

Look-up table   Data stored in memory that when read is the desired translation of the input data. Also referred to as a translation table.

LSB   Least significant bit. The rightmost bit in a binary number.

LSD   Least significant digit. The rightmost digit in a number.

LSI   Large-scale integration. ICs with 100 to 1000 gates per chip.

Machine code   The binary bit patterns (usually bytes) that are the actual instructions executed by the CPU.

Macroinstructions   Instructions that constitute a macroprogram.

Macroprogram   A program that is stored in main memory and is to be executed by the CPU.

Magnetic bubble memory (MBM)   A mass storage device that uses magnetic domains on a magnetic medium such as memory cells.

Magnetic tape   A magnetic tape that is used as a medium to store digital data.

Magnitude   The value or size of a quantity.

Main memory   The working memory of the computer system.

Mask   A bit, or bits, that are used to mask out other bits in a binary number.

Mask-byte   A byte that is used as a mask to remove or filter out certain bits.

Mask ROM (MROM)   A semiconductor memory that had its contents programmed when manufactured using a photomasking technique.

Master-slave flip-flop   A pulse-triggered flip-flop. A type of flip-flop that changes states on detecting a complete pulse.

Mass storage device   A device in which programs and data can be stored on a permanent storage medium for later use.

Maxterm   A logical sum that will produce a 0 for one and only one binary combination of the independent variables.

Memory cell   An object that is used to store a bit of data.

Memory map   A rectangle that symbolically represents all memory addresses within the system. It is further subdivided into blocks that identify specific memory chips in the system. *See also* Address map.

Memory-mapped I/O   An addressing scheme that maps I/O addresses into the memory addresses space. That is, a system that does not differentiate between I/O and memory addresses.

Microinstructions   Instructions that constitute a microprogram.

Microprocessor unit (MPU)   A semiconductor IC that is used to implement the CPU, and other possible functions, of a computer.

Microprogram   A program within the control unit of an MPU that instructs it as to how to execute a particular instruction from its instruction set.

Minterm   A logical product that will produce a 1 for one and only one binary combination of the independent variables.

Mnemonic   A memory aid for representing MPU instructions. Mnemonics symbolically indicate the MPU operation to be performed.

Modulus   The number of states of a counter.

Monostable multivibrator   A one-shot. A circuit that has only one stable state.

MSB   Most significant bit. The leftmost bit in a binary number.

MSD   Most significant digit. The leftmost digit in a number.

MSI   Medium-scale integration. ICs with 12 to 99 gates per chip.

Multiplexer   A logic circuit that directs several input channels containing binary information onto a single output channel.

Multiplicand   The number being multiplied in a multiplication.

Multiplier   The number used to multiply the multiplicand in a multiplication.

NAND gate   A logic gate that produces a logic 0 at its output if and only if all its inputs are at the logic 1 state.

NAND logic   Circuits that are made up of *NAND* gates only.

Negative edge   The portion of a pulse that makes a transition from a 1 to a 0.

Nibble   A 4-bit binary number or half a byte.

9's complement An unsigned representation of a negative number in the decimal numbering system.

Node An electrical connection of two or more components.

Noise immunity The susceptibility of a circuit to the effects of noise.

Noise margin The difference between worst case output voltage and worst case input voltage.

Nonvolatile RAM An R/W memory that does not require that power be maintained to it in order for it to retain its contents. *See also* Electrically erasable PROM (EEPROM).

NOR **gate** A logic gate that produces a logic 0 at its output if any or all of its inputs are at the logic 1 state.

NOR **logic** Circuits that are made up of *NOR* gates only.

Octal A numbering system that is based on a set of eight digits.

Odd parity An indication that a binary number has an odd number of 1 bits.

One-dimensional memory cell array An array of memory cells in which the cells are arranged in rows with each row constituting a memory location. Memory locations are addressed by a one-dimensional parameter, the row location.

One-shot Monostable multivibrator. A circuit that has only one stable state.

1's complement The unsigned representation of a negative binary number.

Op-code The part of an instruction that represents an MPU operation.

Open-collector TTL logic that does not have the top transistor of the usual totem pole output.

Operand The part of an instruction that is to be operated on by the op-code.

OR **gate** A logic gate that produces a logic 1 at its output if any or all of its inputs are at the logic 1 state.

Output The signal or collection of signals that come out of a logic circuit.

Page of memory A predefined number of memory locations.

Parallel I/O device An I/O device that receives or transmits data in multiple bits (usually 8 bits).

Parallel register A register in which data is loaded and retrieved in parallel.

Parallel-to-parallel I/O port An I/O port that receives or transmits data bits (usually 8 bits) in parallel.

Parasitic capacitance An inherent capacitance that is present in all semiconductors.

Parity An indication of whether there are an even or odd number of 1 bits in a binary number.

Parity checker A circuit that checks a binary number for even or odd parity.

Parity generator A circuit that generates a parity bit for a binary number.

Partial linear addressing A memory addressing scheme that uses linear addressing to select (enable) the semiconductor, and address decoding to select the memory location within the selected semiconductor.

Period The interval of time between periodic clock pulses.

Periodic An event that repeats itself at regular intervals.

Pipelining Using a queue to prefetch instructions so that the CPU may use the data bus and execute instructions at the same time.

Positive edge The portion of a pulse that makes the transition from a 0 to a 1.

Preset The state of a flip-flop when its output is at the logic 1 state.

Priority encoder A logic circuit that produces a prioritized binary code when one or more of its inputs are activated.

Product The result of multiplication.

Product-of-sums (POS) One of the forms that a logic equation can take in which logical sums (maxterms) are *AND*ed together.

Program A set of instructions to the CPU that are executed in a particular order to achieve a task.

Programmable Modifying something through software.

Programmable array logic (PAL) A PLD with a programmable *AND* gate array and a preprogrammed output *OR* gate array.

Programmable logic array (PLA) A PLD with programmable *AND* gate and *OR* gate array.

Programmable logic device (PLD) Semiconductor devices that can be programmed to function like various combinational logic circuits. PLDs include other programmable devices (PLEs, PLAs, and PALs).

Programmable logic element (PLE) A PLD that has a nonprogrammable *AND* gate array with a programmable output *OR* gate array.

Programmable ROM (PROM) A semiconductor memory that has the ability to be irreversibly programmed by the user using a PROM programmer.

Propagation delay The amount of delay that occurs from the time the input of a gate changes states to the time the output changes states.

Pull-up resistor A resistor used to reference (pull-up) the collector of an open-collector circuit to V_{CC} volts.

Pulse A very quick change in state from one logic level to another followed by a return to the original state.

Pulse duration Pulse width. The time interval between the edges of a pulse.

Pulse-triggered flip-flop *See* Master-slave flip-flop.

Pulse width *See* Pulse duration.

Quantization error The inherent error of an ADC due to truncation.

Queue A FIFO that is used to sequentially store prefetched instructions. *See also* Pipelining.

Quotient The result of division.

Random access memory (RAM) An acronym that is descriptive of the ability to access a memory location in a random manner with a constant access time. RAM is often erroneously used to mean the ability to read and write to a memory.

Read The process of obtaining or retrieving data from a register or memory location.

Read-only memory (ROM) A memory that cannot have its contents changed (written) but can only have its contents retrieved (read).

Recycle An event that causes a counter to return to its starting (initial) state.

Refresh cycle The operation cycle performed to refresh the memory content of a DRAM.

Register A logic circuit that is capable of storing a single binary number.

Register array An array or collection of registers used to store several binary numbers.

Register file *See* Register array.

Reset The state of a flip-flop when its output is at the logic 0 state.

Resolution The smallest change in a magnitude that can be detected.

Rise time The time it takes for a pulse to change from a low to a high.

Rising edge The portion of a pulse that makes transition from a 0 to a 1.

Schottky technology A fast TTL-type technology.

Scratch pad memory Registers within an MPU or memory locations external to the MPU that are used to temporarily store data or intermediate results from the ALU. *See also* Register array.

Sequential logic Logic circuits whose outputs depend on the various

combinations applied to the inputs and the past history of the outputs.

Serial I/O device An I/O device that receives or transmits data 1 bit at a time.

Serial I/O port An I/O port that transmits and receives serial data to an from the I/O device. *See also* Universal synchronous / asynchronous receiver / transmitter (USART).

Serial register A register in which data is loaded and / or retrieved serially.

Set The state of a flip-flop when its output is at the logic 1 state.

Setup time The time interval before the clock pulse triggers the flip-flop during which the control inputs must be stable.

Seven-segment decoder A logic circuit that converts a binary code to a code that activates the seven segments of an LED.

Shift register A register in which data is loaded and / or retrieved serially.

Signature analysis A dynamic diagnostic tool that counts the number of logic ones that appear at a node over an interval of time for a repetitive sequence of events. From this count a value is calculated, which becomes the signature of that node.

Sink current That current that flows into a logic circuit.

Software The instructions and programs that instruct the CPU as to the tasks to be performed.

Source current The current that flows out of a logic circuit.

SSI Small-scale integration. ICs with 1 to 11 gates per chip.

Stack memory A portion of main memory reserved and defined to constitute the stack. Its memory locations are often accessed in a LIFO manner.

Static RAM A R / W memory that retains its content without the need for refreshing, as long as power is applied to it.

Static testing Testing under DC conditions.

Subroutine A subprogram that is executed by another program.

Subtractor A logic circuit that is used to obtain the difference between two numbers.

Sum The result of addition.

Sum-of-products (SOP) One of the forms that a logic equation can take in which logical products (minterms) are *OR*ed together.

Switch An electronic component that stops or passes the flow of current in a circuit. Also refers to the capability of a logic circuit to change states.

Switching circuit A circuit that has only two states and is capable of switching between these two states.

Synchronous An event that is synchronized with the clock.

Synthesis A procedure that involves putting together a circuit or designing a circuit.

10's complement An unsigned representation of a negative number in the decimal numbering system.

Test node An electrical node that is monitored for test data.

Three-bus architecture A system that has three busses, usually the address, data, and control bus.

Three-dimensional memory cell array An array of memory cells that is constructed of duplicate planes of two-dimensional memory cell arrays. Memory locations are formed by concatenating the identically located memory cell in each plane.

Timing diagram A diagram that illustrates the operation of a sequential logic circuit by displaying the inputs and outputs of the circuit as a function of time.

Toggle The switching of a logic level from a 0 to a 1 and vice versa.

Totem pole Two bipolar transistors connected with their emitter-collectors in series with each other in the output of a TTL logic gate.

Transistor-transistor logic (TTL) A logic family that utilizes bipolar transistors.

Transition A change in logic state.

Translation table *See* Look-up-table.

Trigger A stimulus such as a pulse that is used to change the state of a logic circuit.

Tristate logic Logic circuits or gates whose outputs can exist in one of three possible states—high, low, or high impedance.

Truth table A table that shows the logic state of the dependent variable(s) for all possible combinations of the independent variable(s).

Two-dimensional memory cell array An array of memory cells in which the cells are organized into an array of rows and columns. The address of each cell within the array is an *x-y* coordinate of its row and column location.

2's complement The unsigned representation of a negative binary number.

UL Unit load. A measure of the electrical loading requirements of a logic gate.

Universal synchronous / asynchronous receiver / transmitter (USART) A programmable serial I/O port.

User memory space The area of main memory available for user programs (application programs).

UV-PROM *See* EPROM.

Vector Directed to a specific address.

VLSI Very large-scale integration. ICs with more than 1000 gates per chip.

Volatile memory A memory that loses its content when power is removed from it.

Weight The value of a digit in a number based on its position in the number.

Wired-*AND* logic The logic formed by connecting the outputs of open-collector logic gates.

Word A 16-bit binary number or a double byte. Could also be used to describe the size of binary numbers processed by a digital circuit.

Write The process of storing data into a register or memory location.

ANSWERS TO ODD-NUMBERED PROBLEMS

Chapter 1

1. The number of combinations $= 100,000,000$
Largest value $= 99,999,999$

3. **(a)** 0 1 2 3 4 5 6 7 8 9 A B C
(b) 64 65 66 67 68 69 6A 6B 6C
(c) 19 1A 1B 1C 1D 1E 1F 20 21 22 23
(d) F5 F6 F7 F8 F9 FA FB FC FD FE FF 100 101

5. 17563

7. **(a)** 22161_{10} **(b)** 112297_{10} **(c)** 64206_{10}
(d) 16_{10} **(e)** 65535_{10} **(f)** 92041_{10}

9. 9, 1, 4, 0

11. **(a)** $B64_{16}$. **(b)** $2F37_{16}$.
(c) 100_{16}. **(d)** 10_{16}.

13. **(a)** 5252_8 **(b)** 345_8 **(c)** 26_8 **(d)** 66_8
(e) 1_8 **(f)** 3_8 **(g)** 500_8

15. **(a)** $F59_{16}$ **(b)** $5B_{16}$ **(c)** 5_{16} **(d)** $4BD_{16}$
(e) $3F7F_{16}$ **(f)** 81_{16} **(g)** 152_{16}

17. **(a)** $3F_{16}$ **(b)** 1049_{16} **(c)** $1D520_{16}$
(d) $193A8B_{16}$ **(e)** $10D1_{16}$ **(f)** A_{16}

19. **(a)** 1001 0000 0000 0001 0111
(b) 0110 0101
(c) 0110 0011 0101 0001 0101
(d) 0001 0000 0001 0000 **(e)** 0101
(f) 0010 0011 0110 0000 0010 0001

21. **(a)** 147_{10} **(b)** 223_8 **(c)** 93_{16}
(d) 10010011_2

23. **(a)** 10000 **(b)** 10010111 **(c)** 100000101
(d) 111100001

25. **(a)** 5 **(b)** 87 **(c)** 3 **(d)** 359 **(e)** 1
(f) 5342

27. **(a)** 7 **(b)** 4 **(c)** 36 **(d)** 837
(e) 357

29. **(a)** -328761 **(b)** -1 **(c)** -78
(d) -180264979

31. **(a)** 100 **(b)** 1000111 **(c)** 01111010
(d) 1000101

33. Quotient (tally) $= 9$, remainder $= 2$

35. Quotient (tally) $= 5$, remainder $= 0$

37. **(a)** 0111 1001 **(b)** 0110 0000 **(c)** 1 0110 0111
(d) 1000 0110 0000 0010

39. **(a)** No parity, 101 0111
Even parity, 1101 0111
Odd parity, 0101 0111
(b) No parity, 011 1001
Even parity, 0011 1001
Odd parity, 1011 1001
(c) No parity, 100 0000
Even parity, 1100 0000
Odd parity, 0100 0000
(d) No parity, 000 0011
Even parity, 0000 0011
Odd parity, 1000 0011

Chapter 2

1. **(a)**

A	B	C	D	L
OFF	OFF	OFF	OFF	OFF
OFF	OFF	OFF	ON	OFF
OFF	OFF	ON	OFF	OFF
OFF	OFF	ON	ON	OFF
OFF	ON	OFF	OFF	OFF
OFF	ON	OFF	ON	ON
OFF	ON	ON	OFF	OFF
OFF	ON	ON	ON	ON
ON	OFF	OFF	OFF	OFF
ON	OFF	OFF	ON	OFF
ON	OFF	ON	OFF	ON
ON	OFF	ON	ON	ON
ON	ON	OFF	OFF	OFF
ON	ON	OFF	ON	ON
ON	ON	ON	OFF	ON
ON	ON	ON	ON	ON

(b)

A	B	C	D	L
OFF	OFF	OFF	OFF	OFF
OFF	OFF	OFF	ON	OFF
OFF	OFF	ON	OFF	OFF
OFF	OFF	ON	ON	OFF
OFF	ON	OFF	OFF	OFF
OFF	ON	OFF	ON	OFF
OFF	ON	ON	OFF	OFF
OFF	ON	ON	ON	OFF
ON	OFF	OFF	OFF	OFF
ON	OFF	OFF	ON	OFF
ON	OFF	ON	OFF	OFF
ON	OFF	ON	ON	OFF
ON	ON	OFF	OFF	OFF
ON	ON	OFF	ON	ON
ON	ON	ON	OFF	ON
ON	ON	ON	ON	ON

(a) $L = (A \cdot C) + (B \cdot D)$
(b) $L = A \cdot B \cdot (C + D)$

3. Switch A and C must be *ON* or switch B and D must be *ON*.

5.

A	B	C	D	E	F	G	H
0	0	1	0	1	0	1	0
1	1	0	1	0	1	0	1

7.

P	Q	R	S	W
0	0	0	0	0
0	0	0	1	0
0	0	1	0	0
0	0	1	1	0
0	1	0	0	0
0	1	0	1	0
0	1	1	0	0
0	1	1	1	0
1	0	0	0	0
1	0	0	1	0
1	0	1	0	0
1	0	1	1	0
1	1	0	0	0
1	1	0	1	0
1	1	1	0	0
1	1	1	1	1

Logic equation: $W = P \cdot Q \cdot R \cdot S$
Conclusion: The circuit functions like a four-input *AND* gate.

9.

P	Q	R	S	W
0	0	0	0	0
0	0	0	1	1
0	0	1	0	1
0	0	1	1	1
0	1	0	0	1
0	1	0	1	1
0	1	1	0	1
0	1	1	1	1
1	0	0	0	1
1	0	0	1	1
1	0	1	0	1
1	0	1	1	1
1	1	0	0	1
1	1	0	1	1
1	1	1	0	1
1	1	1	1	1

Logic equation: $W = P + Q + R + S$
Conclusion: The circuit functions like a four-input *OR* gate.

11. (a)

A	B	C
0	0	1
0	1	1
1	0	1
1	1	0

Conclusion: The circuit functions like a *NAND* gate.

(b)

A	B	C
0	0	1
0	1	0
1	0	0
1	1	0

Conclusion: The circuit functions like a *NOR* gate.

13. **(a)**

A	B	C	D	X
0	0	0	0	0
0	0	0	1	0
0	0	1	0	0
0	0	1	1	1
0	1	0	0	0
0	1	0	1	0
0	1	1	0	0
0	1	1	1	1
1	0	0	0	0
1	0	0	1	0
1	0	1	0	0
1	0	1	1	1
1	1	0	0	1
1	1	0	1	1
1	1	1	0	1
1	1	1	1	1

Logic equation: $X = \overline{\overline{A \cdot B} \cdot \overline{C \cdot D}}$

(b)

A	B	C	D	X
0	0	0	0	0
0	0	0	1	0
0	0	1	0	0
0	0	1	1	0
0	1	0	0	0
0	1	0	1	1
0	1	1	0	1
0	1	1	1	1
1	0	0	0	0
1	0	0	1	1
1	0	1	0	1
1	0	1	1	1
1	1	0	0	0
1	1	0	1	1
1	1	1	0	1
1	1	1	1	1

Logic equation: $X = \overline{\overline{A + B} + \overline{C + D}}$

15.

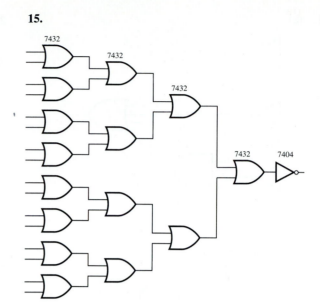

17. **(a)**

W	X	Y	Z
0	0	0	1
0	0	1	1
0	1	0	1
0	1	1	1
1	0	0	1
1	0	1	1
1	1	0	1
1	1	1	0

(b)

W	X	Y	Z
0	0	0	1
0	0	1	0
0	1	0	0
0	1	1	0
1	0	0	0
1	0	1	0
1	1	0	0
1	1	1	0

19. **(a)** $P = (\overline{W \cdot X}) + \overline{Y}$
(b) $P = (\overline{W + X}) \cdot \overline{Y}$

21. (a)

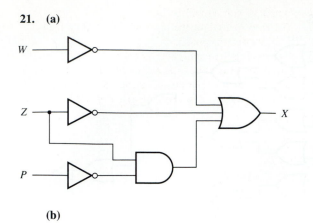

(b)

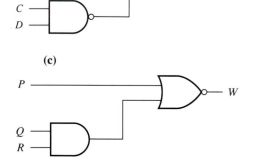

(c)

(d)

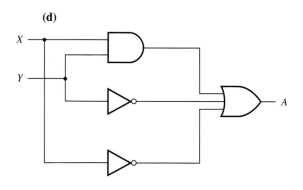

23.

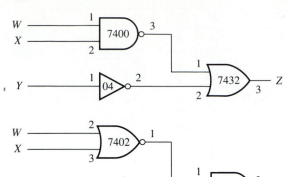

25.

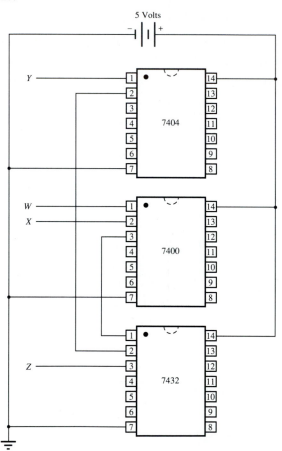

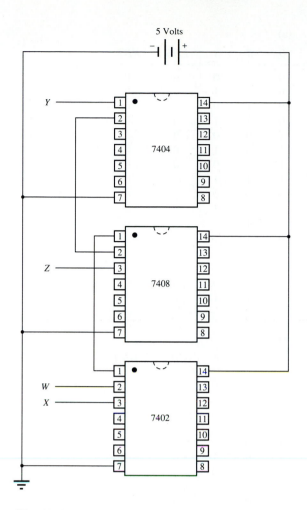

27. 42.5 mW

29. $V_{OH} = 3$ volts, $V_{OL} = 0.5$ volts, $V_{IH} = 2$ volts, $V_{IL} = 0.8$ volts

31. $I_{OH} = 2$ mA, $I_{OL} = 20$ mA, $I_{IH} = 20$ μA, $I_{IL} = 500$ μA

33. FO (HIGH) = 100, FO (LOW) = 40

35. 0.5 UL (HIGH), 0.3 UL (LOW)

37. $V_{NH} = 1$ volt, $V_{NL} = 0.3$ volts

39. $t_{pd} = 23$ ns

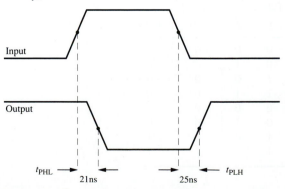

41. Output of $G3$ (Z) will be LOW.

43. The output of $G3$ or the input of $G2$ could be internally shorted to ground.

45. By isolating the connection between $G3$ and $G2$, the output of $G3$ could be measured with the input A set to the *LOW* state. If the probe indicates a *LOW* then $G3$ is defective, but if the probe indicates a *HIGH* then $G2$ is defective.

Chapter 3

1. (a) $X = (A + B) \cdot \overline{C}$
   (b) $Z = PQ + Q + RS$

3. (a)

A	B	C	X
0	0	0	0
0	0	1	0
0	1	0	1
0	1	1	0
1	0	0	1
1	0	1	0
1	1	0	1
1	1	1	0

(b)

P	Q	R	S	Z
0	0	0	0	0
0	0	0	1	0
0	0	1	0	0
0	0	1	1	1
0	1	0	0	1
0	1	0	1	1
0	1	1	0	1
0	1	1	1	1
1	0	0	0	0
1	0	0	1	0
1	0	1	0	0
1	0	1	1	1
1	1	0	0	1
1	1	0	1	1
1	1	1	0	1
1	1	1	1	1

5. (a) Same as 3.3 (a).
   (b) Same as 3.3 (b).

7. $Z = \overline{X}\,\overline{Y} + \overline{X}\,Y$
$S = \overline{PQ}\,R + P\overline{QR} + P\overline{Q}R + PQR$

9. $Z = (\overline{X} + Y) \cdot (\overline{X} + \overline{Y})$
$S = (P + Q + R) \cdot (P + \overline{Q} + R)$
    $\cdot (P + \overline{Q} + \overline{R}) \cdot (\overline{P} + \overline{Q} + R)$

11.

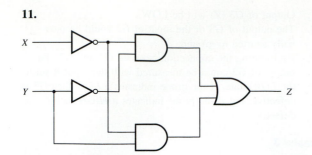

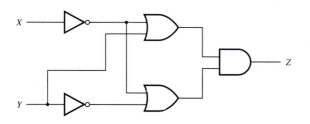

13.

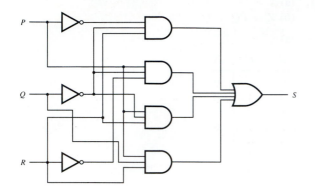

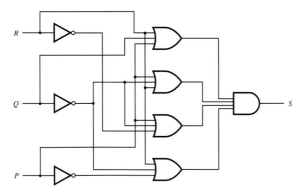

15. $S = 1$
      $S = 0$

17. $Z = \Sigma_m \, 0, 1$
      $S = \Sigma_m \, 1, 4, 5, 7$

19. $Z = \Pi_M \, 2, 3$
      $S = \Pi_M \, 0, 2, 3, 6$

21.   **(a)** 0   **(b)** 1   **(c)** 0   **(d)** 1   **(e)** 1

23.   **(a)**

P	Q	R	S	LHS	RHS
0	0	0	0	0	0
0	0	0	1	0	0
0	0	1	0	0	0
0	0	1	1	0	0
0	1	0	0	0	0
0	1	0	1	0	0
0	1	1	0	0	0
0	1	1	1	0	0
1	0	0	0	0	0
1	0	0	1	0	0
1	0	1	0	0	0
1	0	1	1	0	0
1	1	0	0	0	0
1	1	0	1	1	1
1	1	1	0	0	0
1	1	1	1	1	1

(b)

A	B	C	D	LHS	RHS
0	0	0	0	1	1
0	0	0	1	1	1
0	0	1	0	1	1
0	0	1	1	0	0
0	1	0	0	1	1
0	1	0	1	1	1
0	1	1	0	1	1
0	1	1	1	0	0
1	0	0	0	1	1
1	0	0	1	1	1
1	0	1	0	1	1
1	0	1	1	0	0
1	1	0	0	0	0
1	1	0	1	0	0
1	1	1	0	0	0
1	1	1	1	0	0

(c)

X	Y	Z	LHS	RHS
0	0	0	0	0
0	0	1	0	0
0	1	0	0	0
0	1	1	0	0
1	0	0	0	0
1	0	1	0	0
1	1	0	1	1
1	1	1	1	1

(d)

P	Q	LHS	RHS
0	0	1	1
0	1	1	1
1	0	1	1
1	1	1	1

25. $Z = \overline{X}$ (simplified = one gate, unsimplified = five gates)
$S = \overline{Q}R + \overline{Q}P + PR$ (simplified = five gates, unsimplified = eight gates)

27. $Z = \overline{X}$ (simplified = one gate, unsimplified = five gates)
$S = (P + R) \cdot (P + \overline{Q}) \cdot (\overline{Q} + R)$ (simplified = five gates, unsimplified = eight gates)

29. **(a)** $A + B$ **(b)** 1   **(c)** $\overline{P} + Q\overline{R}$
(d) $\overline{R}$ **(e)** $A\overline{B} + AD + \overline{A}B$ **(f)** $\overline{Q}\,\overline{S}$

31. $Z = \overline{\overline{X \cdot X}}$
$S = \overline{\overline{\overline{QR} \cdot \overline{QP} \cdot \overline{PR}}}$

33. *NAND* logic = one IC, one gate
AND-OR logic = one IC, one gate

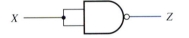

NAND logic = two ICs, five gates
AND-OR logic = three ICs, five gates

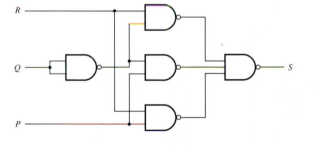

35. **(a)** $Z = \overline{\overline{(\overline{P} + Q)} + \overline{(P + \overline{R})}}$

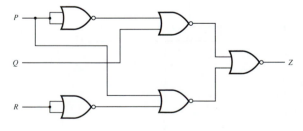

(b) $T = \overline{\overline{(\overline{X} + Y)} + \overline{(\overline{X} + Z)}}$

(c) $S = \overline{\overline{\overline{A} + \overline{(B + \overline{D})}} + \overline{(B + C + D)}}$

(d) $W = \overline{\overline{(P + \overline{Q} + R)} + \overline{(\overline{P} + Q + \overline{S})}}$

(e) $Z = \overline{\overline{\overline{(\overline{A} + D)} + \overline{(B + \overline{C})}} + \overline{(\overline{C} + D)}}$

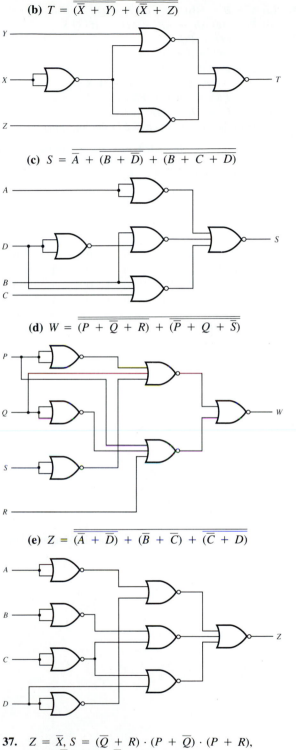

37. $Z = \overline{X}$, $S = (\overline{Q} + R) \cdot (P + \overline{Q}) \cdot (P + R)$,
$X = (\overline{A} + B) \cdot (\overline{B} + D)$
$S = (\overline{Q} + \overline{R})$

39. **(a)** $\overline{A} + \overline{B}$ **(b)** $\overline{P} + \overline{Q}$ **(c)** $P + R$
(d) $(\overline{A} + B) \cdot (\overline{A} + C) \cdot (B + C) \cdot (A + \overline{B} + \overline{C})$
(e) $X + Y$ **(f)** $(\overline{R} + \overline{S} \cdot (\overline{Q} + R) \cdot (\overline{P} + \overline{S}) \cdot (\overline{P} + \overline{Q})$

41. (a) $A + B$ (b) 1 (c) $\bar{P} + Q\bar{R}$
(d) $\bar{R}$ (e) $\bar{A}B + A\bar{B} + AD$ (f) $\bar{Q}\,\bar{S}$

43. (a) $Z = P\bar{Q} + \bar{P}Q$
(b) $X = B\bar{C} + \bar{A}\,\bar{C} + A\bar{B}C$
(c) $S = P\,\bar{Q}\,\bar{R} + \bar{P}QR$
(d) $A = X\bar{Z} + \bar{X}Z + \bar{W}\,\bar{X}\,Y + W\,\bar{X}\,\bar{Y}$
(e) $T = PRS + QR\bar{S} + \bar{P}R\bar{S}$
(f) $J = C + AB + AD$

45. Faults in the circuit are circled in the following table.

P	Q	R	Recorded			Actual		
			G1	G2	G3	G1	G2	G3
0	0	0	1	0	0	0	0	0
0	0	1	1	⓪	0	1	1	0
0	1	0	1	0	0	1	0	0
0	1	1	1	⓪	⓪	1	1	1
1	0	0	0	1	0	0	1	0
1	0	1	0	1	0	0	1	0
1	1	0	0	1	0	0	1	0
1	1	1	0	1	0	0	1	0

Chapter 4

1.

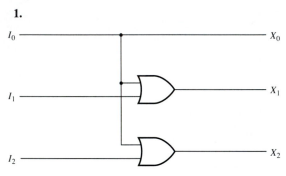

3. Convert to *AND-OR* logic, then apply DeMorgan's law and simplify.

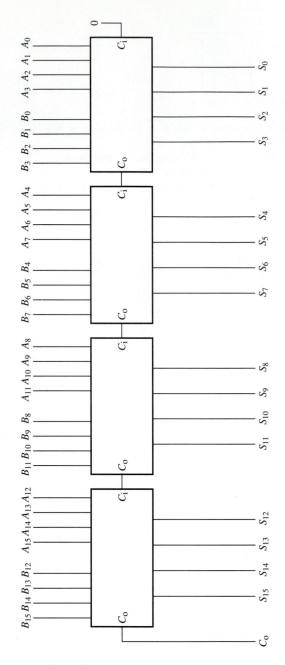

7.

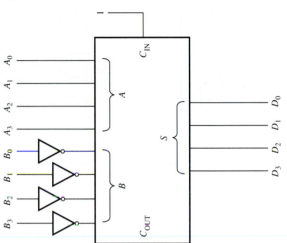

11.

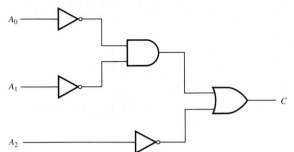

9.

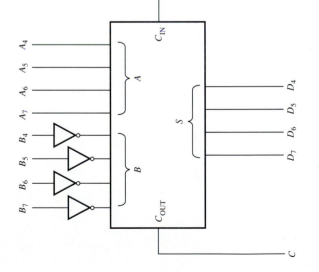

13.

I_0	I_1	I_2	I_3	X_1	X_0
0	0	0	0	1	1
0	0	0	1	1	1
0	0	1	0	1	0
0	0	1	1	1	0
0	1	0	0	0	1
0	1	0	1	0	1
0	1	1	0	0	0
0	1	1	1	0	0
1	0	0	0	0	0
1	0	0	1	0	0
1	0	1	0	0	0
1	0	1	1	0	0
1	1	0	0	0	0
1	1	0	1	0	0
1	1	1	0	0	0
1	1	1	1	0	0

15.

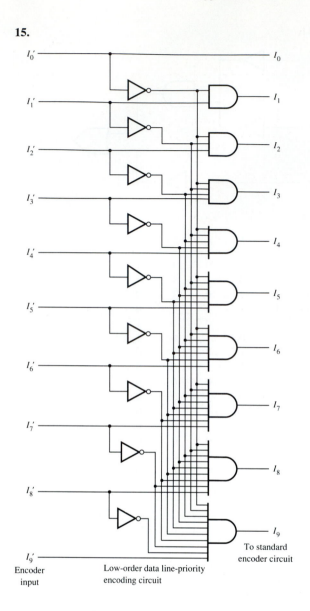

Encoder input

Low-order data line-priority encoding circuit

To standard encoder circuit

17.

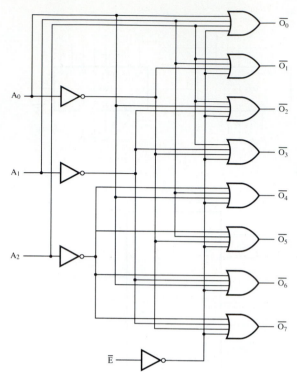

19. Priorities: Highest $\overline{7}\ \overline{6}\ \overline{5}\ \overline{4}\ \overline{3}\ \overline{2}\ \overline{1}\ \overline{0}$ F E D C B $\overline{A}$ 9 $\overline{8}$ Lowest

21.

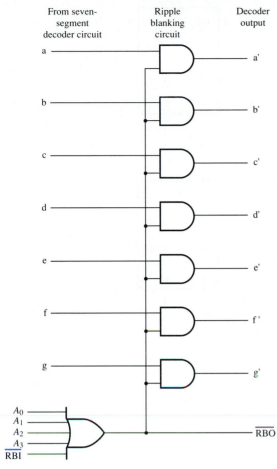

23.

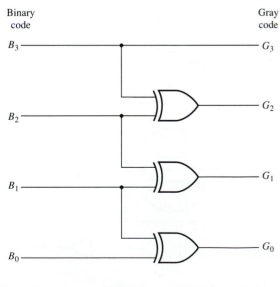

25.

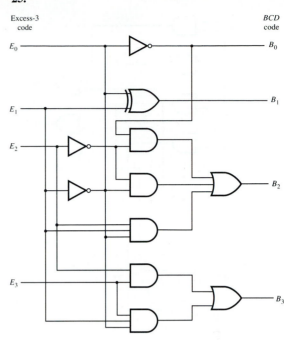

27.

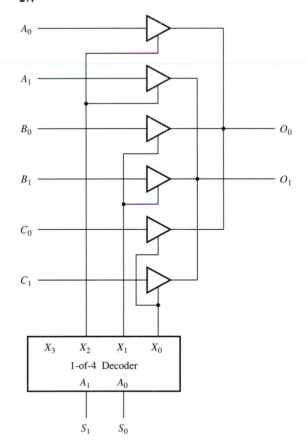

29.

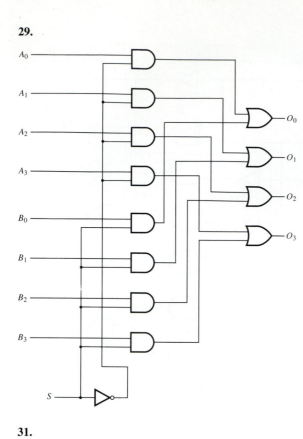

35.

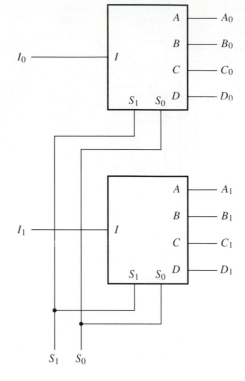

31.

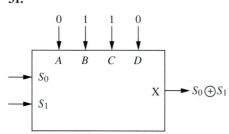

33.

S_1	S_0	X
0	0	1
0	1	1
1	0	0
1	1	1

37.

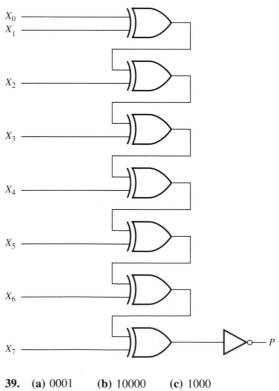

39.   **(a)** 0001   **(b)** 10000   **(c)** 1000
         (d) 1001

41. $\overline{EI}$ is permanently held high due to an internal short to V_{CC} in IC 1 or IC 2 — most probable fault.

Chapter 5

1.

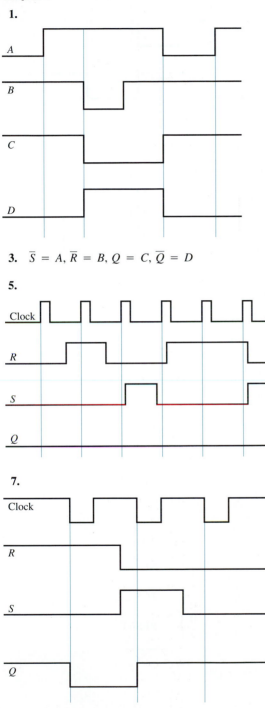

3. $\overline{S} = A, \overline{R} = B, Q = C, \overline{Q} = D$

5.

7.

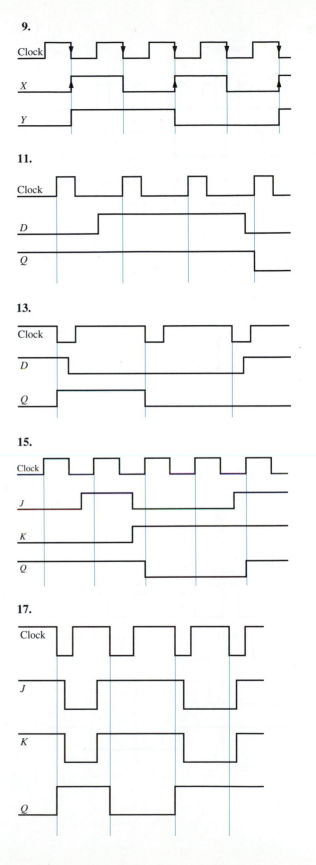

9.

11.

13.

15.

17.

19.

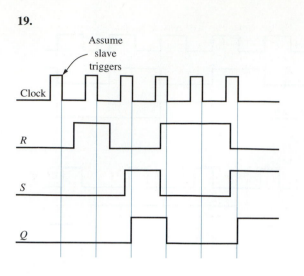

21.

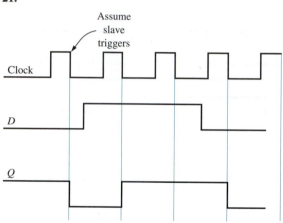

23.

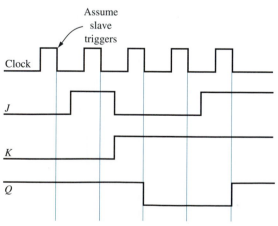

25.

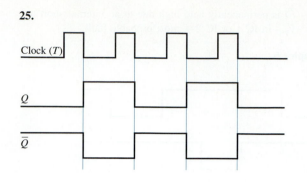

27.

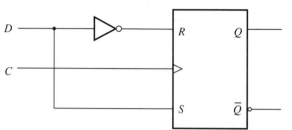

29.

Q^n	S	R	D	Q^{n+1}
0	0	0	0	0
0	0	1	0	0
0	1	0	1	1
0	1	1	1	1
1	0	0	1	1
1	0	1	0	0
1	1	0	1	1
1	1	1	1	1

31.

X	Y	Q^n	Q^{n+1}	
0	0	0	1	Set
0	0	1	1	
0	1	0	0	NC
0	1	1	1	
1	0	0	1	TOGGLE
1	0	1	0	
1	1	0	0	RESET
1	1	1	0	

$$Q^{n+1} = \overline{Y}\,\overline{Q}^n + \overline{X}Q^n$$

33.

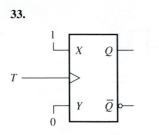

35.

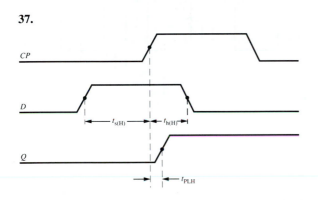

37.

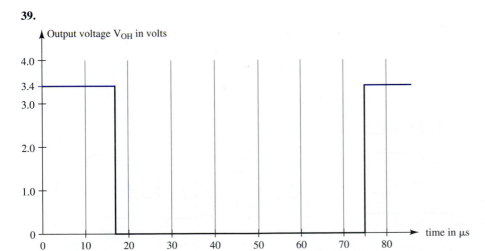

39.

41. By the capacitor that holds the input voltage low.
43. $R_x = 1.6$ K, $C_x = 1\mu$F
45. $C = 0.072\mu$F
$R_A = 500$, $R_B = 10$ K
47. 20 MHz

Chapter 6

1.

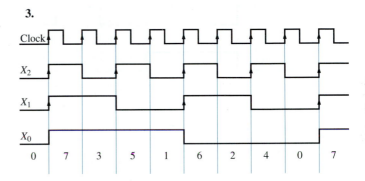

3.

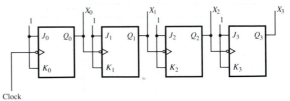

5.

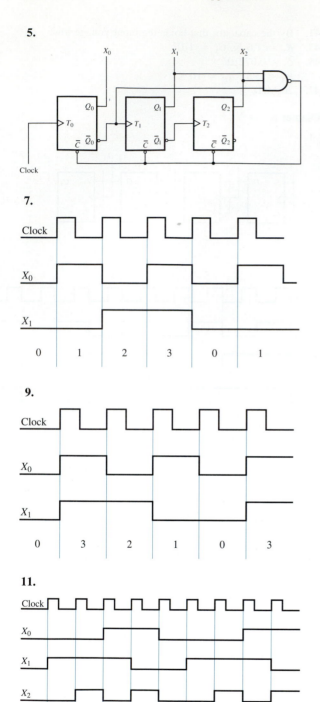

13.

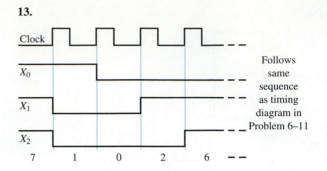

Follows same sequence as timing diagram in Problem 6–11

7.

9.

11.

15.

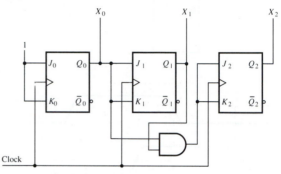

17.

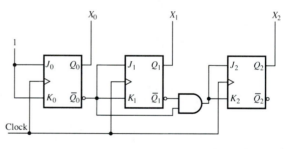

19.

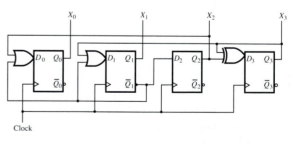

21.

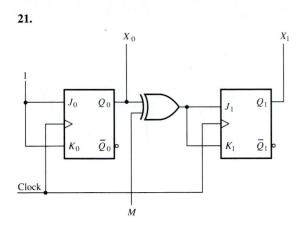

23.

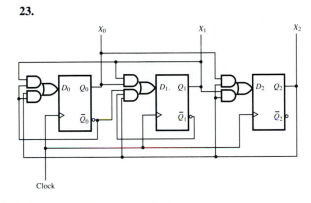

25.

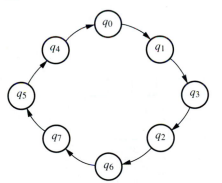

27. q_1 to q_7, q_2 to q_0, q_4 to q_F, q_5 to q_F, q_6 to q_9, q_8 to q_F, q_A to q_A, q_B to q_A, q_C to q_7, q_D to q_7, q_E to q_3

29.

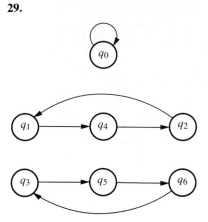

31.

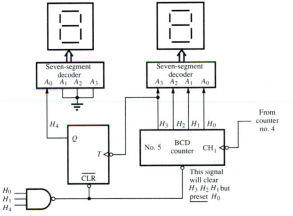

33. Output frequency $= 20$ Hz

35.

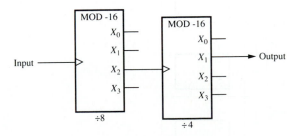

37.

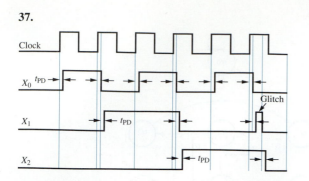

39. Possible remedies: (a) replace asynchronous counter with a synchronous MOD-8 counter, (b) enable the decoder on the negative edge of the clock pulse (*see* Fig. 6.62)

Chapter 7

1.

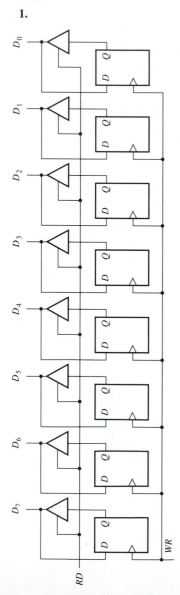

3.

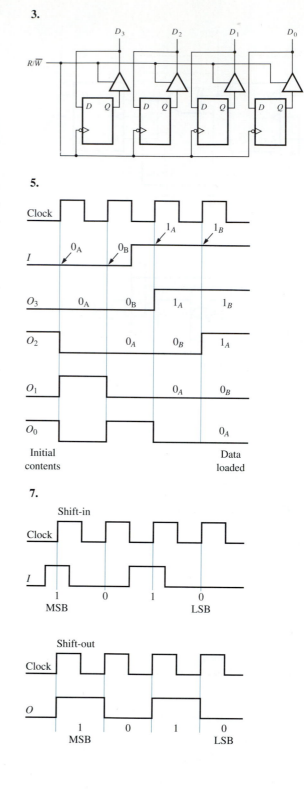

5.

7.

9.

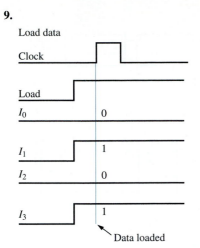

Load data

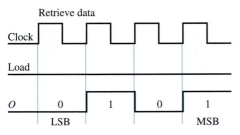

11.

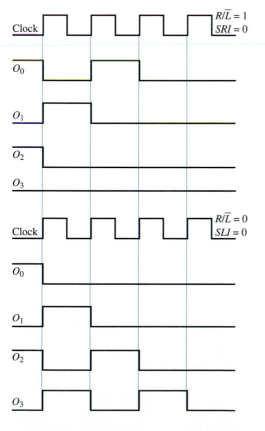

13.

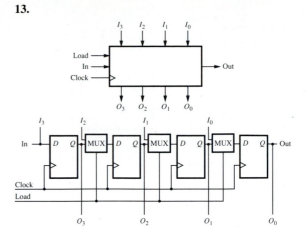

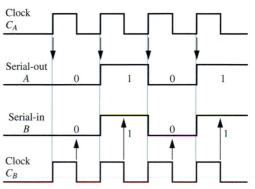

15.

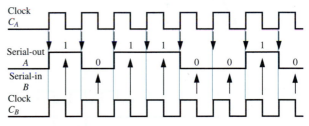

17.

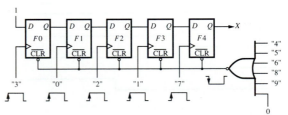

19.

21.

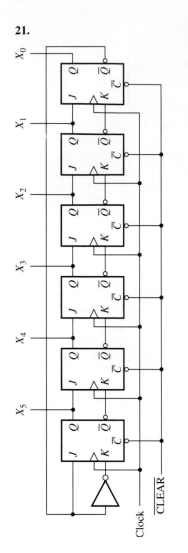

27.

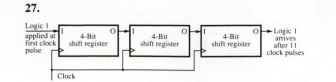

29.

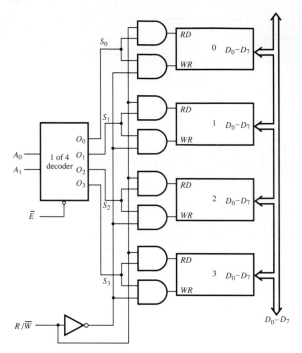

23.

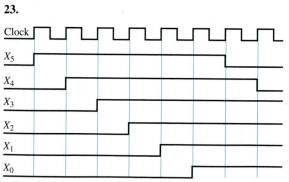

25. $0.3 \ \mu s$

31.

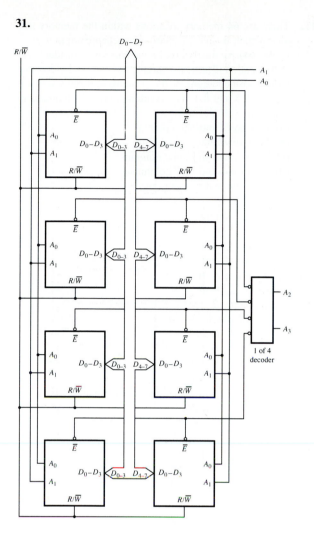

33.

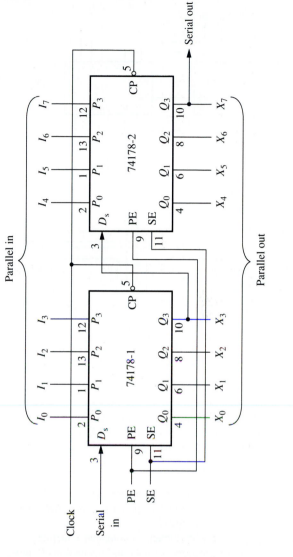

35.

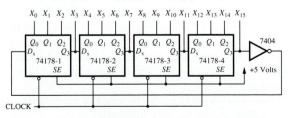

37.

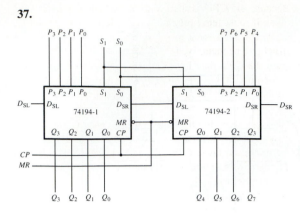

39.

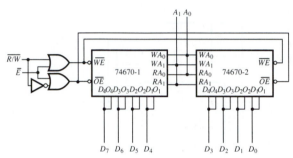

41. Problem: The 74178 is triggered by the negative edge of the clock pulses whereas the 74379 is triggered by the positive edge of the clock pulses. This leads to an error of one half a clock period or 0.25 μs. Solution: Invert the clock connected to the 74379 IC.

Chapter 8

1. Program instructions are interpreted by the computer and tell it which computer operation (*AND, OR, ADD,* etc.) is to be performed.

3. As is the case with humans, a computer's memory serves as a depository for data and instructions (a program) to achieve a task. A computer refers to this memory to retrieve and store instructions and data.

5. Main memory is the working memory of the system in that it is main memory from which the CPU works. Mass storage memory functions as a memory that can be used to store programs and data currently not being operated on.

7. Main memory locations can be accessed much faster than mass storage memory locations.

9. The five basic categories for classifying memories are: (1) organization, (2) manner of accessing, (3) alterability of memory contents, (4) contents of memory retained when power is removed, (5) contents of memory retained even with power applied.

11. There are 64 memory locations within the memory device, which for an *x-y* coordinate scheme requires six ($2^6 = 64$) parameters (columns and rows) to identify each location. These six parameters can be divided into various combinations of the columns and rows seen in Figures 8.3 and 8.4. The combinations of columns and rows are

(a) Five rows and one column
(b) Four rows and two columns
(c) Three rows and three columns
(d) Two rows and four columns
(e) One row and five columns

13. For Problem 11

Rows + Columns = Sum

5	+ 1	=	6
4	+ 2	=	6
3	+ 3	=	6
2	+ 4	=	6
1	+ 5	=	6

For Problem 12

Rows + Columns = Sum

4	+ 1	=	5
3	+ 2	=	5
2	+ 3	=	5
1	+ 4	=	5

From the above patterns, the possible patterns for N memory locations are:

Rows	+	Columns	= Sum
$N-1$	+	1	= N
$N-2$	+	2	= N
$N-3$	+	$N-(N-3)$	= N
$N-4$	+	$N-(N-4)$	= N
.	+	$\cdots$	= .
.	+	$\cdots$	= .
.	+	$\cdots$	= .
$N-(N-1)$	+	$N-[N-(N-1)]$	= N

15.

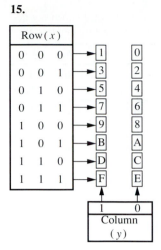

The number of memory cells (bits) per location does not affect the technique used to address memory locations. That is, the address of a memory location is the same regardless of the number of storage cells at that location.

17. Because to address any row-column combination within the memory cell array is the same ($2\ \mu s$) then every location within the array has the same access time.

19. $2 \times 0.2\ \mu s = 0.4\ \mu s$

21. $1.2\ \mu s$

23. Yes. RAM is an acronym for random access memory and all semiconductor ROMs are accessed in this manner.

25. MROMs

27. Either PROMs or EPROMs, it would probably depend on if the company decided to purchase just an EPROM programmer or both an EPROM and PROM programmer (an engineering firm will likely have an EPROM programmer but maybe not a PROM programmer).

29. Volatility describes a memory that cannot retain its contents without power.

31. DRAMs have a much larger memory capacity than SRAMs, hence the increased memory size justifies the additional expense for the refresh control circuitry.

33. Yes. Both lose their contents when power is removed.

35. The CPU controls read and write operations via its read (RD) and write (WR) control signals.

37. (a) For eight address pins there will be four row address pins and four column address pins.
(b) Five row address pins and five column address pins.
(c) Six row address pins and five column address pins.
(d) Six row address pins and six column address pins.

39. Six row address pins and six column address pins. A total of 4096 locations ($64 \times 64 = 4096$, or $2^6 \times 2^6 = 2^{12} = 4096$).

41. Because the CPU is the program execution function-block it must fetch program instructions from memory, hence the need for it to supply the address of the instructions in memory. In the course of executing a program the CPU may have to read and write data from or to memory and I/O, which again requires that the CPU furnish the address of the accessed memory location or I/O device.

43. The row and column select lines must be activated, for then and only then can a read or write operation be performed on those cells. To activate these select lines the memory location address must be input to the memory device.

45. Only the accessed memory cell in each plane is active and therefore only it has control of output lines.

47.

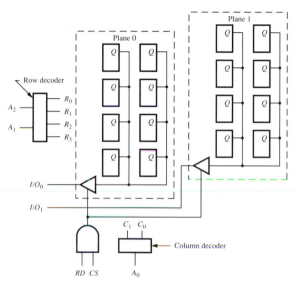

49. A page of memory is a group of memory locations with sequential addresses that have been grouped together to form a block of memory.

One page of memory = 2^N memory locations,

where N is either 10 (if a page of memory is defined to be 1 K) or the number of address pins of the memory IC with the least number of memory locations.

51. (a) $2^{11} = 2048$
(b) One page = 2048 memory locations
(c) $2^{13} = 8192$. Therefore, 13 address bus lines are required.
(d) A total of 15 ($11 + 4$) address bus lines
(e) The control signal $IO/\overline{M}$.

53. A_{14} through A_{11} are used to select memory devices.

55.

Decoding the linear addresses of Problem 55

Address bus lines		A_{14}	A_{13}	A_{12}	A_{11}	A_{10}	A_9	A_8	A_7	A_6	A_5	A_4	A_3	A_2	A_1	A_0
ROM-0	0800H	0	0	0	1	0	0	0	0	0	0	0	0	0	0	0
	0FFFH	0	0	0	1	1	1	1	1	1	1	1	1	1	1	1
RAM-1	1000H	0	0	1	0	0	0	0	0	0	0	0	0	0	0	0
	17FFH	0	0	1	0	1	1	1	1	1	1	1	1	1	1	1
RAM-2	2000H	0	1	0	0	0	0	0	0	0	0	0	0	0	0	0
	27FFH	0	1	0	0	1	1	1	1	1	1	1	1	1	1	1
RAM-3	4000H	1	0	0	0	0	0	0	0	0	0	0	0	0	0	0
	47FFH	1	0	0	0	1	1	1	1	1	1	1	1	1	1	1

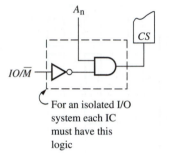

For an isolated I/O system each IC must have this logic

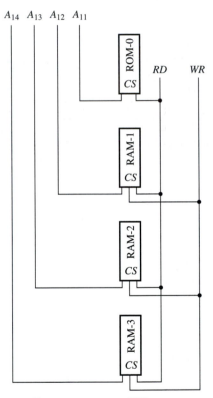

For a memory mapped I/O system

57.

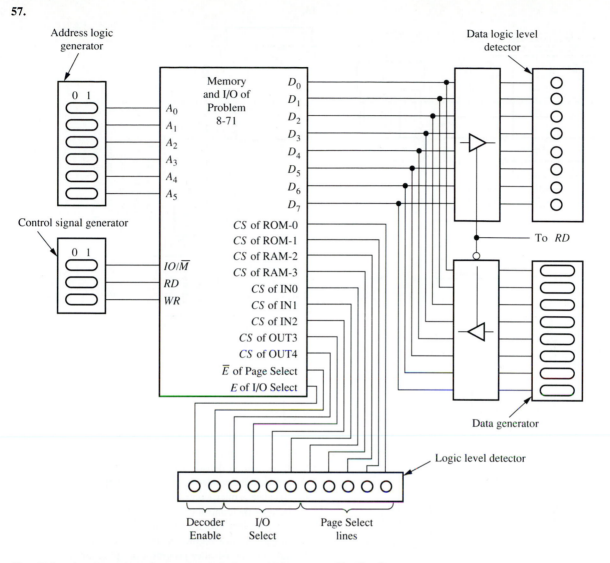

Address logic generator

Control signal generator

Data logic level detector

To *RD*

Data generator

Logic level detector

Decoder Enable   I/O Select   Page Select lines

59. Using the discussion of control and timing and Figure 8.19 as guides. Control signal *RD* is used as an external trigger source for the logic analyzer. Hence, storing just the data on the data bus at the sample times (the leading edge of *RD*) we have the timing diagram and FIFO as shown.

Chapter 9

1. To store temporary data and intermediate data resulting from ALU operations.

3.

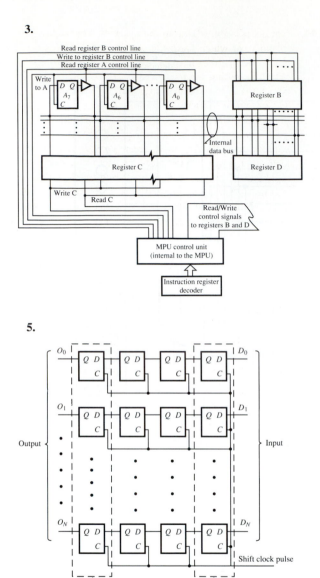

5.

A 4 × N buffer memory

7. Yes

9. The MPU can prefetch instructions from memory and store them in its queue. It will fetch these instructions when the data bus is not used for instruction executions, conserving use of the data bus. It also saves time when the MPU is ready to fetch and execute an instruction because the fetch is internal (from the queue) and the MPU will not have to wait for use of the data bus.

11. Whenever its use would save MPU time by reducing program execution time, such as in a program requiring long, complex calculations.

13.

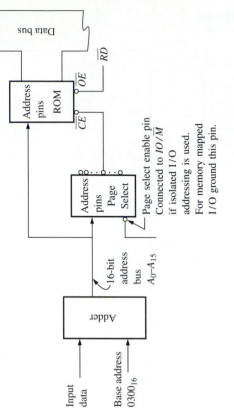

15. Table 9.2 shows that ECL is the fastest technology.

17. It has a high density with fairly fast operating characteristics and moderate power consumption.

19. Because a microcontroller performs very specific tasks, such as the control of some mechanism, its programs are also very specific and do not change. Also, these programs must be present from a cold start. As a result, microcontroller programs are programmed in ROM and usually are relatively small. RAM is usually also very small because if any large quantities of data are to be collected, for example, if a data base was being established, then the collected data would be down loaded into mass storage. Some microcontroller systems are contained on a single chip.

21. Number of rows = 2^7 = 128 (agrees)
Number of columns = 2^4 = 16 (disagrees)
The disagreement as to the number of columns can be accounted for by considering the planes of memory cells, which are eight in number. The manufacturer considers that for each column selected there are eight memory cells (the z coordinate), hence

$$\text{Number of columns} = 16 \times 8 = 128$$

23. Three MPU initiated events occur for every read and write operation and these are (1) the address being output on the address bus, (2) the MPU generated control signal being output on the control bus, (3) the data byte appearing on the data bus. Because the address being output on the address bus is the first event, all others are referenced from it.

25. The chip enable is used to enable the chip for both read and write operations whereas the output enable is only for read operations. Refer to the architecture of Figure 9.7.

27. Figure 9.9 states that the number of columns includes the number of memory planes (eight). Hence, the number of columns would have been stated as 128 (16×8) in architectures such as that shown in Figure 9.7.

29. Figure 9.10b shows that before the write enable pin is active its logic level is a high (inactive for a write), which is the logic level for a read. Because the chip select is active (not shown) the datum is read from the addressed memory location and the datum is output during this time.

31. It is the access time parameter. The value of this parameter is compared to t_{AD} of MPU AC parameters for compatibility, where $t_{AD} > t_{ACC}$. That is, the 2764 must be fast enough to respond to MPU read commands.

33. Refer to the answer to Problem 23.

35. (a) Compared to t_{AD} of Figure 9.15a it is too large (too slow). Not compatible.
(b) $t_{DW} > t_{CC}$ of Figure 9.15b, not compatible.
(c) $t_{OE} > t_{AD} - t_{AC}$ of Figure 9.15a, therefore they are not compatible.
(d) $t_{DS} > t_{CC}$ of Figure 9.15b, not compatible.
See Table 9.6 for MPU values (Fig. 9.15).

37. These enable pins ($\overline{E}_1$, $\overline{E}_2$, and E_3) serve the same function as a logic gate connected to address lines when detecting a specific logic condition, which is the process of decoding those three address bus lines. Hence, utilizing the enable pins for decoding three address bus lines results in as many as eight 74LS138s being paralleled, which allows expansion up to 64 pages of addresses.

39. The page selector is used to select pages of memory. It does this by enabling the chip select pin of the memory IC containing the page being addressed. Once the chip is enabled then the selected memory IC decodes the address at its address pins, which accesses the addressed memory location within the addressed page. Partial linear addressing uses address lines from the address bus to enable the chip select pins of memory ICs rather than the page select lines of the page selector. Hence, the address bus lines replace the function of the page selector.

41. Not necessarily, it's up to the designer. For a small memory, of which hardware cost is of prime importance, then neither have to be prevented. It really is based on getting the most addresses from memory and convenience of programming, relative to addressing.

43. The output of the page selector is used to detect a page of memory within a memory device by enabling that memory device. When multiple pages of memory exist within a memory device, those outputs of the page selector that address those pages are input to a single logic gate. The type of logic gate used (*AND*, *OR*, etc.) will be such that the required logic level to enable that memory device will be output by it whenever any one of the page select outputs is active. For the systems we have designed (active-lows) the logic gate used is an *AND*-gate.

45. Whenever the address space requirements are relatively small, that is, the number of required addresses are small enough that the designer need not be concerned with using all possible addresses, as obtained with a fully decoded addressing system. Partial linear addressing also simplifies the design and reduces cost (a decoder is not required).

47.

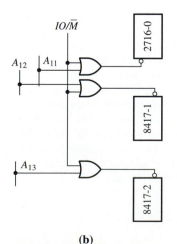

| 2716-0 |
| 3000H–33FFH | ← Note the beginning address of the 2716 |
| 8417-1 |
| 2800H–2FFFH |
| 8417-2 |
| 1800H–1FFFH |

Memory map
(a)

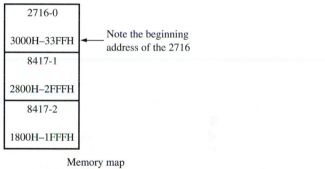

(b)

Address table. Address bus table for Problem 47

Address bus lines	A_{15}	A_{14}	A_{13}	A_{12}	A_{11}	A_{10}	A_9	A_8	A_7	A_6	A_5	A_4	A_3	A_2	A_1	A_0
2716-0						$\overline{CE}$	A_9 ←								→	A_0
8417-1					$\overline{CE}$	A_{10} ←									→	A_0
8417-2				$\overline{CE}$		$\overline{A}_{10}$ ←									→	A_0
2716-0	0	0	1	1	0	0	0 ←								→	0
3000H–33FFH	0	0	1	1	0	0	1 ←								→	1
8417-1	0	0	1	0	1	0 ←									→	0
2800H–2FFFH	0	0	1	0	1	1 ←									→	1
8417-2	0	0	0	1	1	0 ←									→	0
1800H–1FFFH	0	0	0	1	1	1 ←									→	1

49.

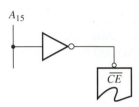

A_{15}

$\overline{CE}$

51. Because the two are pin compatible there is no physical difference. Tables 9.4 and 9.5 show some timing differences, but because all read and write operations will be performed according to the MPU's timing of Figure 9.15 the resultant read and write cycle time will be the same (both are AC compatible). There will be a reduction in power consumption according to Table 9.2.

53. Apply various logic level combinations to address A_0 through A_2 of the decoder and verify the output logic levels.

55. Ground the chip enable and output enable pins. Apply an address to the address pins and detect the corresponding datum output on the data bus. Because the chip enable pin was stuck high (internally) the data bus would be floating (Hi-Z) rather than having the content of the addressed memory location.

57. Write a program that sequentially reads memory locations within 2764-0 of Figure 9.17. Monitor the address, enable, and output pins via the data bus using the logic analyzer. When the enable pins are low check the output pins to verify that the proper datum appeared on the data bus. If the data bus was continuously floating we would know that the enable pins were not functioning. Do not be concerned with which enable pin is internally malfunctioning because if either fails the chip must be replaced.

Chapter 10

1. Address bus: to transmit memory and I/O addresses output by the MPU. It also transmits addresses to main memory from an I/O device having DMA capability.
Data bus: to provide a path for the transmission and receiving of data from or to those devices connected to it.
Control bus: the bus over which control signals are transmitted.

3. The two selectors of Figure 10.1 are the page selector and the I/O selector. The use of a page selector and an I/O selector indicates that this is a fully decoded, isolated addressing scheme. The page selector enables (selects) the memory IC containing the addressed page of memory whereas the I/O selector enables the I/O (port) being addressed.

(a) If there were a common selector (decoder) selecting both memory and I/O, then the addressing scheme would be fully decoded and memory mapped (memory mapped because control signal $IO/\overline{M}$ is not being used).

(b) If no selectors (decoders) were present, but rather address bus lines were used to select memory and I/O, this would be a partial linear addressing scheme that is memory mapped because $IO/\overline{M}$ is not being used.

5. (a) A DMA request is made to the MPU by the I/O device with DMA capability. The MPU grants request by activating its *DMA ACK* pin and at the same time putting its address, data, and control signal pins in the Hi-Z state. The I/O device then takes control of the three busses and begins the DMA process.

(b) The I/O device wishing to be served by the MPU will make the interrupt request to the MPU. When the MPU can service the I/O device it so indicates by activating its *INT REQ* pin.

7. **(a)** The MPU can address memory via its
PC: For fetching instructions from memory.
SP: For accessing that portion of main memory
designated as the stack.
DC: For accessing memory data locations
(registers D and C form the DC).
(b) To address I/O, register pair *I/OPH* and *I/OPL* are
used.

9. Those locations addressed by the PC contain program
instructions whereas the content of those addressed by
the SP is data (return address or register data).

11. The CU loads the address of the instruction to be
fetched from the PC into the address latch (*see* Problem 8). The CU then loads address of the instruction
on to the address bus by sending a signal to the address latch, activating its output buffers connected to
the address bus (refer to Fig. 10.4). This address selects
the memory IC containing the page of memory within
which the instruction resides, via the page selector
(refer to Fig. 10.1). The address also selects the specific memory location within that page of memory.
When the MPU is ready to read the addressed location,
containing the instruction to be fetched, it generates
(via the CU) and outputs the read control signal $\overline{RD}$.
The read control signal activates the output pin of the
addressed memory IC, which loads the content of the
addressed location (the instruction) on to the data bus.
The CU then sends a strobe pulse to the data bus
buffer/latch (refer to Fig. 10.3), causing the data bus
buffer/latch to latch the instruction from the data bus.
From there the CU will cause the data bus buffer/latch
to load the instruction on to the internal data bus from
where the CU will cause the instruction queue (if present) to latch it. The instruction queue will move the
instruction to the IR as instructions in the queue are
executed. Once the instruction is in the IR the IR decoder decodes the instruction, sending signals to the
CU instructing it as to which instruction to be executed. After the instruction has been executed the CU
moves the next instruction in the queue from the queue
and into the IR. It then increments the PC in prepara-
tion to fetch the next instruction. Basically this cycle is
repeated continuously.

13. The CU is a special purpose computer that is dedicated
to the execution of instructions.

15. A microprogram is part of the CU and instructs the CU
as to the execution of instructions. A macroprogram is
stored in main memory and is executed by the CU.

17. The ALU has two inputs, *ACC*1 and *ACC*2.

19. An interrupt will cause the MPU to stop the execution
of one portion of a program and jump (vector) to another and begin program execution there. A reset does
the same thing. It is a reset that is singled out because
a reset always jumps back to the beginning program.

21.

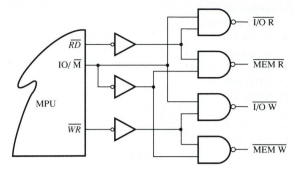

23. The MPU will output address 102CH on the address
bus. At time t_{AC} the read control signal $\overline{RD}$ goes low,
which enables memory. The content of memory loca-
tion 102CH (*MVI B*) is loaded on the data bus before
time t_{AD}. At time $t_{AC} + t_{CC}$ the MPU latches the binary instruction code for *MVI B* from the data bus and
loads it in the queue, via the internal data bus. The CU
increments the PC, which is 102DH (the address of the
second byte of the instruction *MVI B*,50H). The second
byte is fetched and loaded in the queue and again the
CU increments the PC (PC = 102EH). This address is
used to fetch the instruction *ADD ACC1,B*, which is
also loaded in the queue. The queue now has 3 bytes,
and they are the binary codes (machine codes) for *MVI
B*,50H and *ADD ACC1,B*. The CU will move those
bytes that are op-codes from the queue to the IR to be
executed. When byte *MVI B* is loaded in the IR and
decoded by the IR decoder, the CU will execute *MVI
B*, which will cause the next byte (50H) to be loaded
from the queue on to the internal data bus from which
register B will latch it. When byte *ADD ACC1,B* is
moved in the IR the CU will add the contents of register *ACC*1 and *B* and store the result in register *ACC*1,
as well as setting the ALU flags accordingly.

25. From Figure 10.7, by driving the write control signal
$\overline{WR}$ low.

27. Mnemonics are acronyms that indicate the operation of
an instruction and serve as memory aids.

29. An assembler is a program that translates mnemonics
into machine code (object code).

31. The designer will determine what must be coded in an
instruction, such as operation and register codes, and
the pattern with which they should appear for purposes
of being decoded by the IR decoder.

33. **(a)** 00000110 with $X = 0$.
    (b) 00000011
    (c) 00100110 with $X = 0$.
    (d) 01100000
        00000000 = 00H
        00000101 = 05H
    (e) 10011100
    (f) 10101011
    (g) 10111100
    (h) 10100111
    (i) 10110000
        01111010 = 7AH (immediate data)
    (j) 10100000
        00111011

35. Three bits are needed to identify a second register and because only 1 bit is available (T_3) it cannot be programmed with these templates.

37. An I/O port is the interface that enables an I/O device to be connected to the data bus.

39.

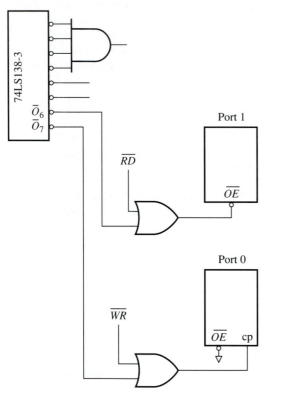

41. The *NAND* gate provides the read/write timing as well as isolation from memory.

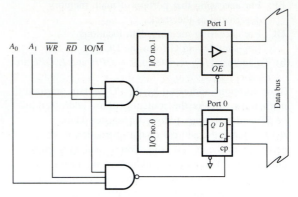

43.

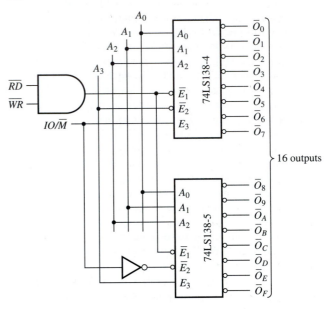

45.

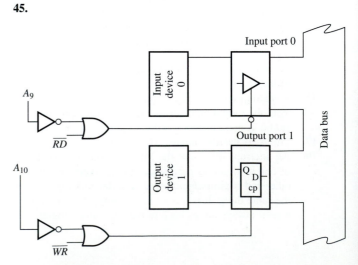

47. When the MPU is reset its PC is reset to 0000H. This address is to address memory location zero in page zero (the first instruction of the operating system). Without the inverters address bus lines A_8 to A_{11} would be active low and then these ports would also be selected (addressed) upon a system reset. This would result in bus contention.

49. The control signal will be one that differentiates between I/O and memory addresses loaded on the address bus by the MPU. This signal ($IO/\overline{M}$) makes the differentiation via its logic level. There are two I/O related instructions, *IN* and *OUT*.

51. The 6800 and 8085 are 8-bit MPUs whereas the 8086 is a 16-bit MPU (refer to its internal data bus size). The 8086 has an instruction queue, whereas neither the 6800 nor 8085 have queues. The 8086 has two architectures in one: the EU and the BIU. The BIU interfaces with the external data bus and therefore to memory and I/O. The EU executes instructions. Memory addresses are segmented by the 8086 via the segment register *CS, DS, SS,* and *ES*. The EU architecture is very similar to the 8085 architecture. The function-block identified as the EU control system is basically composed of 8085 function-blocks, IR, IR decoder, and the CU.

53. Test the I/O select decoder (74LS138-4) by applying the I/O address for port 1 via the MPU (program the MPU to read port 1). Using a logic analyzer verify that the MPU has output the appropriate address and control signal logic levels. If any of these logic levels are not correct (but the program is correct), the MPU has failed. If these logic levels are correct, test the *AND* gate on $\overline{E}_1$ and then test the decoder by verifying the proper output logic level of the decoder $\overline{O}_1$. If it is high then the decoder has failed, whereas if it is low, the port 1 (buffer) has failed.

55. Program the MPU to read from memory chip 2716-0 where the contents of those memory locations are known. Then, using a logic analyzer, monitor the logic levels of the address bus, control signals, and data bus. If these logic levels are correct test the page select decoder by testing for proper input logic levels. This requires that the *NOR* gate be tested first. If the logic levels being input to the decoder are correct, then check the four outputs of 74LS138-3 ($\overline{O}_0$ through $\overline{O}_3$) to verify that the correct output is low. If all are high then the decoder has failed. If any one of these output is low the output of the *AND* gate must be low, and if not, it has failed. If the output of the *AND* gate is low and the output enable pin is also low but the data on the data bus is not correct, the 2716-0 has failed or there is a wiring error on one or more of its address pins.

Chapter 11

1. Fuse-links allow a connection to remain (not blown) or to be broken (blown). As a result it is the fuse-link that enables a user to program a PLD.

3.

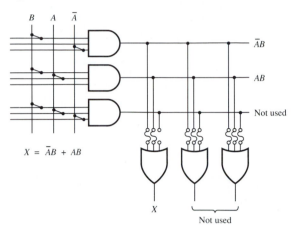

$$X = \overline{A}B + A\overline{B}$$

5.

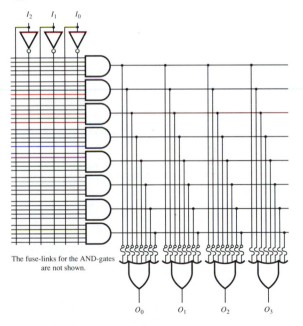

The fuse-links for the AND-gates are not shown.

7. There must be one *AND* gate for each minterm, which is 2^N, where N equals the number of independent variables (inputs).

9. $2^6 = 64$ *AND* gates.

11. One input for each minterm. This allows each output to possibly have each minterm as part of its descriptive Boolean expression.

13.

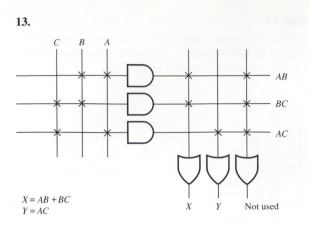

$X = AB + BC$
$Y = AC$

X   Y   Not used

15.

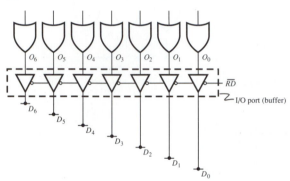

23.

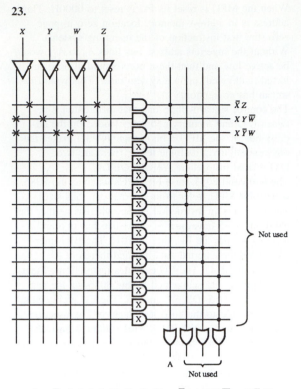

A   Not used

$A = \Sigma_m\ 1,\ 3,\ 5,\ 7,\ 10,\ 11,\ 12,\ 13 = \overline{X}Z + X\,Y\,\overline{W} + X\,\overline{Y}\,W$

Let $X \equiv$ MSB and $Z \equiv$ LSB in the K–map

17. The *AND* gate array is programmed by the manufacturer.
19. (a) PLE4P7 (rather than P it would really be an H, which represents an active high.)
    (b) PLE4R7
21. It can be programmed to detect sum-of-product terms other than just minterms.

25.

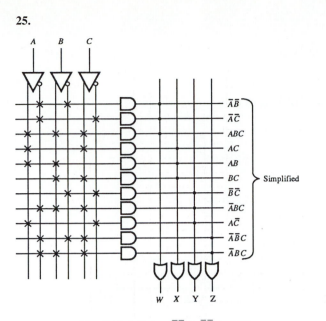

Simplified

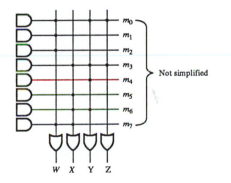

$W = \Sigma_m\ 0, 1, 2, 7 = \overline{AB} + \overline{AC} + ABC$

$X = \Sigma_m\ 3, 5, 6, 7 = AC + AB + BC$

$Y = \Sigma_m\ 0, 3, 4, 6 = \overline{BC} + \overline{A}BC + A\overline{C}$

$Z = \Sigma_m\ 0, 3 = \overline{A}\overline{B}\overline{C} + \overline{A}BC$

Not simplified

27. If the input fuse-links and *AND* gate 1 are left intact, its output would be a logic 0, which would force the output buffer into the Hi-Z state. Then O_n would be an input pin.

29.

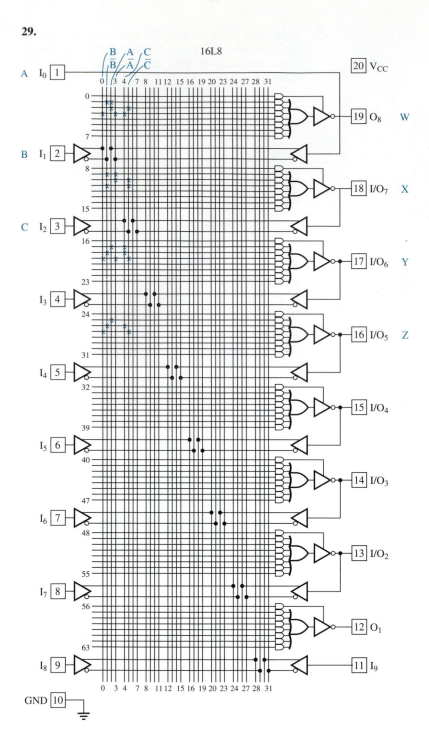

31.

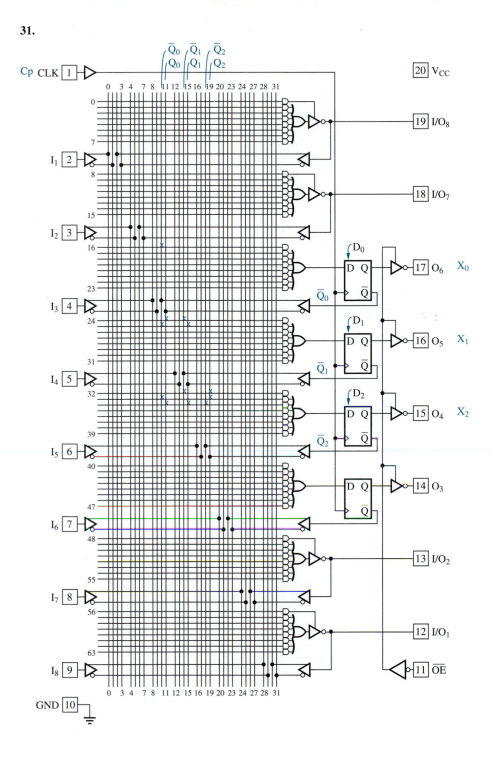

33.

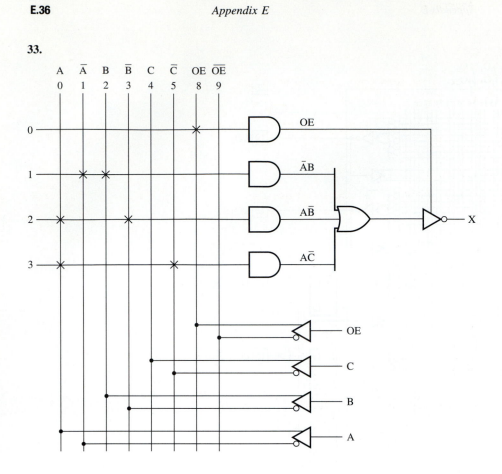

35.

```
;PALASM Design Description

;-------------------------------- Declaration Segment ------------
TITLE      problem 35
PATTERN    prob35.pds (design file)
REVISION   a
AUTHOR     Ismail & Rooney
COMPANY    University of Dayton
DATE       03/21/93

CHIP   prob35   PAL16L8

;-------------------------------- PIN Declarations ---------------

PIN   1         A                      ;INDEPENDENT VARIABLES (INPUTS)
PIN   2         B
PIN   3         C
PIN   4         OE
PIN   5         NC
PIN   6         NC
PIN   7         NC
PIN   8         NC
PIN   9         NC
PIN  10         GND
;
PIN  11         NC
PIN  12         NC
PIN  13         NC
PIN  14         NC
PIN  15         NC
PIN  16         Z              ; MINTERMS: 0,3
PIN  17         Y              ; MINTERMS: 0,3,4,6
PIN  18         X              ; MINTERMS: 3,5,6,7
PIN  19         W              ; MINTERMS: 0,1,2,7
PIN  20         VCC

;-------------------------------- Boolean Equation Segment ------

EQUATIONS

;UNSIMPLIFIED SOP EQUATIONS (NON-INVERTED)

W = /A * /B * /C  +  /A * /B * C  +  /A * B * /C  +  A * B * C

X = /A * B * C  +  A * /B * C  +  A * B * /C  +  A * B * C

Y = /A * /B * /C  +  /A * B * C  +  A * /B * /C  +  A * B * /C

Z = /A * /B * /C  +  /A * B * C

W.TRST = VCC

X.TRST = VCC

Y.TRST = VCC

Z.TRST = VCC
```

37.

```
TITLE      2-bit Magnitude Comparator (Fig 4-28)
PATTERN    prob37.pds (design file)
REVISION   A
AUTHOR     Ismail & Rooney
COMPANY    University of Dayton
DATE       03/21/93

CHIP  prob37  PAL20V8

PIN  1          B1              ; 2-BIT INPUT B
PIN  2          BO
PIN  3          A1              ; 2-BIT INPUT A
PIN  4          AO
PIN  5          NC
PIN  6          NC
PIN  7          NC
PIN  8          NC
PIN  9          NC
PIN  10         NC
PIN  11         NC
PIN  12         GND
;
PIN  13         NC
PIN  14         NC
PIN  15         NC
PIN  16         NC
PIN  17         NC
PIN  18         NC
PIN  19         L               ; A < B
PIN  20         G               ; A > B
PIN  21         E               ; A = B
PIN  22         NC
PIN  23         NC
PIN  24         VCC

EQUATIONS

E = /A1 * /AO * /B1 * /BO  +  /A1 * AO * /B1 * BO  +  A1 * AO * B1 * BO
  + A1 * /AO * B1 * /BO

G = A1 * /B1  +  AO * /B1 * /BO  +  A1 * AO * /BO

L = /A1 * B1  +  /A1 * /AO * BO  +  /AO * B1 * BO

SIMULATION

TRACE_ON A1 AO B1 BO  E G L

SETF /A1 /AO /B1 /BO
SETF /A1 /AO /B1  BO
SETF /A1 /AO  B1 /BO
SETF /A1 /AO  B1  BO
SETF /A1  AO /B1 /BO
SETF /A1  AO /B1  BO
SETF /A1  AO  B1 /BO
SETF /A1  AO  B1  BO
SETF  A1 /AO /B1 /BO
SETF  A1 /AO /B1  BO
SETF  A1 /AO  B1 /BO
SETF  A1 /AO  B1  BO
SETF  A1  AO /B1 /BO
SETF  A1  AO /B1  BO
SETF  A1  AO  B1 /BO
SETF  A1  AO  B1  BO

TRACE_OFF

;-----------------------------------------------------------

PALASM SIMULATION SELECTIVE TRACE LISTING

Title     : 2-bit Magnitude Comparator    Author   : Ismail & Rooney
Pattern   : prob37.pds (design file)      Company  : University of Dayton
Revision  : A                             Date     : 03/21/93

PAL20V8
Page : 1

       gggggggggggggggg
  A1   LLLLLLLLHHHHHHHH
  AO   LLLLHHHHLLLLHHHH
  B1   LLHHLLHHLLHHLLHH
  BO   LHLHLHLHLHLHLHLH
  E    HLLLLHLLLLHLLLLH
  G    LLLLHLLLHHLLHHHL
  L    LHHHLLHHLLLLHLLLL
```

39.

a. $X3 = /X1 * X0 + X1 * X3$

b. $X0 = X0 : + : (X1 : + : X0)$

c. $X1 = X2 * (X1 + /X3)$

d. $X2 = /X1$

41.

```
TITLE      2-bit UP counter
PATTERN    prob41.pds (design file)
REVISION A
AUTHOR     Ismail & Rooney
COMPANY    University of Dayton
DATE       03/21/93

CHIP   prob41  PAL16R4

PIN  1          CP         ;Clock input
PIN  2          NC
PIN  3          NC
PIN  4          NC
PIN  5          NC
PIN  6          NC
PIN  7          NC
PIN  8          NC
PIN  9          NC
PIN  10         GND
;
PIN  11         OE         ;Output enable
PIN  12         NC
PIN  13         NC
PIN  14         NC
PIN  15         NC
PIN  16         X1         ;Counter output
PIN  17         X0         ;Counter output
PIN  18         NC
PIN  19         NC
PIN  20         VCC

EQUATIONS

X0 := /X0
X1 := X1 :+: X0

SIMULATION

TRACE_ON CP OE X1 X0

SETF /OE
CLOCKF CP
CLOCKF CP
CLOCKF CP
CLOCKF CP
CLOCKF CP
CLOCKF CP
CLOCKF CP
CLOCKF CP

TRACE_OFF
;------------------------------------------------------------
```

```
PALASM SIMULATION SELECTIVE TRACE LISTING

Title    : 2-bit UP counter          Author   : Ismail & Rooney
Pattern  : prob41.pds (design file)  Company  : University of Dayton
Revision : A                         Date     : 03/21/93

PAL16R4
Page : 1

      gc c  c  c  c  c  c  c  c
CP    LHHLHHLHHLHHLHHLHHLHHLHHL
OE    LLLLLLLLLLLLLLLLLLLLLLLLL
X1    HHLLLLLLHHHHHHLLLLLLHHHHH
X0    HHLLLHHHLLLHHHLLLHHHLLLHH
```

43.

```
TITLE      7-5-3-1 Counter (State Machine)
PATTERN    prob43.pds (design file)
REVISION A
AUTHOR     Ismail & Rooney
COMPANY    University of Dayton
DATE       03/21/93

CHIP   prob43 PAL20V8

PIN   1           CP              ;CLOCK INPUT
PIN   2           NC
PIN   3           NC
PIN   4           NC
PIN   5           NC
PIN   6           NC
PIN   7           NC
PIN   8           NC
PIN   9           NC
PIN  10           NC
PIN  11           NC
PIN  12           GND
;
PIN  13           NC
PIN  14           NC
PIN  15           NC
PIN  16           NC
PIN  17           NC
PIN  18           NC
PIN  19           X2              ;OUTPUT
PIN  20           X1              ;OUTPUT
PIN  21           X0              ;OUTPUT
PIN  22           NC
PIN  23           NC
PIN  24           VCC

STATE

MOORE_MACHINE

S1 =   /X2 * /X1 * X0
S3 =   /X2 *  X1 * X0
S5 =    X2 * /X1 * X0
S7 =    X2 *  X1 * X0

S1 := VCC -> S7
S3 := VCC -> S1
S5 := VCC -> S3
S7 := VCC -> S5

CONDITIONS

SIMULATION

TRACE_ON CP X0 X1 X2

FOR J := 1 TO 8 DO
   BEGIN
     CLOCKF CP
   END

TRACE_OFF
;------------------------------------------------------

PALASM SIMULATION SELECTIVE TRACE LISTING

Title     : 7-5-3-1 Counter (State Machine)    Author  : Ismail & Rooney
Pattern   : prob43.pds (design file)           Company : University of Dayton
Revision  : A                                  Date    : 03/21/93

PAL20V8
Page : 1

        c c c c c c c c
CP    HHLHHLHHLHHLHHLHHLHHLHHLHHL
X0    HHHHHHHHHHHHHHHHHHHHHHHHHHH
X1    HLLLHHHLLLHHHLLLHHHLLLHH
X2    HHHHLLLLLLHHHHHHHLLLLLLHH
```

45. There need not be any relationship because a PLA designer may choose the number of product-terms and/or minterms that can be programmed. However, as a minimum 2^N (N is the number of inputs) is a guideline. Also, it is an engineering judgment as to the number of outputs (*OR* gates) to have available on a PLA IC.

47. Because for PLDs a user does not have access to any internal connections, only external pins can be monitored. As a result, apply static logic levels to input pins (A_0, B_0, etc.) and generate the logic levels indicated in the truth table of Example 11.4. For each of those input levels detect the resultant output level for outputs C, S_0, and S_1. Also compare these values to the logic levels of the truth table. If any disagree either the PLE was not properly programmed or it has failed.

49.

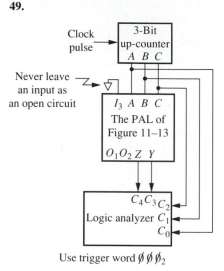

Use trigger word $\emptyset\,\emptyset\,\emptyset_2$

51. Because $Z = \overline{\overline{GH} + GHJ}$, logic levels for G, H, and J are required.

Chapter 12

1. V_{IL} is a maximum voltage for a logic 0 and it just is not of sufficient magnitude.

3. An *AND*.

5. To control the state of Q_2. It does this by directing current either to the base of Q_2 (Q_2 on) or away from it (Q_2 off). When a logic 1 is input to the emitter of Q_1 its base-emitter junction is off but its base collector is on and current is directed to the base of Q_2. When a logic 0 is input to the emitter of Q_1 its base emitter is forward biased and the current from the power supply (V_{CC}) is diverted to ground rather than the base of Q_2.

7. A totem pole is two transistors (BJT in this case) connected in series. Q_2 always turns one of the totem pole transistors on while turning the other off. This keeps the current flow from power supply to ground at a minimum. The electronic switch has just the collector resistor to limit current.

9. Assume an ideal BJT ($V_{be} = 0$), then $I = V_{CC}/R_1$, or $I = 5/4 \times 10^3 = 1.25$ mA. This provides a close approximation (UL = 1.6 mA).

11. Their outputs can be wired together and the result is as if these outputs had been input to an *AND* gate (wired-*AND*).

13. If a negative voltage were accidentally input to the emitter of Q_1 this would possibly cause more current to flow through the base-emitter junction than the transistor was designed to conduct, thus destroying the base-emitter junction. These diodes would be forward biased if, in error, a negative voltage was applied to any of the inputs. Any of these diodes being forward biased would short that input to ground and thus protect Q_1 from drawing too much current.

15. They cannot be driven into saturation due to the base collector Schottky diode, hence they do not build up as much charge on the collector-to-emitter parasitic capacitor, which enables them to be turned off faster.

17. TTL logic has resistors integrated on the IC wafer, and because resistance is proportional to length they require a major portion of the IC wafer area. MOS resistors are MOS transistors (*see* Fig. 12.10) and require very little area.

19. The gate is separated from the channel by an insulator, which ideally draws no current. Hence, any gate current is due to leakage.

21.

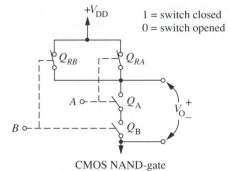

CMOS NAND-gate

23. When the transistors are in series a negated *AND* (*NAND*) exists and when in parallel a negated *OR* (*NOR*) is the resultant logic.

25. Refer to Figures 12.12 and 12.13. By making the logic level of the control element (base or gate) dependent on the complement output element (collector or drain) logic level.

27. The row and column lines must be driven high and the sense lines pulled low for a read. For a read the sense lines are also pulled low, which provides for the on transistor to remain on by providing a path for the emitter current to flow. Because a differential amplifier is connected to both sense lines, it detects which sense line is conducting and outputs the correct voltage logic level. To write to the cell, again, both row and column lines must be driven high and then the appropriate sense line is driven high while the other is pulled low. For a logic 0 ($Q_1 = 0$ and $Q_2 = 1$) to be written into the cell, sense line for Q_1 is pulled low and the sense line for Q_2 is driven high.

29. To provide isolation (buffering) between the data pin D_n and the read and write lines of the cell.

31. The fuse-link remains if a logic 0 is to be programmed in the cell, otherwise it is blown. When a row line goes high all switches in that row with fuse-links intact are turned on and those with blown fuse-links are off. Those transistors that are turned on will short their data lines to ground, which results in a logic 0 on those data lines. Those transistors with blown fuse-links will remain off when their row line is drive high, resulting in their data lines being pulled high by resistor R.

33. 74L (low power). Increase the value of R_3 in Figure 12.7.
    74H (high speed or high power). Decrease the value of R_3 in Figure 12.7.
    74S (Schottky). Use Schottky transistors.
    74LS (low-power Schottky). Increase R_3 and use Schottky transistors.

35. It is a relationship between two logic circuits. The driver is the circuit whose output is connected to an input, which is the driven circuit.

37. The parameters of this table must be such that they are compatible in sign (direction or polarity) and magnitude. Voltage polarities must be the same and current directions must be the opposite. Also, $V_{OL} < V_{IL}$, $V_{OH} > V_{IH}$ and the sink current magnitude must be large enough to handle any source current requirements.

39. Current direction is indicated by the sign that precedes the magnitude of the current. If the sign is positive then the current flows into the gate (the gate sinks the current), and if negative, the current flows away from the gate (the gate sources the current).

41. **(a)** 74L driven by 74C
   $V_{IL} = 0.7$ volts > 0.33 volts $= V_{OL}$: OK
   $V_{IH} = 2.0$ volts < 2.7 volts $= V_{OH}$: OK
   $I_{IL} = -0.18$ mA and $I_{OL} = 3.6$ mA: OK
   $I_{IH} = 1.0$ μA and $I_{OH} = -0.36$ mA: OK
   (b) 74S driven by 74HC
   $V_{IL} = 0.8$ volts $> V_{OL} = 0.33$ volts: OK
   $V_{IH} = 2.0$ volts $< V_{OH} = 3.84$ volts: OK
   $I_{IH} = 50$ μA and $I_{OH} = -4.0$ mA: OK
   $I_{IL} = -2$ mA and $I_{OL} = 4$ mA: OK
   (c) 74H driven by 74AC: they are compatible.
   (d) 74HC driven by 74S: not compatible.
   (e) 74C driven by 74: $V_{IH} = 3.5$ volts $> V_{OH} = 2.4$ volts, hence they are not compatible.

43. $I_{OL} = 8.0$ mA and $I_{IH} = \pm 10$ μA: OK
   $I_{OH} = -0.4$ mA and $I_{IH} = \pm 10$ μA: OK
   They are current compatible.

45. Internal parasitic capacitance that must be charged and discharged.

47. Because each of the chip enable pins is connected directly to address bus lines there is no propagation delay time, as measured from when the address is loaded on the address bus to when a memory IC is enabled. However, there will be a delay from when a memory IC is enabled to when the addressed location outputs its data to the data pins. For memories this delay is known as access time. The output enable pins do not have a propagation delay time, as measured from when the control signal $\overline{RD}$ goes active, because they are directly connected to $\overline{RD}$. As mentioned above, there is a delay from when the output enable is activated to when the data of the addressed location is loaded on the data bus but the effects of this delay are known as access time.

49. Some or all of the output voltage and/or current specifications (V_{OL}, V_{OH}, I_{OL}, and I_{OH}) will not be within the required limits.

51. See the answers to Problems 49 and 50. Measure logic levels and propagation delay times.

53. If too much voltage is dropped across the resistor, V_{OH} of the output will not be large enough to be within specification.

Chapter 13

1. The data sheets of Appendix C show that the control pin of a 74125 requires a low for the buffer to be active, whereas a 74126 requires a high. Then the chip enable pin has an active logic level for a 74125 that is low and for a 74126 that is high.

3.

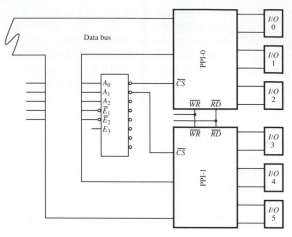

5.

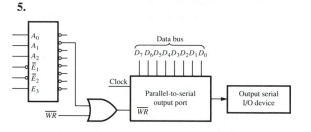

7. The serial data must be shifted, which requires a clock.

9. $e(t) < 0$ then $v_o = -V_{SAT}$
   otherwise $v_o = V_{SAT}$

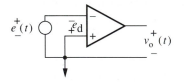

11. The maximum magnitude is V_{SAT}, which can be either positive or negative polarity if both positive and negative power supplies are connected.

13. $e_d = 19.5 / 10^5 = 195$ μV (For $V = 20$ volts), which is approximately 0 volts.

15. $R_f = 183.96$ K ~ 184 K
   $R_0 = R_f / 2.5 = 73.58$ K ~ 74 K
   $R_1 = R_f / 5 = 36.79$ K ~ 37 K
   $R_2 = 18.96$ K ~ 19 K

17. **(a)** $v_o = 1.50$ volts
   (b) $v_o = 1.625$ volts
   (c) $v_o = 1.406$ volts
   (d) $v_o = 1.414$ volts

19. For 0000: $v_o = 0.309$ volts
    For 1000: $v_o = 2.394$ volts
    For 1100: $v_o = 3.958$ volts
    For 1111: $v_o = 4.219$ volts

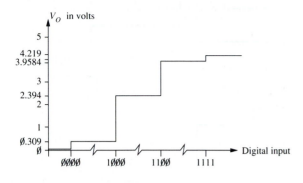

21. $R = 145 \text{ K}/11 = 13.18 \text{ K}$
    $R_f = 10R = 131.8 \text{ K}$
23. $I_f = 0.12 \text{ mA}$

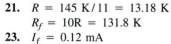

$e(t)$ is the output voltage of the
$R - 2R$ ladder

25. (a) v_o max $= -12.65$ volts
    (b) v_o max $= -13.44$ volts
27. (a) QE $= 0.078$ volts
    (b) QE $= 4.882$ mV
29. (a) $v_o = -0.5859375$ volts
    (b) $v_o = -9.9609375$ volts
31. Use the negative reference voltage input at pin 15.
33. For a 4-bit output there would be 15 op-amps (comparators) and 16 resistors.
35. $V_{15} = 18.75$ volts
37. $n V_{\text{REF}}/N = V_n < e_a$
    $V_{\text{REF}} < (N/n)e_a$ where e_a is its maximum value.
39. For a 4-bit ADC there are 16 counts, hence 16 clock pulses must occur between each S/H.

41.

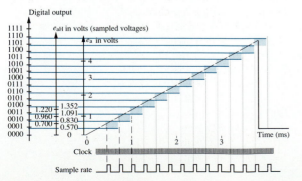

43. With a successive approximation the ADC will approximate the output in a successive fashion. Because the analog ramp voltage begins at 0 volts but the ADC begins its approximation in the middle, the ADC will have to reduce the count by resetting the higher-weighted bits until it is approximately equal to the sample/hold value of the analog signal. When the 16th clock pulse within the sample interval occurs it triggers the EOC signal and the next approximation would begin.

45. (a) 00000001
    (b) 00000101
    (c) 00010100
    (d) 00011001

47. First program the port. This requires that static logic generators be used to put the address of the control register on the address bus, the proper control word on the data bus, and then drive the control signals to the correct logic levels ($IO/\overline{M} = 1$, $\overline{WR} = 0$, and $\overline{RD} = 1$). Now that the port is programmed verify its operation by reading and writing data from and to the I/O devices. Of course this requires that the address, data (for a write), and control bus have the proper logic levels applied to them.

49. Program the MPU to (1) write a byte to the serial output device and (2) read a byte from the serial input device. The MPU would generate the dynamic logic levels required ($\overline{RD}$, $\overline{WE}$, C/D, and $\overline{CS}$ via the address and control busses). The clock signals TXC and RXC would be generated by some pulse-generating circuitry. Thus, for a write operation the MPU would address the USART, load the data on the data bus, and then drive $\overline{WR}$ low. Use a logic analyzer to verify that these events occurred. A channel of the logic analyzer would monitor the TXD to verify that the data being written is transmitted to the serial output device. For a read, load known data into the serial input device and then read that device by again programming the MPU to do so. Monitor the RXD line and busses to verify proper operation.

51. For a DAC with three inputs a 3-bit counter is used, which yields eight counts. We would have manual control over this counter by being able to advance its count when desired (we would pulse the clock input). One trace of the scope would monitor the output voltage whereas the other could be used to monitor the input logic levels (one input at a time).

53.

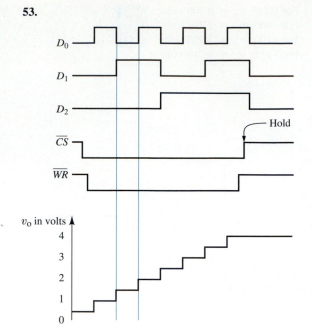

INDEX